PROCESS-INDUCED FOOD TOXICANTS

PROCESS-INDUCED FOOD TOXICANTS

Occurrence, Formation, Mitigation, and Health Risks

RICHARD H. STADLER
Nestle Product Technology Centre

DAVID R. LINEBACK
Joint Institute for Food Safety and Applied Nutrition
University of Maryland

A JOHN WILEY & SONS, INC., PUBLICATION

Published by John Wiley & Sons, Inc., Hoboken, New Jersey
Published simultaneously in Canada

For general information on our other products and services or for technical support, please contact our Customer Care Department within the United States at (800) 762-2974, outside the United States at (317) 572-3993 or fax (317) 572-4002.

Wiley also publishes its books in a variety of electronic formats. Some content that appears in print may not be available in electronic formats. For more information about Wiley products, visit our web site at www.wiley.com.

Library of Congress Cataloging-in-Publication Data:

ISBN: 978-0-470-07475-6

Printed in the United States of America

10 9 8 7 6 5 4 3 2 1

CONTENTS

PREFACE ix

CONTRIBUTORS xiii

PART I SPECIFIC TOXICANTS RELATED TO PROCESSING TECHNOLOGY 1

1 **Introduction to Food Process Toxicants** 3
David R. Lineback and Richard H. Stadler

2 **Thermal Treatment** 21

2.1 **Acrylamide** 23
Craig Mills, Donald S. Mottram, and Bronislaw L. Wedzicha

2.2 **Acrolein** 51
Takayuki Shibamoto

2.3 **Heterocyclic Aromatic Amines** 75
Robert J. Turesky

2.4 **Hazards of Dietary Furan** 117
P. Michael Bolger, Shirley S-H. Tao, and Michael Dinovi

2.5 **Hydroxymethylfurfural (HMF) and Related Compounds** 135
Francisco J. Morales

2.6 **Chloropropanols and Chloroesters** 175
Colin G. Hamlet and Peter A. Sadd

2.7 **Maillard Reaction of Proteins and Advanced Glycation End Products (AGEs) in Food** 215
Thomas Henle

2.8 **Polyaromatic Hydrocarbons** 243
Jong-Heum Park and Trevor M. Penning

3 **Fermentation** 283

3.1 **Ethyl Carbamate (Urethane)** 285
Colin G. Hamlet

3.2 **Biogenic Amines** 321
Livia Simon Sarkadi

4 **Preservation** 363

4.1 ***N*-Nitrosamines, Including *N*-Nitrosoaminoacids and Potential Further Nonvolatiles** 365
Michael Habermeyer and Gerhard Eisenbrand

4.2 **Food Irradiation** 387
Eileen M. Stewart

4.3 **Benzene** 413
Adam Becalski and Patricia Nyman

5 **High-Pressure Processing** 445
Alexander Mathys and Dietrich Knorr

6 **Alkali and/or Acid Treatment** 473

6.1 **Dietary Significance of Processing-Induced Lysinoalanine in Food** 475
Mendel Friedman

6.2 **Dietary Significance of Processing-Induced D-Amino Acids** 509
Mendel Friedman

6.3 **Chloropropanols** 539
Jan Velíšek

PART II GENERAL CONSIDERATIONS 563

7 **Application of the HACCP Approach for the Management of Processing Contaminants** 565
Yasmine Motarjemi, Richard H. Stadler, Alfred Studer, and Valeria Damiano

8 **Emerging Food Technologies** 621
Fanbin Kong and R. Paul Singh

9 **Food Processing and Nutritional Aspects** 645
Josef Burri, Constantin Bertoli, and Richard H. Stadler

10 **Risk Communication** 679
David Schmidt and Danielle Schor

11 **Risk/Risk and Risk/Benefit Considerations** 695
Leif Busk

INDEX 711

PREFACE

Diets are developed to ensure adequate nutrition with the desire to maintain "good" health. This, in turn, directs attention to the quality and safety of individual foods comprising the diet. Whether justified or not, food safety concerns have increased over the past two decades, primarily due to food- and water-borne disease outbreaks (microbiological causation) and some issues of a chemical nature. This has also been accompanied by consumers becoming more aware—and indeed better informed—concerning the risks that consumption of certain foods and food components have with a potential link to degenerative diseases such as diabetes and cancer. Not only is the choice and balance of the "right" foods an essential part of a healthy lifestyle, but how food is prepared and processed can be an important factor.

It has been known for a long time that formation of certain chemicals during food processing or preparation may pose a risk to human health. Examples are polycyclic aromatic amines (PAHs) in grilled/barbecued meat, N-nitrosamines in cured meats and fish, and heterocyclic aromatic amines (HAAs) in overheated meats and fish, which were identified as food-borne carcinogens and consequently a potential human health concern since the early 1970s. In the past few years, the formation of chemicals, particularly those with a potentially adverse effect on human health, has received increased attention. This is illustrated by the report that surfaced unexpectedly in 2002 of the occurrence of acrylamide in commonly consumed foods. The acrylamide issue has become worldwide in scope.

Several more chemicals have been added to the list of undesired process-induced chemicals, also termed process contaminants or process toxicants. With the increasing sensitivity of analytical methodologies and knowledge concerning the formation of both beneficial and potential toxicants during the complex reactions, such as the Maillard reaction, occurring during the processing of foods, the numbers of these chemicals of potential concern will continue to increase. Often these are present in very small amounts in foods consumed in normal diets. An important question then becomes: at what amount does the presence of these toxicants in foods consumed by humans become a potential health problem? Answering such a question normally involves carrying out a quantitative risk assessment.

Readers will notice that the two terms, "contaminant" and "toxicant," are used interchangeably in this book. This is because today there is no clear distinction between a process contaminant and toxicant. It is, however, the editors' view that the term "contaminant" may encompass a broader definition not necessarily restricted to toxic substances.

This book is divided into two parts and presents a comprehensive update of the major toxicants that may be generated during food processing and food preparation. Part 1 considers the different processes used in the manufacture of foods, including food prepared in the home, and the risk of formation of food-borne toxicants linked to the different technologies and processes. Common methods encountered in a typical industrial or home cooking environment are included. For each of these, the aspects addressed include (1) occurrence in food, (2) methods of analysis, (3) routes of formation, (4) mitigation options, (5) human exposure, (6) health risks, and (7) risk management.

Information regarding the first five aspects is critical to accomplishing a quantitative risk assessment that will indicate the likelihood of occurrence of a potentially adverse health effect. Thorough evaluation of the toxicological risks of food-borne toxicants is pivotal for developing risk management approaches. From the consumer's perspective, risk management of the potentially adverse risks is an important issue.

An important step in managing such risks initially occurs in the food manufacturing environment. This begins with the application of well-established food safety procedures such as HACCP (Hazard Analysis Critical Control Point). In terms of microbiological and allergen risks, these are usually relatively well controlled in an industrial setting through the use of such procedures. However, knowledge is, in general, lacking on how to effectively deal with processing contaminants. This involves the need for innovative technologies with both a high level of food safety while yielding the functionalities that maintain the nutritional quality of the food *vis-à-vis* traditional techniques.

Furthermore, it has become increasingly important to recognize that health beneficial chemicals are also formed during the processing of foods, as indicated by several scientific studies linking cocoa, tea, and coffee consumption to certain health benefits. Mitigation measures may in some cases lead to changes in the nutritional profile of foods; for example, acrylamide reduction in some bakery wares is achieved through the replacement of the baking agent ammonium bicarbonate with the corresponding sodium salt. This raises the question of the potential impact of the mitigation measure on sodium intake and possible health risks, particularly in susceptible subgroups of the population suffering from elevated blood pressure.

To perform a transparent and understandable quantitative assessment of this type, there is a need for a common or at least comparable denominator of risks versus benefits. This is where a major challenge lies today. Finally, communicating food safety risks is a key component of the risk analysis framework and cannot be over-emphasized. All these important topics are dealt with in Part 2 of this book.

The editors of this book have been very fortunate in attracting contributions from leading experts in their field. These individuals come from industry, academia, and regulatory/government bodies, bringing the diversity of backgrounds and viewpoints necessary to adequately address the wide scope of issues inherent in dealing with food process toxicants. We have also attempted to provide both European and North American approaches to dealing with food process contaminants, selecting experts from both geographical regions.

However, many challenges remain. Today, in most cases, clear-cut answers concerning the impact of food process toxicants on human health cannot be given. It is hoped that this book will help clarify important issues and caveats involved in the safety assessment of such substances, and will highlight the endeavors and commitments of all stakeholders at an international level to ensure that the foods we prepare are safe to eat.

Richard H. Stadler
David R. Lineback

CONTRIBUTORS

Adam Becalski, Food Research Division, Bureau of Chemical Safety, Health Products and Food Branch, Health Canada, Address Locator 2203D, 2 51 Sir F. Banting driveway, Ottawa, Ontario K1A 0L2, Canada; E-mail: Adam_Becalski@hc-sc.gc.ca

Constantin Bertoli, Nestlé Product Technology Center Konolfingen, Nestlé Str. 3, Konolfingen 3510, Switzerland; E-mail: constantin.bertoli@rdko.nestle.com

P. Michael Bolger, Center for Food Safety and Applied Nutrition, U.S. Food and Drug Administration, 5100 Paint Branch Parkway, College Park, MD 20740, USA; E-mail: Mike.Bolger@fda.hhs.gov

Josef Burri, Nestlé Product Technology Centre Orbe, CH-1350 Orbe, Switzerland; E-mail: josef.burri@rdor.nestle.com

Leif Busk, Department of Research and Development, National Food Administration, Box 622, Uppsala 75126, Sweden; E-mail: leif.busk@slv.se

Valeria Damiano, Nestlé Product Technology Center Orbe, CH-1350 Orbe, Switzerland; E-mail: valeria.damiano@nestle.com

Michael Dinovi, Center for Food Safety and Applied Nutrition, U.S. Food and Drug Administration, 5100 Paint Branch Parkway, College Park, MD 20740, USA; E-mail: michael.dinovi@fda.hhs.gov

Gerhard Eisenbrand, Department of Chemistry, Division of Food Chemistry and Toxicology, University of Kaiserslautern, Erwin-Schroedinger-Str. 52, Kaiserslautern 67663, Germany; E-mail: eisenbra@rhrk.uni-kl.de

Mendel Friedman, Western Regional Research Center, Agricultural Research Service, United States Department of Agriculture, Albany, CA 94710, USA; E-mail: mfried@pw.usda.gov

Michael Habermeyer, Department of Chemistry, Division of Food Chemistry and Toxicology, University of Kaiserslautern, Erwin-Schroedinger-Str. 52, Kaiserslautern 67663, Germany; E-mail: habermey@rhrk.uni-kl.de

Colin G. Hamlet, RHM Technology, RHM Group Ltd., Lord Rank Centre, Lincoln Road, High Wycombe, Bucks HP12 3QR, UK; E-mail: colin.g.hamlet@rhm.com

Thomas Henle, Institute of Food Chemistry, Technische Universität Dresden, D-01062 Dresden, Germany; E-mail: Thomas.Henle@chemie.tu-dresden.de

Dietrich Knorr, Berlin University of Technology, Department of Food Biotechnology and Food Process Engineering, Koenigin-Luise-Str. 22, Berlin D-14195, Germany; E-mail: dietrich.knorr@tu-berlin.de

Fanbin Kong, Department of Biological and Agricultural Engineering, University of California, Davis, CA 95616, USA; E-mail: fanbin_kong@yahoo.com

David R. Lineback, Joint Institute for Food Safety and Applied Nutrition (JIFSAN), University of Maryland, College Park, MD 20742, USA; E-mail: lineback@umd.edu

Alexander Mathys, Berlin University of Technology, Department of Food Biotechnology and Food Process Engineering, Koenigin-Luise-Str. 22, Berlin D-14195, Germany; E-mail: alexander.mathys@tu-berlin.de

Craig Mills, UK Food Standards Agency, 125 Kingsway, London WC2B 6NH, UK; E-mail: mills@physchem.ox.ac.uk

Francisco J. Morales, Consejo Superior de Investigaciones Científicas (CSIC), Instituto del Frío José Antonio Novais 10, Madrid 28040, Spain; E-mail: fjmorales@if.csic.es

Yasmine Motarjemi, Quality Management, Nestlé, 55 Avenue Nestlé, Vevey CH-1800, Switzerland; E-mail: Yasmine.Motarjemi2@nestle.com

Donald S. Mottram, University of Reading, Department of Food Biosciences, Whiteknights, Reading RG6 6AP, UK; E-mail: d.s.mottram@reading.ac.uk

Patricia Nyman, Center for Food Safety and Applied Nutrition, U.S. Food and Drug Administration, College Park, Maryland 20740, USA; E-mail: patricia.nyman@fda.hhs.gov

Jong-Heum Park, Center of Excellence in Environmental Toxicology, Department of Pharmacology, University of Pennsylvania, School of Medicine, Philadelphia, PA 19104-6084, USA; E-mail: hmpark@mail.med.upenn.edu

Trevor M. Penning, Department of Pharmacology, 130C John Morgan Bldg., 3620 Hamilton Walk, Philadelphia, PA 19064, USA; E-mail: penning@pharm.med.upenn.edu

Peter A. Sadd, RHM Technology, RHM Group Ltd., Lord Rank Centre, Lincoln Road, High Wycombe, Bucks HP12 3QR, UK; E-mail: Peter.A.Sadd@rhm.com

Livia Simon Sarkadi, Budapest University of Technology and Economics, 1111 Budapest, Müegyetem rkp 3, Hungary; E-mail: sarkadi@mail.bme.hu

David Schmidt, President, International Food Information Council, 1100 Connecticut Ave. NW, Suite 430, Washington, DC 20036, USA; E-mail: Schmidt@ific.org

Danielle Schor, International Food Information Council, 1100 Connecticut Ave. NW, Suite 430, Washington, DC 20036, USA; E-mail: Schor@ific.org

Takayuki Shibamoto, Department of Environmental Toxicology, University of California, Davis, CA 95616, USA; E-mail: tshibamoto@ucdavis.edu

R. Paul Singh, Department of Biological and Agricultural Engineering, University of California, Davis, CA 95616, USA; E-mail: rpsingh@ucdavis.edu

Richard H. Stadler, Nestlé Product Technology Centre Orbe, CH-1350 Orbe, Switzerland; E-mail: richard.stadler@rdor.nestle.com

Eileen M. Stewart, Agriculture, Food and Environmental Science Division, Agri-Food and Biosciences Institute (AFBI), Newforge Lane, Belfast BT9 5PX, UK; E-mail: eileen.stewart@afbini.gov.uk

Alfred Studer, Nestlé Research Centre, Vers-chez-les-Blanc, CH-1000 Lausanne 26, Switzerland; E-mail: alfred.studer@rdls.nestle.com

Shirley S-H. Tao, Center for Food Safety and Applied Nutrition, U.S. Food and Drug Administration, 5100 Paint Branch Parkway, College Park, MD 20740, USA; E-mail: shyyhwa.tao@fda.hhs.gov

Robert J. Turesky, Division of Environmental Disease Prevention, Wadsworth Center, NYS Department of Health, Albany, NY 12201, USA; E-mail: rxt07@health.state.ny.us

Jan Velíšek, Institute of Chemical Technology, Department of Food Chemistry and Analysis, Technická 1905, Prague 166 28, Czech Republic; E-mail: Jan.Velisek@vscht.cz

Bronislaw L. Wedzicha, University of Leeds, Procter Department of Food Science, Leeds LS2 9JT, UK; E-mail: BronekWedzicha@food.leeds.ac.uk

PART I

SPECIFIC TOXICANTS RELATED TO PROCESSING TECHNOLOGY

1

INTRODUCTION TO FOOD PROCESS TOXICANTS

DAVID R. LINEBACK[1] AND RICHARD H. STADLER[2]

[1] *Joint Institute for Food Safety and Applied Nutrition (JIFSAN), University of Maryland, College Park, MD 20742, USA*
[2] *Nestlé Product Technology Centre Orbe, CH-1350 Orbe, Switzerland*

1.1 HISTORY AND ROLE OF FOOD PROCESSING

Food processing and preservation, the traditional focus of food science and technology, have played, and continue to play, important roles in achieving food sufficiency (availability, quality, and preservation) for the human race. These practices originated in recognition of a need to improve the edibility of many food sources and to maintain food supplies for longer periods of time than their seasonal availability. With the transition from a hunter–gatherer society to life in villages and early agriculture, this need became even greater and emphasis on food preservation became increasingly important. This, of course, was paralleled by the development of processes/processing of animal, vegetable, and marine raw materials into usually more palatable, portable, and nutritionally dense foods. In many cases, if not most, this occurred in a fortuitous, rather than planned, manner as natural causes of food processing and preservation were observed and adapted to human use.

Food processing involves the actions taken from the time a raw product (crop, animal, fish) is harvested, slaughtered, or caught until it is sold to the consumer. By this process, the parts regarded as most valued are separated from by-products or waste. Equally enhanced is the palatability/digestibility

Process-Induced Food Toxicants: Occurrence, Formation, Mitigation, and Health Risks,
Edited by Richard H. Stadler and David R. Lineback

of foods, illustrated in the transformation of baking flour into bread, to maintain or increase quality attributes and to ensure safety. Increasing understanding of the science involved in food loss, deterioration of quality, and means of improving the palatability of foods has resulted in development of the sophisticated methods of food processing and preservation now in use. The work of Pasteur, resulting in identification of the role of microorganisms in food spoilage and development of technology leading to canning by Nicolas Appert in 1809, can be considered initial steps in the development of modern food processing and preservation (1). As the world population continues to grow, resulting in increasing requirements and demands for food availability and safety, new and improved methods of food processing and preservation are needed and in development.

The term "minimal processing" is frequently used to describe foods, such as vegetables, that are harvested, sorted, and washed (or similar minimal invasive procedures) before distribution and sale. This is done to distinguish these more "natural" products from those that undergo more extensive processing procedures. Over the last years the development and distribution of minimally processed foods has been increasing steadily. This trend has been triggered by the demand for fresh and convenient products as well as for more natural products, i.e., less processed or containing less salt, sugar, or preservatives.

Such foods range from fruits and vegetables, which are usually only submitted to washing (with or without biocides), trimming, slicing, or shredding, to prepared foods processed by applying minimal bactericidal treatments in combination with different physicochemical hurdles to ensure their stability and safety. These foods represent certainly a challenge to manufacturers since no or only minimal killing steps are applied, and, at the same time, requirements for more global availability and longer shelf life are increasing. The fact that these challenges are frequently underestimated or not mastered sufficiently is illustrated by the occurrence of numerous incidents linked to a variety of products involving different pathogens. Outbreaks related to minimally processed foods often encompass chilled foods such as *sous-vide* products, pasteurized vegetables, and baked potatoes, which have frequently been linked to *Clostridium botulinum* intoxication (2–4).

Early types of processing/preservation evolved from observations of natural processes, e.g., drying, curing (such as salting), smoking, fermentation, and reducing storage temperature (refrigeration or freezing). Salting and smoke processing originated at the beginning of human civilization, mainly employed to preserve meat and fish. In fact, salting, pickling, and drying continued as the primary means of preserving foods until the twentieth century and the advent of mechanical refrigeration (5). More modern means of preservation precluded the use of copious amounts of salt, exemplified by the far reduced concentrations of salt in ham today ($<2\%$) versus that in hams produced in the first half of the twentieth century ($>6\%$). Changes to technologies were also introduced a few decades ago with regard to cured meats and residual nitrite content. Nitrite, used to cure meat, acts as a preservative against

Clostridium botulinum and other spoilage bacteria. However, during the 1970s, concern arose due to the role of nitrites in the formation of carcinogenic nitrosamines (Chapter 4.1), as well as its contribution to the body burden. In modern cured meats, the nitrite amounts have decreased and are typically one-fifth of those found some 30 years ago. Moreover, the use of ascorbate—an effective inhibitor of nitrosamine formation—is an additional mitigation measure introduced in the production of most cured meats.

Smoke processing is still used today to preserve meat, especially in tropical countries. Smoke imparts appealing organoleptic properties, with concomitant preservation of nutrients. However, concern has been raised about the presence of both polycyclic aromatic hydrocarbons (PAHs) and nitrosamines in smoked foods. PAHs are covered in Chapter 2.8, with special attention to their formation, mitigation, and toxicology. Although the exposure risks in modern manufacture of meats and fish are considered minimal, alternatives to traditional smoking have been developed. Liquid smoke flavorings have gained popularity as they provide the same traits, i.e., desirable organoleptic properties, and preservation through antioxidation and bacteriostasis. Additional benefits include increased product consistency and absence of detectable animal carcinogens. In fact, approximately 75% of hot dogs produced in the United States contain aqueous liquid smoke flavorings (5).

Food preservation can be considered part of or an extension of food processing, since it involves the use of procedures to prevent or reduce spoilage of foods. Examples include the inactivation of enzymes and microorganisms by heating or reduction of moisture content, use of antimicrobial compounds, pasteurization (heat or irradiation), freezing, modified atmospheric packaging, and fermentation.

Techniques that have been used in food processing and preservation include:

- Drying/dehydration
- Curing
- Smoking
- Fermentation
- Canning
- Pasteurization (heat or irradiation)
- Freezing and refrigeration
- Additives
- Controlled atmosphere storage
- Aseptic packaging

Until the last quarter of the twentieth century, canning was widely used in homes throughout the rural United States. Inadequate heat treatment during the canning process occasionally resulted in severe illness or death caused by *Clostridium botulinum* that was not inactivated during the heating process and resulted in subsequent formation of the toxin. Commercial canning, while having some outbreaks of botulinum poisoning, became used much more widely due to improved quality, safety, and increased urban populations.

1.2 GENERAL APPROACHES TO FOOD PROCESSING

The rapid growth and development of commercial food processing in the twentieth century has continued and now dominates food processing, particularly in developed countries. However, food processing in the home, such as canning, decreased with increasing urbanization. Although some aspects of food processing still occur frequently in home situations, particularly in developing nations. Food preparation/processing in the home is primarily related to heat treatment, which plays an important role in the formation of desirable flavors, colors, aromas, and textures. In fact, exposure of food to heat can be considered the most used processing step in modern society, involving frying, baking, grilling, roasting, toasting, microwaving, and broiling, using ovens (convection, microwave), stoves, toasters, grills (gas, wood, and charcoal), and fat-based fryers.

There are, however, considerable differences between practices in the home and in commercial industrial settings. Home appliances tend to have less accurate temperature controls, resulting in actual oven temperatures differing considerably from what is indicated by the oven setting or temperature gauge. In general, there is also less rigid timing due to interruptions and delays in homes as contrasted to an industrial environment. High-quality standards are pivotal for industrialized processes, and food manufacturers have identified early on the need for quality control tools and stringent targets to achieve consumer preference in terms of nutritional quality, shelf life, and organoleptic properties at all times. Ideally, quality is addressed early on in the product and process design phase, identifying those process steps that impact (key) quality parameters. For this purpose, modern industrial lines are equipped with appropriate measuring systems/sensors (temperature profiles, moisture content, texture, color, pH, etc.) to deal with raw material variability and process complexity.

1.3 CONCERNS ABOUT FOOD SAFETY DURING FOOD PROCESSING

1.3.1 Types of Hazards

The major hazards considered in food safety are allergens, and those that are microbiological, physical, and chemical in nature.

Microbiological contamination with pathogens such as enterohemorrhagic *Escherichia coli* strains, *Listeria monocytogenes*, and *Salmonella* spp. represents a major problem in modern food safety (pathogen identification, control, and prevention). It is considered the most important aspect of improving food safety globally, displacing the emphasis on chemical contaminants of previous decades. For further reading, various books and reviews on this topic can be consulted (see References 6 and 7).

Physical hazards are considered acute hazards if not adequately addressed and controlled and may pose a serious threat to human health (e.g., glass, hard plastic and metal pieces, bones, wood, stones). There are different sources of physical hazards, and the origins of the potential risks must be clearly understood (raw materials/ingredients or the operations in the manufacturing of food per se may be a source of a physical hazard, e.g., potential glass breakage along a glass filling line). Within the frame of Hazard Analysis Critical Control Points (HACCP), measures are identified that remove or reduce such hazards to an acceptable level in the final product (e.g., filtration, sieving, centrifugation). Procedures must be put in place by the manufacturer to verify that the measures to control such hazards are indeed effective (e.g., metal detectors, X-ray machines).

Food allergens are generally recognized as a serious food safety issue and manufacturers are responsible for controlling them and providing concise information to consumers. Through good manufacturing practice (GMP), identifying possible sources of cross contact, integration of allergen hazards into HACCP studies, and appropriate ingredient labeling, the health risks can be minimized.

The chemical hazards in foods can be multiple, and as depicted in Fig. 1.1 may enter the food and feed supply chain at many different points. Traditionally, the environment has been thought to be the origin of many chemical food hazards, such as heavy metals and persistent organic pollutants (POPs). An increased risk of pathogenic microorganisms may also be attributed to the contamination of the agricultural water supply caused by human and animal waste, and use of manure as fertilizer.

Essentially, the potential chemical contaminants in food can be broadly classified into:

(1) natural toxins, e.g., mycotoxins, higher plant toxicants, and marine biotoxins;
(2) environmental contaminants, e.g., heavy metals, dioxins, and radionuclides;
(3) chemicals used as aids in food manufacture and, in the event of a failure, which may contaminate food, e.g., through leakage, spillage, or misuse of lubricants, cleansing agents, or disinfectants;
(4) agrochemical residues, e.g., fungicides, pesticides, and veterinary drugs;
(5) packaging migrants, e.g., isopropylthioxanthone, semicarbazide, and styrene;
(6) processing toxicants, e.g., heterocyclic aromatic amines, acrylamide, and furan.

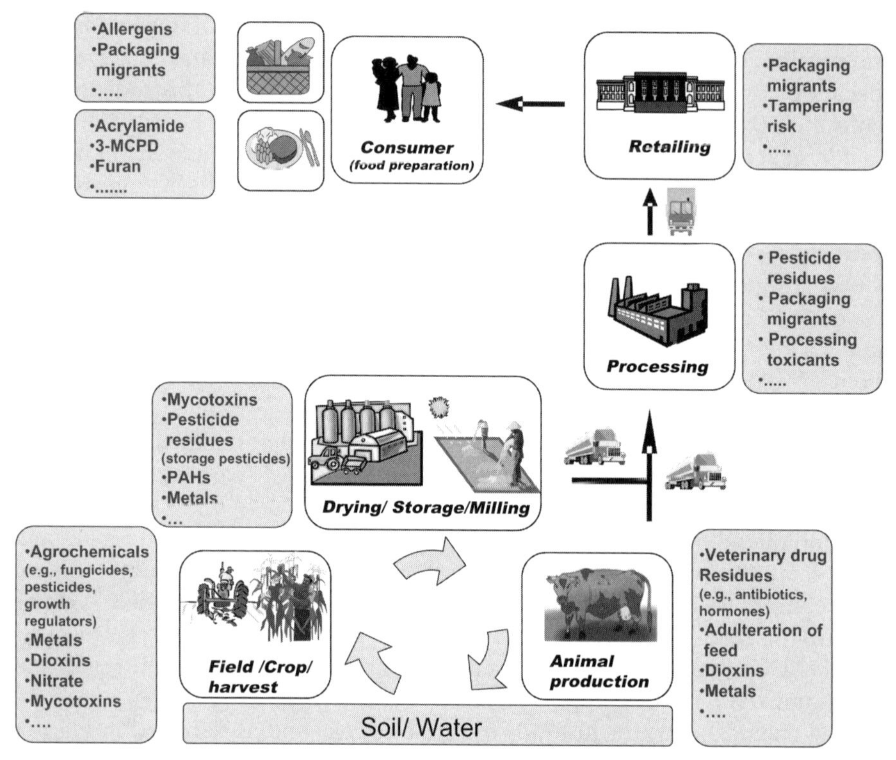

Figure 1.1 Overview of possible routes of chemical contamination of food (simplified). See color insert.

The latter class of substances is the focus of this book, and the reader is referred to further sources of information on general chemical risks in food (see References 8 and 9).

1.3.2 Definition of a Process Toxicant

Processing toxicants (process-induced toxicants, process-formed toxicants) as used in this book are defined as those substances present in food as a result of food processing/preparation that are considered to exert adverse physiological (toxicological) effects in humans, i.e., substances that create a potential or real risk to human health. Food in this definition also includes beverages and nonalcoholic drinks such as coffee and tea, and thus both parts of the diet are included.

Ingredients commonly occurring in food formulations (recipes) are excellent substrates for chemical reactions occurring under the conditions encountered in food processing. The reaction products formed depend on the processes and conditions used, such as fermentation, irradiation, and heat processing.

Products from such reactions can have beneficial properties and/or adverse physiological effects on consumers. Examples of the former include compounds such as antioxidants, anticarcinogens, and those resulting in or contributing to nutritional properties, desirable flavor, aroma, texture, and color in food products. Examples of the latter include carcinogens, genotoxins, neurotoxins, anti-nutrients, and undesirable flavors or aromas. Many of these coexist as a result of being formed during common food processing technologies, particularly those involving heating, e.g., toasting, roasting, frying, broiling, baking, grilling and microwaving.

1.3.3 Progress in Technological Developments

The development of new food processing technologies continues at a rapid pace, with some of these, such as high-pressure processing (HPP), already in commercial use (see Chapters 5 and 8). In fact, the application of HPP is not a new concept, and was already described in certain foods in the late nineteenth century (10). The use of HPP is not uncommon in foods such as whole shell oysters, salsa, ready-to-eat (RTE) meats, and jams. New technologies are aimed at delivering products with superior organolceptic quality, minimal changes to nutrients, safety, and shelf life (product life, preservation of quality), and ideally the minimal formation of undesirable compounds. Pulsed electric field, ohmic heating, jet impingement, infrared radiation, and new biotechnological applications are just a few that can be considered new processing techniques. Innovative nonthermal processing technologies (photosensitization, pulsed electric field technologies, high-pressure homogenization, and HPP coupled to packaging under inert atmosphere) to improve the quality and safety of RTE meals are being investigated within the European "HighQ RTE" project, with the goal to improve the safety and quality of three representative categories of European RTE foods, i.e., salads, fluid foods, and vegetable-based meals (11).

Only very few studies have been performed on the use of alternative or new processing technologies to mitigate process toxicants. Work has recently been reported on the application of infrared radiation to baking with the goal to reduce the amount of acrylamide (12) while maintaining the sensorial properties of the food. Steam baking and steam roasting have also been assessed to reduce acrylamide, and in the case of coffee beans the steam roast had a major impact on the sensorial properties of the coffee, with only a marginal reduction in acrylamide (13).

1.4 FOOD-BORNE PROCESSING TOXICANTS: SETTING PRIORITIES

The processing of ingredients into food products can lead to the formation of a number of chemical compounds having properties desired in the flavor, aroma, and color of the food. One of the most important sources of these compounds is the Maillard reaction (14). This complex reaction, involving reducing sugars and amino acids, has been, and continues to be, the subject of intensive research for many years due to its importance in the formation of characteristic flavors, aromas, and colors (browning) in foods prepared by heating. A major emphasis has been on the identification of the compounds involved in these attributes as formed during the Maillard cascade and in understanding the chemical pathways involved. The Maillard reaction is known to produce more than 550 volatile compounds of which more than 330 have been identified in the volatiles of cooked foods (15). Many of these contribute to the flavors and aromas in these foods. Nonvolatile products such as the melanoidins contribute to the browning colors.

However, compounds having adverse physiological effects or potential health risks are often formed also. The number of studies involving detection, identification, and measurement of such compounds continues to increase as more sensitive analytical methodologies become available and are applied to foods. One example is the discovery in early 2002 of acrylamide in foodstuffs (16), present in the μg/kg (part per billion) range in a wide variety of common foods that are heated at temperatures >120 °C, such as potato chips, French fries, bread, cereal-based products, and coffee (17). Acrylamide has received unprecedented attention since it was first reported in food, with several books and reviews summarizing the research efforts across the disciplines, and reflected by more than 600 research publications to date (15, 18).

Development of new analytical methods to detect and determine acrylamide was required for the very low concentrations encountered in foods. As analytical methodologies continue to become even more sensitive, more of such compounds undoubtedly will be found in foods. When these compounds are determined or known to have adverse effects or potential health risks, toxicological studies can become difficult since these usually are accomplished in animals at concentrations much higher than may be observed in foods. This has become an issue with acrylamide, since definitive toxicological studies are not yet available (anticipated to be complete and reported beginning 2009).

A key question raised by food safety authorities, by academics in the field, and in particular by food producers, is which toxicants are of greatest concern in foods from a dietary health perspective? Numerous compounds have been identified over the past years in foods that show carcinogenic, mutagenic (genotoxic), or neurotoxic properties at high doses in animal studies. Such toxicants can be classified by chemical (structural) type or by the processing methods in which they occur. There will be some overlap in either case. In this book, the issue is approached from the processing method involved and includes (i) thermal treatments, e.g., frying, baking, grilling, roasting, broiling, toasting, and microwaving (Chapter 2); (ii) fermentation (Chapter 3); (iii) preservation (Chapter 4); (iv) high hydrostatic pressure (Chapter 5); and (v) other selected processes such as acid and base treatment (Chapter 6).

When one begins to seek substances that have the potential to be toxic when present in foods, an entirely different set of issues is raised. The HEATOX

(Heat-generated Food Toxicants, Identification, Characterization and Risk Minimization) project had as one of its emphases identification of heat-formed toxicants, other than acrylamide, in food. Within the frame of this European project, supported within the European Commission's 6th Framework Programme on research, nearly 800 volatile compounds have been identified and listed in two databases (19), one of which contains approximately 570 formed through the Maillard reaction and the second which contains about 200 compounds from heated lipid systems. Information on toxicity and carcinogenicity of these compounds is scarce, and therefore computer-assisted toxicity prediction systems, i.e., Topkat and Derek, were employed. These make use of molecular attributes to construct quantitative structure–activity relationship (QSAR) models, which are useful tools for prescreening and setting priorities (20, 21). Of the total list, about 50 of these substances were identified as potential carcinogens and mutagens (19).

The HEATOX inventories for Maillard and heated lipid reaction products are in a spreadsheet format and list compounds that may be formed in model systems and/or are known to occur in food, the latter also featuring the food where the concentration has been reported to be higher than 1 mg/kg (different literature sources).

A key question, however, is the ranking of these compounds in terms of risk level. One approach that is employed to help prioritize risks of food-borne genotoxic carcinogens is the margin of exposure (MoE). The MoE is usually calculated as a range, taking in most cases the $BMDL_{10}$ value (the lower confidence limit on the benchmark dose associated with a 10% cancer incidence) and the upper- and lower-bound human exposure estimate. Table 1.1 illustrates the principle, and shows selected food-borne toxicants and an approach to estimate the margin of safety or MoE. A MoE band >10,000 is interpreted as unlikely to be of concern. Such a procedure would provide a first indication of the degree of risk. However, the interpretation of any MoE is complex and comparisons are not straightforward without knowledge of the methodologies used to analyze the data and data quality (for both the animal carcinogenicity data and dietary exposure estimates).

For a majority of the compounds, however, toxicity data may simply be lacking. In this case, a first screening will be required using existing data (considering also the limitations and uncertainties of the predictive toxicity models), and preferably probabilistic modeling, followed where warranted by in-depth research including method development, analytical measurement, and chronic animal studies. Other databases such as the carcinogenicity potency database (CPDB), which contains information on the toxicity of more than 1400 chemicals, some of which are naturally present in foods such as coffee, may also be a useful source of data (22). This, combined with occurrence databases or other published data on amounts of the selected chemicals in food, may allow a first ranking based on the margin of safety, provided that the compounds are retrievable in the database.

When considering process-induced toxicants and their potential health risks, a number of additional factors come into play in establishing the context

TABLE 1.1 Risk assessment of selected food-borne toxicants.

Food-borne toxicant	Estimated dietary uptake, ng/kg bw/d*	Tox. dose: point of departure, ng/kg bw/d	MoE/Safety factor
1,3-DCP/2,3-DCP (**)	3–200	6,300,000¶	2,100,000–32,000
Heterocyclic aromatic amines (PhIP)	4.8–7.6 (***)	1,250,000§	260,000–164,000
Polyaromatic hydrocarbons [Benzo(a)pyrene]	4[(a)]–10[(b)]	100,000§	25,000–10,000
N-nitrosamines (NDMA)	3.3–5.0	60,000§	18,200–12,000
Ethylcarbamate	33[(c)]–55[(d)]	300,000§	9,000–5,460
Furan	260[(e)]–610[(f)]	1,000,000¶	3,900–1,600
3-MCPD[(g)]	360[(h)]–1,380[(i)]	1,100,000¶	3,055–800
Acrylamide	1,000[(j)]–4,000[(k)]	300,000§	300–75

*Data sources vary and are shown here for illustrative purposes only; a = mean intake; b = high-level intake; c = lower-bound mean; d = upper-bound mean; e = mean for 2+-year-old children; f = 90th percentile for 2+-year-old children; g = non-genotoxic mode of action; h = highest mean of participating country for adult population (23); i = highest 95th percentile of participating country for adult population (23); j = average intake for general population; k = high consumers.
** (24).
*** (25).
¶LOAEL.
§$BMDL_{10}$ (lower conservative end of the range was chosen).
DCP = dichloropropanol; NDMA = *N*-nitrosodimethylamine; PhIP = 2-Amino-1-methyl-6-phenylimidazo- [4,5-b]pyridine.

within which concerns are raised and decisions made, particularly those relating to risk analysis. In this book, these are included in Part II: General Considerations. One of the most commonly used approaches to food safety during food processing is the HACCP concept. Some countries mandate application of this approach, which recognizes that safety should be built into a product during the early development phase, rather than depending on final product testing to detect safety defects. This subject is presented in Chapter 7. A pivotal factor in food processing, including development of new process technologies, is the impact of the processing technology on nutritional aspects (gain/loss during processing, bioavailability, formation of allergens). Chapter 9 deals with this very important area.

In recent years it has become apparent that decisions, particularly those involving regulatory action, need to be risk-based. This is often encompassed in the concept that the amount of regulation should be in direct proportion to the health risk in the food product. As a result, risk analysis (risk assessment, risk communication, risk management) is playing an increasingly important role in this general area. Importance of risk communication is addressed in

Chapter 10 and the concepts of risk–risk and risk–benefit are presented in Chapter 11.

1.5 ISSUES OF PROCESS TOXICANTS PRESENT IN SMALL AMOUNTS IN FOODS

As analytical methodologies increase in sensitivity, allowing detection and determination of process toxicants formed during food processing, an increasing number of significant issues will arise. The analytical instrumentation required for accurate determination of these toxicants will become more costly, prohibitively so in many cases. Already, this has been encountered in the case of acrylamide, for which the preferred analytical procedures for determining the content in foods involve mass spectrometry combined with other techniques, such as liquid chromatography-tandem mass spectrometry (LC-MS/MS). The analytical instrumentation for measurement has been beyond affordability in many institutions, particularly in developing countries.

As a result, this has limited accurate determination of acrylamide in the food supply in many countries, resulting in an inadequacy of data needed for a quantitative risk assessment (26) of potential health implications from its consumption in the diet. Also, this results in an inadequate amount of data to determine exposures to acrylamide in different countries, data also required for risk assessments. Lacking the data needed for quantitative risk assessment, how does one make judgments concerning the reality of a potential public health risk from the chemical involved?

Large amounts of resources have been committed to acrylamide investigations worldwide. The requisite information/data needed to make public health decisions is, however, still not available six years after the announcement of finding it in foods common to diets globally. This raises an important question: Are there criteria that can be established to enable a relatively rapid decision to be made regarding whether a large investment of resources should occur for further investigation of any new toxicant identified at the low concentrations observed for acrylamide, perhaps even when adverse effects have been noted in animal studies?

1.5.1 Risk Assessment Strategies for Food-Borne Toxicants

A risk assessment is a scientific evaluation of a chemical substance and is used to determine whether a particular chemical poses a significant risk to human health and/or the environment. Chemical risk assessments follow a set paradigm comprised of four steps, i.e.:

(1) *Hazard identification*: reviews research into any potential problems that the chemical may cause.

(2) *Hazard characterization*: describes and evaluates dose–response and dose–effect, mode of action, extrapolation of animal to man with ideally an acceptable daily intake (ADI) or tolerable daily intake (TDI) established.
(3) *Exposure assessment*: the amount in food, duration, and pattern of exposure are estimated, encompassing average, medium, and maximum figures for intake.
(4) *Risk characterization*: assesses the risk for the substance to cause cancer or other illnesses in a given population as well as the seriousness of any health risk.

As described by Tritscher (27), traditionally for chemicals in food two approaches to risk assessment are discerned depending on the presumed mode of action of the compound. For non-genotoxic substances, a threshold of action is assumed, from which a no-effect level can be extrapolated experimentally (NOEL or NOAEL). From this safe level, a TDI or reference dose (RfD) can be derived for humans by applying safety factors to the equation. For genotoxic carcinogens, the assumption taken is a non-threshold mechanism and hence no safe level of intake can be derived. In the case of food-borne contaminants, it will be difficult or almost impossible to conduct formal risk assessments for each compound, and regulators may apply the ALARA (as low as reasonably achievable) or ALARP (as low as reasonably practicable) principle aimed at reducing exposures as far as possible. This leaves the decision to the risk managers on the extent of reduction that is technologically and financially practicable and justifiable. However, ALARA does not give a quantitative dimension on which to prioritize the risk. Moreover, the number of undesired substances to be assessed in foods is very high and more compounds are continuously added to the list as detection techniques achieve better performance, setting immense constraints on the limited resources at hand.

One pragmatic approach that is being investigated and discussed, particularly in Europe and North America, is the threshold of toxicological concern (TTC). The TTC is a concept that a threshold of human exposure can be established for all chemicals such that there would be no appreciable risk to health below this value (28, 29). This is an interesting concept, with considerable interest and potential, particularly as an initial screening tool. Regulators undoubtedly will find considerable difficulty with the concept that a threshold can be set for all chemicals below which no appreciable health risk will exist. The possibility is much higher that some criteria can be developed to guide the decisions required for different compounds in a timely manner.

The question as to whether criteria can be established to allow rapid initial decision making, particularly commitment of resources, is being actively discussed. A workshop was held (30) in the United States (i) to develop recommendations on approaches or criteria useful for prioritizing potential risks of chemical and microbial contaminants in foods, as potential tools for resource allocation and decision making; and (ii) to develop recommendations on next steps to advance the use of prioritization tools, including identification of critical knowledge gaps and research needs. Several common themes and conclusions were identified during the workshop with a publication to be prepared on the risk prioritization framework concepts that emerged from the workshop discussions. A second workshop was held (November 2008) to bring together experts who are working on tools for dietary exposure estimates to discuss approaches and develop recommendations for using exposure estimates in a framework for prioritizing risks associated with chemical and microbial contaminants in food.

The outcomes of these efforts and discussions will be of critical importance to how the results of detecting and determining very low concentrations (i.e., ppb and less) of process contaminants in foods will be interpreted in terms of risks to human health and how they are potentially regulated. It must be remembered that just because a toxicant is present in low amounts in foods does not mean that it will not be a significant contributor to exposure in an individual's diet. As an example, acrylamide is present in low amounts in coffee or bread. However, since both of these are consumed in significant amounts in the daily diet of many individuals in some countries, they end up being in the top five or so foods/drinks in terms of daily sources of exposure to (intake of) acrylamide in those nations.

1.5.2 Risk–Benefit Considerations

A further important aspect to be considered also in the mitigation of process toxicants is risk–benefit. Changes made to the nutritional profile of a product may negatively impact the inherent beneficial properties of the food. For example, the removal of whole grain accepted to deliver a nutritional benefit will result in a product with relatively lower amounts of acrylamide. However, the current assessment of such measures is mainly subjective, and lacking is an appropriate risk–benefit framework that allows legitimate comparisons to be made. As noted by the European Food Safety Authority (EFSA), a risk–benefit analysis should not be performed as a routine procedure but only applied in those cases where an impact on public health outcomes can be expected (31). Such an activity has been started by ILSI Europe and models have already been established, comparing for example, the benefits of eating fish that are rich in omega-3-polyunsaturated fatty acids (associated with a lower risk of coronary heart disease and a beneficial role in inflammatory conditions such as arthritis) versus dioxin and methylmercury that may accumulate in certain fish (associated with risk of cancer development and developmental changes in the fetus) (32). Similarly, a recent project comparing the benefits associated with the heat processing of foods (reduced microbial spoilage, extended product shelf life) versus the risks (formation of acrylamide and other process toxicants) has been initiated within the frame of the BRAFO project (Benefit-Risk Analysis of Foods). The main goal is to develop a framework that quantitatively compares human health risks and benefits of foods and food components, using common scales of measurement such as the indices Disability Adjusted Life Years (DALYs) or Quality Adjusted Life

Years (QUALYs) (see Chapter 11). Recently, a case study on risk–benefit considerations of mitigation measures of acrylamide in potatoes, cereals, and coffee has been completed and published (33). Such studies have the ultimate goal of facilitating the decision-making process in public health nutrition.

1.6 OUTLOOK

1.6.1 Research Needs

Research needs in process toxicants were recently highlighted in several international scientific meetings. Worthwhile mentioning is the symposium organized by the German DFG (Deutsche Forschungsgemeinschaft, German Research Foundation) Senate Commission on Food Safety (SKLM) in September 2005 on the potential health benefits and risks of thermal processing of foods, which summarized the major research needs. The SKLM emphasized the importance of further study of the risk–benefits of foods as well as specific needs related to the substances per se. In this context, four compounds or compound classes were selected, i.e., acrylamide, heterocyclic aromatic amines, furan, and 3-MCPD (34).

Similarly, the UK Food Standards Agency convened a stakeholder meeting to discuss the research commissioned by the Agency on food process contaminants in April 2005 (35) and to share the progress made in the different projects on surveys and minimization of risk. The substances of concern were chloropropanols (including 3-MCPD), ethyl carbamate, and acrylamide. Recently, EC-DG SANCO and the CIAA (European Food and Drink Federation) organized a joint workshop on acrylamide in March 2006, pulling together the latest findings from academic and industry research (36), and integrating this into the CIAA "Acrylamide Toolbox," which is *the* reference for the most promising avenues and achievements to reduce acrylamide in the main concerned food categories (37). Soon thereafter, a joint workshop on "furan in food" was hosted by the EC-DG SANCO/EFSA/DG JRC to update the latest knowledge pertaining to analysis and formation/mitigation of furan in food (36). Research in all these areas is progressing at a rapid pace and these selected examples show that process toxicants have in the past few years gained significant attention on a global scale in terms of potential human health risk.

1.6.2 Interdisciplinary Efforts of All Stakeholders

To effectively address food chemical toxicants, the pooling and coordination of knowledge coming from a variety of disciplines and areas of expertise is crucial (agronomy, toxicology, epidemiology, analytical chemistry, food chemistry, food technology, nutrition, consumer research), as well as committed efforts of all stakeholders. Hence, interdisciplinary research projects and partnerships are a more effective means of improving the science–policy interface and ensuring that the work has both scientific and societal relevance. This requires a strategic assessment of research trends in food safety, encompassing the complete chains of food and feed production. Food safety networks have been established at the European level, such as the European Association for Food Safety (SAFE consortium). Its member organizations work together to develop and promote interdisciplinary research projects and partnerships. The output of the consortium as a whole can be consulting activities, larger projects within the Framework Programmes, and participation in the ERA Net Scheme (38) and in Technology Platforms such as the European Technology Platform (ETP) "Food for Life," which consists of more than 90 projects (ongoing or completed) that are related to food safety (39). Cost Action 927 is a further endeavor to coordinate research and foster exchange of scientists working primarily on the health effects of thermally processed foods. The initiatives of the Action are reported on their web site and updated on a regular basis (40). The occurrence of processing toxicants in low amounts in foods will continue to be of increasing concern in the human diet. Many are not new risks, but the occurrence in the diet has raised concern, e.g., furan in baby foods in jars (Chapter 2.4), acrylamide in food (Chapter 2.1), and 3-MCPD esters in refined oils (Chapter 2.6).

That many of these will have adverse toxicological properties/effects will draw attention from consumers and regulators. The fact that it will not be possible to reduce or remove many of these to a significant extent from certain foods, without changing taste, odor, texture, nutrient profiles, or forming other compounds that may increase safety concerns, will have to be carefully and fully explained to consumers in a manner they can understand and accept. Thus, responsible risk communication and management are highly important factors in these issues. Aspects of food processing and preservation, with the attendant potential relationships to adverse effects on public health, are the subject of this book.

REFERENCES

1. Connor, J.M., Schiek, W.A., Food, P. (1997). *An Industrial Powerhouse in Transition*, John Wiley & Sons, Ltd, New York, pp. 8–22.
2. Angulo, F.J.J., Getz, J., Taylor, J.P., Hendricks, K.A., Hatheway, C.L., Barth, S.S., Solomon, H.S., Larson, A.E., Johnson, E.A., Nickey, L.N., Reis, A.A. (1998). A large outbreak of botulism: the hazardous baked potato. *Journal of Infectious Diseases*, *178*, 172–177.
3. Doan, C.H., Davidson, P.M. (2000). Microbiology of potatoes and potato products: a review. *Journal of Food Protection*, *63*, 668–683.
4. Peck, M.W. (2006). *Clostridium botulinum* and the safety of minimally heated, chilled foods: an emerging issue? *Journal of Applied Microbiology*, *101*, 556–570.
5. Pariza, M.W. (1997) CAST Issue Paper Number 8, November 1997, Examination of Dietary Recommendations for Salt-Cured, Smoked, and Nitrite-Preserved Foods.

6. Lund, B.M., Baird-Parker, A.C., Gould, G.W. (2000). *The Microbiological Safety and Quality of Foods* (eds B.M. Lund, A.C. Baird-Parker, G.W. Gould), Aspen Publication.
7. ICMSF (International Commission of Microbiological Specifications for Foods) (2005). *Micro-Organisms in Foods Volume 7 "Microbiological Testing in Food Safety Management"*, Springer Verlag.
8. Watson, D.H. (ed.) (1993). *Safety of Chemicals in Food: Chemical Contaminants*, Ellis Horwood, New York.
9. D'Mello, J.P.F. (ed.) (2003). *Food Safety: Contaminants and Toxins*, CABI Publishing.
10. Hoover, G. (2008). High pressure processing of foods. *New Food*, *1*, 25–27.
11. http://www.highqrte.eu/ (accessed 12 March 2008).
12. Final Leaflet (HEATOX Project) (2007). http://www.slv.se/upload/heatox/documents/D62_final_project_leaflet_.pdf (accessed 16 January 2008). http://www.slv.se/upload/heatox/documents/Heatox_Final%20_report.pdf (accessed 14 September 2008).
13. Theurillat, V., Leloup, V., Liardon, R., Heijmans, R., Bussmann, P. (2007). Impact of roasting conditions on acrylamide formation in coffee, in 21st ASIC Colloquium.
14. Maillard, L.C. (1912). Action des acides amine sur les sucres: formation des melanoidines par voie methodique. *Compte-rendu de l'Académie des Sciences*, *154*, 66–68.
15. Mottram, D.S., Low, M.Y., Elmore, J.S. (2006). The Maillard reaction and its role in the formation of acrylamide and other potentially hazardous compounds in foods, in *Acrylamide and Other Hazardous Compounds in Heat-Treated Foods* (eds K. Skog, J. Alexander), CRC Press, Boca Raton, FL, pp. 1–22.
16. Tareke, E.P., Rydberg, P., Karsson, P., Ericksson, S., Törnqvist, M. (2002). Analysis of acrylamide, a carcinogen formed in heated foodstuffs. *Journal of Agriculture and Food Chemistry*, *50*, 4994–5006.
17. Friedman, M. (2003). Chemistry, biochemistry and safety of acrylamide. A review. *Journal of Agriculture and Food Chemistry*, *51*, 4504–4526.
18. Skog, K., Alexander, J. (2006). *Acrylamide and Other Hazardous Compounds in Heat-Treated Foods*, CRC Press, Boca Raton, FL.
19. Assessed Compounds in Maillard and Lipid Reactions. 2007. http://www.slv.se/templates/SLV_Page.aspx?id=20211&epslanguage=EN-GB (accessed 16 January 2008).
20. Cronin, M.T.D., Livingstone, D. (2004). *Predicting Chemical Toxicity and Fate*, CRC Press.
21. Sanderson, D.M., Earnshaw, C.G. (1991). Computer prediction of possible toxic action from chemical structure; the DEREK system. *Human & Experimental Toxicology*, *10*, 261–273.
22. Gold, L.S., Manley, N.B., Slone, T.H., Rohrbach, L., Garfinkel, G.B. (2005). Supplement to the carcinogenic potency database (CPDB): results of animal bioassays in the general literature through 1997 and by the National Toxicology Program in 1997–1998. *Toxicological Sciences*, *85*, 747–808.
23. European Commission (2004). Report of experts participating in Task 3.2.9. Collection and collation of data on levels of 3-monochloropropanediol (3-MCPD) and

related substances in foodstuffs, June 2004. Brussels: Directorate-General Health and Consumer Protection.

24. Fresenius Conference, Köln (2004). Presentation Dr. A.J. Baars, RIVM.
25. O'Brien, J., Renwick, A.G., Constable, A., Dybing, E., Müller, D.J.G., Schlatter, J., Slob, W., Tueting, W., van Benthem, J., Williams, G.M., Wolfreys, A. (2006). Approaches to the risk assessment of genotoxic carcinogens in food: a critical appraisal. *Food and Chemical Toxicology*, *44*, 1613–1635.
26. WHO Joint FAO/WHO Expert Committee (2005). WHO Joint FAO/WHO Expert Committee on Food Additives sixty-fourth meeting, Rome, 8–17 February 2005. http://www.who.int/ipcs/food/jecfa/summaries/summary_report_64_final.pdf (accessed 16 January 2008).
27. Tritscher, A. (2004). Human health risk assessment of processing-related compounds in food. *Toxicology Letters*, *149*, 177–186.
28. Kroes, R., Kleiner, J., Renwick, A. (2005). The threshold of toxicological concern concept in risk assessment. *Toxicological Sciences*, *86*, 226–230.
29. Barlow, S. (2005). Threshold of Toxicological Concern (TTC)—A Tool for Accessing Substances of Unknown Toxicity Present at Low Levels in the Diet. ILSI Europe Concise Monograph Series, Brussels, Belgium.
30. Institute for Food Safety and Applied Nutrition l (JIFSAN) (2007). Tools for Prioritizing Food Safety Concerns Workshop, 4–6 June 2007, University of Maryland, Greenbelt, Maryland. http://www.jifsan.umd.edu/Tools/Workshop%20Summary.pdf (accessed 16 January 2008).
31. Summary Report EFSA Scientific Colloquium 6, 13–14 July 2006, Tabiano, Italy.
32. http://europe.ilsi.org/activities/ecprojects/BRAFO/ (accessed 3 March 2008).
33. Seal, C.J., de Mul, A., Eisenbrand, G., Haverkort, A.J., Franke, K., Lalljie, S.P.D., Mykkänen, H., Reimerdes, E., Scholz, G., Somoza, V., Tuijtelaars, S., van Boekel, M., van Klaveren, J., Wilcockson S.J., and Wilms L. (2008). Risk–Benefit Considerations of Mitigation Measures on Acrylamide Content of Foods—A Case Study on Potatoes, Cereals and Coffee. *British Journal of Nutrition*, *99* (S2), S1–S46.
34. Eisenbrand, G. (ed.) (2007). *Symposium Proceedings of the DFG: Thermal Processing of Food: Potential Health Benefits and Risks*, Wiley-VCH Verlag GmbH, Weinheim.
35. UK Food Standards Agency Research Programme on Food Contaminants: Process Contaminants. Stakeholder Meeting, 19 April 2005, London, UK.
36. Stadler, R., Anklam, E. (2007). Update on the progress in acrylamide and furan research. *Food Additives and Contaminants*, *S1*, 1–2.
37. CIAA (2007). The CIAA "Acrylamide Toolbox". http://www.ciaa.eu/documents/brochures/toolbox%20rev11%20nov%202007final.pdf (accessed 8 March 2008).
38. http://www.safeconsortium.org/index.asp and http://cordis.europa.eu/coordination/era-net.htm (accessed 3 March 2008).
39. http://etp.ciaa.be/asp/home/welcome.asp (accessed 3 March 2008).
40. http://www.cost.esf.org/index.php?id=182&action_number=927 (accessed 12 March 2008).

2

THERMAL TREATMENT

2.1

ACRYLAMIDE

Craig Mills,[1] Donald S. Mottram,[2] and Bronislaw L. Wedzicha[3]

[1] *UK Food Standards Agency, 125 Kingsway, London WC2B 6NH, UK*
[2] *University of Reading, Department of Food Biosciences, Whiteknights, Reading RG6 6AP, UK*
[3] *University of Leeds, Procter Department of Food Science, Leeds LS2 9JT, UK*

2.1.1 INTRODUCTION

Acrylamide is a reactive unsaturated amide. As an α,β-unsaturated carbonyl it is electrophilic and thus can react with nucleophilic groups such as amines, carboxylates, and thiols that are commonly found on biological molecules such as DNA. *In vivo*, acrylamide undergoes oxidative metabolism to form glycidamide, which is also electrophilic and hence can also react with biological molecules.

Acrylamide has been used as an industrial chemical since the 1950s. It can be readily polymerized to form polyacrylamides, which have found a range of uses such as strengthening aids during the manufacture of paper, as flocculants during water treatment, as soil conditioning agents, and as grouting agents for the construction of dam foundations, tunnels and sewers. It was the latter use that led to the discovery of acrylamide in foods.

During the construction of a railway tunnel in Sweden, a large quantity of an acrylamide polymer was used as a sealant; some of this sealant entered a nearby stream leading to contamination of the local ground and water systems. Dead fish were found downstream of the tunnel as well as paralyzed cows from herds that drank from the contaminated stream; tunnel workers also reported neurotoxic symptoms (1–3). Researchers investigating the effects of the grouting leakage made some interesting discoveries: tunnel workers were

Process-Induced Food Toxicants: Occurrence, Formation, Mitigation, and Health Risks,
Edited by Richard H. Stadler and David R. Lineback

Acrylamide: At a Glance	
Historical	Acrylamide has been used as an industrial chemical since the 1950s in several applications (e.g., flocculant during water treatment, as grouting agent in tunnel construction). In April 2002, acrylamide was reported for the first time in a wide range of foods that are cooked (baked, fried, grilled), but not in boiled foods.
Analysis	Almost all methods use either GC/MS or LC/MS for separation and analyte identification. Methods today can be considered reliable for all pertinent foods down to the low μg/kg level.
Occurrence in Food	Acrylamide is found in a wide range of heat-treated foods, prepared commercially or cooked in the home, including bread, crisp bread, bakery wares, breakfast cereals, potato products (crisps, French fries), chocolate and coffee.
Main Formation Pathways	The Maillard reaction is the major route of acrylamide formation and there is conclusive evidence that the free amino acid asparagine furnishes the backbone of the acrylamide molecule.
Mitigation in Food	The Food & Drink Federation of the EU (CIAA) have published a catalogue of measures in the different food categories (potato products, cereal products, coffee), which are continuously updated as the science progresses.
Dietary Intake	Several exposure estimates have been published by different National Health Authorities ranging from 0.3 to 3.2 μg/kg bw/day. Considering the current data sets, an average mean intake is ca. 0.4 μg/kg bw/day and an average intake for a high level consumer is estimated to be ca. 1 μg/kg bw/day.
Health Risks	Acrylamide is classified by IARC as a probable human carcinogen (Group 2A). Glycidamide, its main metabolite (epoxide), is thought to be responsible for the genotoxicity of acrylamide. JECFA calculated a Benchmark Dose Lower Confidence Interval (BMDL) for mammary tumor formation from rodent studies and then determined the Margin of Exposure (MoE). For mean and high-level consumers, MoEs of 300 and 75, respectively, were calculated. These MoEs are considered to be low for a compound that is genotoxic and carcinogenic.
Regulatory Information or Industry Standards	Regulatory bodies have so far not opted to set maximum limits for acrylamide. Industry has implemented voluntary measures to reduce the amount of acrylamide in foods. Many authorities have chosen to work closely with industry, focusing on effective mitigation strategies and following-up that these are also being applied.
Knowledge Gaps	More in-depth understanding of the toxicology of acrylamide is needed. Several studies investigating long-term carcinogenicity and developmental neurotoxicity in rodents are expected to be completed by end 2008.
Useful Information (web) & Dietary Recommendations	www.ciaa.be http://www.slv.se/upload/heatox/documents/Heatox_Final%20_report.pdf http://acrylamide-food.org/ Consumers should follow on-pack advice when preparing commercial products. Avoid over-heating foods, i.e., bake or toast to a golden yellow color.

found to have increased amounts of acrylamide–hemoglobin adducts (4), as were the poisoned cattle and fish (1). However, following a series of further investigations, it became apparent that unexpected amounts of acrylamide–hemoglobin adducts could be found in unexposed people, living outside the contaminated area (5). The finding that adduct concentrations in wild animals and cattle potentially exposed to the leak were lower than the background amounts of the adduct in unexposed humans led some researchers to examine the possibility that acrylamide exposure in humans resulted from another source: cooked foods.

The results of an experiment examining the effects of feeding rats a diet of fried standard animal feed showed that rats consuming the fried feed had higher amounts of the acrylamide–hemoglobin adduct (6). In April 2002, the Swedish National Food Administration reported concentrations of acrylamide in a variety of fried and baked foods (7). This report was quickly followed by further publications, reporting that up to mg/kg quantities of acrylamide could be formed in carbohydrate-rich foods during high-temperature cooking (8–10). These findings were rapidly confirmed by other researchers; so began the international efforts to understand, and ultimately reduce, the risk of acrylamide contamination of foods.

2.1.2 OCCURRENCE IN FOODS

Since the discovery of acrylamide in cooked foods, the scientific community has made extensive efforts to collect data on the concentrations of acrylamide found in foods. Several large databases of occurrence data have been compiled. European Union Member States, together with the European food industry, have produced a large database of acrylamide concentrations in foods (11). This live database (maintained by the Institute for Reference Materials and Measurements) represents the largest dataset of the occurrence of acrylamide, with concentrations of acrylamide being reported in a wide range of foods originating from across Europe. Acrylamide concentrations in over 7000 samples of different foods are reported. In 2002, the United States Food and Drugs Administration (US FDA) announced its action plan for acrylamide in food (12). One of the US FDA's aims was to monitor amounts of acrylamide in foods. Occurrence data were collected on an annual basis for a range of foods between 2002 and 2006. Other datasets include the World Health Organization's Summary Information and Global Health Trends database for acrylamide (13) and a number of smaller surveys by government organizations and independent researchers.

Acrylamide has been detected in a wide range of heat-treated foods. It is found in both foods processed by manufacturers and foods that are cooked in the home. Generally, acrylamide is found in carbohydrate-rich foods that have been cooked above 120 °C; however, there are exceptions (see Table 2.1.1). It is interesting to note that studies have shown acrylamide is not formed in

TABLE 2.1.1 Summary of reported concentrations of acrylamide in foods.

Food group	Product	Acrylamide ($\mu g\,kg^{-1}$)	
		Minimum	Maximum
Potatoes	Potato crisps[a]	117	4215
	Chips/French fries[b]	59	5200
	Potato fritters/rösti potatoes	42	2779
	Potatoes (raw)	<10	<50
Cereal products	Bakery products and biscuits	18	3324
	Gingerbread	<10	7834
	Bread	<10	397
	Bread (toast)	25	1430
	Breakfast cereals	<10	1649
	Crisp bread	<10	2838
Fruits and vegetables	Olives	<10	1925
	Bottled prune juice	53	267
Cocoa-based products	Chocolate products	<2	826
	Cocoa powder[c]	<10	909
Beverages	Roasted coffee[d]	45	935
	Coffee substitute	80	5399
	Coffee extract/powder	87	1188
	Roasted tea	<9	567
	Beer	<6	<30
Infant foods	Infant biscuits/rusks	<10	1060
	Jarred/canned baby foods	<10	121

[a]Potato snack product that is thinly sliced and fried.
[b]Potato products that are more thickly sliced.
[c]Cocoa powder for baking.
[d]Analyzed as sold.

foods that have been boiled or microwaved (5, 6). Table 2.1.1 lists some of the foods that have been found to contain acrylamide; it is clear that acrylamide can be formed in a wide range of foods, including dietary staples such as potatoes, cereals, and their products. The wide occurrence of acrylamide in foods presents a problem (that will be discussed further later): how can intake of acrylamide be minimized when it occurs in nutritionally important foods?

One of the intriguing characteristics of acrylamide contamination of foods is the variability in acrylamide concentrations; it is not unusual to find a large variation in the amounts of acrylamide found in samples of the same products, and even between samples originating from the same batch. Figure 2.1.1 shows the variability in acrylamide concentrations in different samples of French fries. This variability becomes important when considering dietary exposure to acrylamide (see Section 2.1.6).

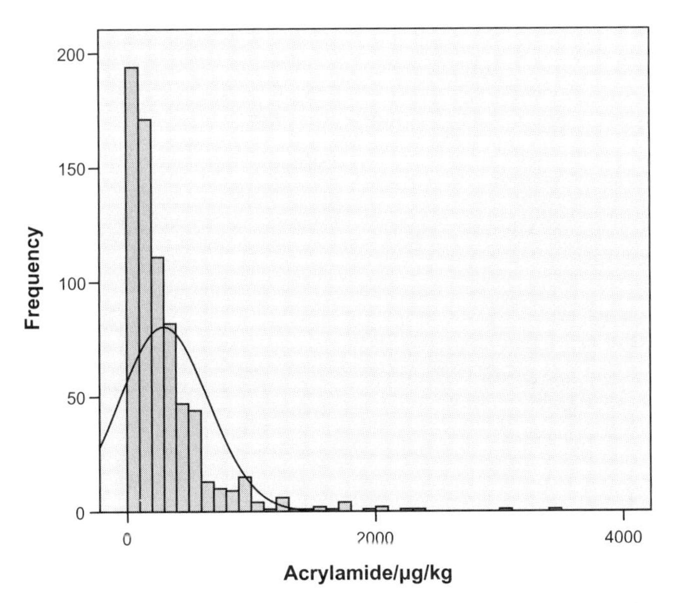

Figure 2.1.1 Concentrations of acrylamide in French fries (data taken from the European Union acrylamide monitoring database).

2.1.3 MECHANISM OF FORMATION

Shortly after acrylamide was first reported in fried and baked foods (5), the Maillard reaction was proposed as the major route for acrylamide formation involving the thermal degradation of free asparagine in the presence of sugars (5, 14, 15). Experiments using ^{13}C and ^{15}N-labeled asparagine confirmed that the carbon skeleton of acrylamide and the nitrogen of the amide group derived from asparagine (16).

The Maillard reaction is a complex reaction between amino compounds (principally amino acids) and reducing sugars, which provides much of the flavor and color characteristics of heated foods (see References 17 and 18). The reaction has implications in other areas of the food industry, including the deterioration of food during processing and storage as well as the protective effect of the antioxidant properties of some Maillard reaction products. Thus, acrylamide formation in heated foods is closely linked to the generation of many of the essential characteristics of cooked foods.

The initial step in the Maillard reaction is the formation of a Schiff base from the reaction of the amino nitrogen on the amino acid with the carbonyl group of a reducing sugar. This Schiff base can rearrange to form an Amadori compound, whose subsequent degradation yields important intermediates for the formation of flavor and color compounds (Fig. 2.1.2). These intermediates include dicarbonyls, hydroxycarbonyls, and furans; further interaction with amino acids yields a wide range of oxygen, nitrogen, and sulfur heterocycles

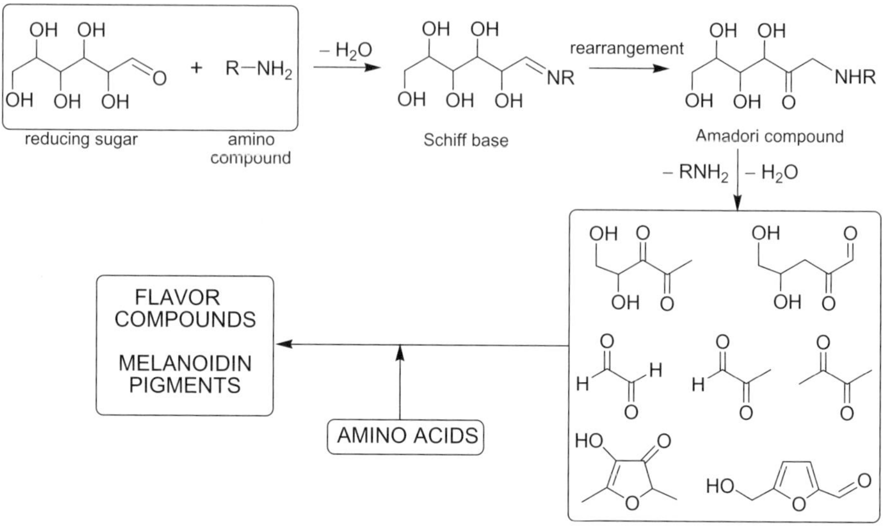

Figure 2.1.2 Simplified scheme showing the formation of flavor and color in the Maillard reaction.

whose nature depends on the composition of the amino acid and sugar composition of the food (19, 20).

As well as providing conclusive evidence that asparagine is the amino acid precursor for acrylamide, Zyzak *et al.* (16) also were able to show the presence of 3-aminopropionamide in heated glucose–asparagine model systems using liquid chromatograph–mass spectrometry (LC-MS). A general reaction scheme for acrylamide formation from asparagine is shown in Fig. 2.1.3. It involves the formation of a Schiff base from the reaction of a carbonyl compound with asparagine, followed by decarboxylation to give an unstable intermediate. Rearrangement of the Schiff base to an Amadori compound is not a prerequisite of this mechanism for acrylamide formation. Hydrolysis of this decarboxylated Schiff base gives 3-aminopropionamide, which yields acrylamide on elimination of ammonia. The ready thermal degradation of 3-aminopropionamide to acrylamide under aqueous or low-water conditions at temperatures between 100 and 180 °C confirmed that it is a very effective precursor of acrylamide (24) and it has been found in heated foods (25, 26). An alternative pathway to acrylamide would be elimination of an imine from the decarboxylated Schiff base.

The mechanism for acrylamide formation via a Schiff base is essentially a Strecker degradation. In food, such reactions explain the formation of aldehydes (Strecker aldehydes) from amino acids by decarboxylation and deamination. α-Dicarbonyls, such as pyruvaldehyde or butanedione, are usually considered the reagents that bring about Strecker degradation, but hydroxy carbonyls, reducing sugars, and conjugated enals could also act as Strecker reagents. The Strecker aldehyde of asparagine (3-oxopropanamide) has not been found in any asparagine–sugar model systems, probably because of its instability.

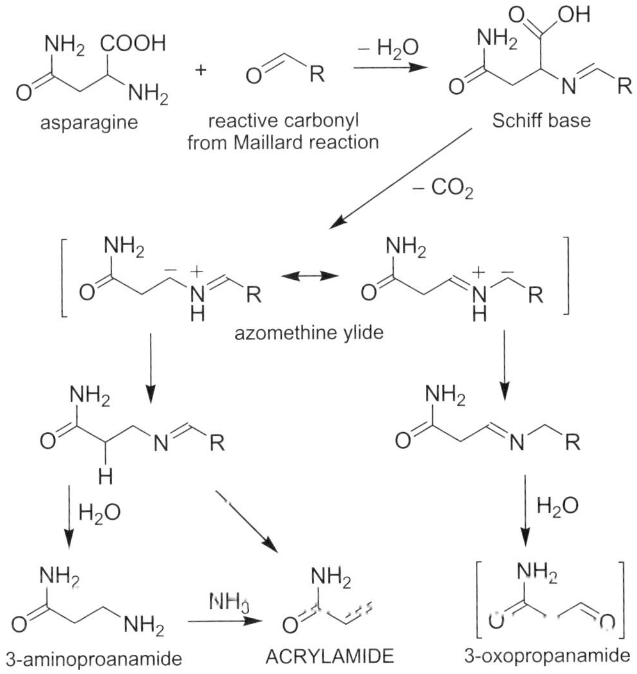

Figure 2.1.3 Proposed mechanism for the formation of acrylamide (21, 22, 23).

The detailed mechanism of each step in the formation of acrylamide may depend on the species involved (i.e., the nature of the carbonyl compound) and on conditions of temperature, water content, and pH. The mechanism has been studied in some detail (21, 22, 27, 28). The initial Maillard intermediate *N*-(D-glucos-1-yl)-L-asparagine, formed from the addition of asparagine to glucose prior to dehydration to the Schiff base, has been shown to release significant quantities of acrylamide upon heating at 180 °C (21). In simple model systems containing only asparagine and a reducing sugar, the Schiff base of asparagine is probably a major precursor of acrylamide. It has been suggested that the hydroxyl group in the β-position to the nitrogen atom will favor decomposition of the Schiff base (28, 29). However, the Maillard reaction produces other reactive carbonyls (Fig. 2.1.2) all of which are capable of forming Schiff bases with asparagine and thus producing acrylamide. Glyoxal, methylglyoxal, and butanedione have all been isolated from asparagine–glucose model systems (30, 31). In heated foods, other free amino acids will react with sugars thus increasing the pool of reactive carbonyl intermediates; consequently the asparagine–sugar Schiff base may assume less importance as the route to acrylamide.

The effectiveness of different carbohydrate moieties in forming acrylamide has been investigated by a number of research groups. Zyzak *et al.* (16) investigated the ability of different carbonyls to generate acrylamide in a potato snack model system and found that a variety of carbohydrate sources could generate acrylamide from asparagine, including glucose, 2-deoxyglucose, ribose, glyceraldehyde, glyoxal, and decanal. Schieberle *et al.* (27) heated equimolar mixtures of asparagine and five different monosaccharides and two disaccharides at 170 °C for 30 min in closed glass vials in the presence of 10% water. Yields of acrylamide ranged from 0.8 to 1.3 mol%, with glucose being the most effective. It is interesting to note that sucrose, a nonreducing sugar, produced almost as much acrylamide as some of the reducing sugars did. This finding suggests that during thermal processing of foods, sucrose may undergo hydrolysis to glucose and fructose. Stadler *et al.* (15) also noted that glucose, fructose, galactose, lactose, and sucrose all gave comparable yields of acrylamide when heated with asparagine at 180 °C. Other carbonyls have been evaluated for their effectiveness in producing acrylamide. Stadler and coworkers (21, 29) compared dicarbonyls and hydroxycarbonyls with glucose, using model systems containing equimolar quantities of carbonyl and asparagine heated, in the presence of water, in sealed tubes at 180 °C for 5 min. The highest yields of acrylamide were for hydroxycarbonyls, followed by glucose, with α-dicarbonyls, such as butanedione and glyoxal, giving yields much lower than that of glucose. Glyoxal (ethanedial) is a very reactive dicarbonyl and these results contrast markedly with a recent report by Amrein *et al.* (30), which showed 300-fold higher yields of acrylamide from asparagine–glyoxal model systems than from similar models with glucose or fructose. Other recent work has confirmed that glyoxal is very reactive toward asparagine in producing acrylamide (31).

2.1.4 ANALYSIS OF ACRYLAMIDE IN FOODS

Products of the Maillard reaction that contribute to color and flavor in both model systems and foods have been studied for several decades and, therefore, it is perhaps surprising that it was not until 2002 that acrylamide, a Maillard reaction product, was found in heated foods. However, it has a low molecular weight (71 daltons) and is highly soluble in water. Analytical methods for Maillard reaction products as flavor compounds have utilized their volatility while the nonvolatile products studied have had higher molecular weights. Such approaches are not likely to isolate acrylamide.

Methods for the analysis of acrylamide in water supplies and in crops had been established well before acrylamide was discovered in heated foods (32, 33). The methods involved bromination to give a dibromo-derivative and separation by gas chromatography.

Since acrylamide is very soluble in water, it can be easily separated from the food using aqueous extraction. Water can be used alone or in conjunction with methanol. In some foods the starch gelatinizes in water, making the aqueous layer difficult to separate from the substrate; the use of aqueous methanol (25%) overcomes this problem (34). Polar organic solvents, such as acetonitrile, ethanol, or propanol have also been used (35). All methods use an internal standard, added at the start of the analysis, for quantification. The most commonly used standard is 1,2,3-$^{13}C_3$-acrylamide although some methods have used methacrylamide.

Almost all published analytical methods for acrylamide use either gas chromatography–mass spectrometry (GC-MS) or LC-MS for the separation and identification of acrylamide. Direct injection of acrylamide onto a GC column is not widely used because the compound does not give a characteristic mass spectrum, acrylamide can be unstable in the heated injection port, and substantial extract cleanup is required. Most GC-MS methods involve the derivatization of acrylamide to 2,3-dibromopropanamide. This derivatization is easily achieved by treating an aqueous extract of the food with a saturated solution of bromine in water overnight and then extracting the 2,3-dibromopropanamide into ethyl acetate (33). LC-MS allows analysis of acrylamide without prior derivatization. However, LC separation does not have as high a resolving power as GC so careful cleanup of the extract is necessary prior to chromatographic separation. Unequivocal identification of the acrylamide chromatographic peak is improved using LC-MS-MS, which provides a protonated molecular ion at *m/z* 72 and a collision-induced ion at *m/z* 55 (36).

A comprehensive review of advantages and disadvantages of the methods available for the analysis of acrylamide in food products has been published by Castle (35).

2.1.5 MITIGATION STRATEGIES

The discovery of acrylamide in heated foods caused considerable concern to legislators and food manufacturers and, consequently, considerable efforts have been made by the food industry and academia to provide strategies by which acrylamide could be reduced in food products. Several reviews of the published research have appeared including those by Taeymans *et al.* (37), Friedman and Levin (38), Zhang and Zhang (39), Foot *et al.* (40), Guenther *et al.* (41), and Konings *et al.* (42). Food manufacturers in Europe have worked with researchers through the Confederation of the Food and Drink Industries of the EU (CIAA) to produce a series of guidelines (CIAA Acrylamide Toolbox) for reduction of acrylamide in a range of foods (43). The first consideration in reducing acrylamide is to examine those factors that are known to affect the Maillard reaction. These include reactant concentrations (i.e., the reducing sugar and free amino acid content of food), time–temperature conditions during processing, moisture levels, pH, and the presence of additives.

Reactant levels are influenced not only by the type of food but also by cultivar, soil conditions, harvesting times, and storage conditions of the raw food. Selection of crops with low concentrations of asparagine and reducing sugars would be an obvious strategy to mitigate acrylamide (44–47). However, since free asparagine is essential for plant growth and biosynthesis of sugars is part of normal plant metabolism, such an approach must have limitations. In potatoes the ratio of reducing sugars to free amino acids is relatively low compared with cereals and the sugar concentration may be a limiting factor in acrylamide formation (40). This is not the case with cereals, where the molar concentrations of sugars are much higher than that of the free amino acids and, therefore, asparagine concentrations in cereals will have a direct impact on the acrylamide-forming potential (48).

Fertilizers have been shown to affect the amounts of asparagine and sugars in potatoes and cereals. Addition of nitrogen increased the amounts of asparagine in wheat and rye flours (49), while the absence of sulfur caused a dramatic increase in asparagine in wheat (26, 50) but the effect in potatoes depended on variety (51). It has been known for many years that sugar concentrations increase in potatoes during storage and this is exacerbated during low-temperature storage (ca. 4 °C). While storage at 12 °C may prevent large increases in sugar content at this temperature, the use of chemical sprout suppressants is necessary for long-term storage.

Proposals for lowering acrylamide formation during food processing include reducing cooking times and temperatures (52, 53), and lowering the pH (54, 55) but these will also reduce color and flavor alongside acrylamide. These approaches are discussed for a wide range of different products in the CIAA Acrylamide Toolbox. Kinetic studies have clearly shown relationships between acrylamide and time, and temperature and moisture content, in keeping with the known mechanisms for the Maillard reaction (23, 56). The reaction is also known to be pH dependent; the initial stages of the Maillard reaction involve the reaction of an amino group and the carbonyl of a reducing sugar and as the pH is lowered, amino groups are protonated thus becoming unavailable for the reaction. Prewashes of raw materials, especially potatoes, can lead to reduction in the concentrations of the reactants available for acrylamide production. The use of the enzyme asparaginase to convert asparagine to aspartic acid has been shown to reduce acrylamide in processed foods (16, 57), and this has been the subject of a number of patents. Two products are now commercially available: Acrylaway (Novozymes, Denmark) and Preventase (DSM Food Specialities, the Netherlands). Both are asparaginases, but they stem from different production strains: Acrylaway from *Aspergillus oryzae* and Preventase from *Aspergillus niger*. The process is being applied to biscuits at commercial scale. It could be applicable to other food products that involve dough or batter processes, such as bread, breakfast cereals, and cakes (42) but would be more difficult to apply to fried potato products.

Addition of specific ingredients to reduce acrylamide has also been investigated. These include other amino acids, such as glycine, which will compete with asparagine for available sugars (58–60). Natural antioxidants, including extracts of bamboo leaves, which are high in polyphenolics, have been shown to reduce amounts of acrylamide in certain food products and in model systems (61, 62). Soaking raw materials (e.g., potato slices) in aqueous solutions containing divalent cations, such as Ca^{2+}, has also been reported to reduce acrylamide in the processed products (63, 64). The effect seems to be greater than that which is achieved by soaking in water.

Most of the strategies for reducing the quantities of acrylamide in heated potato and cereal products affect the extent to which the Maillard reaction occurs. Since the Maillard reaction is also responsible for the generation of desirable flavors and colors in heated food, these quality attributes could be affected alongside acrylamide reduction (60). Both genetic factors and agronomy, which could affect asparagine accumulation, would appear to offer the best long-term potential for a reduction in acrylamide (48). However, if such factors have a major affect on the other free amino acids present in the raw material, the flavor profile of the heated product could be changed (65).

2.1.6 EXPOSURE

Following the announcement of the presence of acrylamide in foods, a consultation was convened by the Food and Agriculture Organization of the United Nations (FAO) and the World Health Organization (WHO) to review and evaluate existing data on acrylamide, and to provide interim advice to governments, industry, and consumers (66). Several preliminary exposure estimates were presented and the committee concluded that long-term acrylamide exposures would be in the range 0.3–0.8 $\mu g\,kg\,bw^{-1}\,day^{-1}$. The committee stressed that the available data were sparse and that further work should be undertaken to produce more robust exposure estimates, taking into account other dietary sources of acrylamide.

The European Scientific Committee on Food also assessed the implications for food safety posed by acrylamide. The Committee considered research conducted across Europe and recommended that amounts of acrylamide in food should be as low as reasonably achievable (ALARA) and endorsed the recommendations of the FAO and WHO.

Two approaches may be used to assess dietary exposure to a chemical contaminant. The first involves a direct method for estimating exposure: diets can be monitored for a full day, with every food that is consumed during the day being prepared in duplicate and analyzed for the chemical(s) of interest. It is usual to composite the analytical samples so that foods of similar types are homogenized and the concentration of the chemical contaminant measured in individual food groups. The exposure estimates are then derived by combining the actual amount of food eaten for each food group with the respective contaminant concentration detected for those food groups. The duplicate method is very burdensome for the consumer and does not allow any investigation of which individual foods contribute most to dietary intake.

The second approach is an indirect method. Exposure estimates are calculated by combining the concentrations of contaminants in individual foods (occurrence data) with the amounts of those foods that are eaten (consumption data). Occurrence data may be collected from a range of monitoring programs. Consumption data may be collected through food composition surveys at an individual consumer or household level. Alternatively, consumption data may be estimated through food purchase statistics. There are obvious disadvantages to each method of consumption estimation: food purchase statistics will only provide data on the quantities of food that are purchased, they do not allow for food wastage. Food consumption surveys involve food frequency questionnaires—consumers may be asked to keep a record of their consumption or be interviewed and asked to recall their consumption. Dietary recall interviews can be flawed because consumers do not remember precisely what was eaten; similarly, consumers may not always wish to state exactly how much of a so-called "bad" food was consumed! Because our diets are extremely complicated and highly variable, it is possible that some foods may not be consumed during the monitoring period. When comparing exposure estimates made using the indirect method, regard must be paid to the method of collection of consumption data: data collected over a longer monitoring period are more likely to reduce some of the possible error from variation in dietary habits.

Table 2.1.2 summarizes the results of the Swiss Federal Office of Public Health investigation into dietary exposure to acrylamide. A duplicate diet study was used to investigate the acrylamide intake of 27 Swiss adults over a two-day period. Duplicate samples of foods were collected and homogenized to yield composite samples of different meals (breakfast, lunch, dinner, snacks, and coffee). A mean daily intake of 0.277 μg kgbw^{-1} day^{-1} was estimated (67). It is interesting to note that coffee was shown to contribute the most acrylamide to the total intake. On reviewing the results of the duplicate diet study, the Swiss Federal Office of Public Health found that consumption of deep fried, baked, and roast potatoes, and crisps reported was below average. Owing to the relatively high concentrations of acrylamide commonly found in these foods, it was decided to correct the possible underestimation. By taking into account import/export data, the estimated intake was recalculated as 0.46 μg kgbw^{-1} day^{-1}; however, even at this higher level, coffee still contributed 22% of the daily intake.

TABLE 2.1.2 Contribution of different meals and beverages to the daily intake of acrylamide by Swiss adults.[a]

Meal category	Intake, μg kgbw^{-1} day^{-1}	Exposure, %
Breakfast	0.022	8
Lunch	0.057	21
Dinner	0.061	22
Snacks	0.037	13
Coffee	0.100	36
Total	*0.277*	*100*

[a]Reference 67.

TABLE 2.1.3 Summary of some exposure estimates for acrylamide.

		Estimated dietary intake, μg kgbw^{-1} day^{-1}	
Organization, country	Population/Sex (age)	Mean	High-level
BfR, Germany (68)	All, 15–18	1.1	3.2[a]
NFAS, Norway (69)	Males	0.49	1.04[b]
	Females	0.46	0.86[b]
	Males (13)	0.52	1.35[b]
	Females (13)	0.49	1.2[b]
AFSSA, France (70)	All (>15)	0.5	0.98[a]
	All (3–14)	1.25	2.54[a]
SNFA, Sweden (71)	All (18–74)	0.45	1.03[a]
WVA, the Netherlands (72)	All (1–97)	0.48	0.60[a]
	All (1–6)	1.04	1.1[a]
	All (7–18)	0.71	0.9[a]
FSA, UK (73)	Males (19–64)	0.4	0.6[c]
	Females (19–64)	0.3	0.6[c]
	All (1.5–4.5)	1.0	1.8[c]
FDA, USA (74)	All (2+)	0.44	0.95[b]
	All (2–5)	1.06	2.33[b]

The UK estimate is for consumers only; other estimates are mostly for the entire population.
[a]95th percentile.
[a]90th percentile.
[a]97.5th percentile.

There have been many estimates of acrylamide intake, some of which are detailed in Table 2.1.3 (68–74). When comparing dietary exposure estimates, it is important to remember their limitations. Exposure may be calculated using deterministic models (actual acrylamide concentrations are combined with consumption data) or by using probabilistic models (here the distribution of acrylamide concentrations is combined with the distribution of consumption data). When using deterministic models to estimate dietary exposure (where a mean, or possibly a maximum, contaminant concentration is assumed), one must be aware of how the distribution of contaminant concentrations may affect the estimated exposure (cf. Fig. 2.1.1 which shows the wide spread of acrylamide concentrations found in samples of French fries recorded in the European Union acrylamide monitoring database).

There are many limitations to exposure estimates. The main limitations focus around data collection and the size of occurrence and consumption datasets. Any estimate will only be valid for those members of the population whose eating habits have been studied. If small datasets for occurrence and consumption data are used, the resulting exposure estimate will not necessarily be representative for people outside of the studied population. Any exposure estimate also makes a number of assumptions concerned with what foods are consumed and how contaminant concentrations might be affected by processing, and a number of other assumptions used to simplify calculations. In short, it is prudent to remember that exposure estimates are always only estimates.

Direct comparison of exposure estimates is not always possible. Differences between estimates can occur owing to different methods being used to calculate exposure (including data collection and estimate calculation) and because different statistical approaches are used (exposure can be calculated for entire populations, or for those parts of the population that actually consume the foods of interest). Such differences must be taken into account before considering how differences between national diets might relate to differences in actual exposures.

Table 2.1.3 shows that acrylamide exposures range from 0.3 to 3.2 μg kgbw^{-1} day^{-1}. There is considerable variation in the estimated exposures. Of course, there is considerable error involved when estimating an average exposure across all populations: dietary habits vary greatly from country to country. In addition, individual country estimates will have been performed in different ways. However, an average mean intake can be taken to be ca. 0.4 μg kgbw^{-1} day^{-1} and an average intake for a high-level consumer to be ca. 1 μg kgbw^{-1} day^{-1}.

Those foods that contribute most to dietary intake will differ from country to country, according to the national diet and the way in which foods are prepared. Generally, the foods that contribute the most to dietary intake of acrylamide are fried potato products (such as French fries and crisps), ready-to-eat breakfast cereals, bread and bakery products, and coffee. Closer inspection of national intake estimates reveals how overall dietary intake of acrylamide varies from region to region. For example, potatoes (and potato products) are estimated to contribute 38% to the total dietary intake of acrylamide for US consumers; contributions from bread (toast and soft bread) and coffee are 9% and 7%, respectively (75). Conversely, the contribution by bread and coffee is higher in Europe. Potatoes (fried, crisps, and French fries) have been estimated to contribute 33% of total intake in Norway, while coffee and bread were shown to contribute 28% and 20.1%, respectively (69). This trend for higher dietary intakes from consumption of coffee and bread was also observed in Sweden (71). However, higher contribution from potatoes was observed in the Netherlands: potatoes (French fries and comparable products, and crisps) were estimated to contribute an astonishing 52% of the total acrylamide intake (entire population, 1–97 years of age); the contributions were found to be even higher for children: 58% for 1-year-olds and 69% for 7- to 18-year-olds (72).

Several studies have been undertaken to investigate the effects of reducing the concentrations of acrylamide in food on dietary intake. In 2004 the US FDA presented work on the potential impact of reducing acrylamide levels in different food groups (76). The 2004 US exposure assessment for the entire population was rerun several times; each rerun assumed that one of the major food groups contributing to acrylamide exposure contained no acrylamide. Their investigations demonstrated that reducing acrylamide levels in just one food group would have little impact on the overall dietary intake of acrylamide

TABLE 2.1.4 Effects of reducing acrylamide levels in different food groups in the United States.[a]

	Population exposure, μg kgbw^{-1} day^{-1}	
	Mean	90th percentile
Original estimate	0.43	0.92
Removal of acrylamide from:		
French fries	0.37	0.78
Snack foods	0.38	0.85
Breakfast cereals	0.38	0.84
Coffee	0.40	0.88

[a]Reference 75.

(see Table 2.1.4). Researchers at the RIKILT Institute of Food Safety in the Netherlands have also investigated the effect of reducing acrylamide levels in certain foods on overall acrylamide intake (77). Several studies have indicated that amounts of acrylamide in potato products can be reduced by frying at lower temperatures. Assuming that cooking potato products at lower temperatures resulted in a decrease of acrylamide levels by 35%, the researchers were able to show that this reduction in acrylamide levels would result in a lowering of overall acrylamide intake in the Netherlands by ca. 13%. Reductions in the levels of other foods were also investigated; however, the resulting decreases in total acrylamide intake were minimal.

Another way to reduce acrylamide intake could be to advise consumers to avoid certain foods. However, many of the foods that contain acrylamide are dietary staples and significantly contribute to nutrient intake. Petersen and Tran (78) looked at the macronutrient and micronutrient contributions of those foods reported to contain acrylamide by the 2003 US FDA acrylamide exposure assessment (79). They were able to show that those foods identified as containing acrylamide contributed 38% of the total daily energy intake and 36% of the total daily intake of fiber. Similarly high contributions were observed for micronutrients, with those foods that contain acrylamide contributing over 30% of the total daily intake of some micronutrients such as iron, selenium, and vitamin C. For these reasons, the advice from the WHO and governments around the world remains that consumers should not make changes to their diet in order to reduce acrylamide intake.

2.1.7 HEALTH RISKS

Acrylamide is rapidly absorbed from the gastrointestinal tract and is widely distributed throughout the body (80, 81); it has been found in fetuses and breast milk (82). It is rapidly metabolized *in vivo* to form glycidamide (83), or can be metabolized by conjugation with glutathione (84). Glycidamide is a reactive epoxide that is thought to be responsible for the genotoxic effects of acrylamide. Both acrylamide and glycidamide are electrophilic and so can react via Michael addition to form adducts on DNA (glycidamide–purine base adducts) and proteins (both acrylamide and glycidamide will bind covalently to amino acids). Acrylamide and glycidamide adducts at the *N*-terminal valine residue of hemoglobin are not toxic, but have been demonstrated to be useful biomarkers of exposure to acrylamide (66, 85). Once in the body, acrylamide and its metabolites are rapidly excreted, in urine, from the body (86).

Single oral doses of acrylamide only produce acute toxic effects in animals at doses greater than $100\,mg\,kgbw^{-1}\,day^{-1}$, with lethal doses reported to be higher than $150\,mg\,kgbw^{-1}\,day^{-1}$. Repeated low-dose exposure to acrylamide in rats has been shown also to cause degenerative peripheral nerve damage (87). Prolonged exposure to acrylamide has been shown to induce degeneration of nerve terminals in brain areas that are critical for learning, memory, and other cognitive functions. For nervous system effects, the Joint FAO/WHO Expert Committee on Food Additives (JECFA) estimated a no observed adverse effect level (NOAEL) of $0.2\,mg\,kgbw^{-1}\,day^{-1}$ for morphological changes in nerves (88). Reproductive and developmental effects have also been demonstrated, with a NOAEL of $2.0\,mg\,kgbw^{-1}\,day^{-1}$ estimated for effects on the testes and male fertility in rodents (89–91).

Acrylamide is both clastogenic (can cause damage to chromosomes) and mutagenic (can cause genetic mutations) in mammalian cells *in vitro* and *in vivo* (92). Metabolism of acrylamide to glycidamide appears to be a prerequisite for the genotoxicity observed *in vitro* for acrylamide. Acrylamide has been shown to induce tumors in rats that have been given drinking water containing acrylamide (93, 94). Many of the tumor sites identified in these studies are hormone responsive, suggesting that acrylamide may act via a neuroendocrine mechanism; however, during its evaluation of acrylamide, JECFA noted that such a mechanism is not proven and as such a genotoxic mode of action could not be excluded.

Evidence in humans suggests that acrylamide acts principally on the nervous system. Occupational exposure to acrylamide has not been linked to overall cancer mortality; studies that have investigated this link are limited in size, and potential cofounders such as tobacco smoking and dietary intake of acrylamide were not considered.

During its review of acrylamide, JECFA also assessed the risk of cancer from the consumption of acrylamide-contaminated food (95). JECFA calculated a benchmark dose lower limit (BMDL) for mammary tumor formation from rodent studies and then calculated the margin of exposure (MoE) between the BDML and human acrylamide intake. The Committee decided to use a mean acrylamide intake for the total population of $1\,\mu g\,kgbw^{-1}\,day^{-1}$ and a high-level consumer acrylamide intake of $4\,\mu g\,kgbw^{-1}\,day^{-1}$ to calculate MoEs of 300 and 75 for mean and high-level consumers, respectively. These MoEs were considered to be low for a compound that is genotoxic and carcinogenic. MoEs were also calculated for morphological nerve changes using a

NOAEL of 0.2 mg $kgbw^{-1}$ day^{-1} calculated for rats; MoEs of 200 and 50 were calculated for mean and high-level consumers. JECFA calculated MoEs of 2000 and 500 for mean and high-level consumers for reproductive, developmental, and non-neoplastic effects. Based on these MoEs, it was concluded that adverse neurological, reproductive, and developmental effects are unlikely for the average consumer, but that morphological changes in nerves cannot be excluded for individuals with very high intake of acrylamide.

To date, a number of epidemiology studies have investigated possible associations between dietary intake of acrylamide and the incidence of several types of cancer in humans (96–100). The majority of these studies have failed to show an association; however, a recent study in Dutch women showed an increase in ovarian cancer incidence that reached statistical significance (101). Many of the negative studies lack sufficient power to detect effects on cancer incidence, partly due to the relatively small difference between high- and low-level intake. The uncertainty surrounding dietary estimates of acrylamide, due to the large variability between individual foods, is a further limitation in these studies.

A recent epidemiology study in postmenopausal women (102) used hemoglobin adducts in blood taken when the women joined the study. The use of hemoglobin adducts as a biomarker of total exposure to acrylamide removes the uncertainty associated with dietary intake estimations, but only represents a limited snapshot of exposure, which may not reflect the whole duration of cancer development. This study appears to show an increase in incidence of breast cancer associated with acrylamide exposure, although required complex adjustment for cigarette smoking, which is an additional source of acrylamide exposure. Clearly, further work is required.

A number of epidemiology studies are ongoing. The US FDA and associated agencies are carrying out a suite of studies on acrylamide and glycidamide as part of the FDA Action Plan for Acrylamide. Several short-term studies have been completed (80, 103–107); however, the results of long-term studies that will give a clearer picture of the situation are not expected until 2008.

2.1.8 RISK MANAGEMENT

The issue of acrylamide in food is a developing area. Owing to the lack of understanding concerning the mechanisms of formation of acrylamide and the mechanism of toxicological effect, regulators have not opted to set maximum levels for acrylamide in foods. Instead, many government organizations/authorities have opted to focus on effective mitigation strategies and monitoring activities, often working with industry toward solving the problem. In Europe, regulators have worked with researchers and manufacturers to develop the Acrylamide Toolbox (43) and a series of brochures (108) that describe current strategies for reducing acrylamide concentrations in a variety of foods and beverages. In 2002, the US FDA published its Draft Action Plan for Acrylamide in Food (12). This plan outlined how the US FDA planned to tackle the issue of acrylamide in food; it has been updated, with the latest version published in 2004 (109). Other countries have conducted research in order to inform risk assessment and risk management measures. The European Union and its member states have compiled an online database containing information on those research projects being undertaken in the European Union (110). A similar database is also maintained by the Joint Institute for Food Safety and Applied Nutrition (111). Recently, the European Commission issued European member states with a recommendation to monitor acrylamide in a variety of foods known to contain high levels, with the aim of obtaining a clear picture of the current levels of acrylamide across Europe (112). The US FDA has undertaken annual acrylamide monitoring, using the Total Diet Study format. Exposure estimates have also been calculated on a regular basis, although no apparent changes in acrylamide intake over time have been observed (75).

2.1.8.1 HEATOX Project

The European Commission funded the HEATOX Project, a large multidisciplinary project, from 2003 to 2007. The project involved 24 partners from 14 countries and aimed to estimate the health risks associated with hazardous compounds found in heat-treated foods and to establish cooking/processing methods that could minimize the occurrence of these compounds in foods. The project focused mostly on acrylamide, but other heat-generated toxicants such as furan and hydroxymethylfurfural (HMF) were also studied. The project produced outputs in the areas of formation, analysis, hazard characterization, risk assessment, and risk characterization. Some advice was also issued from the project board to European industry and governments. Further details of the HEATOX project are available at www.heatox.org.

2.1.8.2 CIAA Acrylamide Toolbox

The latest version of the CIAA Acrylamide Toolbox (see Section 2.1.5) was published in November 2007 (43). The toolbox was developed over several years and reflects the results of industry cooperation. The document describes a number of mitigation strategies that can be employed to reduce acrylamide levels in a variety of different foods. Primarily aimed to assist individual manufacturers with technical knowledge but limited research and development capability, the Toolbox describes parameters that can be manipulated to control acrylamide levels; these parameters are divided into four groups: agronomy, recipe, processing, and final preparation.

Information for the toolbox was taken from a variety of research papers and presentations from various international meetings, as well as some activities coordinated by various trade associations. CIAA states that the Toolbox is a living document, which should be treated as a catalogue of tried and tested methods for reducing acrylamide, and which will continue to be updated as more information becomes available.

Following the success of the Acrylamide Toolbox, the European Commission, together with European Union member states and the CIAA, has developed a series of brochures (108). These brochures are aimed at smaller businesses and manufacturers with limited technical expertise. Five food sectors are covered by the brochures (French fries, Crisps, Bread, Biscuits, and Cereals), which present the detail of the Acrylamide Toolbox in lay terms; the brochures are published on the Commission's web site in 19 European languages.

2.1.8.3 Codex Alimentarius Acrylamide Code of Practice

The issue of acrylamide in foods has also been under discussion in the Codex Alimentarius. The Codex Alimentarius is an international body that was established in 1963 by the FAO and WHO with the aim of protecting the health of consumers and ensuring fair practices in the food trade. Members of the Codex Committee on Food Additives and Contaminants (now the Codex Committee on Contaminants in Food) produced a discussion paper (113) and have recently presented a first draft of a Code of Practice for the reduction of acrylamide in foods (114). The Codex Code of Practice is being developed in the international arena as a means of disseminating strategies that will help reduce acrylamide concentrations in internationally traded foods and assist those governments who are unable to investigate the issue themselves. The information in the draft Code of Practice consists of established minimization techniques that have been demonstrated to be effective in a commercial setting. Over the next year the draft will be elaborated, in order to take into account recent developments in the use of asparaginase to reduce acrylamide levels in certain foods.

2.1.8.4 Minimization Strategies and Consumer Information

Many national governments and government organizations have issued advice to consumers concerning acrylamide. The US FDA, Health Canada, and the UK Food Standards Agency have advised consumers not to make major dietary changes, but to continue to eat a balanced diet. Health Canada has also made suggestions on how consumers could reduce acrylamide exposure; consumers were advised to eat deep-fried foods and snacks less often, cook French fries at lower temperatures and to a lighter color, to soak potato slices before cooking, and to toast bread to a light color (115).

Germany introduced the "minimization concept" in August 2002 (116). This concept involved establishing signal levels for certain food groups. When monitoring of these food groups shows that the signal levels are being exceeded, manufacturers are requested to reduce acrylamide concentrations in their products.

Following recommendations made by the Swiss Federal Authorities, Swiss food retailers have introduced specific packaging labeling for fresh potatoes, indicating their most appropriate forms of preparation (e.g., roasting, frying, and baking). In Zurich, caterers were trained to reduce acrylamide concentrations in French fries; following the training, participants were invited to send in samples of good-quality French fries to the Cantonal authorities for analysis. The results of the exercise were published and resulted in the catering industry being able to produce low-acrylamide French fries. The Swiss Federal Authorities also produced a leaflet for consumers, describing how to reduce acrylamide formation in French fries and fried potato products.

In California, USA, the Safe Drinking Water and Toxic Enforcement Act of 1986 requires the State of California to maintain a list of substances that are known to cause cancer; acrylamide has appeared on this list since 1990. Under this legislation, any business that knowingly exposes individuals to significant amounts of a substance on the list must provide clear and reasonable warning to those individuals. The State of California has recently prosecuted nine major food companies. The popular fast-food chain KFC has agreed to post warnings in its Californian outlets stating that acrylamide is formed in their potato products during cooking. The company was also fined $341,000. The California Attorney General is also in discussions with other fast-food chains.

The future for regulation of acrylamide in foods is still relatively unclear. Until the expert scientific committees understand the mechanisms of formation and toxicology more clearly, regulators will have difficulty in being able to set appropriate limits for acrylamide. Several US toxicology studies investigating long-term carcinogenicity assays and developmental neurotoxicity assessments are expected to be completed in 2008. Once these results, and the results of other researchers investigating similar effects, are made available, it is expected that JECFA and EFSA will re-evaluate acrylamide. The outcome of this second evaluation will undoubtedly shape the future for acrylamide regulation and risk management.

REFERENCES

1. Godin, A.C., Bengtsson, B., Niskanen, R., Tareke, E., Tornqvist, M., Forslund, K. (2002). Acrylamide and N-methylolacrylamide poisoning in a herd of Charolais crossbreed cattle. *Veterinary Record*, *151*, 724–728.
2. Reynolds, T. (2002). Acrylamide and cancer: tunnel leak in Sweden prompted studies. *Journal of the National Cancer Institute*, *94*, 876–878.
3. Tornqvist, M. (2005). Acrylamide in food: the discovery and its implications, in *Chemistry and Safety of Acrylamide in Food* (eds M. Friedman, D.S. Mottram), Springer, New York, pp. 1–19.
4. Hagmar, L., Tornqvist, M., Nordander, C., Rosen, I., Bruze, M., Kautiainen, A., Magnusson, A.L., Malmberg, B., Aprea, P., Granath, F., Axmon, A. (2001). Health

effects of occupational exposure to acrylamide using hemoglobin adducts as biomarkers of internal dose. *Scandinavian Journal of Work Environment & Health*, *27*, 219–226.

5. Tareke, E., Rydberg, P., Karlsson, P., Eriksson, S., Törnqvist, M. (2002). Analysis of acrylamide, a carcinogen formed in heated foodstuffs. *Journal of Agricultural and Food Chemistry*, *50*, 4998–5006.
6. Tareke, E., Rydberg, P., Karlsson, P., Eriksson, S., Tornqvist, M. (2000). Acrylamide: a cooking carcinogen? *Chemical Research in Toxicology*, *13*, 517–522.
7. Swedish National Food Administration (2002). Analysis of Acrylamide in Food, http://www.slv.se/acrylamide (accessed 11 January 2008).
8. Ahn, J.S., Castle, L., Clarke, D.B., Lloyd, A.S., Philo, M.R., Speck, D.R. (2002). Verification of the findings of acrylamide in heated foods. *Food Additives and Contaminants*, *19*, 1116–1124.
9. Food Standards Agency (2002). Study Confirms Acrylamide in Food, http://www.food.gov.uk/news/newsarchive/2002/may/65268 (accessed 11 January 2008).
10. United States Food and Drug Administration (2002). Survey Data on Acrylamide in Food: Individual Food Products, http://www.cfsan.fda.gov/~dms/acrydata.html (accessed 11 January 2008).
11. European Commission (2006). European Union Acrylamide Monitoring Database, http://irmm.jrc.be/html/activities/acrylamide/database.htm (accessed 11 January 2008).
12. United States Food and Drug Administration (2002). FDA Draft Action Plan for Acrylamide in Food, http://www.cfsan.fda.gov/~dms/acryplan.html (accessed 11 January 2008).
13. World Health Organization (2007). Summary Information and World Health Trends.
14. Mottram, D.S., Wedzicha, B.L., Dodson, A.T. (2002). Acrylamide is formed in the Maillard reaction. *Nature*, *419*, 448–449.
15. Stadler, R.H., Blank, I., Varga, N., Robert, F., Hau, J., Guy, P.A., Robert, M.-C., Riediker, S. (2002). Acrylamide from Maillard reaction products. *Nature*, *419*, 449–450.
16. Zyzak, D.V., Sanders, R.A., Stojanovic, M., Tallmadge, D.H., Eberhart, B.L., Ewald, D.K., Gruber, D.C., Morsch, T.R., Strothers, M.A., Rizzi, G.P., Villagran, M.D. (2003). Acrylamide formation mechanism in heated foods. *Journal of Agricultural and Food Chemistry*, *51*, 4782–4787.
17. Nursten, H.E. (2005). *The Maillard Reaction*, Royal Society of Chemistry, Cambridge, p. 214.
18. Mottram, D.S., Low, M.Y., Elmore, J.S. (2006). The Maillard reaction and its role in the formation of acrylamide and other potentially hazardous compounds in foods, in *Acrylamide and Other Hazardous Compounds in Heat-Treated Foods* (eds K. Skog, J. Alexander), Woodhead Publishing, Cambridge, pp. 3–22.
19. Mottram, D.S. (2007). The Maillard reaction: source of flavour in thermally processed foods, in *Flavours and Fragrances: Chemistry, Bioprocessing and Sustainability* (ed. R.G. Berger), Springer-Verlag, Berlin, pp. 269–284.
20. Mottram, D.S., Mottram, H.R. (2002). An overview of the contribution of sulfur-containing compounds to aroma in cooked foods, in *Heteroatomic Aroma*

Compounds (eds G.A. Reineccius, T.A. Reineccius), American Chemical Society, Washington, DC, pp. 73–92.

21. Stadler, R.H., Robert, F., Riediker, S., Varga, N., Davidek, T., Devaud, S., Goldmann, T., Hau, J., Blank, I. (2004). In-depth mechanistic study on the formation of acrylamide and other vinylogous compounds by the Maillard reaction. *Journal of Agricultural and Food Chemistry*, *52*, 5550–5558.
22. Yaylayan, V.A., Stadler, R.H. (2005). Acrylamide formation in food: a mechanistic perspective. *Journal of AOAC International*, *88*, 262–267.
23. Wedzicha, B.L., Mottram, D.S., Elmore, J.S., Koutsidis, G., Dodson, A.T. (2005). Kinetic models as a route to control acrylamide formation in food, in *Chemistry and Safety of Acrylamide in Food* (eds M. Friedman, D.S. Mottram), Springer, New York, pp. 235–253.
24. Granvogl, M., Jezussek, M., Koehler, P., Schieberle, P. (2004). Quantitation of 3-aminopropionamide in potatoes—a minor but potent precursor in acrylamide formation. *Journal of Agricultural and Food Chemistry*, *52*, 4751–4757.
25. Granvogl, M., Schieberle, P. (2007). Quantification of 3-aminopropionamide in cocoa, coffee and cereal products. *European Food Research and Technology*, *225*, 857–863.
26. Granvogl, M., Wieser, H., Koehler, P., Von Tucher, S., Schieberle, P. (2007). Influence of sulfur fertilization on the amounts of free amino acids in wheat. Correlation with baking properties as well as with 3-aminopropionamide and acrylamide generation during baking. *Journal of Agricultural and Food Chemistry*, *55*, 4271–4277.
27. Schieberle, P., Köhler, P., Granvogl, M. (2005). New aspects on the formation and analysis of acrylamide, in *Chemistry and Safety of Acrylamide in Food* (eds M. Friedman, D.S. Mottram), Springer, New York, pp. 205–222.
28. Yaylayan, V.A., Wnorowski, A., Locas, C.P. (2003). Why asparagine needs carbohydrates to generate acrylamide. *Journal of Agricultural and Food Chemistry*, *51*, 1753–1757.
29. Blank, I., Robert, F., Goldmann, T., Pollien, P., Varga, N., Devaud, S., Saucy, F., Hyunh-Ba, T., Stadler, R.H. (2005). Mechanisms of acrylamide formation: Maillard-induced transformations of asparagine, in *Chemistry and Safety of Acrylamide in Food* (eds M. Friedman, D.S. Mottram), Springer, New York, pp. 171–189.
30. Amrein, T.M., Andres, L., Manzardo, G.G.G., Amado, R. (2006). Investigations on the promoting effect of ammonium hydrogencarbonate on the formation of acrylamide in model systems. *Journal of Agricultural and Food Chemistry*, *54*, 10253–10261.
31. Koutsidis, G., De la Fuente, A., Dimitriou, C., Kakoulli, A., Wedzicha, B.L., Mottram, D.S. (2008). Acrylamide and pyrazine formation in model systems containing asparagine. *Journal of Agricultural and Food Chemistry*, (in press).
32. Bologna, L.S., Andrawes, F.F., Barvenik, F.W., Lentz, R.D., Sojka, R.E. (1999). Analysis of residual acrylamide in field crops. *Journal of Chromatographic Science*, *37*, 240–244.
33. Castle, L. (1993). Determination of acrylamide monomer in mushrooms grown on polyacrylamide gel. *Journal of Agricultural and Food Chemistry*, *41*, 1261–1263.

34. Elmore, J.S., Koutsidis, G., Dodson, A.T., Mottram, D.S., Wedzicha, B.L. (2005). Measurement of acrylamide and its precursors in potato, wheat, and rye model systems. *Journal of Agricultural and Food Chemistry*, *53*, 1286–1293.

35. Castle, L. (2006). Analysis for acrylamide in foods, in *Acrylamide and Other Hazardous Compounds in Heat-treated Foods* (eds K. Skog, J. Alexander), Woodhead Publishing, Cambridge, pp. 115–131.

36. Hoenicke, K., Gatermann, R., Harder, W., Hartig, L. (2004). Analysis of acrylamide in different foodstuffs using liquid chromatography-tandem mass spectrometry and gas chromatography-tandem mass spectrometry. *Analytica Chimica Acta*, *520*, 207–215.

37. Taeymans, D., Wood, J., Ashby, P., Blank, I., Studer, A., Stadler, R.H., Gonde, P., Van Eijck, P., Lalljie, S., Lingnert, H., Lindblom, M., Matissek, R., Muller, D., Tallmadge, D., O'Brien, J., Thompson, S., Silvani, D., Whitmore, T. (2004). A review of acrylamide: an industry perspective on research, analysis, formation and control. *Critical Reviews in Food Science and Nutrition*, *44*, 323–347.

38. Friedman, M., Levin, C.E. (2008). Reduction of the acrylamide content of the diet. A review. *Journal of Agricultural and Food Chemistry*, (in press).

39. Zhang, Y., Zhang, Y. (2007). Formation and reduction of acrylamide in Maillard reaction: a review based on the current state of knowledge. *Critical Reviews in Food Science and Nutrition*, *47*, 521–542.

40. Foot, R.J., Haase, N.U., Grob, K., Gonde, P. (2007). Acrylamide in fried and roasted potato products: a review on progress in mitigation. *Food Additives and Contaminants*, *24* (Suppl. 1), 37–46.

41. Guenther, H., Anklam, E., Wenzl, T., Stadler, R.H. (2007). Acrylamide in coffee: review of progress in analysis, formation and level reduction. *Food Additives and Contaminants*, *24* (Suppl. 1), 60–70.

42. Konings, E.J.M., Ashby, P., Hamlet, C.G., Thompson, G.A.K. (2007). Acrylamide in cereal and cereal products: a review on progress in level reduction. *Food Additives and Contaminants*, *24* (Suppl. 1), 47–59.

43. Confederation of the Food and Drink Industries of the EU (CIAA) (2006). The CIAA Acrylamide Toolbox, http://www.ciaa.eu/documents/brochures/toolbox_rev11_nov_2007final.pdf (accessed 11 January 2008).

44. Amrein, T.M., Bachmann, S., Noti, A., Biedermann, M., Barbosa, M.F., Biedermann-Brem, S., Grob, K., Keiser, A., Realini, P., Escher, F., Amado, R. (2003). Potential of acrylamide formation, sugars, and free asparagine in potatoes: a comparison of cultivars and farming systems. *Journal of Agricultural and Food Chemistry*, *51*, 5556–5560.

45. Biedermann-Brem, S., Noti, A., Grob, K., Imhof, D., Bazzocco, D., Pfefferle, A. (2003). How much reducing sugar may potatoes contain to avoid excessive acrylamide formation during roasting and baking? *European Food Research and Technology*, *217*, 369–373.

46. Grob, K., Biedermann, M., Biedermann-Brem, S., Noti, A., Imhof, D., Amrein, T., Pfefferle, A., Bazzocco, D. (2003). French fries with less than 100 µg/kg acrylamide. A collaboration between cooks and analysts. *European Food Research and Technology*, *217*, 185–194.

47. Haase, N.U., Matthaus, B., Vosmann, K. (2003). Acrylamide formation in foodstuffs—minimising strategies for potato crisps. *Deutsche Lebensmittel-Rundschau*, *99*, 87–90.
48. Halford, N.G., Muttucumaru, N., Curtis, T.Y., Parry, M.A.J. (2007). Genetic and agronomic approaches to decreasing acrylamide precursors in crop plants. *Food Additives and Contaminants*, *24* (Suppl. 1), 26–36.
49. Claus, A., Schreiter, P., Weber, A., Graeff, S., Herrmann, W., Claupein, W., Schieber, A., Carle, R. (2006). Influence of agronomic factors and extraction rate on the acrylamide contents in yeast-leavened breads. *Journal of Agricultural and Food Chemistry*, *54*, 8968–8976.
50. Muttucumaru, N., Halford, N.G., Elmore, J.S., Dodson, A.T., Parry, M., Shewry, P.R., Mottram, D.S. (2006). Formation of high levels of acrylamide during the processing of flour derived from sulfate-deprived wheat. *Journal of Agricultural and Food Chemistry*, *54*, 8951–8955.
51. Elmore, J.S., Mottram, D.S., Muttucumaru, N., Dodson, A.T., Parry, M.A.J., Halford, N.G. (2007). Changes in free amino acids and sugars in potatoes due to sulfate fertilization, and the effect on acrylamide formation. *Journal of Agricultural and Food Chemistry*, *55*, 5363–5366.
52. Surdyk, N., Rosen, J., Andersson, R., Aman, P. (2004). Effects of asparagine, fructose, and baking conditions on acrylamide content in yeast-leavened wheat bread. *Journal of Agricultural and Food Chemistry*, *52*, 2047–2051.
53. Taubert, D., Harlfinger, S., Henkes, L., Berkels, R., Schomig, E. (2004). Influence of processing parameters on acrylamide formation during frying of potatoes. *Journal of Agricultural and Food Chemistry*, *52*, 2735–2739.
54. Jung, M.Y., Choi, D.S., Ju, J.W. (2003). A novel technique for limitation of acrylamide formation in fried and baked corn chips and in French fries. *Journal of Food Science*, *68*, 1287–1290.
55. Rydberg, P., Eriksson, S., Tareke, E., Karlsson, P., Ehrenberg, L., Tornqvist, M. (2003). Investigations of factors that influence the acrylamide content of heated foodstuffs. *Journal of Agricultural and Food Chemistry*, *51*, 7012–7018.
56. Knol, J.J., Van Loon, W.A.M., Linssen, J.P.H., Ruck, A.L., Van Boekel, M., Voragen, A.G.J. (2005). Toward a kinetic model for acrylamide formation in a glucose-asparagine reaction system. *Journal of Agricultural and Food Chemistry*, *53*, 6133–6139.
57. Ciesarova, Z., Kiss, E., Boegl, P. (2006). Impact of L-asparaginase on acrylamide content in potato products. *Journal of Food and Nutrition Research*, *45*, 141–146.
58. Fink, M., Andersson, R., Rosen, J., Aman, P. (2006). Effect of added asparagine and glycine on acrylamide content in yeast-leavened bread. *Cereal Chemistry*, *83*, 218–222.
59. Gokmen, V., Senyuva, H.Z. (2006). A simplified approach for the kinetic characterization of acrylamide formation in fructose-asparagine model system. *Food Additives and Contaminants*, *23*, 348–354.
60. Low, M.Y., Koutsidis, G., Parker, J.K., Elmore, J.S., Dodson, A.T., Mottram, D.S. (2006). Effect of citric acid and glycine addition on acrylamide and flavour in a potato model system. *Journal of Agricultural and Food Chemistry*, *54*, 5976–5983.

61. Zhang, Y., Chen, J., Zhang, X.L., Wu, X.Q., Zhang, Y. (2007). Addition of antioxidant of bamboo leaves (AOB) effectively reduces acrylamide formation in potato crisps and French fries. *Journal of Agricultural and Food Chemistry*, *55*, 523–528.
62. Zhang, Y., Zhang, Y. (2008). Effect of natural antioxidants on kinetic behaviour of acrylamide formation and elimination in low-moisture asparagine-glucose model system. *Journal of Food Engineering*, *85*, 105–115.
63. Gokmen, V., Senyuva, H.Z. (2007). Acrylamide formation is prevented by divalent cations during the Maillard reaction. *Food Chemistry*, *103*, 196–203.
64. Park, Y.W., Yang, H.W., Storkson, J.M., Albright, K.J., Liu, W., Lindsay, R.C., Pariza, M.W. (2005). Controlling acrylamide in French fry and potato chip models and a mathematical model of acrylamide formation, in *Chemistry and Safety of Acrylamide in Food* (eds M. Friedman, D.S. Mottram), Springer, New York, pp. 343–356.
65. Elmore, J.S., Parker, J.K., Halford, N.G., Mottram, D.S. (2008). The effects of plant sulfur nutrition on acrylamide and aroma compounds in cooked wheat. *Journal of Agricultural and Food Chemistry*, (in press).
66. FAO/WHO (2002). Health Implications of Acrylamide in Food: Report of a Joint FAO/WHO Consultation, http://wwwint/foodsafety/publications/chem/en/acrylamide_full.pdf (accessed 11 January 2008).
67. Swiss Federal Office of Public Health (2002). Preliminary Communication: Assessment of Acrylamide Intake by Duplicate Diet Study, http://www.bag.admin.ch/index.html (accessed 11 January 2008).
68. Bundesinstitut für Risikobewertung (2003). Abschätzung der Acrylamid-Aufnahme durch hochbelastete Nahrungsmittel in Deutschland, http://www.bfr.bund.de/cm/208/abschaetzung_der_acrylamid_aufnahme_durch_hochbelastete_nahrungsmittel_in_deutschland_studie.pdf (accessed 11 January 2008).
69. Dybing, E., Sanner, T. (2003). Risk assessment of acrylamide in foods. *Toxicological Sciences*, *75*, 7–15.
70. Agence Française de Securite Sanitaire des Aliments (2005). Acrylamide: point d'information No 3, http://www.afssa.fr/Documents/RCCP2002sa0300b.pdf (accessed 11 January 2008).
71. Svensson, K., Abramsson, L., Becker, W., Glynn, A., Hellenas, K.E., Lind, Y., Rosen, J. (2003). Dietary intake of acrylamide in Sweden. *Food and Chemical Toxicology*, *41*, 1581–1586.
72. Konings, E.J.M., Baars, A.J., van Klaveren, J.D., Spanjer, M.C., Rensen, P.M., Hiemstra, M., van Kooij, J.A., Peters, P.W.J. (2003). Acrylamide exposure from foods of the Dutch population and an assessment of the consequent risks. *Food and Chemical Toxicology*, *41*, 1569–1579.
73. Food Standards Agency (2005). Analysis of Total Diet Study Samples for Acrylamide. Food Survey Information Sheet 71/05. http://www.food.gov.uk/multimedia/pdfs/fsis712005.pdf (accessed 11 January 2008).
74. United States Food and Drug Administration (2006). The 2006 exposure assessment for acrylamide. http://www.cfsan.fda.gov/~dms/acryexpo/acryex1.htm (accessed 11 January 2008).
75. United States Food and Drug Administration. (2006). The 2006 Exposure Assessment for Acrylamide. http://www.cfsan.fda.gov/~dms/acryexpo.html (accessed 11 January 2008).

76. United States Food and Drug Administration. (2004). The Updated Exposure Assessment for Acrylamide. http://www.cfsan.fda.gov/~dms/acrydino.html (accessed 11 January 2008).
77. Boon, P.E., de Mul, A., van der Voet, H., van Donkersgoed, G., Brette, M., van Klaveren, J.D. (2005). Calculations of dietary exposure to acrylamide. *Mutation Research-Genetic Toxicology and Environmental Mutagenesis*, *580*, 143–155.
78. Petersen, B.J., Tran, N. (2005). Exposure to acrylamide, in *Chemistry and Safety of Acrylamide in Food* (eds M. Friedman, D.S. Mottram), Springer, New York, pp. 63–76.
79. United States Food and Drug Administration (2003). The exposure assessment for acrylamide. http://www.cfsan.fda.gov/~dms/acryrob2.html (accessed 11 January 2008).
80. Doerge, D.R., Young, J.F., McDaniel, L.P., Twaddle, N.C., Churchwell, M.I. (2005). Toxicokinetics of acrylamide and glycidamide in B6C3F(1) mice. *Toxicology and Applied Pharmacology*, *202*, 258–267.
81. Doerge, D.R., Young, J.F., McDaniel, L.P., Twaddle, N.C., Churchwell, M.I. (2005). Toxicokinetics of acrylamide and glycidamide in Fischer 344 rats. *Toxicology and Applied Pharmacology*, *208*, 199–209.
82. Joint FAO/WHO Expert Committee on Food Additives (2005). Summary and Conclusions of the 64th Meeting. ftp://ftp.fao.org/es/esn/jecfa/jecfa64_summary.pdf (accessed 11 January 2008).
83. Calleman, C.J., Bergmark, E., Costa, L.G. (1990). Acrylamide is metabolized to glycidamide in the rat—evidence from hemoglobin adduct formation. *Chemical Research in Toxicology*, *3*, 406–412.
84. Friedman, M. (2003). Chemistry, biochemistry, and safety of acrylamide. A review. *Journal of Agricultural and Food Chemistry*, *51*, 4504–4526.
85. European Union Scientific Committee on Food (2002). Opinion of the Scientific Committee on Food on New Findings Regarding the Presence of Acrylamide in Food.
86. Miller, M.J., Carter, D.E., Sipes, I.G. (1982). Pharmacokinetics of acrylamide in Fisher-334 rats. *Toxicology and Applied Pharmacology*, *63*, 36–44.
87. Joint FAO/WHO Expert Committee on Food Additives (2006). Evaluation of certain food contaminants. WHO Technical Report Series, No. 930, WHO.
88. Burek, J.D., Albee, R.R., Beyer, J.E., Bell, T.J., Carreon, R.M., Morden, D.C., Wade, C.E., Hermann, E.A., Gorzinski, S.J. (1980). Sub-chronic toxicity of acrylamide administered to rats in the drinking water followed by up to 144 days of recovery. *Journal of Environmental Pathology and Toxicology*, *4*, 157–182.
89. Tyl, R.W., Friedman, M.A., Losco, P.E., Fisher, L.C., Johnson, K.A., Strother, D.E., Wolf, C.H. (2000). Rat two-generation reproduction and dominant lethal study of acrylamide in drinking water. *Reproductive Toxicology*, *14*, 385–401.
90. Chapin, R.E., Fail, P.A., George, J.D., Grizzle, T.B., Heindel, J.J., Harry, G.J., Collins, B.J., Teague, J. (1995). The reproductive and neural toxicities of acrylamide and 3 analogs in Swiss mice, evaluated using the continuous breeding protocol. *Fundamental and Applied Toxicology*, *27*, 9–24.
91. Wise, L.D., Gordon, L.R., Soper, K.A., Duchai, D.M., Morrissey, R.E. (1995). Developmental neurotoxicity evaluation of acrylamide in Sprague-Dawley rats. *Neurotoxicology and Teratology*, *17*, 189–198.

92. European Chemicals Bureau (2002). European Union Risk Assessment Report: Acrylamide. Office for Official Publications of the European Communities. http://ecb.jrc.it/DOCUMENTS/Existing-Chemicals/RISK_ASSESSMENT/REPORT/acrylamidereport011.pdf (accessed 11 January 2008).

93. Friedman, M.A., Dulak, L.H., Stedham, M.A. (1995). A lifetime oncogenicity study in rats with acrylamide. *Fundamental and Applied Toxicology*, *27*, 95–105.

94. Johnson, K.A., Gorzinski, S.J., Bodner, K.M., Campbell, R.A., Wolf, C.H., Friedman, M.A., Mast, R.W. (1986). Chronic toxicity and oncogenicity study on acrylamide incorporated in the drinking water of Fischer 344 rats. *Toxicology and Applied Pharmacology*, *85*, 154–168.

95. Joint FAO/WHO Expert Committee on Food Additives (2006). Evaluation of certain food contaminants, WHO Technical Report Series, No. 930, WHO. http://whqlibdocint/trs/WHO_TRS_930_eng.pdf (accessed 11 January 2008).

96. Mucci, L.A., Adami, H.O., Wolk, A. (2006). Prospective study of dietary acrylamide and risk of colorectal cancer among women. *International Journal of Cancer*, *118*, 169–173.

97. Mucci, L.A., Dickman, P.W., Steineck, G., Adami, H.O., Augustsson, K. (2003). Dietary acrylamide and cancer of the large bowel, kidney, and bladder: absence of an association in a population-based study in Sweden. *British Journal of Cancer*, *88*, 84–89.

98. Mucci, L.A., Lindblad, P., Steineck, G., Adami, H.O. (2004). Dietary acrylamide and risk of renal cell cancer. *International Journal of Cancer*, *109*, 774–776.

99. Mucci, L.A., Sandin, S., Balter, K., Adami, H.O., Magnusson, C., Weiderpass, E. (2005). Acrylamide intake and breast cancer risk in Swedish women. *Journal of the American Medical Association*, *293*, 1326–1327.

100. Pelucchi, C., Franceschi, S., Levi, F., Trichopoulos, D., Bosetti, C., Negri, E., La Vecchia, C. (2003). Fried potatoes and human cancer. *International Journal of Cancer*, *105*, 558–560.

101. Hogervorst, J.G., Schouten, L.J., Konings, E.J., Goldbohm, R.A., van de Brandt, P.A. (2007). A prospective study of dietary acrylamide intake and the risk of endometrial ovarian and breast cancer. *Cancer Epidemiology Biomarkers & Prevention*, *16*, 2304–2313.

102. Olesen, P.T., Olsen, A., Frandsen, H., Frederiksen, K., Overvad, K., Tjønneland, A. (2008). Acrylamide exposure and incidence of breast cancer among postmenopausal women in the Danish Diet, Cancer and Health Study. *International Journal of Cancer*, *122*, 2094–2100.

103. da Costa, G.G., Churchwell, M.I., Hamilton, L.P., Von Tungeln, L.S., Beland, F.A., Marques, A.M., Doerge, D.R. (2003). DNA adduct formation from acrylamide via conversion to glycidamide in adult and neonatal mice. *Chemical Research in Toxicology*, *16*, 1328–1337.

104. Maniere, I., Godard, T., Doerge, D.R., Churchwell, M.I., Guffroy, M., Laurentie, M., Poul, J.M. (2005). DNA damage and DNA adduct formation in rat tissues following oral administration of acrylamide. *Mutation Research-Genetic Toxicology and Environmental Mutagenesis*, *580*, 119–129.

105. Tareke, E., Twaddle, N.C., McDaniel, L.P., Churchwell, M.I., Young, J.F., Doerge, D.R. (2006) Relationships between biomarkers of exposure and toxicokinetics in

Fischer 344 rats and B6C3F(1) mice administered single doses of acrylamide and glycidamide and multiple doses of acrylamide. *Toxicology and Applied Pharmacology*, *217*, 63–75.

106. Twaddle, N.C., Churchwell, M.I., McDaniel, L.P., Doerge, D.R. (2004). Autoclave sterilization produces acrylamide in rodent diets: implications for toxicity testing. *Journal of Agricultural and Food Chemistry*, *52*, 4344–4349.
107. Twaddle, N.C., McDaniel, L.P., da Costa, G.G., Churchwell, M.I., Belanda, F.A., Doerge, D.R. (2004). Determination of acrylamide and glycidamide serum toxicokinetics in B6C3F(1) mice using LC-ES/MS/MS. *Cancer Letters*, *207*, 9–17.
108. European Commission (2007). Acrylamide Brochures. http://ec.europa.eu/food/food/chemicalsafety/contaminants/acrylamide_en.htm (accessed 11 January 2008).
109. United States Food and Drug Administration. (2004). FDA Draft Action Plan for Acrylamide in Food, http://www.cfsan.fda.gov/~dms/acrypla3.html (accessed 11 January 2008).
110. European Commission. (2007). Acrylamide Information Base. http://ec.europa.eu/food/food/chemicalsafety/contaminants/acryl_database_en.htm (accessed 11 January 2008).
111. Joint Institute for Food Safety and Applied Nutrition (2007). Acrylamide Infonet. http://www.acrylamide-food.org/research_database.htm (accessed 11 January 2008).
112. European Commission (2007). Recommendation on the monitoring of acrylamide levels in foods. *Official Journal of the European Union, L*, 33–40.
113. Codex Alimentarius Commission (2006). Discussion Paper on Acrylamide in Food. ftp://ftp.fao.org/codex/ccfac38/fa38_35e.pdf (accessed 11 January 2008).
114. Codex Alimentarius Commission (2007). Proposed Draft Code of Practice for the Reduction of Acrylamide in Food.
115. Health Canada. (2005). Acrylamide—What You Can Do to Reduce Exposure.
116. Kliemant, A., Göbel, A. (2007). The acrylamide minimisation concept—a risk management tool, in *Thermal Processing of Food: Potential Health Benefits and Risks* (ed. Senate Commission on Food Safety of Deutsche Forschungsgemeinschaft), Wiley VCH, pp. 197–207.

2.2

ACROLEIN

TAKAYUKI SHIBAMOTO
University of California, Department of Environmental Toxicology, Davis, CA 95616, USA

2.2.1 INTRODUCTION

Some highly volatile aldehydes, such as formaldehyde and acrolein, are considerably toxic. In particular, acrolein, which is the simplest α,β-unsaturated aldehyde, is known as a lachrymator or tear gas causing serious irritation to the eyes, nose, and throat (1). Acrolein has various names including acrylaldehyde, allyl aldehyde, ethylene aldehyde, 2-propenal, acrylic aldehyde, and pro-2-en-1-al (2). It has been used as an aquatic herbicide and algicide in various water systems such as irrigation canals, water-cooling towers, and water treatment ponds, as well as a slimicide in the manufacture of paper (3). In addition, acrolein is formed from many anthropogenic activities including combustion of fossil fuels, smoking cigarettes, and pyrolyzing fats. Consequently, acrolein has come to be known as one of the major toxic and volatile organic contaminants in the ambient air (3). Acute toxicity is not a serious issue because one can easily avoid being exposed. However, once acrolein is deposited in the environment, it is extremely difficult to prevent exposure. In 1998, the State of California had estimated that the total acrolein emission from industrial facilities was over 50,000 lb (4). The toxicity of acrolein found as an air pollutant has also been frequently reported. In 2003, the Environmental Protection Agency (EPA) published a 99-page comprehensive toxicological review of acrolein (5). Discussion on the air contamination, however, lies outside the scope of this review.

Process-Induced Food Toxicants: Occurrence, Formation, Mitigation, and Health Risks,
Edited by Richard H. Stadler and David R. Lineback

Acrolein: At a Glance	
Historical	The presence of acrolein in heat-processed foods was first reported in the early 1960s. Acrolein is, however, also formed from many anthropogenic activities and considered one of the most abundant toxic volatile organic compounds in ambient air. It has thus received more attention in the past as an environmental pollutant rather than a food process contaminant.
Analysis	Acrolein is highly volatile and tends to polymerize rapidly in aqueous solution. Most methods of analysis encompass the use of derivatizing agents such as 2,4-dinitrophenylhydrazine to form the more stable corresponding hydrazone. The derivative can be analyzed reliably and at trace amounts in foods using GC or HPLC techniques. The use of headspace GC/MS, albeit without derivatization, has also been described.
Occurrence in Food	Reported in various processed foods as a volatile aroma constituent, including sugarcane molasses, soured salted pork, heated vegetable and animal fats.
Main Formation Pathways	A major source is through thermally induced decomposition of animal and vegetable fats. The degradation of certain amino acids such as methionine, homoserine, and homocysteine upon heating at temperatures >100°C may also contribute to the formation of acrolein.
Mitigation in Food	Only general recommendations valid for essentially most heat-process contaminants can be given, i.e., control of thermal input during heating foods, in particular lipid-rich foods.
Health Risks	Most of the studies conducted into the health effects of acrolein are related to the inhalation toxicity of the compound. The NOEL and ADI values reported in different studies vary considerably. For example, EPA has set an oral reference dose for acrolein at 0.0005 mg/kg bw/d based on an NOEL of 0.05 (safety factor 100).
Regulatory Information or Industry Standards	Several countries have established regulations and/or guidelines regarding acrolein in air, water, and other media.
Knowledge Gaps	No information is currently available regarding the systemic toxicity of acrolein in humans. A more reliable database on exposures is required to assess the risk of sensitive populations such as children, and consequently to set adequate standards that cover all possible sources.
Useful Information (web) & Dietary Recommendations	www.atsdr.cdc.gov/tfacts124.html www.epa.gov/iris/subst/0364.htm Avoiding the over-heating of food; fry/barbecue foods to the appropriate "doneness."

Contamination in foods originating from herbicide may be a major concern with respect to human exposure to acrolein toxicity. The formation and presence of acrolein in foods, in particular, heat-processed foods, were first reported in the early 1960s. Therefore, the main object of this review is discussion of various aspects of acrolein found in processed foods.

2.2.2 OCCURRENCE IN FOODS

The formation of acrolein in foods has been known since the early 1960s when the presence of acrolein was reported in a sample obtained from pyrolysis or the destructive distillation of fats (6). In volatile flavor analysis studies, the presence of acrolein in volatiles from concord grapes was recognized for the first time in foods (7). Subsequently, acrolein was reported in various processed foods, such as sugarcane molasses (8), souring salted pork (9), and cooked horse mackerel (10). Later, acrolein was reported as a volatile aroma constituent in various foods (11).

Around the same time, the generation of significant amounts of acrolein was reported from various amino acids and polyamines under the physicochemical conditions commonly used in food processing (neutral pH and 100 °C) (12). The decarboxylation and deamination of amino acids included methionine, homoserine, homocysteine, and cystathionine produced 3-substituted propanals, which were readily decomposed to yield acrolein (12).

Since the 1970s, there have been numerous studies on the formation of cooked-flavor chemicals using the Maillard reaction system consisting of an amino acid and a sugar (13). As a result, various Maillard reaction model systems consisting of amines/carbonyls including *L*-leucine/furfural (14), *L*-rhamnose/ammonia (15), maltol/ammonia (16), starch/glycine (17), beef fat/glycine (18), and corn oil/glycine (19), have been used to investigate the formation of cooked flavor. Among substances used as a carbonyl reactant, dietary or cooking oils such as corn oil, peanut oil, beef fat, and butter, contributed significantly to the formation of many volatile compounds including acrolein (20). Furthermore, common cooking practices such as deep-fat frying, as well as processing of lipid-rich foods have been reported to exhibit the formation of acrolein (21–23), especially in the air. Detected at concentrations between 2.5 and 30 mg/m^3, acrolein was found in the air 15 cm above the surface of heated oil (24). Therefore, acrolein in vapor formed from oils during cooking has come to receive much attention as one of the major indoor air contaminations.

Table 2.2.1 shows the amounts of acrolein formed from typical cooking oils upon heat treatment. Significant amounts of acrolein were detected in a vapor formed from cooking oils and beef fat when heated at 300 °C for 2 h (25), a temperature level similar to those found during deep-fat frying (200 to 300 °C) (27). The amount of acrolein formed from 120 g of oil samples was 58 mg from sunflower oil, 75 mg from beef fat, 76 mg from soybean oil, 81 mg from corn oil, 86 mg from sesame oil, and 104 mg from olive oil. The acrolein concentration in the headspace of corn oil heated at 240 °C was 583 mg/m^3. On the other hand, the formation of acrolein was not observed when corn oil was heated at 180 °C for 2 h (25). When a headspace was collected from 25 g of oil samples during heating at 180 °C, acrolein was found in lard (109 μg/L), corn oil (164 μg/L), cotton seed oil (5.16 μg/L), and sunflower oil (163 μg/L) (26). Acrolein was not detected in vegetable oils (corn oil, sunflower oil, peanut oil,

TABLE 2.2.1 Acrolein formed from various lipids upon heat treatment.

Lipids	Temperature, °C/time	Collected from	Amount	Reference no.
Corn oil	300/2 h	Vapor	0.45 mg/g (N_2)[a], 0.68 mg/g (air)[b]	25
Soybean oil	300/2 h	Vapor	0.25 mg/g (N_2), 0.63 mg/g (air)	25
Sunflower oil	300/2 h	Vapor	0.31 mg/g (N_2), 0.48 mg/g (air)	25
Liver oil	300/2 h	Vapor	0.60 mg/g (N_2), 0.86 mg/g (air)	25
Sesame oil	300/2 h	Vapor	0.49 mg/g (N_2), 0.71 mg/g (air)	25
Beef oil	300/2 h	Vapor	0.38 mg/g (N_2), 0.63 mg/g (air)	25
Methyl linolenate	80/20 h	Vapor	2258.6 μg/g	22
Methyl linoleate	80/20 h	Vapor	1134.6 μg/g	22
Lard	180	Vapor	109 μg/L[b]	26
Corn oil	180	Vapor	164 μg/L	26
Cotton seed oil	180	Vapor	5.16 μg/L	26
Sunflower oil	180	Vapor	163 μg/L	26
Soybean oil	150~400	Fume	0.02~2.43 mg/kg	27
Peanut oil	150~400	Fume	0.03~1.65 mg/kg	27
Lard	150~400	Fume	0.02~10.93 mg/kg	27
Corn oil	145	Solution	4.3 μM	27
Sunflower oil (1)	145	Solution	1.1 μM	27
Sunflower oil (2)	145	Solution	2.9 μM	27
Peanut oil	145	Solution	2.7 μM	27
Olive oil (1)	145	Solution	5.6 μM	27
Olive oil (2)	145	Solution	59.3 μM	27

[a]Purged with nitrogen then collected.
[b]Purged with air then collected.

and olive oil) that were not subjected to thermal treatment as typical in food preparation, whereas it was found in corn oil (4.3 μM), in sunflower oil (2.9 μM), in peanut oil (2.7 μM), and in olive oil (5.6–9.3 μM) after 10 mL each of oil was heated at 145 °C for 2 h (28). When codfish fillets were fried at 182 and 204 °C for 4 min in used oil, 0.1 mg/kg of acrolein was formed from the oil heated at 204 °C, whereas only trace amounts of acrolein were detected in oil heated at 182 °C (23). Considerable increases of acrolein formation from three dietary oils were observed when the heating temperature was raised from 150 to 400 °C—0.02 to 2.43 mg/kg from soybean, 0.03 to 1.65 mg/kg from peanut oil, and 0.02 to 10.93 mg/kg from lard (27). These results suggest that acrolein formation from cooking oils requires high temperatures.

There are only a few reports on the presence of acrolein in fresh foods. The possible presence of acrolein, which might be a residue from the acrolein herbicide, was reported in Concord grape essence in the 1960s (29). Only trace concentrations of acrolein were determined in leaf lettuce after application of herbicide (active ingredient; 92% minimum acrolein) (30). Acrolein is known to be a harmful by-product of fermentation processes in agricultural distilleries. Among 516 grain spirit samples, the highest concentration of acrolein was >0.05 mg/mL (1 sample) followed by 0.006–0.05 mg/mL (3 samples), 0.002–0.006 mg/mL (17 samples), 0.0003–0.002 mg/mL (284 samples), and <0.0003 mg/mL (211 samples). Also, acrolein concentrations in potato starch spirit samples ranged from 0.003 to >0.05 mg/mL (31). Acrolein analysis of eight exemplary real raw spirits samples using a sodium bisulfite derivative followed by capillary isotachophoresis resulted in values of 0.34–2.5 mg/dm^3 (32). Compared with the amount of acrolein formed by a heat treatment, the presence of acrolein as naturally occurring or as herbicide residue in food is significantly lower.

2.2.3 ANALYSIS OF ACROLEIN IN FOODS AND BEVERAGES

Analysis of acrolein is extremely difficult because it is highly volatile and water soluble as shown in Table 2.2.2. In addition, acrolein tends to polymerize readily in an aqueous solution (34). Therefore, direct analysis of low amounts of acrolein is almost unfeasible. It is necessary to prepare an appropriate derivative for trace analysis of acrolein. Once a derivative is prepared, analysis of the derivative is important and various methods have been developed. Figure 2.2.1 shows mechanisms of the reactions used to prepare acrolein derivatives. Before gas chromatography or high-performance liquid chromatography (HPLC) became readily available, some photometric methods were used. For example, acrolein was reacted with *m*-hydroxyaniline to yield 7-hydroxyquinoline (**A**), which was subsequently determined with fluorescence spectrometry at $\lambda = 510$ nm. The limit of sensitivity was 0.002 μg/mL water (35). Later, formation of acrolein (3.46 μmol) from methional (5 μmol) degraded by heat treatment at 100 °C for 30 min was determined using the same method (12). More accurate and sensitive methods to determine 7-hydroquinoline derived from acrolein became possible using mass spectroscopy. Using this method, 2.26 and 1.13 mg/g of acrolein formed from methyl linolenate and methyl linoleate heated at 80 °C for 20 h were determined, respectively (22).

Since the 1950s, 2,4-dinitrophenylhydrazine (DNPH) has been most widely and commonly used to derivatize carbonyl compounds, including acrolein, to more stable corresponding 2,4-dinitrophenylhydrazones (**B**) (36). In the early 1960s, paper chromatography with acetylated paper was successfully used to detect the presence of acrolein after derivatization to a corresponding 2,4-dinitrophenylhydrazones (37). The monocarbonyl compounds (six alkanals, three 2-alkenals, two 2,4-alkadienals, and three methyl ketones) were identi-

TABLE 2.2.2 Chemical and physical properties of acrolein.

Description	Colorless or yellow mobile liquid with pungent and irritating odor
Odor threshold	160 ppb (370 μg/m^3) (Reference 33)
Molecular formula	C_3H_4O
Chemical Structure	$H_2C=CH-CHO$ (H, H on C = C, with H and C = O)
Molecular weight	56.06 g
Specific gravity	0.843 at 20 °C, 0.862 at 0 °C
Boiling point	52.5 °C
Melting point	–88 °C
Vapor pressure	29.3~36.5 kPa at 20 °C
Solubility	206~270 g/L in water at 20 °C, soluble in ethanol and diethyl ether
Henry's low constant	0.446~19.6 Pa·m^3/mol at 20 °C
Henry's low constant	7.8~180 dimensionless at 25 °C
log Kow	–1.1~1.02
log Koc	–0.219~2.43
Conversion factor	1 ppm = 2.3 mg/m^3 at 25 °C

fied as 2,4-dinitrophenhylhydrazones in bone grease using thin-layer chromatography (TLC) and ultraviolet (UV) spectra. Among the monocarbonyl compounds identified, 4.2 μg/g of acrolein was found in bone grease (38). Later, more sensitive tests with higher-resolution gas chromatography and HPLC were beginning to be used to isolate and identify various 2,4-dinitrophenylhydrazones. Acrolein contents (mg/kg) in yeast-raised and cake doughnuts fried at 182 °C were determined using HPLC. Identity of the acrolein derived hydrazone was confirmed by GC/MS (23). Acrolein formed in fumes from heated cooking oils was trapped in the glass fiber filter incorporated with the Sp-Pak DNPH-silica cartridge containing 1,4-DNPH. After the sample was extracted with acetonitrile, acrolein was analyzed as a hydrazone derivative using HPLC with a ultraviolet/visual (UV/VIS) detector (27). When the analytical results of low-molecular-weight aldehydes including acrolein were examined using the 2,4-DNPH/HPLC method from 14 different laboratories, this technique was usually comparable among the different laboratories within ±2 times the coefficient of variation (~6–15%) (39). Even though many derivatives, in addition to hydrazones, have been developed for analysis of low-molecular-weight carbonyl compounds, 2,4-DNPH remains the most widely used derivatizing agent for carbonyl compounds. Above all, the recent rapid advancement of liquid chromatography/tandem mass spectrometry (LC-MS/MS) may promote the use of this method. In fact, the analysis of the γ-aminobutyric acid (GABA) analogue succinic semialdehyde in urine and cerebrospinal fluid was recently performed using 2,4-DNPH derivatization and liquid chromatography/mass spectrometry (LC/MS) (40).

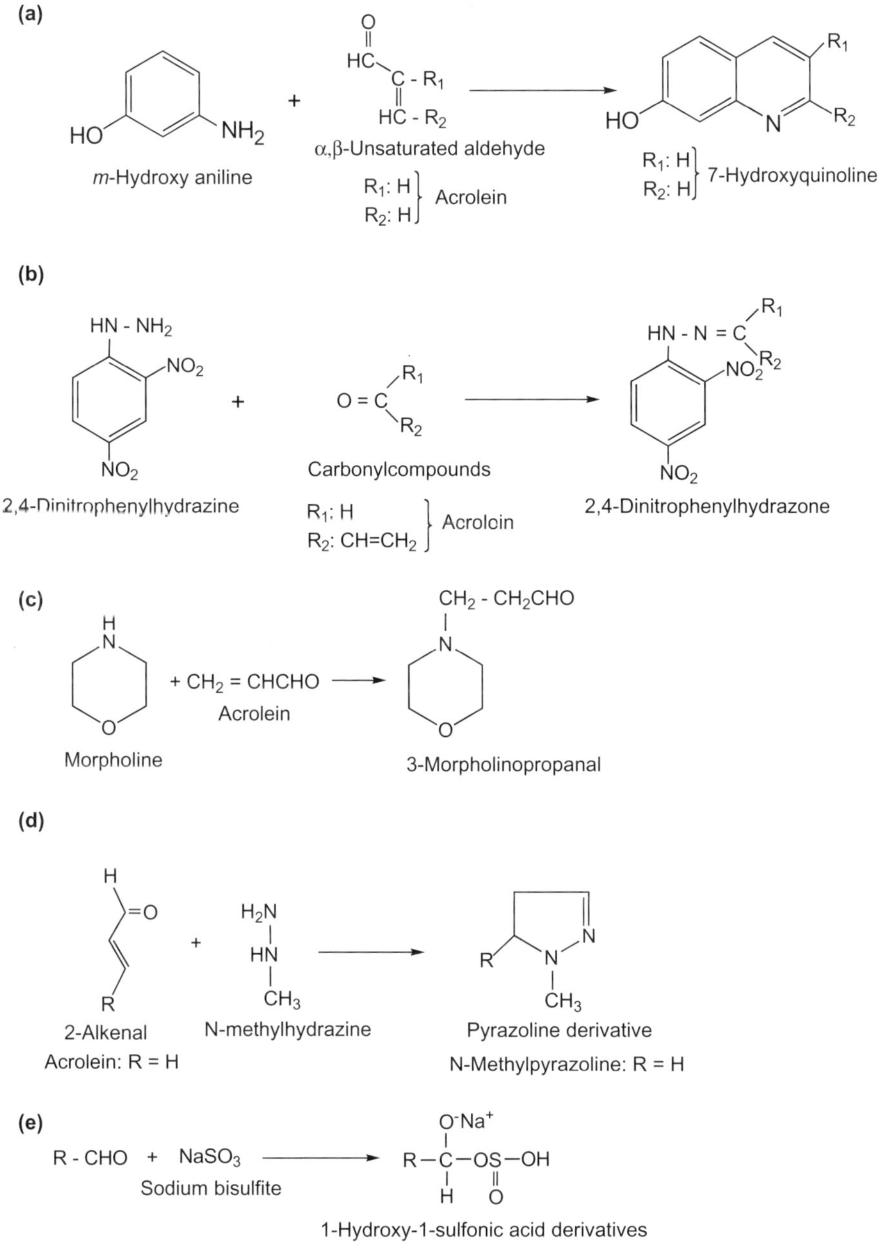

(a)

m-Hydroxy aniline + α,β-Unsaturated aldehyde (R_1: H, R_2: H: Acrolein) → 7-Hydroxyquinoline (R_1: H, R_2: H)

(b)

2,4-Dinitrophenylhydrazine + Carbonylcompounds (R_1: H, R_2: CH=CH$_2$: Acrolein) → 2,4-Dinitrophenylhydrazone

(c)

Morpholine + CH_2 = CHCHO (Acrolein) → 3-Morpholinopropanal

(d)

2-Alkenal (Acrolein: R = H) + N-methylhydrazine → Pyrazoline derivative (N-Methylpyrazoline: R = H)

(e)

R - CHO + $NaSO_3$ (Sodium bisulfite) → 1-Hydroxy-1-sulfonic acid derivatives

Figure 2.2.1 Mechanisms of reactions used to prepare acrolein derivatives. (a) Reaction of α,β-unsaturated aldehydes with *m*-hydroxy aniline to form 7-hydroxyquinolines. (b) Reaction of carbonyl compounds with 2,4-DNPH to form 2,4-dinitrophenylhydrazones. (c) Reaction of acrolein with morpholine to form 3-morpholinopropanal. (d) Reaction of 2-alkenals with *N*-methylhydrazine to form pyrazoline derivatives. (e) Addition of sodium bisulfite to aldehydes to form 1-hydroxy-1-sulfonic acid derivatives. (f) Reaction of aldehydes with PFBHA to form oxime derivatives. (g) Reaction of 2-alkenal with PFPH to form PFPH adducts.

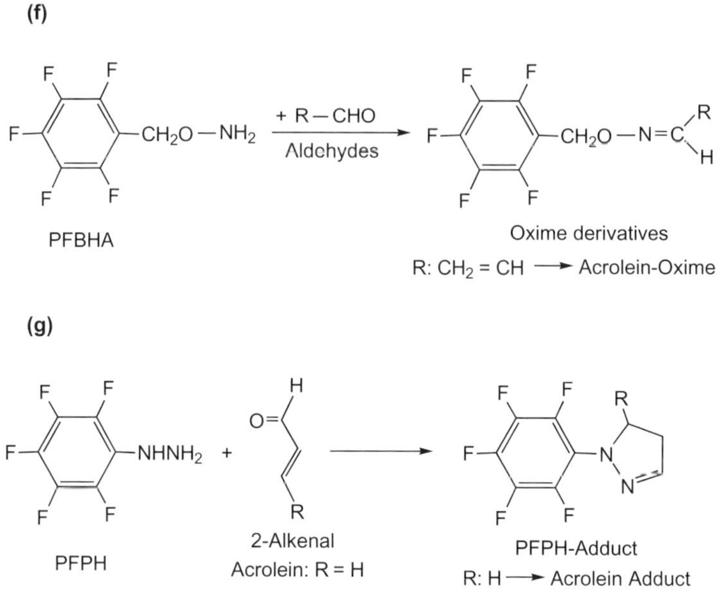

(f)

F F F CH₂O—NH₂ + R—CHO Aldehydes

$$\text{PFBHA} + \text{R-CHO} \xrightarrow{\text{Aldehydes}} \text{C}_6\text{F}_5\text{-CH}_2\text{O-N=C(R)H}$$

PFBHA

Oxime derivatives

R: $CH_2 = CH$ → Acrolein-Oxime

(g)

PFPH + 2-Alkenal → PFPH-Adduct

PFPH

2-Alkenal
Acrolein: R = H

PFPH-Adduct
R: H → Acrolein Adduct

Figure 2.2.1 *Continued*

As mentioned earlier, many methods for trace acrolein analysis have been developed and used successfully for analysis in foods. A unique derivative 3-morpholinopropanal (**C**), produced from acrolein and morpholine, was applied to analyze acrolein in the headspace of heated cooking oils and beef fat (25). After the pyrazoline derivative of acrolein with *N*-methylhydrazine (NMH) was prepared (**D**) for trace analysis of acrolein in 1990 (41), many analyses of acrolein in photo-irradiated or heated lipid samples were conducted using this method. This derivatizing agent, NMH was originally used to measure malonaldehyde as a pyrazole derivative (42, 43). It was found that NMH reacted with saturated monocarbonyl compounds, 2-alkenals (e.g., acrolein), and β-dicarbonyl compounds (e.g., malonaldehyde) to form hydrazones, pyrazolines, and pyrazoles, respectively. Consequently, acrolein was analyzed in various samples: fatty acids and cod liver oil oxidized with Fenton's reagent (44, 45); UV irradiated fatty acids (41), cod liver oil (46), and triolein (47); heated lard, corn oil, sunflower oil, and cotton seed oil (48); kitchen air (49); and cigarette smoke (50, 51). The detection limit of acrolein as *N*-methylpyrazoline was 5.9 pg (49).

Acrolein in raw spirits samples was analyzed by the isotachophoretic method after it was derivatized to 1-hydroxy-1-sulfonic acid derivative with sodium bisulfite (**E**) (32). A limit of detection at 0.03 μg/mL 100% ethanol was achieved by this method. Acrolein in spirits and alcoholic beverages was also determined as *O*-(2,3,4,5,6-pentafluorobenzyl)hydroxylamine (PFBHA) derivative (**F**). The resulting oxime was extracted using polydimethylsiloxane (PDMS) fiber by direct immersion into solution or from headspace. The analyte was thermally desorbed from the fiber and directly introduced into the GC injector and then detected by an electron capture detector (ECD). Using this method, 0.114 mg/L vodka (40% ethanol) of acrolein was determined (52, 53). A similar derivatizing agent, 2,3,4,5,6-pentafluorophenylhyrazine (PFPH), was also used successfully to analyze trace amounts of acrolein (**G**). Acrolein–PFPH adduct was analyzed using a gas chromatograph equipped with ECD or gas chromatography/mass spectrometry (GC/MS) (54). In this study, the detection limits of PFPH–acrolein adduct by ECD and mass spectrometry/single-ion monitoring (MS/SIM) were 16.60 fmol/mL and 1.31 pm/mL, respectively. Later, the same researchers determined acrolein formed from 2 g of sunflower oil heated at 220 °C for 0–120 min using solid-phase microextraction with on-fiber derivatization (55). PFPH was absorbed onto a poly(dimethylsiloxane)/divinylbenzene-coated fiber and exposed to headspace of the heated sunflower oil. The derivative formed on the fiber was subsequently desorbed into the GC with ECD for analysis of acrolein as PFPH adduct. The amounts of acrolein recovered by this method were 7.3–23.6 pmol.

Since the recent developments in the sensitivity of GC/MS and headspace trapping devices, the use of a headspace/solid-phase microextraction/gas chromatography/mass spectrometry (HS/SPME/GC/MS) method without derivatization became common for acrolein analysis. Using this HS/SPME/GC/MS method, direct analysis of acrolein has been conducted in various foods and beverages such as ewes' milk cheeses (56), virgin olive oils (57), and sunflower oil (58). Furthermore, trace amounts of acrylamide, for which acrolein was hypothesized as one of the precursors (59), formed in asparagine/D-glucose Maillard reaction model systems were also successfully determined using the HS/SPME/GC/MS method (60). Acrolein is one of the precursors of genotoxic acrylamide, which is also formed in heat-processed foods.

2.2.4 MECHANISMS OF FORMATION IN FOODS AND BEVERAGES

As previously mentioned, the presence of acrolein in foods and beverages has been known for many years. However, acrolein had received much more attention as one of the hazardous air contaminants in urban and suburban areas than as a food component (61). Acrolein has been known to be released from certain manufacturing plants, such as acrylic acid and cornstarch manufacturing (62, 63), as well as in exhaust gases from both gasoline engines and diesel engines in significantly large amounts (64). However, acrolein was first recognized as one of the volatile flavor chemicals formed in foods. The formation of acrolein was observed in the degradation of amino acids, including methionine, homoserine, homocysteine, and cystathionine by heating at 100 °C and neutral pH (12). It was hypothesized that methional was formed from methionine via Strecker degradation. Subsequently it was determined that methional formed acrolein by further oxidation. This step from methional to acrolein formation was accelerated significantly by the addition of air, or air plus ascorbic acid or hydrogen peroxide (Fig. 2.2.2). This report suggests that amino acids are one of the sources of acrolein in foods.

Major sources of acrolein in foods are heated lipids, which consist mainly of triglycerides of various fatty acids (65). Acrolein was proposed to form from the dehydration of glycerol when animal or vegetable fats were heated to high temperatures (2). As listed in Table 2.2.1, the major sources of acrolein are dietary oils such as corn oil, peanut oil, olive oil, and sunflower oil. Also, formation of various volatile chemicals, including acrolein, from lipids upon oxidation has been known (66).

Hypothesized formation mechanisms of acrolein from triglycerides upon heat treatment are shown in Fig. 2.2.3. As shown in Fig. 2.2.3, the major source of acrolein has been known for many years as the dehydration of glycerol (67). Acrolein is also formed from formaldehyde and acetaldehyde, which are final products of oxidative degradation of lipids, via aldol condensation and croton condensation although this is a minor pathway (68). Acrolein may also be formed from oxidative degradation of various fatty acids formed from triglycerides (66). The formation of acrolein through free radical mechanisms involving homolytic fission of R–O bonds more likely occurs at high-temperature treatment in food processing (25, 27) as well as through glycerol dehydration. The proposed radical mechanisms to yield acrolein from triglyceride are shown in Fig. 2.2.4.

Recently, acrolein has begun to receive a great deal of attention as one of the minor precursors of acrylamide formed during food processing (69–71). For example, when acrolein and ammonia were heated at 180 °C in the vapor phase, 753 mg/g ammonia of acrylamide was formed (59). A proposed formation pathway of acrylamide from acrolein is shown in Fig. 2.2.5. In this pathway,

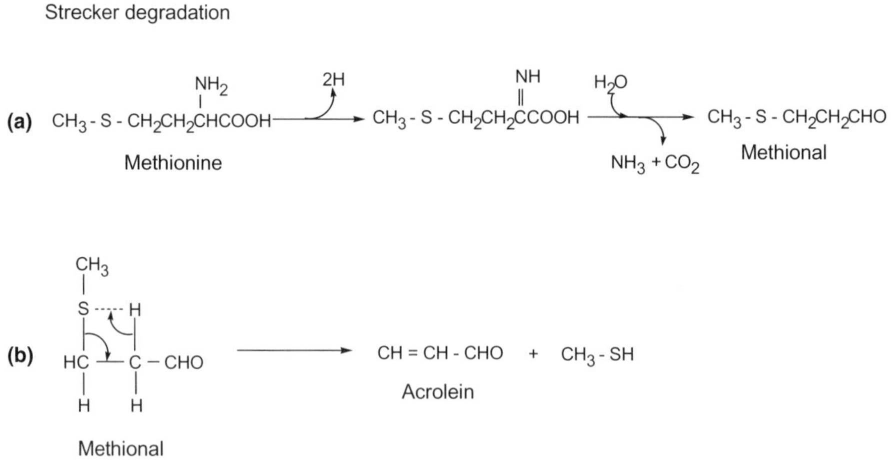

Strecker degradation

(a) CH_3-S-$CH_2CH_2CH(NH_2)COOH$ (Methionine) —(2H)→ CH_3-S-$CH_2CH_2C(=NH)COOH$ —(H_2O; $NH_3 + CO_2$)→ CH_3-S-CH_2CH_2CHO (Methional)

(b) CH_3-S-CH_2-CH_2-CHO (Methional) ——→ CH = CH - CHO (Acrolein) + CH_3-SH

Figure 2.2.2 (a) Formation of methional from methionine via Strecker degradation. (b) Thermal degradation of methional to form acrolein.

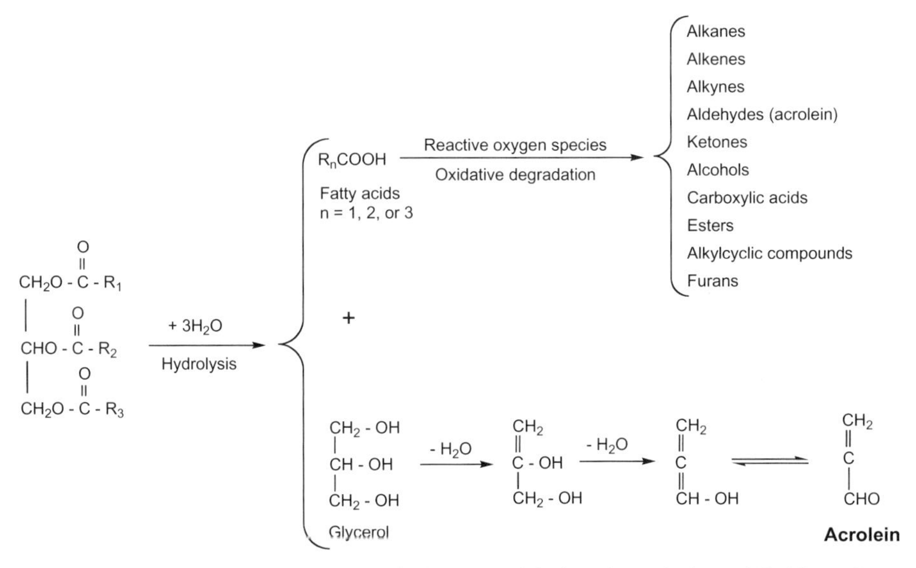

Figure 2.2.3 Typical thermal degradation or oxidative degradation of lipids to form various secondary products, including acrolein.

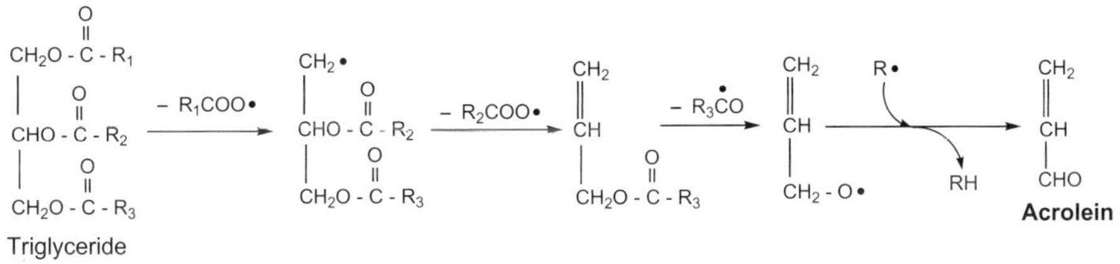

CH_2O-C-R_1
$CHO-C-R_2$
CH_2O-C-R_3
Triglyceride
$-R_1COO\bullet$
$CH_2\bullet$
$CHO-C-R_2$
CH_2O-C-R_3
$-R_2COO\bullet$
CH_2
CH
CH_2O-C-R_3
$-R_3\dot{C}O$
CH_2
CH
$CH_2-O\bullet$
$R\bullet$
RH
CH_2
CH
CHO
Acrolein

Figure 2.2.4 Hypothesized free radical mechanisms of acrolein formation from triglycerides.

acrolein is oxidized to acrylic acid, which subsequently reacts with ammonia to form acrylamide. This pathway may be the major route of acrolein formation because significant amounts of acrolein were formed from the reaction of acrylic acid and ammonia (190 mg/g ammonia) at 180 °C (59). Ammonia is readily formed from amino acids via Strecker degradation and subsequently reacts with acrylic acid to yield acrylamide. Also, radical mechanism under high-temperature processing is proposed.

2.2.5 TOXICITY AND HEALTH RISKS OF ACROLEIN

There have been many reports on the toxicity and health risks of acrolein. Acrolein originally received attention as a hazardous air pollutant because the

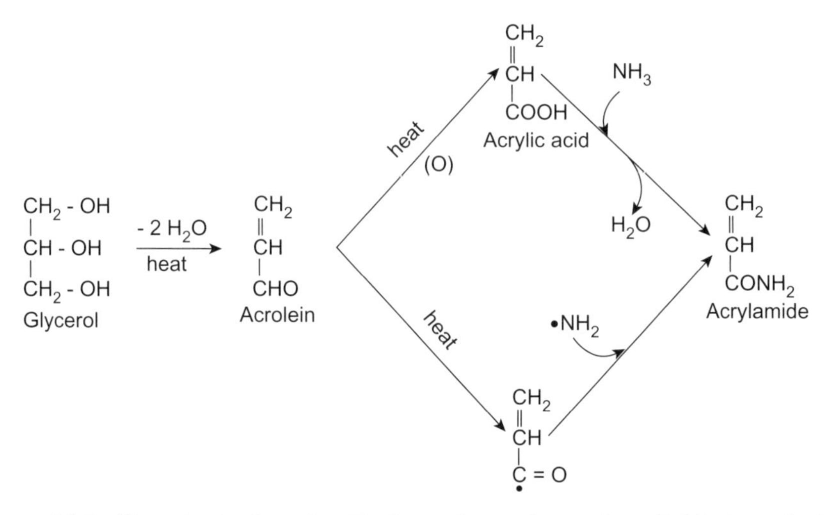

Figure 2.2.5 Hypothesized acrylamide formation pathways from lipids through glycerol and acrolein.

largest source of human exposure to acrolein comes from the incomplete combustion of the many organic materials occurring during fires in urban areas and forests (72, 73). Moreover, acrolein had been widely used as an intermediate of various organic chemicals such as acrylic acid, methionine, and glutaraldehyde released into ambient air during chemical processing. Therefore, numerous comprehensive review articles on toxicities and health risks of acrolein as well as risk assessment have been published. Table 2.2.3 shows a summary of typical reviews on acrolein.

As mentioned earlier, most investigations on acrolein toxicity have been focused on inhalation toxicity because acrolein is commonly present in the outdoor air. However, indoor air pollution by acrolein is also important because it forms directly from lipids such as cooking oils, and lipid-rich foods such as beef and pork, upon high-temperature processing as shown in Table 2.2.1. For example, acrolein was detected at concentrations between 2.5 and 30 mg/m^3 in the air 15 cm above the surface of a heated oil (24).

Because many excellent comprehensive reviews on acrolein are available today (Table 2.2.3), only typical toxicities of acrolein are introduced in the present review. A clinical study on acrolein exposure using human volunteers concluded that the average threshold of sensation ranged from 0.09 (eye irritation) to 0.30 mg/kg (respiration rate, throat irritation) and nasal irritation occurred at 0.15 mg/kg (0.35 mg/m^3) (82). However, there are no current longitudinal studies of humans chronically exposed to acrolein.

A typical acute oral LD_{50} of Sprague-Dawley rats reportedly ranged from 10.3 (males) to 33 mg/kg (females) (83, 84). The acute dermal LD_{50} in New Zealand white rabbits was 223 mg/kg in females and 240 mg/kg in males (83). A median respiratory rate depression (RD_{50}) in various experimental rats and

TABLE 2.2.3 Published reviews on acrolein toxicity and health assessment.

Title	Main focused subjects	Number of references	Reference no.
Draft toxicological profile for acrolein	Public health, health effects, genotoxicity, human exposure, analytical methods, and regulations	942	74
A critical review of the literature on acrolein toxicity	Reactivity and metabolism, pathophysiology, carcinogenicity, teratogenicity, reproductive toxicity, genotoxicity, human toxicity, and environmental effect	520	21
Toxicological review of acrolein	Hazard identification, reproductive/developmental studies, and dose–response assessments	276	75
Acrolein	Analytical methods, human and environmental exposure, environmental transport, distribution, and transformation, metabolism in animals and human, and effects on human	250	76
Fate and effects of acrolein	Chemistry, environmental occurrence, environmental fate, toxicology, and biochemical action mechanisms	225	1
Priority data needs for acrolein	Legislative issues, identification of data needs: exposure, toxicity, and prioritization of data needs	163	77
The molecular effects of acrolein	GSH and apoptosis, cell proliferation, gene expression, thiol redox state, NF-kB, AP-1, *Fos*, *Jun*, *p53*, *c-myc*	99	78
Acrolein	Genetic and toxic effects in various systems: bacteria, yeast, algae, higher plants, *Drosophila*, and mammals	98	2
Acrolein: a respiratory toxin that suppresses pulmonary host defense	Respiratory tract injury and pulmonary host defense	75	79
Mechanisms of nasal toxicity induced by formaldehyde and acrolein	Mechanism of toxic effects in rodents	30	80
Acrolein—its formation, occurrence and properties	Physicochemical properties, industrial production, and presence in spirit products	30	81

mice by acrolein ranged from 1.03 mg/kg (2.4 mg/m^3) in male Swiss Webster mice to 6.0 mg/kg (13.7 mg/m^3) in male F-344 rats (85, 86). When Syrian golden hamsters, SPF Wistar rats, and Dutch rabbits were exposed to acrolein vapor at 4.9 mg/kg (11.3 mg/m^3), six rats died. Also significant physiological changes, including ocular and nasal irritation, growth depression, and histopathological changes of the respiratory tract, were observed in all animals examined (87).

When primary cultured rat hepatocytes were treated with 100 mM of acrolein, gradual accumulation of cellular lipid peroxide was observed (88). Oxidative stress associated with increase of lipid peroxidation is known to have a role in the pathogenesis of Alzheimer's disease. A protein-bound acrolein, such as acrolein-modified keyhole impet hemocyanin, was reportedly a powerful marker of oxidative damage in the formation of neurofibrillary tangles, which subsequently leads to neuronal death in Alzheimer's disease (89).

The exposure of rat liver epithelial RL34 cells to acrolein resulted in a significant accumulation of the acrolein–2′-deoxyadenosine adduct, suggesting that the formation of a carcinogenic aldehyde through lipid peroxidation is involved in the pathophysiological effects associated with oxidative stress (90). A recent study demonstrated that exposure of purified brain mitochondria to acrolein resulted in a dose-dependent increase of reactive oxygen species, which are known to induce lipid peroxidation. The same study reported that acrolein inhibited mitochondrial adenine nucleotide translocase (91). Significant age-dependent increases of acrolein were observed in aged Fisher-344 rats with traumatic brain injury caused by oxidative damage (92). These studies suggested that acrolein, a by-product of lipid peroxidation, also plays an important role in oxidative stress.

Most health assessments of acrolein have been focused on the acrolein contamination in the ambient air. The review articles shown in Table 2.2.3 also contain the reports of acrolein health assessments. Due to public concern for hazardous organic volatiles, including acrolein, present in the ambient air, most health assessments of acrolein have been conducted by Federal or State government offices rather than by laboratories for basic research. Two decades ago, the United States Environmental Protection Agency (USEPA) summarized and evaluated information relevant to a preliminary interim assessment of adverse health effects associated with specific chemicals including acrolein (93). Emission of toxic chemicals such as acrolein from various anthropogenic sources, such as wastewater treatment plants (94), large combustion turbine power projects and power plants (95, 96), as well as from specific sites such as cities in Minnesota (97), along the Arizona–Sonora Border (98), and the Oakland–San Francisco Bay Bridge Toll Plaza (99), has been monitored and assessed for hazardous effects on human and ecosystems. The most recent report indicated that estimated ambient concentrations of acrolein are greater than the USEPA reference concentration throughout the United States, making it a concern for human health. However, there is no method for assessing the extent of risk under the USEPA noncancer risk assessment framework (100). Therefore, establishment of a method for risk assessment of acrolein either present in ambient air or foods is a pressing need.

Also, there have been numerous efforts to establish regulations and guidelines for acrolein. One of the most serious difficulties was to determine specific acrolein concentrations associated with various chronic toxicities, including ADI (acceptable daily intakes), LOAEL (lowest-observed-adverse-effect level), NOEL (no-observable-effect level), and NOAEL (no-observed-adverse-effect level). For example, ADI is determined based on toxicological studies associated with exposure by the oral route. However, because acrolein is highly volatile, it is released into the environment, and consequently, humans are also exposed to acrolein in vapor phase. Therefore, the exposure by the respiratory route (inhalation) may be more significant than by the oral route in the case of acrolein (75). Also, NOAEL and ADI values reported by different studies vary considerably (74). The NOEL and ADI values set for acrolein exposure by the Department of Health and Ageing Office of Chemical Safety of the Australian Government are 0.0005 and 0.05 mg/kg-bw/day, respectively (101). The safety factor used was 100. The EPA sets the oral reference dose for acrolein at 5×10^{-4} based on a NOAEL of 0.05 mg/kg/day (oral gavage for 2 years). Also, the EPA sets the inhalation reference concentration for acrolein at 2×10^{-5} mg/m^3 (102). A comprehensive list of the international and national regulations and guidelines regarding acrolein in air, water, and other media is well summarized in Table 8.1 of the documents published by the Agency for Toxic Substances and Disease Registry (ATSDR) in 2005 (103). For example, the ATSDR derived an acute-duration oral minimal risk level (MRL) of 0.008 mg/kg/day based on a NOAEL of 0.75 mg/kg/day for forestomach squamous epithelial hyperplasia in rats in 13–22 k gavage study (104) and an uncertainty factor of 100 (10 for species extrapolation and 10 for human variability).

2.2.6 CONCLUSIONS

Acrolein is one of the most abundant toxic volatile organic compounds in ambient air. Acrolein has been emitted in large amounts from various sources, including fields treated with herbicide and algicide, paper manufacture, combustion of fuels, cigarette smoke, food plants, and power plants as well as accidental fires of housing and forests. For example, the annual industrial acrolein emissions from facilities in the State of California were estimated to be £54,565 (4). Therefore, most public concerns about acrolein contamination focus on inhalation toxicity. It is critical, however, to note that acrolein forms from triglycerides, which are major components of lipids, upon heat processing (such as cooking). Thus, people are exposed to acrolein not only via ambient air but also from cooked foods. Acrolein can be an intermediate in the formation of genotoxic acrylamide, which is also formed in heat-processed foods. Analysis of acrolein is extremely difficult but advanced technology has permitted detection of acrolein at concentrations of mg/kg to μg/kg in complex matrices such as cooked foods. Consequently, numerous reports on acrolein concentrations in ambient air and various foods, as well as its risk assessment, are available today.

It is important to note that other well-known volatile aldehydes, including formaldehyde, acetaldehyde, propanal, glyoxal, methylglyoxal, and malonaldehyde in addition to acrolein, are formed from lipids and lipid-rich foods during heat processing. In particular, formaldehyde and acetaldehyde may have received more attention than acrolein as toxic volatile chemicals found in foods and beverages. The present review does not cover either of these volatile aldehydes. However, many comprehensive reviews are currently available for formaldehyde (104–112) and for acetaldehyde (113–117).

As mentioned earlier, there are a tremendous number of reports on acrolein (Table 2.2.3). However, it should be noted that additional research is still needed to obtain data adequate to set satisfactory regulations and guidelines for acrolein. The ATSDR also published a comprehensive summary of priority data needs for acrolein (77). This report emphasizes, in particular, the data needed to be able to set standards for the safety of children, including exposure to acrolein through inhalation of ambient and indoor air, particularly in indoor air containing environmental tobacco smoke, as well as simply the ingestion of food containing acrolein. This report also indicated that major gaps exist in the currently available data for assessment of adverse effects of acrolein to human because no information is available regarding the systemic toxicity of acrolein to humans. Therefore, studies on how to integrate the data from animal studies and human health are pressing needs.

REFERENCES

1. Ghilarducci, D.P., Tjeerdema, R.S. (1995). Fats and effects of acrolein. *Rev. Environ. Contam. Toxicol.*, *144*, 95–146.
2. Izard, C., Libermann, C. (1978). *Acrolein. Mutat. Res.*, *47*, 115–138.
3. IARC (International Agency for Research on Cancer) (1985). Allyl compounds, aldehydes, epoxide and peroxides, in *IARC Monograph on the Evaluation of the Carcinogenic Risk of Chemicals to Humans*, *36*, IARC, Lyon, pp. 133–161.
4. CARB (California Air Resources Board) (2000). *California Emissions Inventory Development and Reporting System (CEIDARS)*, CARB, Sacramento, CA, February 12.
5. EPA (2003). *Toxicological Review of Acrolein*, US Environmental Protection Agency, Washington, DC, May.
6. Geyer, B.P. (1962). Methods of preparation, in *Acrolein* (ed. C.W. Smith), John Wiley & Sons, Inc., New York, pp. 4–6.
7. Neudoerffer, T.S., Sandler, S. (1965). Detection of an undesirable anomaly in Concord grape by gas chromatography. *J. Agric. Food Chem.*, *6*, 584–588.
8. Hrdlicka, J., Janicek, G. (1968). Volatile carbonyl compounds isolated from sugarcane molasses. *Sb. Vys. Sk. Chem. Technol. Praze. Potraviny*, E*21*, 77–80.
9. Cantoni, C., Bianchi, M.A., Renon, P., Calcinardi, C. (1969). Bacterial and chemical alterations during souring in salted pork. *Atti. Soc. Ital. Sci. Vet.*, *23*, 752–755.
10. Shinomura, M., Yoshimatsu, F., Matsumoto, F. (1971). Fish odour of cooked horse mackerel. *Kasseigaku Zasshi*, *22*, 106–110.

11. Maier, H.G. (1973). Sorption of volatile aroma constituents by foods VII. Aliphatic aldehydes. *Z. Lebensm.-Untersuchung Forschung*, *151*, 384–386.
12. Alarcon, R.A. (1976). Formation of acrolein from various amino-acids and polyamines under degradation at 100 °C. *Environ. Res.*, *12*, 317–326.
13. Shibamoto, T. (1983). Heterocyclic compounds in browning and browning/nitrite model systems: occurrence, formation mechanisms, flavour characteristics and mutagenic activity, in *Instrumental Analysis of Foods* (eds G. Charalambous, G. Inglett) Academic Press, New York, pp. 229–278.
14. Rizzi, G.P. (1974). Formation of N-alkyl-2-acylpyrroles and aliphatic aldimines in model nonenzymic browning reactions. *J. Agric. Food Chem.*, *22*, 279–282.
15. Shibamoto, T., Bernhard, R.A. (1978). Formation of heterocyclic compounds from the reaction of L-rhamnose with ammonia. *J. Agric. Food Chem.*, *26*, 183–187.
16. Shibamoto, T., Nishimura, O., Mihara, S. (1981). Mutagenicity of products obtained from a maltol-ammonia browning model system. *J. Agric. Food Chem.*, *29*, 643–646.
17. Umano, K., Shibamoto, T. (1987). Analysis of acrolein from heated cooking oils and beef fat. *J. Agric. Food Chem.*, *35*, 909–912.
18. Ohnishi, S., Shibamoto, T. (1984). Volatile compounds from heated beef fat and beef fat with glycine. *J. Agric. Food Chem.*, *32*, 987–992.
19. Macku, C., Shibamoto, T. (1991). Headspace volatile compounds formed from heated corn oil and corn oil with glycine. *J. Agric. Food Chem.*, *39*, 1265–1269.
20. Shibamoto, T.A. (1996). Role of lipids in the formation of cooked flavor, in *The Contribution of Low-and Nonvolatile Materials to the Flavor of Foods* (eds W. Pickenhagen, C.T. Ho, A.M. Spanier), Allured Pub. Corp., Carol Stream, Illinois, pp. 183–192.
21. Beauchamp, R., Andjelkovich, D., Klingerman, A., Morgan, K., Heck, H. (1985). A critical review of the literature on acrolein toxicity. *CRC Crit. Rev. Toxicol.*, *14*, 309–380.
22. Hirayama, T., Yamaguchi, M., Nakata, T., Okumura, M., Yamazaki, T., Watanabe, T., Fukui, S. (1989). Formation of acrolein by the autooxidation of unsaturated fatty acid methyl esters. *Eisei Kagaku*, *35*, 303–306.
23. Lane, R., Smathers, J. (1991). Monitoring aldehyde production during frying by reversed-phase liquid chromatography. *J. AOAC Int.*, *74*, 957–960.
24. Kishi, M. (1975). Effect of inhalation of the vapour from heated edible oil of the circulatory and respiratory systems in rabbits. *Shokuhin Eiseigaku Zasshi*, *16*, 318–322.
25. Umano, K., Dennis, K.J., Shibamoto, T. (1988). Analysis of free malondialdehyde in photoirradiated corn oil and beef fat via a pyrazole derivative. *Lipids*, *23*, 811–814.
26. Yasuhara, A., Shibamoto, T. (1989). Analysis of aldehydes and ketones in the headspace of heated pork fat. *J. Food Sci.*, *54*, 1471–1472, 1484.
27. Lin, J.-M., Liou, S.-J. (2000). Aliphatic aldehydes produced by heating Chinese cooking. *Bull. Environ. Contam. Toxicol.*, *64*, 817–824.
28. Casella, I.G., Contursi, M. (2004). Quantitative analysis of acrolein in heated vegetable oils by liquid chromatography with pulsed electrochemical detection. *J. Agric. Food Chem.*, *52*, 5816–5821.

29. Stern, D.J., Lee, A., McFadden, H., Stevens, K.L. (1967). Volatile from grapes: identification of volatiles from Concord essence. *J. Agric. Food Chem.*, *15*, 1100–1103.
30. Nordone, A.J., Kovacs, M.F., Doane, R. (1997). [^{14}]C Acrolein accumulation and metabolism in leaf lettuce. *Bull. Environ. Contam. Toxicol.*, *58*, 787–792.
31. Szpak, E. (1997). Acrolein in raw grain and potato spirits from 1995/1996 season. *Przemysl Fermentacyjny I Owocowo-Warzywny*, *41*, 16–18.
32. Curylo, J., Wardencki, W. (2005). Determination of acetaldehyde and acrolein in raw spirits by capillary isotachophoresis after derivatization. *Anal. Lett.*, *38*, 1659–1669.
33. Amoore, J.E., Hautala, E. (1983). Odour as an aid to chemical safety: odour thresholds compared with threshold limit values and volatilities for 214 chemicals in air and water dilution. *J. Appl. Toxicol.*, *3*, 272–290.
34. Hess, L.G., Kurtz, A.N., Stanton, D.B. (1978). Acrolein and derivatives. In *Kirk-Othmer, Encyclopedia of Chemical Technology*, 3rd ed., Vol. 1 (ed. H.F. Mark), John Wiley & Sons, Inc., New York, 277–290.
35. Alarcon, R.A. (1968). Fluorometric determination of acrolein and related compounds with *m*-aminophenol. *Anal. Chem.*, *40*, 1704–1708.
36. Shriner, R.L., Hermann, C.K.F., Morrill, T.C., Curtin, D.Y., Fuson, R.C. (1997). *The Systematic Identification of Organic Compounds*, 7th edn, John Wiley & Sons, Inc., New York.
37. Forss, D.A., Ramshaw, E.H. (1963). The chromatography of 2,4-dinitrophenylhydrazones on acetylated paper. *J. Chromatogr.*, *10*, 268–271.
38. Maslowska, J., Bazylak, G. (1985). Thermofractographic determination of monocarbonyl compounds in animal waste fats used as feed fat. *Anim. Feed Sci. Technol.*, *13*, 227–236.
39. Hafkenscheid, T.L., van Oosten, J.A. (2002). Results from interlaboratory comparisons of aldehyde-2,4-dinitrophenylhydrazone analysis. *Anal. Bioanal. Chem.*, *372*, 658–663.
40. Struys, E.A., Jansen, E.E.W., Gibson, K.M., Jakobs, C. (2005). Determination of the GABA analogue succinic semialdehyde in urine and cerebrospinal fluid by dinitrophenylhydrazine derivatization and liquid chromatography-tandem mass spectrometry: application to SSADH deficiency. *J. Inherit. Metab. Dis.*, *28*, 913–920.
41. Dennis, K.J., Shibamoto, T. (1990). Gas chromatographic analysis of reactive carbonyl compounds formed from lipids upon UV-irradiation. *Lipids*, *25*, 460–464.
42. Umano, K., Shibamoto, T. (1984). Chemical studies on heated starch/glycine model systems. *Agric. Biol. Chem.*, *48*, 1387–1393.
43. Ichinose, T., Miller, M.G., Shibamoto, T. (1989). Gas chromatographic analysis of free and bound malonaldehyde in rat liver homogenates. *Lipid*, *24*, 895–898.
44. Miyake, T., Shibamoto, T. (1996). Simultaneous determination of acrolein, malonaldehyde, and 4-hydroxy-2-nonenal produced from lipids oxidized with Fenton's reagent. *Food Chem. Toxicol.*, *34*, 1009–1011.
45. Tamura, H., Kitta, K., Shibamoto, T. (1991). Formation of reactive aldehydes from fatty acids in a Fe2+/H_2O_2 oxidation system. *J. Agric. Food Chem.*, *39*, 439–442.
46. Niyati-Shirkhodaee, F., Shibamoto, T. (1992). Formation of toxic aldehydes in cod liver oil after ultraviolet irradiation. *J. Am. Oil Chem. Soc.*, *69*, 1254–1256.

47. Niyati-Shirkhodaee, F., Shibamoto, T. (1992). *In vitro* determination of toxic aldehydes formed from the skin lipid, triolein, upon ultraviolet irradiation: formaldehyde and acrolein. *J. Toxicol. Cut. Ocular Toxicol.*, *11*, 285–292.
48. Yasuhara, A., Shibamoto, T. (1991). Determination of volatile aliphatic aldehydes in the headspace of heated food oils by derivatization with 2-aminoethanethiol. *J. Chromatogr.*, *547*, 291–298.
49. Yasuhara, A., Shibamoto, T. (1991). Determination of acrolein evolved from heated vegetable oil by N-methylhydrazine conversion. *Agric. Boil. Chem.*, *55*, 2639–2640.
50. Miyake, T., Yasuhara, A., Shibamoto, T. (1995). Gas chromatographic analysis of acrolein as 1-methyl-2-pyrazoline in cigarette smoke. *J. Environ. Chem.*, *5*, 569–573.
51. Fujioka, K., Shibamoto, T. (2006). Determination of toxic carbonyl compounds in cigarette smoke. *Environ. Toxicol.*, *21*, 47–54.
52. Wardencki, W., Sowinski, P., Curylo, J. (2003). Evaluation of headspace solid-phase microextraction for the analysis of volatile carbonyl compounds in spirits and alcoholic beverages. *J. Chromatogr. A*, *984*, 89–96.
53. Sowinski, P., Wardencki, W., Partyka, M. (2005). Development and evaluation of headspace gas chromatography method for the analysis of carbonyl compounds in spirits and vodkas. *Anal. Chim. Acta*, *539*, 17–22.
54. Stashenko, E.E., Derreira, M.C., Sequeda, G., Martinez, J.R., Wong, J.W. (1997). Comparison of extraction methods and detection systems in the gas chromatographic analysis of volatile carbonyl compounds. *J. Chromatogr. A*, *779*, 360–369.
55. Stashenko, E.E., Puertas, M.A., Salgar, W., Delgado, W., Martinez, J.R. (2000). Solid-phase microextraction with on-fibre derivatisation applied to the analysis of volatile carbonyl compounds. *J. Chromatogr. A*, *886*, 175–181.
56. Larrayoz, P., Ibanez, F.C., Ordonez, A.I., Torre, P., Barcina, Y. (2001). Evaluation of supercritical fluid extraction as sample preparation method for the study of Roncal cheese aroma. *Int. Dairy J.*, *10*, 755–759.
57. Vichi, S., Castellote, A.I., Pizzale, L., Conte, L.S. (2003). Analysis of virgin olive volatile compounds by headspace solid-phase microextraction coupled to gas chromatography with mass spectrometric and flame ionization detection. *J. Chromatogr. A*, *983*, 19–33.
58. Biswas, S., Heindselmen, K., Wohltjen, H., Staff, C. (2004). Differentiation of vegetable oils and determination of sunflower oil oxidation using a surface acoustic wave sensing device. *Food Control*, *15*, 19–26.
59. Yasuhara, A., Tanaka, Y., Hengel, M., Shibamoto, T. (2003). Gas chromatographic investigation of acrylamide formation in browning model systems. *J. Agric. Food Chem.*, *51*, 3999–4003.
60. El-Ghorab, A.H., Fujioka, K., Shibamoto, T. (2006). Determination of acrylamide formed in asparagine/D-glucose Maillard model systems by using gas chromatography with headspace solid-phase microextraction. *J. AOAC Int.*, *89*, 149–153.
61. Brodzinsky, R., Singh, H.B. (1982). Volatile organic chemicals in the atmosphere: an assessment of available data, in *Final Report, U.S. Environmental Protection Agency, Contract No 68-02-3452*, SRI International, Menlo Park, pp. 3–4.

62. Serth, R.W., Tierney, D.R., Hughes, T.W. (1978). Sources of Human and Environmental Exposure. Source assessment, acrylic acid manufacture, state-of-the-art, in *Report EPA-600/2-78-004w, Industrial Environmental Research Laboratory*, U.S. Environmental Protection Agency, Cincinnati.
63. Hoshika, Y., Nihei, Y., Muto, G. (1981). Pattern display for characterization of trace amounts of odorants discharged from nine odour sources. *Analyst*, *106*, 1187–1202.
64. IARC (1995). Acrolein. *IARC Monogr. Eval. Carcinog. Risks Hum.*, *3*, 337–372.
65. Gunstone, F.D., Harwood, J.L., Padley, F.B. (1986). *The Lipid Handbook*, Chapman & Hall, New York.
66. Frankel, E.N. (1982). Volatile lipid oxidation products. *Prog. Lipid Res.*, *22*, 1–33.
67. Adkins, H., Hartung, W.H. (1935). Acrolein, in *Synthesis Organics*, Masson, Paris, pp. 1–4.
68. Fishbein, L. (1972). Pesticidal, industrial, food additive and drug mutagens, in *Mutagenic Effects of Environmental Contaminants* (eds H.E. Sutton, M.I. Harris), Academic Press, New York, pp. 129–170.
69. Yaylayan, V.A., Stadler, R.H. (2005). Acrylamide formation in food: a mechanistic perspective. *J. AOAC Int.*, *88*, 262–267.
70. Ehling, S., Hengel, M., Shibamoto, T. (2005). Formation of acrylamide from lipids. *Adv. Exp. Medicine Biol.*, *561*, 223–233.
71. Taeymans, D., Wood, J., Ashby, P., Blank, I., Studer, A., Stadler, R., Gonde, P., Eijck, P., Lalljie, S., Lingnert, H., Lindblom, M., Matissek, R., Mueller, D., Tallmadge, D., O'Brien, J., Thompson, S., Silvani, D., Ahitmore, T. (2004). A review of acrylamide: an industry perspective on research, analysis, formation, and control. *Crit. Rev. Food Sci. Nutr.*, *44*, 323–347.
72. Friedli, H.R., Atlas, E., Stroud, V.R., Giovanni, L., Campos, T., Radke, L.F. (2001). Volatile organic trace gases emitted from North American wildfires. *Global Biogeochem. Cycles*, *15*, 435–452.
73. National Academy of Sciences (1977). Acrolein, in *Drinking Water and Health*, National Academy of Sciences, Washington, DC, pp. 553–556.
74. ATSDR (2005). *Draft Toxicological Profile for Acrolein. U.S. Department of Health and Human Services, Public Health Service*, Agency for Toxic Substance and Disease Registry, September 2005.
75. USEPA (2003). *Toxicological Review of Acrolein*, United States Environmental Protection Agency, Washington, DC (EPA/635/R-03/003).
76. Gomes, R., Meek, M.E. (2002). Acrolein, in *Concise International Chemical Assessment Document 43*, World Health Organization, Geneva. http://www.inchem.org/documents/cicads/cicads/cicad43.htm (accessed on 23 May 2007).
77. ATSDR (2006). *Priority Data Needs for Acrolein*, Syracuse Research Corporation. U.S. Department of Health and Human Services. Public Health Service, Agency for Toxic Substance and Disease Registry, August 2006.
78. Kehrer, J.P., Biswal, S.S. (2000). The molecular effects of acrolein. *Toxicol. Sci.*, *57*, 6–15.
79. Li, L., Holian, A. (1998). Acrolein: a respiratory toxin that suppresses pulmonary host defence. *Rev. Environ. Health*, *13*, 99–108.

80. Heck, H.A., Casanova, M., McNulty, M.J., Lam, C.W. (1984). Mechanisms of nasal toxicity induced by formaldehyde and acrolein, in *Toxicol. Nasal Passages [CIIT Conf. Toxicol.]*, 7th edn (ed. C.S. Barrow), Hemisphere, Washington, DC, pp. 234–247.
81. Wasiak, M. (2000). Acrolein–its formation, occurrence and properties. *Przemysl Fermentacyjny I Owocowo-Warzywny*, *44*, 22–24.
82. Weber-Tschopp, A., Fischer, T., Gierer, R., Grandjean, E. (1977). Experimental irritating effects of acrolein on man. *Int. Arch. Occup. Environ. Health*, *40*, 117–130.
83. Bioassay Systems Corporation (1981). Acute oral toxicity (LD50) of acrolein in rats. Project #10258.
84. Microbiological Associates (1989). Acute oral toxicity study of acrolein, inhibited in rats. Summary of final report (#G-7230.220).
85. Steinhagen, W.H., Barrow, C.S. (1984). Sensory irritation structure-activity study of inhaled aldehydes in B6C#F1 and Swiss-Webster mice. *Toxicol. Appl. Pharmacol.*, *72*, 495–503.
86. Babiuk, C., Steinhagen, W.H., Barrow, C.S. (1985). Sensory irritation response to inhaled aldehydes after formaldehyde pre-treatment. *Toxicol. Appl. Pharmacol.*, *79*, 143–149.
87. Feron, V.J., Kruyses, A., Til, H.P., Immel, H.R. (1978). Repeated exposure to acrolein vapour: subacute studies in hamsters, rats and rabbits. *Toxicol.*, *9*, 47–57.
88. Watanabe, M., Sugimoto, M., Ito, K. (1992). The acrolein cytotoxicity and cytoprotective action of α-tocopherol in primary cultured rat hepatocytes. *Gastroenterologia Japonica*, *27*, 199–205.
89. Calingasan, N.Y., Uchida, K. Gibson, G. (1999). Protein-bound acrolein: a novel marker of oxidative stress in Alzheimer's disease. *J. Neurochem.*, *72*, 751–756.
90. Kawai, Y., Furuhata, A., Toyokuni, S., Aratani, Y., Uchida, K. (2003). Formation of acrolein-derived 2′-deoxyadenosine adduct in an iron-induced carcinogenesis model. *J. Biol. Chem.*, *278*, 50346–50354.
91. Luo, J., Shi, R. (2005). Acrolein induces oxidative stress in brain mitochondria. *Neurochem. Int.*, *46*, 243–252.
92. Shao, C.-X., Roberts, K.N., Markesbery, W.R., Scheff, S.W., Lovell, M.A. (2006). Oxidative stress in head trauma in aging. *Free Radic. Biol. Med.*, *41*, 77–85.
93. EPA (1987). *Health Effects Assessment for Acrolein. Report 1987*, US Environmental Protection Agency, Cincinnati.
94. Ahn, T., Kogan, V., Torres, E.M. (1995). Development of health risk assessment of toxic air emissions from wastewater treatment plants. In Proceedings of the Waste Environment Federation Annual Conference & Exposition, 68th, Miami Beach, Oct. 21–25, Water Environment Federation: Alexandria, 1995, 5, pp. 347–358.
95. Macak, J.J. III, Greidanus, B.E., Torosan, J. (1998). Inhalation health risk assessment of air toxic emissions from large combustion turbine power projects. In Proceedings, Annual Meeting—Air & Waste Management Association, Air & Waste Management Association.
96. Koehler, J., Stein, D.A. (2002). California experience with air toxics health risk assessments for proposed new power plants. In Proceedings of the Air & Waste Management Association's Annual Conference and Exhibition, 95th, Baltimore, Maryland, June 23–27, 2002, pp. 3678–3687.

97. Pratt, G.C., Palmer, D., Wu, C.-Y., Oliaei, F., Hollerbach, C., Fenske, M.J. (2000). An assessment of air toxics in Minnesota. *Environ. Health Perspect.*, *108*, 815–825.
98. Monroy, G.J., Keene, F.E. (1999). Binational air quality studies along the Arizona-Sonora border: ambos Nogales and Douglas-Agua Prieta. In Proceedings of Annual Meeting & Exhibition, Air & Waste Management Association, 92nd, St. Louis, Missouri, June 20–24, Air & Waste Management Association: Pittsburgh, pp. 4951–4962.
99. Destaillats, H., Spaulding, R.S., Charles, M.J. (2002). Ambient air measurement of acrolein and other carbonyls at the Oakland-San Francisco bay bridge toll plaza. *Environ. Sci. Technol.*, *36*, 2227–2235.
100. Woodruff, T.J., Wells, E.M., Holt, E.W., Burgin, D.E., Axelrad, D.A. (2007). Estimating risk from ambient concentrations of acrolein across the United States. *Environ. Health Perspect.*, *115*, 410–415.
101. ADI LIST (2006). *Acceptable Daily Intakes for Agricultural and Veterinary Chemicals*, Office of Chemical Safety, Department of Health and Ageing, Canberra, Australia.
102. Acrolein (CASRIN 107-02-8) (2007). Integrated Risk Information System. IRIS/US EPA. http://www.epa.gov/iris/subst/0364.htm (accessed 2 September 2007).
103. National Toxicology Program (1995). 13-Week gavage toxicity studies of ally acetate, ally alcohol, and acrolein in Fisher 344 rats and B6C3F1 mice (Tox report #48). Research Triangle Park, NC: U.S. Department of Health and Human Services, Public Health Service. National Toxicology Program.
104. Binetti, R., Costamanga, F.M., Marcello, I. (2006). Development of carcinogenicity classifications and evaluations: the case of formaldehyde. *Ann. Ist. Super. Sanita*, *42*, 132–143.
105. Speit, G., Schmid, O. (2006). Local genotoxic effects of formaldehyde in humans measured by the micronucleus test with exfoliated epithelial cells. *Mutat. Res.*, *613*, 1–9.
106. Golden, R. (2005). Formaldehyde: overview of current issues and challenges for the future. Proceedings of International Nonwovens Technical Conference, St. Louis Missouri.
107. Golden, R., Pyatt, D., Shields, P.G. (2006). Formaldehyde as a potential human leukemogen: an assessment of biological plausibility. *Crit. Rev. Toxicol.*, *36*, 135–153.
108. d'A. Heck, H., Casanova, M. (2004). The implausibility of leukaemia induction by formaldehyde: a critical review of the biological evidence on distant-site toxicity. *Regul. Toxicol. Pharmacol.*, *40*, 92–106.
109. Ngokere, A.A., Ofordile, P.M. (2003). The toxicity, mutagenicity and carcinogenicity of formaldehyde used in histology and histochemistry: a review. *Biomed. Res.*, *14*, 166–170.
110. Luttrell, W.E. (2003). Toxic tips: formaldehyde. *Chem. Health Safety*, *10*, 29–300.
111. Chenier, R. (2003). An ecological risk assessment of formaldehyde. *Human Ecol. Risk Assess.*, *9*, 483–509.
112. Liteplo, R.G., Meek, M.E. (2003). Inhaled formaldehyde: exposure estimation, hazard characterization, and exposure-response analysis. *J. Toxicol. Environ. Health, B: Crit. Rev.*, *6*, 85–114.

113. Koskinas, J. (2002). Acetaldehyde adducts: role in ethanol-induced liver disease. *Ethanol Liver*, 130–149.

114. Quertemont, E., Tambour, S. (2004). Is ethanol a pro-drug? The role of acetaldehyde in the central effects of ethanol. *Trends Pharmacol. Sci.*, *25*, 130–134.

115. Salaspuro, M.P. (2004). Alcohol, acetaldehyde, and digestive tract cancer, in *Nutrition and Alcohol*, CRC Press, Boca Raton, pp. 393–411.

116. Quertemont, E., Grant, K.A., Correa, M., Arizzi, M.N., Salamone, J.D., Tambour, S., Aragon, C.M.G., McBride, W.J., Rodd, Z.A., Goldstein, A., Zaffaroni, A., Li, T.-K., Pisano, M., Diana, M. (2005). The role of acetaldehyde in the central effects of ethanol. *Clin. Exp. Res.*, *29*, 221–234.

117. Brooks, P.J., Theruvathu, J.A. (2005). DNA adducts from acetaldehyde: implications for alcohol related carcinogenesis. *Alcohol*, *35*, 187–193.

2.3

HETEROCYCLIC AROMATIC AMINES

ROBERT J. TURESKY

Division of Environmental Disease Prevention, Wadsworth Center, NYS Department of Health, Albany, NY 12201, USA

2.3.1 INTRODUCTION

Heterocyclic aromatic amines (HAAs) are a class of hazardous chemicals in the diet that are receiving attention as a risk factor for human cancer. These chemicals were discovered 30 years ago by Professor Sugimura and colleagues in Japan (1), who showed that the charred parts and smoke generated from broiled fish and beef contained potent activity in *Salmonella typhimurium*-based mutagenicity assays. Since that discovery, more than 20 HAAs have been identified in cooked meats, fish, and poultry (2–5). Several HAAs have also been identified in cigarette smoke condensate and diesel exhaust (6, 7). Many HAAs are potent mutagens in bacteria, genotoxic to mammalian cells, and carcinogenic in experimental laboratory animals (4). The recent Report on Carcinogens, Eleventh Edition, of the National Toxicology Program, concluded that several prevalent HAAs are "reasonably anticipated" to be human carcinogens (8). Therefore, questions have been raised about the safety of foods containing HAAs and much research has been devoted to the biochemical toxicology of HAAs and their potential role in the etiology of human cancer.

Process-Induced Food Toxicants: Occurrence, Formation, Mitigation, and Health Risks,
Edited by Richard H. Stadler and David R. Lineback

Heterocyclic Aromatic Amines (HAAs): At a Glance	
Historical	HAAs were described some 30 years ago as potent *in vitro* mutagens present in smoke generated from broiled fish and beef. Since that discovery, >20 HAAs have been identified in cooked meats, fish, and poultry.
Analysis	Tandem SPE coupled with LC/MS provides a robust and selective method to detect and quantify HAAs in complex food matrices.
Occurrence in Food	Formed in many types of foods (meats, fish) and under diverse cooking conditions.
Main Formation Pathways	Two major classes of HAAs can be distinguished, namely, the "pyrolytic" HAAs and the aminoimidazoarenes. The former are formed at elevated temperatures (>250°C) during the pyrolysis of individual amino acids (e.g., tryptophan, glutamic acid, phenylalanine, and ornithine). The latter compounds occur in meats cooked at temperatures as commonly encountered in domestic practices (150–250°C), and the Maillard reaction is thought to play an important role in their formation (creatine, amino acids, and hexoses).
Mitigation in Food	Lower pan temperatures to cook meat and frequent turns as the meat cooks. However, it is critical to attain an internal temperature of 70°C in the cooked meat to ensure microbial inactivation. Addition of ingredients that may react with the precursors/intermediates could also reduce the amounts of HAAs. For thermal process flavors, optimization of the reaction conditions and the use of high-quality ingredients will minimize formation.
Dietary Intake	The major source of human exposure to HAAs is through consumption of household-cooked meats and fish. The range of HAAs detected in foods is highly variable and dependent on cooking preferences, i.e., pan-frying or barbecuing of meats at high temperature produce the greatest amounts of HAAs.
Health Risks	Many HAAs are potent bacterial mutagens, genotoxic to mammalian cells, and carcinogenic in experimental laboratory animals, inducing tumors in rodents in multiple organs (e.g., oral cavity, liver, stomach, lung, colorectum, mammary glands, and prostate). Human cancer risk factor estimates range widely. An upper limit is proposed at 1 cancer case per 1000 individuals, and a lower limit as 50 cases per million individuals.
Regulatory Information or Industry Standards	The European Commission has proposed regulations for maximum levels of PhIP and 4,8-DiMeIQx of 50 µg/kg in thermal process flavorings.
Knowledge Gaps	Sensitive quantitative methods are needed to measure human biomarkers of ideally multiple HAA exposure (urinary, protein, and DNA adducts), as the extent of exposure to various HAAs can vary in the diet. With better analytical techniques, it may be feasible to more reliably assess the exposure to HAAs, and to determine the biologically effective dose of each HAA and the resultant potential genetic damage. Another uncertainty in the risk assessment is the bioavailability of HAAs in cooked foods.
Useful Information (web) & Dietary Recommendations	http://eur-lex.europa.eu/LexUriServ/site/en/com/2006/com2006_0427en01.pdf Avoiding eating pan residues and scrapings of fried meat and fish. Fry/barbecue foods to the appropriate "doneness."

2.3.2 MECHANISMS OF HAA FORMATION

There are two major classes of HAAs. One class of compounds is formed at elevated temperatures and is known as "pyrolytic HAAs." These compounds arise during the pyrolysis (>250 °C) of individual amino acids that include tryptophan, glutamic acid, phenylalanine, and ornithine. Several HAAs are also formed during the pyrolysis of proteins such as soybean globulin and casein (1, 9). The reactions of these amino acids or proteins at high temperature produce deaminated and decarboxylated products, and reactive radical fragments, which combine to form heterocyclic ring structures. The pyrolytic HAAs are comprised of five structurally distinct groups that contain pyridoindoles, pyridoimidazoles, phenylpyridine, tetraazofluoranthene, or benzmidazole moieties (Fig. 2.3.1a). The tryptophan pyrolysate mutagens 2-amino-1,4-

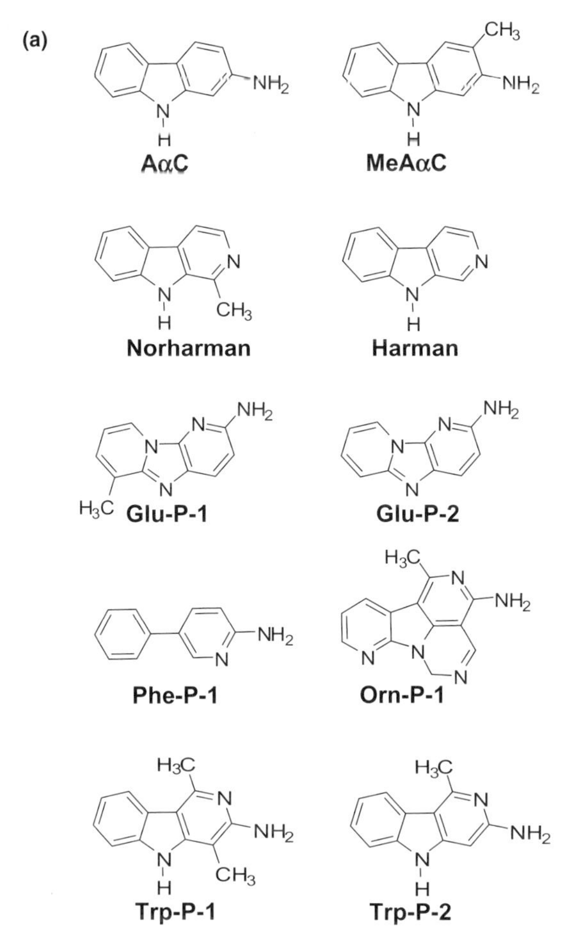

Figure 2.3.1 Chemical structures of HAAs: (a) pyrolysis HAAs and (b) aminoimidazoarene HAAs.

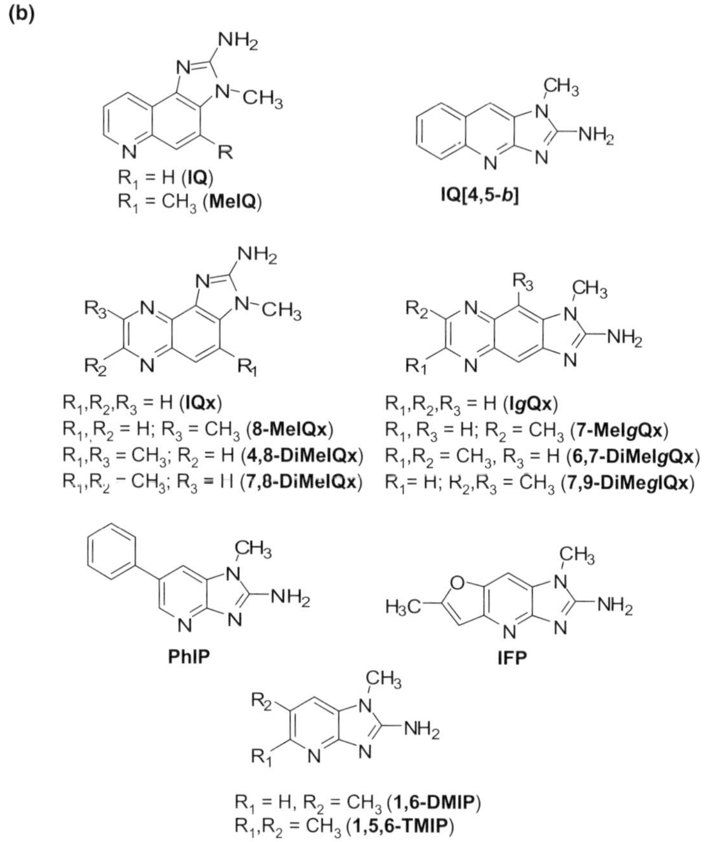

(b)

R_1 = H (**IQ**)
R_1 = CH_3 (**MeIQ**)

IQ[4,5-*b*]

R_1,R_2,R_3 = H (**IQx**)
R_1,R_2 = H; R_3 = CH_3 (**8-MeIQx**)
R_1,R_3 = CH_3; R_2 = H (**4,8-DiMeIQx**)
R_1,R_2 = CH_3; R_3 = H (**7,8-DiMeIQx**)

R_1,R_2,R_3 = H (**I*g*Qx**)
R_1, R_3 = H; R_2 = CH_3 (**7-MeI*g*Qx**)
R_1,R_2 = CH_3, R_3 = H (**6,7-DiMeI*g*Qx**)
R_1= H; R_2,R_3 = CH_3 (**7,9-DiMe*g*IQx**)

PhIP

IFP

R_1 = H, R_2 = CH_3 (**1,6-DMIP**)
R_1,R_2 = CH_3 (**1,5,6-TMIP**)

Figure 2.3.1 *Continued*

dimethyl-5*H*-pyrido[4,3-*b*]indole (Trp-P-1), 2-amino-4-methyl-5*H*-pyrido[4,3-*b*]indole (Trp-P-2), and the soybean globulin pyrolysate mutagens 2-amino-9*H*-pyrido[2,3-*b*]indole and 2-amino-3-methyl-9*H*-pyrido[2,3-*b*]indole (MeAαC), each contain an indole moiety as a part of their structure, which may be derived from tryptophan. The high temperature of burning cigarettes can catalyze the formation of several HAAs (10). AαC and MeAαC are two of the most abundant HAAs that arise in mainstream cigarette smoke (11), with levels reported at up to 258 and 37 ng/cigarette, respectively. These levels are considerably higher than those reported for many polycyclic aromatic hydrocarbons (PAHs) and arylamines (AAs), which are established human carcinogens (12).

The second class of HAAs is the aminoimidazoarenes (AIAs) (Fig. 2.3.1b). These compounds occur in meats cooked at temperatures commonly used in the household kitchen (150–250 °C). The Maillard reaction is thought to play an important role in the formation of AIAs. Jägerstad and coworkers proposed that creatine, free amino acids, and hexoses, which are present in uncooked meats, are the precursors of IQ and IQx compounds (13). The proposed pathways of AIA formation are presented in Fig. 2.3.2. The 2-amino-*N*-methylimidazo portion of the molecule is derived from creatine, and the remaining parts

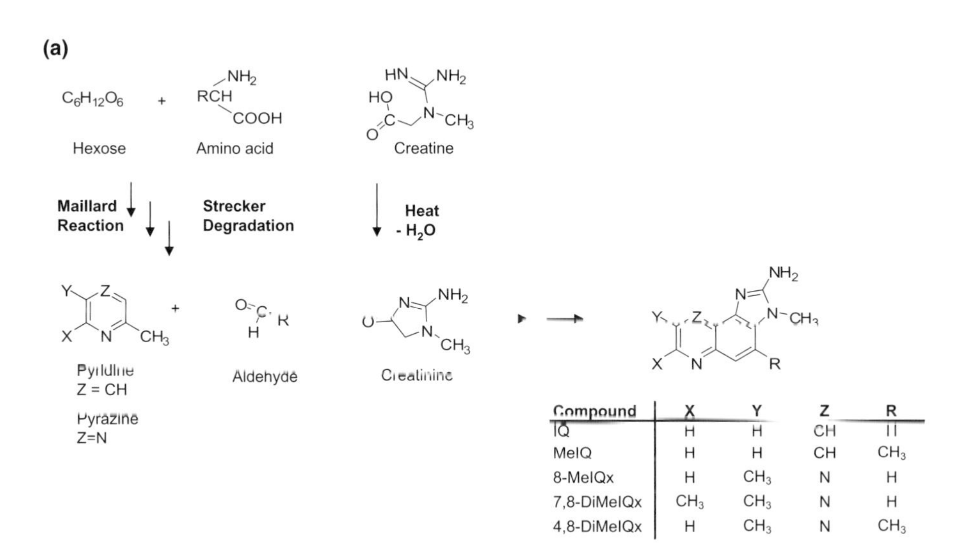

Compound	X	Y	Z	R
IQ	H	H	CH	H
MeIQ	H	H	CH	CH_3
8-MeIQx	H	CH_3	N	H
7,8-DiMeIQx	CH_3	CH_3	N	H
4,8-DiMeIQx	H	CH_3	N	CH_3

Figure 2.3.2 Proposed pathways of HAA formation: (a) formation of IQ and IQx-type compounds; and (b) formation of PhIP via phenylacetaldehyde as an intermediate (adapted from References 13 and 14, respectively, with permission).

of the 2-amino-3-methylimidazo[4,5-*f*]quinoline (IQ) and 2-amino-3-methylimidazo[4,5-*f*]quinoxaline (IQx) skeleton are assumed to arise from Strecker degradation products (e.g., pyridines or pyrazines), formed in the Maillard reaction between hexoses and amino acids (13, 15). An aldol condensation is thought to link the two molecules through an aldehyde or related Schiff base, to form IQ- and IQx-ring-structured HAAs.

Model systems composed of mixtures of creatine, amino acids, and glucose in aqueous diethylene glycol refluxed at temperatures ≥130 °C give rise to IQx, 2-amino-3,8-dimethylimidazo[4,5-*f*]quinoxaline (8-MeIQx) and 2-amino-3,4,8-trimethylimidazo[4,5-*f*]quinoxaline (4,8-DiMeIQx) (13, 16). In experiments using ^{14}C-labeled glucose in a model system with threonine and creatinine, the radioactive C-atoms of glucose were shown to be incorporated into the pyridine or pyrazine moiety of the AIAs (16). 2-Amino-1-methyl-6-phenylimidazo[4,5-*b*]pyridine (PhIP) was shown to form in a model system containing phenylalanine, creatinine, and glucose (17); however, PhIP can also form in the absence of sugar. Phenylalanine and creatine were shown, by dry heating of ^{13}C-labeled phenylalanine and creatine, to be precursors of PhIP (2); the Strecker aldehyde phenylacetaldehyde was identified as a critical intermediate (14) (Fig. 2.3.2b).

Free radicals appear to be involved in AIA formation in model systems, and *N,N*1-dialkylpyrazinium free radicals have been proposed as intermediates (15, 18, 19). IQx-type compounds arise in model systems containing glucose, glycine, and creatine through formation of pyrazine cation radical and carbon-centered radicals (19); evidence for this pathway was obtained by heating creatinine with 2,5-dimethylpyrazine or 2-methylpyridine and acetaldehyde to produce, respectively, 4,8-DiMeIQx or IQ (15). Moreover, the amounts of these AIA compounds formed in model systems were reduced in the presence of the phenolic antioxidants epigallocatechin gallate and flavonoids, suggesting that radical intermediates are involved in the formation of HAAs (19–21).

The kinetics and temperature dependence of HAA formation have been investigated in model systems containing HAA precursors (22, 23) and in fried ground beef (24). The strong temperature dependence of HAA formation for PhIP in particular, has been reported for cooked beef and poultry (25–27). In model systems, appreciable formation of HAAs generally only occurs at temperatures above 130 °C; however, pmol concentrations of PhIP were formed in a mixture containing 100-mM creatinine and 100-mM L-phenylalanine in the presence of sugar, heated at 37 or 60 °C for up to 7 days (28).

2.3.3 ENDOGENOUS FORMATION OF NOVEL HAAs

The β-carboline compounds 9*H*-pyrido[3,4-*b*]indole (norharman) and 1-methyl-9*H*-pyrido[3,4-*b*]indole (harman) are formed during the pyrolysis of tryptophan and are present at much higher amounts in tobacco condensates and cooked foods than are other pyrolytic or AIA HAAs (Fig. 2.3.1) (29). Norharman and harman are not mutagenic in *S. typhimurium* in the presence or absence of liver S9 fraction mixture; however, these β-carbolines become mutagenic when incubated with non-mutagenic aniline or *o*-toluidine in the presence of S9 fraction mixture (30). The co-mutagenic effect was attributed to the formation of novel HAAs (31). The structures of these compounds have been determined as 9-(4′-aminophenyl)-9*H*-pyrido[3,4-*b*]indole (aminophenylnorharman, APNH), 9-(4′-amino-3-methyl-phenyl)-9*H*-pyrido [3,4-*b*]indole (aminomethyl-phenylnorharman, AMPNH), and 9-(4′-aminophenyl)-1-methyl-9*H*-pyrido[3,4-*b*]indole (aminophenylharman, APH); the compounds are produced by the respective reactions of norharman with aniline, norharman and *o*-toluidine, and harman with aniline (32). The formation of APH derivatives is proposed to occur via a P450 complex, where an ipso attack on the aniline (or toluidine) can occur by norharman or harman to produce these compounds (Fig. 2.3.3a). Human P450s 1A2 and 3A4 are the most active of the enzyme isoforms that catalyze this process (33). APNH is a liver and colon carcinogen in F344 rats (34). Moreover, APNH is present at comparable concentrations in the urine of smokers, nonsmokers, and patients

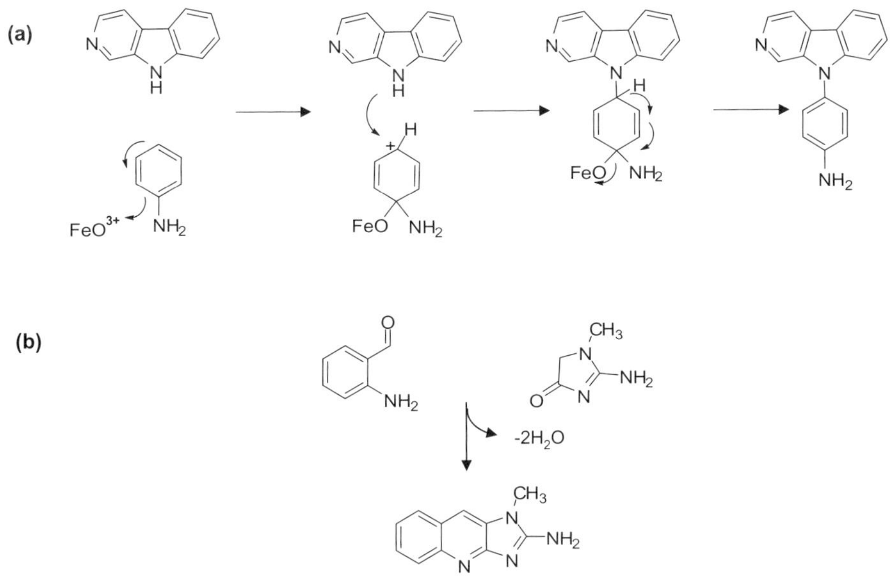

Figure 2.3.3 Endogenous formation of HAAs: (a) formation of 9-(4′-aminophenyl)-1-methyl-9*H*-pyrido[3,4-*b*]indole (APH) via a P450 complex with aniline (adapted from Reference 32 with permission); (b) formation of IQ[4,5-b] by reaction of 2-aminobenzaldehyde with creatinine.

receiving parenteral alimentation (35). These results suggest that APNH is a novel mutagen/carcinogen that is produced endogenously.

2-Amino-1-methylimidazo[4,5-*b*]quinoline (IQ[4,5-*b*]) is a weak bacterial mutagen (36, 37) and an isomer of IQ, a powerful experimental animal carcinogen (4). The amounts of IQ[4,5-*b*] measured in the urine of human volunteers who consumed grilled beef ranged from 15% to 135% of the ingested dose (38). Base treatment of urine at 70 °C increased the amount of IQ[4,5-*b*] by more than 100-fold in carnivores and also in the urine of vegetarians. Moreover, IQ[4,5-*b*], but not IQ, 8-MeIQx, or PhIP, arose in the urine incubated for 3 h at 37 °C: creatinine and 2-aminobenzaldehyde are the likely precursors of IQ[4,5-*b*] (Fig. 2.3.3b). These findings suggest that IQ[4,5-*b*] is present in nonmeat staples, or alternatively that this HAA occurs endogenously within the urine or other biological fluids.

2.3.4 ANALYTICAL METHODS TO MEASURE HAAs

Early investigations that sought to identify HAAs in cooked meat employed multiple chromatography steps, and the compounds were monitored, by the Ames bacterial mutagenesis *S. typhimurium* assay, at each step of the purification (39, 40). The final purification step was done by HPLC. IQ, MeIQ, and 8-MeIQx, the first AIAs discovered, were originally isolated from grilled or broiled fish and meat (41–44). The purified mutagenic fractions were characterized by ^{1}H NMR and mass spectrometry (MS) for structural elucidation, followed by chemical synthesis for corroboration of the structure (45, 46). These methods were extremely labor-intensive, and kg quantities of grilled meat were required to produce amounts of HAAs sufficient for spectroscopic measurements (41–44). Thereafter, a number of other AIAs were identified in cooked meats and poultry (3, 39, 47). The identification of PhIP was first reported in 1986 (48). Although the mutagenic potency of PhIP is not as great as that of IQ, MeIQ, or 8-MeIQx, PhIP was estimated to account for 75% of the mass of genotoxic material attributed to characterized HAAs in fried ground beef. Because the mutagenic potencies of HAAs vary over a range of >10,000-fold in bacterial assays (4), only HAAs possessing high mutagenic activity or present in great abundance were successfully isolated and characterized from cooked meats when bacterial mutagenicity assays were employed for screening.

The analysis of HAAs in cooked foods has been simplified through the advances in solid-phase extraction (SPE) methods, such as blue cotton and diatomaceous earth (silica). These SPE techniques permit the rapid isolation of many different HAAs from cooked meats and provide a cost-effective method of chemical analysis. Hayatsu discovered that copper phthalocyanine trisulfonate, a blue pigment commonly used in the dye industry, had a high affinity for aromatic compounds with three or more fused rings in their structure. This pigment, bound to a cotton support matrix, served as a selective means to isolate HAAs from cooked meats (49). Since then, more robust supports have been devised that employ rayon or chitin (50).

The tandem SPE technique, developed by Gross and colleagues (51, 52), employs diatomaceous earth, which is placed in series with a propylsulfonic acid silica (PRS)-based resin, to purify HAAs from cooked meats. The HAAs are selectively desorbed from the resins as a function of polarity and basicity. A final cleanup is achieved with a C_{18} resin. The technique was recently modified by Turesky *et al.* (27); replacement of the PRS with an Oasis® MCX (Waters) (mixed cation exchange/hydrophobic resin) cartridge, enabled HAAs of diverse polarities and basicities to be collected in one fraction. The tandem SPE technique has been widely applied to measure HAAs in cooked meat samples (53).

The determination of HAAs has been carried out by numerous techniques: HPLC with ultraviolet (41, 51, 54, 55), electrochemical (56) or fluorescence detection (51, 53), liquid chromatography/mass spectrometry (LC/MS) (57–60), gas chromatography/mass spectrometry (GC/MS) (61–63), capillary zone electrophoresis (64, 65), enzyme-linked immunosorbent assays (ELISA) (66), and immunoaffinity purification followed by HPLC-UV detection (67) (see References 53 and 64 and references therein).

LC/MS is the most robust and selective method to detect HAAs in complex matrices. The first LC/MS analyses of HAAs in cooked meat were done by thermospray ionization (57, 59), monitoring the protonated molecules $[M + H]^+$ in the selected ion monitoring (SIM) scan mode. With the advent of atmospheric pressure ionization (API) techniques, tandem MS methods were established to measure HAAs. Both atmospheric pressure chemical ionization (APCI) (58, 68, 69) and electrospray ionization (ESI) (27, 70) have been employed; these ionization techniques provide robust methods of identification and quantification of HAAs, when stable isotopically labeled internal standards are employed in the assay. Both triple quadrupole (58, 69) and ion trap (71) mass spectrometers have been used to measure HAAs in cooked foods. When operated in the selected reaction monitoring (SRM) scan mode, tandem MS instruments can detect HAAs at amounts of <1 picogram (27). The tandem SPE method devised by Gross (51, 52), followed by LC-ESI-MS/MS, provides the most sensitive method of quantification of HAAs. A limit of quantification (LOQ) of ≈ 30 parts per trillion (ppt or pg/g) was reported for a number of HAAs in cooked meats, when 1- to 2-g quantities of cooked meat was assayed by this method (27). The HPLC-based fluorescence or UV detection is generally 3- to 10-fold less sensitive than that of the LC-ESI-MS/MS assay (53, 64).

The API product ion spectra of HAAs have been characterized. In the positive ionization mode, HAAs of the AIA series undergo a principal fragmentation, by APCI or ESI, at the *N*-methylimidazole moiety ($[M + H]^+ \rightarrow [M + H - CH_3]^{+\cdot}$) to produce radical cation species (5, 27, 58, 69, 72). These species can then undergo further fragmentation of the pyrazine or pyridine ring systems (27, 69, 72). For pyrolytic HAAs, such as the carbolines, the most important fragmentation occurs through the loss of ammonia ($[M + H]^+ \rightarrow [M + H - NH_3]^+$), followed by fragmentation of the heterocyclic rings (27, 72). The pyrolytic HAAs that contain a methyl group can also undergo fragmentation with the loss of the methyl group, to form radical cation species that undergo further fragmentation (27, 72). The product ion spectra scan mode has been used to corroborate the identities of many HAAs, and also to detect previously unknown HAA compounds that arise in cooked meats; the latter novel HAAs contain the 2-amino-1-methylimidazo[4,5-*g*]quinoxaline (I*g*Qx) or IQx skeletons, as judged from their product ion spectra (27).

2.3.5 HAAs IN COOKED MEATS, FISH, AND POULTRY

The concentrations of HAAs formed in many types of meats and under diverse cooking conditions have been summarized in several review articles (2, 47, 73). The amounts of individual HAAs in cooked meats range from 0.01 to several hundred ng/g cooked meat; the concentrations are dependent upon the type of meat and method of cooking. In general, panfrying or barbecuing of meats at high temperatures produces the greatest concentrations of HAAs. Roasting or broiling of meats generates lower amounts of HAAs, perhaps attributable to less efficient heat transfer and migration of HAA precursors to the meat surface, at which HAA formation occurs (26, 74, 75). Several pyrolytic HAAs have been reported to form at low ng/g concentrations in some broiled or grilled fish and meats cooked very well-done or burnt (27, 52, 62, 76–79). In contrast to pyrolytic HAAs, the AIAs arise in appreciable concentrations in meats cooked at a lower temperature: concentrations can range from <0.03 to ≈15 ng/g in meats that are cooked from medium to very well-done (25, 27, 74, 75, 80–82). One notable exception is PhIP: it can arise to concentrations of ≈500 ng/g in very well-done barbecued chicken (26).

The importance of the cooking temperature surface and the duration of cooking on the formation of HAAs in fried hamburgers are shown in Fig. 2.3.4. A cooking temperature of 160 °C gives rise to low concentrations of HAAs over time, but both the concentrations and numbers of individual HAAs increase significantly at higher temperatures. These kinetic data are in agreement with results from other studies (25–27). It is noteworthy that AIAs containing the I*g*Qx skeleton, linear tricyclic ring isomers of the angular tricyclic ring-structured IQx compounds, have recently been discovered in cooked meats (5, 83). One of these newly discovered compounds is 2-amino-1,7-dimethylimidazo[4,5-*g*]quinoxaline (7-MeI*g*Qx). 7-MeI*g*Qx is a weaker bacterial mutagen than many other HAAs present in fried beef, but it is the most abundant HAA formed in fried ground beef and steak, and occurs at up to 30 ng/g (84).

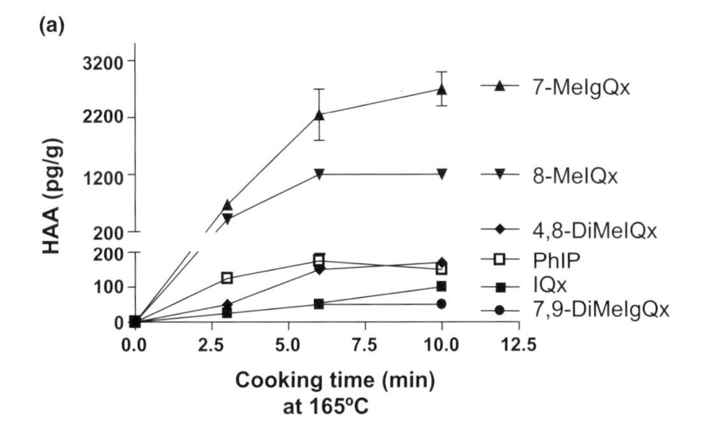

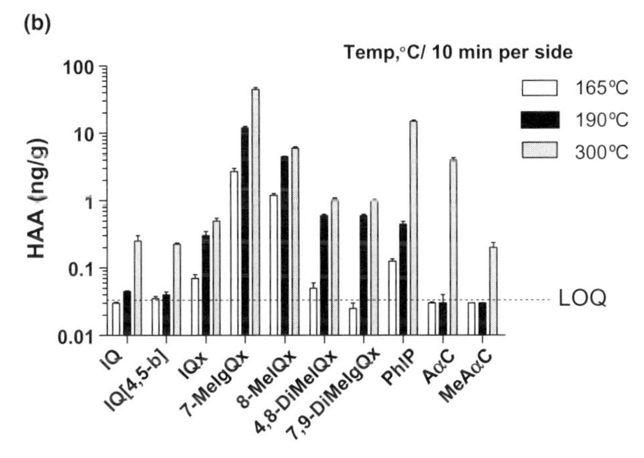

Figure 2.3.4 HAA formation as a function of (a) cooking time and (b) temperature (adapted from Reference 5 with permission). LOQ = Limit of quantification.

2.3.6 HAAs IN PAN-RESIDUE SCRAPINGS, PROCESSED FLAVORINGS, AND COMMERCIAL MEAT PRODUCTS

The pan residues and scrapings of cooked meats prepared in the household are often used as flavoring agents and sauce bases: such extracts have been reported to contain very high concentrations of HAAs measured on a per weight basis (up to 25–100 ng individual HAA/g) (27, 80, 85, 86). There are few reports on the amounts of HAAs formed in processed beef, chicken, and fish flavorings, and in commercial industrial beef extracts, which contain some of the same precursors as those used in model systems and cooked meats that generate HAAs. Different brands of commercial meat extracts contain variable levels of mutagenic activity in *S. typhimurium*-based assays (87). The brand of meat extract with the greatest activity contained approximately

70 ng/g of total IQ and IQx compounds, while the brand of meat extract with the weakest activity contained less than 10 ng/g of these HAAs (59). Samples of individual commercial meat flavorings also showed variable levels of mutagenic potency and were related to the specific creatinine content (88). IQ, 8-MeIQx, 4,8-DiMeIQx, and AαC have been identified in some batches of processed flavorings at concentrations ranging from 7 to 21 ng of HAA/g dry weight (89, 90). The processed flavorings or meat extract are added to foods, usually in the range of 0.1% to 2% by weight, particularly soups, gravies, sauces, snacks, and convenience foods. The concentrations of these HAAs in final packaged products, such as bouillon cubes, are generally below 1 ng/g (69, 91). Compared with published literature values of the HAA content of meats cooked under laboratory or household cooking conditions, industrial processed flavoring additives in final products, and also commercial (fast-food) restaurant hamburgers and other meat products, show low values overall, and appear to contribute only a small percentage of the estimated daily dietary intake of HAAs (92).

2.3.7 CONDITIONS AND FACTORS THAT CAN BE USED TO REDUCE THE FORMATION OF HAAs IN COOKED MEAT

The most effective and practical means by which to reduce the HAA content in fried hamburger or steak is to heat the meat at only a moderate temperature and to frequently turn the meat as it cooks (93). It is critical that an internal temperature of 70 °C be attained in the cooked meat, since a final minimum internal temperature of 70 °C is required to thermally inactivate *Escherichia coli*, in fried ground beef (93). Surprisingly, however, the time needed to reach the 70 °C internal temperature in fried hamburger patties was not dissimilar between frying beef at 250 °C (7 min) and frying at 160 °C (9 min): such similarity may be due to the limit of the slow transfer of heat through the meat (93). Use of lower pan temperatures to cook meat is probably the most practical and effective method to reduce HAA formation.

Removal of precursors is another method that can diminish HAA formation, although this approach may not be readily adapted for the average household. Microwave pretreatment was reported to reduce the amount of HAAs formed during the frying of ground beef (94). The meat juice released by the microwave pretreatment contained creatine, creatinine, amino acids, glucose, water, and fat: discarding these precursors resulted in lower amounts of HAAs. The sum of the HAAs decreased threefold following microwaving and frying at 200 °C or ninefold following microwaving and frying at 250 °C, relative to the HAA content of non-microwave-pretreated beef patties fried under identical conditions (94).

The amounts of HAAs formed in cooked meats can also be reduced if ingredients that can react with precursors or intermediates of the Maillard reaction are incorporated. An effective inhibitor (i) must be capable of causing significant reduction in the total HAA content; (ii) must not lead to formation of new HAAs; (iii) must not induce the formation of novel genotoxicants; and (iv) must not alter the organoleptic quality of the cooked meat (95). For many candidate ingredients, the efficacy of inhibition over variable temperatures and cooking time has not been thoroughly evaluated. Few reported studies have fully addressed the effectiveness and safety of inhibitors (see Reference 95 and references therein). Free radicals are believed to be involved in the formation of HAAs (19, 20), and antioxidants, such as butylated hydroxyanisole (BHA), butylated hydroxytoluene (BHT), propyl gallate (PG), and tert-butylhydroquinone (TBHQ), have been investigated as inhibitors of HAA formation in model systems containing HAA precursors (19, 20, 96). The efficacies of these antioxidants in reducing HAA formation widely varied. Natural antioxidants and some spices have been reported to effectively reduce HAA formation (97, 98). Among 14 antioxidants tested, green tea catechins and the major component ((–)-epigallocatechin gallate), two flavonoids (luteolin and quercetin) and caffeic acid were found to suppress the formation of both 8-MeIQx and PhIP in a model system, at levels of 3.2–75% of the amounts of the controls without additives (96). Theaflavin 3,3′-digallate, epicatechin gallate, rosmarinic acid, and naringenin were capable of simultaneously reducing the concentrations of PhIP, MeIQx, and 4,8-DiMeIQx in fried ground beef patties (99). Tea polyphenols were also found to diminish the formation of mutagens attributed to HAAs in fried ground beef (100). Minced garlic containing organosulfur compounds, such as diallyl sulfide, could inhibit HAA formation in model systems (101) and in assays using pork juice meat (102).

Marinades, which coat the outer surface of the meat where HAAs arise, have been shown to affect the amounts of HAAs formed during grilling or barbecuing of meat. Marinades containing sugar, soy sauce, and other spices applied to the meat greatly reduced the PhIP content in grilled chicken breast, but they increased the amount of 8-MeIQx (103). However, another study reported little change in the amounts of HAA formed, between marinated and non-marinated grilled chicken (104). Unexpectedly, a reduction of HAAs was seen when various types of carbohydrates were mixed with ground beef prior to panfrying (105); in contrast, experiments with pork showed that boiling of pork juice with sugar and soy sauce increased the content of several HAAs severalfold (106).

The fat content of the meat and the type of oil used for panfrying are two other factors that influence HAA formation. The fat content can either increase or decrease the formation of individual HAAs, depending on the method of cooking (86, 107–109). The total amount of HAAs formed in fried ground beef patties and pan residue combined was significantly lower with the use of sunflower seed oil or margarine than with the use of butter, margarine fat phase, liquid margarine, or rapeseed oil (108).

Oligosaccharides, soy proteins, dietary fiber, or sodium chloride/sodium tripolyphosphate, which are used as ingredients in some industrial meat products to improve the textural properties and to increase water-retentive capacity, can affect the transport of water-soluble precursors of HAAs from the inner part of the meat to the surface. Several of these dietary components have been shown to diminish HAA formation in fried beef. Addition of small amounts of complex carbohydrates to ground beef patties was reported to be an effective method of reducing the content of known HAAs (see Reference 109 and references therein).

2.3.8 BACTERIAL MUTAGENESIS ASSAYS, *IN VIVO* SHORT-TERM ASSAYS, AND CARCINOGENESIS OF HAAs

The mutagenic potency of an HAA is dependent upon the chemical structure and also upon the ability of the molecule to undergo *N*-oxidation to form the reactive nitrenium ion (110). This oxidation reaction is usually carried out by the cytochrome P450 enzymes (111). MeIQ, IQ, and 8-MeIQx rank among the most potent mutagens ever tested in the Ames bacterial reversion assay (4, 112, 113), while the other HAAs: PhIP, AαC, and 7-MeIgQx, are respectively ~200-, 1000- and 10,000-fold weaker in potency. The high propensity of some HAAs to induce frameshift revertant mutations in *S. typhimurium* TA98 and TA1538 tester strains is attributed to a preference by these compounds to react at a site about nine base pairs upstream of the original CG deletion in the *his*D$^+$ gene, within a run of GC repeats (114). Several HAAs also induce strong genotoxic effects in strain TA100, which reverts to the wild type through point mutations. However, many of these HAA–DNA lesions can be repaired, since the mutagenic potencies of several HAAs are 100-fold less active in the uvrB$^+$ proficient *S. typhimurium* strain (115). A correspondingly wide range of genotoxic potencies of HAAs is not generally seen in mammalian cell assays, in which many HAAs induce effects more comparable in magnitude (115–118). In some instances, the mutagenic potencies of even the weaker HAAs can be increased by up to 250-fold in *S. typhimurium* TA1538/1,8-DNP-derived strains engineered to express *N*-acetyltransferase (NATs) or sulfotransferase (SULTs) proteins (117, 118), or mammalian cells (117, 119, 120), thereby demonstrating the importance of xenobiotic metabolism enzymes (XMEs) in biological properties of these genotoxicants.

The transgene Muta mouse, Big Blue mouse, and rat models have been used to assess the mutagenicity of various classes of genotoxicants (121), including the HAAs IQ (122, 123), MeIQ (124), MeIQx (125), PhIP (126), and AαC (127) in various organs of transgene animals with either *lacZ* or *lacI* genes. In studies on MeIQ, IQ, and PhIP, the nature of the mutations in the *lacI* gene was consistent with that of known mutations in *Ha-ras* and *Apc* genes in HAA-induced tumors of rodents (128, 129). The induction of *in vivo* mutations in the *cII* transgene of liver and colon of IQ-treated rats (123), and small and large intestines (130) of mice treated with MeIQ, PhIP, and 2-AαC, has also been reported (127, 130). The highest mutation frequencies in the *cII* gene have been reported for PhIP: a high-fold induction of both G/C to AT transitions and G/C to T/A transversions in this gene occurred in the colon of male rats and mice (131, 132). The latter mutation is also frequently induced by PhIP in endogenous loci such as the human *HPRT* gene (133, 134), an event consistent with the near-exclusive reaction of PhIP with dG to form the adduct dG-C8–PhIP (135, 136), which can form a miscode base pair with adenine (137). Additionally, –G frameshifts in homopolymeric runs of guanine bases around the nucleotide 179 –(G:C) of the *cII* gene have been seen in the colon following PhIP treatment (131, 132). A–G frameshift mutation in the GGGA sequence is a frequent signature mutation of PhIP in the *APC* tumor suppressor (138) and *HPRT* genes (133, 134); however, this sequence occurs only once in the *cII* gene, and no mutations were observed at this position in this transgene. While reporter transgenes may be useful predictors of HAA-induced carcinogenesis and represent characteristic mutations that can also occur in cancer-related genes, relationships among DNA adduct formation, mutation frequencies, and cancer incidences of several HAAs have not shown quantitative correlations. Moreover, mutations in transgenes also occur in organs that do not develop tumors. Some of these discrepancies can be attributed to differences in cell-proliferation rates among different organs that affect mutation frequencies. Alternatively, the numbers of mutations and types of genetic alterations required for cancer development likely vary among different organs (124).

HAAs are carcinogenic in rodents and induce tumors in the oral cavity, liver, stomach, lung, colorectum, prostate, and mammary glands, following long-term feeding studies (4, 34). The total dose required to induce tumor formation (TD_{50}) varies for each HAA and is host species-dependent. The TD_{50} values of the most prevalent dietary HAAs range from 0.1 to 64.6 mg/kg/day in rodents (4). IQ is also a powerful liver carcinogen in nonhuman primates: tumor induction occurs within several years, making this compound one of the most powerful carcinogens assayed in nonhuman primates (4). Summaries of the genetic alterations of target genes of HAAs in experimental animal carcinogenicity studies are available (4, 129, 139).

2.3.9 METABOLISM OF HAAs AND IMPLEMENTATION OF HAA BIOMARKERS IN EPIDEMIOLOGY STUDIES

Reliable data on exposures to and bioavailability of HAAs in epidemiological studies are required if we are to assess the actual role of these compounds in human cancers (140). A tabulation of 30 human cancer epidemiology studies that related the consumption of well-done meat to various tumor sites has indicated that in about 80% of the studies, a positive correlation was seen between cancer incidence and consumption of well-done meat (141). However, the extent of HAA exposure was unknown in the study populations and a causal role for HAAs or other genotoxicants in cooked meat and cancer risk could not be established. A food frequency questionnaire (FFQ) containing a meat-cooking practice module was established, as a means to gauge an individual's preference of meat doneness, and to estimate the HAA intake (140), and the FFQ has been incorporated into several epidemiological studies. The findings revealed that red meat consumption, a preference for well-done cooked meat, and putative HAA exposure correlate with increased risk of cancer of the colon, prostate, and female mammary gland (142–145).

Ideally, long-lived biomarkers of HAA exposure and genetic damage should be incorporated into population-based studies, to enable reliable assessment of exposure to health risks of these genotoxicants. The chemical modification of DNA by a genotoxicant is believed to be the initiating event that ultimately leads to cancer (146). However, the acquisition of target tissues is generally not feasible in population-based studies of healthy subjects. Consequently, surrogate markers that measure genetic damage and the biologically effective dose have been developed for tissues and fluids that can be obtained noninvasively. Potential surrogate markers include HAA–DNA adducts in lymphocytes and HAA metabolites bound to circulating blood proteins, such as hemoglobin (Hb) and serum albumin (SA). Although quantitation of these biomarkers can provide an estimate of exposure and the biologically effective dose (147), the markers do not provide a direct measure of genetic damage occurring in the target tissue.

The metabolic activation of HAAs occurs via P450-mediated *N*-oxidation of the exocyclic amino group to form the *N*-hydroxy-HAAs (148, 149); these metabolites can react with DNA or can undergo further transformation with SULTs or NATs, to produce unstable esters that also form adducts with DNA (150, 151). HAAs are principally metabolized by hepatic P450 1A2 and by P450s 1A1 and 1B1 in extrahepatic tissues of rodents and humans (149, 152–156). The principal HAA–DNA adducts arise from reaction between the C8 atom of dG and the exocyclic amino groups of the HAAs, to produce dG-C8–HAA adducts (150, 151, 157). For IQ and MeIQx, adducts also occur between the N^2 group of dG and the C-5 atom of IQ and 8-MeIQx, indicating charge delocalization of the nitrenium ion over the heteronucleus of the HAA (Fig. 2.3.5) (159). These adducts are believed to be responsible for the genotoxic effects of HAAs.

There are reports of putative HAA–DNA adducts in human colon, breast, and lymphocytes, when assayed by ^{32}P-postlabeling, immunohistochemical or accelerator mass spectrometric (AMS) methods (160–164). dG-C8-MeIQx was detected in colon and kidney of some individuals at levels of several adducts per 10^9 DNA bases, when assayed by ^{32}P-postlabeling (165). A GC/MS assay, based upon alkaline hydrolysis of putative dG-C8–HAA adducts back to the parent HAAs, revealed the presence of PhIP in the colorectal mucosa of individuals at several adducts per 10^8 DNA bases (166). Magagnotti and coworkers detected a base-labile adduct of PhIP, presumably dG-C8–PhIP, in long-lived lymphocytes of colorectal cancer subjects at levels of several adducts per 10^8 DNA bases (162). The adduct was detected in about 30% of subjects, and its level varied by over a 10-fold range, suggesting a variable PhIP intake

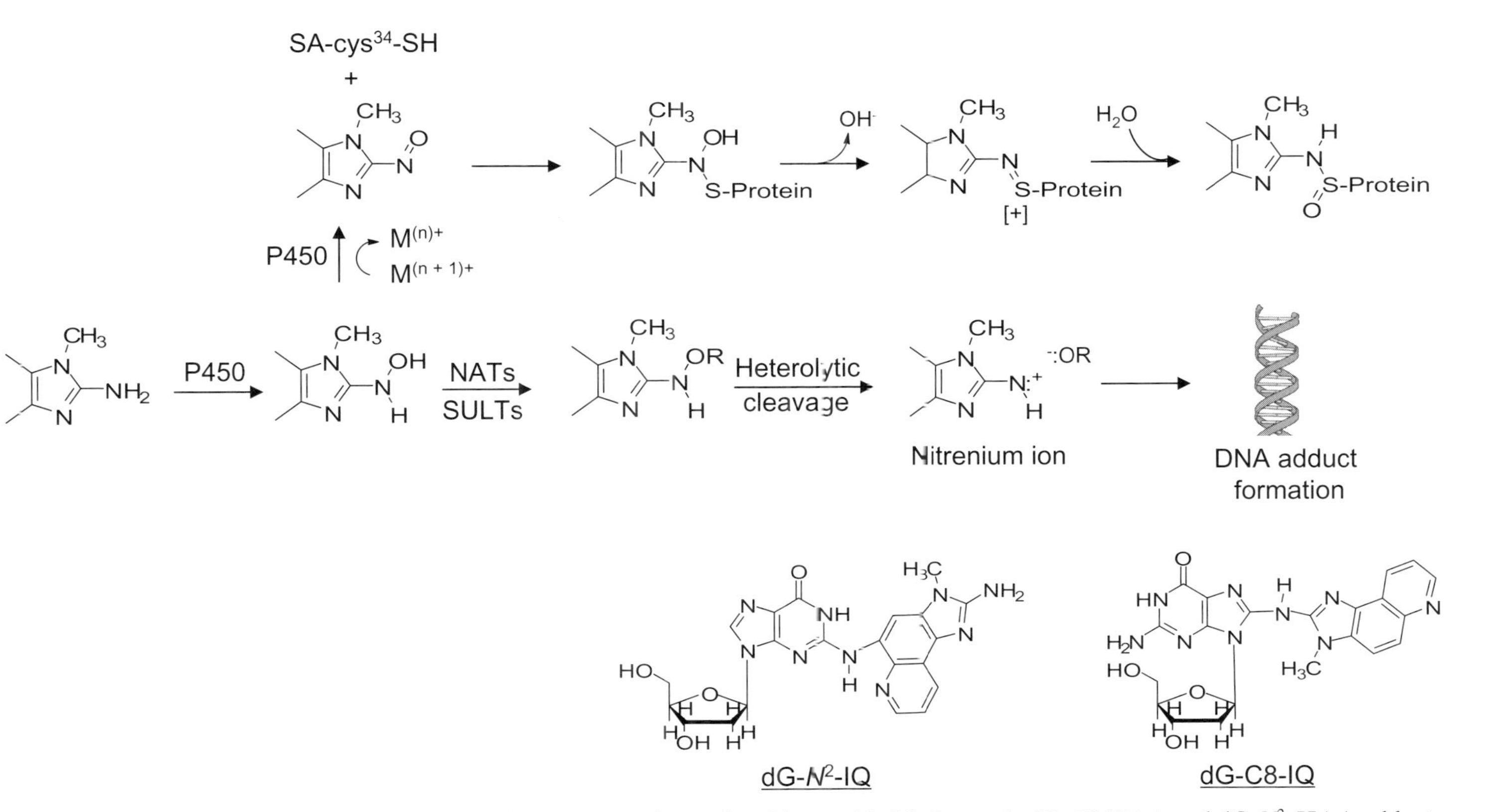

Figure 2.3.5 Metabolic activation of HAAs to form DNA and protein adducts with SA. Isomeric dG-C8-HAA and dG-N^2–HAA adducts have been reported for IQ and MeIQx. The *N*-hydroxy-HAA metabolite also may react with the cys^{34} sulfhydryl group of SA following oxidation to the nitroso-HAA, where formation is catalyzed by P450 1A2 catalysis or metal oxidation (158).

among individuals. The adduct levels were not significantly higher either in smokers or in high meat consumers compared with individuals who ate meat less frequently (≤5 servings per week). Relationships were not observed between PhIP–DNA adduct formation and alleles of *NAT1*, *NAT2*, or *SULT1A1*, genes of XMEs that are involved in bioactivation of PhIP (167). However, a subset of younger individuals carrying two mutated *GSTA1* alleles had higher adduct levels than did homozygous wild-type or heterozygous subjects (162). This observation is consistent with the activity of the GSTA1 protein in the detoxication of *N*-acetoxy-PhIP (168, 169). In two other studies, a DNA adduct of PhIP, presumably, dG-C8–PhIP, was detected by immunohistochemistry or ^{32}P-postlabeling in human breast tissue or epithelial cells at high frequency: levels were ~1 adduct per 10^7 bases (170, 171). Collectively, the data suggest that HAA–DNA adducts are formed in human tissues even though the concentrations of HAAs in the diet are low. However, the demonstration of the adduct structures by the methods employed in all of these studies was ambiguous: the ^{32}P-postlabeling and immunohistochemical methods lack specificity and are not quantitative (157); AMS does not provide confirmatory spectral data. Further, it requires treatment of subjects with radiolabeled isotopes, which is generally not feasible in large population-based studies, although postlabeling of DNA with radioactive derivatization reagents may circumvent this problem (172). Robust and sensitive LC-ESI tandem MS-based methods have emerged as critical tools for biomonitoring of intact-DNA adducts at levels approaching one adduct per 10^8 DNA levels, in experimental animal models and in humans (157, 173–177). Thus, the limit of detection (LOD) of some adducts measured by tandem MS approach the LOD determined by ^{32}P-postlabeling, and the ambiguities of analyte identity and measurement are circumvented by MS techniques, particularly when stable, isotopically labeled internal standards are used for quantification.

Protein adduct formation also occurs after metabolic activation of the HAAs to the *N*-hydroxy-HAA species. On the basis of experimental animal model studies and pilot human studies, the levels of IQ, 8-MeIQx, and PhIP bound to Hb are low (~0.01% of the dose) (161, 178, 179). This low level of protein binding will probably preclude the development of Hb–HAA adducts as biomarkers, because the LOD values for the adducts are beyond the limits of current MS instruments. However, the reaction product(s) of PhIP with SA shows promise as biomarker(s). Human SA is of 585 amino acids in length and is the most abundant protein in plasma (~45 mg/mL) (180). Its roles include maintenance of osmotic pressure and transport of endogenous (i.e., fatty acids, bilirubin, and steroids) and exogenous (drugs) chemicals. The cys^{34} is one of 35 conserved cysteine residues in SA across species (180). Thirty-four of these cysteines are involved in 17 disulfide bonds. The single unpaired cys^{34} is present either as a free thiol or in an oxidized form: this residue is present partially as disulfide linkages with low-molecular-weight thiols. Albumin functions as an antioxidant because of its scavenging of reactive oxygen and nitrogen species that are generated by basal aerobic metabolism. The cys^{34} of SA is thought to be responsible for many of the antioxidant properties of SA and accounts for ~80% of the net free thiols in plasma (181, 182), and it is a major transporter of NO in blood (183). The scavenging properties of this cys^{34} residue to reactive carcinogenic and some toxic electrophiles are well documented, and adducts at the cys^{34} have been identified with reactive metabolites of various toxicants in rodents or human SA, including IQ (184), 8-MeIQx (185), PhIP (186, 187), acrylamide (188), sulfur mustard (189), benzene (190), and acetaminophen (191).

An adduct formed between IQ and SA in the rat has been characterized by MS, ^{1}H NMR, and amino acid analysis as a sulfinamide adduct. Adduct formation occurred through the sulfhydryl group of the cys^{34} residue of SA and nitroso-IQ, accounting for about 10% of the total adducts (184). This adduct is labile toward acid and undergoes hydrolysis with quantitative regeneration of the parent amine. The mechanism of adduct formation is shown in Fig. 2.3.5. Other studies have reported that 8-MeIQx and PhIP form acid-labile adducts with SA in experimental animal models; these structures may be sulfinamide linkages at the cys^{34} (185, 192). Acid-labile PhIP–SA adduct(s) was detected in human subjects on a noncontrolled diet; levels were 10-fold higher in meat eaters than in vegetarians (6.7 ± 1.6 fmol vs. 0.7 ± 0.3 fmol PhIP-SA/mg protein; mean ± SE) (192). The structure(s) of the adduct attributed to the acid-labile lesion remains to be determined. It seems likely that some portion of the acid-labile PhIP adduction products was formed at the cys^{34} residue in human SA. Because the chemical stability of the adduct is unknown, additional studies are required to validate this biomarker prior to its use in population-based studies.

Urine is a useful biological fluid for the measurement of exposure to various classes of carcinogens and their metabolites, since large quantities can be obtained noninvasively (193). Although measurements of HAAs or their metabolites in urine do not shed light on DNA damage, they can assess the capacity of an individual to bioactivate and detoxicate HAAs. HAAs are rapidly absorbed from the gastrointestinal tract and are eliminated in urine as multiple metabolites within 24 h of consumption of grilled meats (194–197). Widely ranging concentrations of unmetabolized HAAs have been detected in urine of individuals evaluated worldwide; such differences are probably attributable to variability in the concentrations of HAAs in the diet. 8-MeIQx, PhIP, and the tryptophan pyrolysate mutagens Trp-P-1 and Trp-P-2 have been detected in the urine of healthy volunteers on a normal diet, but they were not found in the urine of patients receiving parenteral alimentation (198).

8-MeIQx and PhIP undergo extensive metabolism *in vivo*, so that only a percentage of the ingested dose is eliminated in urine as the unaltered compound. The excretion of unchanged 8-MeIQx and PhIP in the urine of male subjects in Western Europe and the United States was reported to be ~2% to 5% (8-MeIQx) and 0.5% to 2% (PhIP) of the dose within 10 h after consumption of well-done fried beef. Comparable percentages of unchanged AαC were eliminated in the urine of subjects during the same time frame, following consumption of very well-done fried beef (199). In the United States, the urinary excretion of 8-MeIQx and total acid-labile phase II conjugates of 8-MeIQx in male African-American subjects was 1.3- and 3.0-fold higher than in Asians and Caucasians, respectively (200). These differences in urinary 8-MeIQx content are probably due to racial dietary preferences for consuming fried beef or bacon rather than racial genetic differences. The urinary concentrations of 8-MelQx were positively associated with intake frequencies of bacon, pork/ham, and sausage/luncheon meats among study subjects. However, the urinary excretion concentrations of PhIP did not correlate with intake frequencies of any cooked meat, when the self-administered dietary questionnaire responses of the same group of subjects were analyzed (201). Therefore, urinary excretion levels of a single HAA may not serve as a reliable predictor of other levels of HAAs, in estimates of exposure to these compounds, for humans consuming unrestricted diets.

Individual expression of P450 1A2 can affect the metabolism and genotoxic potency of HAAs (202). The constitutive P450 1A2 mRNA expression levels in human liver can vary by as much as 15-fold (203, 204), and interindividual expression of P450 1A2 protein may vary by 60-fold in human liver (205–208). The variable expression of P450 1A2 in humans is attributed to environmental and dietary factors, such as smoking (209–211) and frequent consumption of cruciferous vegetables (212, 213) and grilled meats (214), all of which induce P450 1A2 expression. Several genetic polymorphisms (215, 216) and a variation in the extent of CpG methylation (204) have been detected in the upstream 5′-regulatory region of the P450 1A2 gene; these can affect the level of P4501A2 mRNA expression, and can lead to variation in the level of protein expression. In the case of 8-MeIQx, lower concentrations of the unchanged compound were found in the urine of individuals with high P450 1A2 activity, indicating that P450 1A2 is an important enzyme in the metabolism of 8-MeIQx *in vivo* (217). The contribution of P450 1A2 to the metabolism of PhIP in humans was reported to be less important than the enzyme's contribution to the metabolism of 8-MeIQx (218). However, glutathione *S*-transferases (GSTs) are able to reduce *N*-hydroxy-PhIP and *N*-acetoxy-PhIP back to the parent amine, and the occurrence of this "reverse" reaction can obscure the relationship between rapid P450 1A2 activity and PhIP metabolism (168). The importance of P450 1A2 in the metabolism of MeIQx and PhIP in humans was demonstrated in a pharmacokinetic study that used furafylline, a mechanism-based inhibitor of P450 1A2 (219). In that study, as much as 91% of the MeIQx and 70% of the PhIP following consumption of grilled meat were estimated to undergo metabolism by the enzyme (195).

The major pathways of metabolism of 8-MeIQx and PhIP in humans have been characterized (196, 220–222). Direct glucuronidation of these HAAs and their genotoxic *N*-hydroxylated metabolites was reported to be the major pathways of metabolism (Fig. 2.3.6). The levels of glucuronide conjugates of *N*-hydroxy-MeIQx and *N*-hydroxy-PhIP in urine were significantly higher than the levels measured in the urine of rodents (196, 197, 213), a finding consistent with the superior catalytic efficiency of human P450 1A2, relative

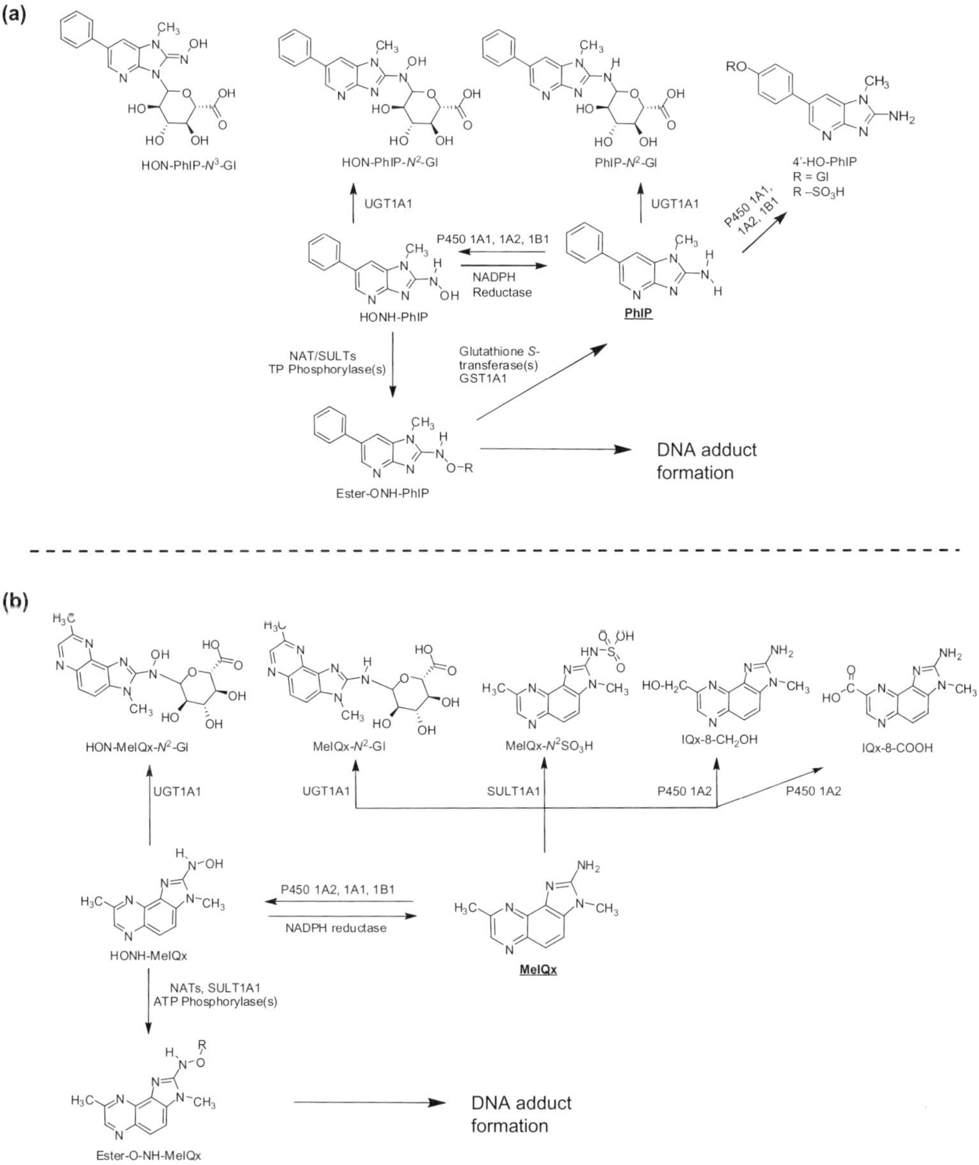

Figure 2.3.6 Major pathways of (a) PhIP and (b) MeIQx metabolism in humans.

to the rat orthologue, in *N*-oxidation of these HAAs (202, 207). In the case of MeIQx, oxidation of the C^8-methyl group was the major pathway of transformation to form the detoxicated metabolite, 2-amino-3-methylimidazo[4,5-*f*]quinoxaline-8-carboxylic acid (IQx-8-COOH) (223), which accounted for more than 50% of the dose excreted in the urine of subjects (196). All of the oxidation steps to form the carboxylic acid are catalyzed by P450 1A2 (224), a fact that underscores the prominent role of P450 1A2 in the metabolism of 8-MeIQx in humans (195). AIAs, unlike primary arylamines (225, 226), do not undergo direct metabolism by NATs to form detoxicated *N*-acetylated products.

2.3.10 HEALTH RISKS OF HAAs AND UNCERTAINTIES IN ASSESSMENT

Human cancer risk factor estimates for HAAs have ranged widely. An upper limit was proposed as one cancer case per 1000 individuals (227), and a lower limit as 50 cases per 10^6 individuals (228). The spread among the estimates can be attributed to inter-study differences in the assumptions used to calculate risk factors, including differing estimates of daily HAA intake, and the usage of TD_{50} values from various animal carcinogen bioassays, in which differences are seen in the HAA carcinogenic potency (229–231). As noted earlier, the actual exposure to HAAs can vary by more than 100-fold in cooked meats (2, 47, 80, 232, 233). Further complicating the extrapolation of the TD_{50} values, the carcinogen bioassays have used doses of HAAs at amounts exceeding the daily human exposure by 10^4- to 10^6-fold (231). Such high levels of HAA exposure may trigger metabolic pathways that preferentially lead to formation of chemically reactive metabolites normally not arising under low-dose treatments, or they may cause saturation of enzymatic detoxication systems and result in an enhanced HAA toxicity (234).

In contrast to many experimental animal models, humans show large inter-individual variations in the expression of cytochrome P450 enzymes and phase II enzymes that metabolize HAAs (207, 235 and references therein). The resultant differences in the expression of these enzymes may lead to different susceptibilities among individuals, which must be considered in risk assessment. Indeed, several epidemiological studies have shown a markedly increased risk of these cancers in subjects who frequently consume meats cooked well-done, and who are both rapid CYP1A2-mediated *N*-oxidizers and rapid acetylators (236–238): these enzymes bioactivate HAAs. There are important interspecies differences in P450 catalysis and regioselectivity of HAA oxidation; these differences affect the genotoxic potency of HAAs, and so must be taken into account when health risks are assessed (239).

Diet is another obvious variable that obscures the interspecies extrapolation of toxicity data from laboratory animals to humans. In contrast to the standardized diet fed to the experimental animals, the human diet is highly diverse and complex; numerous constituents of it may enhance or diminish the genotoxic potency of HAAs. Components in grilled meats can increase the expression of P450 1A2 and the bioactivation of HAAs in humans (214). Fortunately, dietary constituents can diminish the genotoxic potential of HAAs by inhibition of P450 1A2-mediated bioactivation of HAAs (240, 241) and/or induction of the expression of GSTs (168) and UDP-glucuronosyltransferases

(UGTs) (242), phase II enzymes involved in the detoxication of HAAs. Apiaceous vegetables, fruit juices, and beverages contain high levels of dietary components such as methoxypsoralens, apigenin, resveratrol, prenylflavonoids, and furanocoumarins, all of which are potent inhibitors of P450 1A2 activity (212, 243–246).

Another uncertainty in risk assessment of HAAs is the bioavailability of HAAs in cooked foods. The bioavailability of HAAs may be reduced by dietary constituents, such as chlorophyll, which bind to HAAs (247), or by the cooked meat matrix, where increasing the doneness of the meat appears to decrease the amount of HAA accessible from the meat matrix (248). Cooked meats contain a variety of carcinogens at low concentrations, which include polycyclic aromatic hydrocarbons, *N*-nitroso compounds, lipid peroxides, and other pro-oxidative agents, and fungal products, in addition to HAAs. The carcinogenic potency of grilled meats and health risk may be related not only to HAAs, but also to this complex mixture of genotoxic compounds (249–251).

2.3.11 FUTURE PROSPECTS ON RESEARCH OF HAAs IN HUMAN HEALTH RISK

The analysis of HAA biomarkers in humans remains a challenging analytical task, because HAAs are present at the ppb levels in the diet. However, as the sensitivity of MS instrumentation continues to improve, the establishment of LC/MS-based methods to measure biomarkers such as urinary HAA metabolites, and HAA–protein and HAA–DNA adducts in human populations will become more facile. Future studies will require MS analysis of biomarkers of multiple HAAs, since the extent of exposure to various HAAs can vary in the diet. Through a combination of analyses for multiple urinary metabolites, HAA–protein and HAA–DNA adducts, and other biomarkers of longer-term exposures, such as HAA accumulation in hair (252, 253), it may be feasible to assess more reliably the exposure to HAAs and to determine the biologically effective dose of each HAA and the resultant potential genetic damage. With the identification of such biomarkers, the interactive effects of genetic polymorphisms of XMEs involved in HAA metabolism (activation and detoxication) will be able to be correlated with the levels of adduction products and cancer risk in human population studies. Such analyses should clarify the role of HAAs as a critical dietary factor in the initiation of colorectal and other common human cancers.

REFERENCES

1. Sugimura, T., Nagao, N., Kawachi, T., Honda, M., Yahagi, T., Seino, Y., Stao, S., Matsukura, N., Matsushima, T., Shirai, A., Sawamura, M., Matsumoto, H. (1977).

Mutagen-carcinogens in food, with special reference to highly mutagenic pyrolytic products in broiled foods, in *Origins of Human Cancer, Book C* (eds H.H. Hiatt, J.D. Watson, J.A. Winstein), Cold Spring Harbour Laboratory, Cold Spring Harbour, New York, 1561–1577.

2. Felton, J.S., Jägerstad, M., Knize, M.G., Skog, K., Wakabayashi, K. (2000). Contents in foods, beverages and tobacco, in *Food Borne Carcinogens Heterocyclic Amines* (eds M., Nagao, T., Sugimura), John Wiley & Sons, Ltd, Chichester, England, pp. 31–71.
3. Skog, K.I., Johansson, M.A., Jägerstad, M.I. (1998). Carcinogenic heterocyclic amines in model systems and cooked foods: a review on formation, occurrence and intake. *Food Chem. Toxicol.*, *36*, 879–896.
4. Sugimura, T., Wakabayashi, K., Nakagama, H., Nagao, M. (2004). Heterocyclic amines: mutagens/carcinogens produced during cooking of meat and fish. *Cancer Sci.*, *95*, 290–299.
5. Turesky, R.J., Goodenough, A.K., Ni, W., McNaughton, L., LeMaster, D.M., Holland, R.D., Wu, R.W., Felton, J.S. (2007). Identification of 2-Amino-1,7-dimethylimidazo[4,5-g]quinoxaline: an abundant mutagenic heterocyclic aromatic amine formed in cooked beef. *Chem. Res. Toxicol*, *20*, 520–530.
6. Manabe, S., Izumikawa, S., Asakuno, K., Wada, O., Kanai, Y. (1991). Detection of carcinogenic amino-alpha-carbolines and amino-gamma-carbolines in diesel-exhaust particles. *Environ. Pollut.*, *70*, 255–265.
7. Manabe, S., Tohyama, K., Wada, O., Aramaki, T. (1991). Detection of a carcinogen, 2-amino-1-methyl-6-phenylimidazo[4,5-b]pyridine, in cigarette smoke condensate. *Carcinogenesis*, *12*, 1945–1947.
8. National Toxicology Program (2005) Report on Carcinogenesis, 11th edn, U.S. Department of Health and Human Services, Public Health Service, Research Triangle Park, NC.
9. Sugimura, T. (1992). Multistep carcinogenesis: a 1992 perspective. *Science*, *258*, 603–607.
10. Manabe, S., Wada, O., Kanai, Y. (1990). Simultaneous determination of amino-alpha-carbolines and amino-gamma-carbolines in cigarette smoke condensate by high-performance liquid chromatography. *J. Chromatogr.*, *529*, 125–133.
11. Yoshida, D., Matsumoto, T. (1980). Amino-alpha-carbolines as mutagenic agents in cigarette smoke condensate. *Cancer Lett.*, *10*, 141–149.
12. Hecht, S.S. (2003). Tobacco carcinogens, their biomarkers and tobacco-induced cancer. *Nat. Rev. Cancer*, *3*, 733–744.
13. Jägerstad, M., Skog, K., Grivas, S., Olsson, K. (1991). Formation of heterocyclic amines using model systems. *Mutat. Res.*, *259*, 219–233.
14. Murkovic, M. (2004). Formation of heterocyclic aromatic amines in model systems. *J. Chromatogr., B*, *802*, 3–10.
15. Milic, B.L., Djilas, S.M., Candadanoic-Brunet, J.M. (1993). Synthesis of some heterocyclic aminoimidazoarenes. *Food Chem.*, *46*, 273–276.
16. Skog, K., Jägerstad, M. (1993). Incorporation of carbon atoms from glucose into the food mutagens MeIQx and 4,8-DiMeIQx using 14C-labelled glucose in a model system. *Carcinogenesis*, *14*, 2027–2031.
17. Shioya, M., Wakabayashi, K., Sato, S., Nagao, M., Sugimura, T. (1987). Formation of a mutagen, 2-amino-1-methyl-6-phenylimidazo[4,5-b]-pyridine (PhIP) in

cooked beef, by heating a mixture containing creatinine, phenylalanine and glucose. *Mutat. Res.*, *191*, 133–138.

18. Pearson, A.M., Chen, C., Gray, J.I., Aust, S.D. (1992). Mechanism(s) involved in meat mutagen formation and inhibition. *Free Radic. Biol. Med.*, *13*, 161–167.
19. Kato, T., Harashima, T., Moriya, N., Kikugawa, K., Hiramoto, K. (1996). Formation of the mutagenic/carcinogenic imidazoquinoxaline-type heterocyclic amines through the unstable free radical Maillard intermediates and its inhibition by phenolic antioxidants. *Carcinogenesis*, *17*, 2469–2476.
20. Kikugawa, K., Kato, T., Hiramoto, K., Takada, C., Tanaka, M., Maeda, Y., Ishihara, T. (1999). Participation of the pyrazine cation radical in the formation of mutagens in the reaction of glucose/glycine/creatinine. *Mutat. Res.*, *444*, 133–144.
21. Kikugawa, K. (1999). Involvement of free radicals in the formation of heterocyclic amines and prevention by antioxidants. *Cancer Lett.*, *143*, 123–126.
22. Arvidsson, P., Van Boekel, M.A.J.S., Skog, K., Jägerstad, M. (1997). Kinetics of polar heterocyclic amines in a meat model system. *J. Food Sci.*, *62*, 911–916.
23. Ahn, J., Grun, I.U. (2005). Heterocyclic amines: 1. Kinetics of formation of polar and nonpolar heterocyclic amines as a function of time and temperature. *J. Food Sci.*, *70*, C173–C179.
24. Tran, N.L., Salmon, C.P., Knize, M.G., Colvin, M.E. (2002). Experimental and simulation studies of heat flow and heterocyclic amine mutagen/carcinogen formation in pan-fried meat patties. *Food Chem. Toxicol.*, *40*, 673–681.
25. Knize, M.G., Dolbeare, F.A., Carroll, K.L., Moore, D.H., Felton, J.S. (1994). Effect of cooking time and temperature on the heterocyclic amine content of fried beef patties. *Food Chem. Toxicol.*, *32*, 595–603.
26. Sinha, R., Rothman, N., Brown, E.D., Salmon, C.P., Knize, M.G., Swanson, C.S., Rossi, S.C., Mark, S.D., Levander, O.A., Felton, J.S. (1995). High concentrations of the carcinogen 2-amino-1-methyl-6-phenylimidazo[4,5-*b*]pyridine (PhIP) occur in chicken but are dependent on the cooking method. *Cancer Res.*, *55*, 4516–4519.
27. Turesky, R.J., Taylor, J., Schnackenberg, L., Freeman, J.P., Holland, R.D. (2005). Quantitation of carcinogenic heterocyclic aromatic amines and detection of novel heterocyclic aromatic amines in cooked meats and grill scrapings by HPLC/ESI-MS. *J. Agric. Food Chem.*, *53*, 3248–3258.
28. Manabe, S., Kurihara, N., Wada, O., Tohyama, K., Aramaki, T. (1992). Formation of PhIP in a mixture of creatinine, phenylalanine and sugar or aldehyde by aqueous heating. *Carcinogenesis*, *13*, 827–830.
29. Totsuka, Y., Ushiyama, H., Ishihara, J., Sinha, R., Goto, S., Sugimura, T., Wakabayashi, K. (1999). Quantification of the co-mutagenic beta-carbolines, norharman and harman, in cigarette smoke condensates and cooked foods. *Cancer Lett.*, *143*, 139–143.
30. Totsuka, T., Nishigaki, R., Sugimura, T., Wakabayashi, K. (2006). The possible involvement of mutagenic and carcinogenic heterocyclic amines in human cancer, in *Acrylamide and Other Hazardous Compounds in Heat-Treated Foods* (eds K. Skog, J. Alexander), Woodhead Publisher, Boca Raton, FL, pp. 296–515.
31. Hada, N., Totsuka, Y., Enya, T., Tsurumaki, K., Nakazawa, M., Kawahara, N., Murakami, Y., Yokoyama, Y., Sugimura, T., Wakabayashi, K. (2001). Structures of mutagens produced by the co-mutagen norharman with o- and m-toluidine isomers. *Mutat. Res.*, *493*, 115–126.

32. Oda, Y., Totsuka, Y., Wakabayashi, K., Guengerich, F.P., Shimada, T. (2006). Activation of aminophenylnorharman, aminomethylphenylnorharman and aminophenylharman to genotoxic metabolites by human N-acetyltransferases and cytochrome P450 enzymes expressed in Salmonella typhimurium umu tester strains. *Mutagenesis*, *21*, 411–416.

33. Nishigaki, R., Totsuka, Y., Takamura-Enya, T., Sugimura, T., Wakabayashi, K. (2004). Identification of cytochrome P-450s involved in the formation of APNH from norharman with aniline. *Mutat. Res.*, *562*, 19–25.

34. Kawamori, T., Totsuka, Y., Uchiya, N., Kitamura, T., Shibata, H., Sugimura, T., Wakabayashi, K. (2004). Carcinogenicity of aminophenylnorharman, a possible novel endogenous mutagen, formed from norharman and aniline, in F344 rats. *Carcinogenesis*, *25*, 1967–1972.

35. Nishigaki, R., Totsuka, Y., Kataoka, H., Ushiyama, H., Goto, S., Akasu, T., Watanabe, T., Sugimura, T., Wakabayashi, K. (2007). Detection of aminophenylnorharman, a possible endogenous mutagenic and carcinogenic compound, in human urine samples. *Cancer Epidemiol. Biomarkers Prev.*, *16*, 151–156.

36. Ronne, E., Olsson, K., Grivas, S. (1994) One-step synthesis of 2-amino-1-methylimidazo[4,5-*b*]quinoline. *Synth. Commun.*, *24*, 1363–1366.

37. Vikse, R., Hatch, F.T., Winter, N.W., Knize, M.G., Grivas, S., Felton, J.S. (1995). Structure-mutagenicity relationships of four amino-imidazonaphthyridines and imidazoquinolines. *Environ. Mol. Mutagen.*, *26*, 79–85.

38. Holland, R.D., Gehring, T., Taylor, J., Lake, B.G., Gooderham, N.J., Turesky, R.J. (2005). Formation of a mutagenic heterocyclic aromatic amine from creatinine in urine of meat eaters and vegetarians. *Chem. Res. Toxicol.*, *18*, 579–590.

39. Felton, J.S., Knize, M.G., Shen, N.H., Andresen, B.D., Bjeldanes, L.F., Hatch, F.T. (1986). Identification of the mutagens in cooked beef. *Environ. Health Perspect.*, *67*, 17–24.

40. Wakabayashi, K., Kim, I.S., Kurosaka, R., Yamaizumi, Z., Ushiyama, H., Takahashi, M., Koyota, S., Tada, A., Nukaya, H., Goto, S. (1995). Identification of new mutagenic heterocyclic amines and quantification of known heterocyclic amines, in *Heterocyclic Aromatic Amines: Possible Human Carcinogens. Proceedings of the 23rd International Princess Takamatsu Symposium* (eds R.H. Adamson, D.R. Gustafson, N. Ito, M. Nagao, T. Sugimura, K. Wakabayashi, Y. Yamazoe), Princeton Scientific Publishing Co., Inc, Princeton, pp. 39–49.

41. Kasai, H., Nishimura, S., Nagao, M., Takahashi, Y., Sugimura, T. (1979). Fractionation of a mutagenic principle from broiled fish by high-pressure liquid chromatography. *Cancer Lett.*, *7*, 343–348.

42. Kasai, H., Nishimura, K., Wakabayashi, K., Nagao, M., Sugimura, T. (1980) Chemical synthesis of 2-amino-3-methylimidazo[4,5-f]quinoline, a potent mutagen isolated from broiled fish. *Proc. Jpn. Acad.*, *56B*, 382–384.

43. Kasai, H., Yamaizumi, K., Wakabayashi, K., Nagao, M., Sugimura, T., Yokoyama, T., Miyazawa, T., Nishimura, S. (1980) Structure and chemical synthesis of Me-IQ, a potent mutagen isolated from broiled fish. *Chem. Lett.*, 1391–1394.

44. Kasai, H., Yamaizumi, Z., Shiomi, T., Yokoyama, T., Miyagawa, K., Wakabayashi, K., Nagao, M., Sugimura, T., Nishimura, S. (1981) Structure of a potent mutagen isolated from fried beef. *Chem. Lett.*, 488.

45. Sugimura, T., Sato, S. (1983). Mutagens-carcinogens in foods. *Cancer Res.*, *43*, 2415s–2421s.
46. Grivas, S. (1995). Synthetic routes to the food carcinogen 2 amino-3,8-dimethylimidazo[4,5-f]quinoxaline (8-MeIQx) and related compounds, in *Heterocyclic Aromatic Amines: Possible Human Carcinogens. Proceedings of the 23rd International Princess Takamatsu Symposium* (eds R.H. Adamson, J.-A. Gusatfsson, N. Ito, M. Nagao, T. Sugimura, K. Wakabayashi, Y. Yamazoe), Princeton Scientific Publishing Co., Inc, Princeton, pp. 1–8.
47. Skog, K., Solyakov, A. (2002). Heterocyclic amines in poultry products: a literature review. *Food Chem. Toxicol.*, *40*, 1213–1221.
48. Felton, J.S., Knize, M.G., Shen, N.H., Lewis, P.R., Andresen, B.D., Happe, J., Hatch, F.T. (1986). The isolation and identification of a new mutagen from fried ground beef: 2-amino-1-methyl-6-phenylimidazo[4,5-b]pyridine (PhIP). *Carcinogenesis*, *7*, 1081–1086.
49. Hayatsu, H. (1995). Complex formation of heterocyclic amines with porphyrins: its use in detection and prevention. *Princess Takamatsu Symp.*, *23*, 172–180.
50. Skog, K. (2004). Blue cotton, Blue Rayon and Blue Chitin in the analysis of heterocyclic aromatic amines—a review. *J. Chromatogr., B*, *802*, 39–44.
51. Gross, G.A. (1990). Simple methods for quantifying mutagenic heterocyclic aromatic amines in food products. *Carcinogenesis*, *11*, 1597–1603.
52. Gross, G.A., Grüter, A. (1992). Quantitation of mutagenic/carcinogenic heterocyclic aromatic amines in food products. *J. Chromatogr.*, *592*, 271–278.
53. Pais, P., Knize, M.G. (2000). Chromatographic and related techniques for the determination of aromatic heterocyclic amines in foods. *J. Chromatogr., B*, *747*, 139–169.
54. Spingarn, N.E., Kasai, H., Vuolo, L.L., Nishimura, S., Yamaizumi, Z., Sugimura, T., Matsushima, T., Weisburger, J.H. (1980). Formation of mutagens in cooked foods. III. Isolation of a potent mutagen from beef. *Cancer Lett.*, *9*, 177–183.
55. Turesky, R.J., Wishnok, J.S., Tannenbaum, S.R., Pfund, R.A., Buchi, G.H. (1983). Qualitative and quantitative characterization of mutagens in commercial beef extract. *Carcinogenesis*, *4*, 863–866.
56. Takahashi, M., Wakabayashi, K., Nagao, M., Yamamoto, M., Masui, T., Goto, T., Kinae, N., Tomita, I., Sugimura, T. (1985). Quantification of 2-amino-3-methylimidazo[4,5-f]quinoline (IQ) and 2-amino-3,8-dimethylimidazo[4,5-f]quinoxaline (MeIQx) in beef extracts by liquid chromatography with electrochemical detection (LCEC). *Carcinogenesis*, *6*, 1195–1199.
57. Yamaizumi, Z., Kasai, H., Nishimura, S., Edmonds, C.G., McCloskey, J.A. (1986). Stable isotope dilution quantification of mutagens in cooked foods by combined liquid chromatography-thermospray mass spectrometry. *Mutat. Res.*, *173*, 1–7.
58. Holder, C.L., Preece, S.W., Conway, S.C., Pu, Y.M., Doerge, D.R. (1997). Quantification of heterocyclic amine carcinogens in cooked meats using isotope dilution liquid chromatography/atmospheric pressure chemical ionization tandem mass spectrometry. *Rapid Commun. Mass Spectrom.*, *11*, 1667–1672.
59. Turesky, R.J., Bur, H., Huynh-Ba, T., Aeschbacher, H.U., Milon, H. (1988). Analysis of mutagenic heterocyclic amines in cooked beef products by high-performance liquid chromatography in combination with mass spectrometry. *Food Chem. Toxicol.*, *26*, 501–509.

60. Pais, P., Moyano, E., Puignou, L., Galceran, M.T. (1997). Liquid chromatography-atmospheric-pressure chemical ionization mass spectrometry as a routine method for the analysis of mutagenic amines in beef extracts. *J. Chromatogr., A*, *778*, 207–218.

61. Murray, S., Gooderham, N.J., Boobis, A.R., Davies, D.S. (1988). Measurement of MeIQx and DiMeIQx in fried beef by capillary column gas chromatography electron capture negative ion chemical ionisation mass spectrometry. *Carcinogenesis*, *9*, 321–325.

62. Skog, K., Solyakov, A., Arvidsson, P., Jägerstad, M. (1998). Analysis of nonpolar heterocyclic amines in cooked foods and meat extracts using gas chromatography-mass spectrometry. *J. Chromatogr., A*, *803*, 227–233.

63. Reistad, R., Rossland, O.J., Latva-Kala, K.J., Rasmussen, T., Vikse, R., Becher, G., Alexander, J. (1997). Heterocyclic aromatic amines in human urine following a fried meat meal. *Food Chem. Toxicol.*, *35*, 945–955.

64. Kataoka, H. (1997). Methods for the determination of mutagenic heterocyclic amines and their applications in environmental analysis. *J. Chromatogr., A*, *774*, 121–142.

65. Sentellas, S., Moyano, E., Puignou, L., Galceran, M.T. (2003). Determination of heterocyclic aromatic amines by capillary electrophoresis coupled to mass spectrometry using in-line preconcentration. *Electrophoresis*, *24*, 3075–3082.

66. Vanderlaan, M., Watkins, B.E., Hwang, M., Knize, M.G., Felton, J.S. (1988). Monoclonal antibodies for the immunoassay of mutagenic compounds produced by cooking beef. *Carcinogenesis*, *9*, 153–160.

67. Turesky, R.J., Forster, C.M., Aeschbacher, H.U., Würzner, H.P., Skipper, P.L., Trudel, L.J., Tannenbaum, S.R. (1989). Purification of the food-borne carcinogens 2-amino-3-methylimidazo[4,5-f]quinoline and 2-amino-3,8-dimethlyimidazo[4,5-f]quinoxaline in heated meat products by immunoaffinity chromatography. *Carcinogenesis*, *10*, 151–156.

68. Toribio, F., Moyano, E., Puignou, L., Galceran, M.T. (2000). Determination of heterocyclic aromatic amines in meat extracts by liquid chromatography-ion-trap atmospheric pressure chemical ionization mass spectrometry. *J. Chromatogr., A*, *869*, 307–317.

69. Guy, P.A., Gremaud, E., Richoz, J., Turesky, R.J. (2000). Quantitative analysis of mutagenic heterocyclic aromatic amines in cooked meat using liquid chromatography-atmospheric pressure chemical ionisation tandem mass spectrometry. *J. Chromatogr., A*, *883*, 89–102.

70. Busquets, R., Bordas, M., Toribio, F., Puignou, L., Galceran, M.T. (2004). Occurrence of heterocyclic amines in several home-cooked meat dishes of the Spanish diet. *J. Chromatogr., B*, *802*, 79–86.

71. Toribio, F., Moyano, E., Puignou, L., Galceran, M.T. (2002). Ion-trap tandem mass spectrometry for the determination of heterocyclic amines in food. *J. Chromatogr., A*, *948*, 267–281.

72. Toribio, F., Moyano, E., Puignou, L., Galceran, M.T. (2002). Multistep mass spectrometry of heterocyclic amines in a quadrupole ion trap mass analyser. *J. Mass Spectrom.*, *37*, 812–828.

73. Skog, K. (1993). Cooking procedures and food mutagens: a literature review. *Food Chem. Toxicol.*, *31*, 655–675.

74. Sinha, R., Rothman, N., Salmon, C.P., Knize, M.G., Brown, E.D., Swanson, C.A., Rhodes, D., Rossi, S., Felton, J.S., Levander, O.A. (1998). Heterocyclic amine content in beef cooked by different methods to varying degrees of doneness and gravy made from meat drippings. *Food Chem. Toxicol.*, *36*, 279–287.
75. Sinha, R., Knize, M.G., Salmon, C.P., Brown, E.D., Rhodes, D., Felton, J.S., Levander, O.A., Rothman, N. (1998). Heterocyclic amine content of pork products cooked by different methods and to varying degrees of doneness. *Food Chem. Toxicol.*, *36*, 289–297.
76. Yamaizumi, Z., Shiomi, T., Kasai, H., Nishimura, S., Takahashi, Y., Nagao, M., Sugimura, T. (1980). Detection of potent mutagens, Trp-P-1 and Trp-P-2, in broiled fish. *Cancer Lett.*, *9*, 75–83.
77. Matsumoto, T., Yoshida, D., Tomita, H. (1981). Determination of mutagens, amino-alpha-carbolines in grilled foods and cigarette smoke condensate. *Cancer Lett.*, *12*, 105–110.
78. Brockstedt, U., Pfau, W. (1998). Formation of 2-amino-α-carbolines in pan-fried poultry and ^{32}P-postlabelling analysis of DNA adducts. *Z. Lebensm. Unters. Forsch. A*, *207*, 472–476.
79. Skog, K., Augustsson, K., Steineck, G., Stenberg, M., Jägerstad, M. (1997). Polar and non-polar heterocyclic amines in cooked fish and meat products and their corresponding pan residues. *Food Chem. Toxicol.*, *35*, 555–565.
80. Skog, K., Steineck, G., Augustsson, K., Jägerstad, M. (1995). Effect of cooking temperature on the formation of heterocyclic amines in fried meat products and pan residues. *Carcinogenesis*, *16*, 861–867.
81. Felton, J.S., Knize, M.G. (1990). Heterocyclic amine mutagens/carcinogens in foods, in *Handbook of Experimental Pharmacology* (eds C.S. Cooper, P.L. Grover), Springer-Verlag, Berlin Heidelberg, pp. 471–502.
82. Solyakov, A., Skog, K., Jägerstad, M. (1999). Heterocyclic amines in process flavours, process flavour ingredients, bouillon concentrates and a pan residue. *Food Chem. Toxicol.*, *37*, 1–11.
83. Nukaya, H., Koyota, S., Jinno, F., Ishida, H., Wakabayashi, K., Kurosaka, R., Kim, I.S., Yamaizumi, Z., Ushiyama, H., Sugimura, T. (1994). Structural determination of a new mutagenic heterocyclic amine, 2-amino-1,7,9-trimethylimidazo[4,5-g]quinoxaline (7,9-DiMeIgQx), present in beef extract. *Carcinogenesis*, *15*, 1151–1154.
84. Ni, W., McNaughton, L., LeMaster, D., Sinha, R., Turesky, R.J. (2008). Quantitation of thirteen heterocyclic aromatic amines in cooked beef, pork and chicken by liquid chromatography-electrospray ionization/tandem mass spectrometry. *J. Agric. Food Chem.*, *56*, 68–79.
85. Gross, G.A., Turesky, R.J., Fay, L.B., Stillwell, W.G., Skipper, P.L., Tannenbaum, S.R. (1993). Heterocyclic aromatic amine formation in grilled bacon, beef and fish and in grill scrapings. *Carcinogenesis*, *14*, 2313–2318.
86. Johansson, M.A., Jägerstad, M. (1994). Occurrence of mutagenic/carcinogenic heterocyclic amines in meat and fish products, including pan residues, prepared under domestic conditions. *Carcinogenesis*, *15*, 1511–1518.
87. Aeschbacher, H.U., Turesky, R.J., Wolleb, U., Wurzner, H.P., Tannenbaum, S.R. (1987). Comparison of the mutagenic activity of various brands of food grade beef extracts. *Cancer Lett.*, *38*, 87–93.

88. Laser Reutersward, A., Skog, K., Jägerstad, M. (1987). Effects of creatine and creatinine content on the mutagenic activity of meat extracts, bouillons and gravies from different sources. *Food Chem. Toxicol.*, *25*, 747–754.
89. Jackson, L.S., Hargraves, W.A., Stroup, W.H., Diachenko, G.W. (1994). Heterocyclic aromatic amine content of selected beef flavours. *Mutat. Res.*, *320*, 113–124.
90. Stavric, B., Lau, B.P., Matula, T.I., Klassen, R., Lewis, D., Downie, R.H. (1997). Mutagenic heterocyclic aromatic amines (HAAs) in "processed food flavour" samples. *Food Chem. Toxicol.*, *35*, 185–197.
91. Solyakov, A., Skog, K., Jägerstad, M. (1999). Heterocyclic amines in process flavours, process flavour ingredients, bouillon concentrates and a pan residue. *Food Chem. Toxicol.*, *37*, 1–11.
92. Knize, M.G., Sinha, R., Rothman, N., Brown, E.D., Salmon, C.P., Levander, O.A., Cunningham, P.L., Felton, J.S. (1995). Heterocyclic amine content in fast-food meat products. *Food Chem. Toxicol.*, *33*, 545–551.
93. Salmon, C.P., Knize, M.G., Panteleakos, F.N., Wu, R.W., Nelson, D.O., Felton, J.S. (2000). Minimization of heterocyclic amines and thermal inactivation of Escherichia coli in fried ground beef. *J. Natl. Cancer Inst.*, *92*, 1773–1778.
94. Felton, J.S., Fultz, E., Dolbeare, F.A., Knize, M.G. (1994). Effect of microwave pretreatment on heterocyclic aromatic amine mutagens/carcinogens in fried beef patties. *Food Chem. Toxicol.*, *32*, 897–903.
95. Cheng, K.W., Chen, F., Wang, M. (2006). Heterocyclic amines: chemistry and health. *Mol. Nutr. Food Res.*, *50*, 1150–1170.
96. Oguri, A., Suda, M., Totsuka, Y., Sugimura, T., Wakabayashi, K. (1998). Inhibitory effects of antioxidants on formation of heterocyclic amines. *Mutat. Res.*, *402*, 237–245.
97. Murkovic, M., Steomberger, D., Pfannhauser, W. (1998). Antioxidant spices reduce the formation of heterocyclic amines in fried meat. *Z. Lebensm. Unters. Forsch. A/Food Res. Technol.*, *207*, 477–480.
98. Balogh, Z., Gray, J.I., Gomaa, E.A., Booren, A.M. (2000). Formation and inhibition of heterocyclic aromatic amines in fried ground beef patties. *Food Chem. Toxicol.*, *38*, 395–401.
99. Cheng, K.W., Chen, F., Wang, M. (2007). Inhibitory activities of dietary phenolic compounds on heterocyclic amine formation in both chemical model system and beef patties. *Mol. Nutr. Food Res.*, *51*, 969–976.
100. Weisburger, J.H., Veliath, E., Larios, E., Pittman, B., Zang, E., Hara, Y. (2002). Tea polyphenols inhibit the formation of mutagens during the cooking of meat. *Mutat. Res.*, *516*, 19–22.
101. Shin, H.S., Strasburg, G.M., Gray, J.I. (2002). A model system study of the inhibition of heterocyclic aromatic amine formation by organosulfur compounds. *J. Agric. Food Chem.*, *50*, 7684–7690.
102. Tsai, S.J., Jenq, S.N., Lee, H. (1996). Naturally occurring diallyl disulfide inhibits the formation of carcinogenic heterocyclic aromatic amines in boiled pork juice. *Mutagenesis*, *11*, 235–240.
103. Salmon, C.P., Knize, M.G., Felton, J.S. (1997). Effects of marinating on heterocyclic amine carcinogen formation in grilled chicken. *Food Chem. Toxicol.*, *35*, 433–441.

104. Tikkanen, L.M., Latva-Kala, K.J., Heinio, R.L. (1996). Effect of commercial marinades on the mutagenic activity, sensory quality and amount of heterocyclic amines in chicken grilled under different conditions. *Food Chem. Toxicol.*, *34*, 725–730.

105. Skog, K., Jägerstad, M., Reutersward, A.L. (1992). Inhibitory effect of carbohydrates on the formation of mutagens in fried beef patties. *Food Chem. Toxicol.*, *30*, 681–688.

106. Lan, C.M., Chen, B.H. (2002). Effects of soy sauce and sugar on the formation of heterocyclic amines in marinated foods. *Food Chem. Toxicol.*, *40*, 989–1000.

107. Johansson, M., Skog, K., Jägerstad, M. (1993). Effects of edible oils and fatty acids on the formation of mutagenic heterocyclic amines in a model system. *Carcinogenesis*, *14*, 89–94.

108. Johansson, M.A., Fredholm, L., Bjerne, I., Jägerstad, M. (1995). Influence of frying fat on the formation of heterocyclic amines in fried beefburgers and pan residues. *Food Chem. Toxicol.*, *33*, 993–1004.

109. Skog, K., Jägerstad, M. (2006). Modifying cooking conditions and ingredients to reduce the formation of heterocyclic amines, in *Acrylamide and Other Hazardous Compounds in Heat-Treated Foods* (eds K. Skog, J. Alexander), Woodhead Publishing Ltd, Cambridge, England, pp. 407–424.

110. Hatch, F.T., Knize, M.G., Colvin, M.E. (2001). Extended quantitative structure-activity relationships for 80 aromatic and heterocyclic amines: structural, electronic, and hydropathic factors affecting mutagenic potency. *Environ. Mol. Mutagen.*, *38*, 268–291.

111. Guengerich, F.P. (2002). N-hydroxyarylamines. *Drug Metab. Rev.*, *34*, 607–623.

112. Maron, D.M., Ames, B.N. (1983). Revised methods for the Salmonella mutagenicity test. *Mutat. Res.*, *113*, 173–215.

113. Sugimura, T. (1988). Successful use of short-term tests for academic purposes: their use in identification of new environmental carcinogens with possible risk for humans. *Mutat. Res.*, *205*, 33–39.

114. Fuscoe, J.C., Wu, R., Shen, N.H., Healy, S.K., Felton, J.S. (1988). Base-change analysis of revertants of the hisD3052 allele in *Salmonella typhimurium*. *Mutat. Res.*, *201*, 241–251.

115. Felton, J.S., Knize, M.G., Dolbeare, F.A., Wu, R. (1994). Mutagenic activity of heterocyclic amines in cooked foods. *Environ. Health Perspect.*, *102* (Suppl. 6), 201–204.

116. Thompson, L.H., Tucker, J.D., Stewart, S.A., Christensen, M.L., Salazar, E.P., Carrano, A.V., Felton, J.S. (1987). Genotoxicity of compounds from cooked beef in repair-deficient CHO cells versus Salmonella mutagenicity. *Mutagenesis*, *2*, 483–487.

117. Wu, R.W., Tucker, J.D., Sorensen, K.J., Thompson, L.H., Felton, J.S. (1997). Differential effect of acetyltransferase expression on the genotoxicity of heterocyclic amines in CHO cells. *Mutat. Res.*, *390*, 93–103.

118. Glatt, H., Pabel, U., Meinl, W., Frederiksen, H., Frandsen, H., Muckel, E. (2004). Bioactivation of the heterocyclic aromatic amine 2-amino-3-methyl-9H-pyrido [2,3-b]indole (MeAalphaC) in recombinant test systems expressing human xenobiotic-metabolizing enzymes. *Carcinogenesis*, *25*, 801–807.

119. Bendaly, J., Zhao, S., Neale, J.R., Metry, K.J., Doll, M.A., States, J.C., Pierce, W.M., Jr, Hein, D.W. (2007). 2-Amino-3,8-dimethylimidazo-[4,5-f]quinoxaline-induced DNA adduct formation and mutagenesis in DNA repair-deficient Chinese hamster ovary cells expressing human cytochrome P4501A1 and rapid or slow acetylator N-acetyltransferase 2. *Cancer Epidemiol. Biomarkers Prev.*, *16*, 1503–1509.

120. Metry, K.J., Zhao, S., Neale, J.R., Doll, M.A., States, J.C., McGregor, W.G., Pierce, W.M., Jr, Hein, D.W. (2007). 2-amino-1-methyl-6-phenylimidazo [4,5-b] pyridine-induced DNA adducts and genotoxicity in Chinese hamster ovary (CHO) cells expressing human CYP1A2 and rapid or slow acetylator N-acetyltransferase 2. *Mol. Carcinog.*, *46*, 553–563.

121. Lambert, I.B., Singer, T.M., Boucher, S.E., Douglas, G.R. (2005). Detailed review of transgenic rodent mutation assays. *Mutat. Res.*, *590*, 1–280.

122. Bol, S.A., Horlbeck, J., Markovic, J., de Boer, J.G., Turesky, R.J., Constable, A. (2000). Mutational analysis of the liver, colon and kidney of Big Blue rats treated with 2-amino-3-methylimidazo[4,5-f]quinoline. *Carcinogenesis*, *21*, 1–6.

123. Moller, P., Wallin, H., Vogel, U., Autrup, H., Risom, L., Hald, M.T., Daneshvar, B., Dragsted, L.O., Poulsen, H.E., Loft, S. (2002). Mutagenicity of 2-amino-3-methylimidazo[4,5-f]quinoline in colon and liver of Big Blue rats: role of DNA adducts, strand breaks, DNA repair and oxidative stress. *Carcinogenesis*, *23*, 1379–1385.

124. Nagao, M., Ochiai, M., Okochi, E., Ushijima, T., Sugimura, T. (2001). LacI transgenic animal study: relationships among DNA-adduct levels, mutant frequencies and cancer incidences. *Mutat. Res.*, *477*, 119–124.

125. Itoh1, T., Suzuki, T., Nishikawa, A., Furukawa, F., Takahashi, M., Xue, W., Sofuni, T., Hayashi, M. (2000). In vivo genotoxicity of 2-amino-3,8-dimethylimidazo[4, 5-f]quinoxaline in lacI transgenic (Big Blue) mice. *Mutat. Res.*, *468*, 19–25.

126. Lynch, A.M., Gooderham, N.J., Boobis, A.R. (1996). Organ distinctive mutagenicity in MutaMouse after short-term exposure to PhIP. *Mutagenesis*, *11*, 505–509.

127. Zhang, X.B., Felton, J.S., Tucker, J.D., Urlando, C., Heddle, J.A. (1996). Intestinal mutagenicity of two carcinogenic food mutagens in transgenic mice: 2-amino-1-methyl-6-phenylimidazo[4,5-b]pyridine and amino(alpha)carboline. *Carcinogenesis*, *17*, 2259–2265.

128. Okonogi, H., Ushijima, T., Zhang, X.B., Heddle, J.A., Suzuki, T., Sofuni, T., Felton, J.S., Tucker, J.D., Sugimura, T., Nagao, M. (1997). Agreement of mutational characteristics of heterocyclic amines in lacI of the Big Blue mouse with those in tumour related genes in rodents. *Carcinogenesis*, *18*, 745–748.

129. Nagao, M., Ushijima, T., Toyota, M., Inoue, R., Sugimura, T. (1997). Genetic changes induced by heterocyclic amines. *Mutat. Res.*, *376*, 161–167.

130. Itoh, T., Kuwahara, T., Suzuki, T., Hayashi, M., Ohnishi, Y. (2003). Regional mutagenicity of heterocyclic amines in the intestine: mutation analysis of the cII gene in lambda/lacZ transgenic mice. *Mutat. Res.*, *539*, 99–108.

131. Stuart, G.R., Thorleifson, E., Okochi, E., de Boer, J.G., Ushijima, T., Nagao, M., Glickman, B.W. (2000). Interpretation of mutational spectra from different genes: analyses of PhIP-induced mutational specificity in the lacI and cII transgenes from colon of Big Blue rats. *Mutat. Res.*, *452*, 101–121.

132. Smith-Roe, S.L., Hegan, D.C., Glazer, P.M., Buermeyer, A.B. (2005). Mlh1-dependent suppression of specific mutations induced in vivo by the food-borne

carcinogen 2-amino-1-methyl-6-phenylimidazo [4,5-b] pyridine (PhIP). *Mutat. Res.*, *596*, 101–112.

133. Yadollahi-Farsani, M., Gooderham, N.J., Davies, D.S., Boobis, A.R. (1996). Mutational spectra of the dietary carcinogen 2-amino-1-methyl-6-phenylimidazo[4,5-b]pyridine(PhIP) at the Chinese hamsters hprt locus. *Carcinogenesis*, *17*, 617–624.
134. Glaab, W.E., Kort, K.L., Skopek, T.R. (2000). Specificity of mutations induced by the food-associated heterocyclic amine 2-amino-1-methyl-6-phenylimidazo-[4,5-b]-pyridine in colon cancer cell lines defective in mismatch repair. *Cancer Res.*, *60*, 4921–4925.
135. Frandsen, H., Grivas, S., Andersson, R., Dragsted, L., Larsen, J.C. (1992). Reaction of the *N*2-acetoxy derivative of 2-amino-1-methyl-6-phenylimidazo[4,5-*b*]pyridine (PhIP) with 2′-deoxyguanosine and DNA. Synthesis and identification of N2-(2′-deoxyguanosin-8-yl)-PhIP. *Carcinogenesis*, *13*, 629–635.
136. Lin, D.-X., Kaderlik, K.R., Turesky, R.J., Miller, D.W., Lay, O.J. Jr. (1992). Identification of *N*-(Deoxyguanosin-8-yl)-2-amino-1-methyl-6-phenylimidazo[4,5-*b*]pyridine as the major adduct formed by the food-borne carcinogen 2-amino-1-methyl-6-phenylimidazo[4,5-b]pyridine, with DNA. *Chem. Res. Toxicol.*, *5*, 691–697.
137. Shibutani, S., Fernandes, A., Suzuki, N., Zhou, L., Johnson, F., Grollman, A.P. (1999). Mutagenesis of the N-(deoxyguanosin-8-yl)-2-amino-1-methyl-6-phenylimidazo[4,5-*b*]pyridine DNA adduct in mammalian cells. Sequence context effects. *J. Biol. Chem.*, *274*, 27433–27438.
138. Kakiuchi, H., Watanabe, M., Ushijima, T., Toyota, M., Imai, K., Weisburger, J.H., Sugimura, T., Nagao, N. (1995) Specific 5′-GGGA-3′ → 5′-GGA-3′ mutation of the APC gene in rat colon tumours induced by 2-amino-1-methyl-6-phenylimidazo[4,5-b]pyridine. *Proc. Natl. Acad. Sci. U.S.A.*, *92*, 910–914.
139. Nagao, M. (2000). Mutagenicity, in *Food Borne Carcinogens Heterocyclic Amines* (eds M. Nagao, T. Sugimura), John Wiley & Sons, Ltd, Chichester, England, pp. 163–195.
140. Sinha, R. (2002). An epidemiologic approach to studying heterocyclic amines. *Mutat. Res.*, *506–507*, 197–204.
141. Knize, M.G., Felton, J.S. (2005). Formation and human risk of carcinogenic heterocyclic amines formed from natural precursors in meat. *Nutr. Rev.*, *63*, 158–165.
142. Zheng, W., Gustafson, D.R., Sinha, R., Cerhan, J.R., Moore, D., Hong, C.P., Anderson, K.E., Kushi, L.H., Sellers, T.A., Folsom, A.R. (1998). Well-done meat intake and the risk of breast cancer. *J. Natl. Cancer Inst.*, *90*, 1724–1729.
143. Sinha, R., Chow, W.H., Kulldorff, M., Denobile, J., Butler, J., Garcia-Closas, M., Weil, R., Hoover, R.N., Rothman, N. (1999). Well-done, grilled red meat increases the risk of colorectal adenomas. *Cancer Res.*, *59*, 4320–4324.
144. Cross, A.J., Sinha, R. (2004). Meat-related mutagens/carcinogens in the etiology of colorectal cancer. *Environ. Mol. Mutagen.*, *44*, 44–55.
145. Cross, A.J., Peters, U., Kirsh, V.A., Andriole, G.L., Reding, D., Hayes, R.B., Sinha, R. (2005). A prospective study of meat and meat mutagens and prostate cancer risk. *Cancer Res.*, *65*, 11779–11784.

146. Miller, E.C. (1978). Some current perspectives on chemical carcinogenesis in humans and experimental animals: presidential address. *Cancer Res.*, *38*, 1479–1496.

147. Skipper, P.L., Tannenbaum, S.R. (1990). Protein adducts in the molecular dosimetry of chemical carcinogens. *Carcinogenesis*, *11*, 507–518.

148. Kato, R., Yamazoe, Y. (1987). Metabolic activation and covalent binding to nucleic acids of carcinogenic heterocyclic amines from cooked foods and amino acid pyrolysates. *Jpn. J. Cancer Res.*, *78*, 297–311.

149. Yamazoe, Y., Abu-Zeid, M., Manabe, S., Toyama, S., Kato, R. (1988). Metabolic activation of a protein pyrolysate promutagen 2-amino-3,8-dimethylimidazo[4,5-f]quinoxaline by rat liver microsomes and purified cytochrome P-450. *Carcinogenesis*, *9*, 105–109.

150. Turesky, R.J. (1994). DNA adducts of heterocyclic aromatic amines, arylazides and 4-nitroquinoline 1-oxide, in *DNA Adducts: Identification and Biological Significance* (eds K. Hemminki, A. Dipple, D.E.G. Shuker, F.F. Kadlubar, D. Segerbäck, H. Bartsch), International Agency for Research on Cancer, Lyon, pp. 217–228.

151. Schut, H.A., Snyderwine, E.G. (1999). DNA adducts of heterocyclic amine food mutagens: implications for mutagenesis and carcinogenesis. *Carcinogenesis*, *20*, 353–368.

152. Shimada, T., Guengerich, F.P. (1991). Activation of amino-α-carboline, 2-amino-1-methyl-6-phenylimidazo[4,5-b]pyridine, and a copper phthalocyanine cellulose extract of cigarette smoke condensate by cytochrome P-450 enzymes in rat and human liver microsomes. *Cancer Res.*, *51*, 5284–5291.

153. Wallin, H., Mikalsen, A., Guengerich, F.P., Ingelman-Sundberg, I., Solberg, K.E., Rossland, O.J., Alexander, J. (1990). Differential rates of metabolic activation and detoxification of the food mutagen 2-amino-1-methyl-6-phenylimidazo[4,5-b]pyridine by different cytochrome P450 enzymes. *Carcinogenesis*, *11*, 489–492.

154. Turesky, R.J., Lang, N.P., Butler, M.A., Teitel, C.H., Kadlubar, F.F. (1991). Metabolic activation of carcinogenic heterocyclic aromatic amines by human liver and colon. *Carcinogenesis*, *12*, 1839–1845.

155. Crofts, F.G., Strickland, P.T., Hayes, C.L., Sutter, T.R. (1997). Metabolism of 2-amino-1-methyl-6-phenylimidazo[4,5-b]pyridine (PhIP) by human cytochrome P4501B1. *Carcinogenesis*, *18*, 1793–1798.

156. Crofts, F.G., Sutter, T.R., Strickland, P.T. (1998). Metabolism of 2-amino-1-methyl-6-phenylimidazo[4,5-b]pyridine by human cytochrome P4501A1, P4501A2 and P4501B1. *Carcinogenesis*, *19*, 1969–1973.

157. Turesky, R.J., Vouros, P. (2004). Formation and analysis of heterocyclic aromatic amine-DNA adducts in vitro and in vivo. *J. Chromatogr., B*, *802*, 155–166.

158. Kim, D., Kadlubar, F.F., Teitel, C.H., Guengerich, F.P. (2004). Formation and reduction of aryl and heterocyclic nitroso compounds and significance in the flux of hydroxylamines. *Chem. Res. Toxicol.*, *17*, 529–536.

159. Turesky, R.J., Rossi, S.C., Welti, D.H., Lay, O.J., Jr, Kadlubar, F.F. (1992). Characterization of DNA adducts formed in vitro by reaction of N-hydroxy-2-amino-3-methylimidazo[4,5-f]quinoline and N-hydroxy-2-amino-3,8-dimethylimidazo[4,5-f]quinoxaline at the C-8 and N2 atoms of guanine. *Chem. Res. Toxicol.*, *5*, 479–490.

160. Turteltaub, K.W., Mauthe, R.J., Dingley, K.H., Vogel, J.S., Frantz, C.E., Garner, R.C., Shen, N. (1997). MeIQx-DNA adduct formation in rodent and human tissues at low doses. *Mutat. Res.*, *376*, 243–252.

161. Dingley, K.H., Curtis, K.D., Nowell, S., Felton, J.S., Lang, N.P., Turteltaub, K.W. (1999). DNA and protein adduct formation in the colon and blood of humans after exposure to a dietary-relevant dose of 2-amino-1-methyl-6-phenylimidazo[4,5-b]pyridine. *Cancer Epidemiol. Biomarkers Prev.*, *8*, 507–512.

162. Magagnotti, C., Pastorelli, R., Pozzi, S., Andreoni, B., Fanelli, R., Airoldi, L. (2003). Genetic polymorphisms and modulation of 2-amino-1-methyl-6-phenylimidazo[4,5-b]pyridine (PhIP)-DNA adducts in human lymphocytes. *Int. J. Cancer*, *107*, 878–884.

163. Turteltaub, K.W., Dingley, K.H., Curtis, K.D., Malfatti, M.A., Turesky, R.J., Garner, R.C., Felton, J.S., Lang, N.P. (1999). Macromolecular adduct formation and metabolism of heterocyclic amines in humans and rodents at low doses. *Cancer Lett.*, *143*, 149–155.

164. Lightfoot, T.J., Coxhead, J.M., Cupid, B.C., Nicholson, S., Garner, R.C. (2000). Analysis of DNA adducts by accelerator mass spectrometry in human breast tissue after administration of 2-amino-1-methyl-6-phenylimidazo[4,5-b]pyridine and benzo[a]pyrene. *Mutat. Res.*, *472*, 119–127.

165. Totsuka, Y., Fukutome, K., Takahashi, M., Takashi, S., Tada, A., Sugimura, T., Wakabayashi, K. (1996). Presence of N^2-(deoxyguanosin-8-yl)-2-amino-3,8-dimethylimidazo[4,5-f]quinoxaline (dG-C8-MeIQx) in human tissues. *Carcinogenesis*, *17*, 1029–1034.

166. Friesen, M.D., Kaderlik, K., Lin, D., Garren, L., Bartsch, H., Lang, N.P., Kadlubar, F.F. (1994). Analysis of DNA adducts of 2-amino-1-methyl-6-phenylimidazo[4,5-b]pyridine in rat and human tissues by alkaline hydrolysis and gas chromatography/electron capture mass spectrometry: validation by comparison with ^{32}P-postlabeling. *Chem. Res. Toxicol.*, *7*, 733–739.

167. Turesky, R.J. (2004). The role of genetic polymorphisms in metabolism of carcinogenic heterocyclic aromatic amines. *Curr. Drug Metab.*, *5*, 169–180.

168. Lin, D.-X., Meyer, D.J., Ketterer, B., Lang, N.P., Kadlubar, F.F. (1994). Effects of human and rat glutathione-S-transferase on the covalent binding of the N-acetoxy derivatives of heterocyclic amine carcinogens *in vitro*: a possible mechanism of organ specificity in their carcinogenesis. *Cancer Res.*, *54*, 4920–4926.

169. Coles, B.F., Kadlubar, F.F. (2003). Detoxification of electrophilic compounds by glutathione S-transferase catalysis: determinants of individual response to chemical carcinogens and chemotherapeutic drugs? *Biofactors*, *17*, 115–130.

170. Zhu, J., Chang, P., Bondy, M.L., Sahin, A.A., Singletary, S.E., Takahashi, S., Shirai, T., Li, D. (2003). Detection of 2-amino-1-methyl-6-phenylimidazo[4,5-b]-pyridine-DNA adducts in normal breast tissues and risk of breast cancer. *Cancer Epidemiol. Biomarkers Prev.*, *12*, 830–837.

171. Gorlewska-Roberts, K., Green, B., Fares, M., Ambrosone, C.B., Kadlubar, F.F. (2002). Carcinogen-DNA adducts in human breast epithelial cells. *Environ. Mol. Mutagen.*, *39*, 184–192.

172. Farmer, P.B., Brown, K., Tompkins, E., Emms, V.L., Jones, D.J., Singh, R., Phillips, D.H. (2005). DNA adducts: mass spectrometry methods and future prospects. *Toxicol. Appl. Pharmacol.*, *207*, 293–301.

173. Singh, R., Farmer, P.B. (2006). Liquid chromatography-electrospray ionization-mass spectrometry: the future of DNA adduct detection. *Carcinogenesis*, *27*, 178–196.

174. Ricicki, E.M., Soglia, J.R., Teitel, C., Kane, R., Kadlubar, F., Vouros, P. (2005). Detection and quantification of N-(deoxyguanosin-8-yl)-4-aminobiphenyl adducts in human pancreas tissue using capillary liquid chromatography-microelectrospray mass spectrometry. *Chem. Res. Toxicol.*, *18*, 692–699.

175. Beland, F.A., Churchwell, M.I., Von Tungeln, L.S., Chen, S., Fu, P.P., Culp, S.J., Schoket, B., Gyorffy, E., Minarovits, J., Poirier, M.C., Bowman, E.D., Weston, A., Doerge, D.R. (2005). High-performance liquid chromatography electrospray ionization tandem mass spectrometry for the detection and quantitation of Benzo[a]pyrene-DNA adducts. *Chem. Res. Toxicol.*, *18*, 1306–1315.

176. Grollman, A.P., Shibutani, S., Moriya, M., Miller, F., Wu, L., Moll, U., Suzuki, N., Fernandes, A., Rosenquist, T., Medverec, Z., Jakovina, K., Brdar, B., Slade, N., Turesky, R.J., Goodenough, A.K., Rieger, R., Vukelic, M., Jelakovic, B. (2007). Aristolochic acid and the etiology of endemic (Balkan) nephropathy. *Proc. Natl. Acad. Sci. U.S.A.*, *104*, 12129–12134.

177. Chen, L., Wang, M., Villalta, P.W., Luo, X., Feuer, R., Jensen, J., Hatsukami, D.K., Hecht, S.S. (2007). Quantitation of an acetaldehyde adduct in human leukocyte DNA and the effect of smoking cessation. *Chem. Res. Toxicol.*, *20*, 108–113.

178. Skipper, P.L., Peng, X., SooHoo, C.K., Tannenbaum, S.R. (1994). Protein adducts as biomarkers of human carcinogen exposure. *Drug Metab. Rev.*, *26*, 111–124.

179. Dingley, K.H., Freeman, S.P., Nelson, D.O., Garner, R.C., Turteltaub, K.W. (1998). Covalent binding of 2-amino-3,8-dimethylimidazo[4,5-f]quinoxaline to albumin and hemoglobin at environmentally relevant doses. Comparison of human subjects and F344 rats. *Drug Metab. Dispos.*, *26*, 825–828.

180. Peters, T., Jr (1996). *All about Albumin. Biochemistry, Genetics, and Medical Applications*, Academic Press, San Diego, CA.

181. Carballal, S., Radi, R., Kirk, M.C., Barnes, S., Freeman, B.A., Alvarez, B. (2003). Sulfenic acid formation in human serum albumin by hydrogen peroxide and peroxynitrite. *Biochemistry*, *42*, 9906–9914.

182. Beck, J.L., Ambahera, S., Yong, S.R., Sheil, M.M., de Jersey, J., Ralph, S.F. (2004). Direct observation of covalent adducts with Cys34 of human serum albumin using mass spectrometry. *Anal. Biochem.*, *325*, 326–336.

183. Keaney, J.F., Jr, Simon, D.I., Stamler, J.S., Jaraki, O., Scharfstein, J., Vita, J.A., Loscalzo, J. (1993). NO forms an adduct with serum albumin that has endothelium-derived relaxing factor-like properties. *J. Clin. Invest.*, *91*, 1582–1589.

184. Turesky, R.J., Skipper, P.L., Tannenbaum, S.R. (1987). Binding of 2-amino-3-methylimidazo[4,5-f]quinoline to hemoglobin and albumin in the rat. Identification of an adduct suitable for dosimetry. *Carcinogenesis*, *8*, 1537–1542.

185. Lynch, A.M., Murray, S., Zhao, K., Gooderham, N.J., Boobis, A.R., Davies, D.S. (1993). Molecular dosimetry of the food-borne carcinogen MeIQx using adducts of serum albumin. *Carcinogenesis*, *14*, 191–194.

186. Reistad, R., Frandsen, H., Grivas, S., Alexander, J. (1994). In vitro formation and degradation of 2-amino-1-methyl-6- phenylimidazo[4,5-*b*]pyridine (PhIP) protein adducts. *Carcinogenesis*, *15*, 2547–2552.

187. Chepanoske, C.L., Brown, K., Turteltaub, K.W., Dingley, K.H. (2004). Characterization of a peptide adduct formed by N-acetoxy-2-amino-1-methyl-6-phenylimidazo[4,5-b]pyridine (PhIP), a reactive intermediate of the food carcinogen PhIP. *Food Chem. Toxicol.*, *42*, 1367–1372.

188. Noort, D., Fidder, A., Hulst, A.G. (2003). Modification of human serum albumin by acrylamide at cysteine-34: a basis for a rapid biomonitoring procedure. *Arch. Toxicol*, *77*, 543–545.

189. Noort, D., Hulst, A.G., de Jong, L.P., Benschop, H.P. (1999). Alkylation of human serum albumin by sulfur mustard in vitro and in vivo: mass spectrometric analysis of a cysteine adduct as a sensitive biomarker of exposure. *Chem. Res. Toxicol.*, *12*, 715–721.

190. Bechtold, W.E., Willis, J.K., Sun, J.D., Griffith, W.C., Reddy, T.V. (1992). Biological markers of exposure to benzene: S-phenylcysteine in albumin. *Carcinogenesis*, *13*, 1217–1220.

191. Hoffmann, K.J., Streeter, A.J., Axworthy, D.B., Baillie, T.A. (1985). Structural characterization of the major covalent adduct formed in vitro between acetaminophen and bovine serum albumin. *Chem. Biol. Interact.*, *53*, 155–172.

192. Magagnotti, C., Orsi, F., Bagnati, R., Celli, N., Rotilio, D., Fanelli, R., Airoldi, L. (2000). Effect of diet on serum albumin and hemoglobin adducts of 2-amino-1-methyl-6-phenylimidazo[4,5-b]pyridine (PhIP) in humans. *Int. J. Cancer*, *88*, 1–6.

193. Hecht, S.S. (2002). Human urinary carcinogen metabolites: biomarkers for investigating tobacco and cancer. *Carcinogenesis*, *23*, 907–922.

194. Lynch, A.M., Knize, M.G., Boobis, A.R., Gooderham, N., Davies, D.S., Murray, S. (1992). Intra- and interindividual variability in systemic exposure in humans to 2-amino-3,8-dimethylimidazo[4,5-f]quinoxaline and 2-amino-1-methyl-6-phenylimidazo[4,5-b]pyridine, carcinogens present in food. *Cancer Res.*, *52*, 6216–6223.

195. Boobis, A.R., Lynch, A.M., Murray, S., de la Torre, R., Solans, A., Farré, M., Segura, J., Gooderham, N.J., Davies, D.S. (1994). CYP1A2-catalyzed conversion of dietary heterocyclic amines to their proximate carcinogens is their major route of metabolism in humans. *Cancer Res.*, *54*, 89–94.

196. Turesky, R.J., Garner, R.C., Welti, D.H., Richoz, J., Leveson, S.H., Dingley, K.H., Turteltaub, K.W., Fay, L.B. (1998). Metabolism of the food-borne mutagen 2-amino-3,8-dimethylimidazo[4,5-f]quinoxaline in humans. *Chem. Res. Toxicol.*, *11*, 217–225.

197. Kulp, K.S., Knize, M.G., Fowler, N.D., Salmon, C.P., Felton, J.S. (2004). PhIP metabolites in human urine after consumption of well-cooked chicken. *J. Chromatogr., B*, *802*, 143–153.

198. Ushiyama, H., Wakabayashi, K., Hirose, M., Itoh, H., Sugimura, T., Nagao, M. (1991). Presence of carcinogenic heterocyclic amines in urine of healthy volunteers eating normal diet, but not of inpatients receiving parenteral alimentation. *Carcinogenesis*, *12*, 1417–1422.

199. Holland, R.D., Taylor, J., Schoenbachler, L., Jones, R.C., Freeman, J.P., Miller, D.W., Lake, B.G., Gooderham, N.J., Turesky, R.J. (2004). Rapid biomonitoring of heterocyclic aromatic amines in human urine by tandem solvent solid phase extraction liquid chromatography electrospray ionization mass spectrometry. *Chem. Res. Toxicol.*, *17*, 1121–1136.

200. Ji, H., Yu, M.C., Stillwell, W.G., Skipper, P.L., Ross, R.K., Henderson, B.E., Tannenbaum, S.R. (1994). Urinary excretion of 2-amino-3,8-dimethylimidazo[4,5-f]quinoxaline in white, black, and Asian men in Los Angeles County. *Cancer Epidemiol. Biomarkers Prev.*, *3*, 407–411.

201. Kidd, L.C., Stillwell, W.G., Yu, M.C., Wishnok, J.S., Skipper, P.L., Ross, R.K., Henderson, B.E., Tannenbaum, S.R. (1999). Urinary excretion of 2-amino-1-methyl-6-phenylimidazo[4,5-b]pyridine (PhIP) in White, African-American, and Asian-American men in Los Angeles County. *Cancer Epidemiol. Biomarkers Prev.*, *8*, 439–445.

202. Turesky, R.J., Constable, A., Fay, L.B., Guengerich, F.P. (1999). Interspecies differences in metabolism of heterocyclic aromatic amines by rat and human P450 1A2. *Cancer Lett.*, *143*, 109–112.

203. Schweikl, H., Taylor, J.A., Kitareewan, S., Linko, P., Nagorney, D., Goldstein, J.A. (1993). Expression of CYP1A1 and CYP1A2 genes in human liver. *Pharmacogenetics*, *3*, 239–249.

204. Hammons, G.J., Yan-Sanders, Y., Jin, B., Blann, E., Kadlubar, F.F., Lyn-Cook, B.D. (2001). Specific site methylation in the 5′-flanking region of CYP1A2 interindividual differences in human livers. *Life Sci.*, *69*, 839–845.

205. Shimada, T., Yamazaki, H., Mimura, M., Inui, Y., Guengerich, F.P. (1994). Interindividual variations in human liver cytochrome P-450 enzymes involved in the oxidation of drugs, carcinogens and toxic chemicals: studies with liver microsomes of 30 Japanese and 30 Caucasians. *J. Pharmacol. Exp. Ther.*, *270*, 414–423.

206. Eaton, D.L., Gallagher, E.P., Bammler, T.K., Kunze, K.L. (1995). Role of cytochrome P450 1A2 in chemical carcinogenesis: implications for human variability in expression and enzyme activity. *Pharmacogenetics*, *5*, 259–274.

207. Turesky, R.J., Constable, A., Richoz, J., Varga, N., Markovic, J., Martin, M.V., Guengerich, F.P. (1998). Activation of heterocyclic aromatic amines by rat and human liver microsomes and by purified rat and human cytochrome P450 1A2. *Chem. Res. Toxicol.*, *11*, 925–936.

208. Belloc, C., Baird, S., Cosme, J., Lecoeur, S., Gautier, J.-C., Challine, D., de Waziers, I., Flinois, J.-P., Beaune, P.H. (1996). Human cytochrome P450 expressed in *Escherichia coli*: production of specific antibodies. *Toxicology*, *106*, 207–219.

209. Conney, A.H. (1982). Induction of microsomal enzymes by foreign chemicals and carcinogenesis by polycyclic aromatic hydrocarbons: G. H. A. Clowes Memorial Lecture. *Cancer Res.*, *42*, 4875–4917.

210. Sesardic, D., Boobis, A.R., Edwards, R.J., Davies, D.S. (1988). A form of cytochrome P450 in man, orthologous to form d in the rat, catalyses the O-deethylation of phenacetin and is inducible by cigarette smoking. *Br. J. Clin. Pharmacol.*, *26*, 363–372.

211. Schrenk, D., Brockmeier, D., Morike, K., Bock, K.W., Eichelbaum, M. (1998). A distribution study of CYP1A2 phenotypes among smokers and non-smokers in a cohort of healthy Caucasian volunteers. *Eur. J. Clin. Pharmacol.*, *53*, 361–367.

212. Lampe, J.W., King, I.B., Li, S., Grate, M.T., Barale, K.V., Chen, C., Feng, Z., Potter, J.D. (2000). Brassica vegetables increase and apiaceous vegetables decrease cytochrome P450 1A2 activity in humans: changes in caffeine metabolite ratios in response to controlled vegetable diets. *Carcinogenesis*, *21*, 1157–1162.

213. Murray, S., Lake, B.G., Gray, S., Edwards, A.J., Springall, C., Bowey, E.A., Williamson, G., Boobis, A.R., Gooderham, N.J. (2001). Effect of cruciferous vegetable consumption on heterocyclic aromatic amine metabolism in man. *Carcinogenesis*, *22*, 1413–1420.
214. Sinha, R., Rothman, N., Brown, E.D., Mark, S.D., Hoover, R.N., Caporaso, N.E., Levander, O.A., Knize, M.G., Lang, N.P., Kadlubar, F.F. (1994). Pan-fried meat containing high levels of heterocyclic aromatic amines but low levels of polycyclic aromatic hydrocarbons induces cytochrome P4501A2 activity in humans. *Cancer Res.*, *54*, 6154–6159.
215. Nakajima, M., Yokoi, T., Mizutani, M., Kinoshita, M., Funayama, M., Kamataki, T. (1999). Genetic polymorphism in the 5′-flanking region of human CYP1A2 gene: effect on the CYP1A2 inducibility in humans. *J. Biochem. (Tokyo)*, *125*, 803–808.
216. Sachse, C., Brockmoller, J., Bauer, S., Roots, I. (1999). Functional significance of a C→A polymorphism in intron 1 of the cytochrome P450 CYP1A2 gene tested with caffeine. *Br. J. Clin. Pharmacol.*, *47*, 445–449.
217. Sinha, R., Rothman, N., Mark, S.D., Murray, S., Brown, E.D., Levander, O.A., Davies, D.S., Lang, N.P., Kadlubar, F.F., Hoover, R.N. (1995). Lower levels of urinary 2-amino-3,8-dimethylimidazo[4,5-f]-quinoxaline (MeIQx) in humans with higher CYP1A2 activity. *Carcinogenesis*, *16*, 2859–2861.
218. Stillwell, W.G., Kidd, L.-C.K.S.-B., Wishnok, J.W., Tannenbaum, S.R., Sinha, R. (1997). Urinary excretion of unmetabolized and phase II conjugates of 2-amino 1-methyl-6-phenylimidazo[4,5-b]pyridine and 2-amino-3,8-dimethylimidazo[4,5-f]quinoxaline in humans: relationship to cytochrome P450 1A2 and *N*-acetyltransferase activity. *Cancer Res.*, *57*, 3457–3464.
219. Kunze, K.L., Trager, W.F. (1993). Isoform-selective mechanism-based inhibition of human cytochrome P450 1A2 by furafylline. *Chem. Res. Toxicol.*, *6*, 649–656.
220. Turesky, R.J., Guengerich, F.P., Guillouzo, A., Langouet, S. (2002). Metabolism of heterocyclic aromatic amines by human hepatocytes and cytochrome P4501A2. *Mutat. Res.*, *506–507*, 187–195.
221. Malfatti, M.A., Dingley, K.H., Nowell-Kadlubar, S., Ubick, E.A., Mulakken, N., Nelson, D., Lang, N.P., Felton, J.S., Turteltaub, K.W. (2006). The urinary metabolite profile of the dietary carcinogen 2-amino-1-methyl-6-phenylimidazo[4,5-b]pyridine is predictive of colon DNA adducts after a low-dose exposure in humans. *Cancer Res.*, *66*, 10541–10547.
222. Chen, C., Ma, X., Malfatti, M.A., Krausz, K.W., Kimura, S., Felton, J.S., Idle, J.R., Gonzalez, F.J. (2007). A comprehensive investigation of 2-amino-1-methyl-6-phenylimidazo[4,5-b]pyridine (PhIP) metabolism in the mouse using a multivariate data analysis approach. *Chem. Res. Toxicol.*, *20*, 531–542.
223. Langouët, S., Welti, D.H., Kerriguy, N., Fay, L.B., Huynh-Ba, T., Markovic, J., Guengerich, F.P., Guillouzo, A., Turesky, R.J. (2001). Metabolism of 2-amino-3,8-dimethylimidazo[4,5-f]quinoxaline in human hepatocytes: 2-amino-3-methylimidazo[4,5-f]quinoxaline-8-carboxylic acid is a major detoxification pathway catalyzed by cytochrome P450 1A2. *Chem. Res. Toxicol.*, *14*, 211–221.
224. Turesky, R.J., Parisod, V., Huynh-Ba, T., Langouet, S., Guengerich, F.P. (2001). Regioselective differences in C(8)- and N-oxidation of 2-amino-3,8-dimethylimidazo[4,5-f]quinoxaline by human and rat liver microsomes and cytochromes P450 1A2. *Chem. Res. Toxicol.*, *14*, 901–911.

225. Hein, D.W., Doll, M.A., Rustan, T.D., Gray, K., Feng, Y., Ferguson, R.J., Grant, D.M. (1993). Metabolic activation and deactivation of arylamine carcinogens by recombinant human NAT1 and polymorphic NAT2 acetyltransferases. *Carcinogenesis*, *14*, 1633–1638.

226. Hein, D.W. (2002). Molecular genetics and function of NAT1 and NAT2: role in aromatic amine metabolism and carcinogenesis. *Mutat. Res.*, *506–507*, 65–77.

227. Gaylor, D.W., Kadlubar, F.F. (1991). Quantitative risk assessments of heterocyclic amines in cooked foods, in *Mutagens in Foods: Detection and Prevention* (ed. H. Hayatsu), CRC Press, Boca Raton, FL, pp. 229–236.

228. Lutz, W.K., Schlatter, J. (1992). Chemical carcinogens and overnutrition in diet-related cancer. *Carcinogenesis*, *13*, 2211–2216.

229. Bogen, K.T., Keating, G.A. (2001). U.S. dietary exposures to heterocyclic amines. *J. Expo. Anal. Environ. Epidemiol.*, *11*, 155–168.

230. Keating, G.A., Bogen, K.T. (2004). Estimates of heterocyclic amine intake in the US population. *J. Chromatogr., B*, *802*, 127–133.

231. Sugimura, T., Nagao, M., Wakabayashi, K. (2000). How we should deal with unavoidable exposure of man to environmental mutagens: cooked food mutagen discovery, facts and lessons for cancer prevention. *Mutat. Res.*, *447*, 15–25.

232. Skog, K., Johansson, M., Jagerstad, M. (1995). Factors affecting the formation of and yield of heterocylic amines. In: *Heterocyclic aromatic amines in cooked foods: possible human carcinogens. 23rd Proceedings of the Princess Takamatsu Cancer Society*, (eds R.H. Adamson, J.-A. Gustafsson, N. Ito, M. Nagao, T. Sugimura, K. Wakabayashi, Y. Yamazoe), Princeton Scientific Publishing Co., Princeton, NJ, 9–19.

233. Augustsson, K., Skog, K., Jägerstad, M., Steineck, G. (1997). Assessment of the human exposure to heterocyclic amines. *Carcinogenesis*, *18*, 1931–1935.

234. Swenberg, J.A., La, D.K., Scheller, N.A., Wu, K.Y. (1995). Dose-response relationships for carcinogens. *Toxicol. Lett.*, *82–83*, 751–756.

235. MacLeod, S., Nowell, S., Lang, N.P. (2000). Genetic polymorphisms, in *Food Borne Carcinogens Heterocyclic Amines* (eds M. Nagao, T. Sugimura), John Wiley & Sons, Ltd, West Sussex, pp. 112–130.

236. Lang, N.P., Butler, M.A., Massengill, J.P., Lawson, M., Stotts, R.C., Hauer-Jensen, M., Kadlubar, F.F. (1994). Rapid metabolic phenotypes for acetyltransferase and cytochrome P4501A2 and putative exposure to food-borne heterocyclic amines increase the risk for colorectal cancer or polyps. *Cancer Epidemiol. Biomarkers Prev.*, *3*, 675–682.

237. Le Marchand, L., Hankin, J.H., Wilkens, L.R., Pierce, L.M., Franke, A., Kolonel, L.N., Seifried, A., Custer, L.J., Chang, W., Lum-Jones, A., Donlon, T. (2001). Combined effects of well-done red meat, smoking, and rapid N-acetyltransferase 2 and CYP1A2 phenotypes in increasing colorectal cancer risk. *Cancer Epidemiol. Biomarkers Prev.*, *10*, 1259–1266.

238. Chan, A.T., Tranah, G.J., Giovannucci, E.L., Willett, W.C., Hunter, D.J., Fuchs, C.S. (2005). Prospective study of N-acetyltransferase-2 genotypes, meat intake, smoking and risk of colorectal cancer. *Int. J. Cancer*, *115*, 648–652.

239. Guengerich, F.P. (1997). Comparisons of catalytic selectivity of cytochrome P450 subfamily members from different species. *Chem. Biol. Interact.*, *106*, 161–182.

240. Schwab, C.E., Huber, W.W., Parzefall, W., Hietsch, G., Kassie, F., Schulte-Hermann, R., Knasmuller, S. (2000). Search for compounds that inhibit the genotoxic and carcinogenic effects of heterocyclic aromatic amines. *Crit. Rev. Toxicol.*, *30*, 1–69.

241. Schwab, C., Kassie, F., Qin, H.M., Sanyal, R., Uhl, I.M., Hietsch, G., Rabot, S., Darroudi, F., Knasmuller, S. (1999). Development of test systems for the detection of compounds that prevent the genotoxic effects of heterocyclic aromatic amines: preliminary results with constituents of cruciferous vegetables and other dietary constituents. *J. Environ. Pathol. Toxicol. Oncol.*, *18*, 109–118.

242. Malfatti, M.A., Wu, R.W., Felton, J.S. (2005). The effect of UDP-glucuronosyltransferase 1A1 expression on the mutagenicity and metabolism of the cooked-food carcinogen 2-amino-1-methyl-6-phenylimidazo[4,5-b]pyridine in CHO cells. *Mutat. Res.*, *570*, 205–214.

243. Bendriss, E.K., Bechtel, Y., Bendriss, A., Humbert, P.H., Paintaud, G., Magnette, J., Agache, P., Bechtel, P.R. (1996). Inhibition of caffeine metabolism by 5-methoxypsoralen in patients with psoriasis. *Br. J. Clin. Pharmacol.*, *41*, 421–424.

244. Chang, T.K., Chen, J., Lee, W.B. (2001). Differential inhibition and inactivation of human CYP1 enzymes by trans-resveratrol: evidence for mechanism-based inactivation of CYP1A2. *J. Pharmacol. Exp. Ther.*, *299*, 874–882.

245. Miranda, C.L., Yang, Y.H., Henderson, M.C., Stevens, J.F., Santana-Rios, G., Deinzer, M.L., Buhler, D.R. (2000). Prenylflavonoids from hops inhibit the metabolic activation of the carcinogenic heterocyclic amine 2-amino-3-methylimidazo[4, 5-f]quinoline, mediated by cDNA-expressed human CYP1A2. *Drug Metab. Dispos.*, *28*, 1297–1302.

246. Tassaneeyakul, W., Guo, L.Q., Fukuda, K., Ohta, T., Yamazoe, Y. (2000). Inhibition selectivity of grapefruit juice components on human cytochromes P450. *Arch. Biochem. Biophys.*, *378*, 356–363.

247. Dashwood, R.H. (2002). Modulation of heterocyclic amine-induced mutagenicity and carcinogenicity: an "A-to-Z" guide to chemopreventive agents, promoters, and transgenic models. *Mutat. Res.*, *511*, 89–112.

248. Kulp, K.S., Fortson, S.L., Knize, M.G., Felton, J.S. (2003). An in vitro model system to predict the bioaccessibility of heterocyclic amines from a cooked meat matrix. *Food Chem. Toxicol.*, *41*, 1701–1710.

249. Sugimura, T. (2000). Nutrition and dietary carcinogens. *Carcinogenesis*, *21*, 387–395.

250. Bingham, S.A. (1999). High-meat diets and cancer risk. *Proc. Nutr. Soc.*, *58*, 243–248.

251. Sweeney, C., Coles, B.F., Nowell, S., Lang, N.P., Kadlubar, F.F. (2002). Novel markers of susceptibility to carcinogens in diet: associations with colorectal cancer. *Toxicology*, *181–182*, 83–87.

252. Alexander, J., Reistad, R., Hegstad, S., Frandsen, H., Ingebrigtsen, K., Paulsen, J.E., Becher, G. (2002). Biomarkers of exposure to heterocyclic amines: approaches to improve the exposure assessment. *Food Chem. Toxicol.*, *40*, 1131–1137.

253. Hashimoto, H., Hanaoka, T., Kobayashi, M., Tsugane, S. (2004). Analytical method of 2-amino-1-methyl-6-phenylimidazo[4,5-b]pyridine in human hair by column-switching liquid chromatography-mass spectrometry. *J. Chromatogr., B*, *803*, 209–213.

2.4

HAZARDS OF DIETARY FURAN

P. MICHAEL BOLGER, SHIRLEY S-H. TAO, AND MICHAEL DINOVI

Center for Food Safety and Applied Nutrition, U.S. Food and Drug Administration, 5100 Paint Branch Parkway, College Park, MD 20740, USA

2.4.1 INTRODUCTION

Furan is a volatile flammable molecule that is clear, colorless, and water-insoluble. It is used as an intermediate in the organic synthesis and production of compounds such as lacquers, resins, insecticides, and pharmaceuticals. Furan is used primarily as an intermediate in the synthesis and production of tetrahydrofuran, pyrrole, and thiophene. It is also used in the production of agricultural chemicals (insecticides), stabilizers, and pharmaceuticals. Furan is a by-product of high-energy radiation and thermal treatments of food and is found in a variety of foods and beverages, especially in certain processed foods, such as coffee, baked bread, and canned and jarred foods, including baby food. These techniques have long been essential methods of food preparation and preservation. Furan is also found in surface water, industrial effluents, ambient air, human milk samples, and in the breath and urine of healthy humans.

While furan has been known as a flavor volatile for many years, only limited data on furan concentrations in various foods are available and estimates of dietary exposure are limited. For chronic hazards like furan, it is well recognized that the lifetime exposure is the most relevant measure. The best available dietary estimates of exposure to furan in the US population were initially developed by the Center for Food Safety and Applied Nutrition (CFSAN), US Food and Drug Administration (USFDA) in 2004 (1) and later updated in 2007 (2).

Process-Induced Food Toxicants: Occurrence, Formation, Mitigation, and Health Risks,
Edited by Richard H. Stadler and David R. Lineback

Furan: At a Glance	
Historical	Known as a flavor volatile and already tentatively identified in the 1930s in roasted coffee. Some reports published on its occurrence in canned foods in the 1970s, and the primary route of formation was thought to be by thermal decomposition of carbohydrates.
Analysis	Because of its volatility, furan is best measured by a headspace sampling method followed by GC/MS.
Occurrence in Food	Fresh fruits and vegetables show little or no furan. Foods that are heat processed in cans and jars are typically found to contain highest amounts of furan, such as soups, pastas and sauces (gravies) with meat, and baby foods in jars. Coffee powders may contain up to 5 mg/kg on a dry weight basis.
Main Formation Pathways	Several routes of formation and different precursors that may lead to furan in foods when exposed to heat have been identified, such as vitamin C, amino acids, reducing sugars, organic acids, carotenes, and polyunsaturated fatty acids (PUFAs).
Mitigation in Food	So far no specific measures have been identified to mitigate furan formation in food. The reduction of the thermal load may be an avenue to explore in some products, but then within the broader context of food safety.
Health Risks	Furan is an animal carcinogen and classified by IARC as a possible carcinogen to humans. NOAELs based on a 2-year bioassay have been identified for cytotoxicity and hepatocarcinogenicity of 0.5 and 2 mg/kg bw, respectively. Evidence indicates that the metabolite of furan, cis-2-butene-1,4-dial, plays an important role in furan-induced toxicity, including carcinogenesis, probably attributable to a genotoxic mechanism.
Regulatory Information or Industry Standards	No limits have been set to date in foods.
Knowledge Gaps	More exposure data are needed, as well as dose-response data at lower dose levels, knowledge on formation of reactive metabolites and their dependency on dose. Studies on reproductive and developmental toxicity are also needed.
Useful Information (web) & Dietary Recommendations	http://www.cfsan.fda.gov/~dms/furanexp/sld09.htm. http://www.fda.gov/ohrms/dockets/ac/04/slides/2004-4045s2-03-ihalov.pdf. Currently, no specific recommendation can be made other than that there is no justification to make any changes in dietary habits.

The presence of furan in foods is a potential concern because, based on animal tests, furan is considered possibly carcinogenic to humans (3). When administered chronically by gavage, furan induced an increase in the incidence of hepatic cholangiocarcinoma, hepatocellular adenoma and carcinoma, and mononuclear cell leukemia in male and female F344/N rats (4). When chronically administered by gavage to male and female B6C3F1 mice, furan induced a dose-dependent increase in the incidence of hepatocellular adenoma and carcinoma and benign pheochromocytoma when treated up to 2 years (4). However, no human studies of the relationship between exposure to furan and human cancer have been reported.

2.4.2 ANALYSIS

Furan is analyzed using a gas chromatography/mass spectrometry (GC/MS) method. Because furan is volatile, a headspace collection method is employed. Semi-solid and solid foods are weighed into headspace vials, diluted with either water or saturated NaCl solution, and fortified with internal standard (*d4*-furan); the vials are then sealed. Liquid foods are weighed into headspace vials and fortified with internal standard (*d4*-furan), followed by sealing. Automated headspace sampling followed by GC/MS analysis is used to detect furan and *d4*-furan in the scan mode. Furan is quantified by using a standard additions curve, where the concentration of furan in the fortified test portions is plotted versus the furan/*d4*-furan response factors. For most foods, the limit of detection for this method ranges from 0.3 to 1.0 μg/kg. The limit of detection for liquid foods, such as coffee, is as low as 0.2 μg/kg. The complete method of analysis used by USFDA is published on the CFSAN web site (5).

2.4.3 OCCURRENCE

The biogenesis/formation of furan in food products is not understood. Various carbohydrate precursors have been postulated; ascorbic and citric acids have been implicated in furan production in those food types known to contain elevated concentrations. USFDA's program of analysis of foods thought to contain furan was developed by exploring heat-processed foods containing these precursors. Further developments in the understanding of the mechanism of formation for furan in foods could better direct explorations into those food types likely to contain furan at significant concentrations.

Canned soups, pastas, and sauces (gravies) made with meat are typically found to contain the highest concentrations of furan. Due to the volatility of furan, however, heating of these products in open containers will result in lower exposures from the food as consumed. Brewed coffees are also uniformly high, with concentrations up to approximately 100 μg/kg. Jarred baby foods consistently have furan concentrations of 10–100 μg/kg, with mixtures of meat and vegetables having the highest concentrations. Fresh fruits and vegetables show little or no furan. Irradiation of cut, fresh fruits and vegetables produces low concentrations of furan, typically less than 5 μg/kg. Data submitted to the USFDA by the Swiss Federal Office of Public Health were consistent with the published USFDA data (6). Coffee powders had very high concentrations of furan, however, ranging from 2650 to over 5000 μg/kg. The coffee beverages prepared from these powders had furan concentrations ranging from 3 to 25 μg/kg. A recent publication from the Swiss Federal Office of Public Health (7) confirms these findings.

2.4.4 EXPOSURE

Measured furan concentrations in selected (cooked) foods were combined with estimates of food frequency consumption, using Monte Carlo simulation and individual dietary records. For the Monte Carlo analyses, all of the measured furan concentrations were employed as a distributional input. For use with individual dietary records, a mean concentration for each food type analyzed was used. The CFSAN 2004 exposure assessment, using the first set of preliminary data, reported a mean and 90th percentile of 0.3 and 0.6 μg furan/kg body weight (bw)/day for the 2+-year-olds, and 0.4 and 1.0 μg furan/kg bw/day for the 0- to 1-year-olds from the consumption of adult and infant foods, respectively (1). The 2007 CFSAN reassessment derived very similar results of 0.26 and 0.61 μg furan/kg bw/day for the 2+-year-olds, and 0.41 and 0.99 μg furan/kg bw/day for the 0- to 1-year-olds from the consumption of adult and infant foods, respectively. The results using Monte Carlo and individual dietary records were consistent. The contribution to overall furan exposure from 11 food types is shown in Table 2.4.1, in descending order of exposure from each.

TABLE 2.4.1 Exposure to furan from the highest 11 contributing food types according to USFDA.

Food type	Level of furan, μg/kg bw/day
Brewed coffee	0.15
Chili	0.04
Cereals	0.01
Salty snacks	0.01
Soups containing meat	0.01
Pork and beans	0.004
Canned pasta	0.004
Canned string beans	0.004
Pasta sauces	0.001
Juices	0.001
Canned tuna (water-packed)	0.00008

2.4.5 MITIGATION/REDUCTION

Specific mitigation measures for the production of furan in foods are difficult to develop, as the exact process leading to its formation in heat-processed foods is not clear. Further, it has not been determined whether furan can be formed in foods naturally, through an as-yet-unknown enzymatic process. Paradoxically, the furan concentration in some foods has been shown to *increase* when sealed and stored under refrigeration conditions. This may point to endogenous formation, or may simply occur as components of cooked foods react during storage. Once the processes leading to the formation of furan in foods are elucidated, methods for mitigating or preventing the formation in foods may be developed.

Because of the volatility of furan, heating in open containers, or leaving ready-to-eat foods open to the air for a period of time after preparation may reduce the concentration of furan in the foods when consumed. While this has been shown to occur during the development at USFDA of the method of analysis for furan (personal communication to authors), there is currently no way to determine its overall effectiveness in reducing overall exposure to furan from foods.

2.4.6 METABOLISM/TOXICOKINETICS

In rats, a dose of 8mg [^{14}C]furan/kg bw in corn oil is rapidly absorbed after ingestion with the highest concentration in the liver followed by kidney, intestines, blood, and lung. Radioactivity was found to be associated with protein, but not with DNA, in the liver (8). About 80% of radioactivity was eliminated in 24h post dosing: 40% in the expired air (14% as unchanged furan, 26% as CO_2), 22% in feces, and 20% in the urine. Furan was metabolized extensively, but attempts to purify and identify the metabolites were unsuccessful. A physiological-based pharmacokinetic (PBPK) model, which simulated furan metabolism 24h after an oral dose of 8mg/kg bw, predicted that 84% of the dose would be metabolized and that 16% would be expired as furan (9). These predicted values agree fairly well with the data reported by Burka and coworkers (8).

In vitro and *in vivo* studies show that metabolic activation is involved in furan-induced toxicity. Cytochrome P450 enzymes in the target tissues mediate the formation of highly reactive, electrophilic furan metabolites, presumably one of which is cis-2-butene-1,4-dial, which bind covalently to tissue macromolecules (9). Glutathione (GSH) inhibited the covalent binding of reactive furan metabolites to microsomal protein *in vitro* (10), presumably by forming less reactive, water-soluble conjugates with the activated furans.

Similarly, Kedderis and Held (11), using freshly isolated hepatocytes from mice and humans, determined biotransformation kinetics of furan *in vitro* and developed species-specific pharmacokinetic models. In rodents, hepatic blood flow limitation of furan bioactivation was observed following oral bolus administration of a 2 mg/kg bw dose of furan (12). These studies suggest that individual variations in cytochrome P450 enzyme activity observed among human populations would not have a significant effect on the amount of toxic furan metabolite(s) formed in the liver.

2.4.7 TOXICOLOGY

In 16-day gavage studies, male rats were dosed with furan at 0, 5, 10, 20, 40, or 80 mg/kg bw and female rats and mice of both sexes at 0, 10, 20, 40, 80, and 160 mg/kg bw. High mortality rates were observed in rats and mice at doses of 80 or 160 mg/kg bw. Mean body weights were decreased in rats at doses of 20 mg/kg bw (male) and 40 mg/kg bw (male and female). Mottled and enlarged livers were observed in male rats dosed with 20, 40, or 80 mg/kg furan and in female rats dosed with 40 to 160 mg/kg of furan. Mice showed no lesions with similar furan doses. As shown in Tables 2.4.2–2.4.5, other non-neoplastic lesions reported in 13-week and 2-year gavage studies in rats and mice were hyperplasia of bile ducts, which were usually accompanied by fibrosis, inflammation and cysts, liver-cell proliferation, liver cytomegaly, degeneration, nodular hyperplasia, necrosis, and vacuolization. Increased severity of nephropathy associated with increased incidence of parathyroid hyperplasia was also seen in rats. In the 13-week rat study, significantly higher incidences of lesions of the biliary tract and hepatocytes were observed at furan doses equal to or in excess of 4 and 8 mg/kg bw, respectively. Kidney lesions were present in rats dosed with 60 mg/kg bw of furan. Incidences of liver lesions were higher in mice with furan doses of 30 and 60 mg/kg bw. In the 2-year study, increased incidences of these liver and kidney lesions were found in all groups of furan-dosed animals including the lowest dose groups tested of 2 and 8 mg/kg bw in rats and mice, respectively (4).

As shown in Table 2.4.6, furan is not a mutagen in the Ames *Salmonella typhimurium* assay. Furan tested negative with and without S9 in TA98, TA100, TA1535, and TA1537 (4), but was weakly positive in TA100 in another study by the group of Lee (16). Furan is not mutagenic in the germ cells of male Drosophila melanogaster. In contrast, furan tested positive in *in vitro* mammalian systems, such as in mouse lymphoma cells, and causes sister chromatid exchanges (SCE) and chromosomal aberrations in Chinese hamster ovary cells, with and without S9 activation (4). In *in vivo* mammalian systems, it induced chromosomal aberrations, but not sister chromatid exchanges in bone marrow cells in mice, or in hepatocytes in mice and rats (14). It did not cause unscheduled DNA synthesis in mouse or rat hepatocytes (14).

The International Agency for Research on Cancer (IARC) classified furan as possibly carcinogenic to humans (Group 2B) based on inadequate evidence in humans, but sufficient evidence in rats and mice for the carcinogenicity of furan (4). No human studies are available on effects of furan including

TABLE 2.4.2 Incidences of selected non-neoplastic lesions of F344 rats in the NTP 13-week study.[a]

Dose, mg/kg	0		4		8		15		30		60	
Sex	Male	Female	Male	Female	Male	Female	Male	Female	Male	Female	Male	Female
Number of animals examined	10	10	10	10	10	10	10	10	10	10	10	10
Liver												
Biliary tract												
Cholangiofibrosis	0	0	4	1	7	7	10	10	10	10	10	9
Hyperplasia	0	0	4	7	9	10	10	10	10	10	10	9
Hepatocytes												
Cytomegaly	0	0	0	0	0	0	8	10	10	10	10	9
Degeneration	0	0	0	0	7	1	9	10	10	10	10	10
Necrosis	0	0	0	0	0	0	9	8	10	10	10	10
Hyperplasia, nodular	0	0	0	0	0	0	0	0	10	8	10	9
Kupffer cells												
Pigmentation	0	0	4	2	6	8	10	10	10	10	9	9
Kidney												
Renal tubule												
Dilation	0	0	ND	ND	ND	ND	ND	ND	2	0	9	8
Necrosis	0	0	ND	ND	ND	ND	ND	ND	0	0	10	7

[a]Reference 4.
ND = not determined.

TABLE 2.4.3 Incidences of selected non-neoplastic lesions of B6C3F1 mice in the NTP 13-week study.[a]

Dose, mg/kg	0		2		4		8		15		30		60	
Sex	Male	Female	Male	Female	Male	Female	Male	Female	Male	Female	Male	Female	Male	Female
Number of animals examined	10	10	10	0	10	10	10	10	10	10	10	10	0	10
Liver														
Biliary tract														
Cholangiofibrosis	ND	0	ND		ND	ND	ND	0	ND	0	ND	4		10
Hyperplasia	0	0	ND		0	ND	0	0	0	0	2	8		10
Hepatocytes														
Cytomegaly	0	0	ND		0	ND	0	0	0	0	10	10		10
Degeneration	0	0	ND		0	ND	0	0	1	3	10	10		10
Necrosis	0	0	ND		0	ND	1	0	1	0	8	9		10
Kupffer cells														
Pigmentation	0	0	ND		0	ND	0	0	0	0	3	9		10

[a]Reference 4.
ND = not determined.

TABLE 2.4.4 Incidences of selected non-neoplastic lesions of F344 rats in the NTP 2-year study.[a]

Dose, mg/kg	0	2	4	8	0	2	4	8
Sex	Male	Male	Male	Male	Female	Female	Female	Female
Number of animals examined	50	50	50	50	50	50	50	50
Liver								
Biliary tract, %								
Chronic focal inflammation	0	88	96	98	0	98	100	98
Cyst	0	88	94	98	0	98	100	92
Focal fibrosis	0	88	96	98	0	98	100	96
Focal hyperplasia	0	88	96	98	0	98	100	98
Metaplasia	0	88	96	98	0	98	100	98
Hepatocytes, %								
Cytomegaly	0	70	92	98	0	88	100	98
Cytoplasmic vacuolization	2	78	90	98	0	86	98	94
Focal degeneration	0	66	92	98	0	70	98	94
Focal hyperplasia	0	60	92	98	0	64	94	92
Focal necrosis	0	64	92	98	0	36	92	94
Kupffer cells, %								
Focal pigmentation	0	88	96	98	0	98	100	96
Kidney								
Nephropathy, mean severity grade	1.56	2.38	3.16	3.16	0.48	1.84	2.54	3.10

[a]Reference 4.

TABLE 2.4.5 Incidences of selected non-neoplastic liver lesions of B6C3F1 mice in the NTP 2-year study.[a]

Dose, mg/kg	0	8	15	0	8	15
Sex	Male	Male	Male	Female	Female	Female
Number of animals examined	50	50	50	50	50	50
Cytoplasmic vacuolization, %	16	48	72	12	58	72
Focal hyperplasia, %	2	88	98	0	96	96
Mixed cell cellular infiltration, %	4	46	58	16	46	64
Biliary tract, %						
Chronic inflammation	0	88	98	4	96	100
Fibrosis	0	90	98	0	94	100
Hyperplasia	0	92	98	0	94	100
Hepatocytes, %						
Cytomegaly	16	90	100	0	96	100
Degeneration	0	86	86	0	94	96
Necrosis	4	78	82	0	88	94
Kupffer cells, %						
Focal pigmentation	4	86	100	10	96	100
Parenchyma, %						
Focal atrophy	2	90	100	0	96	100

[a]Reference 4.

whether or not it causes carcinogenic effects like those seen in the rodent bioassays.

The National Toxicology Program (NTP) 2-year bioassay (1993) showed that there was clear evidence of carcinogenic activity of furan in male and female F344/N rats based on increased incidences of cholangiocarcinoma and hepatocellular neoplasm of the liver and on increased incidences of mononuclear cell leukemia. There was clear evidence of carcinogenic activity of furan in male and female B6C3F1 mice based on increased incidences of hepatocellular neoplasm of the liver and benign pheochromocytomas of the adrenal gland.

In these bioassays, it is clear that rats are more sensitive than mice to furan-induced neoplasms. Rats were dosed with 0, 2, 4, or 8 mg/kg for 5 days/week for 2 years. Mice were dosed with 0, 8, or 15 mg/kg bw for 5 days/week for 2 years. The results are summarized as follows:

1. *Cholangiocarcinomas*: The incidence was found high in rats receiving 2 mg/kg for 9 months (40–50%) or 15 months (70–90%). Rats receiving 2 mg/kg bw of furan for 2 years showed 86–98% incidence (Table 2.4.7). While no incidence was found at week 13, a 100% incidence was found at the 9- and 15-month interim evaluations or after in male rats exposed to 30 mg/kg for 13 weeks in a stop-exposure experiment.

TABLE 2.4.6 Genotoxicity studies of furan.

	Reference no.
Negative results	
Nonmammalian systems	
• *Salmonella typhimurium* (±rat and hamster liver S9):	
In strains TA100, TA1535, TA1537, and TA98	4
Drosophila melanogaster, sex-linked recessive lethal mutation	4, 13
Mammalian systems	
In vitro	
• Unscheduled DNA synthesis (UDS) in rat and mouse hepatocytes	14
• Chromosomal aberrations (–rat S9 only) in Chinese hamster ovary cells	15
In vivo	
• UDS in mouse or rat hepatocytes after a single oral dose of 200 (mouse) or 100 (rat) mg/kg bw.	4, 14
• Sister chromatid exchanges (SCE) in mouse bone marrow cells of male B6C3F$_1$ mice (350 mg/kg bw, ip)	4
Positive results	
Nonmammalian systems	
• *Salmonella typhimurium* (±rat and hamster liver S9):	
In strain TA100 (weakly positive)	16
Mammalian systems	
In vitro	
• Mouse L5178Y lymphoma cells (–rat S9)	17
• Mouse L5178Y lymphoma cells (±rat S9)	4
• Chinese hamster ovary cells (CHO):	
Sister chromatid exchanges and chromosomal aberrations (± rat S9)	4
Chromosomal aberrations (+rat S9 only)	15
• DNA double-strand breaks in isolated rat hepatocytes	18, 19
In vivo	
• Chromosome aberrations in mice bone marrow cells of male B6C3F$_1$ mice (250 mg/kg bw, ip)	4
• Mutation in H-*ras* oncogene in mouse liver	20

TABLE 2.4.7 Incidence of cholangiocarcinomas of the liver of F344 rats by furan (NTP 2-year gavage study).

	9 months		15 months		24 months	
Dose[a], mg/kg	Male	Female	Male	Female	Male	Female
0	0/10	0/10	0/10	0/10	0/50 (0%)	0/50 (0%)
2	5/10	4/10	7/10	9/10	43/50 (86%)	49/50 (98%)
4	7/10	9/10	9/10	9/10	48/50 (96%)	50/50 (100%)
8	10/10	10/10	6/10	7/10	49/50 (98%)	48/50 (100%)

[a]By gavage, 5 days a week (4).

TABLE 2.4.8 Incidence of liver neoplasms in the NTP 2-year gavage study in rats and mice.[a]

	Furan, mg/kg				
Neoplasm	0	2	4	8	15
Male rats					
Cholangiocarcinomas	0 (0%)	43 (86%)	45 (96%)	49 (98%)	
Hepatocellular adenomas	1 (2%)	4 (8%)	18 (36%)	27 (54%)	
Hepatocellular carcinomas	0 (0%)	1 (2%)	6 (12%)	18 (36%)	
Hepatocellular adenoma/carcinomas	1 (2%)	5 (10%)	22 (44%)	35 (70%)	
Female rats					
Cholangiocarcinomas	0 (0%)	49 (98%)	50 (100%)	48 (96%)	
Hepatocellular adenomas	0 (0%)	2 (4%)	4 (8%)	7 (14%)	
Hepatocellular carcinomas	0 (0%)	0 (0%)	0 (0%)	1 (2%)	
Hepatocellular adenoma/carcinomas	0 (0%)	2 (4%)	4 (8%)	8 (16%)	
Male mice					
Hepatocellular adenomas	20 (40%)			33 (66%)	42 (84%)
Hepatocellular carcinomas	7 (14%)			32 (64%)	34 (68%)
Hepatocellular adenoma/carcinomas	26 (52%)			44 (88%)	50 (100%)
Female mice					
Hepatocellular adenomas	5 (10%)			31 (62%)	48 (96%)
Hepatocellular carcinomas	2 (4%)			7 (14%)	27 (54%)
Hepatocellular adenoma/carcinomas	7 (14%)			34 (68%)	50 (100%)

Fifty animals were examined per group. Dose was given in corn oil by gavage 5 days per week. Historical incidence of hepatocellular adenoma or carcinoma: 2.5 ± 2.8%, range 0–10% (male rats), 1.2 ± 2.7%, range 0–10% (female rats); 35.1 ± 11.0%, range 14–52% (male mice); 10.1 ± 4.3%, range 2–16% (female mice).

[a]Reference 4.

2. *Hepatocellular adenomas and carcinomas*: Hepatocellular neoplasms were induced in both sexes of rats and mice at furan low-dose levels of 2 and 8 mg/kg bw, respectively (Table 2.4.8). The increase in incidence was dose-related. Hepatocellular neoplasms were not observed at the 9- or 15-month interim evaluation in rats. Background (control mice) incidence of hepatocellular adenoma/carcinoma in male mice was high (52%). Hepatocellular carcinoma was first observed at the 15-month interim evaluation in 2 of the 10 male rats dosed with 30 mg/kg for 13 weeks of a stop-exposure study.
3. *Mononuclear cell leukemia*: Incidence was increased in male and female rats at 4 and 8 mg/kg dose levels (Table 2.4.9).
4. *Benign pheochromocytoma of the adrenal medulla*: Incidence was increased in male mice at 8 and 15 mg/kg bw and in female mice at 15 mg/kg bw (Table 2.4.9).

TABLE 2.4.9 Incidence of mononuclear cell leukemia in rats and lesions of the adrenal medulla in mice in the NTP 2-year gavage study.[a]

	Furan, mg/kg				
Neoplasm/lesion	0	2	4	8	15
Mononuclear cell leukemia					
Male rats, %	16	22	34	50	
Female rats, %	16	18	34	42	
Adrenal medulla					
Male mice					
Focal hyperplasia, %	0			6	18
Benign pheochromocytoma, %	2			12	20
Female mice					
Focal hyperplasia, %	4			2	16
Benign pheochromocytoma, %	4			2	12

Fifty animals were examined per group. Dose was given in corn oil by gavage 5 days per week. Historical incidence of mononuclear cell leukemia of vehicle controls: 21.3 ± 8.9%, range 4–38% (male rats) and 26.8 ± 7.0%, range 16–38% (female rats). Historical incidence of benign pheochromocytoma of vehicle controls: 2.7 ± 1.6%, range 0–4% (male mice) and 1.5 ± 2.4%, range 0–8% (female mice).

[a]Reference 4.

The NTP studies demonstrated that furan induced cytotoxic and carcinogenic effects at all doses tested in rats and mice, and a no observed adverse effect level (NOAEL) was not identified.

A recent study by Moser *et al.* (G. Moser, 2007, personal communication) examined the hypothesis that cytotoxicity and cell proliferation may play a role in the liver tumor formation in female mice treated with furan. Like the NTP bioassay, female B6C3F1 mice were exposed to furan, 5 days a week, by gavage in corn oil for 3 weeks or 2 years. Furan dosages included the carcinogenic doses used in the NTP bioassay and lower dosages of 0, 0.5, 1.0, 2.0, 4.0, and 8.0 mg/kg bw. There was a dose–response relationship between the incidence of liver tumors and the induction of cytotoxicity and cell proliferation. The thresholds (Lowest Observed Adverse Effect Level, LOAEL) for cytotoxicity and tumors were 1.0 and 4.0 mg/kg bw, respectively. An increase in cell proliferation in the 3-week study was significant only at the 8.0 mg/kg bw dosage. This study did identify the NOAELs for both cytotoxicity and hepatocarcinogenicity in female B6C3F1 mice. Furthermore, it demonstrated that furan-induced liver tumorigenicity is a threshold-dependent event, which occurs at doses (≥4.0 mg/kg bw) well above the cytotoxic levels (≥1.0 mg/kg bw).

The most sensitive end point and species/sex combination was the incidence of cholangiocarcinomas (bile duct tumors) in female F344 rats (Table 2.4.7). In this bioassay, there were four dose groups: a control, 2 mg/kg/day, 4 mg/kg/day, and 8 mg/kg/day. The incidence of cholangiocarcinoma was 0% in 50 controls and near 100% in all other dose groups. In the NTP study, interim sacrifices were conducted at both 9 and 15 months. The data at 15 months showed a saturated dose–response similar to that seen at 24 months (lifetime).

Only the data at 9 months showed appreciably less than 100% incidence at any dose group. At 9 months' interim sacrifice, the observed incidence of cholangiocarcinoma was 0/10 (0%), 4/10 (40%), 9/10 (90%), and 10/10 (100%) at 0, 2, 4, and 8mg/kg/day, respectively. The lowest dose group with >10% incidence of tumors and significantly elevated above that in the control group is the 2mg/kg/day female group with an incidence of 40%.

2.4.8 MECHANISM OF ACTION

Furan is both cytotoxic and carcinogenic in rodents. It has been shown that cis-2-butene-1,4-dial is a toxic metabolite (21) that covalently binds to proteins and causes DNA damage. As stated previously, this metabolite was mutagenic in Ames assay at nontoxic levels in TA104, a strain that is sensitive to aldehydes (22), but not in TA97, TA98, TA100, and TA102. In comparison, furan tested negative in Ames assays with strains TA98, TA100, TA1535, and TA1537 (4). Peterson and coworkers also reported that glutathione, which inhibits covalent binding, was able to inhibit both acute toxic and genotoxic activity of cis-2-butene-1,4-dial *in vitro* (22). Previous work by Chen *et al.* showed that cis-2-butene-1,4-dial reacts readily with *N*-acetyl-L-cysteine and *N*-acetyl-L-lysine to form pyrrole and pyrrolin-2-one derivatives, indicating that this furan metabolite is a likely candidate for forming protein adducts (23). Recently, this furan metabolite was shown to react with 2′-deoxycytidine, 2′-deoxyguanosine, and 2′-deoxyadenosine to form diastereomeric adducts (24). These data further support the hypothesis that cis-2-butene-1,4-dial plays an important role in the furan-induced toxicity, including carcinogenesis. It is not clear why a macromolecular binding study with [^{14}C]furan found that the radioactivity was associated only with hepatic proteins, but not with DNA (8). Whether the method used was not sensitive enough for detecting DNA adducts or furan–DNA adducts were unstable needs to be investigated.

Furan induced apoptosis in mice at hepatocarcinogenic doses (8 and 15mg/kg bw) but not at 4mg/kg bw (25). The authors suggest that the apoptosis may have occurred in response to an increased number of DNA-altered cells. This is supported by the fact that furan induces DNA double-strand breaks in isolated rat hepatocytes at a dose of 100μM (18) and that furan at cytotoxic doses was shown in both *in vivo* and *in vitro* to cause irreversible uncoupling of hepatic mitochondrial oxidative phosphorylation, which leads to ATP depletion well before cell death (19).

2.4.9 DISCUSSION/SUMMARY

Furan is an established laboratory animal carcinogen and has been classified by the IARC as a possible carcinogen to humans (Group 2B) on this basis (3). It is also clear that furan-induced carcinogenesis in rodents occurs in conjunction with cytotoxicity. In the NTP bioassays, furan administrated by gavage to F344/N rats and B6C3F1 mice of both sexes for up to 2 years, produced hepatic cholangiocarcinoma and mononuclear cell leukemia in rats, and hepatocellular adenoma and carcinoma in rats and mice. In both species, it also caused non-neoplastic liver lesions, including biliary tract fibrosis, hyperplasia, inflammation, and proliferation as well as hepatocellular cytomegaly, degeneration, hyperplasia, necrosis, and vacuolization. Increased severity of nephropathy was also seen in rats. NOAELs were not identified in these bioassays. However, a recent report of a 2-year bioassay in female B6C3F1 mice identified NOAELs of furan for both cytotoxicity and hepatocarcinogenicity of 0.5 and 2 mg/kg bw, respectively.

Furan is negative in a conventional Ames *S. typhimurium* assay in strains TA98, TA100, TA1535, and TA1537. It does not induce unscheduled DNA repair, but induces liver cell proliferation and apoptosis. Furan is oxidized by cytochrome P450 to cis-2-butene-1,4-dial, a microsomal metabolite, which has been identified and shown to form both protein and nucleoside adducts. In addition, cis-2-butene-1,4-dial, tested positive in the Ames assay in strain TA104, but was not mutagenic in strains TA97, TA98, TA100, and TA102. These data support the hypothesis that cis-2-butene-1,4-dial is an important genotoxic metabolite in furan-induced carcinogenesis.

As is always the case with hazard information derived from animal bioassays, there is significant uncertainty in the projection of risk from laboratory animals to the equivalent risk for humans. This uncertainty arises from two major areas. One is the extrapolation from high-dose laboratory animal studies to low-dose human exposure as seen in the diet and the second from cross-species extrapolation where there can be significant toxicokinetic and dynamic differences between rodent bioassays and humans. In part this is about differences in allometry between rodents and humans, where substantial differences in body size, life span, and basal metabolism exist. The differences in metabolism are particularly important for furan because of the fact that the proximal toxicant is not the parent compound, furan, but a metabolite, cis-2-butene-1,4-dial.

It is well known that "processing" can produce physical and chemical alterations in food. The improved storage conditions and the reduction of the disease burden of harmful microorganisms are two of the basic motivations for processing. Chemical alteration of food from processing produces new molecular entities, which may be associated with their own risks. In the case of furan derived during food processing, the central critical public health issue is the risk versus risk trade-off between microbiological and chemical hazards.

REFERENCES

1. Center for Food Safety and Applied Nutrition (CFSAN, US Food and Drug Administration) (2004). http://www.fda.gov/ohrms/dockets/ac/04/slides/2004-4045s2.htm (accessed 18 September 2008).

2. Center for Food Safety and Applied Nutrition (CFSAN, US Food and Drug Administration) (2007). http://www.cfsan.fda.gov/~dms/furanexp/sld09.htm (accessed 18 September 2008).
3. IARC. (1995). *IARC Monographs on the Evaluation of Carcinogenic Risks to Humans, Volume 63, Dry Cleaning, Some Chlorinated Solvents and Other Industrial Chemicals*, pp. 394–407.
4. National Toxicology Program (NTP) (1993). Toxicology and Carcinogenesis Studies of Furan (CAS No. 110-00-9) in F344/N Rats and B6C3F$_1$ Mice (Gavage Studies): NTP Technical Report No. 402, U.S. Department of Health and Human Services, Public Health Service, National Institutes of Health, Research Triangle Park, NC.
5. Center for Food Safety and Applied Nutrition (CFSAN), US Food and Drug Administration (2006). http://www.cfsan.fda.gov/~dms/furan.html (accessed 18 September 2008).
6. Swiss Federal Office of Public Health (SFOPH) (2004). Document submitted to the US FDA Docket, 2004N-0205: Furan in Food, Thermal Treatment. http://www.fda.gov/ohrms/dockets/default.htm (accessed 18 September 2008).
7. Zoller, O., Sager, F., Reinhard, H. (2007). Furan in food: headspace method and product survey. *Food Addit. Contam.*, *24* (**S1**), 91–107.
8. Burka, L.T., Washburn, K.D., Irwin, R.D. (1991). Disposition of [^{14}C]furan in the male F344 rat. *J. Toxicol. Environ. Health*, *34*, 245–257.
9. Kedderis, G.L., Carfagna, M.A., Held, S.D., Batra, R., Murphy, J.E., Gargas, M.L. (1993). Kinetic analysis of furan biotransformation by F344 rats in vivo and in vitro. *Toxicol. Appl. Pharmacol.*, *123*, 274–282.
10. Parmar, D., Burka, L.T. (1993). Studies on the interaction of furan with hepatic cytochrome P-450. *J. Biochem. Toxicol.*, *8*, 1–9.
11. Kedderis, G.L., Held, S.D. (1996). Prediction of furan pharmacokinetics from hepatocyte studies: comparison of bioactivation and hepatic dosimetry in rats, mice, and humans. *Toxicol. Appl. Pharmacol.*, *140*, 124–130.
12. Kedderis, G.L. (1997). Extrapolation of in vitro enzyme induction data to human in vivo. *Chem. Biol. Interact.*, *107*, 109–121.
13. Foureman, P., Mason, J.M., Valencia, R., Zimmering, S. (1994). Chemical mutagenesis testing in Drosophila. IX. Results of 50 coded compounds tested for the National Toxicology Program. *Environ. Mol. Mutagen.*, *23* (**1**), 51–63.
14. Wilson, D.M., Goldsworthy, T.L., Popp, J.A., Butterworth, B.E. (1992). Evaluation of genotoxicity, pathological lesions, and cell proliferation in livers of rats and mice treated with furan. *Environ. Mol. Mutagen.*, *19*, 209–222.
15. Stich, H.F., Rosin, M.P., Wu, C.H., Powrie, W.D. (1981). Clastogenicity of furans found in food. *Cancer Lett.*, *13*, 89–95.
16. Lee, H., Bian, S.S., Chen, Y.L. (1994). Genotoxicity of 1,3-dithiane and 1,4-dithiane in the CHO/SCE assay and the Salmonella/microsomal test. *Mutat. Res.*, *321*, 213–218.
17. McGregor, D.B., Brown, A., Cattanach, P., Edwards, I., McBride, D., Riach, C., Caspary, W.J. (1988). Responses of the L5178Y tk$^+$/tk$^-$ mouse lymphoma cell forward mutation assay. III: 72 coded chemicals. *Environ. Mol. Mutagen.*, *12*, 85–154.

18. Mugford, C.A., Kedderis, G.L. (1996). Furan-mediated DNA double strand breaks in isolated rat hepatocytes. *Fundam. Appl. Toxicol.*, *30* (**1**, Part 2), 128.
19. Mugford, C.A., Carfagna, M.A., Kedderis, G.L. (1997). Furan-mediated uncoupling of hepatic oxidative phosphorylation in Fisher-344 rats: an early event in cell death. *Toxicol. Appl. Pharmacol.*, *144*, 1–11.
20. Reynolds, S.H., Stowers, S.J., Patterson RM Maronpot, R.R., Aaronson, S.A., Anderson, MW (1987). Activated oncogenes in B6C3F1 mouse liver tumours: implications for risk assessment. *Science*, *237*, 1309–1316.
21. Chen, L-J, Hecht, S.S., Peterson, L.A. (1995). Identification of cis-2-butene-1,4-dial as a microsomal metabolite of furan. *Chem. Res. Toxicol.*, *8*, 903–906.
22. Peterson, L.A., Naruko, K.C., Predercki, D.P. (2000). A reactive metabolite of furan, cis-2-butene-1,4-dial, is mutagenic in the Ames assay. *Chem. Res. Toxicol.*, *13* (**7**), 531–534.
23. Chen, L-J, Hecht, S.S., Peterson, L.A. (1997). Characterization of amino acid and glutathione adducts of cis-2-butene-1,4-dial, a reactive metabolite of furan. *Chem. Res. Toxicol.*, *10*, 866–874.
24. Byrns, M.C., Predecki, D.P., Peterson, L.A. (2002). Characterization of nucleoside adducts of cis-2-butene-1,4-dial, a reactive metabolite of furan. *Chem. Res. Toxicol.*, *15* (**3**), 373–379.
25. Fransson-Steen, R., Goldsworthy, T.L., Kedderis, G.L., Maronpot, R.R. (1997). Furan-induced liver cell proliferation and apoptosis in female B6C3F1 mice. *Toxicology*, *118*, 195–204.

2.5

HYDROXYMETHYLFURFURAL (HMF) AND RELATED COMPOUNDS

Francisco J. Morales
Consejo Superior de Investigaciones Científicas (CSIC), Instituto del Frío José Antonio Novais 10, Madrid 28040, Spain

2.5.1 INTRODUCTION

Hydroxymethylfurfural (5-hydroxymethyl-2-furaldehyde, HMF) is formed naturally as an intermediate in the Maillard reaction (MR) (1) and by dehydration of hexoses under mild acidic conditions (2). The chemical structure and selected physical characteristics of HMF are shown in Fig. 2.5.1. Heat processing is the most common way of preserving food, and making it edible and organoleptically attractive to consumers. However, overprocessing may cause damage to food and consequently a possible loss in the nutritional value. Different chemical indicators for assessing the quality of heat-treated foods have proved to be useful for the control of processes, creating the possibility of optimizing manufacturing conditions. The formation of HMF is directly related to the heat load applied to many foods. Although HMF is not present in fresh or untreated foods, it rapidly accumulates during the heat treatment and storage of carbohydrate-rich products, sometimes exceeding 1 g/kg in certain dried fruits and caramel products (3–5). However, the content of HMF can vary largely in various food products (6a). Another source of HMF in food is related to ingredients used in the product formulation such as the addition of caramel solutions, flavorings, honey, or heat-processed dairy products such as dried whey. The range of HMF found in selected food commodities is shown in Table 2.5.1.

Process-Induced Food Toxicants: Occurrence, Formation, Mitigation, and Health Risks,
Edited by Richard H. Stadler and David R. Lineback

Hydroxymethylfurfural (HMF): At a Glance	
Historical	Applied as a marker of thermal processing, storage, or abuse practices. Colorimetric methods were developed in the 1950s to determine HMF in heated dairy products.
Analysis	Modern techniques are mainly based on HPLC coupled with UV detection. Other approaches such as capillary electrophoresis, ion chromatography, or photometric detection after derivatization have been described.
Occurrence in Food	Present in many different foods (industrial and domestic) that are subjected to thermal treatment. The main foods include honey, coffee, fruit juices, candies, cakes, tomato paste, processed cereal products, heat-treated milk, and alcoholic beverages.
Main Formation Pathways	Formed as an intermediate in the Maillard reaction and during acid-catalyzed dehydration of hexoses. The formation of HMF is dependent on temperature, pH, and water activity.
Mitigation in Food	Raw material and ingredient selection; control of thermal conditions during processing; prolonged heating that under certain conditions may lead to accelerated elimination. Choice of appropriate storage conditions (avoiding excess temperature) over product shelf life where this is product relevant.
Health Risks	HMF at very high concentrations is cytotoxic, causes irritation to the eyes, upper respiratory tract, skin, and mucus membranes. The available *in vitro* data generated in cellular systems suggests that HMF does not pose a serious risk to human health, but some concerns have been raised in conjunction with the potential genotoxic properties of specific metabolites.
Regulatory Information or Industry Standards	European quality standards and Codex Alimentarius allow a maximum of 40 mg/kg HMF in honey. The fruit juice industry has set a standard value of 20 mg/kg as an upper limit for the product.
Knowledge Gaps	Data on human dietary exposure is scarce. More clarity on the toxicity and potential genotoxic mode of action of metabolites, as well as cellular effects *in vivo* are warranted.
Useful Information (web) & Dietary Recommendations	http://www.slv.se/upload/heatox/documents/Heatox_Final%20_report.pdf Avoid the over-heating of food; cook/prepare foods to the appropriate "doneness."

HMF is widely recognized as a marker of quality deterioration, generated as a result of excessive heating or inadequate storage conditions in a wide range of foods containing carbohydrates, apart from spoilage. Generally, HMF could be used as marker of the extent of the thermal treatment applied for any food where the MR and/or hexoses degradation are the main reactions taking place. HMF has been considered a chemical index for a wide range of carbohydrate-rich foods such as processed fruits (4–6b), coffee (27), honey (24, 25), and milk (26–28). HMF is also used for monitoring the heating process applied to cereal products such as pasta (29), cookies, bread (18, 30, 31),

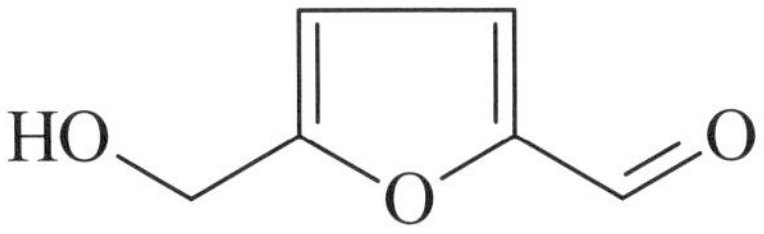

HMF

5-hydroxymethyl-2-furfuraldehyde

5-hydroxymethyl-2-furancarbaldehyde

Hydroxymethylfurfural

$C_6H_6O_3$ 126.11g/mol

CAS number: [67-47-0]

Density: 1.206g/cm^3

Melting point: 32–34°C

Molar absorptivity: 18,000

UV-maximum: 284nm

Refractive index: 1.5627at 18°C

Solubility:

- high: water, methanol, ethanol, ethylacetate
- medium: ether, benzene, chloroformether
- low: petroleum ether

Figure 2.5.1 Chemical structure and chemical/physical parameters for HMF.

extruded baby cereals (32, 33), breakfast cereals (19, 34), and toasted bread (35). The presence of HMF has also been reported in alcoholic beverages such as brandies (36), beer (13), and wine (37, 38). However, the significance of the measurement of HMF is different depending on the type of food where it is employed as marker and its purpose, i.e., adulteration (e.g., honey), overprocessing (e.g., milk or juices), or quality of ageing (e.g., vinegar). In other foods, HMF is frequently reported together with furanic compounds such as furfural. In the following sections, the relevance of HMF in several food commodities is described.

2.5.2 OCCURRENCE IN FOODS

2.5.2.1 HONEY

HMF is naturally present in honey, which is produced by action of the normal honey acidity on reducing sugars and sucrose usually at room temperature. The amount of HMF increases upon thermal treatment and/or storage at improper temperatures (39). HMF was first used as an indicator of adulteration with invert syrups (40, 41), abusive heat treatment, or improper storage conditions (42, 43). Honey is submitted to thermal treatments for two different reasons: (i) to modify its tendency to crystallize or delay its appearance; and (ii) to reduce the microbial load. Crystallization makes honey difficult to handle and pour; also, dosifiers or filling and packaging machines cannot work properly. Moreover, the presentation of the product is changed and loses appeal to most consumers. Thermal treatments applied to honey produce a simultaneous decrease in diastase activity and an increase in HMF content. The temperature and time applied must be limited when pasteurizing and stabilizing it; both diastase activity and HMF content are national and international parameters used as controls so as to limit thermal application.

TABLE 2.5.1 Range of HMF content (mg/kg) in different food commodities.

Food commodities	HMF content	Method	Reference no.
Coffee	100–1,900	LC-UV	7, 8
Coffee (instant)	400–4,100	LC-UV	9
Coffee (decaffeinated)	430–494	LC-UV	7
Chicory	200–22,500	LC-UV	7
Malt	100–6,300	LC-UV	7
Barley	100–1,200	LC-UV	7
Honey	10.4–58.8	DNPH-LC-UV	10
	nd–85.5	LC-UV	11
Honey (fresh)	0–22.8	LC-UV	12
(Nectar)	5.2–151.5	LC-UV	12
(Honeydew)	7.1–18.5	LC-UV	12
Beer	3.0–9.2[a]	DNPH-LC-UV	13, 14
Jam	5.5–37.7	LC-UV	6b
Fruit juices	2.0–22.0	IE-LC-UV	14, 15
Wine (red)	1.0–1.3[a]	LC-UV	14
Vinegar (malt)	1.6–7.3[a]	LC-UV	16
Vinegar (sherry)	13.8–34.8[a]	LC-UV	16
Vinegar (wine)	0–21.5[a]	LC-UV	16
Vinegar (caramel)	0–9.0[a]	LC-UV	16
Vinegar (balsamic)	316.4–35,251.3[a]	LC-UV	16
Cookies	0.5–74.5	LC-UV	17
Bread (white)	3.4–68.8	LC-UV	18
Bread (toast)	11.8–87.7	LC-UV	18
Bread (snacks)	2.2–10.0	LC-UV	18
Breakfast cereals	6.9–240.5	LC-UV	19
Whiskey (blended)	1.4–13.1[a]	LC-UV	20
Whiskey (straight)	2.0–8.2[a]	LC-UV	20
Brandy	1.5–4.8[a]	LC-UV	21
Rum	1.7[a]	LC-UV	21
Baby food (milk-based)	0.18–0.25	LC-MS	22
Baby food (cereal-based)	0–57.18	LC-MS	22
Dried pears	3,500	LC-UV	6a
Dried fruits	25–2,900	LC-UV	6a
Roasted almond	9	LC-UV	6a
Caramel products	110–9,500	LC-UV	6a

[a]Values are in mg/L.
nd = not detected.

The legislation of several countries and the CODEX Alimentarius Commission Standards fix a minimum amount of 8 units of diastatic activity and an upper limit of 40 mg HMF/kg honey to ensure that honey has not been excessively heat treated (44). Very high HMF values (>500 mg/kg) after processing and/or blending demonstrate an adulteration with invert syrup (44, 45). The EC Directive 2001/110/EC1 stipulates a limit <40 mg/kg, with the following exceptions: honey coming from countries or regions with tropical temperatures (80 mg/kg), and honey with a low enzymatic level (15 mg/kg; 8–3 Schade units) (11).

2.5.2.2 Coffee

In roasted products, the HMF content may reach appreciable amounts. Examples are chicory and coffee, with concentrations reported in the 0.2–22.5 g/kg and 0.1–6 g/kg range, respectively (7). For furfural, the highest reported content is 250 and 113 mg/kg in coffee and roasted chicory, respectively. Amounts between 26 and 120 mg/kg in natural coffees and between 500 and 2300 mg/kg have been found in torrefacto coffee (coffee with sugar added before roasting). In coffee, the HMF content has been applied to distinguish between coffee and coffee–chicory mixtures, but this application is limited to instant coffees (9).

2.5.2.3 Fruit Juices

Thermal treatments and unsuitable storage temperatures unfavorably impact the sensorial and nutritional properties of juices. Color quickly fades (46), and the fresh odor is altered both by loss of the most aromatic components and by formation of some detrimental off-flavors (47). The HMF content is practically zero in fresh, untreated fruit juice (48). HMF accumulates in stored juices and has been associated with color and flavor degradation. Logically, a high content of HMF should represent an index of thermal abuse and prolonged or improper storage (49–54). In apple sauce and grape jelly, the content of HMF has been used as indicator of long-term storage quality (55). There are some studies on the presence of HMF in jams (56, 57). HMF formation is judged to be the most useful method for assessing the effectiveness of heat treatment in destroying spoilage organisms in jams and fruit products (58). Although HMF is not present in fresh grapes, it can be formed during the juice production especially as a result of thermal stress during heat processing to which the must is subjected. Grapes seem to have a higher capacity to form HMF than other fruits. This could be due to the sugar content and composition of grape juices. It has been found that most commercial grape juices contained measurable quantities of HMF (59). An increase of the HMF content has been observed during storage time (60). This has been shown for apple juices and other fruit juices such as citrus fruit juices (39, 61, 62). The Association of the Industry of Juices and Nectars from Fruits and Vegetables of the European Economic Community (63) included the amount of HMF among the absolute parameters of quality (max. 20 mg/L in fruit juices and 25 mg/kg in concentrates) in the Code of Practice for the evaluation of fruits and vegetable juices.

2.5.2.4 Milk and Milk-Based Baby Foods

HMF is formed in milk upon treatment at temperatures above 120 °C. The content of HMF in milk is controversial as the terms "potential HMF," "total

HMF," or "HMF value" were widely used in the literature usually resulting in an overestimation of the actual "free HMF" content (64). Total HMF is calculated as the sum of the HMF formed from precursors (potential HMF) plus free HMF. Although the so-called HMF value is useful in a technological sense, this fact should be taken into account before its application for nutritional or consumption studies.

Special attention has been given to the HMF content in infant formulas or baby foods in general. Infant formulas are sometimes the only food given to infants in a certain period of their development, and therefore they must fulfill all their nutritional needs. However, due to the enrichment of infant milks with various compounds including ascorbic acid, iron, and lactose, they are more susceptible to the MR than cow's milk. Generally speaking, the content of HMF in milk and milk-based products is dependent on (i) quality of the raw material used in the recipe (65), (ii) thermal treatment applied (66), and (iii) storage conditions (67).

2.5.2.5 Cereals

The amount of HMF present in cereal-based products is highly variable and ranges between 0.4 and 65.5 mg/kg in infant cereals (33), 3.7 and 194 mg/kg in breakfast cereals (34), 2.2 and 88 mg/kg in bread (68, 30), and 0.1 and 7 mg/kg in dried pasta (31). During the baking of bread, the water content on the surface of the loaf is reduced at a rate faster than in the crumb, with exposure of the dough to higher temperatures at the surface of the loaf (69). Consequently the HMF content in the crumb is relatively lower than in the crust, typically in the range 0.6–2.2 mg/kg and 18.3–176 mg/kg, respectively (18). The fermentation and baking times required to obtain a large bread loaf with a thick crust are approximately double those required for other breads. For this reason, four to eight times more HMF is measured in large versus stick breads (18).

2.5.2.6 Alcoholic Beverages (Wine and Brandy)

It has been shown that the HMF content in white and red wine does not exceed a concentration of about 10 mg/L (37, 70, 71). The mean values were found to be in the range of 0.1–1.5 mg/L. Higher values have only been observed in dessert wine and other sweet wines (72–74a,b). The HMF contents of normally produced white wines and of those that have been aged in oak barrels were not significantly different (38). Sherries were reported with HMF concentrations in the range 20–340 mg/L, sweet sherries ranged from 130 to 1245 mg/L (74a), and dry sherry 0–152 mg/L. HMF was hardly detected in fino wine (73, 75).

Jeurings and Kuppers (44) found that part of the furfural present in brandy originated during the distillation process, so that the presence of this furanic aldehyde in a spirit cannot be taken as an indicator or marker of aging. Profumo *et al.* (76) also expressed doubts about the use of furfural as marker of ageing for evaluation of the length of maturation of the distillate, as there are no appreciable differences between recent distillates and the samples aged in oak. If we also take into account that most of the HMF found in the samples of bottled brandy is the result of the addition of caramel coloring (77), this compound can hardly be taken as a possible aging marker either (78). Straight whiskey is aged in a freshly charred oak barrel for a maximum of 2 years and is not colored with added caramel, which is accepted in blended whiskey. Hexoses present in the syrup may undergo dehydration, forming HMF. Thus, straight whiskeys have a 2:1 higher ratio of furfural to HMF.

2.5.2.7 Tomato Paste

Industrial processing of tomato to a final moisture content of <15% often involves high temperatures (60–110 °C) for a period of 2–10 h in the presence of oxygen, and therefore products may show some degree of oxidative damage (79). During processing and storage, tomato products undergo changes in their nutritional and sensorial quality. HMF is normally used to evaluate the extent of heat damage in processed tomato products (80, 81). Air-drying of tomatoes also leads to an increase in the HMF content, which results in undesirable color and appearance changes of dried tomatoes (79). Drying of tomato halves at 110 °C resulted in a 20-fold increase in HMF compared with 80 °C. An increase in time from 390 to 430 min resulted in a corresponding increase in HMF content from 10 to 36 mg/kg dry matter at 80 °C, and 18 to 512 mg/kg dry matter at 110 °C. Some authors also reported that an HMF value >20 mg/kg dry matter corresponded to a change in color from red to brick red, then to brown. However, others claim that HMF does not represent a valid descriptor of heat damage or thermal history of tomato products (82, 83). The decrease in HMF during storage at room temperature limits its use as an index of heat damage, as the true product damage caused by heat treatment is masked. The HMF content of a tomato paste sample (dry matter of 28.1%), treated, and stored at 25 °C, decreased from 609 to 339 mg/kg after 29 days and to 17 mg/kg after 98 days, while furosine showed a slight increase (about 10%) after the same storage times.

2.5.2.8 Vinegar

Depending on the origin of the food matrix, different types (e.g., wine, apple, sherry, or malt vinegar) can be distinguished. Aqueous ethanol (e.g., industrial quality or distilled spirits) can also be fermented and this gives colorless spirit vinegar. As all food matrices used for the production of vinegar contain sugar, the formation of HMF during either the production process or storage is possible (16). The highest concentrations of HMF were found in the sherry vinegar samples that are produced from sherry wine and subsequently aged in oak barrels. Following the so-called *Solera* aging system leads to unique organoleptic properties in the final product. One possible origin of the high HMF concentrations could be cellulose degradation of the oak barrels. HMF could barely be detected in another sherry vinegar product obtained by a quick acidification process in which the vinegar had not been submitted to aging (84). Another possible source of HMF is the addition of caramel which is well established in vinegar wineries (84). While the values for the samples labeled as "Aceto Balsamico di Modena" were in the range of 300 mg/L to 3.3 g/L, those for the traditionally produced vinegar samples ("Aceto Balsamico di Modena tradizionale") clearly depended on their age and reached values of up to 5.5 g/L after a fermentation time of 25 years. These high values for HMF in balsamic vinegar could be due to several reasons: (i) an elevated level of HMF in the concentrated must, (ii) the long fermentation process, and (iii) the storage in wooden barrels (85). Among other substances, the chemical composition of wood includes lignin, cellulose, and hemicellulose. When an oak barrel is produced, the inner part of the staves is burnt and wetted in order to prevent breakage. This treatment leads to a partial reduction of cellulose, thus generating HMF (86). Therefore, the HMF concentration could be considered a good indicator for the age of vinegar.

2.5.2.9 Food Flavorings and Additives

HMF has been detected in caramel, which is a widely used coloring agent in food and pharmaceutical syrups (87). Caramel is used in different food categories such as processed fruits, confectionery, ready-to-eat savories, cereal, and milk products.

2.5.2.10 Parenteral and Enteral Solutions

Apart from the traditional dietary source, glucose infusions are commonly used as vehicles for administering a variety of drugs. Enteral formulas are products with physicochemical and biological properties that allow them to be administrated through a tube into the gastrointestinal (GI) tract. Rufian-Henares *et al.* (88) reported amounts from 0.05 to 19.1 mg/L of HMF in 19 commercial enteral formulas. During the production of the glucose infusion, the solutions must be sterilized and subsequently HMF may be formed. Ulbricht *et al.* (89) reported the results of several studies that have determined the concentrations of HMF in parenteral solutions. In sterile glucose solutions, HMF concentrations range from 1 to 90 mg HMF/L (90). Some consequences have been observed during continuous ambulatory peritoneal dialysis, where mesothelial cells are continually exposed to glucose degradation products, such as HMF (from the peritoneal dialysis fluids), resulting in a potential pathogenic effect during long-term exposure. The HMF concentration correlated positively with high acidity ($pH < 4$), high sterilization temperature (>110 °C), and a long sterilization time (30 min) (91). HMF has been determined in fructose-containing solutions for intravenous injection. It appears to be formed during sterilization whenever a fructose-containing solution with pH lower than 3.5 to 4.0 is heated at about 110 to 130 °C (92). Murty *et al.* (93)

reported that a 50% dextrose injection showed an HMF concentration of 0.72 mg/L, 24 h after manufacture. In addition, there is little information on the potential interaction of HMF and active drugs with an amino group.

2.5.3 ANALYSIS

The reliable qualitative analysis of HMF in food has been a central objective due to its application as marker of thermal processing or abuse practices, but also in the fields of clinical chemistry and biotechnology. The occurrence of HMF in several food products (e.g., fruit juices, honey, tomato products, milk, and jam) may serve as an indication of quality deterioration. However, the use of different analytical methods for HMF determination and the use of inaccurate analytical methods or inadequate extraction procedures are a drawback to establishing a reliable database for HMF content in processed foodstuffs, pivotal for estimation of exposure and consequently a reliable risk assessment. Some analytical procedures are aimed or developed only for distinguishing food quality or processing conditions, but results on HMF are often not acceptable from a quantitative point of view. An in-house validation of the procedure is mandatory as a first step and many of the published analytical methods lack this.

In the scientific literature, a wide variety of methods are described that include colorimetric, spectrophotometric, and chromatographic approaches. Table 2.5.2 summarizes the most frequently employed analytical approaches to measure HMF in different foodstuffs.

2.5.3.1 Sample Extraction and Cleanup

Many compounds naturally present or formed in foods during processing could also absorb at 280 nm, which is the common wavelength for HMF detection when using chromatographic/UV detection methods. Some cleanup steps are needed depending on the food matrix or potential co-extractives. A number of approaches have been applied to extract HMF from foods. For simple matrices, such as honey, alcoholic beverages, coffee, and fruit juices, extraction in water and subsequent filtration were commonly applied. However, the use of solid-phase extraction (SPE) cartridges is recommended to clarify the extracts, particularly important for coffee and fruit juices (8). In milk-based products, the extract is heated for about 25 min after the addition of oxalic acid (28, 114). In other cases, extraction of HMF is improved by use of methanol or ethyl acetate, with subsequent evaporation and reconstitution of the extract in water (or mobile phase when employing liquid chromatography) prior to analysis (115).

The use of Carrez I (15% $K_4Fe(CN)_6$) and Carrez II (30% $ZnSO_4$) solutions as a clarifying agent instead of classical acids such as trichloroacetic (TCA), metaphosphoric, or sulfosalicylic, is recommended for cereal and tomato products, due to the possible *in situ* production of HMF from glucose present in

TABLE 2.5.2 Summary of some analytical methods applied for HMF analysis in foods.

	Reference no.
Colorimetric	
Resorcinol/HCl	94
Diphenylamine	95
Aniline	49
Thiobarbituric acid	96
p-Toluidine/barbituric acid	
Direct	97
Automated (FIA)	98
Spectrophotometric	
Direct measurement	
Automated	99
Derivative spectra	100
Indirect measurement	
Thiosemicarbazide	73
Polarography	101
Spectroscopy	
^{1}H NMR	102
Chromatographic	
Paper chromatography	
2,4-DNPH derivative	103
Liquid–liquid partition	104
Thin-layer chromatography	105
Gas chromatography	75, 106, 107, 108
Liquid chromatography	
2,4-DNPH derivative	10
Anion exchange	15, 109
Cation exchange	14, 110
Reversed-phase	
Cereals	34
Honey	44
Jams	6b
Dried fruits	8
Vinegar	16
Fruit juices	39
Tomato products	81
Coffee	8, 111
Milk	28, 64
Capillary electrophoresis	
Micellar electrokinetic capillary chromatography	26, 113

the food matrix at low pH (30, 68, 81). Moreover, three consecutive extractions are usually applied for cereals products (30, 116). However, TCA (40% w/v) has been traditionally used in other foods such as milk (26, 32) and fruit preparations (4, 6b). Recently, Ait-Ameur and coworkers showed that the use of TCA during preparation of cereals samples did not lead to artifactual production of HMF (17).

Extensive sample pretreatment with SPE has been reported by several authors. Lee *et al.* (61) used a C-18 cartridge following washing with hexane, to remove contaminants in orange juice extracts, and reported a limit of detection (LOD) of 50 µg/L. Gökmen and Senyuva (22) used Oasis HLB cartridges packed with a macroporous copolymer of the lipophilic divinylbenzene and the hydrophilic *N*-vinylpyrrolidone to clean the aqueous extract prior to LC analysis. HMF present in the extract strongly interacted with the sorbent material, whereas most of the co-extractives did not. HMF retained in the cartridge was eluted with diethyl ether, evaporated to dryness, and redissolved in water prior to LC-MS analysis. The authors commented that traditional LC/UV analyses without an SPE cleanup step overestimate the HMF concentrations in baby foods by about 32%.

2.5.3.2 Colorimetric Methods

Initially, the classical methods for the qualitative identification of carbonyl compounds were based on colorimetric procedures, and were later adapted for quantitative measurement of furanic compounds, and are often the official method for the determination of HMF in certain foods. Colorimetric methods are based on color reactions that can ideally be measured in the visible range. Resorcinol, diphenylamine, aniline, *p*-toluidine, and 2-thiobarbituric acid (TBA) have widely been used for HMF determination. The so-called Winkler method for assessing quality of honey involves the use of *p*-toluidine but has been questioned due to uncertainties in the color measurement (97). The method described by White (117) involves measurement of UV absorbance (284 and 336 nm) of clarified aqueous honey solutions with and without bisulfite, while the Winkler method involves absorbance (550 nm) with added barbituric acid and *p*-toluidine. This reaction and the reaction with TBA have been widely used to quantify HMF in honey (118), wine (71–73, 75), brandy (119, 120), grape syrup, and must (59). In coffee, HMF may be overestimated by around 9% due to the furfural content as both substances form red colors with *p*-toluidine/barbituric acid (7). However, an automated flow injection analysis (FIA) version of the Winkler method was proposed by Salinas *et al.* (121).

The TBA method has been widely applied in dairies since Keeney and Bassette (96) developed a simple spectrophotometric method for determining HMF in dairy products using a TBA reaction product (after boiling in an oxalic acid solution). However, the main drawbacks of the method are the lack of specificity of TBA for HMF, since other aldehydes may take part in the reaction, and the limited stability of the colored complex (112). Morales *et al.* (112)

compared the chromatographic and colorimetric methods for HMF determination and found that about 70% of the HMF measured by the colorimetric method was due to interferences. Hence, the TBA method is not a reliable measurement of HMF content, although it is still used as a quick, cost-effective measurement of the heat load of milk products. Since the development of the Keeney and Bassette method (96), a distinction has been made between free HMF and the measurement of the so-called total HMF content described earlier. To determine the latter, the milk sample is heated in 0.3 N oxalic acid to release HMF. Potential HMF is the sum of the HMF precursors (lactulosyllysine, 1-2 enolized products, etc.), plus free HMF. Free HMF is determined by omitting the heating step. Later, Morales *et al.* (122) described an approach to estimate the potential HMF formed only from the Amadori rearrangement product (ARP) degradation in milk-resembling systems, which could be applied as an indirect determination of the ARP in milk-based products.

2.5.3.3 Spectrophotometric Methods

Spectrophotometric determinations of HMF can be carried out either by direct measurement of the absorption, or by preparing a derivative (indirect methods). White and Siciliano modified the previous Winkler method by recording the absorbance at 284 and 336 nm with a spectrophotometer (45); a plate reader can also be used to expedite measurement (123). However, experimental errors are expected if other compounds that also absorb at 284 nm and react with hydrogensulfite are present at appreciable concentrations. In an indirect method, HMF reacts with thiosemicarbazide to give the corresponding thiosemicarbazones that absorb at 322 nm (73). The use of derivative spectra, first to fourth, and the use of partial least squares multivariate calibration enhance the selectivity and have been applied for the determination of HMF in citrus fruit juices (100, 123).

In summary, both the colorimetric and spectrophotometric methods have several drawbacks. They are often tedious, prone to interference especially in strongly colored foods, and employ toxic or hazardous chemicals. Further, they require a strict control of both reaction time and temperature as the instability of the reaction product may lead to low recoveries and hence a high variability in the results.

2.5.3.4 Chromatographic Methods

Today, LC techniques are preferably used for accurate and reliable measurement of furanic compounds in several food products. These techniques can determine HMF and furfural specifically, and the formation of a colored derivative is not required because of the strong UV absorption of furfurals at approximately 280 nm. For many analyses, a detection wavelength of 280 nm is chosen because it is located between the maxima of HMF (284 nm) and furfural (277 nm). HMF shows a band at 284 nm (18,000 molar absorptivity) and a less intense band at 230 nm. Reversed-phase LC (RP/LC) methods have been widely used to determine the contents of 5-HMF and furfural in many food items, such as apple juices and concentrates (124), commercial brandies and caramels (86), coffee (111), milk (64), infant formulas (20a), breakfast cereals (34), tomato products (81), and jams (6b). HMF is often eluted isocratically with mobile phases containing 5–10% acetonitrile or methanol in water, acidified water (0.1% acetic acid), or sodium-acetate buffers (pH 3.6), although gradient elution was reported for coffee and whiskies (7, 20b).

Some authors reported a derivative of the 2,4-dinitrophenylhydrazone (DNPH) of HMF in order to enhance both the selectivity and sensitivity of the method in fruit juices and beers (62). The reaction between DNPH and carbonyl compounds is highly specific. The UV/Vis spectra of HMF hydrazone reveal three maxima at 287, 331, and 352 nm. DNPH should be in excess, and a ratio 1:20 is considered sufficient. The derivatives are formed in 25 min and are stable at room temperature for several days. Colored or fluorescent hydrazones are detected photometrically in acidic medium or fluorimetrically after an extraction step or after dialysis into methanol (125, 126).

Yuan and Chen (14) described an ion exchange LC/PDA (photodiode array detection) approach for the separation of HMF, 2,5-dimethyl-4-hydroxy-3(2H)-furanone (DMHF), furoic acid (FA), furfural, 2-acetylfuran, and furfuryl alcohol in several fruit juices. The results of HMF from honey correlated well with the Winkler method, but poorly with RP/LC (cation-exchange LC) (110). Recently, Gökmen and Senyuva (22) proposed the use of positive atmospheric pressure chemical ionization/mass spectrometry (APCI/MS) for the analysis of HMF. Characteristic are the precursor ion [M+1] and the analyte-specific ion [$C_6H_5O_2$] due to loss of water from the protonated molecule. Both ions at *m/z* of 127 and 109 were used to monitor HMF in the selected ion monitoring (SIM) mode. Furthermore, the ratio of these ions (responses of ions 127/109 = 1.12) confirmed the purity of the HMF peak. Importantly, when using traditional LC/UV methodologies, it is recommended to determine the purity of the HMF peak (19).

2.5.3.5 Capillary Electrophoresis Methods

Nowadays, capillary electrophoresis (CE) has been proven a powerful and promising technique for food analysis, mainly because of its resolving power, small sample, and buffer requirements. Moreover, CE demands less rigorous sample cleanup, which is advantageous over traditional separation techniques such as LC (113). Electromigration can be conducted in an uncoated fused-silica capillary with phosphate buffer (50 mM, pH 7.5) containing sodium dodecyl sulphate as electrolyte; separation is achieved within 5 min and HMF can be recorded at 280 nm and confirmed by its UV spectrum (26).

2.5.3.6 Other Methods

Reyes-Salas *et al.* (101) described an electrochemical approach for the analysis of HMF in honey. HMF presented a single, well-defined reduction signal at

−1100 mV versus Ag/AgCl by using borate as supporting electrolyte. Proton nuclear magnetic resonance (NMR) analysis has been also used for quantitative determination of HMF in traditional balsamic vinegar (102). Gas chromatography (GC) has rarely been applied for the analysis of HMF in foodstuffs despite its good resolution. In wines, Guerra-Hernandez and coworkers demonstrated the lack of sensitivity of the Winkler method as compared with GC (75).

2.5.4 FORMATION

During thermal treatment of foods, sugars decompose into furfural compounds by caramelization, the MR, or pyrolysis of either free monosaccharides (hexoses) or reducing moieties of disaccharides (1, 127, 128). Figure 2.5.2 summarizes the main reactions involved in the formation of HMF in foods.

2.5.4.1 Caramelization

Caramelization is an example of nonenzymatic browning involving the degradation of reducing sugars without the condensation step. When sugars are heated above their melting points, they darken to form brown color polymers under alkaline or acidic conditions. Polysaccharides are first to decompose into monosaccharides. The formation of HMF from the thermal treatment of hexoses in acidic medium even under mild conditions has been known for a long time. HMF arises from hexoses through an acid-catalyzed dehydration and cyclization mechanism (2, 129). Furfural and methylfurfural are mainly produced from pentoses (130, 131). In foods, predominantly glucose, fructose, sucrose (previous hydrolysis), maltose, maltotriose, and lactose and to some extent reducing pentoses together with amino acids and proteins are involved. HMF is formed from both the degradation of hexoses and is also an intermediate in the MR; its formation is dependent on the temperature, pH, water activity, acidity, and presence of bivalent metals, organic or inorganic acids, or salts in the reaction media (132). HMF is a common product of these two reactions and is formed from 3-deoxyhexosulose, the dehydration product derived from 1,2-enolization of glucose and fructose. During caramelization, the reducing carbohydrates, including maltose and maltotriose directly undergo 1,2 enolization, dehydration, and cyclization reactions. Caramelization requires higher temperatures than the MR does to develop. Similarly, different sugars have a different impact on the formation of HMF; for example, fructose is apparently twice as reactive as glucose.

2.5.4.2 Maillard Reaction

The MR is a complex cascade of reactions that has been recorded in heated, dried, or stored foods as well as *in vivo* in the mammalian organism (1, 133,

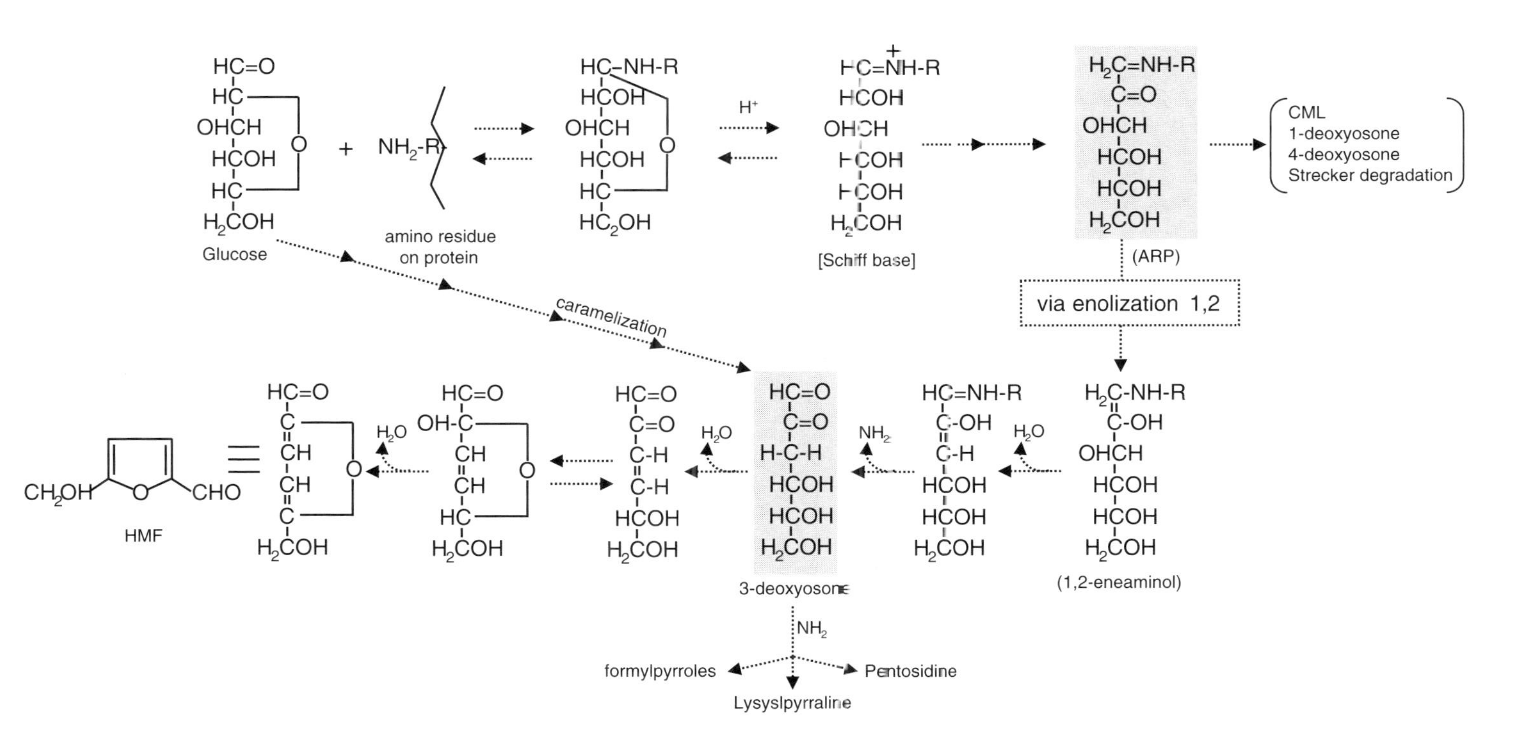

Figure 2.5.2 Scheme of the main routes for HMF formation during heat processing of foods.

134). During the heating of foods, the MR takes place and the resulting products contribute to the characteristic flavor and color of the products. At this point it is important to remind readers that a plethora of foodstuffs that make up the typical Western diet, e.g., bread, snack items, breakfast cereals, roasted meat, cakes, pastries, and baked potatoes, are heat processed. Since the food constituents participating in the MR represent major dietary constituents, it is necessary to define the safety and quality of food products that are consumed, a topic under current investigation (135).

From a mechanistic point of view, the MR can be divided into three stages, i.e., early, advanced, and final. However, in food the stages can be concomitant since many heat-induced reactive intermediary compounds can feedback into the reaction loop. The early stage of the MR involves the condensation of a carbonyl group, for example, a reducing sugar such as glucose, with a free amino group such as the ε-amino group of protein lysine residues. In general, primary amines are more important than secondary amines because the concentration of primary amino acids in foods is usually higher, and in proteins the primary ε-amino groups of the lysine side chains preferentially react with sugars (1, 136).

At this stage, the formation of glycosylamines involves the addition of amine to the acyclic aldehydes or the cation generated from the sugar during mutarotation. The nucleophilic attack of the amine at C_1, followed by the elimination of the hydroxyl ion, results in the formation of the iminium ion of glycosylamines. Later, enolization by elimination of the hydrogen atom at C_2 yields the enol form, 1,2-enaminol. This results in the formation of an unstable Schiff base, which spontaneously rearranges to form the more stable ARP. As nucleophiles, the amines rearrange with aldoses to the amino ketoses (ARP, for instance fructoselysine) and with ketoses to the amino aldoses (Heyns products). The ARP of free amino acids has been detected in various foods such as dried fruits and vegetables (137), beer (138), and honey (12). This reaction also occurs under physiological conditions as demonstrated by the identification of the hemoglobin variant HbA_{IC}, in which the *N*-terminal valine residue has reacted with glucose to *N*-α-fructosylvaline (139).

At the advanced stage of the MR, the ARP—although fairly stable in foods with low water activity—may undergo several degradation reactions during severe heating or prolonged storage, leading to the formation of 1,2-dicarbonyls. Among these, 3-deoxyosone (140), 1-deoxyosone (141, 142), methylglyoxal, and glyoxal (143) are the most important. However, many other reactive compounds are formed as intermediates, such as keto and aldoseamine compounds (136), furans (144), maltol (145), pyrrole derivatives (146), pyrazines (147), pyranones (131), lactones (148), and substituted imidazoles (149). The ARP is degraded via various pathways, leading to the formation of reductones and furfurals, which can react further to afford colored, high-molecular-mass products and melanoidins in the final stage. Dicarbonyl compounds are more reactive than their precursors and can lead to the formation of compounds that contribute to flavor such as Strecker aldehydes, pyrazines, thiophenes, and furans. In addition, the dicarbonyl fragments can act as precursors of acrylamide. Besides degradation of carbohydrates or Amadori products, 1,2-dicarbonyls may also be formed under oxidative conditions from the Schiff base, the initial condensation product of a reducing carbohydrate and an amino compound, via the so-called Namiki pathway (150). At this stage, proteins are modified into colored, fluorescent, and cross-linked molecules leading to the formation of advanced glycation end products (AGEs).

HMF is formed by degradation of the ARP via 1,2-enolization, as under weakly alkaline conditions the 2,3-enolization pathway is favored. Detailed discussions of possible reaction mechanisms, which include enolization, elimination of water as well as retroaldolization reactions, can be found in the literature (151). Briefly, in the 1,2-enolization pathway, the presence of a positively charged amino group assists in shifting the equilibrium to the enol form, which undergoes elimination of the hydroxyl group from C_3 to yield the 2,3-enol, which is readily hydrolyzed at the C_1 Schiff base to the glycosulose-3-ene, which is an unsaturated dicarbonyl compound and undergoes cyclodehydration to form HMF. Furfurals and 1,2-dicarbonyls are also formed upon caramelization (152). The condensation of amino compounds allows enolization and elimination to occur near to neutral pH and at low temperatures, conditions commonly found in food systems.

2.5.5 DECOMPOSITION

HMF is considered a final product of the MR during food processing in a practical sense since it may be employed as a chemical marker. However, the well-known combinations of HMF with some food components, and the chemical instability of the compound, may in some cases diminish its usefulness as an indicator of food quality, particularly when strong thermal or prolonged heating treatments of food are employed. At certain processing and storage conditions, the content of HMF recorded in foods is a balance between formation and degradation reactions (Fig. 2.5.3). On the other hand, HMF is a volatile substance that can be lost during baking or roasting.

In principle, HMF can react further by decarboxylation, oxidation, dehydration, and polycondensation reactions (130, 153–155). Dehydration furnishes levulinic and formic acids. Small amounts of 5-hydroxymethyl-2-furan carboxylic acid (HMFA) are formed upon the oxidation of HMF. Durham *et al.* (90) have reported the formation of 5-hydroxymethylfuroic acid and furan-2,5-dicarboxylic acid by HMF decomposition under autoclaving conditions. 2-(2′hydroxyacetyl)-furan is formed as a side-product of fructose degradation and by condensation of both compounds in the presence of only catalytic amounts of amino compounds, leading to the formation of colored products. Self-condensation of HMF affords the corresponding ether. Studies of the decomposition of glucose, ^{13}C-labeled in positions 1 and 6, under Curie-point pyrolysis conditions at 300 °C, allow the specific incorporation of the label into

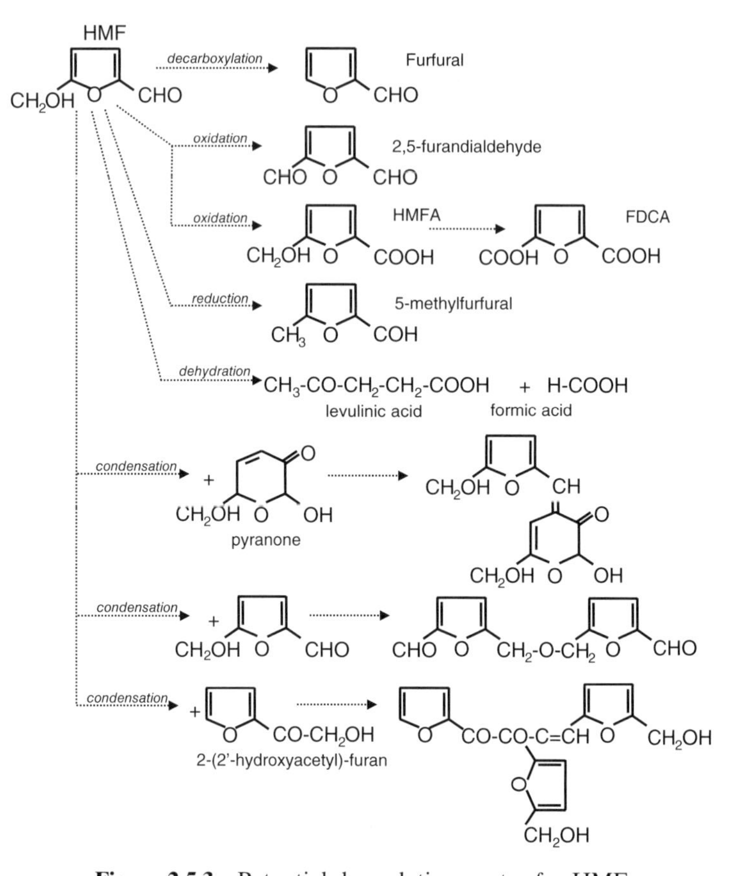

Figure 2.5.3 Potential degradation routes for HMF.

2,5-furandialdehyde, 5-methyfurfural, and furfural (156). Furthermore, *in vivo* experiments have shown the presence of *N*-(5-hydroxymethyl-2-furoyl)-glycine and 5-hydroxymethyl-2-furoic acid when HMF was administered to rats (157) (see Section 2.5.7.2).

2.5.6 MITIGATION

HMF is a substance formed upon thermal treatment and found in cooked food, particularly in plant-derived foods. Strategies could be devised to minimize its formation during the cooking process. In fact, much information is already known through model system studies that have addressed the kinetics of the reactions involved and the activation energy required at different moisture levels. However, in each mitigation approach, it is important to consider the risk/benefit equation and final product quality in terms of consumer acceptability. Moreover, prior to any measures, it would be pivotal to establish a database to know which food categories contribute most to the daily intake, and consider practical strategies to reduce HMF in those priority products. Given the ubiquity of HMF in the diet, complete removal without affecting the overall organoleptic quality and safety of foods is not possible. Possible mitigation strategies are stipulated in the next sections.

2.5.6.1 Selection of Appropriate Raw Materials in the Formulation

Both sugar caramelization and the MR are involved in the formation of HMF. Limiting the content of reducing sugars will reduce the potential formation of HMF. The use of sugar alcohols (i.e., maltitol) instead of fructose or glucose will limit the extent of the nonenzymatic browning reactions (Morales *et al.*, unpublished results). On the other hand, HMF measured in the final product could not be directly related to the thermal processing conditions applied. The use of ingredients with a significant HMF content such as brown cane sugar, sugar syrup, or honey, will increase the amount of HMF in the final product. However, these ingredients will reinforce the palatability and overall consumer acceptance, and a balance must be struck. In some cases, it is possible to vary some components of the recipe (e.g., dough formulation during cereal processing or infant formula formulation) before thermal processing.

2.5.6.2 Selection of Appropriate Thermal Processing Conditions

Certain types of food require the development of the MR to impart flavor, color, and/or textural properties, as typically encountered in roasted coffee, malt, or toasted cereals. In contrast, this reaction is to be minimized in milk and dairy products due to the development of a cooked flavor or brown color, which is not appreciated by consumers. Many complex phenomena, reactions, and interactions of food constituents occur during thermal processing, such as protein denaturation and aggregation, fat melting, and water loss, just to mention a few. Temperature effects are explained by the Arrhenius equation and activation energy, which is necessary to start the reaction and can be used to determine the optimum thermal load. Low water content of cereal-based products as compared with liquid products such as milk favors HMF formation. Ait-Ameur *et al.* (17) reported that HMF formation after baking of cookies starts from an average water activity of 0.40, independent of the temperature in the cookie. The MR and caramelization could be favored as they are highly dependent on the temperature/time profile applied and the conditions of the reaction media (water content, acidity, bivalent cations, or transient ions). However, a side effect could appear since cations (Ca^{2+}, Mg^{2+}) may in some cases reduce the formation of acrylamide, but promote the dehydration of glucose leading to HMF and furfural (158). HMF can also be formed during storage and thus products should be stored at appropriate conditions over their shelf life. Gökmen *et al.* (159) described that sucrose is decomposed very rapidly if ammonium bicarbonate is used as leavening agent for baking of cookies, and subsequently HMF is formed. Replacing ammonium bicarbonate with sodium bicarbonate maintained the pH of cookies in the range 9.0–10.0.

2.5.6.3 Elimination of Formed HMF

Under drastic baking conditions, Ait-Ameur *et al.* (17) observed a decrease in HMF in cookies, indicating that HMF was degraded into secondary products, mostly volatiles. However, these processing conditions will drastically affect the overall product quality and may lead to other undesired compounds such as acrylamide.

2.5.7 HEALTH RISK: TOXICOLOGY

Several toxicological studies have been performed with HMF (and related substances) and these have investigated possible associations with colon cancer (160), skin papillomas (161, 162), lipomatous tumors in kidney (163), or induction of chromosomal aberrations (164). Based on the *in vitro* mutagenic activity of a sulfate conjugate of HMF, 5-sulfoxymethylfurfural (SMF), there appears to be sufficient evidence to raise some concern regarding the genotoxic potential of the metabolite. However, additional data are necessary to corroborate this activity also *in vivo*.

2.5.7.1 Absorption and Elimination

Administration of [U-^{14}C]-HMF via oral or gavage to rodents at different concentrations (0.08–500 mg/kg bw) showed rapid absorption from the GI tract. The difficulties of extrapolating and interpreting rodent data to humans are well known, since humans may not follow the same patterns of HMF absorption. Recently, transportation and uptake of HMF has been studied in the Caco-2 cell lines (165). The authors noticed a clear impact of food composition. Most likely, the fiber content may play a role in HMF bioavailability, with lower absorption at higher fiber level.

Microbial transformation after ingestion may also be expected since enteric bacterial strains are capable of converting furfural to furfuryl alcohol under both aerobic and anaerobic conditions. Both furfuryl alcohol and furfural are expected to be present in the GI tract of animals given furfural. Enteric bacterial strains are also capable of reducing HMF to a compound postulated to be 5-hydroxymethylfurfuryl alcohol under both aerobic and anaerobic conditions during an 8-h incubation period (166). Biotransformation of HMF was accomplished by co-metabolism in the presence of glucose and peptone as main substrates (166).

The distribution of [U-^{14}C-]-HMF and its metabolites after oral dosage by whole-body autoradiography has also been studied (167, 168). HMF-derived radiography is rapidly cleared from all major tissues, with no evidence of accumulation, but some covalent binding could be demonstrated in liver, bladder, kidney, and the GI tract after 1, 8, and 24h. Within 48h, 70–82% of the administrated dose is excreted in the urine of rats; and 8–12% is excreted in the feces. However, 80–100% of the total amount of radioactivity excreted via the urine was recovered with the first 24h post dosing, indicating no accumulation of radioactivity and complete elimination of HMF and its metabolites. The rate of urinary elimination of [U-^{14}C]-HMF is, however, not dependent on the administrated dose and $^{14}CO_2$ (exhaled) was not detected.

2.5.7.2 Metabolism and Detoxification Route of HMF

Four routes have been hypothesized for the biotransformation of HMF: (i) formation of a glycine conjugate, (ii) oxidation, (iii) α-ketoglutaric acid with formation of CO_2, and (iv) sulfonation. Pathways (i)–(iii) comprise as first step the oxidation of the aldehyde group to an FA derivative (5-HMFA). Figure 2.5.4 summarizes the different pathways described in the literature. HMF, as well as furfuryl alcohol, FA, and their derivatives participates in the same pathways as those involved in the detoxification of furfural in rodents. The metabolism of non-sulfur-containing furan derivatives, such as HMF, included the formation of furfural, a known reactive aldehyde that may lead to hepatotoxicity (169).

In the first and major pathway (Fig. 2.5.4, route I), HMF oxidation may be followed by conjugation of the resulting carboxylic acid with glycine, yielding 5-hydroxymethyl-2-furoyl-glycine (HMFG, the glycine conjugate of HMFA) as the main metabolite, which is readily excreted in the urine. The ratio of HMFA/HMFG concentrations increases at higher doses of administrated HMF, indicating that the availability of the free glycine may limit the rate of conjugation, resulting in the excretion of free FA or 2,5-furandicarboxylic acid (FDCA) through the second pathway (Fig. 2.5.4).

Godfrey *et al.* (167) identified and quantified HMFA, HMFG, and FDCA in the urine of rats and mice, accounted for 78–85%, 5–8%, and 2–6% of the administrated dose of rats and mice, respectively (Fig. 2.5.4, route II). Excretion of HMFG was inversely proportional to dose due to glycine depletion. Earlier, Germond *et al.* (168) identified two major HMF urinary metabolites in rats, namely HMFA and HMFG, and a third minor polar metabolite not identified further. In humans, furoylglycine, HMFA, and FDCA are detected in the urine, and it has been demonstrated that these are metabolites of HMF (92, 170).

Recently, Pryor *et al.* (171) identified HMFA, HMFG, CAFG (5-carboxylic-2-furoyl-glycine), and CAFAM (5-carboxylic-2-furoylamino-methane) in urine of subjects that consumed dried plum juice. Total recovery in the urine (% of initial HMF dose) during the first 6h after dried plum juice consumption was

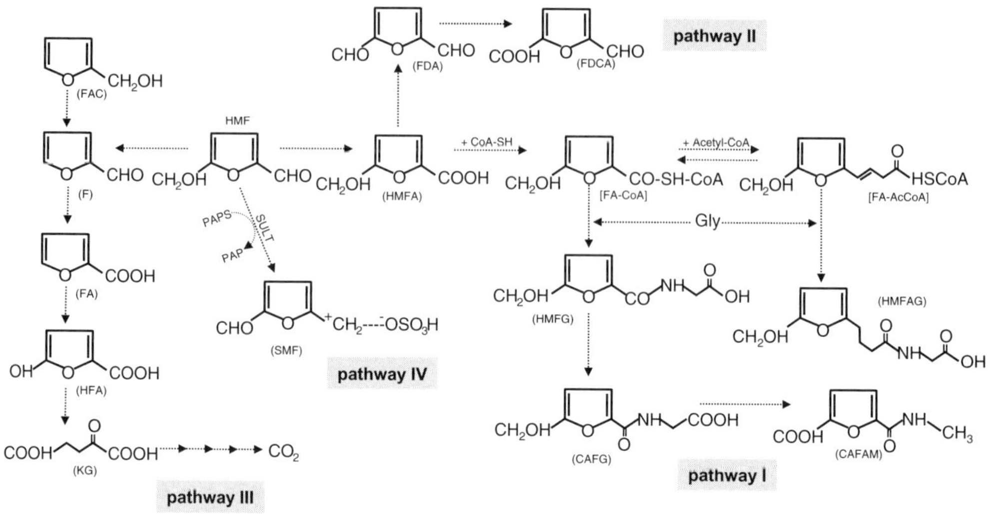

Figure 2.5.4 Scheme of biotransformation routes for HMF.

estimated for HMFA (36.9%), HMFG (3.4%), CAFG (4.2%), and CAFAM (1.8%). The ratio of HMFA/HMFG excreted in the urine was 10.7 in subjects that consumed dried plum juice and 5.8 in subjects that ate dried plums. These results suggest that increased HMF intake reduces the formation of the glycine conjugate, as reported by Godfrey *et al.* (167) and Germond *et al.* (168). HMFA was also rapidly detected in plasma after 30 min of ingestion and began to decline after 60 min (171). In summary, HMF appears to be rapidly metabolized to glycine conjugates and other metabolites and excreted in the urine.

A third possible—albeit minor route—of biotransformation has been described for rodents and involves the complete oxidation of the furan moiety to CO_2 (Fig. 2.5.4, route III). The process requires the opening of the furan ring and formation of reactive intermediates (172). HMF and alpha-ketoglutaric acid can be simultaneously determined in human plasma by their respective derivative hydrazones (173). The formation of labeled CO_2 from [^{14}C]-furfural and [^{14}C]-furfuryl alcohol in rodents may occur via decarboxylation of 2-furoic acid and formation of alpha-ketoglutaric acid that could enter into the TCA cycle.

In addition to the aforementioned pathways, HMF has been shown to be bioactivated *in vitro* to SMF, through sulfonation of its allylic hydroxyl functional group, catalyzed by sulfotransferases (SULTs) (Fig. 2.5.4, route IV). In the resulting ester, the sulfate is a good leaving group, thus producing a highly electrophilic allyl carbocation, which could be stabilized by distribution of charges on the furan ring. The resulting ester has been demonstrated to induce genotoxic and mutagenic effects. The subsequent interaction of this reactive intermediate with critical cellular nucleophiles (i.e., DNA, RNA, and proteins) may result in structural damages to these macromolecules, thereby causing toxicity and mutagenicity, although its occurrence *in vivo* has not been confirmed yet. This sulfate conjugate appears to be too unstable to allow intact excretion as such and thus detection in urine. Indeed, when HMF was incubated with 35S-PAPS, a sulfo-group donor, and liver cytosol, an unstable conjugate was formed, which disappeared within 60 min. The time-dependent decline in the amount of the reaction product appears to be associated with its hydrolysis in an aqueous environment (161). A recent investigation shows that SMF is not present in human urine (8). The authors postulated that one reason may be that SMF is a short-lived metabolite that can react with proteins and DNA and consequently does not appear in the urine.

Furthermore, Pryor *et al.* incubated human urine samples with a mixture of beta-glucuronidase and sulfatase, which did not significantly change the area of HMF metabolic substances (171). Today, there is no evidence that glucuronide or sulfate conjugates of HMF are formed in humans. Therefore, it is concluded that non-sulfur furan derivatives, including HMF, cannot be predicted to be metabolized to potentially toxic compounds (169).

2.5.7.3 Genotoxicity, Mutagenicity, and Clastogenicity Potential

Today, it is not confirmed whether human exposure to HMF represents a potential health risk, although it is known that HMF at high concentrations is cytotoxic, causing irritation to eyes, upper respiratory tract, skin, and mucous membranes. However, the *in vitro* and *in vivo* data available raise some concern with respect to genotoxicity. Furthermore, the mechanisms by which HMF exerts its genotoxic and tumorigenic effects remain unclear and hence the potential health risk is still a matter of debate. HMF is a multifunctional molecule containing a furan ring, a carbonyl group, and an allylic hydroxyl group that may undergo formation of Schiff bases with amino groups and Michael addition reactions. These reactive sites on the molecule will influence the biological activity and fate of HMF in the body.

Data from epidemiological studies or case reports on potential association of HMF with cancer risk in humans are not available. To support the genotoxic potential of HMF, certain indications of tumorigenic activities of HMF have, however, been deduced from rodent bioassays. HMF can act as an initiator and also as a promoter, as shown for the induction of colonic aberrant crypt foci (ACF; a pre-neoplastic lesion) as marker for colon cancer in rats treated with orally applied HMF (0–300 mg/kg bw) (160, 174, 175), chromosomal aberrations (164), induction of skin papillomas after topical application of 10–25 mmol HMF to mice (161, 162), or development of lipomatous tumors in kidney in rats treated subcutaneously (200 mg/kg bw) (163). In another study in mice, the increase of skin tumor rates associated with HMF treatment was not statistically significant (176).

Controversial results have been published on mutagenicity/genotoxicity of HMF *in vitro* by the traditional Ames *Salmonella* test (177, 178). Experiments with *Salmonella typhimurium* strains TA98, TA100, TA104, TA1535, and TS1537, with or without activation with S9, have been carried out. In experiments with TA98 and TA100, both positive (132) and negative (179–181) mutagenicity results have been obtained, albeit only positive in the presence of S9 metabolic activation mix. On the other hand, HMF was effective in reducing the mutagenic activity (in terms of revertants per plate) of heterocyclic amines. This activity is presumed to be a result of the presence of carbonyl groups and their reaction with the amino residues. Later, Kong *et al.* (182) reported a dose effect in the antimutagenic activity of HMF which, like other furan compounds, appeared to show bifunctionality by exhibiting a weak mutagenic effect and a desmutagenic effect. Another example for contradictory results on the biological activity of HMF is the study of Ying *et al.* (183), who reported that HMF and 5-methyl-2-furfuraldehyde potentially inhibit tumor necrosis factor alpha or interleukin-1 beta expression. In addition, it is proposed that HMF derivatives could be used as promising anticancer substances through inhibition of tubulin polymerization (184).

Other publications report that HMF is mutagenic (185) and is transformed by rat and human SULTs to SMF, which was shown to be mutagenic (162, 186). SULTs are a class of enzymes that are frequently involved in bioactivation reactions and are characterized by a high detoxification as well as toxification potential, depending on the chemical involved.

Shinohara *et al.* (132) found HMF to be positive in the Rec-assay with *Bacillus subtilis*, with and without activating system. In Chinese hamster V79 cells, HMF (2 mg/mL) induced chromosomal aberrations without external activation (164). In V79 cells, HMF (up to 80 mM) induced a small (although statistically significant) increase in chromosomal aberrations, a reduction in mitotic index, and HPRT (hypoxanthine-guanine-phosphoribosyl-transferase) mutations (187). In human TK6 lymphoblast cells, mutagenicity was not observed up to 75 mg HMF/mL in the HPRT and TK assay (160, 161). A positive result was obtained also in the *Umu* assay, although only at high concentrations, resulting in reduced cell viability (187). On the other hand, some positive responses reported are still controversial since the purity of the target compound could not be evaluated (185) or experiments were performed only at a single dose (177).

In an Ames test with TA 104 strain upon inclusion of 3′-phosphoadenosine-5′-phosphosulfate (PAPS), a sulfo-group donor and cofactor of cytosolic SULTs (EC. 2.8.2), and rat liver cytosol into the experimental model, HMF gave a positive result suggesting that it can be activated to reactive metabolites following sulfation, with formation of SMF (160, 161). As mentioned before, HMF is an allylic alcohol and as such may be activated via sulfo conjugation. Indeed, the mutagenic effect could be partly suppressed by the addition of SULT inhibitors such as 2,6-dichloro-4-nitrophenol and dehydroepiandrosterone, clearly suggesting that HMF can be metabolically bioactivated to an allylic sulfate with genotoxic potential. In addition, HMF is an α,β-unsaturated aldehyde and as such may undergo Michael addition reactions with cellular macromolecules. The subsequent interaction of such reactive species with critical cellular nucleophiles such as DNA and proteins may cause the cytotoxicity, genotoxicity, and tumorigenicity of HMF (188). In this context, the studies carried out by Mizushina *et al.* (189) concluded that HMF exerts a dose-dependent and selective (very highly structure-dependent) inhibition of mammalian DNA polymerase λ (pol λ) and terminal deoxynucleotidyltransferase (TdT). Inhibition was not reversible, indicating that HMF might bind to or interact with the hydrophilic region of the pol β-like core on the pol λ and TdT molecules. However, HMF had no effect on the activity of pol β from the same family of molecules.

Glatt *et al.* (190) analyzed all 13 human SULT forms known for their ability to convert HMF to SMF. Based on the kinetic parameters, SULT1A1 was the most active form which is expressed in many tissues, including colon (unlike the corresponding mouse and rat forms where expression is low in colon). Colorectal cancer is common in man, but is only induced in experimental animals by very few carcinogens and it has not yet been shown that HMF or SMF may be involved.

SMF is inactivated by glutathione (GSH) in the presence of glutathione transferase (GST) activity. GSH is a soft nucleophile whereas SMF is a strong nucleophile and their interaction would therefore not be favored. GST can facilitate the covalent binding of SMF with GSH by lowering the energy barrier required for this reaction to proceed. In accordance, SMF in TA 104 was genotoxic in the absence of any metabolic system (cytotoxicity not specified); the effect was reduced by addition of GSH and GST and restored when the enzyme was inhibited (188).

Genotoxicity of chemically synthesized SMF was tested in *Salmonella* strain TM677 (8-AG-resistance), without any metabolic activation, giving a clear positive response obtained at concentrations that reduced cell survival to <60%. Genotoxicity was also observed with SMF in human lymphoblasts at the TK and HPRT loci, at concentrations (>40 μg/mL) reducing cell survival to >63%. No genotoxicity was observed with HMF, with its acetate ester or with the sulfation product of 2-methyl furfuryl alcohol, suggesting that the genotoxicity of SMF requires the presence of both a reactive sulfate group and a free aldehyde group.

Carcinogenic activity has been evaluated in rats under chronic administration of HMF from the diet (0–250 mg/kg bw/day) or by gavage (0–160 mg/kg bw/day) for 10 and 11 months, respectively (191, 192). Estimated NOAELs (no observed adverse effects levels) for diet and gavage were 250 and 80 mg/kg/day, respectively. An assay for primary DNA damage (comet assay) did not show an effect of HMF in V79 and Caco-2 cells up to cytotoxic concentrations (80 mM). HMF causes a slight but significant increase in DNA single-strand breaks in primary rat hepatocytes at cytotoxic levels (40–100 mM), whereas in human colon biopsy material the same effect was seen in the absence of cytotoxicity. HMF at non-cytotoxic concentrations induced a substantial concentration-related GSH depletion in V79, Caco-2, and rat liver cells. The effect of sulfate conjugation was not directly studied, but since this activity is present at least in primary hepatocytes, it might have contributed to the depletion of GSH and to induction of DNA strand breaks in these cells (187). However, the former study cannot be considered evidence of a mutagenic activity of SMF in mammalian cells *in vivo*. At present, some research is conducted to replace endogenous SULTs in laboratory animals by human SULTs, e.g., SULT1A1, by genetic engineering. In a previous step, Glatt *et al.* (190) succeeded to construct a Chinese hamster V79-derived cell line that expresses human cytochrome P450 (CYP)2e1 and human SULT1A1.

An investigation conducted recently by NIEHS (The National Institute of Environmental Health Sciences, US) under the National Toxicology Program (NTP, No-TR-554) evaluated the toxicology and carcinogenicity of HMF in F344/N rats and B6C3F1 mice in a long-term (2 years) gavage study (193). It was concluded that there was no evidence of carcinogenic activity of HMF in male/female rats administrated 188 to 750 mg/kg but there was some evidence of carcinogenicity in female mice based on incidences of hepatocellular adenomas. Additionally, administration of HMF was associated with increased incidences of olfactory and respiratory epithelium lesions in rodents.

2.5.8 EXPOSURE

Detailed information regarding the human dietary exposure to HMF is scarce. Apart from the diet, humans are also potentially exposed to HMF through pharmaceutical preparations and cigarette smoke. The intake of HMF from the regular diet has to be estimated at the time of consumption for exposure assessment. Studies have been carried out in laboratory animals; for instance, oral acute toxicity data are available for HMF by oral or gavage administration to rodents. The LD50 ranged between 2.5 and 5.0 g/kg bw (194, 195) (WARF Institute 1977, unpublished result). In another study carried out with rats, the acute oral LD_{50} has been reported at 2.5 g/kg for males between 2.5 and 5.0 g/kg for females (196). Recently, an oral LD_{50} of 3.1 g/kg bw has been determined in rats (187). HMF is abundant in the human diet and it is estimated that humans may ingest up to 150 mg HMF/day, equating to 2.5 mg bw/day for a 60-kg person (as reviewed in Reference 89). This value exceeds the intake of other food-processing contaminants such as acrylamide and furan by several orders of magnitude.

The Scientific Panel on Food Additives, Flavourings, Processing Aids and Materials in Contact with Foods (AFC) estimated the intake of non-sulfur substituted furan derivatives (16 substances, including HMF) as flavoring substances in Europe based on the mTAMDI-approach (modified Theoretical Added Maximum Daily Intake). This method is regarded as a conservative estimate of the actual intake in most consumers because it is based on the assumption that the consumer regularly eats and drinks several food products containing the same substance at the upper level. For HMF the mTAMDI value was 1600 μg/person/day, which is above the threshold of concern of 540 μg/person/day derived from a large database containing data on subchronic and chronic animal studies (197). Finally, a recommendation of additional toxicological data is highlighted by the Scientific Panel although the procedures used in the risk assessment of contaminants are essentially the same as those employed for food additives (169). However, in this case a tolerable daily intake (TDI) is applied instead of the acceptable daily intake (ADI) or mTAMDI. An ADI, using a 40-fold margin of safety of 2 mg/kg bw has been recommended for humans (192). On the other hand, a maximum of 100 mg HMF/L for syrup containing highly inverted sugars was recommended for the preparation of nonalcoholic liquids (194).

Specific studies for HMF should be carried out covering average, medium, and maximum intake from regular foods, specific foods, for the whole population, segments of the population, or certain individuals that are exposed to high amounts through the diet. These studies will estimate the intake of HMF from food at the time of consumption, which is necessary for exposure assessment. Recently, Delgado-Andrade *et al.* (198) monitored HMF among other indicators of the MR in two realistic diets adjusted to the requirements of an adolescent population. The control diet was as regularly consumed by adolescents in a catering service at school and compared with a modified diet similar to the control one but taking food cooked under less severe conditions to minimize the MR. The authors reported that the HMF content was significantly higher (nearly threefold) in the control diet.

2.5.9 FUTURE RESEARCH NEEDS

HMF is naturally formed during food processing or cooking and exposure to this substance is basically unavoidable. The formation of HMF may follow multiple routes and involve different precursors and intermediates. In general, knowledge on the formation of HMF in complex food matrices is lacking, in particular the impact of ingredients, recipes/formulation, processing conditions, and storage of the products until end of shelf life. Little is known about the formation of HMF and related compounds in a typical home cooking environment.

Although the toxicological studies conducted to date show that HMF does not pose a serious health risk, high amounts in certain foods may approach the biologically effective dose as determined in cellular test systems. This leads to the question of exposure assessment, which deserves further attention particularly for specific individuals who may be exposed to significantly higher amounts of HMF through, for example, the regular intake of dried fruits or caramel products. In this context, a reliable database of contents of HMF in food and diets would be useful to conduct a more reliable exposure assessment.

ACKNOWLEDGMENT

The author is thankful to Prof. Vural Gökmen (Department of Food Engineering, Hacettepe University, Ankara, Turkey) for the critical review of the manuscript.

REFERENCES

1. Ames, J.M. (1992). *Biochemistry of Food Proteins* (ed. B.J.F. Hudson), Elsevier Applied Science, London, pp. 99, 153.
2. Kroh, L.W. (1994). Caramelisation in food and beverages. *Food Chemistry*, *51*, 373–379.
3. Akkan, A.A., Özdemir, Y., Ekiz, H.L. (2001). Derivative spectrophotometric determination of 5(hydroxymethyl)-2-furaldehyde HMF and furfural in Locust bean extract. *Nahrung/Food*, *45* (**1**), 43–46.
4. Ibarz, A., Pagán, J., Garza, S. (2000). Kinetic models of nonenzymatic browning in apple puree. *Journal of Science and Food Agriculture*, *80* (**8**), 1162–1168.
5. Rada-Mendoza, M., Luz Sanz, M., Olano, A., Villamiel, M. (2004). Formation of hydroxymethylfurfural and furosine during the storage of jams and fruit-based infant foods. *Food Chemistry*, *85*, 605–609.

6a. Bachmann, S., Meier, M., Kaenzig, A. (1997). 5-Hydroxymethyl-2-furfural (HMF) in Lebensmitteln. *Lebensmittelchemie*, *51*, 49–50.

6b. Rada-Mendoza, M., Olano, A., Villamiel, M. (2002). Determination of hydroxymethylfurfural in commercial jams and in fruit-based infant foods. *Food Chemistry*, *79*, 513–516.

7. Kanjahn, D., Jarms, U., Maier, H.G. (1996). Hydroxymethylfurfural and furfural in coffee and related beverages. *Deutsche Lebensmittel-Rundschau*, *92*, 328–331.
8. Murkovic, M., Pichler, N. (2006). Analysis of 5-hydroxymethylfurfural in coffee, dried fruits and urine. *Molecular Nutrition Food Research*, *50*, 842–846.
9. Smith, R.M. (1981). Determination of 5-hydroxymethylfurfural and caffeine in coffee and chicory extracts by high performance liquid chromatography. *Food Chemistry*, *6*, 41–45.
10. Lo-Coco, F., Valentini, C., Novelli, V., Ceccon, L. (1996). High performance liquid chromatographic determination of 2-furaldehyde and 5-hydroxymethyl-2-furaldehyde in honey. *Journal of Liquid Chromatography*, *749*, 95–102.
11. Zappala, M., Fallico, B., Arena, E., Verzera, A. (2005). Methods for the determination of HMF in honey: a comparison. *Food Control*, *16*, 273–277.
12. Sanz, M.L., del Castillo, M.D., Corzo, N., Olano, A. (2003). 2-Furoylmethyl amino acids and hydroxymethylfurfural as indicators of honey quality. *Journal of Agricultural and Food Chemistry*, *51*, 4278–4283.
13. Lo-Coco, F.L., Valentini, C., Novelli, V., Ceccon, L. (1995). Liquid chromatographic determination of 2-furaldehyde in beer. *Analytica Chimica Acta*, *306*, 57–64.
14. Yuan, J.P., Chen, F. (1998). Separation and identification of furanic compounds in fruit juices and drinks by high-performance liquid chromatography photodiode array detection. *Journal of Agricultural and Food Chemistry*, *46*, 289–1291.

15. Kim, H-J, Richardson, M. (1992). Determination of 5-hydroxymethylfurfural by ion-exclusion chromatography with UV detection. *Journal of Chromatography, A*, *593*, 153–156.
16. Theobald, A., Müller, A., E. (1998). Determination of 5-hydroxymethylfurfural in vinegar samples by HPLC. *Journal of Agricultural and Food Chemistry*, *46*, 1850–1854.
17. Ait-Ameur, L.A., Trystam, G., Birlouez-Aragon, I. (2006). Accumulation of 5-HMF in cookies during the baking process: validation of an extraction method. *Food Chemistry*, *98*, 790–796.
18. Ramírez-Jiménez, A., Guerra-Hernández, E., García-Villanova, B. (2000). Browning indicators in bread. *Journal of Agricultural and Food Chemistry*, *48*, 4176–4181.
19. Rufian-Henares, J.A., Delgado-Andrade, C., Morales, F.J. (2006). Analysis of heat-damage indices in breakfast cereals: influence of composition. *Journal of Cereal Science*, *43*, 63–69.
20a. Albala-Hurtado, S., Veciana-Nogués, M.T., Izquierdo-Pulido, M., Vidal-Carou, M.C. (1997). Determination of free and total furfural compounds in infant milk formulas by high-performance liquid chromatography. *Journal of Agricultural and Food Chemistry*, *45*, 2128–2133.
20b. Jaganathan, J., Dugar, S. (1999). Authentication of straight whiskey by determination of the ratio of furfural to 5-hydroxymethylfurfural. *Journal of the AOAC International*, *82* (**4**), 997–1001.
21. Goldberg, D.M., Hoffman, B., Yang, J., Soleas, G.J. (1999). Phenolic constituents, furans, and total oxidant status of distilled spirits. *Journal of Agricultural and Food Chemistry*, *47*, 3978–3985.
22. Gökmen, V., Senyuva, H.Z. (2006). Improved method for the determination of hydroxymethylfurfural in baby foods using liquid chromatography-mass spectrometry. *Journal of Agricultural and Food Chemistry*, *54*, 2845–2849.
23. Dauberte, B., Estienne, J., Guerre, M., Guerra, N. (1990). Contribution à l'étude de la formation d'hydromethylfurfural dans les boissons a base da jus de fruits et dans les cafés Torrefacto. *Annales des Falsifications de L'Expertise Chimique*, *889*, 231–253.
24. Fallico, B., Zappalà, M., Arena, E., Verzera, A. (2004). Effects of conditioning on HMF content in unifloral honeys. *Food Chemistry*, *85* (**2**), 305–313.
25. Tosi, E., Ciappini, M., Ré, E., Lucero, H. (2002). Honey thermal treatment effects on hydroxymethylfurfural content. *Food Chemistry*, *77* (**1**), 71–74.
26. Morales, F.J., Jiménez-Pérez, S. (2001). Hydroxymethylfurfural determination in infant milk-based formulas by micellar electrokinetic capillary chromatography. *Food Chemistry*, *72*, 525–531.
27. Van Boekel, M.A.J.S. (1998). Effect of heating on Maillard reactions in milk. *Food Chemistry*, *62*, 403–414.
28. Van Boekel, M.A.J.S., Rehman, Z. (1987). Determination of hydroxymethylfurfural in heated milk by high-performance liquid chromatography. *Netherlands Milk and Dairy Journal*, *41*, 297–306.
29. Resmini, P., Pellegrino, L., Pagani, M.A., De Noni, I. (1993). Formation of 2-acetyl-3-D-glucopyranosylfuran (glucosylisomaltol) from nonenzymatic browning in pasta drying. *Italian Journal of Food Science*, *4*, 341–353.

30. Ramírez-Jiménez, A., García-Villanova, B., Guerra-Hernández, E. (2000). Hydroxymethylfurfural and methylfurfural content of selected bakery products. *Food Research International*, *33*, 833–838.
31. Sensidoni, A., Peressini, D., Pollini, C.M. (1999). Study of the Maillard reaction in model systems under conditions related to the industrial process of pasta thermal VHT treatment. *Journal of Science and Food Agriculture*, *79*, 317–322.
32. Ferrer, E., Alegría, A., Farré, R., Abellán, P., Romero, F. (2002). High-performance liquid chromatographic determination of furfural compounds in infant formulas, changes during heat treatment and storage. *Journal of Chromatography, A*, *947*, 85–95.
33. Ramírez-Jiménez, A., Guerra-Hernández, E., García-Villanova, B. (2003). Evolution of non enzymatic browning during storage of infant rice cereal. *Food Chemistry*, *83*, 219–225.
34. García–Villanova, B., Guerra-Hernández, E., Martínez Gómez, E., Montilla, J. (1993). Liquid chromatography for the determination of 5-(hydroxymethyl)-2-furaldehyde in breakfast cereals. *Journal of Agricultural and Food Chemistry*, *41*, 1254–1255.
35. Ramírez-Jiménez, A.J. (1998). *Indicadores de las reacciones de pardeamiento químico en productos panarios. Memory of pharmacy degree of licenciate*, The University of Granada, Spain.
36. Frischkorn, H.E., Wanderly Casado, M., Frischkorn, C.G. (1982). Schnelle Bestimmung von Furfural und Hydromethylfurfural in alkoholischen Getränken mit Hilfe der Umkehrphasen-Chromatographie. *Zaitung Lebensmittel Untersuchung und Forschung*, *174*, 117–121.
37. Laszlavik, M., Gal, L., Misik, S., Erdei, L. (1995). Phenolic compounds in two Hungarian red wines matured in *Quercus robur* and *Quercus petrea* barrels: HPLC analysis and diode array detection. *American Journal of Enology and Viticulture*, *46*, 67–74.
38. Chatonnet, P., Dubourdieu, D., Boidron, J.N. (1992). Incidence des conditions de fermentation et d'élevage des vins blancs secs en barriques sur leur composition en substances cedées par le bois de chêne. *Sciences des Aliments*, *12*, 665–685.
39. Mijares, R.M., Park, G.L., Nelson, D.B., McIver, R.C. (1986). HPLC analysis of HMF in orange juice. *Journal of Food Science*, *51*, 843–844.
40. Swallow, K.W., Low, N.H. (1994). Determination of honey authenticity by anion-exchange liquid chromatography. *Journal of AOAC International*, *77*, 695–702.
41. B.O.E. (Boletín Oficial del Estado) (1983). *Norma de Calidad para la Miel Destinada al Mercado Interior*, Madrid, Spain, pp. 22384–22386.
42. Doner, L.W. (1977). The sugars of honey—a review. *Journal of the Science Food Agriculture*, *28*, 443–456.
43. Bath, P.K., Singh, N.A. (1999). Comparison between *Helianthus annus* and *Eucalyptus lanceolatus* honey. *Food Chemistry*, *67*, 389–397.
44. Jeurings, H.J., Kuppers, F.J.E.M. (1980). High performance liquid chromatography of furfural and hydroxymethylfurfural in spirits and honey. *Journal of the Association of Official Analytical Chemists*, *63*, 1215–1218.
45. White, J.W., Siciliano, J. (1980). Hydroxymethylfurfural and honey adulteration. *Journal of the Association of Official Analytical Chemists*, *63*, 7–10.

46. Maccarone, E., Maccarrone, A., Rapisarda, P. (1985). Stabilization of anthocyanins of blood orange fruit juice. *Journal of Food Science*, *50*, 901–904.
47. Fallico, B., Lanza, M.C., Maccarone, E. (1996). Role of hydroxycinnamic acids and vinylphenols in the flavor alteration of blood orange juices. *Journal of Agricultural and Food Chemistry*, *44*, 2654–2657.
48. Askar, A. (1984). Flavour changes during production and storage of fruit juices. *Flüssiges Obst*, *51*, 564–569.
49. Dinsmore, H.L., Nagy, S. (1972). Colorimetric furfural measurement as an index of deterioration in stored citrus juices. *Journal of Food Science*, *37*, 768–770.
50. Nagy, S., Randall, V. (1973). Use of furfural content as an index of storage temperature abuse in commercially processed orange juice. *Journal of Agricultural and Food Chemistry*, *21*, 272–275.
51. Nagy, S., Dinsmore, H.L. (1974). Relationship of furfural to temperature abuse and flavor change in commercially canned single-strength orange juice. *Journal of Food Science*, *39*, 1116–1119.
52. Lee, H.S., Nagy, S. (1988). Quality changes and nonenzymic browning intermediates in grapefruit juice during storage. *Journal of Food Science*, *53*, 168–172.
53. Kaanane, A., Kane, D., Labuza, T.P. (1988). Time and temperature effect on stability of Moroccan processed orange juice during storage. *Journal of Food Science*, *53*, 1470–1473.
54. Arena, E., Fallicio, B., Maccarone, E. (2001). Thermal damage in blood orange juice: kinetics of 5-hydroxymethyl-2-furancarboxaldehyde formation. *International Journal of Food Science and Technology*, *36*, 145–151.
55. Shaw, C.P., Roche, C., Dunne, C.P. (1996). Changes in the Hydroxymethylfurfural and the furfural content of applesauce and grape jellying long-term storage (pp. 91–92). Institute of Food Technologists Annual Meeting: book of abstracts (ISSN 1082–1236).
56. Corradini, C., Nicoletti, I., Cannarsa, G., Corradini, D., Pizzoferrato, L., Vivanti, V. (1995). Microbore liquid chromatography and capillary electrophoresis in food analysis. Current status and future trends. Proceedings of EURO FOOD CHEM VIII, Vienna, Austria, 18–20 September, 2, pp. 299–302.
57. Simonyan, T.A. (1971). Hygienic assessment of the hydroxymethylfurfural level in a daily food ration of man. *Voprosy Pitaniya*, *30*, 50–53.
58. Steber, F., Klostermeyer, H. (1987). Heat treatment of fruit preparations and jams, and monitoring its efficacy. *Molkerei Zeitung Welt der Milch*, *41*, 289–290; 292–295.
59. Malik, F., Drak, M., Crhova, K. (1981). 5-Hydromethylfurfural in Produkten der Weinerzeugung. *Die Wein-Wissenschaft*, *36*, 360–365.
60. Wucherpfennig, K., Burkardt, D. (1983). Die Bedeutung der Endtemperatur bei der Rückkühlung heissgefüllter. *Fruchtsäfte. Flüssiges Obst*, *9*, 416–422.
61. Lee, H.S., Rouseff, R.L., Nagy, S. (1986). HPLC determination of furfural and 5-hydroxymethylfurfural in citrus juices. *Journal of Food Science*, *51*, 1075–1076.
62. Lo-Coco, F., Valentini, C., Novelli, V., Ceccon, L. (1994). High performance liquid chromatographic determination of 2-furaldehyde and 5-hydroxymethyl-2-furaldehyde in processed citrus juices. *Journal of Liquid Chromatography*, *17*, 603–617.

63. Association of the Industry of Juices and Nectars from Fruits and Vegetables (AIJN) (1996). Association of the Industry of Juices and Nectars of the European Economic Community Code of Practice for Evaluation of Fruit and Vegetable Juices. Brussels: AIJN.
64. Morales, F.J., Romero, C., Jiménez-Pérez, S. (1992). An enhanced liquid chromatographic method for 5-hydroxymethylfurfural determination in UHT milk. *Chromatographia*, *33*, 45–48.
65. Hewedy, M.M., Kiesner, C., Meissner, K., Hartkopf, J., Erbersdobler, H.F. (1994). Effects of UHT heating of milk in an experimental plant on several indicators of heat treatment. *Journal of Dairy Research*, *61*, 305–309.
66. Morales, F.J., Romero, C., Jiménez-Pérez, S. (1995). New methodologies for kinetic study of 5-(hydroxymethyl)furfural formation and reactive lysine blockage in heat-treated milk and model systems. *Journal of Food Protection*, *58*, 310–315.
67. Ferrer, E., Alegría, A., Farré, R., Abellán, P., Romero, F. (2000). Effects of thermal processing and storage on available lysine and furfural compounds contents of infant formulas. *Journal of Agricultural and Food Chemistry*, *48*, 1817–1822.
68. Cárdenas-Ruiz, J., Guerra-Hernández, E., García-Villanova, B. (2004). Furosine is a useful indicator in pre-baked breads. *Journal of Science Food and Agriculture*, *84*, 366–370.
69. Thorvaldsson, K., Kjjöledbrand, C. (1998). Water diffusion in bread during baking. *Lebensmittel-Wissenschaft und Technologie*, *31*, 658–663.
70. Malik, F., Rudicka, L., Drak, M. (1983). Gehalt und Eigenschaften des 5-Hydroxymethylfurfurals in Produkten der Weinerzeugung. *Wein-Wiss*, *38*, 51–57.
71. Malik, F., Navara, A., Minarik, E. (1985). 5-Hydroxymethylfurfural-Gehalte in Traubenweinen. *Mitteilungen Klosterneuburg*, *35*, 45–47.
72. Navara, A., Malik, F., Minarik, E. (1986). 5-Hydroxymerhylfurfural in Weinen aus dem ostslowakischen und Tokayer Weinbaugebiet der CSSR. *Mitteilungen Klosterneuburg*, *36*, 28–33.
73. Montilla-Gomez, J., Olea-Serrano, F., García-Villanova, R. (1988). Determinación espectrofotométrica de furfural y 5-hidroximetilfurfural al estado de tiosemicarbazonas en vinos tipo Malaga. *Química Analítica*, *7*, 77–84.
74a. Meidell, E., Filipello, F. (1969). Quantitative determination of hydroxymethylfurfural in sherries and grape concentrate. *American Journal of Enology and Viticulture*, *20* (**3**), 164–168.
74b. Sigler, J. (1977). 5-Hydroxymethylfurfural (HMF) in Likörwein. *Lebensmittelchemie*, *51*, 13–14.
75. Guerra-Hernández, E.J., Montilla-Gómez, J., García-Villanova, R. (1988). Determinación espectrofotométrica y cromatográfica de hidroximetil-furfural en vino: estudio comparativo. *Química Analítica*, *7*, 100–106.
76. Profumo, A., Riolo, C., Pesavento, M., Francoli, A. (1988). Evolution of the Italian distillate "grappa" during aging in wood: a gas chromatographic and high performance liquid chromatographic study. *American Journal of Enology and Viticulture*, *39*, 273–278.
77. Pons, I., Garrault, C., Jaubert, J.N., Morel, J. (1991). Analysis of aromatic caramel. *Food Chemistry*, *39*, 311–320.
78. Puech, J.L., Robert, A., Rabier, P., Moutounet, M. (1988). Caractéristiques et degradation physico-chimique de la lignine du bois de chêne (Characteristics and

physicochemical degradation of lignin in oak wood). *Bulletin Liaison Groupe Polyphenols*, *14*, 157–160.

79. Zanoni, B., Peri, C., Nani, R., Lavelli, V. (1999). Oxidative heat damage of tomato halves as affected by drying. *Food Research International*, *31*, 395–401.
80. Allen, B.H., Chin, H.B. (1980). Rapid HPLC determination of HMF in tomato paste. *Journal of the Association of Official Analytical Chemists*, *63*, 1074–1076.
81. Porretta, S., Sandei, L. (1991). Determination of 5-(hydroxymethyl)-2-furfural (HMF) in tomato products: proposal of a rapid HPLC method and its comparison with the colorimetric method. *Food Chemistry*, *39*, 51–57.
82. Hidalgo, A., Pompei, C., Zambuto, R. (1998). Heat damage evaluation during tomato products processing. *Journal of Agricultural and Food Chemistry*, *46*, 4387–4390.
83. Hidalgo, A., Pompe, C. (2000). Hydroxymethylfurfural and furosine reaction kinetics in tomato products. *Journal of Agricultural and Food Chemistry*, *48*, 78–82.
84. Garcia-Parrilla, C., Heredia, F.J., Troncoso, A.M. (1996). Phenols HPLC analysis by direct injection of sherry wine vinegar. *Journal of Liquid Chromatography and Related Technologies*, *19*, 247–258.
85. Galletti, G.C., Carnacini, A. (1995). Chemical composition of wood casks for wine aging as determined by pyrolysis/GC/MS. *Rapid Communications in Mass Spectrometry*, *9*, 1331–1334
86. Villalon Mir, M., Quesada Granados, J., Lopez, G., De la Serrana, H., Lopez Martinez, M.C. (1992). High performance liquid chromatography determination of furanic compounds in commercial brandies and caramels. *Journal of Liquid Chromatography*, *15*, 513–524.
87. Hewala, I.I., Blaih, S.M., Zoweil, A.M., Onsi, S.M. (1993). Detection and determination of interfering 5-hydroxymethylfurfural in the analysis of caramel-coloured pharmaceutical syrups. *Journal of Clinical Pharmacy and Therapeutics*, *18*, 49–53.
88. Rufian-Henares, J.A., Garcia-Villanova, B., Guerra-Hernandez, E. (2001). Determination of furfural compounds in enteral formula. *Journal of Liquid Chromatography and Related Technologies*, *24*, 3049–3061.
89. Ulbricht, R.J., Northup, S.J., Thomas, J.A. (1984). A review of 5-hydroxymethylfurfural (HMF) in parenteral solutions. *Fundamental and Applied Toxicology*, *4*, 843–853.
90. Durham, D.G., Hung, C.T., Taylor, R.B. (1982). Identification of some acids produced during autoclaving of D-glucose solutions using HPLC. *International Journal Pharmaceutics*, *12*, 31–40.
91. Cook, A.P., Macleod, T.M., Appleton, J.D., Fell, A.F. (1989). Reversed-phase high-performance liquid chromatographic method for the quantification of 5-hydroxymethylfurfural as the major degradation product of glucose in infusion liquids. *Journal of Chromatography*, *467*, 395–401.
92. Jellum, E., Borrenssen, H.C., Eldjarn, L. (1973). The presence of furan derivatives in patients receiving fructose containing solutions intravenously. *Clinica Chimica Acta*, *47*, 191–201.
93. Murty, B.S.R., Kapoor, J.N., Smith, F.X. (1977). Levels of 5-hydroxymethylfurfural in dextrose injection. *American Journal of Hospital Pharmacy*, *34*, 205–206.

94. Hadorn, H., Kovacs, A.S. (1960). On the examination and evaluation of foreign bee honey with regards to its hydroxymethyl furfural and diastase content. *Travaux de Chimie Alimentaire et d'Hygiène*, *51*, 373–375.
95. Garoglio, P.G. (1961). *Il Corriere Vinicolo*, *27*, 42.
96. Keeney, M., Bassette, R. (1959). Detection of intermediate compounds in the early stages of browning reaction in milk products. *Journal of Dairy Science*, *42*, 945–960.
97. Winkler, O. (1955). Beitrag zum Nachwals und zur Bestimmung von Oxymethylfurfural in Honig und Kunsthonig. *Zeitschrift fur Lebensmittel-Untersuchung und-Forschung*, *102*, 161–167.
98. De la Iglesia, F., Lázaro, F., Puchades, R., Maquieira, A. (1997). Automatic determination of 5-hydroxymethylfurfural (5-HMF) by a flow injection method. *Food Chemistry*, *60*, 245–250.
99. Dattatreya, A., Rankin, S.A. (2006). Moderately acidic pH potentiates browning of sweet whey powder. *International Dairy Journal*, *16*, 822–828.
100. Espinosa-Mansilla, A., Muñoz-delaPeña, A., Salinas, F. (1993). Semiautomatic determination of furanic aldehydes in food and pharmaceutical samples by a stopped-flow injection analysis method. *Journal of AOAC International*, *76*, 255–1261.
101. Reyes-Salas, E.O., Manzanilla-Cano, J.A., Barceló-Quintal, M.H., Juárez-Mendoza, D., Reyes-Salas, M. (2006). Direct electrochemical determination of hydroxymethylfurfural (HMF) and its application to honey samples. *Analytical Letters*, *39*, 161–171.
102. Consonni, R., Gatti, A. (2004). ^{1}H NMR studies for quantitative determination of HMF in traditional balsamic vinegar. *Journal of Agricultural and Food Chemistry*, *52*, 3446–3450.
103. Rice, R.G., Keller, G.J., Kirchner, J.G. (1951). Separation and identification of 2,4-dinitrophenylhydrazones of aldehydes and ketones, and 3,5-dinitrobenzoates of alcohols by filter-paper chromatography. *Analytical Chemistry*, *23*, 194–195.
104. Van Duin, H. (1957). Chromatography by liquid–liquid partition and liquid–liquid interface adsorption. *Nature*, *180*, 1473.
105. Schuck, D.F., Pavlina, T.M. (1994). Rapid detection of 5-(hydroxymethyl)-2-furfural in parenteral solutions with high, performance thin, layer chromatography. *Journal of Planar Chromatography*, *7*, 242–246.
106. Ralls, J.W. (1960). Correction. *Analytical Chemistry*, *32*, 332–332.
107. Cerrutti, P., Resnik, S.L., Seldes, A., Ferro-Fontan, C.F. (1985). Kinetics of deteriorative reactions in model food systems of high water activity. Glucose loss, HMF accumulation and fluorescent development due to non-enzymatic browning. *Journal of Food Science*, *50*, 627–630.
108. Teixidó, E., Santos, F.J., Puignou, L., Galceran, M.T. (2006). Analysis of 5-hydroxymethylfurfural in foods by gas chromatography–mass spectrometry. *Journal of Chromatography, A*, *1135*, 85–90.
109. Bouchard, J., Chornet, E., Overnet, R.P. (1988). High-performance liquid chromatographic monitoring of carbohydrate fractions in partially hydrolyzed corn starch. *Journal of Agricultural and Food Chemistry*, *36*, 1188–1192.
110. Riesner, C.H., Kiser, M.J., Dube, M.E. (2006). An aqueous high-performance liquid chromatographic procedure for the determination of 5-hydroxymethylfur-

fural in honey and other sugar-containing materials. *Journal of Food Science*, *71*, 179–184.

111. Chambel, P., Oliveira, M.B., Andrade, P.B., Seabra, R.M., Ferreira, M.A. (1997). Development of an HPLC/diode-array detector method for simultaneous determination of 5-HMF, furfural, 5-O-caffeoylquinic acid and caffeine in coffee. *Journal of Liquid Chromatography and Related Technologies*, *20*, 2949–2957.
112. Morales, F.J., Romero, C., Jiménez-Pérez, S. (1996). Study on 5-hydroxymethylfurfural formation in milk during UHT treatment measured by two analytical procedures, in *Heat Treatments and Alternative Methods*, International Dairy Federation, Brussels, pp. 354–357.
113. Corradini, D., Corradini, C. (1992). Separation and determination of 5-hydroxymethyl-2-fulfuraldehyde in fruit juices by micellar electrokinetic capillary chromatography with direct sample injection. *Journal of Chromatography*, *624*, 503–509.
114. Chávez-Servín, J.L., Castellote, A.I., López-Sabater, M.C. (2005). Analysis of potential and free furfural compounds in milk-based formulae by high-performance liquid chromatography. Evolution during storage. *Journal of Chromatography, A*, *1076*, 133–140.
115. Fallico, B., Arena, E., Zappala, M. (2003). Roasting of hazelnuts. Role of oil in colour development and hydroxymethylfurfural formation. *Food Chemistry*, *81*, 569–573.
116. Rufian-Henares, J.A., Delgado-Andrade, C., Morales, F.J. (2006). Application of a reverse phase high-performance liquid chromatography method for simultaneous determination of furanic compounds and glucosylisomaltol in breakfast cereals. *Journal of AOAC International*, *89*, 1–5.
117. White, J. (1979). Spectrophotometric method for hydroxymethylfurfural in honey. *Journal of the Association of Official Analytical Chemists*, *62*, 509–514.
118. Wood, R.M. (1993). A HMF validated method for the analysis of foodstuffs. *J. Assoc. Publ. Anal.*, *28*, 195–199.
119. Quesada-Granados, J., Villalon, M., Lopez-Serrana, H., Lopez-Martinez, M.C. (1992). Comparison of spectrophotometric and chromatographic methods of determination of furanic aldehydes in wine distillates. *Food Chemistry*, *52*, 203–208.
120. Duran-Meras, I., Espinosa-Mansilla, A., Salinas, F. (1995). Simultaneous kinetic spectrophotometric determination of 2-furfuraldehyde and 5-hydroxymethyl-2-fulfuraldehyde by application of a modified Winkler's method and partial least squares calibration. *Analyst*, *120*, 2567–2571.
121. Salinas, F., Espinosa, A., Berzas, J.J. (1991). Flow-injection determination of HMF in honey by the Winkler method. *Fresenius Journal of Analytical Chemistry*, *340*, 250–252.
122. Morales, F.J., Romero, C., Jiménez-Pérez, S. (1997). Chromatographic determination of bound hydroxymethylfurfural as an index of milk protein glycosylation. *Journal of Agricultural and Food Chemistry*, *45*, 1570–1573.
123. Espinosa-Mansilla, A., Salinas, F., Berzas Navado, J.J. (1992). Differential determination of furfural and hydroxymethylfurfural by derivative spectrophotometry. *Journal of AOAC International*, *75*, 678–684.

124. Blanco-Gomis, D., Gutierrez-Alvarez, M.D., Sopena-Naredo, L., Mangas-Alonso, J.J. (1991). High-performance liquid chromatographic determination of furfural and hydroxymethylfurfural in apple juices and concentrates. *Chromatographia*, *32*, 45–48.
125. Michael, L., May, P.C., André, C.M. (1985). Automated fluorimetric determination of furfurals. *Analytical Biochemistry*, *144*, 6–14.
126. Lang, M., Malyusz, M. (1994). Fast method for the simultaneous determination of 2-oxo acids in biological fluids by high-performance liquid chromatography. *Journal of Chromatography, B. Biomedical Applications*, *662*, 97–102.
127. Antal, M.J.Jr, Mok, W.S., Richards, G.N. (1990). Mechanism of formation of 5-(hydroxymethyl)-2-furaldehyde from D-fructose and sucrose. *Carbohydrate Research*, *199*, 91–109.
128. Feather, M.S., Harris, J.F. (1973). Dehydration reactions of carbohydrates. *Advances in Carbohydrate Chemistry*, *28*, 161–224.
129. Wolfrom, M.L., Schuetz, R.D., Calvalieri, L.F. (1948). Discoloration of sugar solutions and 5-(hydroxymethyl)furfural. *Journal of the American Chemist Society*, *70*, 514.
130. Ledl, F., Fritsch, G., Hiebl, J., Pachmayr, O., Severin, T. (1986). Degradation of Maillard products, in *Amino-Carbonyl Reactions in Food and Biological Systems* (eds M. Fujimaki, M. Namiki, H. Kato), Kadansha Ltd., Tokyo; Elsevier Science Publishers, B.V. Amsterdam, p. 173.
131. Ledl, F., Hiebl, J., Severin, T. (1983). Formation of coloured 8-pyranones from hexoses and pentoses. *Zeitschrift fur Lebensmittel-Untersuchung und-Forschung*, *177*, 353–355.
132. Shinohara, K., Kim, E.-H., Omura, H. (1986). Furans as the mutagens formed by amino-carbonyl reactions. *Developments in Food Science*, *13*, 353–362.
133. Friedman, M. (1996). Food browning and its prevention: an overview. *Journal of Agricultural and Food Chemistry*, *44*, 631–653.
134. Yaylayan, V.A. (1997). Classification of the Maillard reaction: a conceptual approach. *Trends in Food Science and Technology*, *8*, 12–18.
135. COST Action 927 (2004). Thermally processed food: possible health implications. European Science Foundation, Brussels, http://www.cost.esf.org/index.php?id=181&action_number=927 (accessed 10 June 2007).
136. Hodge, J.E. (1953). Chemistry of browning reactions in model systems. *Journal of Agricultural and Food Chemistry*, *1*, 928–943.
137. Eichner, K. (1982). Analytical detection of thermally induced qualitative changes in vegetables and fruits. *Zaitung Lebensmittel Untersuchung und Forschung*, *36*, 101–104.
138. Wittmann, R., Eichner, K. (1989). Detection of Maillard products in malts, beers, and brewing colorants. *Zeitschrift fur Lebensmittel Untersuchung und Forschung*, *188*, 212–220.
139. Rahbar, S., Blumenfeld, O., Ranney, H.M. (1969). Studies of unusual hemoglobin in patients with diabetes mellitus. *Biochemical and Biophysical Research Communications*, *36*, 838–843.
140. Madson, M., Feather, M.S. (1981). An improved preparation of 3-deoxy-D-erythro-hexo-2-ulose via the bis(benzohydrazone) and some related constitutional studies. *Carbohydrate Research*, *94*, 183–191.

141. Ishizu, A., Lindberg, B., Theander, O. (1967). 1-Deoxy-D-erythro-2,3-hexodiulose, an intermediate in the formation of D-glucosaccharinic acid. *Carbohydrate Research*, *5*, 329–334.
142. Glomb, M.A., Pfahler, C. (2000). Synthesis of 1-deoxy-D-erythro-hexo-2,3-diulose, a major hexose Maillard intermediate. *Carbohydrate Research*, *329*, 515–523.
143. Thornalley, P.J. (1996). Pharmacology of methylglyoxal: formation, modification of proteins and nucleic acids, and enzymatic detoxification—a role in pathogenesis and antiproliferative chemotherapy. *General Pharmacology*, *27*, 565–573.
144. Shaw, P.E., Berry, R.E. (1977). Hexose-amino acid degradation involving formation of pyrroles, furans and other low molecular weight products. *Journal of Agricultural and Food Chemistry*, *25* (**3**), 641–644.
145. Potter, P.E., Patton, S. (1956). Evidence of maltol and HMF in evaporated milk as shown by paper chromatography. *Journal of Dairy Science*, *39*, 978–982.
146. Klein, E., Ledl, F., Bergmueller, W., Severin, T. (1992). Reactivity of Maillard products with a pyrrole structure. *Zaitung Lebensmittel Untersuchung und Forschung*, *194* (**6**), 556–560.
147. Milic, B.L., Piletic, M.V. (1984). The mechanism of pyrrole pyrazine and pyridine formation in non-enzymic browning reaction. *Food Chemistry*, *13* (**3**), 165–180.
148. Shigematsu, H., Kurata, T., Kato, H., Fujimaki, M. (1971). Formation of 2-(5-hydroxymethyl-2-fomrylpyrrol-a-yl)alkyl acid lactones on roasting alkyl-a-amino acid with D-glucose. *Agricultural and Biological Chemistry*, *35*, 2105–2097.
149. Davidek, T., Velisek, J., Davidek, J., Pech, P. (1992). Amino acids derived 13-disubstituted imidazoles in nonenzymatic browning reactions. *Sbornik*, *54*, 165–183.
150. Hayashi, T., Namiki, M. (1980). Formation of two-carbon sugar fragment at an early stage of the browning reaction of sugar with amine. *Agricultural and Biological Chemistry*, *44*, 2575–2580.
151. Weenan, H. (1998). Reactive intermediates and carbohydrate fragmentation in Maillard chemistry. *Food Chemistry*, *62*, 393–401.
152. Hoellnagel, A., Kroh, L.W. (1998). Formation of alpha-dicarbonyl fragments from mono- and disaccharides under caramelization and Maillard reaction conditions. *Zaitung Lebensmittel Untersuchung und Forschung*, *207*, 50–54.
153. Olsson, K., Pernemalm, P.A., Theander, O. (1981). Maillard reaction in food. Chemical, physiological and technological aspects, in *Progress in Food Nutrition and Science*, Vol. 5 (ed. C. Eriksson), Pergamon Press, Oxford, New York, p. 47.
154. Miller, R.E., Cantor, S.M. (1952). 2-Hydroxyacetylfuran from sugars. *Journal of the American Chemical Society*, *74*, 5236–5237.
155. Theander, O. (1981). Maillard reaction in food. Chemical, physiological and technological aspects, in *Progress in Food Nutrition and Science*, Vol. 5 (ed. C. Eriksson), Pergamon Press, Oxford, New York, p. 471.
156. Schrödter, R. (1992). Modelluntersuchungen zur strukturaufklarung von modellmelanoidinen mittels hochfrequenzpyrolyse. Dissertation, Technische Universitat Berlin.
157. Germond, J.E., Philippossian, G., Richli, U., Bracco, I., Arnaud, M.J. (1987). Rapid and complete urinary elimination of [14C]-5-(hydroxymethyl)-2-furaldehyde. *Journal of Toxicology and Environmental Health*, *22* (**1**), 790–789.

158. Gökmen, V., Senyuva, H.Z. (2007). Effects of some cations on the formation of acrylamide and furfurals in glucose-asparagine model systems. *European Food Research and Technology*, *225*, 815–820.
159. Gökmen, V., Acar, O.C., Serpen, A., Morales, F.J. (2008). Effect of leavening agents and sugars on the formation of hydroxymethylfurfural in cookies during baking. *European Food Research and Technology*, *226*, 1031–1037.
160. Archer, M.C., Bruce, W.R., Chan, C.C., Corpet, D.E., Medline, A., Roncucci, L., Stamp, D., Zhang, X.M. (1992). Aberrant crypt foci and microadenoma as markers for colon cancer. *Environmental Health Perspectives*, *98*, 195–197.
161. Surh, Y.J., Tannenbaum, S.R. (1994). Activation of the Maillard reaction product 5-hydroxymethyl)furfural to strong mutagens via allylic sulfonation and chlorination. *Chemical Research in Toxicology*, *7*, 313–318.
162. Surh, Y.J., Liem, A., Miller, J.A., Tannenbaum, S.R. (1994). 5-Sulfooxymethylfurfural as a possible ultimate mutagenic and carcinogenic metabolite of the Maillard reaction product, 5-hydroxymethylfurfural. *Carcinogenesis*, *15*, 2375–2377.
163. Schoental, R., Hard, G.C., Gibbard, S. (1971). Histopathology of renal lipomatous tumours in rats treated with the natural products, pyrrolizidine alkaloids and a,ß-unsaturated aldehydes. *Journal of the National Cancer Institute*, *47*, 1037–1044.
164. Nishi, Y., Miyakawa, Y., Kato, K. (1989). Chromosome aberrations induced by pyrolysates of carbohydrates in Chinese hamster V79 cells. *Mutation Research*, *227*, 117–123.
165. Delgado-Andrade, C., Seiquer, I., Navarro, M.P., Morales, F.J. (2008). Estimation of hydroxymethylfurfural availability in breakfast cereals. Studies in Caco-2 cells. *Food and Chemical Toxicology*, *46* (**5**), 1600–1607.
166. Boopathy, R., Bokang, H., Daniels, L. (1993). Biotransformation of furfural and 5-hydroxymethyl furfural by enteric bacteria. *Journal of Industrial Microbiology*, *11*, 147–150.
167. Godfrey, V.B., Chen, L.J., Griffin, R.J., Lebetkin, E.H., Burka, L.T. (1999). Distribution and metabolism of (5-hydroxymethyl)furfural in male F344 rats and B6C3F1 mice after oral administration. *Journal of Toxicology and Environmental Health, Part A*, *57*, 199–210.
168. Germond, J.E., Philippossian, G., Richli, U., Bracco, I., Arnaud, M.J. (1987). Rapid and complete urinary elimination of (14C)-5-hydroxymethyl-2-furaldehyde administered orally or intravenously to rats. *Journal of Toxicology and Environmental Health*, *22*, 79–89.
169. AFC (2005). Opinion of the Scientific Panel on food additives, flavourings, processing aids and materials in contact with foods (AFC) on a request from the Commission related to flavouring group evaluation 13: furfuryl and furan derivatives with and without additional side-chain substituents and heteroatoms from chemical group 14. *The EFSA Journal*, *215*, 1–73.
170. Pettersen, J.E., Jellum, E. (1972). The identification and metabolic origin of 2-furoylglycine and 2,5-furandicarboxylic acid in human urine. *Clinica Chimica Acta*, *41*, 199–207.
171. Pryor, R.L., Wu, X., Gu, L. (2006). Identification of urinary excretion of metabolites of 5-(hydroxymethyl)-2-furfural in human subjects following consumption of dried plums or dried plum juice. *Journal of Agricultural and Food Chemistry*, *54*, 3744–3749.

172. Nomeir, A.A., Silvena, D.M., McComish, M.F., Chadwick, M. (1992). Comparative metabolism and disposition of furfural and furfuryl alcohol in rats. *Drug Metabolism and Disposition*, *20*, 198–204.
173. Michail, K., Juan, H., Maier, A., Matzi, V., Greiberger, J., Wintersteiger, R. (2007). Development and validation of liquid chromatographic method for the determination of hydroxymethylfurfural and alpha-ketoglutaric acid in human plasma. *Analytica Chimica Acta*, *581*, 287–297.
174. Bruce, W.R., Archer, M.C., Corpet, D.E., Medline, A., Minkin, S., Stamp, D., Yin, Y., Zhang, X.-M. (1993). Diet, aberrant crypt foci and colorectal cancer. *Mutation Research*, *290*, 111–118.
175. Zhang, X.-M., Chan, C.C., Stamp, D., Minkin, S., Archer, M.C., Bruce, W.R. (1993). Initiation and promotion of colonic aberrant crypt foci in rats by 5-hydroxymethyl-2-furaldehyde in thermolyzed sucrose. *Carcinogenesis*, *14*, 773–775.
176. Miyakawa, Y., Nishi, Y., Kato, K., Sato, H., Takahashi, M., Hayashi, Y. (1991). Initiating activity of eight pyrolysates of carbohydrates in a two stage mouse skin tumourigenesis model. *Carcinogenesis*, *12*, 1169–1173.
177. Kim, S.B., Hayase, F., Kato, H. (1987). Desmutagenic effect of a-dicarbonyl and a-hydroxycarbonyl compounds against mutagenic heterocyclic amines. *Mutation Research*, *177*, 9–15.
178. Kong, Z.L., Mitsuiki, M., Nonaka, M., Omura, H. (1988). Mutagenic activities of furfurals and the effects of Cu2+. *Mutation Research*, *203*, 376.
179. Aeschbacher, H.U., Chappuis, C., Manganel, M., Aeschbach, R. (1981). Investigation of Maillard products in bacterial mutagenicity test systems. *Progress in Food and Nutrition Science*, *5*, 279–294.
180. Florin, I., Rutberg, L., Curvall, M., Enzell, C.R. (1980). Screening of tobacco smoke constituents for mutagenicity using the Ames test. *Toxicology*, *18*, 219–232.
181. Kasai, H., Kumeno, K., Yamaizumi, Z., Nishimura, S., Nagao, M., Fujita, Y., Sugimura, T., Nukaya, H., Kim, S.B., Hayase, F., Kato, H. (1987). Desmutagenic effect of a-dicarbonyl and a-hydroxycarbonyl compounds against mutagenic heterocyclic amines. *Mutation Research*, *177*, 9–15.
182. Kong, Z.L., Shinohara, K., Mitsuiki, M., Murakami, H., Omura, H. (1989). Desmutagenicity of furan compounds towards some mutagens. *Agricultural and Biological Chemistry*, *53*, 2073–2079.
183. Ying, D., Kevin, P., Weinhan, Z., Xiaoquiang, Y., Jianrong, H. (2005). US Patent 2005 124648 A1.
184. Uckum, F.M., Tai, S., Jan, M. (2003). US Patent 6,258,841 B1.
185. Omura, H., Jahan, N., Shinohara, K., Muramaki, H. (1983). Formation of mutagens by the Maillard reaction, in *The Maillard Reaction in Foods and Nutrition* (eds G.R. Waller, M.S. Feather), American Chemical Society, Washington, DC, pp. 537–564.
186. Sommer, Y., Hollnagel, H., Schneider, H., Glatt, H.R. (2003). Metabolism of 5-hydroxymethyl-2-furfural (HMF) to the mutagen 5-sulfoxymethyl-2-furfural (SMF) by individual sulfotransferases. *Naunyn Schmiedebergs Archives of Pharmacology*, *367* (Suppl. 1), r166.
187. Janzowski, C., Glaab, V., Samimi, E., Schlatter, J., Eisenbrand, G. (2000). 5-Hydroxymethylfurfural: assessment of mutagenicity, DNA damaging potential and reactivity towards cellular glutathione. *Food Chemical Toxicology*, *38*, 801–809.

188. Lee, Y-C., Shlyankevich, M., Jeong, H-K., Douglas, J.S., Surh, Y.J. (1995). Bioactivation of 5-hydroxymethyl-2-furaldehyde to an electrophilic and mutagenic allylic sulphuric acid ester. *Biochemical and Biophysical Research Communications*, *209*, 996–1002.

189. Mizushina, Y., Yagita, E., Kuramochi, K., Kuriyama, I., Shimazaki, N., Koiwai, O., Uchiyama, Y., Yomezawa, Y., Sugawara, F., Kobayashi, S., Sakaguchi, K., Yoshida, H. (2006). 5-(Hydroxymethyl)-2-furfural: a selective inhibitor of DNA polymerase λ and terminal deoxynucleotidyltransferase. *Archives of Biochemistry and Biophysics*, *446*, 69–76.

190. Glatt, H., Schneider, H., Yungang, L. (2005). V79-h-CYP2E1SULT1A1, a cell line for the sensitive detection of genotoxic effects induced by carbohydrate pyrolysis products and other food-borne chemicals. *Mutation Research*, *580*, 41–52.

191. Lang, K., Kieckebusch, W., Bässler, K.H., Griem, W., Czok, G. (1970). [Tolerance of 5-hydroxymethylfurfural (HMF). 1. Chronic administration of HMF and metabolites of the substance]. *Zeitschrift für Ernährungswissenschaft*, *10*, 97–101.

192. Zaitzev, A.N., Simonyan, T.A., Pozdnyakov, A.L. (1975). [Hygienic standardization of oxymethylfurfurol in food products]. *Voprosy Pitaniya*, *1*, 52–55. (In Russian)

193. NIEHS (2008). Toxicology and Carcinogenesis studies of 5-(Hydroxymethyl)-2-Furfural (CAS no. 67-47-0) in F344/N rats and B6C3F1 mice (gavage studies). Draft technical report TR-554. http://ntp.niehs.nih.gov (accessed 12 March 2007).

194. Simonyan, T.A. (1969). [Toxico-hygienic characteristics of hydroxymethylfurfural]. *Voprosy Pitaniya*, *28*, 54–58.

195. Czok, G. (1970). [Tolerance to 5-hydroxymethylfurfural (HMF). 2. Pharmacological properties]. *Zeitschrift fur Ernahrungwissenschaft*, *10*, 103–110.

196. US EPA. (1992). Initial submission: acute oral LD50 study with 5-hydroxymethylfurfural in rats. Cover letter dated 073192, EPA/OTS. Doc. #88-920005429, Chicago, IL.

197. JECFA (1996). *Toxicological Evaluation of Certain Food Additives. The Forty-Fourth Meeting of the Joint FAO/WHO Expert Committee on Food Additives and Contaminants*. WHO Food Additives Series: 35, IFCS, WHO, Geneva.

198. Delgado-Andrade, C., Seiquer, I., Navarro, M.P., Morales, F.J. (2007). Maillard reaction indicators in diets usually consumed by adolescent population. *Molecular Nutrition and Food Research*, *51*, 341–351.

2.6

CHLOROPROPANOLS AND CHLOROESTERS

Colin G. Hamlet and Peter A. Sadd
RHM Technology, RHM Group Ltd, Lord Rank Centre, Lincoln Road, High Wycombe, Bucks HP12 3QR, UK

2.6.1 INTRODUCTION

Chloropropanols and their fatty acid esters (chloroesters) are contaminants that are formed during the processing and manufacture of certain foods and ingredients. The presence of these compounds in foods is of concern because toxicological studies have shown that they could endanger human health (1).

2.6.1.1 Historical Perspective: 1979–2000

Velíšek *et al.* (2–6) and Davídek *et al.* (7, 8) were the first to demonstrate that chloropropanols and chloroesters could be formed in hydrolyzed vegetable proteins (HVP) produced by hydrochloric acid hydrolysis of proteinaceous by-products from edible oil extraction, such as soybean meal, rapeseed meal, and maize gluten. It was shown that hydrochloric acid could react with residual glycerol and lipids associated with the proteinaceous materials to yield a range of chloropropanols. The main chloropropanol found in HVP was 3-chloropropane-1,2-diol (3-MCPD) together with lesser amounts of 2-chloropropane-1,3-diol (2-MCPD), 1,3-dichloropropanol (1,3-DCP), 2,3-dichloropropanol (2,3-DCP), and 3-chloropropan-1-ol (see Fig. 2.6.1).

Process-Induced Food Toxicants: Occurrence, Formation, Mitigation, and Health Risks,
Edited by Richard H. Stadler and David R. Lineback

3-Monochloropropane-1,2-diol (3-MCPD): At a Glance	
Historical	3-MCPD is the main chloropropanol found in acid-HVP and was first described by Velíšek and co-workers in 1978.
Analysis	The majority of methods are based on isotope dilution GC/MS after derivatization of the analyte. Fully validated analytical methods with a limit of quantification in the low part-per-billion range are available.
Occurrence in Food	Besides HVP and soy sauce produced by acid hydrolysis, the food groups most likely to contain 3-MCPD are retail ready-to-eat processed foods including bread, cakes, biscuits, cheese, cooked/cured fish, and meat; licorice confectionery; and commercial food ingredients such as processed garlic, liquid smokes, and malts.
Main Formation Pathways	Mainly thermally driven reactions that require chloride ion, the backbone of the molecule being furnished by different precursors such as glycerol, monoacylglycerols, and lysophospholipids.
Mitigation in Food	During the late 1980s, acid-HVP manufacturers implemented the necessary procedures to minimize 3-MCPD formation. For example, careful control of the acid hydrolysis step and subsequent neutralization or alternatively decomposition of 3-MCPD by a subsequent alkali treatment stage are known approaches. Further, some manufacturers have changed processes, e.g., employ enzyme hydrolysis; addition of flavoring.
Health Risks	Non-genotoxic carcinogen; TDI for 3-MCPD set at 2 μg kg^{-1} bw (JECFA 2001).
Regulatory Information or Industry Standards	The EU has set a maximum limit for 3-MCPD in HVP and soy sauce of 0.02 mg/kg on a liquid basis (40% solids). This level was set using the principle of "as low as reasonably achievable" (when 3-MCPD was originally considered a genotoxic carcinogen) and exposure estimates derived before any remedial action was taken by manufacturers of HVP and soy sauces. This limit may be increased if the EC is satisfied that exposure to all dietary sources of 3-MCPD (including chloroesters) is low. Other countries have mandatory or recommended limits ranging from 0.02–1.0 mg/kg for soy sauce and HVP. However, limits for other food products may be set in the future.
Knowledge Gaps	There is growing evidence that; 3-MCPD esters (chloroesters) may be widespread in processed foods, in most cases they are present at higher concentrations than 3-MCPD, and upon hydrolysis would consequently contribute to the dietary load of free 3-MCPD. Their occurrence in foods, as well as toxicological significance, still needs to be assessed.
Useful Information (web) & Dietary Recommendations	http://www.foodstandards.gov.au/_srcfiles/3-monochloropropane-1-2-diol_in_foods.pdf http://www.inchem.org/documents/jecfa/jecmono/v48je18.htm http://eur-lex.europa.eu/LexUriServ/LexUriServ.do?uri=CELEX:32006R1881:EN:NOT No specific dietary recommendations but consumers should follow on-pack advice when preparing commercial products. Avoid over-heating foods, i.e., bake or toast to a golden yellow color.

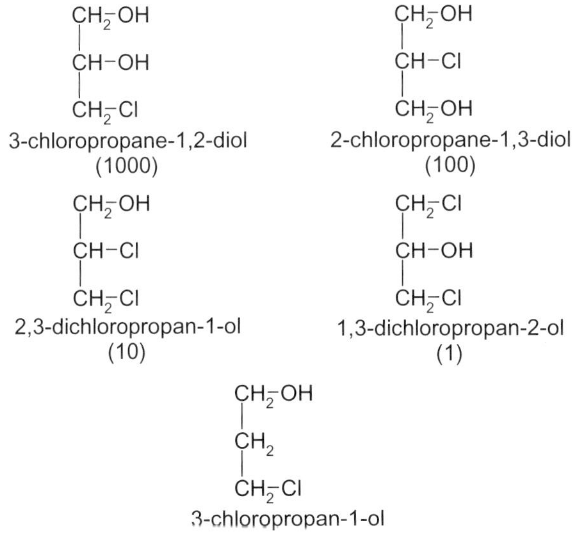

$$\begin{array}{c} CH_2\text{–}OH \\ | \\ CH\text{–}OH \\ | \\ CH_2\text{–}Cl \end{array}$$
3-chloropropane-1,2-diol
(1000)

$$\begin{array}{c} CH_2\text{–}OH \\ | \\ CH\text{–}Cl \\ | \\ CH_2\text{–}OH \end{array}$$
2-chloropropane-1,3-diol
(100)

$$\begin{array}{c} CH_2\text{–}OH \\ | \\ CH\text{–}Cl \\ | \\ CH_2\text{–}Cl \end{array}$$
2,3-dichloropropan-1-ol
(10)

$$\begin{array}{c} CH_2\text{–}Cl \\ | \\ CH\text{–}OH \\ | \\ CH_2\text{–}Cl \end{array}$$
1,3-dichloropropan-2-ol
(1)

$$\begin{array}{c} CH_2\text{–}OH \\ | \\ CH_2 \\ | \\ CH_2\text{–}Cl \end{array}$$
3-chloropropan-1-ol

Figure 2.6.1 Chloropropanols originally found in hydrolyzed vegetable proteins and their relative amounts (in parentheses).

Although fatty acid esters of 1,3-DCP and 3-MCPD (mono- and diesters) were formed in acid HVP (4, 9, 10), it was shown that only traces remained in the final commercial product since the majority were effectively removed during filtration of the hydrolysate. Collier *et al.* (11) later showed that chloroesters were intermediates in the formation of dichloropropanols (DCPs) and monochloropropanediols (MCPDs) from lipids and the relative amounts of each chloropropanol formed in acid HVP were dependent on the composition of the lipid in the raw materials used.

The first reported occurrence of chloroesters in a natural unprocessed food was by Cerbulis *et al.* (12), who identified a small but significant quantity ($<1\%$ of total neutral lipids) of diesters of 3-MCPD in raw milk from several herds of goats. Subsequent investigations (13, 14) appeared to exclude the biosynthesis of these chloroesters *in vivo*; however, it was not clear whether these compounds were formed from dietary substances, e.g., 3-MCPD or its esters, passed through the organism or from anthropogenic chlorine-containing compounds (such as chlorine-based sanitizers used in dairy operations) post secretion. The structure of the fatty acid esters of 3-MCPD and 1,3-DCP is given in Fig. 2.6.2.

In 1994 and 1997, the EU Scientific Committee for Food concluded that 3-MCPD should be regarded as a genotoxic carcinogen and since a safe threshold dose could not be determined, residues in foods should be undetectable by the most sensitive analytical method (15). This spurred intense analytical (16–19), monitoring (20–23), and mitigation activities in many countries. In the case of HVP, progress made by European manufacturers in implementing

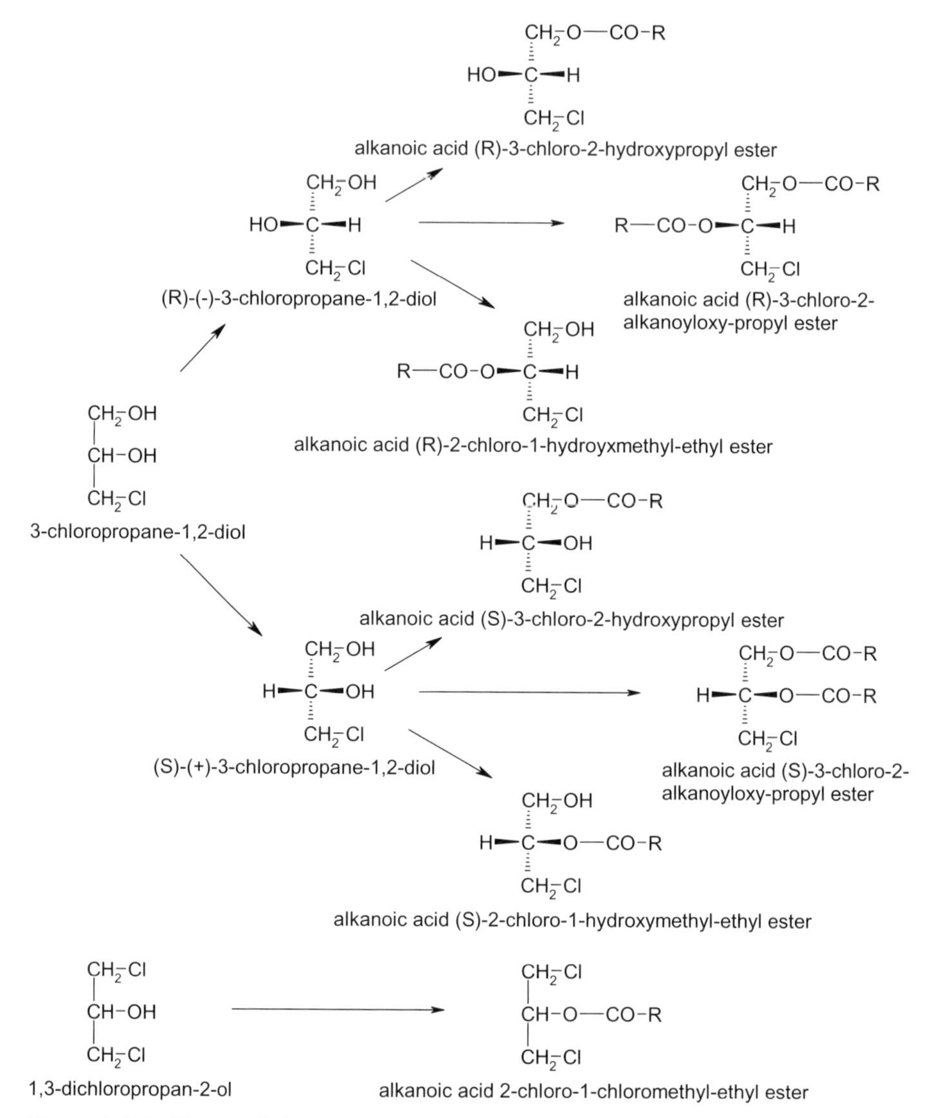

Figure 2.6.2 Esters of chloropropanols found in foodstuffs and their relationship to the corresponding chloropropanol isomers.

initial chloropropanol reduction processes (24–28) has significantly reduced amounts of both MCPDs and DCPs in these products.

2.6.1.2 Scope of Chapter

The subsequent discovery of chloropropanols in food contact materials treated with wet-strength resins (29), in drinking water, and in a range of processed foods that do not use acid-HVP as an ingredient (30), has renewed interest in these contaminants. This chapter brings together the information available (since approximately 2000) on health risks, analysis, occurrence, formation, risk management, exposure, and risk assessment for chloropropanols and chloroesters in foodstuffs excluding HVPs and soy sauces. The latter products are reviewed in Section 6.2.2.

2.6.2 HEALTH RISKS

2.6.2.1 3-MCPD

The chloropropanol 3-MCPD has been investigated in short- and long-term toxicity studies, and the most recent toxicology, mutagenicity, and carcinogenicity data have been summarized previously by the Joint FAO/WHO Expert Committee on Food Additives (JECFA) at its meeting in 2001 (31). In rats and mice, the kidney was the main target organ for toxicity with effects also observed on male fertility. Studies have demonstrated that 3-MCPD has mutagenic activity *in vitro* (32, 33), although negative results reported from a bone marrow micronucleus assay in rats and a rat liver unscheduled DNA synthesis (UDS) assay (34, 35) have provided reassurance that the mutagenic activity seen *in vitro* was not expressed *in vivo* (36). No epidemiological or clinical studies in humans have been reported.

In June 2001, JECFA (37) assigned a provisional maximum tolerable daily intake (PMTDI) for 3-MCPD of $2\,\mu g\,kg^{-1}$ body weight on the basis of the lowest observed effect level (LOEL) and a safety factor of 500. The safety margin included a factor of 5 for extrapolation from a LOEL to a no observed effect level (NOEL) and was considered adequate to account for the effects on male fertility and for inadequacies in the reproductive toxicity data.

2.6.2.2 Dichloropropanols

The available toxicology, mutagenicity, and carcinogenicity data for 1,3-DCP have been summarized previously by the 57th session of JECFA in 2001 (38). JECFA concluded that 1,3-DCP was hepatotoxic, induced a variety of tumors in various organs in the rat, and was genotoxic *in vitro*. The UK Committees on Mutagenicity (COM) and Committees on Carcinogenicity (COC) of Chemicals in Food, Consumer Products and the Environment considered 1,3-DCP in 2003 and 2004, respectively, following the publication of results from *in vivo* rat bone marrow micronucleus and rat liver UDS tests (36, 37). The COM (39) concluded that 1,3-DCP is not genotoxic *in vivo* in the tested tissues. However, the COC (40) concluded that 1,3-DCP should be regarded as a genotoxic carcinogen as it was not possible to exclude a genotoxic mechanism for the induction of tumors of rat tongue observed in the 2-year carcinogenicity study. The Committee also recommended that further investigation regarding the mechanisms of 1,3-DCP carcinogenicity *in vivo* is needed. The most recent assessment of 1,3-DCP (41) concluded that the critical effect of 1,3-dichloro-2-propanol was carcinogenicity and that a genotoxic mode of action could not be excluded. On the basis of these findings, a tolerable daily intake for 1,3-DCP has not been set.

There are very few data on the absorption, distribution, and excretion of 2,3-DCP. Theoretically, 2,3-DCP could be metabolized to produce epichlorohydrin (and subsequently glycidol) and therefore there are structural alerts for genotoxicity and carcinogenicity (42). Limited *in vitro* mutagenicity data indicate that 2,3-DCP is genotoxic with and without metabolic activation in bacterial and mammalian cells (42). Recently published *in vivo* rat bone marrow micronucleus and rat liver UDS assays were negative (34, 35). No appropriate carcinogenicity studies of 2,3-DCP are available. The UK COM (42) considered that 2,3-DCP has no significant genotoxic potential *in vivo* in the tissues evaluated (i.e., bone marrow and liver in the rat). The COC (40) considered that no conclusions regarding carcinogenicity of 2,3-DCP could be reached.

2.6.2.3 Chloroesters

The recent discovery of chloroesters in foods other than acid-HVP has raised questions about the bioavailability of chloropropanols from the dietary intake of these substances. To date, there are insufficient data for expert bodies to evaluate dietary intake or the toxicological significance of chloroesters, and JECFA (41) has recommended that studies be undertaken to address these questions.

The toxicological concern associated with exposures to 3-MCPD-esters is likely to vary according to their metabolic fate in the human gut. Limited preliminary data in model systems indicate that like triacylglycerols (TAG), 3-MCPD-esters may be substrates of lipases, releasing 3-MCPD (43). The efficiency of this enzymatic reaction was found to be higher toward 1-monoesters than toward diesters. In addition, the release of 3-MCPD was significantly reduced when the esters were provided in an oil matrix. Altogether, these data suggest that a limited hydrolysis of 3-MCPD-esters in the gut is probable, although a complete release of bound 3-MCPD is unlikely. Therefore, to address the health significance of exposure to 3-MCPD-esters, it is required to consider both the esters per se, and the potentially released 3-MCPD from this source.

The systemic effects of 3-MCPD-esters will depend on their rate of absorption. Because of their structural similarity and sensitivity to pancreatic lipases, the mechanisms of TAG absorption (44) may provide relevant information for 3-MCPD-esters. Ingested TAGs are not absorbed as such. In the stomach and small intestine, lipases with higher affinity for positions 1 and 3 release free fatty acids, and 2-monoacylglycerol (sn2-MAG) from the TAGs. sn2-MAG is readily taken up by the enterocytes, re-esterified by cellular acyltransferases, then incorporated into TAG-rich lipoprotein particles (i.e., chylomicrons). Chylomicrons are secreted into the lymph, enter the circulation through the thoracic duct, and supply tissues and organs with TAG. Excess of fats is stored in adipose tissue.

Assuming a similar metabolism for 3-MCPD-esters, the de-esterification in position sn1 would be favored by pancreatic lipases. Hence 3-MCPD and 3-MCPD-2-monoacylesters would therefore be released respectively from the 1-monoesters, and the diesters potentially present in food. The 2-monoesters of 3-MCPD would be absorbed, re-esterified in the enterocytes, packaged in the chylomicrons, and then possibly incorporated in cellular membranes and/or adipose tissue.

The biological significance of such a fate for 3-MCPD-esters is unknown. It may however explain the relatively high concentrations of 3-MCPD-esters reported in some samples of mammalian milks, including human breast milk.

2.6.3 ANALYSIS

2.6.3.1 Chloropropanols

The measurement of chloropropanols at trace contaminant concentrations in foodstuffs presents a challenge for the analyst: the absence of suitable chromophores has made approaches based on high-performance liquid chromatography (HPLC) with ultraviolet or fluorescence detection unsuitable. The low volatility and high polarity of, for example, MCPDs give rise to unfavorable interactions with components of gas chromatography (GC) systems that result in poor peak shape and low sensitivity, while the low molecular weights of both MCPDs and DCPs make mass detection difficult since diagnostic ions cannot be reliably distinguished from background chemical noise. Many of these limitations have been overcome and methods based on the formation of stable GC volatile derivatives with mass-selective detection predominate. Restrictions on the commercial availability of some reference standards have necessitated custom syntheses of, for example, 2-MCPD. The latter compound has been quantified to a first approximation by using the response factors from readily available 3-MCPD calibration standards (45).

2.6.3.1.1 Heptafluorobutyryl (HFB) Ester Derivatives Procedures based on the GC/mass spectrometry (MS) detection of HFB esters of chloropropanols have been universally applied to the analysis of a wide range of foodstuffs and have subsequently become the normative reference methods (46, 47). Heptafluorobutyryl imidazole (HFBI) and heptafluorobutyric anhydride (HFBA) are the derivatization reagents of choice, and the formation of the corresponding HFB esters is selective for the hydroxyl function and hence the analysis of all chloropropanols (see Fig. 2.6.3). Since HFBI and HFBA are moisture sensitive, derivatization must be carried out under anhydrous conditions.

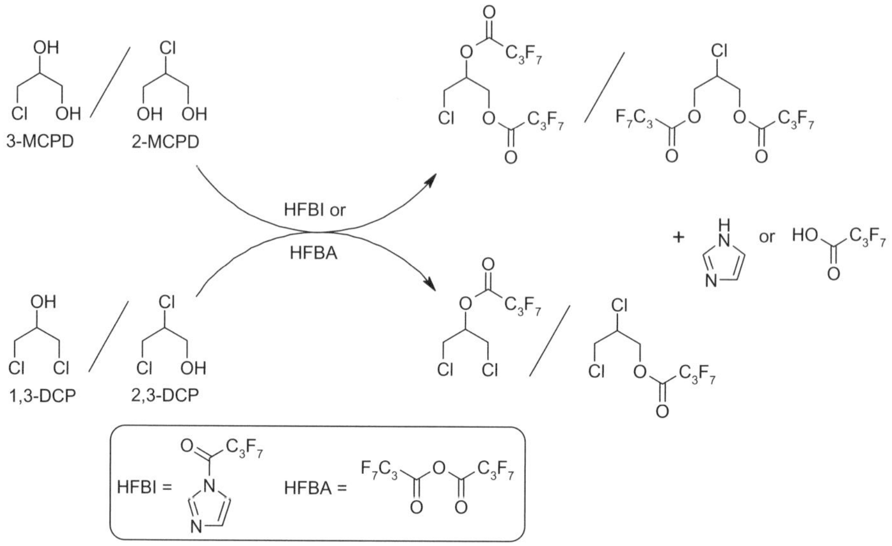

Figure 2.6.3 Derivatization reactions of chloropropanols with HFBI and HFBA.

Brereton *et al.* (46) validated by collaborative trial the procedures developed by Hamlet *et al.* (16, 17) for the determination of 3-MCPD at the low $\mu g\,kg^{-1}$ level in a wide range of foods and ingredients. The water-soluble 3-MCPD was extracted into saline solution and then partitioned into diethyl ether using a solid-phase extraction technique based on diatomaceous earth. Dried and concentrated extracts were treated with HFBI to convert 3-MCPD to the corresponding HFB diester (see Fig. 2.6.3) prior to analysis by GC and MS. Quantification was by a stable isotope internal standard method using deuterium-labeled 3-MCPD (3-MCPD-d_5) added to the sample prior to extraction. The method was suitable for the quantification of 3-MCPD at concentrations of $\geq 10\,\mu g\,kg^{-1}$ and was adopted as an AOAC First Action Official Method. The three main advantages of this approach were: (i) high sensitivity resulting from the formation of a GC volatile 3-MCPD derivative, (ii) high specificity associated with mass spectrometric detection at higher mass, and (iii) accurate quantification from the use of a stable isotope internal standard.

The procedure of Brereton *et al.* (46) has subsequently been modified and extended to the analysis of MCPDs and/or DCPs. Nyman *et al.* (48) optimized a method for the analysis of 1,3-DCP in soy sauces using 10% diethyl ether/hexane to isolate the dichloropropanol at the solid-phase extraction stage. The mean recovery of 1,3-DCP from spiked test samples was 100% with a relative standard deviation of 1.32%: the method limits of quantification (LOQs) was $0.185\,\mu g\,kg^{-1}$. Xu *et al.* (49) added hexane at the initial aqueous extraction stage to remove fat from samples and also reported improved sensitivity and selectivity over electron ionization (EI)-MS when using negative ion chemical ionization MS. The authors also showed that the derivatization of DCPs with HFBA was more effective when a catalytic amount of triethylamine was added. Chung *et al.* (50) used silica gel columns with ethyl acetate as the elution solvent for the solid-phase extraction step in the analysis of soy sauces. Abu-El-Haj *et al.* (51) used dichloromethane as the chloropropanol elution solvent: aliquots of aqueous samples (e.g., soy sauce and soups) were pre-absorbed onto alumina columns while dry samples (e.g., crumbed cereal products) were eluted directly following a lipid removal step with hexane. Chung *et al.* (50) and Abu-El-Haj *et al.* (51) both used HFBA for the derivatization step and the LOQs for these methods were 3–5 $\mu g\, kg^{-1}$.

The EI mass spectra of the HFB esters of 3-MCPD, 1,3- and 2,3-DCP, and the characteristic ions for all chloropropanols are given in Fig. 2.6.4 and Table 2.6.1, respectively.

2.6.3.1.2 Dioxaborolane/Dioxaborinane Derivatives Boronic acids react with 1,2- and 1,3-diols to give cyclic dioxaborolane/dioxaborinane derivatives. Rodman and Ross (52) carried out derivatization under anhydrous conditions while Pesselman and Feit (53) showed that a boronic acid derivative of 3-MCPD could be prepared under aqueous conditions, despite the liberation of water from the formation reaction (see Fig. 2.6.5). Hence, this approach has been adopted for the analysis of liquid seasonings such as HVP and soy sauces (23, 54). Since boronic acids react specifically with diol compounds, other chloropropanols such as 1,2- and 1,3-DCP cannot be determined by this procedure.

Breitling-Utzmann *et al.* (55) and Divinová *et al.* (56) prepared boronic acid derivatives of MCPDs extracted from a wide variety of retail foodstuffs. Breitling-Utzmann *et al.* used saline solution for both extraction and derivatization, while Divinová *et al.* used hexane/acetone (1:1 v/v) for the initial extraction followed by a partition step into saline solution. Derivatization was carried out in the saline solution using phenylboronic acid (PBA) and the reaction products were extracted into hexane. Quantification was by a stable isotope internal standard method using 3-MCPD-d_5 and GC/MS operating in selected ion monitoring (SIM) mode. Typical LOQs ranged from 10 $\mu g\, kg^{-1}$ (e.g., soy sauce) to 50 $\mu g\, kg^{-1}$ (e.g., toasted bread). Velíšek *et al.* (57) used a Chiraldex G-TA fused silica capillary column to separate the optical isomers of 3-MCPD prior to quantification by MS. Kuballa and Ruge (58) showed that lower LOQs could be achieved using GC combined with tandem mass spectrometry (MS/MS) on a triple quadrupole mass spectrometer operating in the selected reaction monitoring mode. In a recent publication, Huang *et al.* (59) demonstrated that the PBA derivative of 3-MCPD formed in liquid HVPs could be extracted and quantified at the low $\mu g\, kg^{-1}$ level using headspace solid-phase microextraction (SPME) combined with GC/MS detection.

The EI mass spectrum of the PBA derivative of 3-MCPD and the characteristic ions for all MCPDs are given in Fig. 2.6.6 and Table 2.6.2, respectively.

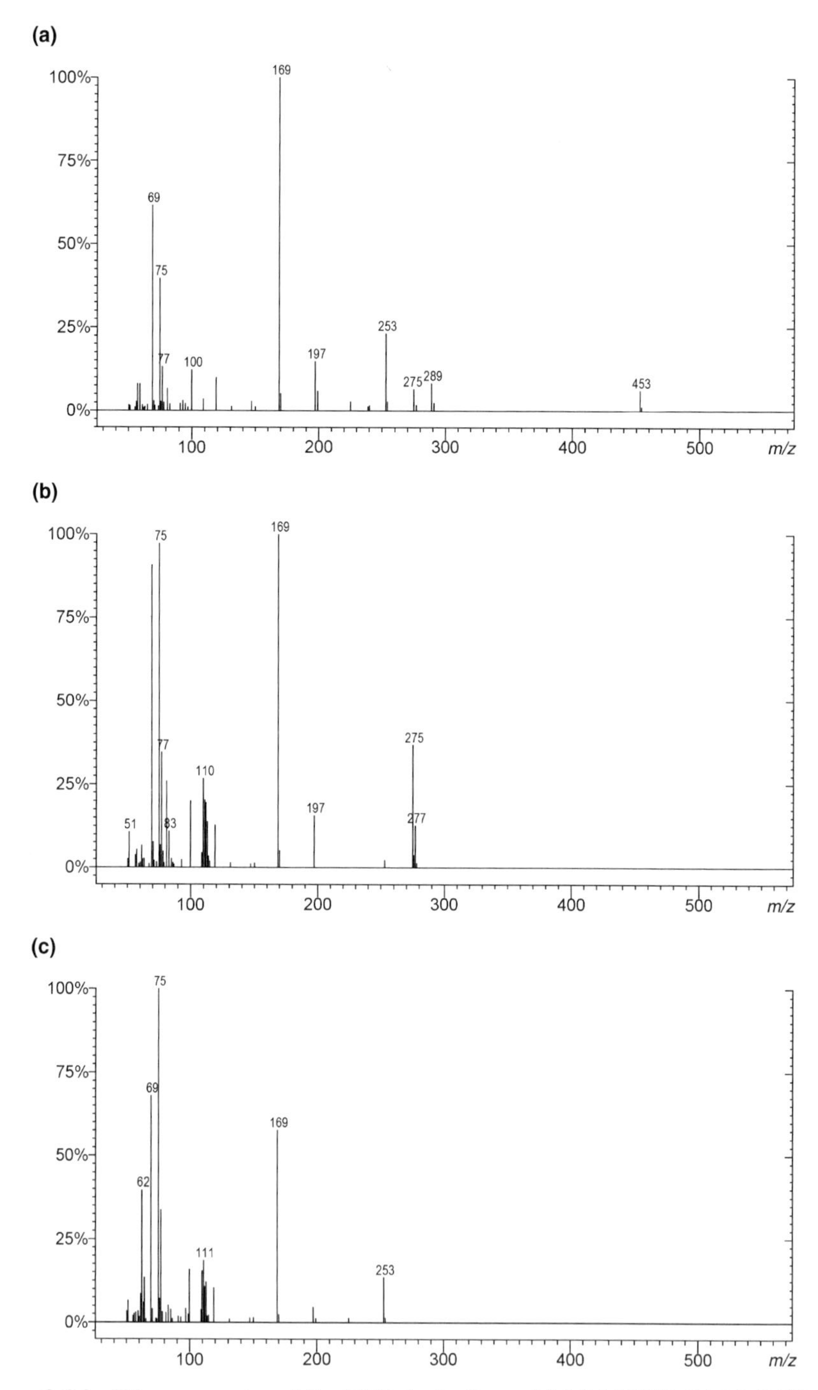

Figure 2.6.4 EI mass spectra of the HFB derivatives of (a) 3-MCPD, (b) 1,3-DCP, and (c) 2,3-DCP (data courtesy of RHM Technology, UK).

TABLE 2.6.1 Characteristic ions (*m/z*) in the mass spectra of the HFB esters of chloropropanols.

(a) EI mass spectra

HFB ester of:	MW	$[M-CH_2Cl]^+$	$[M-C_3F_7CO_2]^+$	$[M-C_3F_7CO_2CH_2]^+$	$[M-C_3F_7CO_2-HCl]^+$	$[M-C_3F_7CO_2-C_3F_7CO_2H]^+$
3-MCPD	502	453	289/291[a]	275/277[a]	253	75/77[a]
3-MCPD-d_5	507	456	294/296[a]	278/280[a]	257	79/81[a]
2-MCPD	502	—	289/291[a]	—	253	75/77[a]
1,3-DCP	324	275/277[a]	111/113/115[a]	—	75/77[a]	—
1,3-DCP-d_5	329	278/280[a]	116/118/120[a]	—	79/81[a]	—
2,3-DCP	324	—	111/113/115[a]	—	75/77[a]	—

(b) NCI mass spectra

HFB ester of:	MW	$[M-HF]^-$	$[M-HF-HCl]^-$	$[M-HF-HCl-HF]^-$
3-MCPD	502	482/484[a]	446	426
3-MCPD-d_5	507	486/488[a]	449	428
2-MCPD	502	482/484[a]	446	426
1,3-DCP	324	304/306/308[a]	268/270[a]	248/250[a]
1,3-DCP-d_5	329	308/310/312[a]	271/273[a]	250/252[a]
2,3-DCP	324	304/306/308[a]	268/270[a]	248/250[a]

[a]Isotopic chlorine cluster ions.

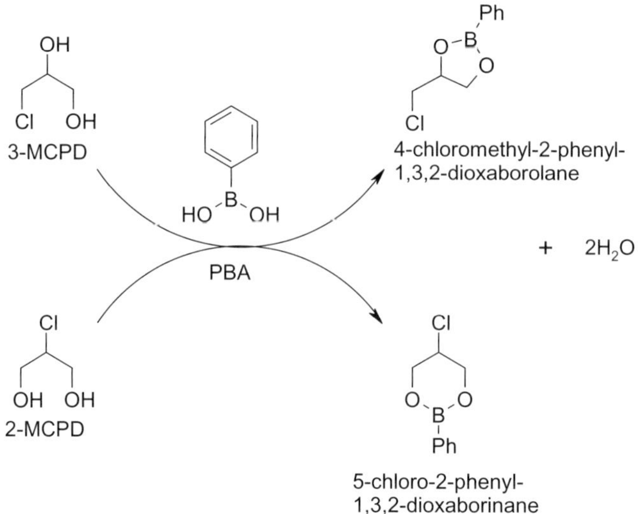

Figure 2.6.5 Derivatization reactions of MCPDs with PBA.

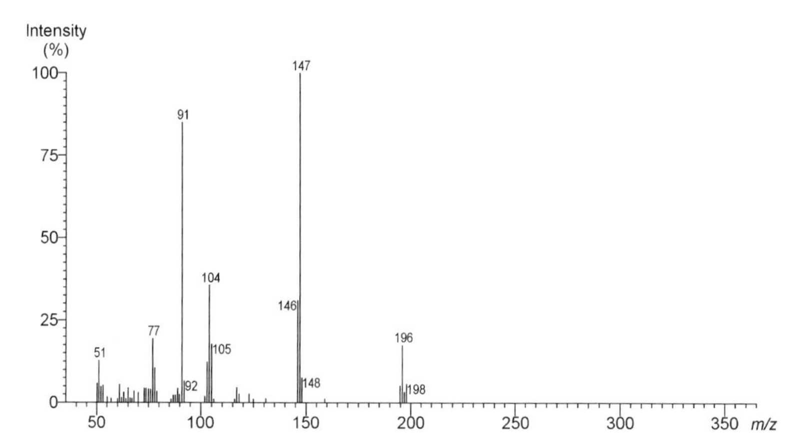

Figure 2.6.6 EI mass spectrum of the PBA derivative of 3-MCPD (data courtesy of RHM Technology, UK).

TABLE 2.6.2 Characteristic ions (*m/z*) in the EI mass spectra of the PBA derivatives of MCPDs.

PBA derivative of:	MW	$[M]^{+\cdot}$	$[M\text{-}CH_2Cl]^+$	Other structurally significant ions
3-MCPD	196	196/198[a]	146/147[b]	103/104[b] $[Ph\text{-}BO]^{+\cdot}$, 91 $[C_7H_7]^+$
3-MCPD-d_5	201	201/203[a]	149/150[b]	103/104[b] $[Ph\text{-}BO]^{+\cdot}$, 93 $[C_7H_5D_2]^+$
2-MCPD	196	196/198[a]	—	103/104[b] $[Ph\text{-}BO]^{+\cdot}$, 91 $[C_7H_7]^+$

[a]Isotopic chlorine cluster ions.
[b]Isotopic boron cluster ions.

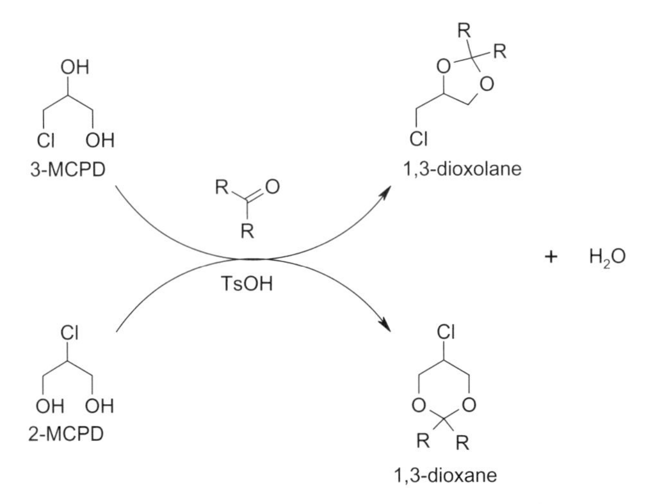

Figure 2.6.7 Derivatization reactions of MCPDs with ketones.

2.6.3.1.3 ***Dioxolane/Dioxane Derivatives*** The reaction of 1,2 and 1,3 diols with aldehydes and ketones to give cyclic acetals and ketals is well known (60), and Meierhans *et al.* (18) were the first to report a quantitative method for 2- and 3-MCPD using dry acetone as the derivatizing reagent in the presence of toluene-4-sulfonic acid (TsOH) according to the reaction scheme given in Fig. 2.6.7. The resulting 1,3-dioxolane and 1,3-dioxane derivatives were characterized by GC/MS. MCPDs were partitioned from aqueous samples/extracts using solid-phase extraction on diatomaceous earth and diethyl ether elution described previously. In keeping with the HFB reagents, anhydrous conditions are required for derivatization.

Subsequent modifications to the procedure of Meierhans *et al.* (18) include the addition of MCPD-d_5 to samples prior to extraction for quantification (61–63), the use of 3-pentanone or 4-heptanone/TsOH for derivatization (61, 63), and the use of ethyl acetate for the solid-phase partition step and additional cleanup post derivatization using basic aluminum oxide cartridges (63). The LOQs for these methods are in the range 1–5 $\mu g\,kg^{-1}$.

Although the occurrence of 2-MCPD has been reported in some foodstuffs (63), no mass spectral data have been given for the acetone derivative (dioxane) of this isomer. These methods have not been used for the determination of dichloropropanols since these compounds do not form cyclic derivatives with ketones. The EI mass spectra of the acetone derivative of 3-MCPD and the characteristic ions for all MCPDs are given in Fig. 2.6.8 and Table 2.6.3, respectively.

2.6.3.1.4 ***Other Methods of Interest*** The higher volatility and lower polarity of dichloropropanols, compared with MCPDs, has permitted the direct determination of 1,3-DCP using GC/MS, i.e., without derivatization (64).

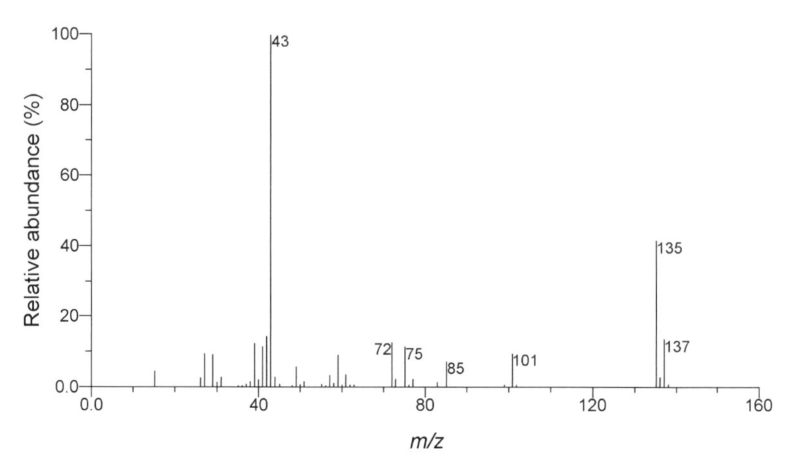

Figure 2.6.8 EI mass spectrum of the acetone derivative of 3-MCPD (data courtesy of RHM Technology, UK).

TABLE 2.6.3 Characteristic ions (*m/z*) in the EI mass spectra of the dioxolane derivatives of MCPDs.

Chloropropanol	Dioxolane of:	MW	$[M\text{-}C_nH_{2n+1}]^+$	Other structurally significant ions
3-MCPD	Acetone	150	135/137[a]	43 $[C_2H_3O]^+$
3-MCPD-d_5	Acetone	155	140/142[a]	43 $[C_2H_3O]^+$
3-MCPD	3-Pentanone	178	149/151[a]	57 $[C_3H_5O]^+$
3-MCPD-d_5	3-Pentanone	183	154/156[a]	57 $[C_3H_5O]^+$
3-MCPD	4-Heptanone	206	163/165[a]	71 $[C_4H_7O]^+$
3-MCPD-d_5	4-Heptanone	211	168/170[a]	71 $[C_4H_7O]^+$

[a]Isotopic chlorine cluster ions.

Schumacher *et al.* (65) extracted 1,3-DCP from river water samples using ethyl acetate, while Crews *et al.* (66) showed that 1,3-DCP was sufficiently volatile to be quantified in the headspace of soy sauces. Both authors used GC/MS to monitor ions at *m/z* 79 and *m/z* 81 for 1,3-DCP and *m/z* 82 for the stable isotope internal standard 1,3-DCP-d_5. The procedure of Crews *et al.* (66) was subsequently validated by a collaborative trial in which participants used both static headspace and SPME, and the method was shown to be rapid, accurate, and fit for purpose (67) with a method limit of detection (LOD) of $\leq 10\,\mu g\,kg^{-1}$.

Although the formation of trimethylsilyl ethers of chloropropanols has previously been used for the analysis of MCPDs and DCPs in HVPs (64) and resin-treated papers (29), recent applications have been limited to the analysis of soy sauces (68).

Leung *et al.* (69) recently demonstrated the feasibility of using molecularly imprinted polymers (MIP) as a qualitative tool for the screening of food products for 3-MCPD. An MIP derived from 4-vinylphenylboronic acid was shown to act as a potentiometric chemosensor via the increase in Lewis acidity of the receptor sites, upon reaction of the aryl boronic acid with 3-MCPD (see Fig. 2.6.5). A simple pH glass electrode was sufficient to monitor the analyte-specific binding, and in water a linear response was obtained over 0–350 mg kg^{-1}.

Xing and Cao (70) described a capillary electrophoresis technique with electrochemical detection for the rapid analysis of 3-MCPD in soy sauces. The linear range for of the method was 6.6–200 mg L^{-1} with an LOD of 0.13 mg L^{-1}.

2.6.3.2 Chloroesters

There are fewer methods for the analysis of chloroesters. Hamlet *et al.* (71), Hamlet and Sadd (72), and Zelinková *et al.* (73) developed methods for the direct analysis of esters of 3-MCPD in cereal products and edible oils, based on an adaptation of the earlier procedure of Davídek *et al.* (7). Fat containing the chloroesters was extracted from samples using either ethyl acetate or diethyl ether and separated by preparative TLC into individual fractions containing the diesters and monoesters of 3-MCPD. The individual fractions were analyzed by GC/MS, and the mono- and diesters of MCPDs were identified by comparison with synthetic 3-MCPD-palmitate esters and reference mass spectral data (7, 74). All MCPD-esters were quantified as 3-MCPD-dipalmitate using 5-α-cholestane as an internal standard. No recent methods for the analysis of monoesters of 1,3- and 2,3-DCP have been reported.

The isolation and measurement of all chloroesters by these methods is a lengthy process due to the many species arising from the different fatty acid combinations associated with each chloropropanol moiety. Chloroesters may be hydrolyzed to the corresponding toxic chloropropanol, either during food processing or *in vivo*. Consequently, methods have been devised to measure the total amount of chloropropanol, i.e., free and ester derived.

Hamlet *et al.* (71) and Hamlet and Sadd (72) measured total MCPDs in dried cereal products, i.e., free and esterified with fatty acids. The MCPDs esterified with fatty acids were released by enzyme hydrolysis following incubation in 0.1 M phosphate buffer (pH 7.0) with a commercial lipase from *Aspergillus oryzae* for 24 h (see Fig. 2.6.9). Total MCPDs were determined as the HFB-esters by GC/MS as described previously (see Section 2.6.3.1.1). The recovery of 3-MCPD from 3-MCPD-dipalmitate reference standard was 106% and 91% at 250 and 750 μg kg^{-1} (sample basis), respectively. The repeatability of the method, expressed as a coefficient of variation, was 3.7% and the LOQ was <10 μg kg^{-1}. Divinová *et al.* (56) developed a method to measure total 3-MCPD in a wide range of foodstuffs. Free and esterified 3-MCPD was isolated from samples as fat using diethyl ether as the extraction solvent. The solvent-free fat extract was then subjected to methanolysis using methanolic sulfuric acid and the 3-MCPD generated was quantified as the PBA derivative using

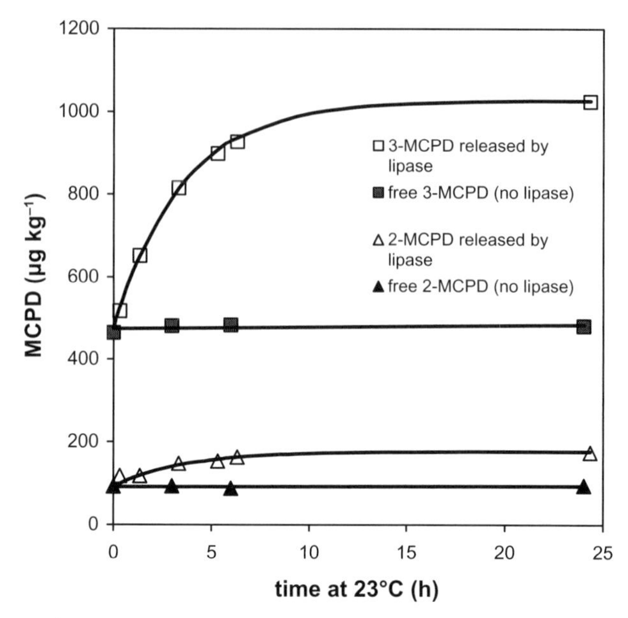

Figure 2.6.9 Release of MCPDs from MCPD-esters in bread crust by enzyme hydrolysis.

GC/MS as described previously (see Section 2.6.3.1.2). The LOD and LOQ were determined to be 1.1 and 3.3 mg kg^{-1}, respectively, on the extracted fat basis. Samples and extracts prepared by these methods were not analyzed for the corresponding DCPs.

2.6.4 OCCURRENCE

2.6.4.1 Chloropropanols

2.6.4.1.1 Retail Foodstuffs The chloropropanol 3-MCPD has been found in a wide range of retail ready-to-eat processed foods including cereal, coffee, cheese, licorice, fish, and meat, suggesting that formation may occur by a range of mechanisms/processes (see Table 2.6.4). Reassuringly, the amounts of 3-MCPD are much lower than those reported for some soy sauces.

Thermally processed products (e.g., cereals) account for the greatest incidence of 3-MCPD with some of the highest amounts found in products attaining high temperatures (e.g., bread crust and toasted bread). The mechanisms of formation of 3-MCPD in thermally processed cereal products are discussed in Section 2.6.5.

Low concentrations of 3-MCPD have also been found in products not subjected to high-temperature treatments, such as cheese, salami, and cold smoked fish. Some possible occurrence routes for 3-MCPD in cheese and salami

TABLE 2.6.4 Retail, ready-to-eat foodstuffs with quantifiable amounts of 3-MCPD.

Foodstuff	Incidence[a]	Mean[b], μg kg^{-1}	Range, μg kg^{-1}	Reference no.
Cereal products				
Breads	14/27	12	<10–49	75
Bread	6/9	23	<10–76	55
Crust	9/9	91	24–275	55
Toast	26/26	214	30–679	55
Toast	10/10	136	20–322	76
Cake, fruit	8/10	78	<10–210	77
Crackers/toasts	30/34	38	<10–134	75
Doughnuts	5/5	18	11–24	75
Rusks	10	21	<10–48	77
Sweet biscuits	5/19	5	<10–32	75
Cheese	4/30	8	<10–31	75
Cheese	11/105	8	<10–95	77
Cheese	4/9	12	<10–37	66
Cheese	3/3	42	13–83	78
Coffee	11/15	12	<9–19	79
Fish				
Anchovies, in olive oil	2/2	48	15–81	75
Crumbed	5/6	37	<5–83	80
Smoked fish	6/8	37	<10–191	77
Licorice	2/2	22	20–23	77
Meats				
Bacon	3/6	11	<5–22	80
Beef burger/hamburger	5/7	25	<10–71	75
Beef burger/hamburger	6/6	15	7–49	77
Salami	9/20	12	<10–69	75
Salami	2/2	31	14–48	78
Sausages	3/6	16	<5–69	80
Smoked meats				
Bacon	2/10	11	<10–47	75
Fermented sausages/ smoked ham	29/33	43	<5–74	81

[a]Number of samples above the limit of detection.
[b]Derived value from reported data (one-half of the reported limit of detection value used to calculate the mean).

include enzymatic release of 3-MCPD from chloroesters and/or migration from epichlorohydrin containing resin-treated food contact materials (see Section 2.6.4.1.3). In the case of smoked products, the occurrence of 3-MCPD in smoke generated from the wood during the process and from liquid smoke ingredients may contribute to the amounts found in these products (81). High concentrations of 3-MCPD, up to 1150 μg kg^{-1}, have been reported in uncooked convenience chicken products, e.g., chicken nuggets and crumb dressed chicken

TABLE 2.6.5 Occurrence data for chloropropanols in raw meats.[a]

Sample	Number of samples	3-MCPD		1,3-DCP	
		Incidence[b]	Range, $\mu g\, kg^{-1}$	Incidence	Range, $\mu g\, kg^{-1}$
Beef, minced	10	0/10	<5	9/10	<3–110
Beef, steak	5	0/5	<5	1/5	70
Sausages	10	3/10	<5–13	7/10	<3–69
Ham, leg	5	2/5	<5–27	2/5	<3–21
Lamb, chops	6	0/6	<5	1/6	91

[a]Adapted from FSANZ (80).
[b]Number of samples above the limit of detection.

breast (81). The authors offered the suggestion that these products may have been treated with highly contaminated protein hydrolysates (to increase water-binding capacity), although no further details were given.

Although information on other chloropropanols in retail foods is scarce at present, data from the recent EC SCOOP task (77) indicate that the incidence and concentrations of 2-MCPD (8/115), 1,3-DCP (0/42), and 2,3-DCP (1/28) in retail foods are likely to be very low. However, results from a recent survey of food products carried out by Food Standards Australia New Zealand (80) revealed that, unlike soy sauces, 1,3-DCP may be present in food without the presence of 3-MCPD or at amounts substantially higher than those of 3-MCPD. These foods were all raw meats (see Table 2.6.5) and the concentration of 1,3-DCP decreased when the samples were cooked. A subsequent survey of 28 raw meats from retail outlets in the United Kingdom (82) did not find any quantifiable amounts of 3-MCPD or 1,3-DCP, and to date, the occurrence route for 1,3-DCP in the raw meats from the 2003 FSANZ survey remains unknown.

2.6.4.1.2 Commercial Food Ingredients Survey data on the occurrence of 3-MCPD in commercial food ingredients are given in Table 2.6.6. Occurrence data for other chloropropanols are very limited at present.

Processed garlic accounted for the highest incidence and concentration, with 20/21 samples containing between 5 and 690 $\mu g\, kg^{-1}$ 3-MCPD. The mechanism of formation of 3-MCPD in these products is discussed in Section 2.6.5.

The presence of 3-MCPD in commercial smoke flavorings has recently been reported by Kuntzer and Weisshaar (81) in the concentration range 200–760 $\mu g\, kg^{-1}$. Based on these findings, the estimated contribution to, for example, smoke flavored sausages was 9 $\mu g\, kg^{-1}$. The occurrence of 3-MCPD was widespread in malt products and could be attributed to the additional heat treatments to which these malts had been subjected to produce the desired colors and flavors (84). Although processing details were not available for the modified starches, the two samples with quantifiable amounts of 3-MCPD were both maize yellow dextrins. These samples can be prepared by heat treatment

TABLE 2.6.6 Commercial food ingredients (excluding HVPs/flavorings) with quantifiable amounts of 3-MCPD.

Food ingredient	Incidence[a]	Mean[b], $\mu g\,kg^{-1}$	Range, $\mu g\,kg^{-1}$	Reference no.
Bread crumbs	1/6	7	<10–14	83
Garlic, processed	20/21	106	<10–690	77
Liquid smokes	6/6	—[c]	200–760	81
Malts	31/63	96	<10–850	77
Meat extracts	1/5	7	<10–14	83
Modified starches	2/9	59	<10–488	77

[a]Number of samples above the limit of detection.
[b]Derived value from reported data (one-half of the reported limit of detection value used to calculate the mean).
[c]Data not available.

in the presence of a mineral acid, a process that can produce chloropropanols by a mechanism analogous to that of acid-HVP by Hamlet *et al.* (83).

2.6.4.1.3 *Other Sources of Chloropropanols in Foodstuffs* Epichlorohydrin copolymers with polyamines and/or polyamides are used to provide wet strength to food contact papers. Typical applications include tea bag paper, coffee filters, absorbents packaged with meats, and cellulose casings (for ground meat products such as sausage). These resins can accumulate chloropropanol-forming species during storage, resulting in contamination of foodstuffs by migration of chloropropanols from treated papers (85). Concentrations of 3-MCPD ranging from 0.005 to 219 $mg\,kg^{-1}$ (mean 13.9 $mg\,kg^{-1}$) were found in 7 of 16 edible sausage casings reported in the EC data compilation on chloropropanols in foods (77).

1,3-DCP up to 1000 $mg\,kg^{-1}$ has also been reported in dimethylamine-epichlorohydrin copolymer (DEC) used at concentrations of up to 150 $ng\,g^{-1}$ by weight of sugar solids in sugar refining (85). DEC is used as a flocculent or decolorizing agent for sugar liquors. It is also used to immobilize glucose isomerase enzymes for production of high-fructose corn syrup (85).

Chloropropanols have been found in epichlorohydrin polyamine polyelectrolytes used in drinking water treatment chemicals as coagulation and flocculation products (85), and low concentrations of 3-MCPD have been found in finished water from flocculent use in the United Kingdom (86). In the United Kingdom, the Drinking Water Inspectorate (87) concluded that limiting the dosing rate of the flocculent to no more than 2.5 $mg\,L^{-1}$ drinking water would indirectly regulate concentrations of the chloropropanols 1,3-DCP, 2,3-DCP, and 3-MCPD.

Kuntzer *et al.* (81) isolated 3-MCPD from smoke generated by heating commercial food grade wood pellets; hence, foods in contact with smoke could become contaminated with 3-MCPD. Dichloropropanols were not detected in the generated smoke.

2.6.4.1.4 Other Compounds of Interest Rétho and Blanchard (63) identified 3-bromopropane-1,2-diol and 2-bromopropane-1,3-diol at concentrations of 35, 45, and 7 μg kg^{-1} in grape seed, rapeseed, and sesame seed oils (total of both isomers), respectively. In all cases, it was a mixture (about 2.5 : 1.0) of the 3-bromo and 2-bromo isomers.

2.6.4.2 Chloroesters

At present, data on chloroesters in foods, other than HVPs (see Chapter 6.2), are scarce and confined to the occurrence of esters of 3-MCPD. However, recent findings indicate that the formation of 3-MCPD-esters (monoesters and diesters with higher fatty acids) may be widespread in processed foods derived from cereals, potatoes, meat, fish, nuts, and oils (71–73, 78, 79). These compounds represent a new class of food contaminants, which might release 3-MCPD into foods during processing and storage or possibly *in vivo* as a result of lipase-catalyzed hydrolysis reactions. Table 2.6.7 shows the amounts of

TABLE 2.6.7 Concentrations of 3-MCPD-esters in foodstuffs quantified as 3-MCPD.

Foodstuff	Number of samples	Mean, μg kg^{-1}	Range, μg kg^{-1}	Reference no.
Bread	1	6.7	–	71, 72
Crumb	1	4.9	–	
Crust	1	547	–	
Toast	7	86[a]	60–160	
Crispbread	1	420	–	78
Coffee	15	140[a]	<100–390	79
Cracker	1	140	–	78
DATEM	1	66	–	71, 72
Doughnut	1	1210	–	78
French fries	1	6100	–	78
Malt, dark	1	580	–	78
Nuts, roasted	3	1370	433–500	73
Oils				73
Virgin seed oils	9	63	<100–<300	
Roasted	1	337	–	
Virgin olive oils	4	75	<100–<300	
Virgin germ oils	2	100	<100–<300	
Refined seed oils	5	524	<300–1234	
Refined olive oils	5	1464	<300–2462	
Pickled herring	1	280	–	78
Salami	1	1760	–	73
Wheat flour	1	<5	–	71, 72

[a]Derived value from reported data (one-half of the reported limit of detection value used to calculate the mean).

3-MCPD-esters, measured as 3-MCPD released by hydrolysis of the esters, in a wide range of foods. In many cases, the level of 3-MCPD-esters exceeded that of the free 3-MCPD.

Hamlet *et al.* (71) and Hamlet and Sadd (72) measured 3-MCPD-esters in bread and toast. The highest amounts were found in regions of the bread that attained the highest temperature (i.e., the crust) and concentrations increased from 60 to 160 $\mu g\,kg^{-1}$ when the bread was toasted over 40–120 s. The highest level of 3-MCPD-esters (6100 $\mu g\,kg^{-1}$) was found in a sample of French fries (78). The level of 3-MCPD-esters in roast coffee was relatively low and varied between 6 $\mu g\,kg^{-1}$ (soluble coffees) and 390 $\mu g\,kg^{-1}$ (decaffeinated coffee), although it exceeded the free 3-MCPD level by a factor of 8 to 33 times (79). The presence of 3-MCPD-esters in bread crumb (71, 72), pickled olives, and herrings suggests that these compounds can also form at relatively low temperatures and even in acid media (78).

Zelinková *et al.* (73) analyzed 25 retail virgin and refined edible oils for the content of free 3-MCPD and 3-MCPD-esters (as 3-MCPD released from its esters with higher fatty acids). The oils contained free 3-MCPD ranging from $<3\,\mu g\,kg^{-1}$ (LOD) to 24 $\mu g\,kg^{-1}$. Surprisingly, the 3-MCPD-ester level was much higher and varied between $<100\,\mu g\,kg^{-1}$ (LOD) and 2462 $\mu g\,kg^{-1}$. Generally, virgin oils had relatively low quantities of 3-MCPD-esters ranging from $<100\,\mu g\,kg^{-1}$ (LOD) to $<300\,\mu g\,kg^{-1}$ (LOQ). Higher amounts of 3-MCPD-esters were found in oils obtained from roasted oilseeds (337 $\mu g\,kg^{-1}$) and in the majority of refined oils (<300 to 2462 $\mu g\,kg^{-1}$), including refined olive oils. In general, it appeared that the formation of 3-MCPD-esters in oils was linked with the preliminary heat treatment of oilseeds and with the process of oil refining. The analysis of crude, degummed, bleached, and deodorized rapeseed oil showed that the level of MCPD-esters decreased during the refining process. However, additional heating of seed oils for 30 min at temperatures ranging from 100 to 280 °C, and heating at 230 and 260 °C for up to 8 h, led to an increase of the level of 3-MCPD-esters. Conversely, heating olive oil resulted in a decrease in the 3-MCPD-ester level.

Analysis of fat isolated from a salami containing 1670 $\mu g\,kg^{-1}$ of 3-MCPD-esters mainly consisted of 3-MCPD-diesters together with lesser amounts of 3-MCPD-monoesters (73). The major types of 3-MCPD-diesters (about 85%) were mixed diesters of palmitic acid with C18 fatty acids (stearic, oleic, linoleic acids), 3-MCPD distearate (11%), and 3-MCPD dipalmitate (4%). The level of 3-MCPD as the free compound in the fat extract was 31 $\mu g\,kg^{-1}$.

2.6.5 FORMATION

2.6.5.1 From Hydrochloric Acid and Glycerol/Acylglycerols

In the laboratory, MCPDs and DCPs can be prepared by the action of concentrated hydrochloric acid or dry hydrogen chloride gas on glycerol alone, or

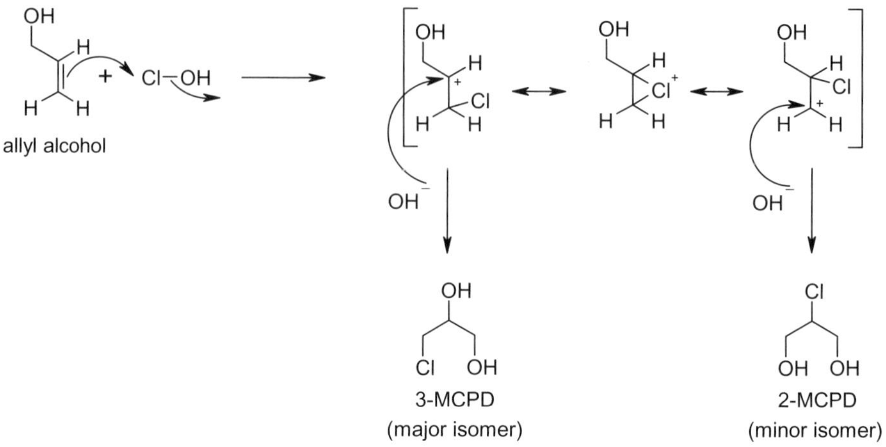

allyl alcohol

3-MCPD
(major isomer)

2-MCPD
(minor isomer)

Figure 2.6.10 Formation of MCPDs from allyl alcohol and hypochlorous acid.

in the presence of glacial acetic acid (88, 89). These reactions require prolonged heating at temperatures of about 100 °C and are applicable to the formation of chloropropanols in acid-HVP (3, 11) discussed in Chapter 6.2.

2.6.5.2 From Hypochlorous Acid and Allyl Alcohol

Hypochlorous acid (HOCl, from chlorine and water) will add to the double bond of allyl alcohol (2-propenol) to give 2- and 3-MCPD in accordance with Markovnikov's rules governing carbenium ion stability (see Fig. 2.6.10). The reaction is reported to proceed rapidly at 50–60 °C with an 88% yield of MCPDs (90). Sources of allyl alcohol include (S)-allyl-L-cysteine sulfoxide (alliin), a cysteine amino acid found in garlic and related species (91) and a soft cheese made from ewe's milk (92). This mechanism is believed to account for the concentrations of 3-MCPD found in processed garlic (45).

2.6.5.3 From Sodium Chloride and Glycerol/Acylglycerols

2.6.5.3.1 Model System Studies Hasnip *et al.* (93), Dolezal *et al.* (94), and Calta *et al.* (95) demonstrated that 3-MCPD could be formed by heating sodium chloride with emulsified (Tween 80) glycerol/acylglycerol precursor mixtures in sealed vials. The generation of 3-MCPD increased with increasing sodium chloride concentration and maximum formation occurred at a water content between 10% and 20%. The relative amounts of 3-MCPD (relative to monoacylglycerols, mol mol^{-1}) formed by heating sealed samples at 200 °C for 30 min were 1.0 (monoacylglycerols) < 1.1 (triacylglycerols) < 1.6 (glycerol) < 1.8 (diacylglycerols) < 3.3 (lecithin).

2.6.5.3.2 Baked Cereal Products Hamlet (96) developed and validated a model system to determine the major precursors, intermediates, and the mechanisms of formation of MCPDs in bread. Free glycerol was shown to be the major precursor of MCPDs in leavened dough (97). This glycerol was generated by yeast during proving and its subsequent reaction with added chloride during baking could account for approximately 70% of the MCPDs formed. The production of glycerol by yeast in dough was dependent upon time and temperature and was limited by available sugar. The addition of glucose promoted MCPD generation in dough but not via increased glycerol production by yeast. The promoting effect of glucose was subsequently shown to be due to the removal of potential amino inhibitors (e.g., amino acids) via the Maillard reaction (98). Breitling-Utzmann *et al.* (99) found that the addition of glucose and sucrose to bread dough promoted 3-MCPD formation during toasting of bread slices, presumably by a similar mechanism. The generation of MCPDs in breads also showed strong moisture dependence: levels increased with decreasing dough moisture to a point where formation was limited by the solubility of chloride and competing reactions involving glycerol and the reaction intermediate, glycidol (97), shown in Fig. 2.6.11.

Minor precursors of MCPDs (together with chloride) in dough were found to be monoacylglycerols, lysophospholipids, and phosphatidylglycerols present at significant concentrations in white flour and germ, together with DATEM (diacetyl tartaric acid esters of monoacyl glycerols), an emulsifier used in bread making (100). These compounds could account quantitatively for the remaining contribution (30%) to MCPD in bread. A characteristic of these precursors was the proximity of —OH with respect to either acyl- or phosphoryl- groups on the glycerol skeleton. It was concluded that (i) the subsequent reaction with chloride ion involved a neighboring group mechanism, and (ii) the removal of the acyl group from the resulting MCPD-ester

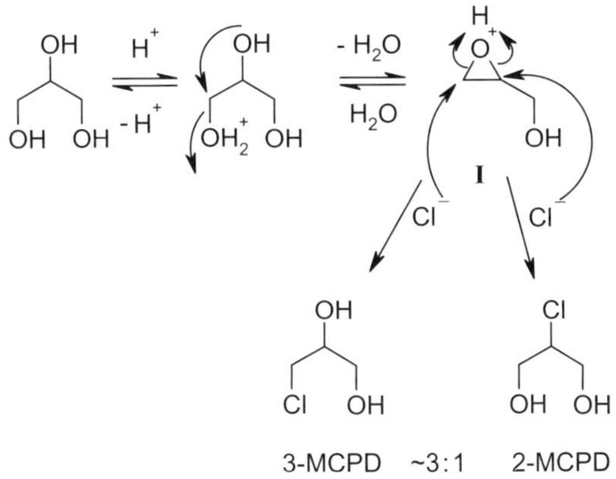

Figure 2.6.11 Formation of MCPDs from glycerol via the intermediate epoxide, glycidol (I).

intermediate was rate determining. These mechanistic features could explain the relatively nonreactive nature of the di- and triacylglycerols with respect to MCPD generation. The generation of MCPDs in bread was pH dependent and increased with decreasing pH. This effect was not due to reduced degradation of MCPDs at reduced pH (see Section 2.6.5.6) but attributed to an increased rate of hydrolysis of intermediate MCPD-esters (98).

Hamlet *et al.* (101) studied the formation of 3-MCPD (and other process contaminants) in bread prepared in domestic bread machines in the United Kingdom. It is usual for such domestic machines to employ extended proof times (e.g., 2–3 h) compared with commercial bread production (e.g., 50 min). This extended proof leads to increased yeast activity and hence higher glycerol precursor concentrations in the dough (97). Despite these conditions, MCPD generation was found to be only slightly higher than that of commercial products. An explanation for this is the lower baking temperatures used in domestic bread machines.

Evidence for the formation of MCPD-ester intermediates in bread was provided by Hamlet *et al.* (71) and Hamlet and Sadd (72), who showed that chloroester formation was correlated with MCPD generation and that amounts of both species increased on heating (see Fig. 2.6.12). The presence of low levels of chloroesters in bread crumb, i.e., at temperatures <100 °C, illustrates that these compounds may form readily from partial acylglycerols, presumably

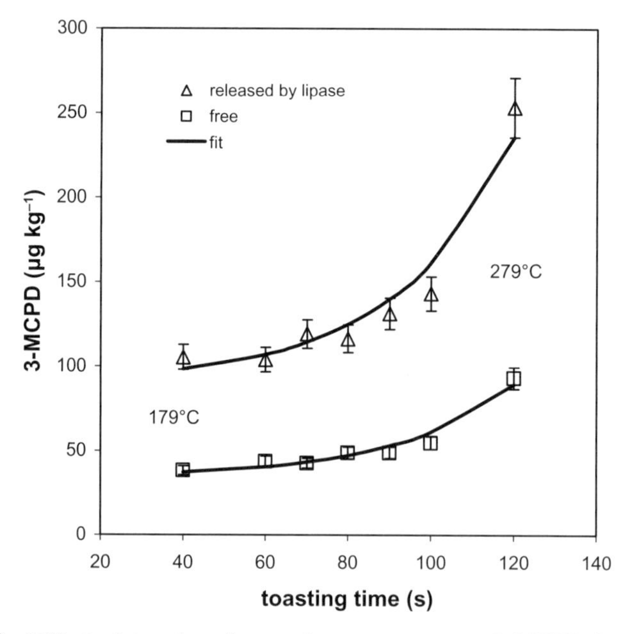

Figure 2.6.12 Effect of toasting time and temperature on 3-MCPD levels in bread: the generation of lipase released 3-MCPD (from chloroesters) and free 3-MCPD was correlated ($r^2 = 0.94$).

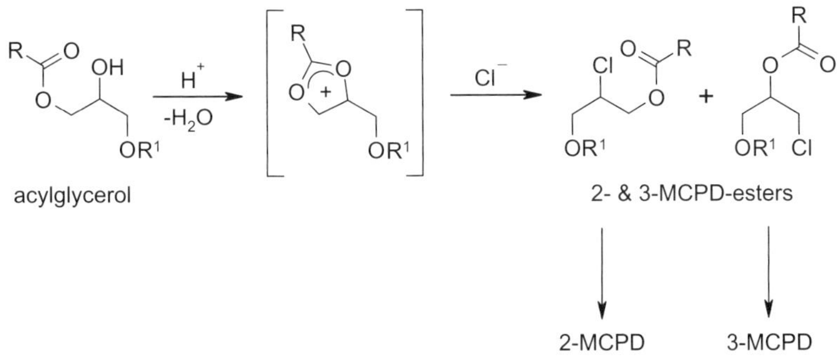

Figure 2.6.13 Proposed mechanism of formation of MCPD-esters from mono- and di-acylglycerols (R = alkyl, R[1] = H or COR).

as a consequence of facile cyclic acyl oxonium ion formation and subsequent ring opening by chloride ion (see Fig. 2.6.13).

2.6.5.3.3 Formation in Malts and Roasted Cereals Studies on the formation of chloropropanols in malts (84) have shown that 3-MCPD can form when malted or unmalted barley is dry roasted (higher than 170 °C), and that endogenous lipids and glycerol/acylglycerols in the grain are sufficient to promote this synthesis. The 3-MCPD produced during roasting was correlated with color development, while extended heating at temperatures of 200 °C or greater gave a reduction in 3-MCPD, presumably as a result of degradation (84).

2.6.5.4 Mechanisms Involving Enzymes

The formation of MCPDs has been demonstrated in model systems comprising pepper, oil, and sodium chloride, and over a period of several days (102). As spices are known to be a source of enzyme activity, this was subsequently shown to be a general reaction of a lipase in the presence of chloride and lipid. The highest yield of 3-MCPD was obtained in reaction mixtures containing lipase from *Rhizopus oryzae*, and all the lipases studied exhibited a high hydrolytic activity toward triglycerides from palm and peanut oil (103). The authors proposed that MCPDs were formed by a lipase-catalyzed reaction between a triacylglycerol and chloride ion. However, it is more likely that MCPDs were formed by the hydrolysis of residual MCPD-esters that are now known to be present in the raw materials used (73).

2.6.5.5 Other Formation Mechanisms of Interest

In a study of the smoking process, Kuntzer and Weisshaar proposed that 3-MCPD generated in smoke from burning wood chips could be derived from cellulose (81). A mechanism was proposed for the reaction of chloride ions with acetol (3-hydroxyacetone), a thermal degradation product of cellulose and wood smoke. It is interesting to note that 3-hydroxyacetone is isomeric with the known 3-MCPD intermediate glycidol.

2.6.5.6 Stability and Reactions of MCPDs

Hamlet *et al.* studied the degradation reactions of MCPDs in pure water (104) and in bread (96, 105). Under aqueous conditions, 3-MCPD decayed to glycerol, via the intermediate epoxide glycidol, according to first-order kinetics. The stability of 3-MCPD was sensitive to both pH and temperature, particularly over a range applicable to baked cereal products. In bread dough, the decay of 2-MCPD and 3-MCPD was slower than that in pure water: the decay reaction was inhibited by a decrease in moisture content and the pH drop seen in cooked dough at elevated temperatures. At high temperatures (i.e., >100 °C), 2- and 3-MCPD exhibited similar stability. At lower temperatures (i.e., <100 °C), this behavior was reversed and 2-MCPD was more stable. It was shown that in bread dough, added 3-MCPD was converted into the isomeric compound 2-MCPD in a mechanism involving the known decay intermediate, glycidol (96, 105).

The structure of 3-MCPD indicates that the compound should undergo reactions characteristic of both alcohols and alkyl chlorides. For example, 3-MCPD is known to react readily with alcohols, aldehydes, amino compounds, ammonia, ketones, organic acids, and thiols (106). Some of these reaction products, such as 3-MCPD-derived amino alcohols, dihydroxypropylamines, and amino acids, have been identified in HVPs (106–108). Model system studies (107, 108) have shown that the reaction of 3-MCPD with either ammonia or amino acids occurs readily and at moderate temperatures (20–90 °C). The reactions of cysteine and glutathione with 3-MCPD have also been reported (109).

2.6.6 MITIGATION

Manufacturers of HVPs and soy sauces and producers of wet-strength resins for food contact applications have made considerable efforts to reduce chloropropanols in, and contamination of foods from, their products. For example, in alkaline media both 2- and 3-MCPD are decomposed to glycerol (110, 111) and hence alkalization is a method that is used commercially to reduce the level of MCPDs in protein hydrolysates (112) and also in wet-strength resins (113). However, strategies to reduce chloropropanols and chloroesters in other products have not yet been fully explored and may not be possible for all foodstuffs. These strategies need to consider whether interventions to reduce the risk of chloropropanols and chloroesters might increase the risk of other process contaminants, such as furan or acrylamide (114). For example,

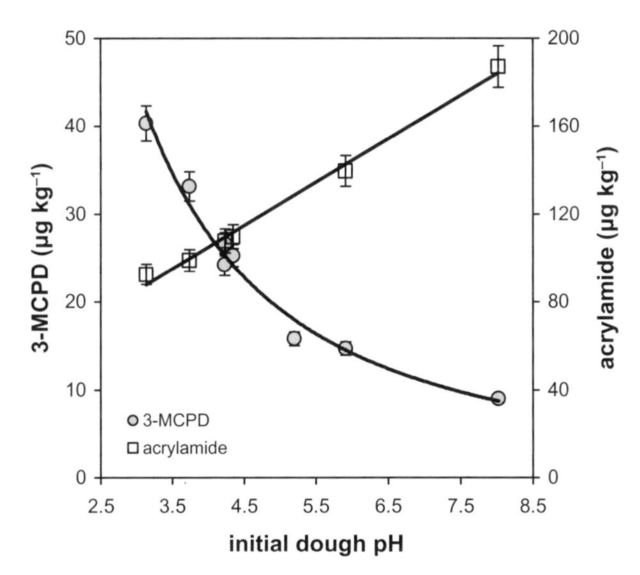

Figure 2.6.14 Effect of pH on 3-MCPD and acrylamide formation in baked cereal products (adapted from References 98 and 115).

Fig. 2.6.14 illustrates how increasing dough pH to reduce 3-MCPD formation in cereal products has the opposite effect on acrylamide generation. A summary of potential mitigation options together with literature references is given in Table 2.6.8.

2.6.7 EXPOSURE

The most recent exposure assessments have considered 1,3-DCP and 3-MCPD and were carried out by JECFA in 2006 (41). Unlike previous JECFA evaluations (37), which mainly considered exposure from soy sauces, the most recent evaluation was based on contributions from all food groups in the diet using data from, for example, the 2003 FSANZ survey (80) and the EC compilation produced under tasks for scientific cooperation (77). Particular consideration was given to groups that might have higher levels of exposure to chloropropanols.

JECFA also noted recent reports that fatty acid esters of 3-MCPD are present in foods, and recommended that studies should be undertaken to enable their intake or toxicological significance to be evaluated.

2.6.7.1 3-MCPD

Estimated exposures at the national level considered a wide range of foods, including soy sauce and soy sauce-related products, and ranged from 1% to

TABLE 2.6.8 Summary of potential mitigation measures for chloropropanols and chloroesters in foods and food contact materials.

Product/ application	Potential control measure	Chloropropanol(s)	Reference no.
Bread/toast	Time/temperature/ minimum fermentation/ recipe management	3 MCPD	96–100
HVP and soy sauces produced by acid hydrolysis	Alkalization/neutralization	3-MCPD	112
	Degradation by *Pseudomonas* sp. OS-K-29	Esters of dichloropropanols	116
Roasted cereals/malts	Time/temperature	3-MCPD	84
Savory foodstuffs	Lipase/esterase inactivation (temperature, pH, water activity)	3-MCPD released from chloroesters	117
Smoke	$CaCO_3$ pretreatment of wood pellets	3-MCPD	81
Unspecified	Degradation by bacterial cultures	1,3-DCP and 3-MCPD	118
		1,3-DCP	119, 120
	Degradation by *Saccharomyces cerevisiae*	3-MCPD	121
	Reactive food additives, e.g., cysteine, disodium carbonate, glutathione, and sodium bicarbonate	3-MCPD	109
Wet-strength resins	Chemical (base)/enzyme/ process treatment of epichlorohydrin copolymers	1,3-DCP	122
		MCPDs	113, 123, 124

35% of the PMTDI for average exposure in the general population. For the consumers at the high percentile (95th), the estimated intakes ranged from 3% to 85% and up to 115% of the PMTDI in young children. The estimates were based on concentrations of 3-MCPD derived before any remedial action had been taken by government or industry. JECFA noted that reduction in the concentration of 3-MCPD in soy sauce and related products made with acid-HVP could substantially reduce the intake of this contaminant by certain consumers of this condiment.

2.6.7.2 1,3-DCP

JECFA (41) concluded that a representative mean intake of 1,3-DCP for the general population was of 0.051 μg kg^{-1} body weight per day and an estimated high-level intake (young children included) was 0.136 μg kg^{-1} body weight per day. Comparison of these mean and high-level intake values for consumers with the lowest calculated dose for incidence data on tumor-bearing animals indicated margins of exposure of approximately 65,000 and 24,000, respectively. Based on these margins of exposure, it was concluded that the estimated intakes of 1,3-dichloro-2-propanol were of low concern for human health.

2.6.8 RISK MANAGEMENT

To date, regulatory controls have been adopted for HVPs and soy sauces (see Table 2.6.9) since chloropropanol intake levels from other foods are very low. With appropriate manufacturing controls, the amounts of chloropropanols can be controlled in HVPs and soy sauces produced by acid hydrolysis. In instances where brand loyalty could result in regular consumption of, for example, contaminated soy sauces, these controls could markedly reduce intake levels (37). It has been demonstrated that 1,3-DCP is typically associated with high con centrations of 3-MCPD. Hence, regulatory control of 3-MCPD should negate

TABLE 2.6.9 International maximum limits/specifications for chloropropanols in foodstuffs.

Country/region	DCPs, mg kg^{-1}	3-MCPD, mg kg^{-1}	Scope	Reference no.
Australia/New Zealand	0.005	0.2	Soy/oyster sauces	125
Canada	—	1	Soy/oyster sauces	126
China	—	1	Acid-HVP	127
European Community	—	0.02	HVP and soy sauces (40% solids)	128
Korea	—	0.3	Soy sauce containing acid-HVP	129
	—	1	HVP	129
Malaysia	—	0.02	Liquid foods with acid-HVP	130
	—	1	Acid-HVP, industrial product	130
Switzerland	0.05	0.2	Savory sauces	131
Thailand	—	1	Hydrolyzed soybean protein	132
United States	0.05	1	Acid-HVP	133

the need for specific limits on 1,3-DCP, although some countries have imposed maximum limits (see Table 2.6.9).

2.6.9 FUTURE PROSPECTS AND CONCLUSIONS

The recent observations of chloropropanols in a wide range of foodstuffs have led to a renewal of research worldwide on methods of analysis, occurrence, and formation mechanisms of these contaminants. Analytical methods for MCPDs and DCPs have been updated significantly and robust procedures are available, covering a wide range of matrices. Considerable progress has been made on occurrence routes, particularly in thermally treated cereal products, where the precursors and mechanisms of MCPD formation have been elucidated. Overall, the contribution from foodstuffs (excluding soy sauces) to the dietary intake of chloropropanols is relatively low, and to date, regulatory controls to minimize exposure have not been deemed necessary. However, higher concentrations of chloropropanols (e.g., 3-MCPD) found in certain soy sauces have necessitated maximum limits for these materials in some countries. With the appropriate manufacturing controls, contamination of soy sauces by chloropropanols can be eliminated.

There is growing evidence that chloroesters may be widespread in processed foods and that amounts of these contaminants are higher than the corresponding chloropropanols. Model studies have shown that lipases can release 3-MCPD from chloroesters. Although the extent to which this may occur in foodstuffs is unknown, the major concern here is the potential release of 3-MCPD from the chloroesters *in vivo* during digestion in the gastrointestinal tract. According to Seefelder *et al.*, chloroesters are hydrolyzed in the same way as the triacylglycerols, i.e., lipases have specificity for the sn-1 and sn-3 positions (43). Hence, the monoesters of 3-MCPD are completely hydrolyzed to 3-MCPD and it is the ratio of monoesters versus diesters that is likely to determine the contribution of the chloroesters to 3-MCPD exposure. Inevitably, further studies will be required to fully understand the bioavailability, toxicokinetics, and toxicology of the chloroesters.

Several potential mitigation measures for chloropropanols and chloroesters in foodstuffs have already been identified. However, it is important that nutritionists, food chemists, and toxicologists jointly consider the wider risks and benefits of these measures. This will ensure that interventions to reduce these contaminants do not have a negative impact on health and nutrition or increase the risk of producing other undesirable compounds.

REFERENCES

1. Lynch, B.S., Bryant, D.W., Hook, G.J., Nestmann, E.R., Munro, I.C. (1998). Carcinogenicity of monochloro-1,2-propanediol (alpha-chlorohydrin, 3-MCPD). *International Journal of Toxicology*, *17*, 47–76.

2. Velíšek, J., Davídek, J., Hajšlová, J., Kubelka, V., Janíček, G., Mánková, B. (1978). Chlorohydrins in protein hydrolysates. *Zeitschrift fur Lebensmittel-Untersuchung Und-Forschung*, *167*, 241–244.
3. Velíšek, J., Davídek, J., Kubelka, V., Bartošová, J., Tuèková, A., Hajšlová, J., Janíček, G. (1979). Formation of volatile chlorohydrins from glycerol (triacetin, tributyrin) and hydrochloric acid. *Lebensmittel-Wissenschaft und-Technologie—Food Science and Technology*, *12*, 234–236.
4. Velíšek, J., Davídek, J., Kubelka, V., Janíček, G., Svobodová, Z., Šimicová, Z. (1980). New chlorine-containing organic compounds in protein hydrolysates. *Journal of Agricultural and Food Chemistry*, *28*, 1142–1144.
5. Velíšek, J., Davídek, J., Šimicová, Z., Svobodová, Z. (1982). Glycerol chlorohydrins and their esters—reaction products of lipids with hydrochloric acid. *Scientific Papers of the Prague Institute of Chemical Technology, Food, E 53*, 55–65.
6. Velíšek, J., Davídek, J. (1985). Lipidy jako prekurzory organických sloučenin chloru v bílkovinných hydrolyzátech. *Potravinárské*, *1*, 1–18.
7. Davídek, J., Velíšek, J., Kubelka, V., Janíček, G., Šimicová, Z. (1980). Glycerol chlorohydrins and their esters as products of the hydrolysis of tripalmitin tristearin and triolein with hydrochloric acid. *Lebensmittel-Untersuchung und-Forschung*, *171*, 14–17.
8. Davídek, J., Velíšek, J., Kubelka, V., Janíček, G. (1982). New chlorine containing compounds in protein hydrolysates, in *Recent Developments in Food Analysis, Proceedings of Euro Food Chem I, Vienna, Austria, 17–20 February, 1981* (eds W. Baltes, P.B. Czedik-Eysenberg, W. Pfannhauser), Weinheim, Deerfield Beach, Florida, pp. 322–325.
9. Velíšek, J. (1989). Organic chlorine compounds in food protein hydrolysates. DSc. Thesis, Institute of Chemical Technology, Prague, Czech Republic.
10. Velíšek, J., Ledahudcová, K. (1993). Problems of organic chlorine compounds in food protein hydrolysates. *Potravinárské vědy*, *11*, 149–159.
11. Collier, P.D., Cromie, D.D.O., Davies, A.P. (1991). Mechanism of formation of chloropropanols present in protein hydrolysates. *Journal of the American Oil Chemists Society*, *68*, 785–790.
12. Cerbulis, J., Parks, O., Liu, R., Piotrowski, G., Farrell, H. (1984). Occurrence of diesters of 3-chloro-1,2-propanediol in the neutral lipid fraction of goats' milk. *Journal of Agricultural and Food Chemistry*, *32*, 474–476.
13. Kuksis, A., Marai, L., Myher, J.J., Cerbulis, J., Farrell Jr, H.M. (1986). Comparative study of the molecular species of chloropropanediol diesters and triacylglycerols in milk fat. *Lipids*, *21*, 183–190.
14. Myher, J.J., Kuksis, A., Marai, L., Cerbulis, J. (1986). Stereospecific analysis of fatty acid esters of chloropropanediol isolated from fresh goat milk. *Lipids*, *21*, 309–314.
15. European Commission (1997). Opinion on 3-monochloro-propane-1,2-diol (3-MCPD), expressed on 16 December 1994, in *Food Science and Techniques: Reports of The Scientific Committee For Food (Thirty-Sixth Series)*, European Commission, Brussels, pp. 31–33.
16. Hamlet, C.G., Sutton, P.G. (1997). Determination of the chloropropanols, 3-chloro-1,2-propandiol and 2-chloro-1,3-propandiol, in hydrolysed vegetable proteins and

seasonings by gas chromatography ion trap tandem mass spectrometry. *Rapid Communications in Mass Spectrometry*, *11*, 1417–1424.

17. Hamlet, C.G. (1998). Analytical methods for the determination of 3-chloro-1,2-propandiol and 2-chloro-1,3-propandiol in hydrolysed vegetable protein, seasonings and food products using gas chromatography ion trap tandem mass spectrometry. *Food Additives and Contaminants*, *15*, 451–465.
18. Meierhans, D.C., Bruehlmann, S., Meili, J., Taeschler, C. (1998). Sensitive method for the determination of 3-chloropropane-1,2-diol and 2-chloropropane-1,3-diol by capillary gas chromatography with mass spectrometric detection. *Journal of Chromatography, A*, *802*, 325–333.
19. Vanbergen, C.A., Collier, P.D., Cromie, D.D.O., Lucas, R.A., Preston, H.D., Sissons, D.J. (1992). Determination of chloropropanols in protein hydrolysates. *Journal of Chromatography*, *589*, 109–119.
20. Ministry of Agriculture Fisheries and Food (1999). Survey of 3-monochloropropane-1,2-diol (3-MCPD) in acid-hydrolysed vegetable protein. Food Surveillance Information Sheet, No. 181. http://archive.food.gov.uk/maff/archive/food/infsheet/1999/no181/181mcpd.htm (accessed September 2006).
21. Ministry of Agriculture Fisheries and Food (1999). Survey of 3-monochloropropane-1,2-diol (3-MCPD) in soy sauce and similar products. Food Surveillance Information Sheet No. 187. http://archive.food.gov.uk/maff/archive/food/infsheet/1999/no187/187soy.htm (accessed September 2006).
22. Macarthur, R., Crews, C., Davies, A., Brereton, P., Hough, P., Harvey, D. (2000). 3-monochloropropane-1,2-diol (3-MCPD) in soy sauces and similar products available from retail outlets in the UK. *Food Additives and Contaminants*, *17*, 903–906.
23. Wu, H., Zhang, G. (1999). Determination of 3-chloropropane-1,2-diol in soy sauce by GC-MS-SIM. *Fenxi Ceshi Xuebao (Journal of Instrumental Analysis)*, *18*, 64–65.
24. Faesi, R.S., Werner, G., Wolfensberger, U. (1987). Procédé de fabrication d'un condiment. European patent EP0226769.
25. Brown, D.A., Van Meeteren, H.W., Simmons, J.D. (1989). Process for preparing improved hydrolysed protein. European patent EP03610595B1.
26. De Rooij, J.F.M., Ward, B.A., Ward, M. (1989). Process for preparing improved hydrolysed protein. European patent EP0361597 (A1).
27. Hirsbrunner, P., Weymuth, H. (1989). Procédé de fabrication d'un condiment. European patent EP0380371B1.
28. Payne, L.S. (1989). Process for preparing improved hydrolysed protein. European patent EP0361596.
29. Boden, L., Lunddgren, M., Stensio, K., Gorzynski, M. (1997). Determination of 1,3-dichloro-2-propanol in papers treated with polyamidoamine-epichlorohydrin wet-strength resins by gas chromatography-mass spectrometry using selective ion monitoring. *Journal of Chromatography, A*, 195–203.
30. Food Advisory Committee (2000). Genotoxicity of 3-monochloropropane-1,2-diol. Food Advisory Committee paper for discussion, FdAC/Contaminants/48. http://archive.food.gov.uk/pdf_files/papers/fac_48.pdf (accessed November 2006).

31. Schlatter, J., Baars, A.J., DiNovi, M., Lawrie, S., Lorentzen, R. (2002). 3-chloro-1,2-propanediol. In: WHO food additives series 48. Safety evaluation of certain food additives and contaminants, prepared by the Fifty-seventh meeting of the Joint FAO/WHO Rome, 2001. http://www.inchem.org/documents/jecfa/jecmono/v48je18.htm (accessed November 2006).
32. Lynch, B.S., Bryant, D.W., Hook, G.J., Nestmann, E.R., Munro, I.C. (1998). Carcinogenicity of monochloro-1,2-propanediol (alpha-chlorohydrin, 3-MCPD). *International Journal of Toxicology*, *17*, 47–76.
33. Robjohns, S., Marshall, R., Fellows, M., Kowalczyk, G. (2003). In vivo genotoxicity studies with 3-monochloropropan-1,2-diol. *Mutagenesis*, *18*, 401–404.
34. Fellows, M. (2000). 3-MCPD: measurement of unscheduled DNA synthesis in rat liver using an in vitro/in vivo procedure. Report No. 1863/1-D5140. York: Covance Laboratories.
35. Marshall, R.M. (2000). 3-MCPD: induction of micronuclei in the bone-marrow of treated rats. Unpublished report No. 1863/2-D5140 from Covance Laboratories.
36. Committee on Mutagenicity of Chemicals in Food, Consumer Products and the Environment (COM) (2000). Mutagenicity of 3-monochloro propane 1,2-diol (3-MCPD), COM statement COM/00/S4, October 2000. http://www.advisorybodies.doh.gov.uk/com/mcpd2.htm (accessed November 2006).
37. Joint FAO/WHO Expert Committee On Food Additives (JECFA) (2002). Evaluation of certain food additives and contaminants: fifty seventh report of the Joint FAO/WHO Expert Committee on Food Additives, Rome, June 5–14, 2001. WHO technical report series; 909, WHO, Geneva.
38. Schlatter, J., Baars, A.J., DiNovi, M., Lawrie, S., Lorentzen, R. (2002). 1,3-dichloro-2-propanol. In: WHO food additives series 48. Safety evaluation of certain food additives and contaminants, prepared by the Fifty-seventh meeting of the Joint FAO/WHO, Rome, 2001. http://www.inchem.org/documents/jecfa/jecmono/v48je19.htm (accessed November 2006).
39. Committee on Mutagenicity of Chemicals in Food, Consumer Products and the Environment COM (2003). Statement on the mutagenicity of 1,3-dichloropropan-2-ol Com/03/S4, October 2003. http://www.advisorybodies.doh.gov.uk/com/1,3-dcp.htm (accessed November 2006).
40. Committee on Carcinogenicity of Chemicals in Food, Consumer Products and the Environment (COC) (2004). Carcinogenicity of 1,3-dichloropropan-2-ol (1,3-DCP) and 2,3-dichloropropan-1-ol (2,3-DCP), COC/04/S2, June 2004. http://www.advisorybodies.doh.gov.uk/coc/1,3-2,3dcp04.htm (accessed November 2006).
41. Joint FAO/WHO Expert Committee on Food Additives (JECFA) (2006). Summary and conclusions from the Sixty-seventh meeting, Rome, 20–29 June 2006, JECFA67/SC. http://who.int/ipcs/food/jecfa/summaries/summary67.pdf (accessed November 206).
42. Committee on Mutagenicity of Chemicals in Food, Consumer Products and the Environment (COM) (2004). Statement on the mutagenicity of 2,3-dichloropropan-1-ol, Com/04/s1, May 2004. http://www.advisorybodies.doh.gov.uk/com/2,3dcp04.htm (accessed November 2006).
43. Seefelder, W., Varga, N., Studer, A., Williamson, G., Scanlan, F.P., Stadler, R.H. (2007). Esters of 3-chloro-1,2-propanediol (3-MCPD) in vegetable oils:

significance in the formation of 3-MCPD. *Food Additives and Contaminants*, *25*, 391–400.

44. Gurr, M.I. (1992). *Role of Fats in Food and Nutrition*, Elsevier Applied Science, London.
45. Hamlet, C.G., Sadd, P.A., Crews, C., Velisek, J., Baxter, D.E. (2002). Occurrence of 3-chloro-propane-1,2-diol (3-MCPD) and related compounds in foods: a review. *Food Additives and Contaminants*, *19*, 619–631.
46. Brereton, P., Kelly, J., Crews, C., Honour, S., Wood, R., Davies, A. (2001). Determination of 3-chloro-1,2-propanediol in foods and food ingredients by gas chromatography with mass spectrometric detection: collaborative study. *Journal of AOAC International*, *84*, 455–465.
47. British Standards Institution (2004). BS EN. 14573, 2004 Foodstuffs. Determination of 3-monochloropropane-1,2-diol by GC/MS.
48. Nyman, P.J., Diachenko, G.W., Perfetti, G.A. (2003). Determination of 1,3-dichloropropanol in soy sauce and related products by using gas chromatography/mass spectrometry. *Food Additives and Contaminants*, *20*, 903–908.
49. Xu, X., Ren, Y., Wu, P., Han, J., Shen, X. (2006). The simultaneous separation and determination of chloropropanols in soy sauce and other flavouring with gas chromatography-mass spectrometry in negative chemical and electron impact ionization modes. *Food Additives and Contaminants*, *23*, 110–119.
50. Chung, W.C., Hui, K.Y., Cheng, S.C. (2002). Sensitive method for the determination of 1,3-dichloropropan-2-ol and 3-chloropropane-1,2-diol in soy sauce by capillary gas chromatography with mass spectrometric detection. *Journal of Chromatography, A*, *952*, 185–192.
51. Abu-El-Haj, S., Bogusz, M.J., Ibrahim, Z., Hassan, H., Al Tufail, M. (2007). Rapid and simple determination of chloropropanols (3-MCPD and 1,3-DCP) in food products using isotope dilution GC–MS. *Food Control*, *18*, 81–90.
52. Rodman, L.E., Ross, R.D. (1986). Gas-liquid chromatography of 3-chloropropanediol. *Journal of Chromatography*, *369*, 97–103.
53. Pesselman, R.L., Feit, M.J. (1988). Determination of residual epichlorohydrin and 3-chloropropanediol in water by gas chromatography with electron capture detection. *Journal of Chromatography*, *439*, 448–452.
54. Plantinga, W.J., Van Toorn, W.G., Van Der Stegen, G.H.D. (1991). Determination of 3-chloropropane-1,2-diol in liquid hydrolysed vegetable proteins by capillary gas chromatography with flame ionisation detection. *Journal of Chromatography*, *555*, 311–314.
55. Breitling-Utzmann, C.M., Kobler, H., Herbolzheimer, D., Maier, A. (2003). 3-MCPD—Occurrence in bread crust and various food groups as well as formation in toast. *Deutsche Lebensmittel-Rundschau*, *99*, 280–285.
56. Divinová, V., Svejkovská, B., Doležal, M., Velíšek, J. (2004). Determination of free and bound 3-chloropropane-1,2-diol by gas chromatography with mass spectrometric detection using deuterated 3-chloropropane-1,2-diol as internal standard. *Czech Journal of Food Sciences*, *22*, 182–189.
57. Velíšek, J., Doležal, M., Crews, C., Dvorák, T. (2002). Optical isomers of chloropropanediols: mechanisms of their formation and decomposition in protein hydrolysates. *Czech Journal of Food Sciences*, *20*, 161–170.

58. Kuballa, T., Ruge, W. (2004). Analysis and detection of 3-monochloropropane-1,2-diol (3-MCPD) in food by GC/MS/MS. http://www.varianinc.com/cgi-bin/nav?applications/gcms&cid=JLQOILQPFJ (accessed November 2006).
59. Huang, M., Jiang, G., He, B., Liu, J., Zhou, Q., Fu, W., Wu, Y. (2005). Determination of 3-chloropropane-1,2-diol in liquid hydrolyzed vegetable proteins and soy sauce by solid-phase microextraction and gas chromatography/mass spectrometry. *Analytical Sciences*, *21*, 1343–1347.
60. March, J. (1983). *Advanced Organic Chemistry: Reactions Mechanisms, and Structure*, 2nd edn, McGraw-Hill, Tokyo, pp. 810–811.
61. Dayritt, F.M., Ninonuevo, M.R. (2004). Development of an analytical method for 3-monochloropropane-1,2-diol in soy sauce using 4-heptanone as derivatizing agent. *Food Additives and Contaminants*, *21*, 204–209.
62. Jin, Q., Zhang, Z., Luo, R., Li, J. (2001). Survey of 3-monochloropropane-1,2-diol (3-MCPD) in soy sauce and similar products. *Wei Shang Yan Jiu*, *30*, 60–61.
63. Rétho, C., Blanchard, F. (2005). Determination of 3-chloropropane-1,2-diol as its 1,3-dioxolane derivative at the μg kg-1 level: application to a wide range of foods. *Food Additives and Contaminants*, *22*, 1189–1197.
64. Wittmann, R. (1991). Determination of dichloropropanols and monochloropropandiols in seasonings and in foodstuffs containing seasonings. *Zeitschrift fur Lebensmittel-Untersuchung Und-Forschung*, *193*, 224–229.
65. Schumacher, R., Nurmi-Legat, J., Oberhauser, A., Kainz, M., Krska, R. (2005). A rapid and sensitive GC-MS method for determination of 1,3-dichloro-2-propanol in water. *Analytical and Bioanalytical Chemistry*, *382*, 366–371.
66. Crews, C., LeBrun, G., Brereton, P.A. (2002). Determination of 1,3-dichloropropanol in soy sauces by automated headspace gas chromatography-mass spectrometry. *Food Additives and Contaminants*, *19*, 343–349.
67. Hasnip, S., Crews, C., Potter, N., Brereton, P. (2005). Determination of 1,3-dichloropropanol in soy sauce and related products by headspace gas chromatography with mass spectrometric detection: interlaboratory study. *Journal of AOAC International*, *88*, 1404–1412.
68. Mingxia, Z., Jianke, Z., Junhong, L. (2003). Determination of chloropropanols in expanded foods by gas chromatography. *Food and Fermentation Industries*, *29*, 52–54.
69. Leung, M.K.P., Chiu, B.K.W., Lam, M.H.W. (2003). Molecular sensing of 3-chloro-1,2-propanediol by molecular imprinting. *Analytica Chimica Acta*, *491*, 15–25.
70. Xing, X., Cao, Y. (2007). Determination of 3-chloro-1,2-propanediol in soy sauces by capillary electrophoresis with electrochemical detection. *Food Control*, *18*, 167–172.
71. Hamlet, C.G., Sadd, P.A., Gray, D.A. (2004). Chloropropanols and their esters in baked cereal products, in *Abstracts of the Division of Agricultural and Food Chemistry 227th Annual Meeting, Anaheim, California, March 28–April 1, 2004* (ed. W. Yokoyama), American Chemical Society, Anaheim.
72. Hamlet, C.G., Sadd, P.A. (2004). Chloropropanols and their esters in cereal products. *Czech Journal of Food Sciences*, *22*, 259–262.
73. Zelinková, Z., Svejkovská, B., Velíšek, J., Doležal, M. (2006). Fatty acid esters of 3-chloropropane-1,2-diol in edible oils. *Food Additives and Contaminants*, *23*, 1290–1298.

74. Kraft, R., Brachwitz, H., Etzold, H.G., Langen, P., Zöpfl, H.-J. (1979). Halogenolipids. I. Mass spectrometric structure investigation of the isomeric halogenopropanediols esterified with fatty acids (deoxyhalogeno-glycerides). *Journal für Praktische Chemie*, *321*, 756–768.
75. Food Standards Agency (2001). Survey of 3-monochloropropane-1,2-diol (3-MCPD) in selected food groups (number 12/01). Food Standards Agency, London. http://www.food.gov.uk/science/surveillance/fsis2001/3-mcpdsel (accessed November 2006).
76. Crews, C., Brereton, P., Davies, A. (2001). The effects of domestic cooking on the levels of 3-monochloropropanediol in foods. *Food Additives and Contaminants*, *18*, 271–280.
77. European Commission (2004). Report of experts participating in Task 3.2.9. Collection and collation of data on concentration of 3-monochloropropanediol (3-MCPD) and related substances in foodstuffs, June 2004. Directorate-General Health and Consumer Protection, Brussels. http://ec.europa.eu/food/food/chemicalsafety/contaminants/mcpd_data_tables_en.htm (accessed November 2006).
78. Svejkovská, B., Novotný, O., Divinová, V., Réblová, Z., Doležal, M., Velíšek, J. (2004). Esters of 3-chloropropane-1,2-diol in foodstuffs. *Czech Journal of Food Sciences*, *22*, 190–196.
79. Doležal, M., Chaloupská, M., Divinová, V., Svejkovská, B., Velíšek, J. (2005). Occurrence of 3-chloropropane-1,2-diol and its esters in coffee. *European Food Research and Technology*, *221*, 221–225.
80. Food Standards Australia New Zealand (FSANZ) (2003). Chloropropanols in food: an analysis of the public health risk. Technical report series no. 15. FSANZ, Canberra. http://www.foodstandards.gov.au/_srcfiles/Chloropropanol%20Report%20(no%20appendices)-%2011%20Sep%2003b-2.pdf#search=%22chloropropanols%22 (accessed 17 September 2008).
81. Kuntzer, J., Weisshaar, R. (2006). The smoking process—a potent source of 3-chloropropane-1,2-diol (3-MCPD) in meat products. *Deutsche Lebensmittel-Rundschau*, *102*, 397–400.
82. Food Standards Agency (2004). Chloropropanols in meat products, press release, Thursday, 20 May 2004. http://www.food.gov.uk/news/newsarchive/2004/may/chloropropanols (accessed November 2006).
83. Hamlet, C.G., Jayaratne, S.M., Matthews, W. (2002). 3-monochloropropane-1,2-diol (3-MCPD) in food ingredients from UK food producers and ingredient suppliers. *Food Additives and Contaminants*, *19* (1), 15–21.
84. Brereton, P., Crews, C., Hasnip, S., Reece, P., Velíšek, J., Doležal, M., Hamlet, C., Sadd, P., Baxter, D., Slaiding, I., Muller, R., 2005, The origin and formation of 3-MCPD in foods and food ingredients, Report No. FD 04/12. A report prepared for the Food Standards Agency. Central Science Laboratory, York.
85. Masten, S.A. (2005). 1,3-Dichloro-2-propanol [CAS No. 96-23-1], review of toxicological literature. A report prepared for National Toxicology Program (NTP), National Institute of Environmental Health Sciences (NIEHS), National Institutes of Health, U.S. Department of Health and Human Services. NTP/NIEHS, North Carolina. http://ntp.niehs.nih.gov/ntp/htdocs/Chem_Background/ExSumPdf/dichloropropanol.pdf (accessed 17 September 2008).

86. Codex Committee on Food Additives and Contaminants (CCFAC) (2001). Position paper on chloropropanols, CX/FAC 01/31, thirty-third Session the Hague, the Netherlands, 12–16 March 2001. http://www.codexalimentarius.net/ccfac33/fa01_01e.htm (accessed November 2006).
87. Drinking Water Inspectorate (DWI) (2003). *The Use of Polyamine Coagulants in Public Water Supplies*, DWI, London. http://www.dwi.gov.uk/cpp/chloro.htm (accessed November 2006).
88. Conant, J.B., Quayle, O.R. (1947). Glycerol, -dichlorohydrin. *Organic Syntheses Collective*, *1*, 292–294.
89. Conant, J.B., Quayle, O.R. (1947). Glycerol, -monochlorohydrin. *Organic Syntheses Collective*, *1*, 294–296.
90. Myszkowski, J., Zielinski, A.Z. (1965). Synthesis of glycerol monochlorohydrins from allyl alcohol. *Przemysl Chemiczny*, *44*, 249–252.
91. Kubec, R., Velíšek, J., Doležal, M., Kubelka, V. (1997). Sulfur-containing volatiles arising by thermal degradation of alliin and deoxyalliin. *Journal of Agricultural and Food Chemistry*, *45*, 3580–3585.
92. Carbonell, M., Nunez, M., Fernandez-Garcia, E. (2002). Evolution of the volatile components of ewe raw milk La Serena cheese during ripening. Correlation with flavour characteristics. *Lait*, *82*, 683–698.
93. Hasnip, S., Crews, C., Brereton, P., Velíšek, J. (2002). Factors affecting the formation of 3-monochloropropanediol in foods. *Polish Journal of Food and Nutrition Sciences*, *11*, 122–124.
94. Doležal, M., Calta, P., Velíšek, J., Hasnip, S. (2003). Study on the formation of 3-chloropropane-1,2-diol from lipids in the presence of sodium chloride, in *Euro Food Chem XII Conference on "Strategies for Safe Food: Analytical Industrial, and Legal Aspects; Challenges in Organisation and Communication", 24–26 September 2003* (eds T. Eklund *et al.*), KVCV, Brugge, Belgium, pp. 251–254.
95. Calta, P., Velíšek, J., Doležal, M., Hasnip, S., Crews, C., Réblová, Z. (2004). Formation of 3-chloropropane-1,2-diol in systems simulating processed foods. *European Food Research and Technology*, *218*, 501–506.
96. Hamlet, C.G. (2004). Monochloropropanediols in bread: model dough systems and kinetic modelling. Ph.D. Thesis, University of Nottingham.
97. Hamlet, C.G., Sadd, P.A., Gray, D.A. (2004). Generation of monochloropropanediols (MCPDs) in model dough systems. 1. Leavened doughs. *Journal of Agricultural and Food Chemistry*, *52*, 2059–2066.
98. Hamlet, C.G., Sadd, P.A. (2005). Effects of yeast stress and pH on 3-MCPD producing reactions in model dough systems. *Food Additives and Contaminants*, *22* (7), 616–623.
99. Breitling-Utzmann, C.M., Hrenn, H., Haase, N.U., Unbehend, G.M. (2005). Influence of dough ingredients on 3-chloropropane-1,2-diol (3-MCPD) formation in toast. *Food Additives and Contaminants*, *22*, 97–103.
100. Hamlet, C.G., Sadd, P.A., Gray, D.A. (2004). Generation of monochloropropanediols in model dough systems. 2. Unleavened doughs. *Journal of Agricultural and Food Chemistry*, *52*, 2067–2072.
101. Hamlet, C.G., Jayaratne, S.M., Morrison, C.M. (2005). Processing contaminants in bread from bread making machines: a continuing project under C03020. A report

prepared for the UK Food Standards Agency. RHM Technology Ltd., High Wycombe.

102. Robert, M.-C., Stadler, R.H. (2003). Studies on the formation of chloropropanols in model systems, in *Euro Food Chem XII Conference on "Strategies for Safe Food: Analytical Industrial, and Legal Aspects; Challenges in Organisation and Communication", 24–26 September 2003* (eds T. Eklund *et al.*), KVCV, Brugge, Belgium, pp. 527–530.

103. Robert, M.-C., Oberson, J.-M., Stadler, R.H. (2004). Model studies on the formation of monochloropropanediols in the presence of lipase. *Journal of Agricultural and Food Chemistry*, *52*, 5102–5108.

104. Hamlet, C.G., Sadd, P.A. (2002). Kinetics of 3-chloropropane-1,2-diol (3-MCPD) degradation in high temperature model systems. *European Food Research and Technology*, *215*, 46–50.

105. Hamlet, C.G., Sadd, P.A., Gray, D.A. (2003). Influence of composition, moisture, pH and temperature on the formation and decay kinetics of monochloropropanediols in wheat flour dough. *European Food Research and Technology*, *216*, 122–128.

106. Velíšek, J., Davídek, T., Davídek, J., Hamburg, A. (1991). 3-Chloro-1,2-propanediol derived amino alcohol in protein hydrolysates. *Journal of Food Science*, *56*, 136–138.

107. Velíšek, J., Davídek, T., Davídek, J., Kubelka, V., Víden, I. (1991). 3-Chloro-1,2-propanediol derived amino acids in protein hydrolysates. *Journal of Food Science*, *56*, 139–142.

108. Velíšek, J., Ledahudcová, K., Hajšlová, J., Pech, P., Kubelka, V., Víden, I. (1992). New 3-chloro-1,2-propanediol derived dihydroxypropylamines in hydrolysed vegetable proteins. *Journal of Agricultural and Food Chemistry*, *56*, 1389–1392.

109. Velíšek, J., Calta, P., Crews, C., Hasnip, S., Doležal, M. (2003). 3-chloropropane-1,2-diol in models simulating processed foods: precursors and agents causing its decomposition. *Czech Journal of Food Sciences*, *21*, 153–161.

110. Doležal, M., Velíšek, J. (1992). Kinetics of 3-chloro-1,2-propanediol degradation in model systems, in *Proceedings of Chemical Reactions in Foods II, Prague, Czech Republic, September 24–26*, pp. 297–302.

111. Doležal, M., Velíšek, J. (1995). Kinetics of 2-chloro-1,3-propanediol degradation in model systems and in protein hydrolysates. *Potravinaoske Vědy*, *2*, 85–91.

112. Sim, C.W., Muhammad, K., Yusof, S., Bakar, J., Hashim, D.M. (2004). The optimization of conditions for the production of acid-hydrolysed winged bean and soybean proteins with reduction of 3-monochloropropane-1,2-diol (3-MCPD). *International Journal of Food Science and Technology*, *39*, 947–958.

113. Riehle, R.J. (Hercules Inc.) (2006). Treatment of resins to lower levels of CPD-producing species and improve gelation stability. U.S. Patent No 7081512. http://www.uspto.gov/patft/index.html (accessed November 2006).

114. Konings, E.J.M., Ashby, P., Hamlet, G.G., Thompson, G.A.K. (2007). Acrylamide in cereal and cereal products: a review on progress in mitigation. *Food Additives and Contaminants*, *24* (S1), 47–59.

115. Hamlet, C.G., Baxter, D.E., Sadd, P.A., Slaiding, I., Liang, L., Muller, R., Jayaratne, S.M., Booer, C. (2005). Exploiting process factors to reduce acrylamide in cereal-

based foods, Report No. C014. A report prepared for the UK Food Standards Agency. RHM Technology Ltd, High Wycombe.

116. Kasai, N., Suzuki, T., Idogaki, H. (2006). Enzymatic degradation of esters of dichloropropanols: removal of chlorinated glycerides from processed foods. *Food Science and Technology*, *39*, 86–90.

117. Stadler, R.H., Theurillat, V., Studer, A., Scanlan, F.P., Seefelder, W. (2007). The formation of 3-monochloropropane-1,2-diol (3-MCPD) in food and potential measures of control, in *Thermal Processing of Food: Potential Health Benefits and Risks. Symposium of the Deutsche Forschungsgemeinschaft (DFG)* (eds G. Eisenbrand *et al.*), Wiley-VCH Verlag GmbH, Weinheim, pp. 141–155.

118. Mamma, D., Papadopoulou, E., Petroutsos, D., Christakopoulos, P., Kekos, D. (2006). Removal of 1,3-dichloro2-propanol and 3-chloro1,2-propanediol by the whole cell system of pseudomonas putida DSM 437. *Journal of Environmental Science and Health Part A-Toxic/Hazardous Substances & Environmental Engineering*, *41*, 303–313.

119. Bastos, F., Bessa, J., Pacheco, C.C., De Marco, P., Castro, P.M.L., Silva, M., Jorge, R.F. (2002). Enrichment of microbial cultures able to degrade 1,3-dichloro-2-propanol: a comparison between batch and continuous methods. *Biodegradation*, *13*, 211–220.

120. Yonetani, R., Ikatsu, H., Miyake-Nakayama, C., Fujiwara, E., Maehara, Y., Miyoshi, S.I., Matsuoka, H., Shinoda, S. (2004). Isolation and characterization of a 1,3-dichloro-2-propanol-degrading bacterium. *Journal of Health Science*, *50*, 605–612.

121. Bel-Rhlid, R., Talmon, J.P., Fay, L.B., Juillerat, M.A. (2004). Biodegradation of 3-chloro-1,2-propanediol with *Saccharomyces cerevisiae*. *Journal of Agricultural and Food Chemistry*, *52*, 6165–6169.

122. Laurent, H., Dreyfus, T., Poulet, C., Quillet, S. (2002). Process for obtaining aminopolyamide-epichlorohydrin resins with a 1,3-dichloro-2-propanol content which is undetectable by ordinary means of vapour-phase chromatography. U.S. Patent No 6342580. http://gb.espacenet.com (accessed November 2006).

123. Riehle, R.J., Busink, R., Berri, M., Stevels, W. (Hercules Inc.) (2003). Reduced byproduct high solids polyamine-epihalohydrin compositions. U.S. Patent Application 20030070783. http://www.uspto.gov/patft/index.html (accessed November 2006).

124. Yamamoto, S., Yoshida, Y., Kurumatani, M., Ota, M., Asano, S. (2001). Water-soluble thermosetting resin and wet-strength agent for paper using the same. U.S. Patent Application 20010034406. http://www.uspto.gov/patft/index.html (accessed November 2006).

125. Food Standards Australia New Zealand (FSANZ) (2006). Australia New Zealand Food Standards Code, 2006. Standard 1.4.1, Contaminants and Natural Toxicants. Anstat Pty Ltd, Canberra. http://www.foodstandards.gov.au/thecode/foodstandardscode.cfm (accessed 17 September 2008).

126. Health Canada (2005). Canadian Guidelines ("Maximum Limits") for Various Chemical Contaminants in Foods. Health Canada, Ottawa, Canada. http://www.hc-sc.gc.ca/fn-an/securit/chem-chim/contaminants-guidelines-directives_e.html (accessed January 2007).

127. China Standards (2000). SB 10338-12000, acid hydrolysed vegetable protein seasoning. http://www.chinagb.org/search/queryok.asp (accessed January 2007).

128. European Commission (2001). Commission Regulation (EC) No 466/2001 of 8 March 2001 setting maximum levels for certain contaminants in foodstuffs (OJ L 77 16.3.2001, p12).

129. World Trade Organization (WTO) (2001). Committee on Sanitary and Phytosanitary Measures Notification, Republic of Korea. Acid-hydrolyzed vegetable protein and soy sauce containing acid, G/SPS/N/KOR/106, 3 December 2001. http://www.ipfsaph.org/En/default.jsp (accessed January 2007).

130. World Trade Organization (WTO) (2001). Committee on Sanitary and Phytosanitary Measures Notification of Emergency Measures, Malaysia. Foods containing acid-hydrolyzed vegetable protein, G/SPS/N/MYS/10, 26 July 2001. http://www.ipfsaph.org/En/default.jsp (accessed January 2007).

131. Département fédéral de l'intérieur (DFI) (2006). Ordonnance du DFI du 26 juin 1995 sur les substances étrangères et les composants dans les denrées alimentaires, RS 817.021.23 (DFI schedule on foreign substances and components in foodstuffs). DFI, Berne. http://www.admin.ch/ch/f/rs/c817_021_23.html (accessed January 2007).

132. World Trade Organization (WTO) (2002). Committee on Sanitary and Phytosanitary Measures Notification, Thailand. Vegetables and derived products, G/SPS/N/THA/88, 26 March 2002. http://www.ipfsaph.org/En/default.jsp (accessed January 2007).

133. Committee on Food Chemicals Codex (1996). Food Chemicals Codex: First Supplement to Fourth Edition. Institute of Medicine of the National Academies, Washington DC. http://www.iom.edu/report.asp?id=4590 (accessed January 2007).

2.7

MAILLARD REACTION OF PROTEINS AND ADVANCED GLYCATION END PRODUCTS (AGEs) IN FOOD

THOMAS HENLE

Institute of Food Chemistry, Technische Universität Dresden, D-01062 Dresden, Germany

2.7.1 INTRODUCTION AND HISTORICAL PERSPECTIVE

The discovery of fire by *Homo erectus* about 400,000 years ago and the controlled use of this phenomenon was definitely one of the most important achievements of humankind (1). One can easily imagine that the benefits of heating foods, resulting in better hygienic quality, prolonged storage stability, improved digestibility, and—maybe the most important—better taste and flavor, were realized soon after. With regard to this, the Maillard reaction, which leads to the formation of numerous flavor and color compounds, can be attributed as the first chemical reaction that was used for a controlled enhancement of food quality. In other words, man and his ancestors have been eating the products resulting from this chemical reaction since several hundred thousand years. Nowadays, nobody can challenge that heating processes used in food industry are an indispensable prerequisite in order to obtain safe and high-quality food products. The chemistry behind this universal reaction, however, remained mysterious until the early years of the twentieth century. Although some speculations about the chemistry of browning reactions in foods were made before (2), it was the merit of Louis Camille Maillard to give the first detailed protocol on the reactions occurring during heating of amino

Process-Induced Food Toxicants: Occurrence, Formation, Mitigation, and Health Risks, Edited by Richard H. Stadler and David R. Lineback

Advanced Glycation End Products (AGEs): At a Glance	
Historical	Realization of the importance of Maillard-type reactions *in vivo* began in the mid-1970s, and around 1980 researchers began recognizing the importance of the late-stage Maillard processes as mediators of the complications of diabetes and aging. Subsequently, proteins bearing Amadori product have come to be referred to as glycated proteins. AGEs are best known as endogenous products of non-enzymatic glucose-protein interactions, which include a multitude of complex structures.
Analysis	Earlier work for the detection of AGEs mainly made use of HPLC methods, and recently MS techniques (e.g., LC/MS and MALDI-TOF MS) have been applied for the analysis of certain AGEs including carboxymethyllysine (CML). Rapid methods such as ELISA have also been described to measure CML.
Occurrence in Food	The Amadori products are the quantitatively dominating protein-bound Maillard reaction products in most foods, and may represent up to 70% of the lysine residues. CML is predominantly formed from the oxidative cleavage of Amadori products and has been reported in several heated foods. It may account for 3 to 10% of the corresponding Amadori products. Other AGEs such as pyrraline can occur in similar concentrations.
Main Formation Pathways	Reactions of 1,2-dicarbonyl compounds with basic amino acid residues or with Amadori intermediates are thought to be involved in many of the known AGE cross-linking structures. In food, more specifically, these lead to compounds such as CML, pyrraline, pentosidine, and more complex lysine dimer structures.
Mitigation in Food	Glycation reactions are the basis for the formation of many of the desired characteristics of cooked food, such as flavor, color, and shelf life. Thus, any mitigation strategies must consider the beneficial aspects of the Maillard reaction.
Dietary Intake	The amount of glycation products ingested depends on the thermal treatment of the food, and even within the same type of food, the amounts of glycation compounds may vary by one order of magnitude. Thus, the dietary intake can only be crudely estimated.
Health Risks	The possible pathophysiological role of dietary glycation products has become a topic of increasing interest over the past years. In diabetics with impaired renal function, food-derived AGEs can accumulate in serum, implying that normal kidney function is important to protect from dietary AGEs. However, no study so far has unambiguously shown a toxicological effect of a glycation compound in healthy subjects.
Regulatory Information or Industry Standards	Glycation products or AGEs in foods are not regulated.
Knowledge Gaps	There are indications that AGEs, formed either endogenously or taken up through the diet, may play a role in pathophysiological processes. However, the evidence available to date is not conclusive with regard to the possible health effects of dietary AGEs. Clearly, far more work is needed to clarify the possible health impact of AGE-rich foods.
Useful Information (web) & Dietary Recommendations	http://imars.case.edu/costimars2ndcirc.html http://www.if.csic.es/proyectos/cost927/PublicationsCOST-92714-07-05.pdf Avoiding the over-heating of food; cook/prepare foods to the appropriate "doneness."

acids and reducing sugars (3), and to realize the interdisciplinary implications of this "nonenzymatic browning" for food technology, physiology, and analytical chemistry (4). However, the consequences of these findings were not recognized until the 1940s, when several papers reported on the decreased nutritional value of long-term stored milk powder as a result of the reaction between proteins and lactose (5). During the following three decades, research on the Maillard reaction mainly focused on two general aspects, namely analytical strategies to assess the modification of the essential amino acid lysine due to browning reactions (6, 7), and the impact of Maillard reactions on flavor generation (8, 9). A principally new scientific aspect was added with the identification of HbA_{Ic}, a nonenzymatic glycosylated variant of hemoglobin, which was first found in the blood of diabetic patients (10). With this discovery, it was demonstrated that the Maillard reaction also occurs *in vivo*, and the term "glycation" was introduced as a synonym for "nonenzymatic glycosylation," in order to distinguish this process from the well-known enzymatic glycosylation of proteins (11). Protein modifications called "advanced glycation end products" (AGEs), which are formed *in vivo* during aging, diabetes, and in renal failure via comparable chemical pathways as described for heated foods, nowadays are generally accepted to play a pivotal pathophysiological role in several diseases (12). Due to the generally accepted importance of glycation reactions for pathophysiological processes, the Maillard reaction now seems to be a domain of biomedical research (Fig. 2.7.1). Since the

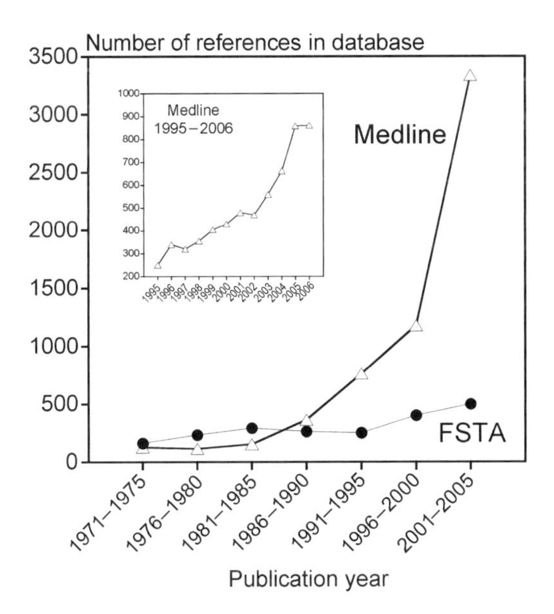

Figure 2.7.1 Number of references found in the databases Medline and FSTA (Food Science and Technology Abstracts) for the keywords "glycation OR Maillard reaction OR nonenzymatic browning OR advanced glycosylation OR furosine OR pentosidine OR carboxymethyllysine OR pyrraline OR methylglyoxal OR fructosamine."

amount of glycation compounds ingested with meals from certain heated foods far exceeds the total amount of AGEs in the human body (13), it was not surprising that questions arose very recently concerning the intake of dietary AGEs via the daily food and possibly related (patho)physiological aspects (14). In the meantime, some authors use the term "glycotoxins" for dietary glycation compounds in order to express that heated foods may represent a risk factor not only for diabetic and uremic patients, but also in "normal" lifestyle (15, 16).

It is the purpose of this chapter to give an insight into the chemistry and formation of individual glycation compounds in foods. Based on a brief review of selected literature, it shall be demonstrated which food items are recognized as the largest source of glycation compounds. Having a view on existing analytical methods on one hand and observed biological effects on the other, a discussion covering the controversial standpoints with respect to a possible "health risk" due to dietary AGEs will follow.

2.7.2 FORMATION PATHWAYS

Since the pioneering review of Hodge (17), most textbooks mention for reasons of clarity three "stages" of the Maillard reaction (Fig. 2.7.2). From a chemical point of view, this breakup into an "early," "intermediate," and "final" or "advanced" stage is somewhat artificial, as all reactions more or less occur simultaneously, with the product spectrum depending on the reaction conditions. Furthermore, it must be kept in mind that in most foods, the Maillard reaction can hardly be distinguished from caramelization reactions, the latter by definition leading to carbohydrate degradation in the absence of amino compounds (18). The formation of 1,2-dicarbonyl compounds such as methylglyoxal may be due mainly to such caramelization rather than to "pure" Maillard reactions. The academic discussion concerning the origin of certain reaction products in foods, however, should not be stressed here. However, it should be pointed out that as a matter of principle, glycation compounds represent amino acid derivatives.

For most food items, the "early stage" might be quantitatively the most important. The first steps of the Maillard reaction are characterized by a nucleophilic addition of amino compounds, which in protein-containing foods predominantly are represented by the ε-amino group of lysine residues and the α-amino group of *N*-terminal amino acids, to the carbonyl moiety of a reducing sugar, which in foods is mainly glucose, fructose, lactose, or maltose. Following the first condensation reaction and a subsequent Amadori rearrangement, peptide-bound ε- or α-deoxyketosyl amino acids or aminoketoses, also referred to as Amadori compounds or "sugar amino acids," are formed (19, 20). If fructose is the reacting sugar, the so-called Heyns compounds or 2-amino-2-deoxy-aldoses are formed, which should not be discussed further, as very little is known about the possible occurrence of these amino acid

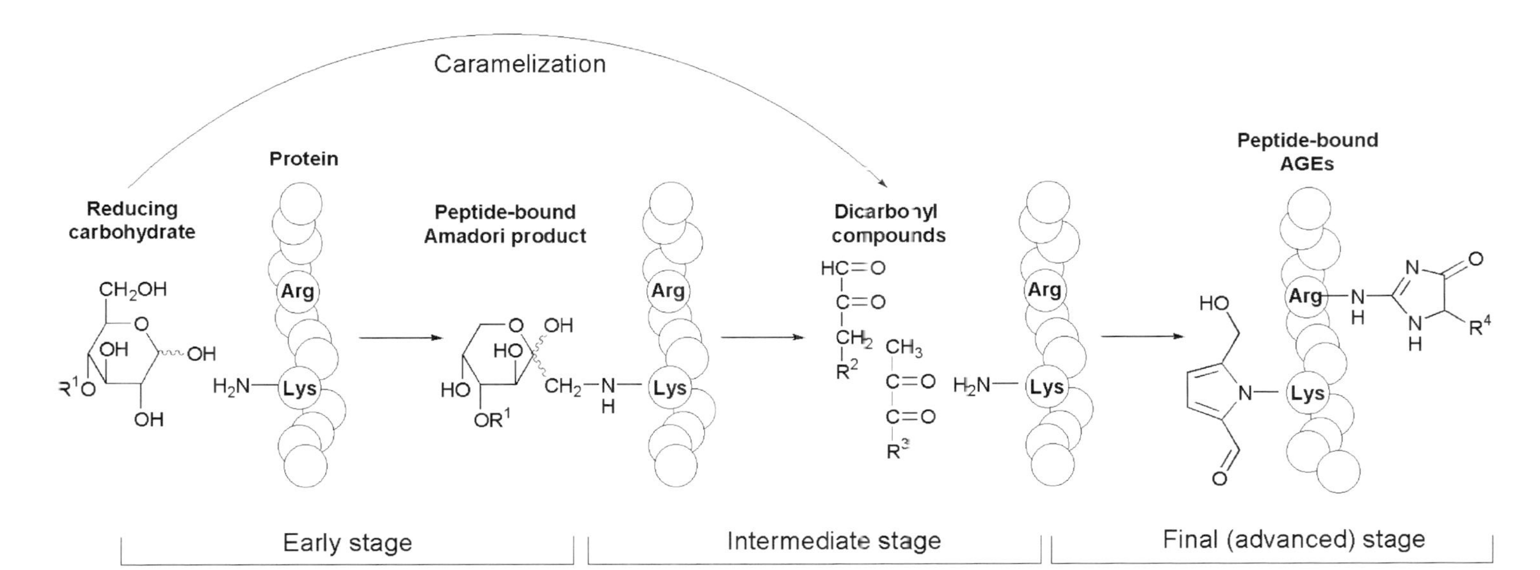

Figure 2.7.2 General scheme of the Maillard reaction (R^1 is H in the case of glucose, a β-galactosyl or α-glucosyl moiety in the case of maltose or lactose, respectively; for explanation of R^2 and R^3 see Fig. 2.7.3; R^4 is a methyl group in the case of MG-H1).

derivatives in foods (21–23). It should also be noted that free amino acids may react to Amadori products, which was demonstrated for foods such as dried fruits and vegetables (24), beer (25), or honey (26). The reactions of free amino acids will not be discussed further; this review will focus on the reaction of peptides and proteins.

The major reaction products of the early Maillard reaction of food proteins are the lysine derivatives *N*-ε-fructosyl-, *N*-ε-maltulosyl-, or *N*-ε-lactulosyllysine (Fig. 2.7.3). For lactose- and peptide-containing foods such as hypoallergenic infant formula, a significant modification of *N*-terminal α-amino acids was reported very recently, resulting in the formation of Amadori products such as *N*-α-lactulosylvaline or *N*-α-lactulosylleucine, respectively (19).

The "intermediate stage" of the Maillard reaction may be characterized by the formation of 1,2-dicarbonyl compounds (Fig. 2.7.2). Mainly due to prolonged storage or severe heating, the amino ketoses may degrade to some extent via enolization and elimination reactions, leading mainly to 3-deoxyglucosulose (27) and 1-deoxyglucodiulose (28) (Fig. 2.7.3). In disaccharide-containing foods, the formation of 1-amino-4-deoxy-2,3-ketodiuloses may be of importance (29, 30). Retroaldolization results in the formation of short-chain carbonyls such as methylglyoxal and glyoxal. For a detailed discussion of possible reaction mechanisms, the reader is referred to general textbooks of food chemistry (31). As mentioned earlier, 1,2-dicarbonyl compounds are also formed from reducing carbohydrates without involvement of amines during

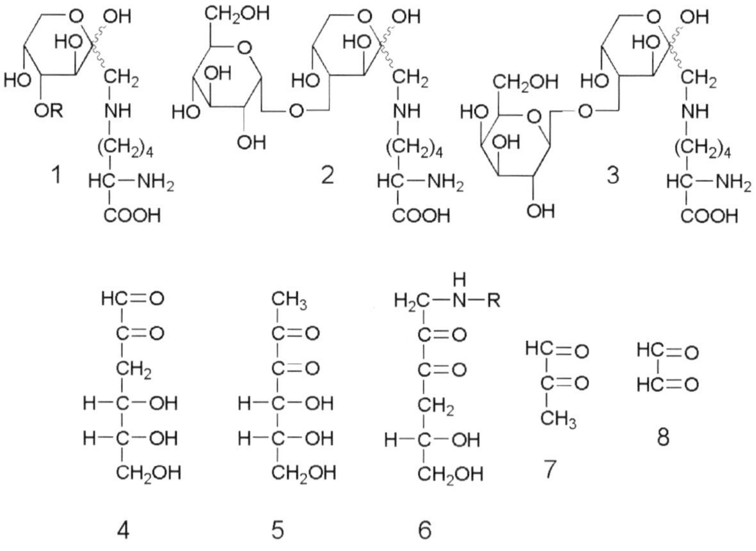

Figure 2.7.3 Examples for Amadori compounds and 1,2-dicarbonyl compounds formed in the early and advanced stages of the Maillard reaction. 1, *N*-ε-fructosyllysine; 2, *N*-ε-maltulosyllysine; 3, *N*-ε-lactulosyllysine; 4, 3-deoxyglucosulose; 5, 1-deoxyglucodiulose; 6, 1-amino-4-deoxy-2,3-ketodiuloses; 7, methylglyoxal; 8, glyoxal.

caramelization (32). The formation of 1,2-dicarbonyls "directly" from reducing carbohydrates should be quantitatively more important compared with their formation via degradation of Amadori products (33).

Although formed in relatively low amounts when compared with the main carbohydrates in foods, the dicarbonyl compounds now represent the direct precursors for the formation of AGEs. The "final" or "advanced" stage" of the Maillard reaction, therefore, can be characterized as the reaction of 1,2-dicarbonyls with side chains of peptides or proteins (Fig. 2.7.2). The side chains of peptide-bound lysine and arginine residues represent main targets for a derivatization by 1,2-dicarbonyl compounds. When screening the literature, it cannot be overlooked that most of the lysine or arginine derivatives known today as glycation compounds have only been described for model mixtures, and have not yet been found in foods or biological systems. For only a handful of compounds, the formation during food processing is reported based on profound analytical characterization and quantification (Fig. 2.7.4). *N*-ε-carboxymethyllysine (CML), initially identified in human lens proteins and collagens (34), was the first amino acid derivative of the advanced Maillard reaction that was quantified in foods, where it may account for 3% to 10% of the corresponding Amadori products (35). CML is predominantly formed from oxidative cleavage of Amadori products (33, 36), but it may also arise from the reaction of the lysine side chain with carbonyl compounds resulting from lipid peroxidation (37). As first representative of heterocyclic

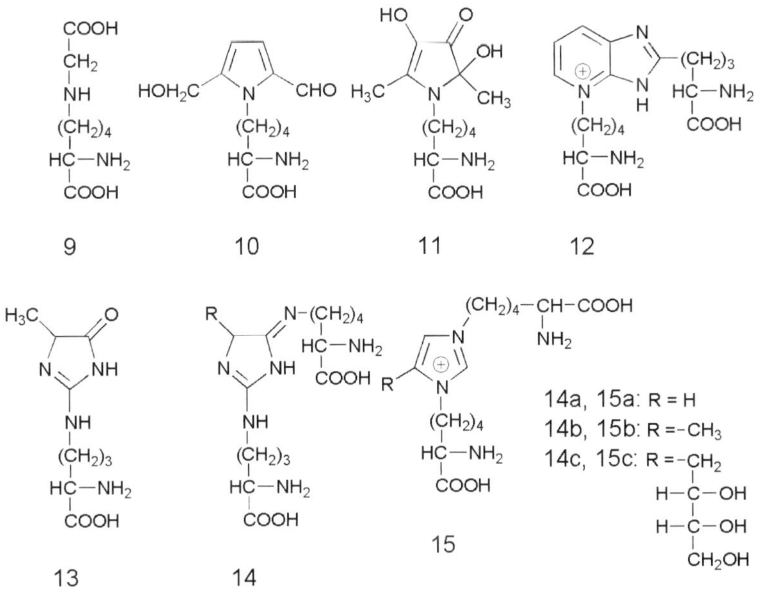

Figure 2.7.4 AGEs detected and quantified in foods. 9, CML; 10, pyrraline; 11, pronyl-lysine; 12, pentosidine; 13, MG-H1; 14a, GODIC; 14b, MODIC; 14c, DODIC; 15a, GOLD; 15b, MOLD; 15c, DOLD.

AGEs, pyrraline, which was initially identified as a minor compound resulting from the reaction of lysine and glucose in heated model mixtures (38), proved to be a quantitatively important glycation compound in several food items (39). Pyrraline is formed by reaction of the ε-amino group of lysine with 3-desoxyglucosulose (40). Due to the fact that the pyrrole compound is labile during acid and alkaline hydrolysis, quantification can only be achieved after complete enzymatic hydrolysis of food proteins (41). A structurally similar lysine derivative, named pronyl-lysine, resulting from lysine side chains and acetylformoin, was recently detected and quantified in bread crust (42). With respect to the well-known phenomenon of protein oligomerization as a consequence of Maillard reactions (43, 44), few cross-linking glycation compounds have been identified in food samples. Among these, the well-known pentosidine, initially identified in hydrolysates of human collagen (45), was the first "Maillard cross-link" that was detected in low amounts in acid hydrolysates of severely heated milk, bakery products, and roasted coffee (46). In this compound, one lysine and one arginine residue are linked together by a C5-precursor resulting from carbohydrate degradation (47). By the identification of pentosidine, it became clear that also peptide-bound arginine moieties are substrates for advanced Maillard reactions. The guanidino side chain of arginine is not involved in Amadori-type reactions, but is predominantly modified by the reaction with 1,2-dicarbonyl compounds. This was proven first with the isolation and identification of an imidazolinone resulting from the reaction of arginine and methylglyoxal (48). This amino acid derivative, meanwhile generally named MG-H1 (49), represents the major form of arginine derivatization in certain bakery products and coffee (46) as well as in physiological samples (50). The formation of further imidazolinones via the reaction of arginine with other dicarbonyl compounds such as glyoxal, 3-deoxypentosulose, or 3-deoxyglucosulose in foods is very likely (29, 51, 52); however, these compounds have not yet been identified and quantified in food samples. The same is true for a structurally different arginine derivative, namely argpyrimidine, which originally was identified as an AGE formed *in vivo* (53). To date, argpyrimidine is only quantified as free amino acid in food samples (54). It remains to be elucidated whether this compound represents a quantitatively important product of protein glycation.

Recently, the occurrence of cross-linking amino acids resulting from the reaction of two lysine side chains and two molecules of glyoxal, methylglyoxal, or 3-deoxyglucosulose in foods was reported (55). These lysine derivatives, GOLD (gyloxal-lysine-dimer), MOLD (methylglyoxal-lysine-dimer), and DOLD (3-deoxyglucosulose-lysine-dimer), previously were identified in biological samples, where they shall form by the reaction of two amino groups of lysine side chains and two dicarbonyl molecules via a Cannizarro-like reaction, yielding formaldehyde during the cyclization reaction (56). Biemel *et al.* (55) also reported on the quantification of the cross-linking amino acids GODIC, MODIC, DODIC, and glucosepan, which are formed by an addition of a lysine side chain to an imidazolinone intermediate such as MG-H1 (57).

In addition to the compounds mentioned here, several other reaction products of advanced stages of the Maillard reaction of proteins have been described, which to date have not been quantified unambiguously in foods, and therefore will not be discussed here.

2.7.3 ANALYSIS AND QUANTITATIVE DATA OF GLYCATION COMPOUNDS IN FOODS

2.7.3.1 Amino Acid Derivatives

Despite the advances in modern instrumental analysis during the last decades, quantification of individual glycation compounds in food still remains an enormous challenge. For most of the compounds mentioned earlier, no standard material is commercially available. Establishing a strategy for analysis, therefore, first of all requires synthesis of pure reference material of the said compound and sufficient characterization of the isolates in terms of identity and purity, based on spectroscopic characterization such as NMR and MS as well as elemental analysis. Next, quantification normally is only possible together with unambiguous identification, which means that the quantitative data generated by an analytical technique are only valid if they are supported by structural information. This is of special importance for methods that do not provide "qualitative" (structural) information together with the "quantitative" signal. Prominent examples, in this context, are immunological methods such as ELISAs (enzyme-linked immunosorbent assays), based on (more or less) specific antibodies. If such methods were used for generating quantitative data, it normally would be an absolute must that the ELISA method used to quantify glycation compounds in certain matrices is validated by cross-checking the results with a more specific method, like HPLC-MS. Especially for immunological methods, it is basic knowledge that the antigen–antibody interaction is widely dependent on the milieu conditions and therefore on the composition of the matrix. An ELISA that works fine in urine may not give reliable results for cheese samples. Keeping this in the back of one's mind while screening the relevant literature, it may be obvious in many cases that data reported for glycation compounds do not fulfill the basic requirements for a credible analysis even if published in "high-impact" journals. It seems to be important to address this issue mainly because of recent publications reporting astonishing data of glycation compounds in foods (58). In the following, therefore, only these reports are mentioned in which valid data are reported based on the use of methods fulfilling the aforementioned criteria.

From the quantitative point of view, the early Maillard reaction, characterized by the formation of peptide-bound Amadori products, is the dominating modification found in foods (Table 2.7.1). It has been known since the early 1980s that heat treatment of milk, in particular sterilization, as well as long-term storage of milk or whey powder may lead to a significant lysine modification, reaching 10% to 20% in certain samples. Lactose-hydrolyzed whey may

TABLE 2.7.1 Range of protein-bound Amadori compounds and AGEs in foods.

	Milk products	Bakery products	Pasta	Coffee	Roasted meat
Amadori	RM/PM: 1–2 UHT: 2–5 SM: up to 50; cheese: up to 10; MP/WP: 50 to 200 (up to 700 in case of lactose-hydrolyzed WP)	Bread crumb up to 200; bread crust up to 800	Up to 400	nd	1–10
CML	nd–10	nd–20	—	—	Up to 0.1
Pyrraline	nd–25	1–10, up to 180 in case of bread crust	nd to 13	—	nd
Pronyllysine	—	Bread crumb: 0.01 Bread crust: 0.1	—	—	—
Pentosidine	UHT: nd–0.01	nd–0.4	—	0.2	nd
MG-H1	—	10–20, up to 700 in case of alkaline-treated products	—	100–350	—

Data are given as mmol/mol lysine or mmol/mol arginine (in the case of pentosidine and MG-H1).
RM, raw milk; PM, pasteurized milk; UHT, UHT-treated milk; SM, sterilized milk; MP, milk powder; WP, whey powder; nd, not detected.

even contain up to 70% modified lysine, which is due to the higher reactivity of monosaccharides compared with lactose (6). It is noteworthy that a lysine modification of 0.1% is found even in raw milk. This low amount of Amadori products is only slightly increased during pasteurization or UHT treatment. In bakery products, *N*-ε-maltuloselysine and *N*-ε-fructoselysine are the dominating lysine derivatives. Quantification of the extent of the early Maillard reaction can be achieved by measuring furosine, a reaction product that is formed from the Amadori products of lysine during acid hydrolysis (Fig. 2.7.5) (59, 60). For the use of furosine as well as corresponding furoylmethyl amino acids (FMAAs) as indirect parameters for Amadori products, it is necessary to have information on to which extent these derivatives are formed during acid hydrolysis (61). If the corresponding conversion factors are known, FMAAs quantified via ion-exchange chromatography (7), RP-HPLC with UV detection (62), or capillary electrophoresis (63), represent valuable analytical tools for sensitive monitoring of Amadori product formation not only in foods but also in biological samples such as blood or urine (64, 65).

The group of Erbersdobler was the first to determine CML in milk products using ion-exchange chromatography with ninhydrin detection (34). Shortly after this initial report, CML was determined in acid hydrolysates of various food samples via gas-liquid chromatography (GLC). For GLC, various

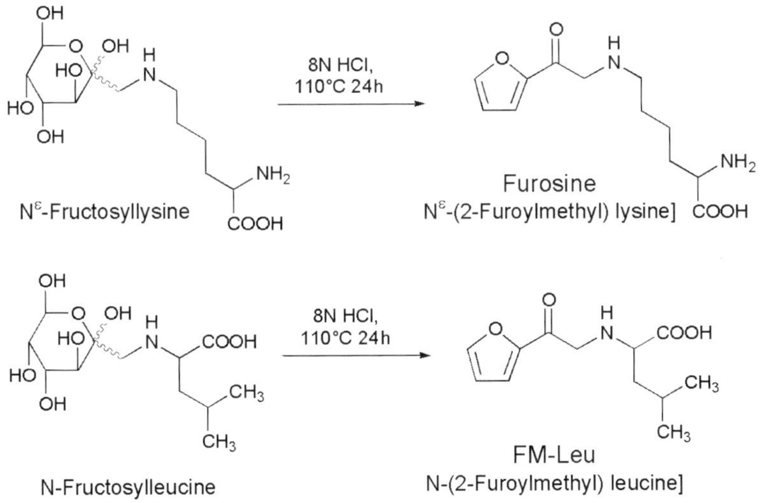

Figure 2.7.5 Formation of *N*-(2-furoylmethyl) amino acids (FMAAs) during acid hydrolysis of Amadori compounds.

derivatization and detection techniques were applied. The corresponding heptafluorobutyryl isobutyl ester of CML was analyzed using phosphorus–nitrogen detection (66). The pentafluoropropionyl isopropyl derivative was measured using mass spectroscopy after chemical or electron impact ionization (67). Alternative methods for quantification include RP-HPLC after derivatization with orthophtaldialdehyde (68) and MALDI-TOF-MS (69). GC-MS using isotopically labeled reference material proved to be most specific method for the quantification of CML (70). At this point, it is noteworthy to comment on the use of immunological methods. Although ELISAs for the quantification of CML in milk products without need for cleanup have been published (71), authors agree that these methods are only suitable for certain types of foods, as long as no validation has been made by comparing data from immunological analysis with those from chromatographic assays. Comparing ELISA with GC-MS, it was recently shown that an ELISA, which was performed on infant formulas, gave satisfactory results in powdered but not in liquid formulas (72). Data for CML published in literature based on the sole use of an ELISA, therefore, must be interpreted with care. Unfortunately, corresponding reports (73) gained much interest in context with nutritional consequences resulting from dietary CML, although many of the published data are somewhat dubious and therefore shall not be discussed further.[1]

[1] For instance, butter and olive oil, respectively, are reported to contain several hundredfold higher "amount" of CML when compared with the crust of whole wheat bread (265,000 U/g or 120,000 U/g versus 730 U), but such high amounts of an amino acid in the lipid samples are virtually impossible. Even if butter would contain a "protein," composed solely of CML, such high concentrations of CML would not be realistic.

Pyrraline, the acid-labile pyrrole derivative of lysine, was quantified after enzymatic hydrolysis in several foods such as milk products, enteral formula, bakery products, and pasta. The methods used were ion-exchange chromatography with photodiode array detection (40) and RP-HPLC with UV detection (74, 75). Very recently, LC-MS was successfully applied for the quantification of pyrraline, and conditions for hydrolysis were carefully evaluated (76). Pyrraline is a suitable indicator for "advanced stages" of the Maillard reaction. Concentrations found ranged from 150 mg/kg protein in sterilized milk up to 3700 mg/kg protein in bread crusts.

For the analysis of pronyllysine, hydrazinolysis with methyl hydrazine and transformation of the glycation compound to a stable derivative, namely 5-acetyl-4-hydroxy-1,3-dimethylpyrazole, had to be performed, which then was analyzed using high-resolution GLC with mass spectroscopic detection. Concentrations of around 60 mg/kg in the crust and 6 mg/kg in the crumb of bread were found (41).

To the best of the author's knowledge, there is only one report demonstrating the occurrence of the cross-linking amino acids GODIC, MODIC, DODIC, and glucosepan in foods (54), in which LC-MS with electrospray ionization (ESI) after enzymatic hydrolysis was used to measure the glycation compounds in bakery products. Concentrations up to 150 mg/kg protein were found. Amounts of GOLD and MOLD were significantly lower, indicating that these imidazolium cross-links may play only a minor role for glycation reactions in foods.

Also for pentosidine, only very low amounts were found in acid hydrolysates of several foods using ion-exchange chromatography with direct fluorescence detection (44). Only for roasted coffee and some bakery products, values up to 35 mg/kg protein were measured. Interestingly, it was found that roasted coffee contains pentosidine in the free form, which is therefore readily absorbed and excreted via the urine (77). High-pressure pasteurization of foods may increase the formation of pentosidine (78).

2.7.3.2 1,2-Dicarbonyl Compounds

It is textbook knowledge that 1,2-dicarbonyl compounds are formed during the late stages of the Maillard reaction, representing important precursors for flavor and color formation in food. However, despite this generally accepted role in food chemistry, only very limited information is available about the amount of individual compounds in foods (Table 2.7.2). Most of the reports deal with the quantification of methylglyoxal. Quantification of methylglyoxal generally requires deproteinization of the food sample, followed by derivatization of methylglyoxal and subsequent chromatography. Various derivatizing agents are reported, and chromatographic methods include HPLC with UV-, fluorescence or MS-detection as well as GLC with ECD or MS (for review, see Reference 79). Data can be found for methylglyoxal in coffee brews, ranging from 23 to 47 mg/L for coffee prepared from roasted beans or for instant coffee, respectively (80). The concentration of methylglyoxal was found to increase in the early phases of the roasting process, followed by a decline.

TABLE 2.7.2 Range of 1,2-dicarbonyl compounds in foods.

	3-DG	MGO	GO
Cheese	—	4–11	4–6
Yogurt	—	0.6–1.3	0.6–0.9
Wine, sherry	—	0.7–1.8	0.6–0.9
Apple juice	—	0.3	nd
Maple syrup	—	2.5	nd
Cocoa	0.5–3.6	0.02	0.9–3.4
Coffee			
Roasted beans	—	20–220	20–130
Brew	—	23–47	—
Honey	79–1451	nd–5.7	nd–4.6
Manuka honey	563–1060	38–761	0.7–7.0

Data are given in mg/kg or mg/L, respectively; nd, not detected.

As light- and medium-roasted coffees had higher contents of glyoxal and methylglyoxal compared with dark roasted coffee (81), it can be speculated that the last mentioned coffee samples may contain significantly higher amounts of peptide bound adducts of methylglyoxal such as the imidazolinones described earlier, which, therefore, may serve as indicators for the control of roasting conditions. Besides glycation reactions, lipid peroxidation may also be a source of methylglyoxal (82). Furthermore, methylglyoxal has also been reported to originate from enzymatic pathways, as up to 1.8 mg/L methylglyoxal has been determined together with trace amounts of glyoxal in fermented foods such as wine, beer (83–86), and dairy products (87). Very recently, honey was reported to contain large amounts of 1,2-dicarbonyl compounds, in particular 3-deoxyglucosulose (88). For a number of commercially available honey samples, amounts of 3-deoxyglucosulose ranged from 79 to 1451 mg/kg, which is up to 100-fold higher compared with 5-hydroxymethylfurfural, a well-known marker for the control of heat treatment of honey. For most honey samples, values for glyoxal and methylglyoxal were in the ranges 2–2.7 mg/kg and 0.4–5.7 mg/kg, respectively, and were not affected by storage. Very recently, however, for samples of New Zealand Manuka honey (*Leptospermum scoparium*), surprisingly high amounts of methylglyoxal were found, ranging from 38 up to 761 mg/kg, around two orders of magnitude higher compared with conventional honeys (89). Methylglyoxal in this concentration proved to be directly responsible for the pronounced antibacterial activity of Manuka honey, which is widely used as a medicinal food (90).

2.7.4 MITIGATION

Due to the fact that glycation reactions are the basis for the benefits of cooking, baking, and roasting, such as flavor, color, and improved storage stability, concrete strategies for a "mitigation" of individual compounds during heat treatment cannot be established without changing the processing conditions, which directly influences product quality. In most cases, formation of glycation compounds is strongly correlated with the intensity and the time of heating. Generally, foods shall be heated high enough, mainly with respect to microbial safety, but heating should not exceed minimal processing conditions in order to keep undesired constituents as low as possible. This can be the basic rule for food processing and storage also with respect to the formation of glycation compounds.

In addition to the control of processing conditions, suggestions for mitigation of the formation of glycation compounds in foods also include additives such as thiol compounds like cysteine or glutathione, which are well-known inhibitors of enzymatic as well as nonenzymatic browning (91). From a mechanistic point of view, thiols may act as scavengers for 1,2-dicarbonyl compounds (92), but at present no information is available concerning mitigation of individual glycation compounds. Due to the complex chemical pathways, which lead to the formation of a myriad of individual compounds on the one hand, and the great demands that are made on high-quality foods, the feasibility of glycation inhibitors for the use in food processing is more than questionable and therefore shall not be stressed further.

2.7.5 CONSIDERATIONS ON EXPOSURE

Based on the given quantitative data, it is obvious that glycation compounds present in various foods represent an important part of our daily diet. The amount of ingested glycation compounds per day depends largely on the heat treatment of the food items present in the diet. Since even within one type of food, say bakery or milk products, the amount of individual glycation compounds may vary within one order of magnitude, the "daily dose" of Maillard compounds can only be estimated. However, some general rules can be drawn. First of all, it must be realized that in practically all protein-containing foods, a well-defined Amadori compound such as *N*-ε-fructosyl-, *N*-ε-maltulosyl-, or *N*-ε-lactulosyllysine, which result from early Maillard reactions, represents the major form of glycation compounds, accounting for more than 90% of the amino acid derivatives present. Compared with this, the concentration of advanced glycation compounds is significantly lower and furthermore is divided up in several individual compounds of varying quantitative importance, among which the lysine derivatives CML and pyrraline, and in some cases the imidazolinones resulting from arginine modification, may be the most important candidates. Furthermore, it is important to notice that bakery products, in particular bread crust, are definitely the main source of glycation compounds, followed by milk products, for which the extent of protein glycation shows the largest variations depending on the processing conditions. Although at present only few data are available, roasted meat contains remarkably low amounts of Maillard compounds. It therefore is noteworthy that vegetarians consuming protein mainly of plant and milk origin will have a higher intake of glycation compounds when compared with carnivores. This may explain why significantly enhanced plasma AGE levels were found in vegetarians in comparison with omnivores (93). In the next section, implications related to this aspect will be discussed in connection with possible physiological consequences.

The intake of glycation compounds in a hypothetical diet consisting of 1 L of heat-heated milk, 500 g of bakery products, and 400 mL of coffee, respectively, was calculated to be between 1500 and 4000 μmol of Amadori compounds (corresponding to 500 to 1200 mg of fructoselysine) and 100 to 300 μmol (or 25 to 75 mg) of AGEs, mainly pyrraline and CML (13).

Estimating the uptake of 1,2-dicarbonyl compounds is even more difficult, as no data are available for staple foods. A rough estimation, taking the data shown in Table 2.7.2 into account, may point to a daily dose of a few milligrams of methylglyoxal. One cup of coffee corresponds roughly to a maximum uptake of 0.5 to 1.0 mg of methylglyoxal. These values, however, significantly increase when Manuka honey rich in methylglyoxal is consumed, which alone may count for 5–10 mg of this 1,2-dicarbonyl compounds per serving of 10 to 15 g of honey. Similar considerations can be made for 3-deoxyglucosulose. Here, with the exception of honey, the supply via daily food is virtually unknown. One serving of conventional honey may account for up to 10–15 mg of 3-desoxyglucosulose. Taking honey for a food item rich in monosaccharides, it must be stated here that other foods containing glucose and/or fructose such as dried fruits and vegetables, fruit juices, and soft drinks may also be candidates containing significant amounts of 1,2-dicarbonyl compounds. Future studies based on the use of reliable methods are needed in order to obtain a more comprehensive view on the supply of 1,2-dicarbonyl compounds via the daily diet.

2.7.6 HEALTH RISKS AND RISK MANAGEMENT

2.7.6.1 From Lysine Modification to the Maillard Reaction *In Vivo*

Evidence for the fact that browning reactions during food processing may have an adverse effect on the nutritional quality of food proteins, which in particular is due to lysine modification, can already be drawn from papers published in the first half of the twentieth century (94, 95). Finot and coworkers (96) were the first to prove that the Amadori product of lysine, namely *N*-ε-fructosyllysine, is not used as a lysine source *in vivo*. The assessment of such "lysine deterioration" caused by food processing first was a domain of dairy science. Several methods for the determination of "blocked" and "available" lysine in milk products have been reported (97–100). Examinations about a possible toxicological role of Amadori products or other peptide-bound Maillard compounds, however, were not undertaken. A virtually new view on possible nutritional consequences resulting from Maillard reactions in foods was introduced following the discovery of the fact that glycation reactions also occur *in vivo*. Starting with the hemoglobin variant HbA_{IC}, in which the *N*-terminal valine of the beta-chain has reacted with glucose to an *N*-terminal fructosylvaline (101), it was soon observed that mainly proteins with a slow turnover *in vivo* are prone to nonenzymatic glycosylation in the human body (102). Increased glycation of proteins *in vivo* was linked to the pathophysiology of diabetes, and corresponding biological disorders such as cataract, joint stiffening, or diabetic nephropathy were discussed to be at least in part due to glucose-modified proteins (103, 104). As consequences of diabetes often were attributed as "accelerated aging," glycation reactions shall also be important for physiological alterations observed during "normal" aging of the human body (105). For nephrologists, AGEs now represent an important class of "uremic toxins" (12), as uremic patients were found to accumulate glycation compounds such as pentosidine or CML in the plasma and tissues, probably due to impaired renal clearance (106). Further strengthening for the hypothesis that the formation of AGEs *in vivo* is detrimental was drawn from the identification of binding proteins for AGEs, among which RAGE, a multiligand receptor for AGEs, became the most popular (107). Binding of ligands to RAGE is discussed as a mechanism to initiate a variety of cellular dysfunctions in inflammatory disorders (108–110). Without going into details, it should be mentioned that in recent publications, serious doubts have arisen concerning the generally accepted role of AGE–RAGE interactions and its pathophysiological consequences (111).

Based on this body of evidence, it may not be a surprise that a possible pathophysiological role of dietary glycation compounds became a topic of scientific interest in recent years. At present, a debate is going on whether dietary glycation compounds represent a risk for the consumer or not (112), and some aspects related to both standpoints shall be mentioned here.

2.7.6.2 Are Dietary AGEs a Risk for the Consumer?

Arguments supporting the motion that an intake of thermally processed foods may be detrimental to human health are based on animal studies as well as on studies with human volunteers (for review, see Reference 113). Some representative reports shall be mentioned here. Papers arguing that dietary glycation compounds are a "risk" for the consumer most often start their argumentation with the observation that the serum levels of circulating AGEs can be raised after eating fructose-modified egg white (15). As for patients with renal failure, the decrease of plasma AGEs was slower when compared with healthy volunteers. Speculations concerning a possible reactivity of AGEs and accumulation *in vivo* were made, which motivated the authors to create the phrase "glycotoxins" for glycation compounds present in foods (15). As mentioned earlier, this "pioneering" study is questionable from an analytical point of view, as an immunological assay based on an antibody against CML was used for quantification, and no information about the reliability of the assay in quantifying CML in plasma, urine, and foods is given. Using the same analytical approach for quantification of AGEs, several papers reported on possible pathophysiological effects of dietary glycation compounds. For rats fed a "high-AGE diet" for 6 months, a decreased glucose and insulin response was found together with 1.5-fold increase in plasma AGEs (114). Improved insulin sensitivity and a prevention of diabetic nephropathy was observed for diabetic mice after a 2-month "low-AGE diet" (115, 116), whereas a diet high in AGEs was made responsible for impaired wound healing in genetically diabetic mice (117). For patients with renal failure, dietary glycation compounds were found to correlate with circulating AGEs (118). For dialysis patients on an AGE-free diet for 4 weeks, a significant decrease of inflammatory markers such as tumor necrosis factor-α and C-reactive protein was observed (119, 120). Following ingestion of a single "high-AGE meal" by patients suffering from type 2 diabetes, impaired micro- and macrovascular function concomitant with an increase of markers for postprandial vascular dysfunction and oxidative stress was found (121).

When reviewing reports presenting evidence for detrimental effect of dietary AGEs, one cannot overlook that most of the studies report biological phenomena without any chemical characterization of the stimulating agent. "AGEs" are generally taken as "anonymous" mixture of compounds, for which no reliable quantitative data are given. Most of the studies refer to the aforementioned database (73) when creating diets high or low in AGEs. In other words, up to now, there is *no* study showing unambiguously any toxicological effect of one single glycation compound based on accepted structure-function assays. Actual studies based on chemically characterized compounds are rare, however, in general report either on low bioavailability of glycation compounds or on quick elimination without accumulation *in vivo*. From peptide-bound Amadori products such as *N*-ε-fructosyllysine or *N*-ε-lactulosyllysine, only 3% to 10% are resorbed and are rapidly excreted via the kidney (122–124). Even in dialysis patients, plasma concentrations of Amadori products did not rise following a diet high in *N*-ε-lactulosyllysine (125). Enzymes produced by the microbiota present in the colon of humans may be responsible for the degradation of Amadori products (126).

For rats fed a diet high in *N*-ε-carboxymethyllysine, about half of the administered CML was recovered in the feces and the urine, whereas most of the compound was discussed to be metabolized by the colonic microflora, as virtually no deposition of CML in organs was found (127). Using positron-emission tomography, rapid elimination without accumulation was found after intravenous injection of radioactively labeled CML (128). *N*-ε-fructoselysine and CML were the only glycation compounds for which a study was performed to investigate the intestinal transepithelial absorption *in vitro* and to determine whether these glycation compounds are substrates for physiologically occurring membrane transport proteins for amino acids and dipeptides. The transepithelial flux measured for these compounds was very low, indicating that resorption most probably occurs by simple diffusion (129). The metabolic fate of dietary pyrraline and pentosidine has been investigated in an intervention study with 18 healthy volunteers. It was found that peptide-bound pyrraline present in bakery products is rapidly proteolyzed and resorbed. Most of the ingested pyrrole derivative (ca. 80%) was found in the urine within 24h after the diet. Similar data were obtained for free pentosidine in coffee, whereas the peptide-bound cross-link seems to be not available during digestion (130). In addition to the studies reporting on bioavailability and rapid elimination of individual compounds, several reports argue against adverse effects of AGEs and even discuss positive aspects. In a cross-sectional study with 312 hemodialysis patients, it was found that high serum AGEs are not linked to increased mortality (131). This observation was confirmed recently for 540 patients suffering from diabetes and nephropathy, for which serum CML content could not be identified as an independent risk factor for cardiovascular or renal outcomes (132). As already mentioned, the generally accepted role of RAGE as a specific receptor for AGEs, which should trigger inflammatory responses on a cellular level, has come under debate. It was found that chemically defined AGEs free of endotoxins do not bind to RAGE, and AGE–RAGE interaction is not sufficient to induce inflammatory signals, thus arguing against a uniform role of AGEs in cellular activation (133–135). In this context, it should also be noted that despite the highly visible number of papers dealing with RAGE, to date, none of them has ever reported on an individual glycation compound as specific ligand of RAGE. A paper in which CML is suggested to specifically interact with RAGE must be handled with care, as binding could also result from unspecific electrostatic interactions between the receptor protein, which due to its low isoelectric point is positively charged under the assay conditions, and negatively charged proteins containing CML residues (136). No controls with other proteins containing, for instance, high amounts of aspartic or glutamic acid were made in order to exclude such unspecific "ion-exchange" mechanisms and to prove selective binding mechanisms.

Finally, it shall be mentioned that more and more papers discuss possible benefits of glycation compounds. A "chemoprotective" role of individual dietary AGEs may result from their antioxidative properties, which is well established for melanoidins from various food sources (137–139). Pronyllysine, which is present in bread crust melanoidins, may be one of the important antioxidants in this context (140). Melanoidins formed during baking of bread or caramelization of fructooligosaccharides may have a prebiotic effect on colonic microorganisms (141, 142). Several Maillard compounds identified as constituents of melanoidins were found to inhibit tumor cell growth (143).

2.7.7 CONCLUSION

A legal system normally presumes the accused as innocent, until guilt is proven. In this context, the potential "health risk" of dietary AGEs must be discussed very carefully. Although there are indications that AGEs, either formed endogenously in the human body or taken up through the daily diet, may play a role in pathophysiological events, it is definitely not proven that AGEs, either in total or one individual glycation compound, are toxic or may be the cause of any adverse effect. On the other hand, several arguments point against adverse effects, and the fact that glycation compounds may also have positive biological effects may support the theory that humans are not only accustomed to consuming cooked foods and their constituents for hundreds of thousands of years, but also that these compounds play a role as potential chemoprotective agents. The data available to date suggest that dietary AGEs are not harmful to healthy individuals. Of course it cannot be excluded that for certain subpopulations, dietary AGEs may pose a greater risk, although to date no studies have unequivocally shown this. Besides diabetics, these groups include for instance uremic patients, due to reduced elimination, or infants, whose gastrointestinal tract is more permeable and therefore may resorb more glycated peptides from the diet.

Reports dealing with the physiological effects of glycation compounds, therefore, must be interpreted with caution. From a chemical viewpoint, only a very small fraction of the glycation compounds present in foods have been adequately characterized. Current data show that the metabolic transit may differ largely. Thus, reference to "AGEs in total" should be avoided and the research focused on individual compounds rather than on crude mixtures. Animal feeding studies or human intervention studies, in which effects of diets "high" or "low" in AGEs are presented, should only be accepted if the amount of individual glycation compounds in the said food items is quantified using reliable chromatographic techniques, ideally calibrated with pure reference standards. "Calculations" of an AGE content on the basis of flawed data collections are out of the question. A risk assessment must be based on a data evaluation for individual Maillard compounds in common foods using reliable chromatographic techniques rather than screening methods such as ELISA. Furthermore, animal experiments and intervention studies must be based on in-depth chemical characterization of the individual compounds. Furthermore, knowledge on the structure–activity relationships and the occurrence of the compounds is a prerequisite for any statement on biological effect. A real risk assessment must be based on sound science, considering all pertinent disciplines and joining the resources of biology, medicine, and chemistry.

REFERENCES

1. Balter, M. (1995). Archaeology—did *Homo erectus* tame fire first? *Science*, *268*, 1570.
2. Ling, A.R. (1908). Malting. *J. Inst. Brewery*, *14*, 494–521.
3. Maillard, L.C. (1912). Action of amino acids on sugars. Formation of melanoidins in a methodical way. *Compt. Rend.*, *154*, 66–68.

4. Finot, P.A. (2006). Historical perspective of the Maillard reaction. *Ann. N. Y. Acad. Sci.*, *1043*, 1–8.
5. Doob, H., Willmann, A., Sharp, P.F. (1942). Influence of moisture on browning of dried whey and skim milk. *Ind. Eng. Chem.*, *34*, 1460–1468.
6. Finot, P.A., Deutsch, R., Bujard, E. (1981). The extent of the Maillard reaction during the processing of milk. *Prog. Food Nutr. Sci.*, *5*, 344–355.
7. Erbersdobler, H.F., Somoza, V. (2007). Forty years of furosine—forty years of using Maillard reaction products as indicators of the nutritional quality of foods. *Mol. Nutr. Food Res.*, *51*, 423–430.
8. Mottram, D.S. (1994). Flavour compounds formed during the Maillard reaction. *ACS Symp. Ser.*, *543*, 104.
9. Van Boekel, M.A. (2006). Formation of flavour compounds in the Maillard reaction. *Biotechnol. Adv.*, *24*, 230–233.
10. Rahbar, S., Blumenfeld, O., Ranney, H.M. (1969). Studies of an unusual haemoglobin in patients with diabetes mellitus. *Biochem. Biophys. Res. Commun.*, *36*, 838–843.
11. Brownlee, M., Vlassara, H., Cerami, A. (1984). Nonenzymatic glycosylation and the pathogenesis of diabetic complications. *Ann. Intern. Med.*, *101*, 527–537.
12. Raj, D.S., Choudhury, D., Welbourne, T.C., Levi, M. (2000). Advanced glycation end products: a nephrologist's perspective. *Am. J. Kidney Dis.*, *35*, 365–380.
13. Henle, T. (2003). AGEs in foods: do they play a role in uremia? *Kidney Int.*, (Suppl. 84), S145–S147.
14. Henle, T., Deppisch, R., Ritz, E. (1996). The Maillard reaction: from food chemistry to uremia research. *Nephrol. Dial. Transplant.*, *11*, 1719–1722.
15. Koschinsky, T., He, C.J., Mitsuhashi, T., Bucala, R., Liu, C., Buenting, C., Heitmann, K., Vlassara, H. (1997). Orally absorbed reactive glycation products (glycotoxins): an environmental risk factor in diabetic nephropathy. *Proc. Natl. Acad. Sci. U.S.A.*, *94*, 6474–6479.
16. Uribarri, J., Cai, W., Peppa, M., Goodman, S., Ferrucci, L., Striker, G., Vlassara, H. (2007). Circulating glycotoxins and dietary advanced glycation endproducts: two links to inflammatory response, oxidative stress, and aging. *J. Gerontol. A Biol. Sci. Med. Sci.*, *62*, 427–433.
17. Hodge, J.E. (1953). Browning reactions in model systems. *J. Agric. Food Chem.*, *1*, 928–943.
18. Kroh, L.W. (1994). Caramelization in food and beverages. *Food Chem.*, *51*, 373–379.
19. Friedman, M. (2003). Nutritional consequences of food processing. *Forum Nutr.*, *56*, 350–352.
20. Penndorf, I., Biedermann, D., Maurer, S.V., Henle, T. (2006). Studies on N-terminal glycation of peptides in hypoallergenic infant formulas: quantification of alpha-N-(2-furoylmethyl) amino acids. *J. Agric. Food Chem.*, *55*, 723–727.
21. Pilkova, L., Pokorny, J., Davidek, J. (1990). Browning reactions of Heyns rearrangement products. *Nahrung/Food*, *34*, 759–764.
22. Brands, C.M.J., Van Boekel, M.A.J.S. (2001). Reactions of monosaccharides during heating of sugar-casein systems: building of a reaction network model. *J. Agric. Food Chem.*, *49*, 4667–4675.

23. Krause, R., Schlegel, K., Schwarzer, E., Henle, T. (2008). Studies on the formation of peptide-bound Heyns compounds. *J. Agric. Food Chem.*, *56*, 2522–2527.
24. Eichner, K. (1982). Analytical detection of thermally induced qualitative changes in vegetables and fruits. *Z. Lebensm. Unters. Forsch.*, *36*, 101–104.
25. Wittmann, R., Eichner, K. (1989). Detection of Maillard products in malts, beers, and brewing colorants. *Z. Lebensm. Unters. Forsch.*, *188*, 212–220.
26. Sanz, M.L., del Castillo, M.D., Corzo, N., Olano, A. (2003). 2-Furoylmethyl amino acids and hydroxymethylfurfural as indicators of honey quality. *J. Agric. Food Chem.*, *51*, 4278–4283.
27. Anet, E.F.L.J. (1960). Degradation of carbohydrates I. Isolation of 3-deoxyhexosones. *Aust. J. Chem.*, *13*, 396–403.
28. Glomb, M.A., Pfahler, C. (2000). Synthesis of 1-deoxy-D-erythro-hexo-2,3-diulose, a major hexose Maillard intermediate. *Carbohydr. Res.*, *329*, 515–523.
29. Pischetsrieder, M., Schroeter, C., Severin, T. (1998). Formation of an aminoreductone during the Maillard reaction of lactose with N-alpha-acetyllysine or proteins. *J. Agric. Food Chem.*, *46*, 928–931.
30. Mavric, E., Kumpf, Y., Schuster, K., Kappenstein, O., Scheller, D., Henle, T. (2004). A new imidazolinone resulting from the reaction of peptide-bound arginine and oligosaccharides with 1,4-glycosidic linkages. *Eur. Food Res. Technol.*, *218*, 213–218.
31. Belitz, H.D., Grosch, W., Schieberle, P. (2004). *Food Chemistry*, 3rd edn, Springer.
32. Hollnagel, A., Kroh, L.W. (2000). Degradation of oligosaccharides in nonenzymatic browning by formation of alpha-dicarbonyl compounds via a "peeling off" mechanism. *J. Agric. Food Chem.*, *48*, 6219–6226.
33. Berg, H.E., van Boekel, M.A.J.S. (1994). Degradation of lactose during heating of milk. *Neth. Milk Dairy J.*, *48*, 157–175.
34. Ahmed, M.U., Dunn, J.A., Walla, M.D., Thorpe, S.R., Baynes, J.W. (1986). Identification of N-epsilon-carboxymethyllysine as a degradation product of fructoselysine in glycated protein. *J. Biol. Chem.*, *263*, 8816–8821.
35. Büser, W., Erbersdobler, H. (1986). Carboxymethyllysine, a new compound of heat damage in milk products. *Milchwissenschaft*, *41*, 780–785.
36. Kasper, M., Schieberle, P. (2005). Labelling studies on the formation pathway of N-epsilon-carboxymethyllysine in Maillard-type reactions. *Ann. N. Y. Acad. Sci.*, *1043*, 59–62.
37. Fu, M.X., Requena, J.R., Jenkins, A.J., Lyons, T.J., Baynes, J.W., Thorpe, S.R. (1996). The advanced glycation end product, N-epsilon-(carboxymethyl)lysine, is a product of both lipid peroxidation and glycoxidation reactions. *J. Biol. Chem.*, *271*, 9982–9986.
38. Nakayama, T., Hayase, F., Kato, H. (1980). Formation of epsilon-(2-formyl-5-hydroxy-methyl-pyrrol-1-yl)-L-norleucin in the Maillard reaction between D-glucose and L-lysine. *Agric. Biol. Chem.*, *44*, 1201–1202.
39. Henle, T., Walter, A.W., Klostermeyer, H. (1994). Simultaneous determination of protein-bound Maillard products by ion-exchange chromatography and photodiode array detection, in *Maillard Reactions in Chemistry, Food and Health* (eds

T.P. Labuza, G.A. Reineccius, V.M. Monnier, J. O'Brien, J.W. Baynes), The Royal Society of Chemistry, pp. 195–200.

40. Henle, T., Bachmann, A. (1996). Synthesis of pyrraline. *Z. Lebensm. Unters. Forsch.*, *202*, 72–75.
41. Henle, T., Klostermeyer, H. (1993). Determination of protein-bound 2-amino-6-(2-formyl-2-hydroxymethyl-1-pyrrolyl)-hexanoic acid ("pyrraline") by ion-exchange chromatography and photodiode array detection. *Z. Lebensm. Unters. Forsch.*, *196*, 1–4.
42. Lindenmeier, M., Faist, V., Hofmann, T. (2002). Structural and functional characterization of pronyl-lysine, a novel protein modification in bread crust melanoidins showing in vitro antioxidative and phase I/II enzyme modulating activity. *J. Agric. Food Chem.*, *50*, 6997–7006.
43. Cho, R.K., Okitani, A., Kato, H. (1986). Polymerization of proteins and impairment of their arginine residues due to intermediate compounds in the Maillard reaction. *Dev. Food Sci.*, *13*, 439–448.
44. Kato, Y., Matsuda, T., Kato, N., Nakamura, R. (1988). Browning and protein polymerization induced by amino-carbonyl reaction of ovalbumin with glucose and lactose. *J. Agric. Food Chem.*, *36*, 806–809.
45. Sell, D.R., Monnier, V.M. (1991). Structure elucidation of a senescence cross-link from human extracellular matrix. Implication of pentoses in the aging process. *J. Biol. Chem.*, *264*, 21597–21602.
46. Henle, T., Schwarzenbolz, U., Klostermeyer, H. (1997). Detection and quantification of pentosidine in foods. *Z. Lebensm. Unters. Forsch.*, *204*, 95–98.
47. Biemel, K.M., Reihl, O., Conrad, J., Lederer, M.O. (2001). Formation pathways for lysine-arginine cross-links derived from hexoses and pentoses by Maillard processes: unraveling the structure of a pentosidine precursor. *J. Biol. Chem.*, *276*, 23405–23412.
48. Henle, T., Walter, A.W., Haessner, R., Klostermeyer, H. (1994). Isolation and identification of a protein-bound imidazolone resulting from the reaction of arginine residues and methylglyoxal. *Z. Lebensm. Unters. Forsch.*, *199*, 55–58.
49. Ahmed, N., Argirov, O.K., Minhas, H.S., Cordeiro, C.A., Thornalley, P.J. (2002). Assay of advanced glycation endproducts (AGEs): surveying AGEs by chromatographic assay with derivatization by 6-aminoquinolyl-N-hydroxysuccinimidyl-carbamate and application to N-epsilon-carboxymethyl-lysine- and N-epsilon-(1-carboxyethyl)lysine-modified albumine. *Biochem. J.*, *364*, 1–14.
50. Thornalley, P.J., Battah, S., Ahmed, N., Karachalias, N., Agalou, S., Babaei-Jadidi, R., Dawnay, A. (2003). Quantitative screening of advanced glycation endproducts in cellular and extracellular proteins by tandem mass spectrometry. *Biochem. J.*, *375*, 581–592.
51. Schwarzenbolz, U., Henle, T., Haessner, R., Klostermeyer, H. (1997). On the reaction of glyoxal with proteins. *Z. Lebensm. Unters. Forsch.*, *205*, 121–124.
52. Konishi, Y., Hayase, F., Kato, H. (1994). Novel imidazolone compound formed by the advanced Maillard reaction of 3-deoxyglucosone and arginine residues in proteins. *Biosci. Biotech. Biochem.*, *58*, 1953–1955.
53. Shipanova, I.N., Glomb, M.A., Nagaraj, R.H. (1997). Protein modification by methylglyoxal: chemical nature and synthetic mechanism of a major fluorescent adduct. *Arch. Biochem. Biophys.*, *344*, 29–36.

54. Glomb, M.A., Rösch, D., Nagaraj, R.H. (2001). N(delta)-(5-hydroxy-4,6-dimethylpyrimidine-2-yl)-l-ornithine, a novel methylglyoxal-arginine modification in beer. *J. Agric. Food Chem.*, *49*, 366–372.

55. Biemel, K.M., Bühler, H.P., Reihl, O., Lederer, M.O. (2001). Identification and quantitative evaluation of the lysine-arginine crosslinks GODIC, MODIC, DODIC and glucosepan in foods. *Nahrung/Food*, *45*, 210–214.

56. Frye, E.B., Degenhardt, T.P., Thorpe, S.R., Baynes, J.W. (1998). Role of the Maillard reaction in aging of tissue proteins. Advanced glycation end product-dependent increase in imidazolium cross-links in human lens proteins. *J. Biol. Chem.*, *273*, 18714–18719.

57. Lederer, M.O., Klaiber, R.G. (1999). Cross-linking of proteins by Maillard processes: characterization and detection of lysine-arginine cross-links derived from glyoxal and methylglyoxal. *Bioorg. Med. Chem.*, *7*, 2499–2507.

58. Goldberg, T., Cai, W., Peppa, M., Dardaine, V., Baliga, B.S., Uribarri, J., Vlassara, H. (2004). Advanced glycoxidation end products in commonly consumed foods. *J. Am. Diet. Assoc.*, *104*, 1287–1291.

59. Resmini, P., Pellegrino, L., Masotti, F., Tirelli, A., Prati, F. (1992). Detection of reconstituted milk powder in raw and in pasteurized milk by HPLC of furosine. *Scienza e Tecnica Lattiero-Casearia*, *43*, 169–186.

60. Henle, T., Zehetner, G., Klostermeyer, H. (1995). Fast and sensitive determination of furosine. *Z. Lebensm. Unters. Forsch.*, *200*, 235–237.

61. Krause, R., Knoll, K., Henle, T. (2003). Studies in the formation of furosine and pyridosine during acid hydrolysis of different Amadori products of lysine. *Eur. Food Res. Technol.*, *216*, 277–283.

62. Rada-Mendoza, M., Olano, A., Villamiel, M. (2002). Furosine as indicator of Maillard reaction in jams and fruit-based infant foods. *J. Agric. Food Chem.*, *50*, 4141–4145.

63. Vallejo-Cordoba, B., Mazorra-Manzano, M.A., González-Córdova, A.F. (2004). New capillary electrophoresis method for the determination of furosine in dairy products. *J. Agric. Food Chem.*, *52*, 5787–5790.

64. Schleicher, E., Scheller, L., Wieland, O.H. (1981). Quantitation of lysine-bound glucose of normal and diabetic erythrocyte membranes by HPLC analysis of furosine [epsilon-N-(L-furoylmethyl)-L-lysine]. *Biochem. Biophys. Res. Commun.*, *99*, 1011–1009.

65. Erbersdobler, H.F., Lohmann, M., Buhl, K. (1991). Utilization of early Maillard reaction products by humans. *Adv. Exp. Med. Biol.*, *289*, 363–370.

66. Bueser, W., Erbersdobler, H.F., Liardon, R. (1987). Identification and determination of N-epsilon-carboxymethyllysine by gas-liquid chromatography. *J. Chromatogr.*, *387*, 515–519.

67. Liardon, R., De Weck Gaudard, D., Philippossian, G., Finot, P.A. (1987). Identification of N-epsilon-carboxymethyllysine—a new Maillard reaction product, in rat urine. *J. Agric. Food Chem.*, *35*, 427–431.

68. Hartkopf, J., Pahlke, C., Lüdemann, G., Erbersdobler, H.F. (1994). Determination of N-epsilon-carboxymethyllysine by a reversed-phase high-performance liquid-chromatography method. *J. Chromatogr., A*, *672*, 242–246.

69. Kislinger, T., Humeny, A., Peich, C.C., Zhang, X., Niwa, T., Pischetsrieder, M., Becker, C.M. (2003). Relative quantification of N-epsilon-(carboxymethyl)lysine, imidazolone A, and the Amadori product in glycated lysozyme by MALDI-TOF mass spectrometry. *J. Agric. Food Chem.*, *51*, 51–57.
70. Delatour, T., Fenaille, F., Parisod, V., Vera, F.A., Buetler, T. (2006). Synthesis, tandem MS- and NMR-based characterization, and quantification of the carbon 13-labeled advanced glycation endproduct, 6-N-carboxymethyllysine. *Amino Acids*, *30*, 25–34.
71. Tauer, A., Hasenkopf, K., Kislinger, T., Frey, I., Pischetsrieder, M. (1999). Determination of N-epsilon-carboxymethyllysine in heated milk products by immunochemical methods. *Eur. Food Res. Technol.*, *209*, 72–76.
72. Charissou, A., Ait-Ameur, L., Birlouez-Aragon, I. (2007). Evaluation of a gas chromatography/mass spectrometry method for the quantification of carboxymethyllysine in food samples. *J. Chromatogr., A*, *1140*, 189–194.
73. Goldberg, T., Cai, W.J., Peppa, M., Dardaine, V., Baliga, B.S., Uribarri, J., Vlassara, H. (2004). Advanced glycoxidation end products in commonly consumed foods. *J. Am. Diet. Assoc.*, *104*, 1287–1291.
74. Resmini, P., Pellegrino, L. (1994). Occurrence of protein-bound lysylpyrrolaldehyde in dried pasta. *Cereal Chem.*, *71*, 254–262.
75. Rufian-Henares, J.A., Guerra-Hernandez, E., Garcia-Villanova, B. (2004). Pyrraline content in enteral formula processing and storage and model systems. *Eur. Food Res. Technol.*, *219*, 42–47.
76. Hegele, J., Parisoda, V., Richoza, J., Förster, A., Maurer, S., Krause, R., Henle, T., Bütler, T., Delatour, T. (2008). Evaluating the extent of protein damage in dairy products by LC-MS/MS: simultaneous determination of early and advanced glycation-induced lysine modifications. *Ann. N. Y. Acad. Sci.*, *1126*, 300–306.
77. Förster, A., Kuhne, Y., Henle, T. (2005). Studies on absorption and elimination of dietary Maillard reaction products. *Ann. N. Y. Acad. Sci.*, *1043*, 474–481.
78. Schwarzenbolz, U., Klostermeyer, H., Henle, T. (2000). Maillard-type reactions under high hydrostatic pressure: formation of pentosidine. *Eur. Food Res. Technol.*, *211*, 208–210.
79. Nemet, I., Varga-Defterdarovic, L., Turk, Z. (2006). Methylgloxal in food and living organisms. *Mol. Nutr. Food Res.*, *50*, 1105–1117.
80. Hayashi, T., Shibamoto, T. (1985). Analysis of methylglyoxal in foods and beverages. *J. Agric. Food Chem.*, *33*, 1090–1093.
81. Daglia, M., Papetti, A., Aceti, C., Sordelli, B., Spini, V., Gazzani, G. (2007). Isolation and determination of α-dicarbonyl compounds by RP-HPLC-DAD in green and roasted coffee. *J. Agric. Food Chem.*, *55*, 8877–8882.
82. Niyati-Shirkhodaee, F., Shibamoto, T. (1993). Gas chromatographic analysis of glyoxal and methylglyoxal formed from lipids and related compounds upon ultraviolet irradiation. *J. Agric. Food Chem.*, *41*, 227–230.
83. de Revel, G., Bertrand, A. (1993). A method for the detection of carbonyl compounds in wine: glyoxal and methylglyoxal. *J. Sci. Food Agric.*, *61*, 267–272.
84. Barros, A., Rodrigues, J.A., Almeida, P.J., Oliva-Teles, M.T. (1999). Determination of glyoxal, methylglyoxal, and diacetyl in selected beer and wine, by HPLC with

UV spectrophotometric detection, after derivatization with o-phenylenediamine. *J. Liquid Chromatogr. Relat. Technol.*, *22*, 2061–2069.

85. de Revel, G., Pripis-Nicolau, L., Barbe, J.-C., Bertrand, A. (2000). The detection of α-dicarbonyl compounds in wine by formation of quinoxaline derivatives. *J. Food Agric.*, *80*, 102–108.
86. Fujioka, K., Shibamoto, T. (2004). Formation of genotoxic dicarbonyl compounds in dietary oils upon oxidation. *Lipids*, *39*, 481–486.
87. Bednarski, W., Jedrychowski, L., Hammond, E.G., Nikolov, Z.L. (1989). A method for the determination of alpha-dicarbonyl compounds. *J. Dairy Sci.*, *72*, 2474–2477.
88. Weigel, K.U., Opitz, T., Henle, T. (2004). Studies on the occurrence and formation of 1,2-dicarbonyls in honey. *Eur. Food Res. Technol.*, *218*, 147–151.
89. Mavric, E., Wittmann, S., Barth, G., Henle, T. (2007). Identification and quantification of methylglyoxal as the dominant antibacterial constituent of Manuka (*Leptospermum scoparium*) honeys from New Zealand. *Mol. Nutr. Food Res.*, *52*, 483–489.
90. Lusby, P.E., Coombes, A.L., Wilkinson, J.M. (2005). Bactericidal activity of different honeys against pathogenic bacteria. *Arch. Med. Res.*, *36*, 464–467.
91. Wagner, K.H., Reichhold, S., Koschutnig, K., Cheriot, S., Billaud, C. (2007). The potential antimutagenic and antioxidant effects of Maillard reaction products used as "natural antibrowning" agents. *Mol. Nutr. Food Res.*, *51*, 496–504.
92. Wondrak, G.T., Cervantes-Laurean, D., Roberts, M.J., Qasem, J.G., Kim, M., Jacobson, E.L., Jacobson, M.K. (2002). Identification of alpha-dicarbonyl scavengers for cellular protection against carbonyl stress. *Biochem. Pharmacol.*, *63*, 361–373.
93. Sebeková, K., Krajcoviová-Kudlácková, M., Schinzel, R., Faist, V., Klvanová, J., Heidland, A. (2001). Plasma levels of advanced glycation end products in healthy, long-term vegetarians and subjects on a western mixed diet. *Eur. J. Nutr.*, *40*, 275–281.
94. McCollum, E.V., Davis, M. (1915). The cause of the loss of nutritive efficiency of heated milk. *J. Biol. Chem.*, *23*, 247–254.
95. Greaves, E.O., Morgan, A.F., Loveen, M.K. (1938). The effect of amino acid supplements and variations in temperature and duration of heating upon the biological value of heated casein. *J. Nutr.*, *6*, 115–128.
96. Finot, P.A., Bujard, E., Mottu, F., Mauron, J. (1977). Availability of the true Schiff's bases of lysine. Chemical evaluation of the Schiff's base between lysine and lactose in milk. *Adv. Exp. Med. Biol.*, *86B*, 343–365.
97. Carpenter, K.J. (1960). The estimation of available lysine in animal-protein foods. *Biochem. J.*, *77*, 604–610.
98. Bujard, E., Handwerck, V., Mauron, J. (1967). The differential determination of lysine in heated milk. I. In vitro methods. *J. Sci. Food Agric.*, *18*, 52–57.
99. Carpenter, K.J., Steinke, F.H., Catignani, G.L., Swaisgood, H.E., Allred, M.C., MacDonald, J.L., Schelstraete, M. (1989). The estimation of "available lysine" in human foods by three chemical procedures. *Plant Foods Hum. Nutr.*, *39*, 129–135.
100. Mauron, J. (1990). Influence of processing on protein quality. *J. Nutr. Sci. Vitaminol. (Tokyo)*, *36*, S57–S69.

101. Bunn, H.F., Haney, D.N., Gabbay, K.H., Gallop, P.M. (1975). Further identification of the nature and linkage of the carbohydrate in haemoglobin A1c. *Biochem. Biophys. Res. Commun.*, *67*, 103–109.
102. Monnier, V.M., Cerami, A. (1981). Nonenzymatic browning in vivo: possible process for aging of long-lived proteins. *Science*, *211*, 491–493.
103. Monnier, V.M., Stevens, V.J., Cerami, A.M. (1981). Maillard reactions involving proteins and carbohydrates in vivo: relevance to diabetes mellitus and aging. *Prog. Food Nutr. Sci.*, *5*, 315–327.
104. Monnier, V.M., Kohn, R.R., Cerami, A. (1984). Accelerated age-related browning of human collagen in diabetes mellitus. *Proc. Natl. Acad. Sci. U.S.A.*, *81*, 583–587.
105. Monnier, V.M. (1989). Toward a Maillard reaction theory of aging. *Prog. Clin. Biol. Res.*, *304*, 1–22.
106. Schwenger, V., Zeier, M., Henle, T., Ritz, E. (2001). Advanced glycation endproducts (AGEs) as uremic toxins. *Nahrung/Food*, *45*, 172–176.
107. Schmidt, A.M., Vianna, M., Gerlach, M., Brett, J., Ryan, J., Kao, J., Esposito, C., Hegarty, H., Hurley, W., Clauss, M. (1992). Isolation and characterization of two binding proteins for advanced glycosylation end products from bovine lung which are present on the endothelial cell surface. *J. Biol. Chem.*, *267*, 14987–14997.
108. Bierhaus, A., Stern, D.M., Nawroth, P.P. (2006). RAGE in inflammation: a new therapeutic target? *Curr. Opin. Investig. Drugs*, *7*, 985–991.
109. Yan, S.F., Barile, G.R., D'Agati, V., Du Yan, S., Ramasamy, R., Schmidt, A.M. (2007). The biology of RAGE and its ligands: uncovering mechanisms at the heart of diabetes and its complications. *Curr. Diab. Rep.*, *7*, 146–153.
110. Ramasamy, R., Yan, S.F., Schmidt, A.M. (2007). Arguing for the motion: yes, RAGE is a receptor for advanced glycation endproducts. *Mol. Nutr. Food Res.*, *51*, 1111–1115.
111. Heizmann, C.W. (2007). The mechanism by which dietary AGEs are a risk to human health is via their interaction with RAGE: arguing against the motion. *Mol. Nutr. Food Res.*, *51*, 1116–1119.
112. Henle, T. (2007). Dietary advanced glycation end products—a risk to human health? A call for an interdisciplinary debate. *Mol. Nutr. Food Res.*, *51*, 1075–1078.
113. Sebekova, K., Somoza, V. (2007). Dietary advanced glycation endproducts (AGEs) and their health effects—PRO. *Mol. Nutr. Food Res.*, *51*, 1079–1084.
114. Sandu, O., Song, K., Cai, W., Zheng, F. *et al.* (2005). Insulin resistance and type 2 diabetes in high-fat-fed mice are linked to high glycotoxin intake. *Diabetes*, *54*, 2314–2319.
115. Hofmann, S.M., Dong, H.J., Li, Z., Cai, W. *et al.* (2002). Improved insulin sensitivity is associated with restricted intake of dietary glycoxidation products in the db/db mouse. *Diabetes*, *51*, 2082–2089.
116. Zheng, F., He, C., Cai, W., Hattori, M. *et al.* (2002). Prevention of diabetic nephropathy in mice by a diet low in glycoxidation products. *Diabetes Metab. Res. Rev.*, *18*, 224–237.
117. Peppa, M., Brem, H., Ehrlich, P., Zhang, J.G. *et al.* (2003). Adverse effects of dietary glycotoxins on wound healing in genetically diabetic mice. *Diabetes*, *52*, 2805–2813.

118. Uribarri, J., Peppa, M., Cai, W., Goldberg, T., Lu, M., Baliga, S., Vassalotti, J.A., Vlassara, H. (2003). Dietary glycotoxins correlate with circulating advanced glycation end product levels in renal failure patients. *Am. J. Kidney Dis.*, *42*, 532–538.

119. Vlassara, H., Cai, W., Crandall, J., Goldberg, T., Oberstein, R., Dardaine, V., Peppa, M., Rayfield, E.J. (2002). Inflammatory mediators are induced by dietary glycotoxins, a major risk factor for diabetic angiopathy. *Proc. Natl. Acad. Sci. U.S.A.*, *99*, 15596–15601.

120. Uribarri, J., Cai, W., Peppa, M., Goodman, S., Ferrucci, L., Striker, G., Vlassara, H. (2007). Circulating glycotoxins and dietary advanced glycation endproducts: two links to inflammatory response, oxidative stress, and aging. *J. Gerontol. A Biol. Sci. Med. Sci.*, *62*, 427–433.

121. Negrean, M., Stirban, A., Stratmann, B., Gawlowski, T., Horstmann, T., Götting, C., Kleesiek, K., Mueller-Roesel, M., Koschinsky, T., Uribarri, J., Vlassara, H., Tschoepe, D. (2007). Effects of low- and high-advanced glycation endproduct meals on macro- and microvascular endothelial function and oxidative stress in patients with type 2 diabetes mellitus. *Am. J. Clin. Nutr.*, *85*, 1236–1243.

122. Finot, P.A., Magnenat, E. (1981). Metabolic transit of early and advanced Maillard products. *Prog. Food Nutr. Sci.*, *5*, 193–207.

123. Erbersdobler, H.F., Lohmann, M., Buhl, K. (1991). Utilization of early Maillard reaction products by humans. *Adv. Exp. Med. Biol.*, *289*, 363–370.

124. Henle, T., Schwenger, V., Ritz, E. (2000). Studies on the renal handling of lactuloselysine from milk products. *Czech J. Food Sci.*, *18* (Spec. Issue), 101–102.

125. Schwenger, V., Morath, C., Schönfelder, K., Klein, W., Weigel, K., Deppisch, R., Henle, T., Ritz, E., Zeier, M. (2006). An oral load of the early glycation compound lactuloselysine fails to accumulate in the serum of uraemic patients. *Nephrol. Dial. Transplant.*, *21*, 383–388.

126. Wiame, E., Delpierre, G., Collard, F., van Schaftingen, E. (2002). Identification of a pathway for the utilization of the Amadori product fructoselysine in *Escherichia coli*. *J. Biol. Chem.*, *277*, 42523–42529.

127. Faist, V., Wenzel, E., Erbersdobler, H.F. (2000). In vitro and in vivo studies on the metabolic transit of N-ε-carboxymethyllysine. *Czech J. Food Sci.*, *18*, 116–119.

128. Bergmann, R., Helling, R., Heichert, C., Scheunemann, M., Mäding, P., Wittrisch, H., Johannsen, B., Henle, T. (2001). Radio fluorination and positron emission tomography (PET) as a new approach to study the in vivo distribution and elimination of the advanced glycation end products N^{ε}-carboxymethyllysine (CML) and N^{ε}-carboxyethyllysine (CEL). *Nahrung/Food*, *45*, 182–188.

129. Grunwald, S., Krause, R., Bruch1, M., Henle, T., Brandsch, M. (2006). Transepithelial flux of early and advanced glycation compounds across Caco-2 cell monolayers and their interaction with intestinal amino acid and peptide transport systems. *Br. J. Nutr.*, *95*, 1221–1228.

130. Förster, A., Kühne, Y., Henle, T. (2005). Studies on absorption and elimination of dietary Maillard reaction products. *Ann. N. Y. Acad. Sci.*, *1043*, 474–481.

131. Schwedler, S.B., Metzger, T., Schinzel, R., Wanner, C. (2002). Advanced glycation end products and mortality in hemodialysis patients. *Kidney Int.*, *62*, 301–310.

132. Busch, M., Franke, S., Wolf, G., Brandstädt, A., Ott, U., Gerth, J., Hunsicker, L.G., Stein, G. (2006). The advanced glycation end product N(epsilon)-

carboxymethyllysine is not a predictor of cardiovascular events and renal outcomes in patients with type 2 diabetic kidney disease and hypertension. *Am. J. Kidney Dis.*, *48*, 571–579.

133. Valencia, J.V., Mone, M., Koehne, C., Rediske, J., Hughes, T.E. (2004). Binding of receptor for advanced glycation end products (RAGE) ligands is not sufficient to induce inflammatory signals: lack of activity of endotoxin-free albumin-derived advanced glycation end products. *Diabetologia*, *47*, 844–852.
134. Demling, N., Ehrhardt, C., Kasper, M., Laue, M., Knels, L., Rieber, E.P. (2006). Promotion of cell adherence and spreading: a novel function of RAGE, the highly selective differentiation marker of human alveolar epithelial type I cells. *Cell Tissue Res.*, *323*, 475–488.
135. Lieuw-a-Fa, M.L., Schalkwijk, C.G., Engelse, M., van Hinsbergh, V.W. (2006). Interaction of N-e-(carboxymethyl)lysine- and methylglyoxal-modified albumin with endothelial cells and macrophages. Splice variants of RAGE may limit the responsiveness of human endothelial cells to AGEs. *Thromb. Haemost.*, *95*, 320–328.
136. Kislinger, T., Fu, C., Huber, B., Qu, W., Taguchi, A., Du Yan, S., Hofmann, M., Yan, S.F., Pischetsrieder, M., Stern, D., Schmidt, A.M. (1999). N(epsilon)-(carboxymethyl)lysine adducts of proteins are ligands for receptor for advanced glycation end products that activate cell signalling pathways and modulate gene expression. *J. Biol. Chem.*, *274*, 31740–31749.
137. Anese, M., Nicoli, M.C. (2003). Antioxidant properties of ready-to-drink coffee brews. *J. Agric. Food Chem.*, *51*, 942–946.
138. Samaras, T.S., Camburn, P.A., Chandra, S.X., Gordon, M.H., Ames, J.M. (2005). Antioxidant properties of kilned and roasted malts. *J. Agric. Food Chem.*, *53*, 8068–8074.
139. Papetti, A., Daglia, M., Aceti, C., Quaglia, M., Gregotti, C., Gazzani, G. (2006). Isolation of an in vitro and ex vivo antiradical melanoidin from roasted barley. *J. Agric. Food Chem.*, *54*, 1209–1216.
140. Somoza, V., Wenzel, E., Lindenmeier, M., Grothe, D., Erbersdobler, H.F., Hofmann, T. (2005). Influence of feeding malt, bread crust, and a pronylated protein on the activity of chemopreventive enzymes and antioxidative defense parameters in vivo. *J. Agric. Food Chem.*, *53*, 8176–8182.
141. Borrelli, R.C., Fogliano, V. (2005). Bread crust melanoidins as potential prebiotic ingredients. *Mol. Nutr. Food Res.*, *49*, 673–678.
142. Böhm, A., Kaiser, I., Trebstein, A., Henle, T. (2005). Heat-induced degradation of inulin. *Eur. Food Res. Technol.*, *220*, 466–471.
143. Marko, D., Habermeyer, M., Kemény, M., Weyand, U., Niederberger, E., Frank, O., Hofmann, T. (2003). Maillard reaction products modulating the growth of human tumour cells in vitro. *Chem. Res. Toxicol.*, *16*, 48–55.

2.8

POLYAROMATIC HYDROCARBONS

JONG-HEUM PARK[1] AND TREVOR M. PENNING[2]

[1] *Center of Excellence in Environmental Toxicology, Department of Pharmacology, University of Pennsylvania, School of Medicine, Philadelphia, PA 19104-6084, USA*

[2] *Department of Pharmacology, 130C John Morgan Bldg., 3620 Hamilton Walk, Philadelphia, PA 19064, USA*

2.8.1 INTRODUCTION

Polycyclic aromatic hydrocarbons (PAHs) are compounds that contain two or more fused aromatic rings. They are ubiquitous environmental pollutants, and comprise the largest class of known chemical carcinogens (1). PAHs are formed as a result of incomplete combustion of fossil fuels (e.g., wood, coal, and oil) used in the generation of power and heat; they are present in car and diesel exhaust, and are components of fine particulate matter (PM2.5). Since they are air pollutants, they can contaminate the soil and ground water and can enter into the food chain. Foods may also be contaminated from the production of PAHs during food processing and cooking, e.g., smoking of meat and fish, and the preparation of charbroiled food, respectively. Since there are multiple sources of human exposure, it is difficult to assess the single contribution that is made from food intake. Nevertheless, there is a need to address and mitigate the overall burden and cumulative exposure to PAHs since representative compounds are complete carcinogens in rodents when administered orally (1, 2). Based on mutagenicity, carcinogenicity, epidemiological data, and mode of action, benzo[*a*]pyrene (B[*a*]P) originally classified as a

Process-Induced Food Toxicants: Occurrence, Formation, Mitigation, and Health Risks,
Edited by Richard H. Stadler and David R. Lineback

Polyaromatic Hydrocarbons (PAHs): At a Glance	
Historical	PAHs are ubiquitous environmental pollutants formed as a result of incomplete combustion of fossil fuels (e.g., wood, coal and oil). The presence of PAHs in food appears to be from several routes including the growth of fruits and vegetables in contaminated soils, ingestion of sea food and fish from oceans contaminated with crude oil, and man-made food preparation techniques that can contaminate food unnecessarily.
Analysis	Early methods for detecting PAHs in foods used HPLC with fluorescence detection. However, GC-MS combined with the use of isotopically labeled internal standards ensures exact quantification and unambiguous structural identification.
Occurrence in Food	The food groups with the highest concentration of PAHs include meat products, smoked fish, smoked and cured cheese, vegetables, tea, and roasted coffee.
Main Formation Pathways	Grilling (broiling) of meat, fish, and other foods over intense heat or by direct contact with flames results in the formation of PAHs. They are formed by the condensation of smaller organic compounds by either pyrolysis or pyrosynthesis. A common reaction may involve Diels-Alder type rearrangements to yield the PAH.
Mitigation in Food	Because exposure is ubiquitous, effective reduction from all sources is not possible. Formation of PAHs during cooking of food can be reduced if excess fat is trimmed from meat and meat is cooked at lower temperatures and an appropriate distance from charcoal when this is used in broiling. Similarly, the type of wood used as a fuel during the smoking of food should be carefully chosen.
Dietary Intake	Benzo[*a*]pyrene (B[*a*]P) is usually used as a surrogate marker of exposure to all PAHs. JECFA reviewed estimates of intake from a number of sources and reported B[*a*]P intakes ranging from <1 to 2 μg/day.
Health Risks	The largest concern associated with PAHs is cancer risk. The mode of action of carcinogenic PAHs appears to be via a genotoxic mechanism. Consequently it is not possible to assume that a safe threshold exists. JECFA also calculated a Benchmark Dose Lower Confidence Interval (BMDL) equivalent of 100 μg B[*a*]P/kg bw/d, derived for PAHs in food on the basis of their carcinogenicity in mice treated orally with mixtures of PAHs. A representative mean and estimated high level intake of B[*a*]P was 0.004 μg/kg bw/d and 0.01 μg/kg bw/d, respectively, equating to a Margin of Exposure (MoE) of 25,000 and 10,000, respectively.
Regulatory Information or Industry Standards	In the EU, maximum amounts of B[*a*]P have been set for several foods and food groups (e.g., oils and fats, baby foods, infant formulas, smoked meats, fish, and other seafoods). Maximum amounts are also defined by WHO and US EPA in drinking water.
Knowledge Gaps	JECFA has identified 13 PAHs that occur in foods as being genotoxic and carcinogenic. However, for certain compounds little or no data exist on their concentrations across the different food groups.
Useful Information (web) & Dietary Recommendations	http://www.who.int/ipcs/food/jecfa/summaries/summary_report_64_final.pdf http://ec.europa.eu/food/fs/sc/scf/out153_en.pdf Avoid the over-heating of food; fry/barbecue foods to the appropriate "doneness."

TABLE 2.8.1 Priority PAH as food toxicants: evidence for carcinogenicity and relative potencies.

Compound	Evidence for carcinogenicity[a]		Ranking[a]	Relative potency[a]	TEF
	Human	Animal			
Naphthalene					
Acenaphthylene					
Acenaphthene					
Fluorene	ND	I	3		
Anthracene	ND	I	3	0.232	0.001
Phenanthrene	ND	I	3		0.001
Fluoranthene	ND	I	3		
Pyrene	ND	I	3	0.81	0.001
Chrysene	ND	L	2B	0.0044	0.01
Benz[*a*]anthracene	ND	S	2B	0.145	0.1
Benzo[*b*]fluoranthene	ND	S	2B	0.141	0.1
Benzo[*k*]fluoranthene	ND	S	2B	0.061	0.1
Benzo[*a*]pyrene	ND	S	1	1.0	1.0
Indeno[1,2,3-*cd*]pyrene	ND	S	2B	0.232	0.1
Benzo[*g,h,i*]perylene	ND	I	3	0.022	0.01
Dibenz[*a,h*]anthracene	ND	S	2A	1.11	5.0

[a]Group 1, the compound is carcinogenic in human; Group 2A, the compound is probably carcinogenic in humans; Group 2B, the compound is possibly carcinogenic in humans; Group 3, the compound is not classifiable as to its carcinogenicity in humans.
TEF = toxicity equivalency factors; ND = no adequate data; I, inadequate evidence; L = limited evidence; S = sufficient evidence.

probable human carcinogen (Group 2A) has been recently reclassified by the WHO International Agency for Research on Cancer (IARC) as a human carcinogen (Group 1 carcinogen; Table 2.8.1) (1,3). Thus, exposure to PAHs is a significant public health problem.

Over 100 different PAHs have been identified and they often occur as complex mixtures as a result of the pyrolytic process. Of these, the US Environmental Protection Agency (EPA) has listed 16 PAHs as priority pollutants (4). These priority pollutants include naphthalene, acenaphthylene, acenaphthene, fluorene, anthracene, phenanthrene, fluoranthene, pyrene, chrysene, benz[*a*]anthracene, benzo[*b*]fluoranthene, benzo[*k*]fluoranthene, B[*a*]P, indeno[1,2,3-*cd*]pyrene, benzo[*g,h,i*]perylene, and dibenz[*a,h*]anthracene (Fig. 2.8.1). This chapter will discuss the occurrence of these PAHs in food, methods of detection and analysis, mechanism of formation in the processing and cooking of food, mitigation, biomonitoring and biomarkers of exposure, health risks, and risk management.

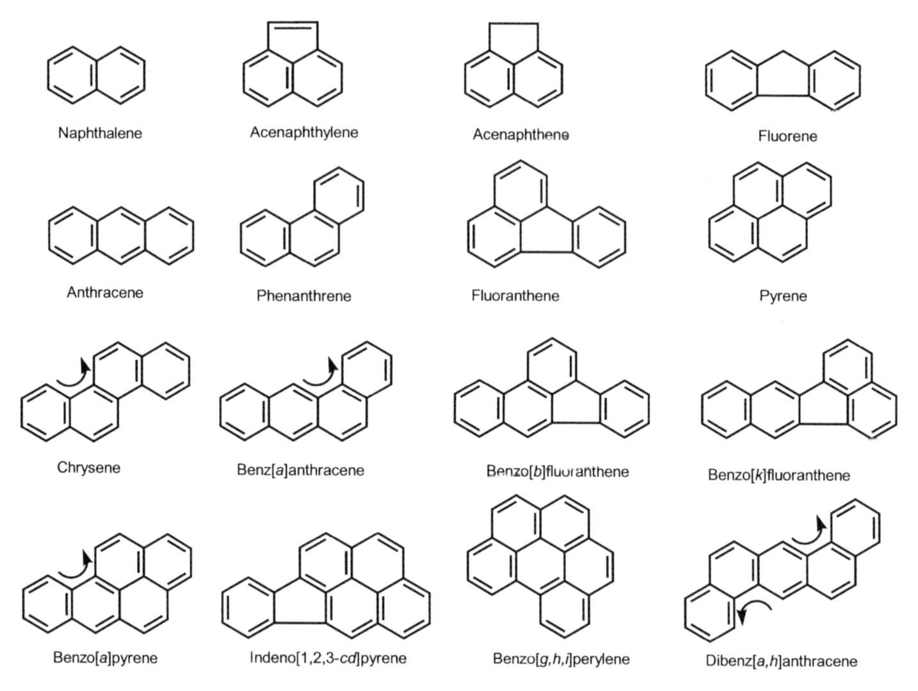

Figure 2.8.1 The 16 PAHs that are priority pollutants listed by the US EPA. The curved arrow shows the presence of a bay region in carcinogenic PAH.

2.8.2 OCCURRENCE

There are numerous reports of detection of the 16 PAH priority pollutants in almost all food sources (5–7). The presence of PAHs in food appears to be from several major routes, including the growth of fruit and vegetables in contaminated soils, the ingestion of seafood from oceans contaminated with crude oil, and man-made food preparation techniques such as smoking, grilling, broiling, and roasting, which can contaminate food unnecessarily. The introduction of PAHs into food by food-processing methods cannot be considered without the recognition that the food itself may already be contaminated with these toxicants prior to processing. To conduct meaningful risk assessment, cross-comparison between food groups and across studies would be ideal. However, this is made almost impossible due to the different analytical protocols that have been applied, and in some instances, state-of-the-art methods (e.g., gas chromatography-mass spectrometry [GC-MS]) have not been used routinely, raising concerns about the robustness of these data. Nevertheless, there appears to be agreement in the trends observed.

2.8.2.1 PAHs in the Food Chain

The natural and anthropogenic sources of PAHs in foods are numerous. PAHs are produced and released into the atmosphere during incomplete combustion or pyrolysis of organic substances by industrial process and human activity. They are emitted into the environment as the result of burning coal, wood, petroleum, and petroleum products, and from coke production, refuse burning, and motor vehicle exhaust fumes (8, 9). They are also formed in coal gasification plants, municipal incinerators, and in aluminum production facilities (10–12), and are thus released as emissions from these plants.

Most PAHs have a low volatility and poor water solubility. Thus, PAHs have a high tendency to absorb onto small organic particulate matter (<2.5 μm) like fly ash and soot. Particles containing PAHs fall from the atmosphere as a primary source of contamination of the soil, a problem that is exacerbated when it rains. Consequently, cereals, grain, and vegetables grown in contaminated soil will absorb the PAHs due to their lipophilic properties. Unfortunately, PAHs in the soil build up in a cumulative fashion. Soil samples collected in Rothamsted Experimental Station in S.E. England over a period of 140 years (1846–1980) were analyzed for PAHs. In the upper 3.0 cm of soil, B[*a*]P concentrations had increased from 18 to 130 μg/kg over this time period (13). As a result of soil contamination in some parts of the world, concentrations as high as 590–2301 μg B[*a*]P/kg can be observed in the peelings of vegetables. Grazing herds that feed in contaminated areas can also have PAHs enter into their milk supply and dairy products can thus become contaminated (14).

Particles containing PAH also fall into surface water; they can enter the nearshore marine environment as a result of urban water drain runoffs, water runoff from highways, effluent from industrial and sewage outlets (15, 16), and from creosote-treated wharfs and pilings (17). They accumulate in river or marine sediments to which they are strongly bound, and these sediments act as a pollution reservoir from which PAHs may be constantly released. Oceans and waterways contaminated with crude oil or engine oil due to either oil tanker disasters or industrial waste will create sediments more heavily contaminated with PAHs. It is estimated that PAHs constitute about 20% of the total hydrocarbon content of crude oil (18). Filter-feeding bivalves such as mussels, clams, and oysters are the first invertebrate targets for exposure to PAHs since they filter large amounts of water and have a very poor metabolic clearance for PAHs. They thus play an important role in the transfer of PAHs to seafood and fish, and to higher animals through the food chain (19, 20).

2.8.2.2 PAH Contamination in Food Preparation and Processing

Food preparation and processing techniques can unavoidably increase the content of PAHs in foods that may or may not be already contaminated.

2.8.2.2.1 Smoking Smoking of food is one of the oldest methods to preserve and flavor food, e.g., meat, fish, and cheese. Smoke is generated in the decomposition of wood. The smoke produced at 650–700°C has the optimal properties for preservation and flavoring (7). The smoke is driven into a kiln and delivered to the food at decreasing temperatures. When the smoke comes into contact with the food, the smoking treatment is of three varieties. Cold smoking, with smoke temperatures between 15 and 20°C, is used for aromatization of uncooked sausage, raw hams, and salami. Warm smoking, with temperatures between 25 and 50°C is used for aromatization and mild pasteurization of frankfurters, sausages, meat pieces, and gammon; and hot smoking with temperatures between 50 and 85°C is used for both aromatization and thermal treatment of hams, salami, and sausages. "Wild" smoking is used to describe the process that would occur in households in developing countries where the process of smoking is uncontrolled, and this can lead to enormous content of PAHs in foods (21).

The presence of PAHs in smoke is influenced by temperature of smoke generation, type of wood, oxygen concentration during smoke generation, and type of smoker (internal or external) (22, 23). For instance, the lowest concentration of PAHs was detected in the smoke from the combustion of tree heather wood in the drum, while the highest concentration of PAHs was reached when rock rose wood was used in the kiln (24). Moisture content can reduce the formation of PAHs during the smoking process, and optimal humidity is 20% and 30% (25).

As an alternative to direct smoking, liquid smoke flavoring (LSF) can be performed. LSF is prepared from smoke condensate but can be quality controlled for its content of PAHs (7). LSFs are used in liquids for spraying and nebulization, as emulsions for injection into food or curing brine, and for use in vegetable oils. The studies using LSF showed that meat products treated with qualified LSFs have a lower content of PAHs than wood-smoked meat products (26–28).

2.8.2.2.2 Roasting and Grilling (Broiling) In these methods, foods are cooked in an oven or in a pan under the source of direct dry heat. The use of high temperatures (150–400°C) is accompanied by the use of edible oil or fat such as butter, margarine, or vegetable oils, which are used to prevent initial excess vaporization of moisture from the surface of foods during cooking. Formation of PAHs is closely linked to the pyrolysis (incomplete combustion) of fat when it drips into the heat source or the pyrolysis of food at high temperature (29). Contact of the food with flames, cooking at high temperatures, and cooking meat with untrimmed fat all increase the formation of PAHs by pyrolysis. The highest concentration of B[*a*]P has been found in fatty beef that is charbroiled (130 μg/kg) (30). Therefore, concentrations of PAHs in foods prepared by these methods are unavoidably increased. These methods of food preparation when combined with smoking foods are among the major sources of ingested PAHs in humans.

PAHs appear especially in cooked meats and meat products. The concentrations of PAHs are dependent upon the fat content of meat, the time of cooking, and the temperature (31). The concentrations can also differ according to which method is preferred for cooking. For example, a comparison of concentrations of PAHs in duck breast steaks undergoing various cooking processes and different cooking times showed that charcoal-grilled duck steak without skin produced the highest concentrations of total PAHs (320 μg/kg), followed by charcoal-grilled steak with skin (300 μg/kg), smoking (210 μg/kg), roasting (130 μg/kg), steaming (8.6 μg/kg), and LSF (0.3 μg/kg) (32).

The roasting method is also applied to coffee bean preparation. The process is typically performed at temperatures of 180–300 °C for periods that vary from a few minutes to about 30 min to give a unique flavor and taste. Two types of roasting process are widely used for manufacturing coffee. One is indirect-fired roasting, in which the burner flame does not contact the coffee beans, although the combustion gases from the burner do contact the beans. Another is direct-fired roasting, in which there is direct contact of the beans with the burner flame and the combustion gases. This process can lead to a high amount of PAHs in coffee (33)

2.8.2.2.3 Frying Frying is a cooking method whereby foods are submerged in hot vegetable oil or fat. This cooking process is much faster than roasting or grilling (broiling) because heat from hot oils or fat is transferred through all surfaces of foods.

The presence of PAHs in vegetable oils used in frying is related to how the oil is processed. Heat treatment and the process used to dry the seeds, in which combustion gases may make contact with the seeds, can lead to contamination by PAHs (34). If oil or fat is overheated or overused, this can promote pyrolysis during the frying process, causing the accumulation of PAHs in their fume or in the oils, thereby resulting in the transfer of PAHs to kitchen air and fried foods (35–37).

2.8.2.2.4 Drying Drying is the oldest method of preserving food, which reduces water content sufficiently to prevent bacterial growth. Most types of meat can be dried, for example, to produce salted ham and beef jerky. This process can be applied to fruits such as apples, pears, bananas, mangoes, and coconuts. It is used as the normal means of preservation for cereal grains such as wheat, barley, and rice.

Tea, which is almost as widely consumed as water in some parts of the world, is prepared by drying. The presence of PAHs in tea occurs via two routes. One is the contamination from air; the other is by infiltration of PAHs formed in the drying process. Currently many tea manufacturers dry tea leaves using combustion gas from burning wood, oil, or coal. Thus, PAHs formed by this drying method can be absorbed into the tea product (38–41).

2.8.2.2.5 Steaming Steaming means to cook foods using steam. This method is based on the concept that steam transfers heat to the food, thus cooking the food. The advantage of steaming over other cooking methods is that it can minimize or avoid pyrolysis (incomplete combustion) of oil or fat during the cooking process. Thus, the exposure of foods to PAHs during steaming is significantly reduced when compared with other food cooking processes. Concentrations of PAHs in duck steaks undergoing various cooking processes showed that steaming gave concentrations of PAHs that were 35, 24, and 15 times lower than observed with grilling, smoking, and roasting, respectively (32).

2.8.2.3 PAH Occurrence in Food Groups

Numerous studies have analyzed different food groups for contamination by PAHs. The results of several studies for each major food group are shown in Table 2.8.2 (5, 6, 62). The food groups with the highest concentration of PAHs include meat products, smoked fish, smoked and cured cheese, vegetables, and tea and coffee. In many studies, the focus has been on a single PAH, namely B[*a*]P, since it is among the most carcinogenic and therefore the most extensively studied.

2.8.2.3.1 Meat and Meat Products Concentrations of B[*a*]P and other PAHs can vary in meat products based on the source of meat and its preparation. Examination of smoked meat in Iceland found that commercial brands had little to none B[*a*]P, whereas home-smoked meat contained considerable amounts in the superficial layers. These results have been seen irrespective of meat type, e.g., smoked sausage, mutton, bologna, and bacon. When charbroiling, it was found that charbroiled sausage gave some of the highest concentrations of priority PAHs (Table 2.8.2, and references therein).

The second National Health and Nutrition Examination Survey (NHANES II) estimated B[*a*]P content in meat samples cooked by different techniques under controlled conditions, as well as B[*a*]P content in meat cooked in restaurants and fast-food chains. The highest concentrations of B[*a*]P were in cooked meat that was grilled or barbecued, yielding values of 4 μg/kg (63).

2.8.2.3.2 Fish and Marine Products Concentrations of B[*a*]P in fish as in meat vary depending on its preparation. Concentrations in smoked fish can vary from being non-detectable to considerable, (8.4 μg/kg) and are largely dependent on the smoking method. In addition, seafoods can be contaminated by the food chain before they are smoked. Some of the highest concentrations of B[*a*]P have been detected in mussels from the Bay-of-Naples, yielding values as high as 130 μg/kg (64).

2.8.2.3.3 Dairy Products: Cheese, Butter, Cream, Milk, and Related Products The most contaminated dairy products are semi-cured and cured cheeses (Table 2.8.2). However, other dairy products formed from cow's milk have lower concentrations of PAHs. The presence in cow's milk is likely due to the grazing of cattle in soil and grass contaminated with airborne PAHs (14, 65).

TABLE 2.8.2 Mean concentration of 16 PAHs (in μg/kg) measured in individual foodstuffs.

Food group	Meat and meat products													
Foodstuff	Smoked pork	Smoked pork	Smoked chicken	Charbroiled hamburger	Charbroiled hamburger	Fried hamburger	Fired hamburger	Charbroiled sausage	Charbroiled sausage	Charbroiled sausage	Charbroiled sausage	Charbroiled sausage	Charbroiled sausage	Charbroiled sausage
Reference no.	42	42	43	44	44	44	44	4	4	4	4	4	4	45
PAH														
Naphthalene	8.6	20.6	**43.8**					**3.4**	0.4	6.6	**30.0**	**15.3**	**26.3**	
Acenaphthyene			7.4					0.3	0.5	0.6	1.1	0.3	13.8	
Acenaphthene		12.4	4.9					0.2	0.4	0.3	0.5	0.9	1.6	
Fluorene		4.4	9.2					0.9	**1.4**	1.0	6.1	3.1	6.0	
Anthracene		28.0	12.5	1.1	0.8			0.2	0.3	0.8	0.7	1.4	1.4	35.4
Phenanthrene	10.2	24.8	32.5					1.4	1.1	**7.0**	8.5	9.2	18.1	**168**
Fluoranthene	7.2	**39.6**	15.8	6.2	5.9	0.1	0.4	0.2	0.3	1.3	1.0	1.8	3.8	119
Pyrene	6.3	33.8	19.6	**14.0**	**19.0**	0.4	0.5	0.2	0.2	1.2	0.9	2.0	4.4	127
Chrysene	9.9	15.4	6.3					0.2		0.5	0.2	0.5	1.5	
Benz[*a*]anthracene	**11.7**		2.1	2.1	2.8	0.1	0.2	0.2		0.2		0.2	0.4	44.5
Benzo[*a*]fluoranthene			4.6							0.2		0.2	0.3	29.8
Benzo[*k*]fluoranthene		5.2	2.6							0.2		0.2	0.5	41.9
Benzo[*a*]pyrene			3.2	1.5	2.9		0.1			0.2			0.3	54.2
Indeno[1,2,3-*cd*]pyrene		8.4	1.7	3.6	5.8								0.3	41.4
Benzo[*g,h,i*]perylene														35.5
Dibenz[*a,h*]anthracene		8.5		2.1	1.3					0.2	0.2		0.5	3.5

TABLE 2.8.2 ***Continued***

Food group	Meat and meat products					Fish and marine products							
Foodstuff	Grilled sausage	Grilled sausage	Grilled sausage	Grilled sausage	Grilled sausage	Tuna	Mackerel	Smoked trout	Smoked herring	Smoked fish	Smoked fish	Smoked fish	Clam
Reference no.	45	45	45	45	45	46	46	47	48	49	50	49	46
PAH													
Naphthalene						**2.3**	1.6						3.9
Acenaphthyene						0.1	0.4						0.5
Acenaphthene						0.4	1.1						0.3
Fluorene						0.4	0.8	38.7					1.1
Anthracene	13.4	13.8	2.3	1.4	0.7	<0.1	0.1	6.8		6.3	14.1	21.0	0.3
Phenanthrene	**76.7**	**85.7**	**21.6**	**9.2**	**4.5**	0.9	**1.7**	27.3		**32.0**	**81.0**	**65.3**	**5.1**
Fluoranthene	31.7	48.6	6.0	3.4	1.9	0.1	0.7	2.3	107	9.1	16.3	26.0	4.9
Pyrene	32.5	55.4	6.5	3.4	1.8	0.4	0.6	1.6	**111**	5.3	10.2	20.5	2.6
Chrysene						<0.1	0.3	<0.1		0.6		2.5	0.9
Benz[*a*]anthracene	10.8	16.7	0.9	0.8	0.3	0.1	0.2		26.7	0.6	1.7	2.5	0.5
Benzo[*b*]fluoranthene	6.4	8.9	0.4	0.1		0.1	0.9	0.2		0.1	0.2	1.2	0.7
Benzo[*k*]fluoranthene	6.8	15.2	0.3	0.2		<0.1	0.3	<0.1		0.1	0.4	0.5	0.2
Benzo[*a*]pyrene	7.7	17.6	0.3	0.2	0.1	<0.1	0.1	0.1	8.4	0.1	0.8	1.2	0.2
Indeno[1,2,3-*cd*]pyrene	4.3	12.9	0.2			<0.1	0.2				<0.1	1.1	0.2
Benzo[*g,h,i*]perylene	4.0	14.3	0.4			<0.1	0.2	<0.1	3.0	<0.1	<0.1	0.7	0.2
Dibenz[*a,h*]anthracene	0.6	1.5				0.1	0.1	<0.1		<0.1		<0.1	<0.1

Food group	Fish and marine products				Dairy products							Vegetable oils and animal fat		
Foodstuff	Mussel	Mussel	Mussel	Squid	Milk	Cheese	Smoked cheese	Semi-cured cheese	Semi-cured cheese	Cured cheese	Cured cheese	Vegetable[a]	Vegetable[a]	Vegetable
Reference no.	44	51	51	46	52	6	53	53	53	53	53	54	54	6
PAH														
Naphthalene	**3.3**			1.0			**40.1**	**176.0**	**18.8**	**9.3**	**105.0**	**7.9**	**7.5**	
Acenaphthyene	0.6			**1.2**			11.0	44.5	1.3	0.4	31.5			
Acenaphthene	0.6			0.4			2.3	10.4	0.9	0.3	7.3	0.4	1.6	
Fluorene	1.1	3.3	2.2	0.2		0.1	3.9	37.9	2.0	0.8	17.5	2.5	3.8	
Anthracene	0.9	0.5	0.4	<0.1	<0.1	**0.8**	4.2	7.0	0.4	0.1	4.1	0.8	0.4	
Phenanthrene	3.2	3.3	6.8	0.5	0.3		18.1	42.9	3.8	1.9	16.0	9.4	6.8	
Fluoranthene	3.5	8.6	12.1	<0.1	0.1		2.5	5.2	0.5	0.4	2.6	7.1	2.8	3.2
Pyrene	2.1			<0.1	0.1		1.8	4.8	0.6	0.8	1.9	8.2	2.9	3.5
Chrysene	1.9	0.5	1.0	<0.1	0.1	0.1	0.5[b]	0.6[b]	0.1[b]	0.1[b]	1.2[b]	0.9	0.6	2.1
Benz[*a*]anthracene	0.9	**15.6**	**14.2**	<0.1	<0.1	0.1	0.3	0.4	<0.1	<0.1	1.2	1.0	0.5	1.9
Benzo[*b*]fluoranthene	1.9		0.5	<0.1	0.4	<0.1	<0.1	0.2			0.5	1.4	0.5	0.7
Benzo[*k*]fluoranthene	0.7	0.2	0.2	<0.1	<0.1	0.1	0.1[c]	0.2[c]			1.6[c]	0.5	0.3	1.0
Benzo[*a*]pyrene	0.4			<0.1	<0.1	<0.1	<0.1	0.2			0.5	1.3	0.5	1.3
Indeno[1,2,3-*cd*]pyrene	0.6			<0.1	<0.1	<0.1		0.1			0.4	1.2	0.3	2.3
Benzo[*g,h,i*]perylene	0.6			<0.1	<0.1	0.1	<0.1	0.1	<0.1	<0.1	0.5	1.2	0.3	1.5
Dibenz[*a,h*]anthracene	0.2			<0.1	<0.1	<0.1					0.3	0.1	0.1	0.2

TABLE 2.8.2 ***Continued***

Food group	Vegetable oils and animal fat							Vegetables						
Foodstuff	Olive	Olive	Corn	Sunflower	Sunflower	Margarine	Butter	Lettuce	Lettuce	Potato (peel)	Potato (core)	Carrot (peel)	Carrot (core)	Chicory[d]
Reference no.	55	55	56	57	57	6	6	58	58	59	59	59	59	60
PAH														
Naphthalene	**15.2**	11.4								0.7	0.23	0.1	<0.1	
Acenaphthyene	0.4	1.1								0.3	<0.1	<0.1	<0.1	
Acenaphthene	1.1	0.4												
Fluorene	2.9	12.2				1.8	0.6			**4.0**	**2.9**	1.4	0.5	
Anthracene		6.3	<0.1	0.3	0.9	2.1	**1.2**	<0.1	<0.1	2.3	0.5	1.5	0.8	32
Phenanthrene	8.3	46.9	1.0	**3.8**				4.3	1.6	3.2	1.1	1.3	0.9	383
Fluoranthene	3.7	111	2.5	3.1	**2.7**			**9.3**	5.5	1.2	0.9	**5.4**	**1.2**	**146**
Pyrene	5.2	142	**2.6**	2.6	**2.7**			6.4	**6.6**	1.6	0.9	2.5	0.1	69
Chrysene	4.7[b]	**340**[b]			1.5	1.9	0.1							
Benz[*a*]anthracene	1.2	124	2.6	0.9	1.0	1.5	0.1			0.4	<0.1	0.7	<0.1	7
Benzo[*b*]fluoranthene	0.7	109			1.6	0.8	<0.1			<0.1	<0.1	0.3	0.1	
Benzo[*k*]fluoranthene	0.8[c]	88.1[c]			0.6	1.1	<0.1			<0.1	<0.1	0.6	0.1	
Benzo[*a*]pyrene	0.4	92.7	1.2	0.7		1.7	0.1	0.5	0.6	0.4	0.2	0.5	0.1	3
Indeno[1,2,3-*cd*]pyrene		42.8	0.8	0.4	0.9	**2.9**	0.2	0.4		0.4	0.2	0.2	<0.1	3
Benzo[*g,h,i*]perylene	0.4	49.7		0.5	1.2	1.8	0.2	0.6	0.4	0.6	0.5	0.5	<0.1	3
Dibenz[*a,h*]anthracene		9.5	<0.1		0.3	0.3	<0.1			<0.1	<0.1	1.1	<0.1	

Food group	Cereals					Coffee and teas									
Foodstuff	Puffed wheat	Wheat bran	Corn bran	Milled wheat	White bread	Ground coffee	Ground coffee	Green tea	Green tea	Green tea	Oolong tea	Oolong tea	Black tea	Black tea	Black tea
Reference no.	6	6	6	61	6	33	33	38	39	40	38	39	38	39	41
PAH															
Naphthalene								51.4	12.3	26.6	81.9	11.0	114	429	80.6
Acenaphthyene								9.0	10.8	20.1	11.7	7.3	80	256	10.9
Acenaphthene				0.7					123	<5.0		**420**		115	3.4
Fluorene	5.4	18	1.9	1.7		10.0	21.5	15.2	28.6	21.4	15.7	22.5	70	491	14.1
Anthracene						1.3	3.5	5.0	8.4	25.1	5.5	4.3	187	795	21.8
Phenanthrene				**10.0**		**6.2**	**20.2**	**168**	**197**	**141.3**	**115**	64.0	**970**	**3460**	**245**
Fluoranthene				1.8	0.7			60.0	93.7	79.6	3.8	25.4	352	1480	173
Pyrene	**5.4**	**21.0**	**2.2**	1.6	0.5	4.0	12.3	4.3	67.9	74.4	5.5	32.8	403	32.8	173
Chrysene					0.2	0.8	2.0		18.2	24.9		6.0		241	45.4
Benz[*a*]anthracene	0.8	3.7	0.6		0.1				6.9	16.3		4.4		175	30.4
Benzo[*b*]fluoranthene	0.2	0.9	0.2		0.1	<0.1	0.5	7.9		10.8	2.9		10.0	37.6	22.0
Benzo[*k*]fluoranthene					0.1	<0.1	0.2			43.7		8.7		31	12.4
Benzo[*a*]pyrene		0.8	0.2	0.2	0.1	<0.1	0.2	6.8		7.4	4.0	5.1	20.1	39.7	5.9
Indeno[1,2,3-*cd*]pyrene	3.0	1.4	0.2		0.2			9.5		5.0	5.8	83.4	5.0	31.4	1.5
Benzo[*g*,*h*,*i*]perylene					0.1					<5.0		62.9		18.5	2.7
Dibenz[*a*,*h*]anthracene	<0.1	3.6			<0.1			0.8		<5.0	0.4		1.6		2.1

[a]Vegetable oils from canned food.
[b]Chrysene + triphenylene.
[c]Benzo[*j*+*k*]fluoranthene.
[d]From contaminated source.
Bold indicates the highest level of a priority PAH reported in the Reference in the column heading.

2.8.2.3.4 Vegetables The composition of PAHs in vegetables mirrors that found in the soil in which they are grown (66, 67). Their presence in potatoes or carrots can be mitigated if they are washed and peeled. But PAHs in leafy vegetables such as lettuce can be difficult to remove by washing since they are trapped in the hydrophobic plant cell wall (58, 68). The soil is likely contaminated with airborne PAHs, and PAHs can be especially high if the vegetables are grown in the vicinity of transportation corridors. With organically farmed vegetables, farmers avoid the use of synthetic fertilizers and pesticides. But if these organically farmed vegetables are grown in soil contaminated with airborne particulates, the PAH hazard will still exist (59).

2.8.2.3.5 Cereals PAH contents in cereals appear to be low compared with other individual foods. The concentrations of PAHs in cereals range from 2.1 to 49.4 μg/kg (21). However, the contribution of cereal to whole food intake of PAHs can be substantial since by weight it may make up a large proportion of the total diet. It is estimated that cereals account for 30% of the uptake of PAHs from the food next to edible oils and fat products (5). Wheat, corn, oats, and barley grown near industrial sites have higher concentrations of PAHs than identical crops found in more remote areas (5, 6).

2.8.2.3.6 Beverages Roasted coffee beans can have a high content of PAHs. Concentrations of B[*a*]P in coffee below 1 μg/kg have been usually reported, but over 20 μg/kg of B[*a*]P has been found in highly roasted coffee (33, 69–71). PAH contents of teas differ according to the type of tea. The concentrations of PAHs in green tea range from 0.8 to 197 μg/kg and those in oolong tea range from 0.4 to 420 μg/kg. Black tea had the highest concentrations of PAHs, ranging from 1.6 to 970 μg/kg in one study and from 18 to 3460 μg/kg in another study (Table 2.8.2). In both instances, the highest contaminant was phenanthrene. This implies that the manufacturing process of tea leaves might be the main source of PAHs in the tea product.

Infusion of coffee or tea can be expected to reduce the uptake of PAHs since they have very low solubility in water. However, recent studies showed that 50% of original PAH content in the tea could be soluble in the infusion, suggesting that the essential oils in tea leaves act as co-solvent for these lipophilic compounds (72). Thus, drinking tea or coffee heavily contaminated with PAHs may impose a health risk to humans.

2.8.2.3.7 Vegetable and Animal Fats and Oils Vegetable oils have been thought of as being free of PAHs. The PAH content of native olive oils can be high (34), but the PAH content of coconut, soybean, maize, and rapeseed oil can be reduced by the use of activated charcoal (73), suggesting a possible remediation strategy. Edible oil and fat products are a substantial source of PAHs in the diet, either directly or indirectly when used for cooking food. Lard and dripping were found to contain concentrations of PAHs ranging from 0.01 μg/kg dibenz[*a*,*h*]anthracene to 6.9 μg/kg fluoranthene (6). High concentrations of PAHs were found in margarine samples from the Netherlands (74) and the United Kingdom (6). Edible oil and fat products can be the main contributors to the uptake of PAHs in the diet (50%) and this is due to the strong lipophilic nature of PAHs.

The occurrence of PAHs in edible oils and fat products appears to be attributed mainly to environmental contamination of vegetable raw material and to contamination coming from some operation performed during its processing like seed drying or solvent extraction. Particularly high concentrations of PAHs in oils can be found when the raw seed material is subjected to direct fire drying.

2.8.3 ANALYSIS

The most common method for detecting PAHs in foods has been to rely upon high-performance liquid chromatography (HPLC) with fluorescence detection (75–77). This approach does not always involve the use of an internal standard to correct for losses in recovery during sample preparation, and identification of the analyte is based on co-elution with an authentic synthetic standard and/or examination of fluorescence properties. Although, excitation and emission wavelengths may be chosen for the detection of different PAHs by fluorescence, this method lacks sufficient specificity and false positives are easily obtained. Often a photodiode array detector (PDA) can be placed either in tandem with the fluorescence detector or used instead of the fluorescence detector. In these instances, the UV-visible properties of the analyte are matched against a spectral library for compound identity. In reality, often the spectral matches are less than perfect and the lower limit of detection (LOD) of the PDA is 20–30 times less sensitive than that of a fluorescence detector.

The field has now moved toward the use of GC-MS combined with isotopically labeled internal standards that ensure exact quantitation and unambiguous structural identification of PAHs (4, 43, 78). GC allows the separation of complex mixtures of PAHs often found in food extracts, and this is best accomplished with stationary phases of methylpolylsiloxanes. Liquid chromatography/mass spectrometry/mass spectrometry (LC/MS/MS) methods are also being developed for PAH metabolites, but are not suitable for measuring the parent hydrocarbon due to poor ionization properties. Studies have revealed that the most exact estimates of PAHs in food are obtained when a perdeuterated internal standard is used for each analyte (4).

Deuterium-labeled internal standard PAH mixtures are available commercially containing the 16 priority PAH pollutants (Cambridge Isotope chemical purity>98% and isotope purity>99%) to permit exact quantitation by GC-MS (4). The typical processing of a cooked meat sample for PAH analysis would involve the following steps: (i) methanolic/KOH for saponification, (ii) extraction into cyclohexane followed by centrifugation, (iii) solid-phase extraction with elution by acetonitrile, and (iv) GC-MS analysis. Samples would be spiked with perdeuterated PAH internal standards after the addition of the methanolic/KOH (4).

GC-MS can be performed on a Hewlett-Packard 6890 series II gas chromatograph system (Hewlett-Packard) interfaced with a 5973 series mass selective detector or equivalent. Separation is typically performed on a Supelco SPB (Sigma-Aldrich) column ($25\,m \times 0.20\,mm$) heated to 250 °C. Helium gas is used as carrier and at the MS interface ionization is by electron impact.

MS analysis is performed by selected ion monitoring (SIM). The molecular ion ($M^{\cdot+}$) of each PAH is used as the target ion for selection, and peak identity and purity is established by the detection of a qualifier ion $(M\text{-}H_2)^{\cdot+}$ and $(M\text{-}D_2)^{\cdot+}$ for the internal standard. Analyte identity is established by the ratio of the target ion to the qualifier ion. Quantitation is achieved by the use of calibration curves. Using GC-MS, the lower LOD is approximately 0.2 µg/kg and the lower limit of quantitation (LOQ) is 0.6 µg/kg (4). These limits should be kept in mind when reviewing the PAH contents in food listed in Table 2.8.2.

2.8.4 MECHANISMS OF FORMATION

Grilling (broiling) meat, fish, and other foods over intense heat or by direct contact with flames results in the formation of PAHs, which adhere to the surface of foods. These PAHs are formed by condensation of smaller organic compounds by either pyrolysis or pyrosynthesis. At high temperatures, organic compounds are easily fragmented (pyrolysis) and the free radicals produced recombine to form stable polynuclear aromatic compounds (pyrosynthesis). A common reaction may involve Diels–Alder-type rearrangements to yield the PAH. The importance of pyrolysis is borne out by the temperature at which cooking occurs. At low temperatures of less than 400 °C, only small amounts of PAHs are formed, but between 400 and 1000 °C, the amounts of PAHs formed increase linearly with temperature. Temperature affects both the structure and diversity of the PAH formed (79).

The PAH content in the smoke from heating model lipids and food lipids was quantitated by GC-MS. All 16 priority PAHs could be detected, and it was found that methyl linolenate produced the highest amount of PAHs, followed by methyl linoleate, methyl oleate, and methyl stearate (79).

2.8.5 MITIGATION

Because the exposure to PAHs is ubiquitous, it is impossible to mitigate effective reduction from all sources. However, knowing the process by which food is contaminated with PAHs can provide recommendations on how to reduce these concentrations. Fruits and vegetables grown in contaminated soil should be thoroughly washed and cleaned of all debris before consumption. Formation of PAHs during the cooking of food can be reduced if excess fat is trimmed from meat and the meat is cooked at lower temperatures and at a distance from charcoal when this is used in broiling. Similarly, cooking and smoking foods using wood fires can lead to highly variable amounts of PAHs depending on the wood used as fuel. Hard woods such as oak and hickory, burn cleanly, while woods such as mesquite generate large quantities of PAHs. When cooking on open log fires, it is recommended that the embers be used. Elimination of PAHs from packaging can be achieved by using cellulose-based wrapping.

An unexpected finding was that LSF can be absorbed by low-density polyethylene. Incubation of PAHs in LSF for 14 days in low-density polyethylene followed by extraction showed that the amount of PAHs that could be recovered fell by two orders of magnitude. The most intense absorption occurs in the first 24-h period (7). Two physiochemical principles determine the outcome, surface absorption, and diffusion into the plastic bulk. The diffusion coefficients for PAH are fluoranthene > pyrene > B[*a*]P. In a side-by-side comparison, polyethylene terephthalate was found to be less effective in absorbing PAHs (7). It is recommended that packaging of smoked food in low-density polyethylene could be an effective remediation strategy (7).

2.8.6 EXPOSURE AND BIOMONITORING

It is generally accepted that human exposure to PAHs is either by inhalation of polluted air, cigarette smoke and environmental tobacco smoke, and occupational exposure. But some studies support the concept that the general population is exposed to more B[*a*]P from food than from smoking. The Total Human Environment Exposure Study (THESS) performed in Phillipsburg, New Jersey, measured B[*a*]P exposure in the home due to food preparation that was attributed to inhalation. Dietary exposures varied from 2 to 500 ng/day depending on the eating habits and food preparation techniques of the study subjects. By contrast, inhalation exposures were of the order of 10–50 ng/day (80, 81).

The most common biomarker used for monitoring human exposure to PAHs has been 1-hydroxypyrene since it is an abundant metabolite of pyrene (82, 83). Furthermore, pyrene is on the list of 16 priority PAHs listed by the EPA. The disadvantage of measuring 1-hydroxypyrene is that it does not measure exposure to a carcinogenic PAH, e.g., B[*a*]P, which are the compounds that pose the greatest health risk. A superior method may therefore lie in the measurement of 3-hydroxy-B[*a*]P. Traditional methods for measuring this analyte have lacked sufficient sensitivity to detect amounts that might result from food consumption, and historically have been only useful for detecting occupational exposures to PAHs (84, 85). The drawback to this approach is that 3-hydroxy-B[*a*]P is not a metabolite derived from a biological intermediate that might be related to the carcinogenic process, e.g., 7*R*,8*R*-dihydroxy-7,8-dihydroB[*a*]P (B[*a*]P-7,8-dihydrodiol) or *r*-7,*t*-8,*t*-9,*c*-10-

tetrahydroxy-7,8,9,10-tetrahydroB[*a*]P (B[*a*]P-tetraol-1). An alternative has been to measure phenanthrene-tetraol amounts by GC/MS since phenanthrene undergoes similar bioactivation to B[*a*]P. Although this method has the requisite sensitivity, phenanthrene is not a carcinogenic PAH (86). Some of these problems may be circumvented by the use of liquid chromatography/atmospheric pressure chemical ionization/mass spectrometry (LC/APCI/MS) in the negative ion mode (87). Ideally these sensitive methods need to be adapted to measure trace carcinogenic PAH metabolites.

The most common biomarkers for monitoring response to PAHs are predicated on the concept that (+)-7α,8β-dihydroxy-7,8-dihydro-9α,10*a*-oxo-B[*a*]P (*anti*-B[*a*]PDE) is an ultimate carcinogenic metabolite of B[*a*]P. This highly reactive diol-epoxide can form stable-DNA adducts, e.g., (+)-*anti*-B[*a*]PDE-N^2-dGuo and stable hemoglobin adducts. Measurement of (+)-*anti*-B[*a*]PDE-N^2-dGuo adducts in human biospecimens can be achieved with sensitive methods that lack specificity. These include [^{32}P]-post-labeling, enzyme-linked immunosorbent assay (ELISA)-based methods (88–90), and HPLC coupled with fluorescence detection to measure the tetraols that are released upon hydrolysis (91).

2.8.7 HEALTH RISKS/EFFECTS

The single largest concern associated with PAH-contaminated food is cancer risk. PAHs are multispecies, multi-organ complete carcinogens (1). PAHs that are carcinogenic contain a terminal benzene ring, which creates a bay region in the structure (Fig. 2.8.1). Of the 16 priority PAHs listed by the EPA, only those that contain a bay region are carcinogenic in experimental animals, e.g., chrysene, benz[*a*]anthracene, B[*a*]P, and dibenz[*a,h*]anthracene (1).

The route of administration of PAH in a single species can determine the end organ in which tumors form. In humans, inhalation toxicity is likely responsible for lung cancer, and skin contact is likely to result in non-melanoma skin cancer. However, these are unlikely to be sites of tumor formation following ingestion in the diet. Based on animal studies, the sites of most relevance are likely to be tumors in the gastrointestinal tract and digestive system.

Human epidemiological studies suggest a probable increased risk in colorectal adenoma and pancreatic cancer. A recent study examined the association between meat and meat-related mutagens, and incidence of colorectal adenoma in the Prostate, Lung, Colorectal and Ovarian (PLCO) cancer screening trial (92). Association between heterocyclic amines (HCA) and B[*a*]P intake based on cooking method, using CHARRED (a computerized database that consists of HCA and PAH data generated for 120 categories of meat; http://charred.cancer.gov/) and incidence of adenoma by sigmoidoscopy was studied. It was concluded that consumption of red meat and meat cooked at high temperature was associated with a high risk of colorectal adenoma (odds ratio 1.21; 95% confidence interval, 1.06–1.37) (92). A similar questionnaire-based study design was used to associate the intake of grilled/barbecued food with a meat-derived

HCA and B[*a*]P mutagen database. The database was built on the HCA and B[*a*]P content of meat samples prepared differently as measured by solid-phase extraction/HPLC methodology and their mutagenic potential in the Ames test. It was found that B[*a*]P from well-done barbecued and panfried meats may be associated with a 2.2 odds ratio for increased pancreatic cancer with a 95% confidence interval of 1.2–4.0 (93).

2.8.7.1 Metabolic Activation

PAHs in food are biologically inert and have to be metabolically activated to exert their mutagenic, carcinogenic, and tumorigenic effects. Three major pathways of PAH activation have been proposed in the literature, but which pathway dominates in a particular organ site in humans is presently unknown (Fig. 2.8.2). These pathways involve the formation of radical cations (P450 peroxidase dependent) (94, 95), the formation of *anti*-diol-epoxides (P4501A1/1B1 monoxygenase dependent) (96, 97), and the formation of reactive and redox active PAH *o*-quinones (aldo-keto reductase [AKR] dependent) (98).

In the radical cation pathway, the PAH acts as a co-reductant of Fe^{IV+}-protoporphyrin cation (compound I) formed in the peroxidase cycle, yielding a radical cation at the most electron-deficient carbon, e.g., C6 of B[*a*]P. The reaction requires a peroxide co-substrate. Once formed, the radical cation is short-lived but can form depurinating DNA adducts leading to abasic sites (99, 100). If unrepaired, the abasic sites can yield G to T transversions, which is a common mutation seen in the tumor protoncogene (e.g., *K*-, *N*-, and *H-ras*) and the tumor suppressor gene (e.g., *p53*).

In the diol-epoxide pathway, the parent PAH undergoes an NADPH-dependent monoxygenation catalyzed by P450 isoforms (P4501A1, 1A2, or 1B1) to yield an arene oxide, e.g., 7*R*,8*S*-B[*a*]P-oxide (96, 97). The arene oxide is then hydrated by epoxide hydrolase (EH) to yield (–)-B[*a*]P-7,8-dihydrodiol, which can then undergo a further monoxygenation to yield the (+)-*anti*-B[*a*]PDE. There is compelling evidence that *anti*-BPDE is an ultimate carcinogen. It forms stable (+)-*anti*-B[*a*]PDE-N^2-dGuo adducts with naked and bulk DNA (101, 102). These adducts can be detected in cell culture models and *in vivo* (103, 104). (+)-*Anti*-B[*a*]PDE is a potent mutagen in bacterial and mammalian cell mutagenicity assays (105). It can cause cellular transformation by mutating codon 12 of H-ras (106); it will form DNA adducts in "hot spots" most mutated in p53 in lung cancer patients (107); and it is tumorigenic in SENCAR mouse skin (108, 109), A/J mouse lung (110), and the murine fore-stomach (111).

In the PAH *o*-quinone pathway, the intermediate *trans*-dihydrodiols, e.g., (–)-B[*a*]P-7,8-dihydrodiol, undergoes an $NADP^+$-dependent oxidation catalyzed by the dihydrodiol dehydrogenase activity of AKRs (98). In humans, five isoforms have been implicated in this transformation: AKR1A1 and AKR1C1 to AKR1C4 (112–114). This reaction results in the formation of a ketol, which spontaneously rearranges to form a catechol. The catechol (*o*-hydroquinone) undergoes two sequential one-electron oxidation steps in the presence of air

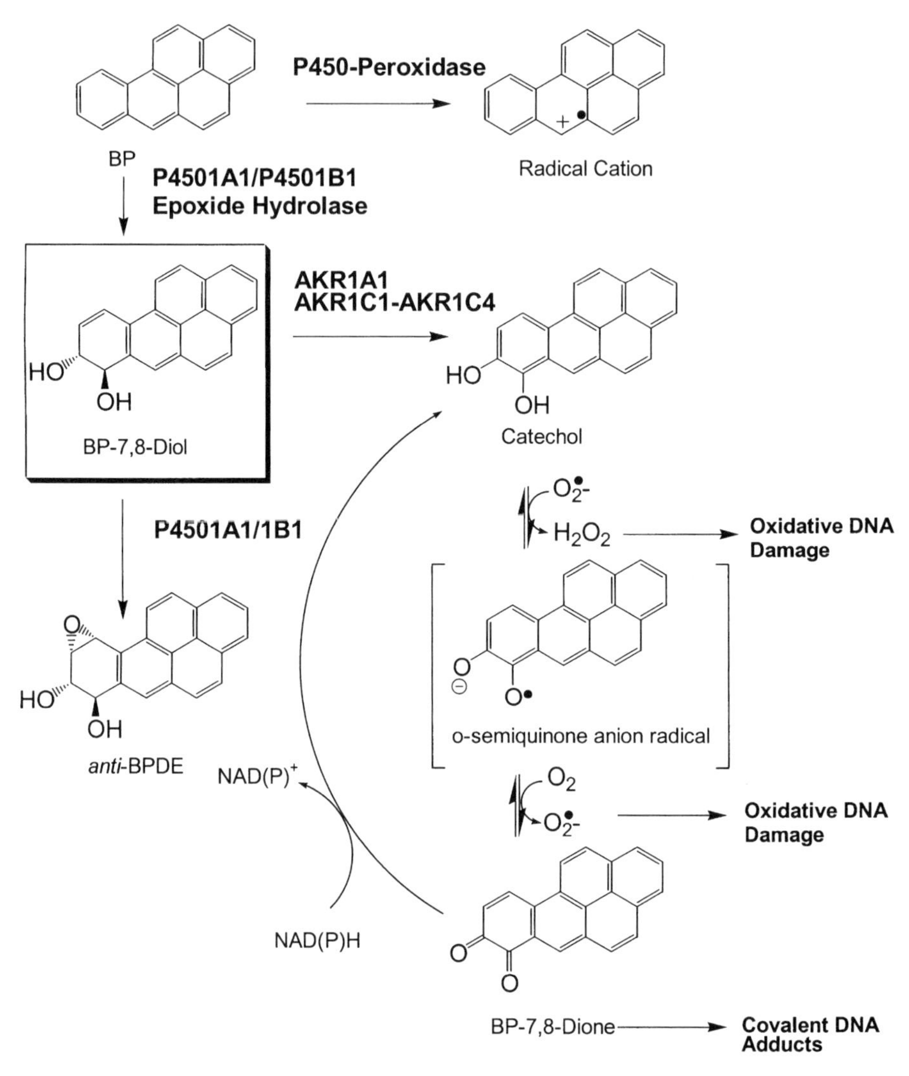

Figure 2.8.2 Major routes of metabolic activation of PAH.

to yield first an *o*-semiquinone anion radical, and second the fully oxidized *o*-quinone (115). As a result of these oxidation events, reactive oxygen species (ROS) are produced (116). Once the *o*-quinone is formed, this reactive Michael acceptor can undergo nucleophilic addition with bases in DNA and RNA. *In vitro*, PAH *o*-quinones can form both stable N^2-deoxyguanosine adducts and depurinating adducts (117–119). The PAH *o*-quinones can also form conjugates with glutathione and amino acid residue side chains in proteins (120, 121). Alternatively, in the presence of cellular reducing equivalent, the *o*-quinone can be reduced back to the catechol for a subsequent round of auto-oxidation. This establishes a futile redox cycle that continues to generate ROS until the reducing equivalent is depleted. This ROS amplification system has been shown to cause a significant amount of 8-oxo-dGuo formation (122), mutation of the *p53* tumor suppressor gene (123), and single-strand DNA breaks (124).

2.8.7.2 Mutagenicity

The assay most commonly performed on the 16 priority PAHs in foods has been the *in vitro Salmonella typhimurium* assay, which scores histidine revertants (Ames test) in tester strains TA97, TA98, TA100, and TA102. These tester strains detect different types of mutations. TA97 and TA98 detect frameshift mutations, TA100 scores predominately point mutations, and TA102 is sensitive to oxidative mutagens. In addition, the mammalian mutagenicity assay, which scores heritable mutations in the hypoxanthine ribosyl transferase (*HPRT*) gene, is scored in V79 cells (Chinese hamster lung fibroblasts). Both assays have the versatility to permit the incorporation of a metabolic activation system (S9, rat liver microsomes, and NADPH-regenerating system) in the cell culture medium. The *HPRT* gene assay can also be modified so that V79 cells are stably transfected with P450 isoforms involved in PAH activation. Each of the 16 priority PAHs has been examined as mutagens in one or both of these assays. The results of these assays are summarized in Table 2.8.3.

- *Naphthalene*—No reliable mutagenicity studies have been reported for naphthalene.
- *Acenaphthylene*—No data available.
- *Acenaphthene*—This scores negative as a mutagen in tester strains TA97, TA98, and TA100 in the absence and presence of rat or hamster S9 fraction following Aroclor induction, at concentrations of 1–200 μg/plate (125).
- *Fluorene*—This scores negative as a mutagen in tester strains TA97, TA98, and TA100 in the absence and presence of rat liver S9 induced with a polychlorinated biphenyl (PCB) at concentrations of 1–250 μg/plate (126). Fluorene scored positively for increases in chromosomal aberrations in Chinese hamster lung cells at 25 μg/mL provided there was metabolic activation (127), but it was not mutagenic in Chinese hamster ovary cells.
- *Anthracene*—This is mutagenic in tester strain TA100 in the absence of a metabolic activation system (128), and is a positive mutagen in tester strain TA97 in the presence of rat liver S9 induced with PCB in the range of 5–250 μg/plate (126).
- *Phenanthrene*—This is mutagenic in tester strains TA100 and TA1535 in the presence of a rat liver S9 induced with Aroclor at concentrations of 1–250 μg/plate (129).
- *Fluoranthene*—This is mutagenic in tester strains TA97, TA98, and TA100 in the presence of a rat liver S9 induced with Aroclor at concentrations of 1–250 μg/plate (130, 131).

TABLE 2.8.3 Mutagenicity of priority PAH.

	Ames test no activation system, μg/plate				Ames test plus activating system, μg/plate				HPRT	
	TA97	TA98	TA100	TA1547	TA97	TA98	TA100	TA1537/8	−Activ	+Activ
Compound										
Naphthalene	ND	ND	ND	ND	ND	ND	ND	ND	ND	ND
Acenaphthylene	ND	ND	ND	ND	ND	ND	ND	ND	ND	ND
Acenaphthene	ND	ND	ND	ND	X 1–200	X 1–200	X 1–200	ND	ND	ND
Fluorene	ND	ND	ND	ND	X 1–250	X 1–250	X 1–250	ND	ND	ND
Anthracene			X 5–250		X 5–250			ND	ND	ND
Phenanthrene	ND	ND	ND	ND			X 1–250	X 1–250		
Fluoranthene	ND	ND	ND	ND	X 1–250	X 1–250	X 1–250	ND	ND	ND
	TA97	TA98	TA100	TA1547	TA97	TA98	TA100	TA1537/8	−Activ	+Activ
Pyrene	ND	ND	ND	ND	X 1–40	X 1–40				
Chrysene	ND	ND	ND	ND	X 1–50		X 1–50		X	
Benz[*a*]anthracene	ND	ND	ND	ND	X 10–100	X 10–100	X 10–100	X 10–100		
Benzo[*b*]fluoranthene	ND	ND	ND	ND		X 0.3–10	X 0.3–10			
Benzo[*k*]fluoranthene	ND	ND	ND	ND			X 2.5–200			
Benzo[*a*]pyrene					X	X 0–1	X 0–1	X	X	
Indeno[1,2,3-*cd*]pyrene	ND	ND	ND	ND	ND	ND	ND	ND	ND	ND
Benzo[*g,h,i*]perylene					X 10–250	X 10–250	X 10–250	X 10–250		
Dibenz[*a,h*]anthracene						X 3–3333	X 0.3–333	X	X	

- *Pyrene*—This is mutagenic in tester strains TA97 and TA100 in the presence of a rat liver S9 induced with Aroclor at concentrations of 1–40 μg/plate (132). It is not mutagenic in mammalian cell-based assays.
- *Chrysene*—Is a positive mutagen in tester strains TA98 and TA100 in the presence of a rat liver S9 induced by either Aroclor or PCB in the range of 1–50 μg/plate (133). In addition, it is mutagenic in the *HPRT* assay in the presence of an activating system (134), and in *in vivo* assays in the *LacZ* transgenic mouse following i.p. administration (135).
- *Benz[*a*]anthracene*—Is positive in the Ames test in tester strains TA98, TA100, and TA1537/TA1538 with a metabolic activating system (rat liver S9 plus phenobarbital, or PCB, or Aroclor) in the range of 10–100 μg/plate (136, 137).
- *Benzo[*b*]fluoranthene*—Positive in the Ames test in tester strains TA98 and TA100 with metabolic activation systems (rat liver or hamster liver S9 induced with Aroclor) in the range of 0.3–10 μg/plate (125, 138, 139).
- *Benzo[*k*]fluoranthene*—Is positive in the Ames test in tester strain TA100 only in the presence of an activation system (rat liver S9 induced with Aroclor) in the range of 2.5–200 μg/plate (140).
- *Benzo[*a*]pyrene*—B[*a*]P has been found to be mutagenic in over 30 separate studies in the Ames test with tester strains TA97, TA98, TA100, and TA1537/8 in every case in the presence of a metabolically activating system was required, e.g., rat liver S9 (126, 136, 141, 142).
- *Indeno[1,2,3-*cd*]pyrene*—Is not mutagenic by itself on TA100 (143). The metabolically activated indeno[1,2,3-*cd*]pyrene is not mutagenic in the standard plate Ames test using tester strain TA100.
- *Benzo[*g,h,i*]perylene*—Is positive in the Ames test in tester strains TA97, TA98, TA100, and TA1538 provided a metabolic activation system is employed (rat liver S9 induced by phenobarbital, Aroclor 1254, and PCB) in the range of 10–250 μg/plate (130, 138, 144).
- *Dibenz[*a,h*]anthracene*—Is positive in the Ames test in tester strains TA98, TA100, and TA1537 in the presence of activating systems (rat S9 induced with Aroclor or 3-methylcholanthrene) (136, 143, 145). It is also mutagenic in V79 cells using rat liver S9 as an activation system (146).

2.8.7.3 Carcinogenicity

The standard tumorigenic assays for PAH involve the use of SENECAR mice, which are an inbred strain with a predisposition to form papillomas when the skin is painted with PAH as an initiating agent with or without repetitive treatment of a tumor promoter (PMA) (108), or the use of the A/J mouse, which has a predisposition to form lung cancer following intra-tracheal or i.p. administration of PAH (110, 147). Other rodent strains used in PAH toxicology studies are F344 male and female rats, B6C3F1 male and female mice, and

NMRI male and female mice, which are routinely used in the National Toxicology Program (NTP), in the United States. Animals used in the NTP permit the toxicity of a wide variety of different chemical entities to be compared within the same species and strain. Published studies on PAH toxicity have also been performed on Swiss albino mice and their derivative strains (CD-1, CFW, and ICF mice), since these represent general all-purpose strains used in safety and efficacy testing. The formation of PAH-induced tumors at specific organ sites can be dependent upon the rodent species and route of administration. Of most relevance to food intake is the evidence for tumor formation following oral administration.

No studies exist regarding cancer in humans following oral administration except those from epidemiological studies. However, data from animal studies exist and these can vary depending on the PAH and whether the exposure is acute, intermediate, or chronic. Each of the 16 priority PAHs will be discussed in turn. The results indicate that benz[*a*]anthracene, B[*a*]P, dibenz[*a*,*h*]anthracene, and possibly other PAHs as well are carcinogenic to rodents following oral administration at high doses.

- *Naphthalene*—Inhalation of naphthalene will cause lung adenoma in female mice B6C3F1 (148), nasal adenoma in F344 male rats, and neuroblastoma in F344 female rats (149). However, inhalation is not the exposure route for food toxicants, and no data exist for the oral administration of naphthalene.
- *Acenaphthylene*—No data available.
- *Acenaphthene*—One study of intermediate duration evaluated the carcinogenic potential of acenaphthene. Male and female mice exposed to 0, 175, 350, or 700 mg/kg/day by gavage for 13 weeks showed no sign of tumorigenesis (148).
- *Fluorene*—No data available.
- *Anthracene*—No data available.
- *Phenathrene*—No data available.
- *Fluoranthene*—Will cause lung adenoma in several mice strains (e.g., ICR, Swiss, and CD-1 male and female mice) following i.p. administration (148).
- *Pyrene*—Will cause lung tumors in mice following i.p. injection (148).
- *Chrysene*—Will cause skin carcinoma following repetitive dermal application in ICR male and female mice (150), and lung carcinoma after intrapulmonary implant to rats (149). No data are available following oral administration.
- *Benz[a]anthracene*—Mice acutely administered 1.5 mg/kg benz[a]-anthracene twice per day for 3 days by oral gavage exhibited an increased incidence of hepatomas and pulmonary adenomas as compared with controls. Mice receiving intermittent gavage doses of 1.5 mg/kg for 5 weeks

showed a 95% incidence of pulmonary adenomas, 46% incidence of hepatomas, and 5% of forestomach papillomas (1, 111).

- *Benzo[*b*]fluoranthene*—No studies on oral administration exist. But dermal application will cause skin cancer in CD-1 mice (151), and lung implants and i.p. exposure will cause lung sarcoma and lung adenoma, in rats and mice, respectively (152, 153).
- *Benzo[*k*]fluoranthene*—No studies on oral administration. But dermal application and will cause skin cancer in mice (153) and lung implants will cause lung carcinoma in rats (154), over their life span.
- *Benzo[*a*]pyrene*—Oral administration of B[*a*]P to male and female CFW mice induced gastric papillomas and squamous cell carcinomas. Tumors also arose in distal sites since there was an increase in pulmonary adenomas, thyomas, lymphomas, and leukemias following oral administration. Gastric tumors were seen in 70% of the mice fed 50–250 ppm B[*a*]P for 4–6 months (155).

Mice fed B[*a*]P at a concentration of 33.3 mg/kg/day exhibit forestomach neoplasms following two or more days of consumption (111). Hamsters will also develop papillomas and carcinomas of the alimentary tract following gavage or dietary exposure to B[*a*]P (111). A 77% incidence of mammary tumors was observed 90 weeks after a single oral dose of 50-mg B[*a*]P (100 mg/kg) was given to rats. Intragastric doses of 67–100 mg/kg of B[*a*]P will elicit pulmonary adenomas and forestomach papillomas in mice. The incidence of forestomach tumors (papillomas and carcinomas) was related to the duration of the oral exposure to B[*a*]P and was dose-dependent. In a similar manner, an association between dietary B[*a*]P intake and the development of leukemia and tumors of the forestomach and lung has been observed in mice (111).

- *Indeno[1,2,3-*cd*]pyrene*—No data available.
- *Benzo[*g,h,i*]perylene*—No studies on oral administration. In the Osborne–Mendel female rat, lung implants with up to 4-mg compound were not tumorigenic (152).
- *Dibenz[*a,h*]anthracene*—Forestomach papillomas were found in 10% mice administered with a single oral dose of 0.05 mg/kg dibenz[*a*, *h*]anthracene, but this incidence increased to 21% if croton oil (tumor promoter) was subsequently administered for 30 weeks. In another study, mice were fed a total of 9–19 mg dibenz[*a*,*h*]anthracene over 5–7 months and tumors in the forestomach were detected (2). After 15 weeks of dosing, 5% of female BALB/c mice had mammary carcinoma but there was no control group. The hydrocarbon will cause skin papillomas in female NMRI/mice following dermal application (156), lung adenomas in NMRI/female mice following subcutaneous administration (156), lung adenoma in A/J mice following i.p. administration (153), and liver adenomas in B6C3F1 mice following i.p. administration (157).

These data suggest that oral intake of benz[*a*]anthracene, B[*a*]P, and dibenz[*a*,*h*]anthracene has to be of greatest concern, since they appear to be multi-organ and multispecies carcinogens when administered by this route at high dose.

2.8.7.4 Oral Bioavailability

B[*a*]P is orally absorbed in humans. B[*a*]P could be detected in the feces of healthy volunteers who ingested broiled meat, but these amounts were similar to those observed in individuals who ate control meat. However, based on animal studies, the extent of B[*a*]P oral absorption may be governed by the presence of oils (158). The extent of oral absorption of B[*a*]P is enhanced in rats when it is dissolved in triolein, soybean oils, and high-fat diets (111, 158). Oral absorption may also be affected by biotransformation in the intestinal flora and fauna, a topic outside the scope of this review.

Five volunteers, who ingested B[*a*]P from grilled beef for 2–3 days, had concentrations of 0.7 to 20 μg/day in their feces. Oral absorption of B[*a*]P in rats is incomplete and can be influenced by oils and fats in the gastrointestinal tract. Radioactivity found in liver, lungs, kidneys, and testis following a low oral dose of [^{3}H]-B[*a*]P to S.D. rats provides supporting evidence for oral absorption (111).

2.8.8 RISK MANAGEMENT

The FAO/WHO Joint Expert Committee on Food Additives (JECFA) concluded that the critical effect of PAHs is their carcinogenicity in animals and potential carcinogenicity in humans (159). Moreover, since the mode of action appears to be via genotoxic metabolites, it is not possible to assume that a safe exposure exists. Their evaluation focused on 13 PAHs that are genotoxic and carcinogenic: benz[*a*]anthracene, benzo[*b*]fluoranthene, benzo[*j*]fluoranthene, benzo[*k*]fluoranthene, B[*a*]P, chrysene, dibenz[*a*,*h*]anthracene, dibenzo[*a*,*e*]pyrene, dibenzo[*a*,*h*]pyrene, dibenzo[*a*,*i*]pyrene, dibenzo[*a*,*l*]pyrene, indeno[1,2,3-*cd*]pyrene, and 5-methylchrysene. Some of these are identical to the 16 priority PAH pollutants listed by the US EPA; the exceptions being the dibenzopyrenes and 5-methylchrysene. The Committee recommended that B[*a*]P be used as a surrogate marker of exposure to all these PAHs and that exposures be limited (159). This seems to be the most prudent course of action since there is considerable debate as to whether there is any safe exposure to compounds that can initiate the carcinogenic process. The Committee calculated a "benchmark dose lower confidence interval (BMDL)" equivalent of 100-μg B[*a*]P/kg of body weight per day, which was derived for PAHs in food on the basis of their carcinogenicity in mice treated orally with mixtures of PAHs. Despite a wide range of estimates of PAH intake, the Committee concluded that a representative mean intake of B[*a*]P was 0.004 μg/kg of body weight per day and an estimated high level of intake of B[*a*]P was 0.01 μg/kg of body weight per day. When compared with the BMDL, these mean and high-level intakes indicate that the margin of exposure (MOE) is 25,000 and 10,000, respectively. Based on these MOEs, the Committee concluded that the estimated intake of PAHs in food was of low concern to human health.

The Committee on Diet Nutrition and Cancer appointed by the National Research Council of the United States stated that of the PAHs that commonly occur in the American diet, only three (B[*a*]P, benz[*a*]anthracene, and dibenz[*a*,*h*]anthracene) have been found to be carcinogenic in animals following oral administration (21).

2.8.8.1 Legislation to Reduce Exposures

Some countries have set legislative limits for PAHs in foods with the aim of minimizing their intake and decreasing risk. In Germany, Poland, and Austria, a limit of 1 mg/kg is currently imposed for B[*a*]P in smoked meat products, while in Italy no limit values exist regarding B[*a*]P and other PAHs in smoked food (111).

In case of edible oils, Spain, Italy, Portugal, and Greece have produced a legislative regulation limiting the concentration of seven PAHs: benz[*a*]anthracene, benzo[*e*]pyrene, benzo[*k*]fluoranthene, B[*a*]P, dibenz[*a*,*h*]anthracene, benzo[*g*,*h*,*i*]perylene, and indeno[1,2,3-*cd*]pyrene. A maximum limit per single PAH is 2 μg/kg but cannot exceed 5 μg/kg as a total (160). The Canadian Food Inspection Agency has also suggested correcting safety recommendations on the basis of toxic equivalency factors (see Table 2.8.1) (161).

For smoke flavoring agents, the EU and United States have allowed the use of these aromas, provided that B[*a*]P residue does not exceed 0.03 μg/kg in the final products. However, to avoid the discrepancies between these and other limits, the EU adopted a new regulation of maximum level for PAHs in foodstuffs in 2005 (Commission Regulation No. 208/2005) and provided a list of 15 PAHs that were of major concern for human health due to their toxic properties (162). This regulation set B[*a*]P as a marker for the occurrence and effect of carcinogenic PAHs in food groups. The maximum concentrations, which can be present in individual foodstuffs according to food group are listed in Table 2.8.4. This regulation limits the content of B[*a*]P to 5 μg/kg in smoked meat and smoked fish products. The European Commission also adopted the 2005/10/EC directive, which set standards for sampling and analysis for B[*a*]P contamination in food groups (163). This directive indicated that validated methods that have an LOD of 0.3 μg/kg and LOQ of 0.9 μg/kg, and which employ methods that yield 50–100% recovery of B[*a*]P should be used. This directive was repealed by European Commission Regulation 333/2007, which made the requirements somewhat more stringent by indicating that the analyte

TABLE 2.8.4 The limit of maximum level of benzo[*a*]pyrene in foodstuffs (Commission Regulation of EU, No. 208/2005).

Product	Maximum level, μg/kg wet weight
Oil and fat	2.0
Foods for infants and young children	1.0
Smoked meats and smoked meat products	5.0
Smoked fish and smoked fishery products	5.0
Muscle meat of fish other than smoked fish	2.0
Crustaceans, cephalopods, other than smoked	5.0
Bivalve mollusks	10.0

should be free of spectral interference and that there should be verification of positive detection (164). Interestingly, neither regulation recommended a method of analysis.

In addition, the European Commission 88/288 directive set standards for the LSF for food (165). Maximum acceptable concentrations were 10μg/kg for B[*a*]P and 20μg/kg for benz[*a*]anthracene. This directive limits maximal residual concentrations of B[*a*]P to 0.03μg/kg in foods prepared using LSF. For international trade purposes, JEFCA tolerates LSF with concentrations that must not exceed 10μg/kg B[*a*]P and 20μg/kg benz[*a*]anthracene.

In the United States, the presence of carcinogens in food is covered by the Delaney Clause of the Federal Food, Drug and Cosmetic Act enforced by the Federal Food and Drug Administration (166). Prior to this 1958 amendment, a substance added to food was presumed safe until it was proven otherwise. The amendment prohibits the FDA approving the use of a food additive found to cause cancer in animals or humans. It has been criticized as being too restrictive by setting a zero level of risk. It applies to only 400 of the 2700 substances that were added to food prior to the amendment since information on the remaining additives is lacking. However, the food-processing practices described in this article are still permitted to take place.

2.8.8.2 Future Directions

PAHs are prevalent in foods and occur either due to contamination from the food chain or from the food preparation technique. It is unlikely that dependence on fossil fuel combustion will be attenuated, indicating that it will be difficult to limit exposure by remediation in the food chain alone. But progress in mitigating amounts of PAHs in man-made food processing is possible, for example, in the use of LSF in place of smoking. Mutagenic and carcinogenic data obtained in mammalian cells and rodents, respectively, support the health hazard associated with PAH exposure, but these data are difficult to extrapolate to human exposure levels. However, the fact that PAHs are multispecies, multi-organ carcinogens in rodents suggests that a cautious approach is required to mitigate exposures. The EU has placed limits on human exposure to PAHs in food products, and other nations could do the same. JECFA has identified 13 PAHs that occur in foods as being genotoxic and carcinogenic. However, among these, little or no analytical data on the concentrations of 5-methyl chrysene, dibenzo[*a*,*i*]pyrene, or dibenzo[*a*,*l*]pyrene exist across food group, and this gap in knowledge needs to be addressed.

REFERENCES

1. IARC (1973). *Monographs on the Evaluation of the Carcinogenic Risk of Chemicals to Humans*, Vol. 3, International Agency for Research on Cancer, Lyon, France.
2. Liaronow, L.F., Soboleva, N.G. (1938). Gastric tumours experimentally produced in mice by means of benzopyrene and dibenz[*a*]anthracene. *Vestnik Rentgenologii i Radiologii*, *20*, 276
3. Straif, K., Baan, R., Grosse, Y., Secretan, B., El Ghissassi, F., Cogliano, V. (2005). Policy watch: carcinogenicity of polycyclic aromatic hydrocarbons. *The Lancet Oncology*, 931–932.
4. Mottier, P., Parisod, V., Turesky, R.J. (2000). Quantitative determination of polycyclic aromatic hydrocarbons in barbecued meat sausages by gas chromatography coupled to mass spectrometry. *Journal of Agriculture and Food Chemistry*, *48*, 1160–1166.
5. Dennis, M.J., Massey, R.C., McWeeny, D.J., Knowles, M.E. (1983). Analysis of polycyclic aromatic hydrocarbons in UK total diets. *Food Chemistry and Toxicology*, *21*, 569–574.
6. Dennis, M.J., Cripps, G., Venn, I., Howarth, N., Lee, G. (1991). Factors affecting the polycyclic aromatic hydrocarbon content of cereals, fats, and other food products. *Food Additives and Contaminants*, *8*, 517–530.
7. Simko, P. (2005). Factors affecting elimination of polycyclic aromatic hydrocarbons from smoked meat foods and liquid smoke flavourings. *Molecular Nutrition and Food Research*, *49*, 637–647.
8. Beak, S.O., Field, R.A., Goldstein, M.E., Kirk, P.W., Lester, J.N., Perry, R. (1991). A review of atmospheric polycyclic aromatic hydrocarbons: sources, fate and behaviour. *Water Air and Soil Pollution*, *60*, 279–300.
9. Marr, L.C., Kirchstetter, T.W., Harley, R.A. (1999). Characterization of polycyclic aromatic hydrocarbons in motor vehicle fuels and exhaust emissions. *Environmental Science and Technology*, *33*, 3091–3099.
10. Boffetta, P., Jourenkova, N., Gustavsson, P. (1997). Cancer risk from occupational and environmental exposure to polycyclic aromatic hydrocarbons. *Cancer Causes and Control*, *8*, 444–472.
11. Stern, A.H., Munshi, A.A., Goodman, A.K. (1989). Potential exposure levels and health effects of neighbourhood exposure to a municipal incinerator bottom ash landfill. *Archives of Environmental Health*, *44*, 40–48.

12. Kriek, E., Van Schooten, F.J., Hillebrand, M.J., Van Leeuwen, F.E., Den Engelse, L., De Looff, A.J., Dijkmans, A.P. (1993). DNA adducts as a measure of lung cancer risk in humans exposed to polycyclic aromatic hydrocarbons. *Environmental Health Perspectives*, *99*, 71–75.
13. Jones, K.C., Stratford, J.A., Waterhouse, K.S., Johnston, A.E. (1987). Polynuclear aromatic hydrocarbons in UK soils: long-term temporal trends and current levels, in Trace Substances in Environmental Health XXI, St. Louis, MO.
14. Grova, N.C., Laurent, C., Feidt, C., Rychen, G., Laurent, F. (2000). Gas-chromatography-mass spectrometry study of polycyclic aromatic hydrocarbons in grass and milk from urban and rural farms. *European Journal of Mass Spectrometry*, *6*, 457–460.
15. Hoffman, E.J., Mills, G.L., Latimer, J.S., Quin, J.G. (1984). Urban runoff as a source of polycyclic aromatic hydrocarbons to coastal waters. *Environmental Science and Technology*, *18*, 580–587.
16. Brown, R.C., Pierce, R.H., Rice, S.A. (1985). Hydrocarbon contamination in sediment from urban storm water runoff. *Marine Pollution Bulletin*, *16*, 236–240.
17. Black, J.J. (1982). Movement and identification of a creosote-derived complex below a river pollution point source. *Archives of Environmental Contamination and Toxicology*, *11*, 161–166.
18. Neff, J.M. (1990). Composition and fate of petroleum and spill-treating agents in the marine environment, in *Sea Mammals and Oil: Confronting the Risks* (eds J. Geraci, D.J. St. Aubin), Academic Press, New York, pp. 1–33.
19. Nendza, M., Herbst, T., Kussatz, C., Gies, C.A. (1997). Potential for secondary poisoning and biomagnification in marine organisms. *Chemosphere*, *35*, 1875–1885.
20. Gardener, G.R., Yevich, P.P., Harshbarger, J.C., Malcolm, A.R. (1991). Carcinogenicity of black rock harbour sediment to the eastern oyster and tropic transfer of black rock harbour carcinogens from the blue mussel to the winter flounder. *Environmental Health Perspectives*, *99*, 71–75.
21. International Programme on Chemical Safety (IPCS) (1998). *Selected Non-Heterocycles Polycyclic Aromatic Hydrocarbons*, World Health Organization, Geneva, pp. 148–156.
22. Howard, J.W., Fazio, T. (1980). Analytical methodology and reported findings of polycyclic aromatic hydrocarbons in foods. *Journal of the Association of Official Analytical Chemists*, *63*, 1077–1104.
23. Garcia-Falcon, M.S., Simal-Gandara, J. (2005). Polycyclic aromatic hydrocarbons in smoke from different woods and their transfer during traditional smoking into chorizo sausages with collagen and tripe casings. *Food Additives and Contaminants*, *22*, 1–8.
24. Conde, F.J., Ayala, J.H., Afonso, A.M., González, V. (2005). Polycyclic aromatic hydrocarbons in smoke used to smoke cheese produced by the combustion of rock rose (*Cistus monspeliensis*) and tree heather (*Erica arborea*) wood. *Journal of Agriculture and Food Chemistry*, *53*, 176–182.
25. Pagliuca, G., Gazzotti, T., Zironi, E., Serrazanetti, G.P., Mollica, D., Rosmini, R. (2003). Determination of high molecular mass polycyclic aromatic hydrocarbons in a typical Italian smoked cheese by HPLC-FL. *Journal of Agriculture and Food Chemistry*, *51*, 5111–5115.

26. Gomaa, E.A., Gray, J.I., Rabie, S., Lopez-Boto, C., Booren, A.M. (1993). Polycyclic aromatic hydrocarbons in smoked food products and commercial liquid smoke flavourings. *Food Additives and Contaminants*, *10*, 503–521.
27. Simko, P., Brunckova, B. (1993). Lowering of polycyclic aromatic hydrocarbons concentration in a liquid smoke flavour by absorption into polyethylene packaging. *Food Additives and Contaminants*, *10*, 257–263.
28. Tilgner, D.J. (1977). The phenomena of quality in the smoke curing process. *Pure and Applied Chemistry*, *49*, 1629–1638.
29. Lijinsky, W., Shubik, P. (1964). Benzo[*a*]pyrene and other polynuclear hydrocarbons in charcoal-broiled meat. *Science*, *145*, 53–55.
30. Lijinsky, W., Ross, A.E. (1967). Production of carcinogenic polynuclear hydrocarbons in the cooking of food. *Food and Cosmetics Toxicology*, *5*, 343–347.
31. Doremire, M.E., Harmon, G.E., Pratt, D.E. (1979). 3,4-Benzopyrene in charcoal grilled meats. *Journal of Food Science*, *44*, 622–623.
32. Chen, B.H., Lin, Y.S. (1997). Formation of polycyclic aromatic hydrocarbons during processing of duck meat. *Journal of Agriculture and Food Chemistry*, *45*, 1394–1403.
33. Lai, J.-P., Niessner, R., Knopp, D. (2004). Benzo[*a*]pyrene imprinted polymers synthesis, characterization and SPE application in water and coffee samples. *Analytica Chimica Acta*, *522*, 137–144
34. Speer, K., Steeg, E., Horstmann, P., Kuehn, T., Montag, A. (1990). Determination and distribution of polycyclic aromatic hydrocarbons in native vegetable oils, smoked fish products, mussels and oysters, and bream from the river Elbe. *Journal of High Resolution Chromatography*, *13*, 104–111.
35. Chen, Y.C., Chen, B.-H. (2003). Determination of polycyclic aromatic hydrocarbons in fumes from fried chicken legs. *Journal of Agriculture and Food Chemistry*, *51*, 4162–4167.
36. Zhu, L., Wang, J. (2003). Sources and patterns of polycyclic aromatic hydrocarbons pollution in kitchen air, China. *Chemosphere*, *50*, 611–618.
37. Li, S. (1994). Analysis of polycyclic aromatic hydrocarbons in cooking oil fumes. *Archives of Environmental Health*, *49*, 119–122.
38. Lin, D., Zhu, L. (2006). Factors affecting transfer of polycyclic aromatic hydrocarbons from made tea to tea-infusion. *Journal of Agriculture and Food Chemistry*, *54*, 4350–4354.
39. Lin, D., Tu, Y., Zhu, L. (2005). Concentrations and health risk of polycyclic aromatic hydrocarbons in tea. *Food Chemistry and Toxicology*, *43*, 41–48.
40. Fiedler, H., Cheung, C.K., Wong, M.H. (2002). PCDD/PCDF, chlorinated pesticides and PAH in Chinese teas. *Chemosphere*, *46*, 1429–1433.
41. Schlemitz, S., Pfannhauser, W. (1997). Supercritical fluid extraction of mononitrated polycyclic aromatic hydrocarbons from tea: correlation with the PAH concentration. *Zeitschrift fur Lebensmittel-Untersuchung und-Forschung. A*, *205*, 305–310.
42. Wang, G., Lee, A.S., Lewis, M., Kamath, B., Archer, R.K. (1999). Accelerated solvent extraction and gas chromatography/mass spectrometry for determination of polycyclic aromatic hydrocarbons in smoked food samples. *Journal of Agriculture and Food Chemistry*, *47*, 1062–1066.

43. Chiu, C.P., Lin, Y.S., Chen, B.H. (1997). Comparison of GC-MS and HPLC for overcoming matrix interferences in the analysis of PAHs in smoked food. *Chromatographia*, *44*, 497–504.

44. Lawrence, J.F., Weber, D.F. (1983). Determination of polycyclic aromatic hydrocarbons in some Canadian commercial fish, shellfish, and meat products by liquid chromatography with confirmation by capillary gas chromatography-mass spectrometry. *Journal of Agriculture and Food Chemistry*, *32*, 789–794.

45. Larsson, B.K., Sahlberg, G.P., Eriksson, A.T., Busk, L.A. (1983). Polycyclic aromatic hydrocarbons in grilled food. *Journal of Agriculture and Food Chemistry*, *31*, 867–873.

46. Llobet, J.M., Falco, G., Bocio, A., Domingo, J.L. (2006). Exposure to polycyclic aromatic hydrocarbons through consumption of edible marine species in Catalonia, Spain. *Journal of Food Protection*, *69*, 2493–2499.

47. Moret, S., Conte, L., Dean, D. (1999). Assessment of polycyclic aromatic hydrocarbon content of smoked fish by means of a fast HPLC/HPLC method. *Journal of Agriculture and Food Chemistry*, *47*, 1367–1371.

48. Dennis, M.J., Massey, R.C., Mcweeny, D.J., Larsson, B., Eriksson, A., Sahlberg, G. (1984). Comparison of a capillary gas chromatographic and a high-performance liquid chromatographic method of analysis for polycyclic aromatic hydrocarbons in food. *Journal of Chromatography*, *285*, 127–133.

49. Karl, H., Leinemann, M. (1996). Determination of polycyclic aromatic hydrocarbons in smoked fishery products from different smoking kilns. *Zeitschrift für Lebensmittel-Untersuchung und-Forschung*, *202*, 458–464.

50. Larsson, B.K. (1982). Polycyclic aromatic hydrocarbons in smoked fish. *Zeitschrift für Lebensmittel-Untersuchung und-Forschung, A*, *174*, 101–107.

51. Perugini, M., Visciano, P., Giammarino, A., Manera, M., Di Nardo, W., Amorena, M. (2007). Polycyclic aromatic hydrocarbons in marine organisms from the Adriatic Sea, Italy. *Chemosphere*, *66*, 1904–1910.

52. Kishikawa, N., Wada, N., Kuroda, N., Akiyama, S., Nakashima, K. (2003). Determination of polycyclic aromatic hydrocarbons in milk samples by high-performance liquid chromatography with fluorescence detection. *Journal of Chromatography, B*, *789*, 257–264.

53. Sopelana, P., Guillen, M.D. (2005). Headspace solid-phase microextraction as a tool to estimate the contamination of smoked cheeses by polycyclic aromatic hydrocarbons. *Journal of Dairy Science*, *88*, 13–20.

54. Moret, S., Purcaro, G., Conte, L.S. (2005). Polycyclic aromatic hydrocarbons in vegetables oils from canned foods. *European Journal of Lipid Science and Technology*, *107*, 488–496.

55. Guillen, M.D., Sopelana, P., Palencia, G. (2004). Polycyclic aromatic hydrocarbons and olive pomace oil. *Journal of Agriculture and Food Chemistry*, *52*, 2123–2132.

56. Speer, K., Montag, A. (1988). Polycyclische aromatische kohlenwasserstoffe in nativen pflanzlichen olen. *Fat Science Technology*, *90*, 163–167.

57. Welling, P., Kaandorp, B. (1986). Determination of polycyclic aromatic hydrocarbons (PAH) in edible vegetable oils by liquid chromatography and programmed fluorescence detection. *Zeitschrift für Lebensmittel-Untersuchung und-Forschung, A*, *183*, 111–115.

58. Wickstrom, K., Pyysalo, H., Plammi-Heikkila, S., Tuominen, J. (1986). Polycyclic aromatic compounds (PAC) in leaf lettuce. *Zeitschrift für Lebensmittel-Untersuchung und-Forschung, A*, *183*, 182–185.
59. Zohair, A., Salim, A-B., Soibo, A.A., Beck, A.J. (2006). Residues of polycyclic aromatic hydrocarbons (PAHs), polychlorinated biphenyls (PCBs) and organochlorine pesticides in organically-farmed vegetables. *Chemosphere*, *63*, 541–553.
60. Corradetti, E., Mazzanti, L., Poli, G., Zucchetti, G. (1990). lla contaminazione delle colture esposte agli idrocarburi policiclici aromatici (IPA) di origine industriale ed autoveicolare. *Bollettino dei Chimici Igienisti*, *41*, 441–479.
61. Tuominen, J.P., Pyysalo, H.S., Sauri, M. (1988). Cereal products as a source of polycyclic aromatic hydrocarbons. *Journal of Agriculture and Food Chemistry*, *36*, 118–120.
62. de Vos, R.H., Van Dokkum, W., Schouten, A., de Jong-Berkhout, P. (1990). Polycyclic aromatic hydrocarbons in Dutch total diet samples (1984–1986). *Food Chemistry and Toxicology*, *28*, 263–268.
63. Kazerouni, N., Sinha, R., Hsu, C.-H., Greenberg, A., Rothman, N. (2001). Analysis of 200 food items for benzo[*a*]pyrene and estimation of its intake in an epidemiologic study. *Food Chemistry and Toxicology*, *39*, 423–436.
64. Bourcart, J., Mallet, L. (1965). Coastal marine pollution in the central region of the Tyrrhenian Sea (Bay of Naples) by polyaromatic hydrocarbons of the benzo-3,4-pyrene type. *Comptes Rendus de l'Academie des Sciences (Paris)*, *260*, 3729–3734.
65. Lutz, S., Fiedt, C., Monteau, F., Rychen, G., Le Bizec, B., Jurjanz, S. (2006). Effect of exposure to soil-bound polycyclic aromatic hydrocarbons on milk contaminations of parent compounds and their monohydroxylated metabolites. *Journal of Agriculture and Food Chemistry*, *54*, 263–268.
66. Fritz, W. (1971). The extent of sources of contamination of our food with carcinogenic hydrocarbons. *Ernahrungsforschung*, *16*, 547–557.
67. Fritz, W. (1983). Analysis and assessment of carcinogenic polycyclic aromatic hydrocarbons from the food hygiene toxicology point of view. *Nahrung*, *27*, 965–973.
68. Larsson, B., Sahlberg, G. (1982). Polycyclic aromatic hydrocarbons in lettuce. Influence of a highway and an aluminium smelter, in *Polynuclear Aromatic Hydrocarbons: Physical and Biological Chemistry* (eds M. Cooke, A.J. Fischer), Battelle Press, Columbus, OH, pp. 417–426.
69. Houessou, J.K., Delteil, C., Camel, V. (2006). Investigation of sample treatment steps for the analysis of polycyclic aromatic hydrocarbons in ground coffee. *Journal of Agriculture and Food Chemistry*, *54*, 7413–7421.
70. Garcia-Falcon, M.S., Cancho-Grande, B., Simal-Gandara, J. (2005). Minimal cleanup and rapid determination of polycyclic aromatic hydrocarbons in instant coffee. *Food Chemistry*, *90*, 643–647.
71. Krujif, N., Schouten, T., Van der Stegen, G.H.D. (1987). Rapid determination of benzo[*a*]pyrene in roasted coffee and coffee brew by high-performance liquid chromatography with fluorescence detection. *Journal of Agriculture and Food Chemistry*, *35*, 545–549.

72. Lin, D., Zhu, L. (2004). Polycyclic aromatic hydrocarbons: pollution and source analysis of a black tea. *Journal of Agriculture and Food Chemistry*, *52*, 8268–8271.
73. Larsson, B., Eriksson, A.T., Cervenka, M. (1987). Polycyclic aromatic hydrocarbons in crude and deodorized vegetable oils. *Journal of American Oil Chemical Society*, *64*, 365–370.
74. Vassen, H.A.M.G., Jekel, A.A., Wilbers, A.A.M.M. (1988). Dietary intake of polycyclic aromatic hydrocarbon. *Toxicology and Environmental Chemistry*, *16*, 281–294.
75. Perfetti, G.A., Nyman, P.J., Fisher, S., Joe, F.L., Diachenko, G.W. (1992). Determination of polynuclear aromatic hydrocarbons in seafood by liquid chromatography with fluorescence detection. *Journal of the Association of Official Analytical Chemists*, *75*, 872–877.
76. Dafflon, O., Gobet, H., Koch, H., Bosset, J.O. (1995). Le dosage des hydrocarbures aromatiques polycycliques dans le poisson le produits carnes at le fromage par chromatographie liquide a haute performance. *Travel Chimica Aliment and Hygiene*, *86*, 534–555.
77. Chen, B.H., Wang, C.Y., Chiu, C.P. (1996). Evaluation of analysis of polycyclic aromatic hydrocarbons in meat products by liquid chromatography. *Journal of Agriculture and Food Chemistry*, *44*, 2244–2251.
78. Nymann, P.J., Perfetti, G.A., Joe, F.L., Diachenko, G.W. (1993). Comparison of two cleanup methodologies for the gas chromatography/mass spectrometric determination of low nanogram/gram levels of polynuclear aromatic hydrocarbons. *Food Additives and Contamination*, *10*, 489–501.
79. Chen, B.H., Chen, Y.C. (2001). Formation of polycyclic aromatic hydrocarbons in the smoke from heated model lipids and food lipids. *Journal of Agriculture and Food Chemistry*, *49*, 5238–5243.
80. Lioy, P.L., Waldman, J.M., Greenberg, A., Harkov, R., Pietarnen, C. (1988). The Total Human Environmental Exposure Study (THEES) to benzo[*a*]pyrene: comparison of the inhalation and food pathways. *Archives of Environmental Health*, *43*, 304–312.
81. Waldman, J.M., Lioy, P.J., Greenberg, A., Butler, J.P. (1991). Analysis of human exposure to benzo[*a*]pyrene via inhalation and food ingestion in the Total Human Environmental Exposure Study (THEES). *Journal of Exposure Analysis and Environmental Epidemiology*, *1*, 193–225.
82. Jongeneelen, F.J. (1994). Biological monitoring of environmental exposure to polycyclic aromatic hydrocarbons: 1-hydroxypyrene in urine of people. *Toxicology Letters*, *72*, 205–211.
83. Jacob, J., Seidel, A. (2002). Biomonitoring of polycyclic aromatic hydrocarbons in human urine. *Journal of Chromatography, B*, *778*, 31–47.
84. Pigini, D., Ciadella, A.M., Faranda, P., Trandfo, G. (2006). Comparison between external and internal standard calibration in the validation of an analytical method for 1-hydroxypyrene in human urine by high-performance liquid chromatography/tandem mass spectrometry. *Rapid Communications in Mass Spectrometry*, *20*, 1013–1018.
85. Fan, R., Dong, Y., Zhang, W., Wang, Y., Yu, Z., Sheng, G., Fu, J. (2006). Fast simultaneous determination of urinary 1-hydroxypyrene and 3-hydroxybenzo[*a*]pyrene

by liquid chromatography-tandem mass spectrometry. *Journal of Chromatography, B*, *836*, 92–97.

86. Carmella, S., Chen, M., Yagi, H., Jerina, D.M., Hecht, S.S. (2004). Analysis of phenanthrols in human urine by gas chromatography-mass spectrometry: potential use in carcinogen metabolite phenotyping. *Cancer Epidemiology Biomarkers & Prevention*, *13*, 2167–2174.
87. Singh, G., Gutierrez, A., Xu, K., Blair, I.A. (2000). Liquid chromatography/electron capture atmospheric pressure chemical ionization/mass spectrometry: analysis of pentafluorobenzyl derivatives of biomolecules and drugs in the attomole range. *Analytical Chemistry*, *72*, 3007–3013.
88. Santella, R.M., Lin, C.D., Cleveland, W.L., Weinstein, I.B. (1984). Monoclonal antibodies to DNA modified by a benzo[*a*]pyrene diol-epoxide. *Carcinogenesis*, *5*, 373–377.
89. Santella, R.M., Yang, X.Y., Hsieh, L.L., Young, T.L., Lu, X.Q., Stefanidid, M., Perera, F.P. (1990). Immunologic methods for the detection of carcinogen adducts in humans. *Basic Life Sciences*, *53*, 33–44.
90. Santella, R.M. (1990). Immunological methods for detection of carcinogen-DNA damage in humans. *Cancer Epidemiology Biomarkers & Prevention*, *8*, 733–739.
91. Pavanello, S., Favretto, D., Bruggnone, F., Mastrangelo, G., Dal Pra, G., Clonfero, E. (1990). HPLC/fluorescence determination of *anti*-BPDE–DNA adducts in mononuclear white blood cells from PAH-exposed humans. *Carcinogenesis*, *20*, 431–435.
92. Sinha, R., Peters, U., Cross, A.J., Kuldordd, M., Weissfeld, J.L., Pinksy, P.F., Rothman, N., Hayes, R.B. and the Prostate, Lung, Colorectal, and Ovarian Cancer Project Team (2005). Meat, meat cooking methods and preservation, and risk for colorectal adenoma. *Cancer Research*, *65*, 8034–8041.
93. Anderson, K.E., Kadulbar, F.F., Kuldorff, M., Hranck, L., Gross, M., Lang, N.P., Barber, C., Sinha, R. (2005). Dietary intake of heterocyclic amines and benzo[*a*]pyrene associations with pancreatic cancer. *Cancer Epidemiology Biomarkers & Prevention*, *14*, 2261–2265.
94. Cavalieri, E.L., Rogan, E.G. (1995). Central role of radical cations in the metabolic activation of polycyclic aromatic hydrocarbons. *Xenobiotica*, *25*, 677–688.
95. Cavalieri, E.L., Rogan, E.G. (2002). Fluoro-substitution of carcinogenic aromatic hydrocarbons: models for understanding mechanisms of metabolic activation and of oxygen transfer catalyzed by cytochrome P450, in *The Handbook of Environmental Chemistry*, Vol. 3 (ed. A.H. Nielson), Springer-Verlag, Berlin, pp. 278–293.
96. Gelboin, H.V. (1980). Benzo[*a*]pyrene metabolism, activation and carcinogenesis: role and regulation of mixed function oxidases and related enzymes. *Physiological Reviews*, *60*, 1107–1166.
97. Conney, A.H. (1982). Induction of microsomal enzymes by foreign chemicals and carcinogenesis by polycyclic aromatic hydrocarbons. G.H.A. Clowes Memorial Lecture. *Cancer Research*, *42*, 4875–4917.
98. Penning, T.M., Burczynski, M.E., Hung, C-F., McCoull, K.D., Palackal, N.T., Tsuruda, L.S. (1999). Dihydrodiol dehydrogenases and polycyclic aromatic hydrocarbon activation: generation of reactive and redox-active *o*-quinones. *Chemical Research in Toxicology*, *12*, 1–17.

99. Devanesan, P.D., RamaKrishna, N.V.S., Todorovic, R., Rogan, E.G., Cavalieri, E.L., Jeong, H., Jankowiak, R., Small, G.J. (1992). Identification and quantitation of benzo[*a*]pyrene-DNA adducts formed by rat liver microsomes in vitro. *Chemical Research in Toxicology*, *5*, 302–309.
100. Chakravarti, D., Pelling, J.C., Cavalieri, E.L., Rogan, E.G. (1995). Relating aromatic hydrocarbon-induced DNA adducts and c-H-*ras*-mutations in mouse skin papillomas: the role of apurinic sites. *Proceedings of the National Academy of Sciences of the United States of America*, *92*, 10422–10426.
101. Jennette, K.W., Jeffery, A.M., Blobstein, S.H., Beland, F.A., Harvey, R.G., Weinstein, I.B. (1977). Nucleoside adducts from the *in vitro* reaction of benzo[*a*]pyrene-7,8-dihydrodiol-9,10-oxide or benzo[*a*]pyrene-4,5-oxide with nucleic acids. *Biochemistry*, *16*, 932–938.
102. Ruan, Q., Kim, H.-Y., Jiang, H., Penning, T.M., Harvey, R.G., Blair, I.A. (2006). Quantification of benzo[*a*]pyrene diol epoxide DNA-adducts by stable isotope dilution liquid chromatography/tandem mass spectrometry. *Rapid Communications in Mass Spectrometry*, *20*, 1369–1380.
103. Koreeda, M., Moore, P.D., Wislocki, P.G., Levin, W., Conney, A.H., Yagi, H., Jerina, D.M. (1978). Binding of benzo[*a*]pyrene-7,8-diol-9,10-epoxides to DNA, RNA and protein of mouse skin occurs with high stereoselectivity. *Science*, *199*, 778–781.
104. Ruan, Q., Gelhaus, S.L., Penning, T.M., Harvey, R.G., Blair, I.A. (2007). Aldo-keto reductase- and cytochrome P450-dependent formation of benzo[*a*]pyrene-derived DNA adducts in human bronchoalveolar cells. *Chemical Research in Toxicology*, *20*, 424–431.
105. Malaveille, C., Kuroki, T., Sims, P., Grover, P.L., Bartsch, H. (1977). Mutagenicity of isomeric diol-epoxides of benzo[*a*]pyrene and benz[*a*]anthracene in *S. typhimurium* TA98 and TA100 and in V79 Chinese hamster cells. *Mutation Research*, *44*, 313–326.
106. Marshall, C.J., Vousden, K.H., Phillips, D.H. (1984). Activation of c-Ha-*ras*-1 proto-oncogene by *in vitro* chemical modification with a chemical carcinogen, benzo[*a*]pyrene diol-epoxide. *Nature*, *310*, 585–589.
107. Denissenko, M.F., Pao, A., Tang, M-S, Pfieifer, G.P. (1996). Preferential formation of benzo[*a*]pyrene adducts at lung cancer mutational hotspots in p53. *Science*, *274*, 430–432.
108. Slaga, T.J., Viaje, A., Berry, D.L., Bracken, W. (1976). Skin tumour initiating ability of benzo[*a*]pyrene 4,5-, 7,5- and 7,8-diol-9,10-epoxides and 7,8-diol. *Cancer Letters*, *2*, 115–122.
109. Slaga, T.J., Bracken, W.M., Viaje, A., Levin, W., Yagi, H., Jerina, D.M., Conney, A.H. (1977). Comparison of the tumour initiating activities of benzo[*a*]pyrene arene oxides and diol-epoxides. *Cancer Research*, *37*, 4130–4133.
110. Nesnow, S., Mass, M.J., Ross, J.A., Galati, A.J., Lambert, G.R., Gennings, C., Carter, W.H., Jr, Stoner, G.A. (1998). Lung tumourigenic interactions in strain A/J mice of five environmental polycyclic aromatic hydrocarbons. *Environmental Health Perspectives*, *106*, 1337–1346.
111. Agency for Toxic Substances & Disease Registry (ASTDR) (1995). *Toxicological Profile for Polycyclic Aromatic Hydrocarbons (PAHs)*, ASTDR, pp. 50–53.
112. Burczynski, M.E., Harvey, R.G., Penning, T.M. (1998). Expression and characterization of four recombinant human dihydrodiol dehydrogenase isoforms:

oxidation of *trans*-7,8-dihydroxy-7,8-dihydrobenzo[*a*]pyrene to the activated *o*-quinone metabolite benzo[*a*]pyrene-7,8-dione. *Biochemistry*, *37*, 6781–6790.

113. Palackal, N.T., Burczynski, M.E., Harvey, R.G., Penning, T.M. (2001). The ubiquitous aldehyde reductase (AKR1A1) oxidizes proximate carcinogen *trans*-dihydrodiols to *o*-quinones: potential role in polycyclic aromatic hydrocarbon activation. *Biochemistry*, *40*, 10901–10910.
114. Palackal, N.T., Lee, S.H., Harvey, R.G., Blair, I.A., Penning, T.M. (2002). Activation of polycyclic aromatic hydrocarbon *trans*-dihydrodiol proximate carcinogens by human aldo-keto reductase (AKR1C) enzymes and their functional overexpression in human lung carcinoma (A549) cells. *Journal of Biological Chemistry*, *277*, 24799–24808.
115. Smithgall, T.E., Harvey, R.G., Penning, T.M. (1988). Spectroscopic identification of *ortho*-quinones as the products of polycyclic aromatic *trans*-dihydrodiol oxidation catalyzed by dihydrodiol dehydrogenase. A potential route of proximate carcinogen metabolism. *Journal of Biological Chemistry*, *263*, 1814–1820.
116. Penning, T.M., Ohnishi, S.T., Ohnishi, T., Harvey, R.G. (1996). Generation of reactive oxygen species during the enzymatic oxidation of polycyclic aromatic hydrocarbon *trans*-dihydrodiols catalyzed by dihydrodiol dehydrogenase. *Chemical Research in Toxicology*, *9*, 84–92.
117. Shou, M., Harvey, R.G., Penning, T.M. (1993). Reactivity of benzo[*a*]pyrene-7,8-dione with DNA. Evidence for the formation of deoxyguanosine adducts. *Carcinogenesis*, *14*, 475–482.
118. McCoull, K.D., Rindgen, D., Blair, I.A., Penning, T.M. (1998). Synthesis and characterization of polycyclic aromatic hydrocarbon *o*-quinone depurinating N^7-guanine adducts. *Chemical Research in Toxicology*, *12*, 237–246.
119. Balu, N., Padgett, W.T., Lambert, G.R., Swank, A.E., Richard, A.M., Nesnow, S. (2004). Identification and characterization of novel stable deoxyguanosine and deoxyadenosine adducts of benzo[*a*]pyrene-7,8-quione from reactions at physiological pH. *Chemical Research in Toxicology*, *17*, 827–838.
120. Murty, V.S., Penning, T.M. (1992). Characterization of mercapturic acid and glutathionyl conjugates of benzo[*a*]pyrene-7,8-dione by two-dimensional NMR. *Bioconjugate Chemistry*, *3*, 218–224.
121. Sridhar, G.R., Murty, V.S., Lee, S.H., Blair, I.A., Penning, T.M. (2001). Amino acid adducts of PAH *o*-quinones: model studies with naphthalene-1,2-dione. *Tetrahedron*, *57*, 407–412.
122. Park, J.-H., Gopishetty, S., Szewczuk, L.M., Troxel, A.B., Harvey, R.G., Penning, T.M. (2005). Formation of 8-oxo-7,8-dihydro-2′-deoxyguanosine (8-oxo-dGuo) by PAH *o*-quinones: involvement of reactive oxygen species and copper (II)/copper(I) redox cycling. *Chemical Research in Toxicology*, *18*, 1026–1037.
123. Shen, Y.M., Troxel, A.B., Vedantam, S., Penning, T.M., Field, J. (2006). Comparison of p53 mutations induced by PAH *o*-quinones with those caused by *anti*-benzo[*a*]pyrene diol epoxide in vitro: role of reactive oxygen and biological selection. *Chemical Research in Toxicology*, *19*, 1441–1450.
124. Flowers, L., Ohnishi, S.T., Penning, T.M. (1997). DNA strand scission by polycyclic aromatic hydrocarbon *o*-quinones: role of reactive oxygen species, Cu(II)/Cu(I) redox-cycling and *o*-semiquinone anion radicals. *Biochemistry*, *36*, 8640–8648.

125. Zeiger, E., Anderson, B., Haworth, S., Lawlor, T., Mortelmans, K. (1992). *Salmonella* mutagenicity tests V. Results from testing of 311 chemicals. *Environmental and Molecular Mutagenesis*, *19* (Suppl. 21), 2–141.
126. Sakai, M., Yooshida, D., Mizusaki, S. (1985). Mutagenicity of polycyclic aromatic hydrocarbons and quinones on *Salmonella typhimurium* TA97. *Mutation Research*, *156*, 61–67.
127. Yuan, J., Liu, S., Cao, J. (2003). Spindle poisons induce TK gene mutation in mouse lymphoma cells. *Di San Junyi Daxue Xuebao*, *25*, 1688–1691.
128. Nagao, M., Takahashi, Y. (1981). Mutagenic activity of 42 coded compounds in the *Salmonella*/microsome assay. *Progress in Mutation Research*, *1*, 302–313.
129. Narbonne, J.F., Cassand, P., Alzieu, P., Grolier, P., Merlina, G., Calmon, J.P. (1987). Structure-activity relationships of the N-methylcarbamate series in *Salmonella typhimurium*. *Mutation Research*, *191*, 21–27.
130. Mosaanda, K., Poncelet, F., Fouassin, A., Mercier, M. (1979). Detection of mutagenic polycyclic aromatic hydrocarbons in African smoked fish. *Food and Cosmetics Toxicology*, *17*, 141–143.
131. Bos, R.P., Pronsen, W.J.C., Van rppy, J.G.M., Jongeneelen, E.J., Theuws, J.L.G., Henderson, P.T. (1987). Fluoranthene, a volatile mutagenic compound, present in creosote and car tar. *Mutation Research*, *187*, 119–125.
132. Brams, A., Buchet, J.P., Crutzen-Fayt, M.C., De Mester, C., Lauwerys, R., Leonard, A. (1987). A comparative study with 40 chemicals of the efficiency of the *Salmonella* assay and the SOS chromotest (kit procedure). *Toxicology Letters*, *38*, 123–133.
133. Glatt, H., Seidel, A., Bochnitschek, W., Marquardt, H., Marquardt, H., Hodgson, R.M., Grover, P.L., Oesch, F. (1986). Mutagenic and cell transforming activities of triol-epoxides as compared to other chrysene metabolites. *Cancer Research*, *46*, 4556–4565.
134. Declos, K.B., Heflich, R.H. (1992). Mutation induction and DNA adduct formation in Chinese hamster ovary cells treated with 6-nitrochrysene, 6-aminochrysene and their metabolites. *Mutation Research*, *279*, 153–164.
135. Yamada, K., Suzuki, T., Hohara, A., Kato, T.A., Hayashi, M., Mizutani, T., Saki, K.I. (2005). Nitrogen-substitution effect on *in vivo* mutagenicity of chrysene. *Mutation Research*, *586*, 1–17.
136. McCann, J., Choi, E., Yamasaki, E., Ames, B.N. (1975) Detection of carcinogens as mutagens in the *Salmonella*/microsome test: assay of 300 chemicals. *Proceedings of the National Academy of Sciences of the United States of America*, *72*, 5135–5139.
137. Commoner, B. (1976) Reliability of bacterial mutagenesis techniques to distinguish carcinogenic and noncarcinogenic chemicals. NTIS PB-259934, EPA-600/1-76-022.
138. Hecht, S.S., La Voie, E., Amin, S., Bedenko, V., Hoffmann, D. (1980) Polynuclear aromatic hydrocarbons: chemistry and biological effects. 4th International Symposium, pp. 417–433.
139. Amin, S., Huie, K., Hecht, S.S. (1985). Mutagenicity and tumour initiating activity of methylated benzo[*b*]fluoranthenes. *Carcinogenesis*, *6*, 1023–1025.
140. Weyand, E.H., Geddie, N., Rice, J.E., Czecg, A., Amin, S., Lavoie, E.J. (1988). Metabolism and mutagenic activity of benzo[*k*]fluoranthene and 3-, 8- and 9-fluo robenzo[*k*]fluoranthene. *Carcinogenesis*, *9*, 1277–1281.

141. Oesch, F., Bentley, P., Glatt, H.R. (1976). Prevention of benzo[*a*]pyrene-induced mutagenicity by homogeneous epoxide hydratase. *International Journal of Cancer*, *18*, 448–452.

142. Dunkel, V.C., Zeigler, E., Brusick, D., McCoy, E., McGregor, D., Moretlmans, K., Rosenkranz, H.S., Simon, V.F. (1984). Reproducibility of microbial mutagenicity assays 1. Tests with *Salmonella typhimurium* and *Escherichia coli* using a standardized protocol. *Environmental Mutagens*, *6* (Suppl. 2), 1–39.

143. Rice, J.E., Coleman, D.T., Hosted, T.J., Lavoie, E.J., McCaustland, D.J., Wiley, J.C. (1985). Identification of mutagenic metabolites of indeno[1,2,3-*cd*]pyrene formed *in vitro* with rat liver enzymes. *Cancer Research*, *45*, 5421–5425.

144. Andrews, A.W., Thibault, L.H., Lijinsky, W. (1978). The relationship between carcinogenicity and mutagenicity of some polynuclear hydrocarbons. *Mutation Research*, *51*, 311–318.

145. Levitt, R.C., Pelkonen, O., Okey, A.B., Nebert, D.W. (1979). Genetic differences in metabolism of polycyclic aromatic carcinogens and aromatic amines by mouse liver microsomes. Detection of DNA binding of metabolites and mutagenicity in histidine dependent *Salmonella typhimurium in vitro*. *Journal of the National Cancer Institute*, *62*, 947–955.

146. Bradley, M.O., Bhuyan, B., Francis, M.C., Langenbach, R., Peterson, A., Hubernan, E. (1981). Mutagenesis by chemical agents in V79 Chinese hamster cells: a review and analysis of the literature. *Mutation Research*, *87*, 81–142.

147. Buening, M.K., Wilsocki, P.G., Levin, W., Yagi, H., Thakker, D.R., Akagi, H., Koreeda, M., Jerina, D.M., Conney, A.H. (1978) Tumourigenicity of the optical enantiomers of the diastereomeric benzo[*a*]pyrene-7,8-diol-9,10-epoxides in newborn mice: exceptional activity of (+)-7β,8α-dihydroxy-9α,10α-epoxy-7,8,9,10-tetrahydrobenzo[*a*]pyrene. *Proceedings of the National Academy of Sciences of the United States of America*, *75*, 5358–5361.

148. Busby, W.F., Stevens, E.K., Martin, C.N., Chow, F.L., Garner, R.C. (1989). Comparative lung tumourigenicity of parent and mononitro-polynuclear aromatic hydrocarbons in the Blu:HA new born mouse assay. *Toxicology and Applied Pharmacology*, *99*, 555–563.

149. Wenzel-Hartung, R., Brune, H., Grimmer, G., Germann, P., Timm, J., Wosniok, W. (1990). Evaluation of the carcinogenic potency of 4 environmental polycyclic aromatic compounds following intrapulmonary application in rats. *Experimental Pathology*, *40*, 221–227.

150. Wood, A.W., Chang, R.L., Levin, W., Ryan, D.E., Thomas, P.E., Mah, H.D., Karie, J., Yagi, H., Jerina, D.M., Conney, A.H. (1979). Mutagenicity and tumourigenicity of phenanthrene and chrysene epoxides and diol-epoxides. *Cancer Research*, *39*, 4069–4077.

151. Lavoie, E.J., Ami, S., Hecht, S.S., Furuya, K., Hoffman, D. (1982). Tumour initiating activity of dihydrodiols of benzo[*k*]fluoranthene. *Carcinogenesis*, *3*, 49–52.

152. Deutsch-Wenzel, R.P., Brune, H., Grimmer, G., Dettbarn, G., Misfeld, J. (1983). Experimental studies in rat lungs on the carcinogenicity and dose-response relationships of eight frequently occurring environmental polycyclic aromatic hydrocarbons. *Journal of the National Cancer Institute*, *71*, 539–544.

153. Ross, J.A., Nelson, G.B., Wilson, K.H., Rabinowitz, J.R., Galati, A., Stoner, G.D., Nesnow, S., Mass, M.J. (1995). Adenomas induced by polycyclic aromatic hydro-

carbons in strain A/J mouse lung correlate with time-integrated DNA adduct levels. *Cancer Research*, *55*, 1039–1044.

154. Wynder, E.L., Hoffman, D. (1959). The carcinogenicity of benzofluoranthenes. *Cancer*, *12*, 1194–1199.
155. Rigdon, R.H., Neal, J. (1969). Relationship of leukaemia to lung and stomach tumours in mice fed benzo[*a*]pyrene. *Proceedings of the Society for Experimental Biology and Medicine*, *130*, 146–148.
156. Platt, K.L., Pferffer, E., Petrovic, P., Friesel, H., Beermann, D., Hecker, E., Oesch, F. (1990). Comparative tumorigenicity of picene and dibenz[*a*,*h*]anthracene in the mouse. *Carcinogenesis*, *11*, 1721–1726.
157. Fu, P.P., von Tungelin, L.S., Chiu, L.H., Zhan, D.J., Deck, J., Bucci, T., Wang, J.C. (1998). Structure, tumourigenicity, microsomal metabolism and DNA binding of 7-nitrodibenz[*a*,*h*]anthracene. *Chemical Research in Toxicology*, *11*, 937–945.
158. Starvic, B., Klassen, R. (1994). Dietary effects on the uptake of benzo[*a*]pyrene. *Food Chemistry and Toxicology*, *32*, 727–734.
159. 64th Report Joint FAO/WHO Expert Committee on Food Additives (2006). WHO Technical Report 930, WHO, Geneva, Switzerland.
160. Gazzetta Ufficiale (GU) (2001). Official Statement, Tenori massimi tollerabili di idrocarburi policlici aromatici nell'olio di sansa di oliva enell'olio di sansa di oliva raffinato. n. 225, Sept 27
161. Canadian Food Inspection Agency (2001). Industry Advisory, Olive Pomace Oil, 17 September 2001.
162. No. 2008/2005 of 4th February 2005 amending regulation (EC) No. 466/2001 as regards polycyclic aromatic hydrocarbons, Commission Regulation (EC) (2005).
163. Commission Directive 2005/10/EC of 4th February 2005 laying down the sampling methods and methods of analysis for the official control of levels of benzo[*a*]pyrene in foodstuffs, Commission Directive (2005).
164. No 333/2007 of 28th March 2007 laying down the methods of sampling and analysis for the official control of the levels of lead, cadmium, mercury, inorganic tin, 3-MCPD and benzo[*a*]pyrene in foodstuffs, Commission Regulation (EC) (2007).
165. Council Directive of 22nd June 1988 on the approximation of the laws of the member states relating to flavourings for use in foodstuffs and to source materials for their production (88/288 ECC) (1988).
166. Delaney Clause is part of the 1958 Food Additives Amendment (Section 409) to the 1954 Federal Food, Drug and Cosmetic Act (FFDCA) (1954).

3

FERMENTATION

3.1

ETHYL CARBAMATE (URETHANE)

Colin G. Hamlet
RHM Technology, RHM Group Ltd, Lord Rank Centre, Lincoln Road, High Wycombe, Bucks HP12 3QR, UK

3.1.1 INTRODUCTION

Ethyl carbamate (EC) ($NH_2COOCH_2CH_3$; CAS No. 51-79-6), also known as urethane, is the ethyl ester of carbamic acid (NH_2COOH). Ethyl carbamate has a broad spectrum of biological activities (1) and is a known multisite carcinogen capable of inducing tumors in various animal species (2). On the basis of extensive studies, it must be concluded that exposure to ethyl carbamate represents a potential risk to human health.

3.1.1.1 Historical Perspective

Human exposure to ethyl carbamate predates the 1940s in which it was administered to man as a hypnotic, as well as for the treatment of chronic leukemia, multiple myeloma, and varicose veins (3, 4). In addition to therapeutic applications, other sources of human exposure may occur from commercial applications in which ethyl carbamate is used in the production of cosmetics, fumigants, pesticides, and textiles (4). Ethyl carbamate has also been detected in tobacco leaves and tobacco smoke (5). In 1943, Nettleship *et al.* (6) were the first to describe the carcinogenicity of ethyl carbamate following its use as an anesthetic for laboratory animals, and since 1948 (7) it has been known that ethyl carbamate is mutagenic in *Drosophila melanogaster*. Subsequent studies in several species of animals using a variety of modes of administration of ethyl

Process-Induced Food Toxicants: Occurrence, Formation, Mitigation, and Health Risks, Edited by Richard H. Stadler and David R. Lineback

Ethyl Carbamate (Urethane): At a Glance	
Historical	Ethyl carbamate has been administered to humans as a hypnotic and for medicinal reasons prior to the 1940s. Ethyl carbamate was detected in fruit juices and fermented foods such as bread, beer, soy sauce, and wine in the 1970s. It attracted more attention upon the discovery of higher concentrations in distilled spirit drinks.
Analysis	GC/MS is the method of choice in essentially all food matrices following isolation of ethyl carbamate using, among others, solid-liquid or liquid-liquid extraction.
Occurrence in Foods and Beverages	Data on ethyl carbamate in foods is sparse, and concentrations in alcoholic beverages (e.g., stone-fruit brandies, tequila) are considerably higher. In the food group, highest amounts were found in toasted bread and soy sauces.
Main Formation Pathways	The major precursors of ethyl carbamate in foodstuffs and alcoholic beverages are cyanide (from cyanogenic glycosides), carbamyl phosphate, citrulline, and urea (from arginine metabolism). Cyanate is believed to be the ultimate precursor, and its subsequent reaction with ethanol at low temperatures to furnish ethyl carbamate is well documented.
Mitigation in Food	Efforts have been made by industry and researchers to identify mitigation measures in alcoholic beverages, covering best practices in viticulture, juice nutrient status, yeast strains and lactic acid bacteria, urease application, sur lies aging, distillation, shipment and storage.
Health Risks	The mean intake of ethyl carbamate from food was estimated by JECFA to be approx. 1 μg per person per day. Using the MoE approach, and a BMDL of 0.3 mg/kg bw/d, different exposure scenarios conclude that exposure to ethyl carbamate from foods (excluding alcoholic beverages) was of low health concern (>10,000). For consumers of particular brands of stone-fruit brandy and tequila, MoEs are considerably lower and may be reason for concern.
Regulatory Information or Industry Standards	Canada has introduced maximum limits for ethyl carbamate in a range of alcoholic beverages. The USA has set voluntary targets agreed with the wine producers and exporters to that country. Some European countries have recommended maximum amounts for ethyl carbamate in alcoholic beverages, but no maximum limits for ethyl carbamate in food exist.
Knowledge Gaps	Although the main precursors of ethyl carbamate have been identified, detail on the reaction mechanism, particularly light-induced ethyl carbamate formation in distilled spirits, is still lacking.
Useful Information (web) & Dietary Recommendations	http://www.who.int/ipcs/food/jecfa/summaries/summary_report_64_final.pdf http://monographs.iarc.fr/ENG/Meetings/vol96-summary.pdf http://www.efsa.eu.int/EFSA/efsa_locale-1178620753812_1178655060600.htm There is no specific dietary advice for ethyl carbamate as the main risk is with particular brands of stone-fruit brandy or tequila, with higher than average amounts of ethyl carbamate.

carbamate led to the observation of various benign and malignant tumors in all species tested (4, 8). In the early 1970s, small amounts of ethyl carbamate were detected in fruit juices (9) that had been treated with the antimicrobial agent diethyl dicarbonate [DEDC; $(C_2H_5OCO)_2O$]. Labeling experiments subsequently showed that ethyl carbamate was formed from the reaction of DEDC with residual ammonia and that this mechanism could account for amounts found in orange juice, wine, and beer (10–12).

In 1976, Ough (13) was the first to demonstrate that ethyl carbamate was formed naturally in fermented foods such as bread, beer, soy sauce, and wine. The concentrations found were generally in the low $\mu g\,kg^{-1}$ range although higher amounts (e.g., $150\,\mu g\,L^{-1}$) were found in *sake*. The latter was attributed to the possible use of DEDC to sterilize the mash prior to fermentation (13) and at that time ethyl carbamate was not considered a high priority by food control agencies. In 1985, much higher amounts of ethyl carbamate (low $mg\,kg^{-1}$) were reported in Canadian wines and distilled spirit drinks (14). These higher concentrations probably were first thought to arise from the legitimate use of urea, a yeast nutrient (nitrogen source), since the reaction of ethanol with urea was well known in the laboratory praxis (15). However, this mechanism could not account for the high concentrations of ethyl carbamate found in spirits that used only ammonium phosphate or sulfate as yeast nutrients. Baumann and Zimmerli (16, 17) subsequently showed that ethyl carbamate was formed from natural components, most probably the reaction of ethanol with cyanate (NCO^-), an oxidation product of cyanide that is a well-known constituent of stone fruits (18). These findings led to intense investigations by government agencies and industry, and other precursors (in conjunction with ethanol) such as citrulline, urea, and carbamyl phosphate were later shown to be significant in, for example, wine and soy sauce (for a review, see Reference 19). In some countries such as Canada, the use of urea is prohibited in alcoholic fermentations and maximum limits exist for ethyl carbamate.

3.1.2 TOXICOLOGICAL ASPECTS

There is a vast literature covering the biological activity of ethyl carbamate (for previous reviews, see References 4, 8, and 20). The acute toxicity of ethyl carbamate is relatively low, i.e., of the order of $2500\,mg\,kg^{-1}$ body weight (bw) [LD50 mouse, oral administration (21)], and in the context of low-dose human exposure via the diet, it can be ignored. Despite the large-scale use of ethyl carbamate in patients during 1950–1975, there are no epidemiological data available on its potential carcinogenicity in humans.

3.1.2.1 Toxicokinetics

Ethyl carbamate is rapidly distributed throughout the body regardless of the route of administration and is metabolized by two different pathways. In the

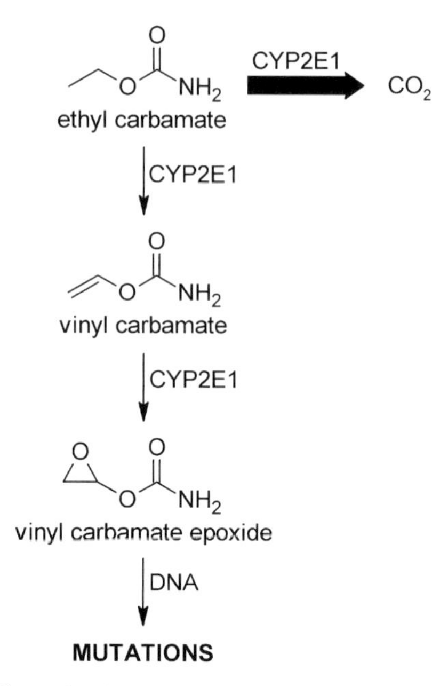

Figure 3.1.1 Metabolic activation and detoxification pathways for ethyl carbamate.

first pathway, studies in rodents have shown that greater than 90% of an administered dose of ethyl carbamate is removed from the body (within 6 h in mice) as carbon dioxide (see Fig. 3.1.1) in reactions that are now believed to be mediated by cytochrome P450 2E1 (CYP2E1), and not by esterase as previously hypothesized (22). Pharmacokinetic modeling revealed that CYP2E1, other cytochromes P450, and esterase contributed 96.4%, 3.2%, and 0.4% to ethyl carbamate catabolism to carbon dioxide, respectively. In the second and minor pathway, ethyl carbamate is oxidized sequentially by CYP2E1 (23, 24) to vinyl carbamate and vinyl carbamate epoxide, the latter of which is considered to be the ultimate carcinogen, forming adducts with DNA, RNA, and proteins (23, 25) (see Fig. 3.1.1). The pathways for the catabolism of ethyl carbamate in rodents and humans are believed to be similar (26).

In rats, mice, and humans, CYP2E1 is induced by ethanol, which suggests that chronic ethanol exposure could increase the oxidation of ethyl carbamate to its epoxide derivative. On the other hand, ethanol has been reported to decrease the catabolism of ethyl carbamate, presumably by acting as a competitive substrate (27–29). The interactions of ethanol with ethyl carbamate are of major interest since intake of the latter is frequently accompanied by alcohol. However, the mechanisms are complex and despite recent studies (30, 31), they are not fully elucidated.

3.1.2.2 Mutagenicity

Various studies have been published concerning the mutagenicity of ethyl carbamate in a wide range of organisms (for reviews, see References 19, 20, 32–41). In bacterial mutagen test systems, the results were mainly negative. However, in *Salmonella typhimurium* in the presence of P450-mediated activation, ethyl carbamate was reported to be mutagenic. Genotoxic potential has also been evaluated using both a standard strain of *D. melanogaster* and a strain in which the genetic control of cytochrome P450-dependent enzyme systems had been altered (constitutively increased P450 enzyme activities) (19). The effects were dose-dependent and the modified strain was more sensitive to ethyl carbamate. These observations were consistent with the involvement of the P450 enzyme system in the activation of ethyl carbamate to the more potent mutagen, vinyl carbamate (37). This mechanism is believed to play a central role in the clastogenicity (chromosome-breaking ability) and mutagenicity of ethyl carbamate (37). In mutagenicity tests using eukaryotic cells, ethyl carbamate was found to induce point mutation, intrachromosomal recombination, chromosomal aberrations, and sister chromatid exchanges in yeast, plant systems, and mammalian cells (20).

3.1.2.3 Carcinogenicity

From the pervious discussion it is evident that ethyl carbamate is genotoxic *in vitro* and *in vivo*. It is therefore not surprising that ethyl carbamate is a well-known multisite animal carcinogen (4). Doses of between 100 and 2000 $mg\,kg^{-1}$ bw using a variety of administration routes have been shown to induce tumors in a wide range of experimental animals. For example, in a recent 2-year drinking water study on mice, exposure-related tumors (neoplasms) were found in the liver, lung, Harderian gland, skin, and forestomach together with rare and invasive cancers (hemangiosarcomas), primarily of the liver and heart (31). Possible mechanisms for the carcinogenicity of ethyl carbamate are induction of DNA damage by its metabolites (40) and an increase in cell proliferation in target tissues. Ethyl carbamate is also teratogenic (causes physical defects in the developing embryo) in mice when administered during gestation. The teratogenic effects were evident in the offspring when either male or female rodents were exposed prior to mating or pregnancy (42). Ethyl carbamate has recently been classified by the International Agency for Research on Cancer as probably carcinogenic to humans (Group 2A) (26).

3.1.3 ANALYSIS

3.1.3.1 Chemistry in General

3.1.3.1.1 Physical Properties Ethyl carbamate ($NH_2COOC_2H_5$; MW 89; CAS 51-79-6) is a colorless and almost odorless white powder that is very

TABLE 3.1.1 Physical properties of ethyl carbamate.

Parameter	Value	Reference no.
Boiling point (°C)	185	44
Melting point (°C)	49	44
Solubility (g 100 mL^{-1})		
in olive oil	3.13	45
in water	48	46
in ether	67.7	45
in chloroform	111	45
in alcohol	125	45
Vapor pressure (mm Hg at 25 °C)	0.262	47
Vapor density (air = 1)	3.07	48
Log K_{ow}	−0.15	49
Density/specific gravity	0.9813	44

soluble both in water and in a wide range of organic solvents (43). This, together with its relatively high vapor pressure of 0.262 mm Hg, can impact on the extraction efficiency and potential losses of ethyl carbamate at trace analytical levels. The physical properties of ethyl carbamate are given in Table 3.1.1.

3.1.3.1.2 Chemical Derivatives As the ester of carbamic acid, ethyl carbamate exhibits reactions typical of both amides and carboxylic acid esters. For example, *N*-acetyl and *N*-alkyl derivatives of ethyl carbamate have been prepared for analysis by gas chromatography/mass spectrometry (GC/MS) and nitrogen selective detectors (50–52), and high-performance liquid chromatography (HPLC) with fluorescence detection (FLD) (53).

3.1.3.1.3 Analytical Precautions Special precautions must be taken when analyzing ethyl carbamate as conditions of sample storage (time, temperature, exposure to light) and extraction (temperature, pH) can have a significant impact on the results obtained.

In wines, long storage times and elevated temperatures (storage and analytical) can increase ethyl carbamate concentrations (from ethanol and urea) (54, 55). In stone-fruit brandies, temperature does not appear to be influential although light can drastically increase ethyl carbamate concentrations, particularly in recently distilled samples (19). In aqueous alkali, ethyl carbamate is reported to decompose to a mixture of the alkali cyanate and carbonate (19).

3.1.3.1.4 Extraction Foodstuffs such as bread are homogenized with water, centrifuged and the aqueous layer separated. Aqueous samples/preparations containing ethyl carbamate are typically extracted into an organic solvent using either liquid–liquid or solid-phase extraction (SPE) prior to GC analysis. Dichloromethane is usually the solvent of choice for liquid–liquid extraction, followed by a cleanup step with Florisil® columns (13), whereas pre-adsorption of aqueous samples onto celite (prior to packing in deactivated alumina/sodium sulfate columns) (56) or the use of diatomaceous earth SPE columns (57) overcomes difficulties with dichloromethane emulsions. The latter procedure is the currently accepted AOAC official method for ethyl carbamate. Recently, the use of hyper-cross-linked hydroxylated polystyrene-divinyl benzene copolymer SPE columns has been reported for the analysis of alcoholic beverages (58). However, removal of ethanol from aqueous samples prior to SPE appears to be a prerequisite for the efficient recovery of ethyl carbamate.

3.1.3.1.5 Mass Spectra The 70-eV electron ionization (EI) mass spectrum of ethyl carbamate (Fig. 3.1.2) gives a weak molecular ion at *m/z* 89 and is dominated by the resonance stabilized even-electron ion at *m/z* 62, i.e., $[M\text{-}C_2H_3]^+$. Labeling experiments have shown that this loss of 27 u to produce

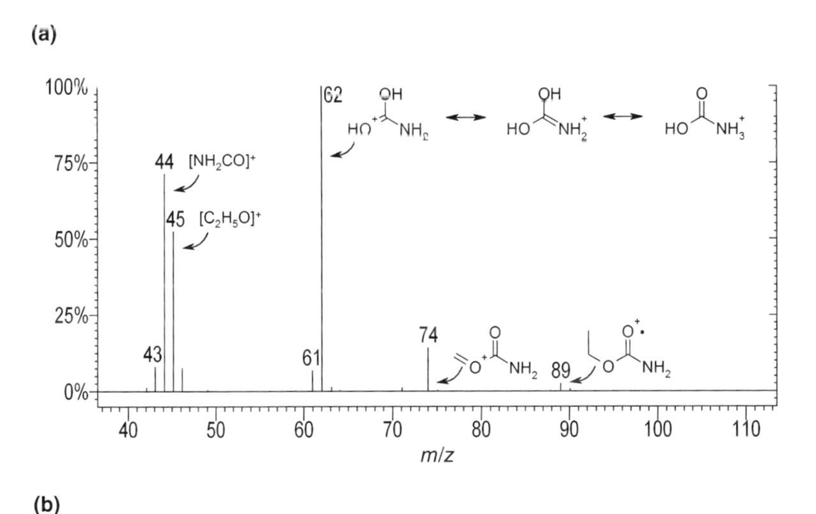

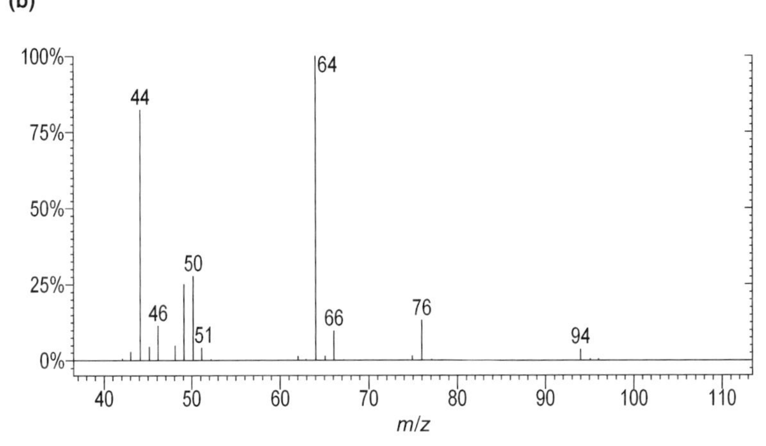

Figure 3.1.2 EI mass spectra of (a) ethyl carbamate and (b) ethyl-d_5 carbamate.

m/z 62 derives from a double hydrogen rearrangement involving the transfer of hydrogen atoms from the methyl and methylene groups (59), while the ions at *m/z* 44 and 45 represent $[NH_2CO]^+$ and $[C_2H_5O]^+$, respectively. In low-resolution MS, interferences at *m/z* 44 and especially *m/z* 74 (contribution from $C_3H_6O_2$ from diethyl succinate (60)) are often observed without efficient extraction cleanup. The ion at *m/z* 62 is chosen for quantification as it provides high selectivity and sensitivity, while the ions at *m/z* 74 and 89 can be used for confirmation. Other carbamates such as *n*-propyl carbamate and *n*-butyl carbamate have been used as internal standards for GC/MS (61–63) and both compounds exhibit the characteristic double hydrogen rearrangement ion at *m/z* 62 ion (for mass spectra, see Reference 64).

When operated in the chemical ionization (CI) mode, ethyl carbamate produces $[M+H]^+$ ions at *m/z* 90 either with methane, isobutene, or ammonia (see Fig. 3.1.3). As expected, methane gives a rearrangement ion at *m/z* 62 ($[M+H\text{-}C_2H_4]^+$), while ammonia produces an abundant adduct ion at *m/z* 107 ($[M+NH_4]^+$) (59). Although methane is preferred over ammonia or isobutane for CI with selected ion monitoring (SIM), the presence of only two ions cannot be considered ideal for unambiguous identification. With the ability to select protonated and/or ammonium adduct ions for collision-induced dissociation (MS/MS), product ions that are structurally related to the precursor ion can be produced, thereby meeting the requirements for confirmation (59).

3.1.3.2 Review of Recent Analytical Methods

Table 3.1.2 gives a summary of analytical methods for ethyl carbamate in a range of matrices published since 2000. GC/MS appears to be the method of choice for all matrices following isolation of ethyl carbamate using solid–liquid extraction (58, 59, 61, 63, 66–70), liquid–liquid extraction (69), or headspace solid-phase microextraction (HS-SPME) (62, 68). Detection limits for these methods range from 0.1 to $30\,\mu g\,L^{-1}$ ($\mu g\,kg^{-1}$) depending on the method used.

In keeping with the increasing popularity of liquid chromatography and tandem mass spectrometry (LC/MS/MS) for the analysis of polar, low-molecular-weight contaminants such as acrylamide (71–73), the first reported use of this technique for ethyl carbamate (63) shows much promise. Other methods based on liquid chromatography using HPLC-FLD (of derivatized ethyl carbamate) have also been reported for the analysis of alcoholic beverages (53) and soy sauce (63).

A rapid screening method using Fourier transform infrared (FTIR) and no sample preparation looks promising (65), although its application appears to be limited to products with potentially higher amounts of EC, such as stone-fruit spirits.

3.1.3.2.1 GC/MS Methods The majority of these procedures use a quadrupole mass spectrometer operating in SIM mode. However, despite the use

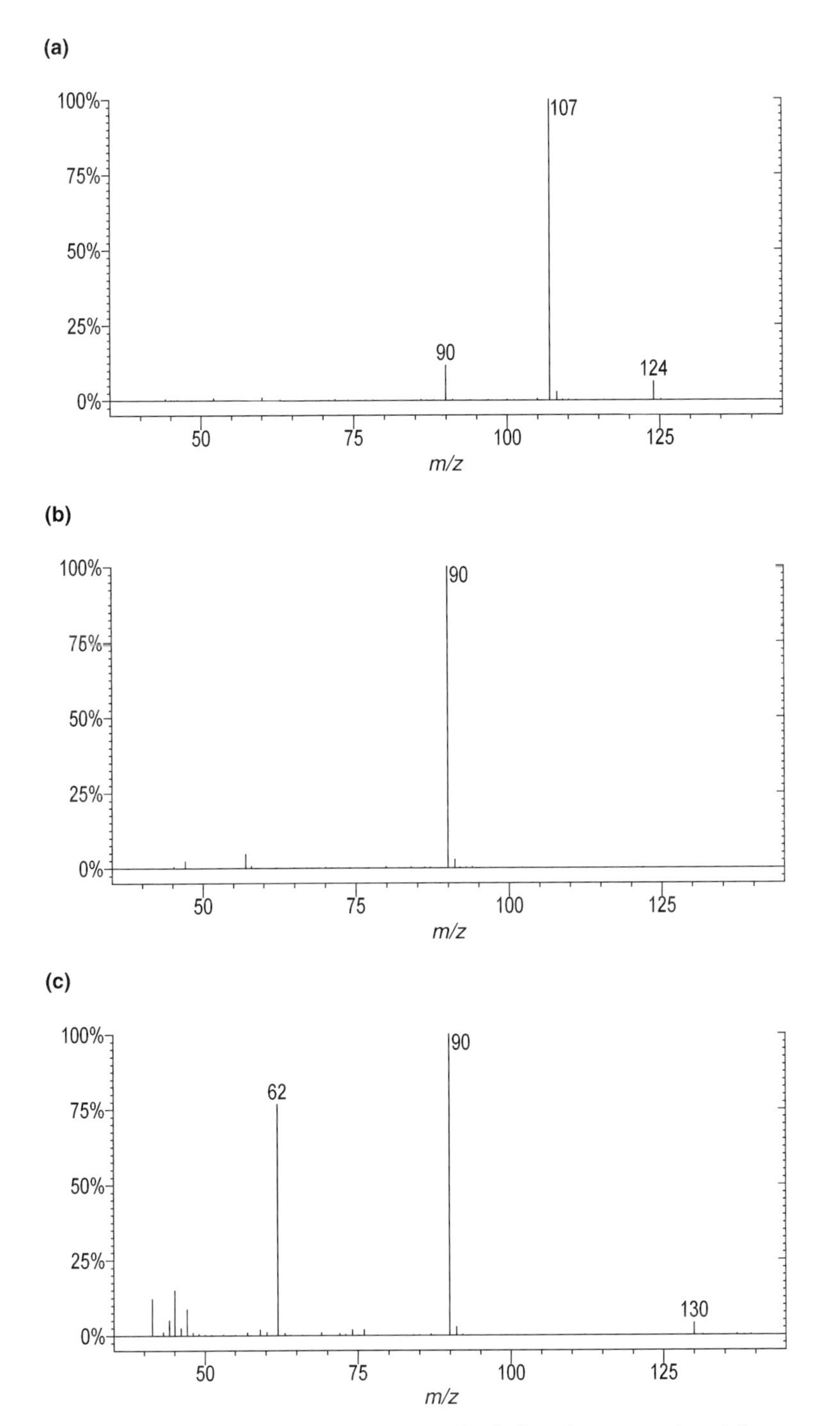

Figure 3.1.3 Positive ion CI mass spectra of ethyl carbamate using (a) ammonia, (b) isobutane, and (c) methane.

TABLE 3.1.2 Summary of analytical methods for ethyl carbamate (since 2000).

Detection system	LOD, $\mu g\,L^{-1}$ or $\mu g\,kg^{-1}$	Matrix	Internal standard	Extraction/preparation	Reference no.
FTIR	600–800	Stone-fruit spirits	None	None	65
GC/MS					
CI-ITMS	1	Bread	Ethyl carbamate-d_5	Solid–liquid	66
CI-MS/MS	0.6	Bread	Ethyl carbamate-d_5	Solid–liquid	59
EI-MS/MS	10	Alcoholic beverages	Ethyl carbamate-d_5	Solid–liquid	67
EI-MS/MS	30	Alcoholic beverages	Ethyl carbamate-d_5	HS-SPME	68
EI-SIM	—[a]	Alcoholic beverages	Propyl carbamate	Liquid–liquid	69
EI-SIM	0.1	Alcoholic beverages	$^{15}N^{13}C$-ethyl carbamate	Solid–liquid	58
EI-SIM	<0.3[b]	Soy sauce	Propyl carbamate	Solid–liquid	61
EI-SIM	0.5	Soy sauce	Butyl carbamate	Solid–liquid	63
EI-SIM	3	Alcoholic beverages	Ethyl carbamate-d_5	Solid–liquid	70
EI-SIM	9.6	Alcoholic beverages	Propyl carbamate	HS-SPME	62
HPLC-FLD	4.2	Alcoholic beverages	None	Derivatization	53
HPLC-FLD	20	Soy sauce	Propyl carbamate	Liquid–liquid, derivatization	63
LC/MS/MS	0.05	Soy sauce	Propyl carbamate	Solid–liquid	63

[a]Value not given by authors.
[b]Estimated.

of mass selective detection, many laboratories still appear to have problems implementing the official method of analysis for wine (74, 75) or lengthy cleanup procedures are required to overcome interferences from matrix components (61). Jagerdeo *et al.* (58) demonstrated that multidimensional GC could be used to overcome the problem of matrix interferences and a limit of detection (LOD) of 1 $\mu g\,kg^{-1}$ could be attained in wines. Although a rapid extraction step using hyper-cross-linked hydroxylated polystyrene-divinylbenzene copolymer SPE (Isolute ENV+[Biotage]) columns could be used, subsequent analysis required specialist equipment comprising a GC with a flame ionization detector and a GC/MS system coupled using a cryogenic trap. More recently, Mirzoian and Mabud (70) combined the ENV+ SPE technique with GC/MS SIM and demonstrated satisfactory method performance for the analysis of selected alcoholic beverages. The ENV+ SPE columns are selective for small polar molecules and avoid the need for harmful chlorinated solvents. However, the recovery of ethyl carbamate is diminished by the presence of ethanol in samples and its removal, using for example, a vacuum centrifuge concentrator, can be time-consuming.

Lachenmeier *et al.* (67) showed that MS/MS in combination with GC could be used to provide a rapid and reliable quantification of ethyl carbamate in stone-fruit spirits with minimal sample cleanup. In EI mode, characteristic transitions of m/z 74 $\rightarrow$ 44 and m/z 62 $\rightarrow$ 44 for ethyl carbamate, and m/z 64 $\rightarrow$ 44 for ethyl carbamate-d_5 were followed using selected reaction monitoring (SRM). Using this technique, ethyl carbamate could be identified and quantified to an LOD of 10 $\mu g\,L^{-1}$. Hamlet *et al.* (59) used CI-MS/MS to measure low concentrations of ethyl carbamate in bread. Using positive ion CI with ammonia reagent gas to generate abundant $[M+H]^+$ and $[M+NH_4]^+$ ions from ethyl carbamate, precursor to product transitions of m/z 107 $\rightarrow$ 90, m/z 107 $\rightarrow$ 62, and m/z 90 $\rightarrow$ 62 for ethyl carbamate and m/z 112 $\rightarrow$ 63 for the ethyl carbamate-d_5 were monitored. This use of CI over EI resulted in a lower LOD for ethyl carbamate of 0.6 $\mu g\,kg^{-1}$.

In the last 10–15 years, solid-phase microextraction (SPME), as discovered by Arthur and Pawliszyn (76), has emerged as a rapid, simple, solvent-free alternative to conventional extraction methods. SPME can dramatically reduce labor, materials, and solvent disposal costs, and is ideally suited to the analysis of aqueous samples such as alcoholic beverages (77). Whiton and Zoecklein (62) optimized an HS-SPME method for the analysis of ethyl carbamate in wine. Using propyl carbamate as internal standard and a carbowax/divinyl benzene SPME fiber exposed to the sample headspace for 30 min, ethyl carbamate could be rapidly and reliably quantified to an LOD of about 10 $\mu g\,L^{-1}$. More recently, Lachenmeier *et al.* (68) used ethyl carbamate-d_5 as internal standard and GC/MS/MS detection to develop an automated HS-SPME method for the confirmatory screening of ethyl carbamate in stone-fruit spirits.

3.1.3.2.2 HPLC Methods It is evident that reliable GC/MS analysis of ethyl carbamate requires either lengthy sample preparation procedures or

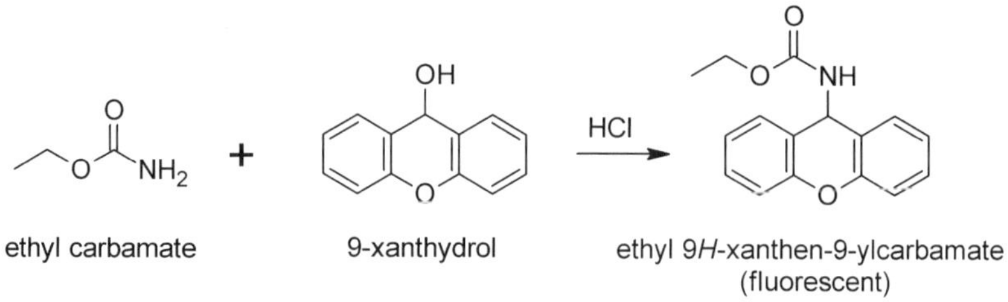

Figure 3.1.4 Formation of ethyl xanthen-9-ylcarbamate (fluorescent) from ethyl carbamate and 9-xanthydrol.

expensive instrumentation such as MS/MS. However, other workers have demonstrated that HPLC-based methods could offer a simpler analytical alternative to the use of GC/MS, particularly for alcoholic beverages. Herbert *et al.* (53) developed a simple, rapid, and cost-effective method for the analysis of ethyl carbamate in wine brandy, fortified wines, and table wines, using HPLC-FLD. To achieve the required selectivity, a fluorescent derivative of ethyl carbamate was prepared from the reagent 9-xanthydrol according to the scheme given in Fig. 3.1.4. Although 9-xanthydrol can form derivatives with, for example, amines and amino acids present naturally in alcoholic beverage such as wine, these did not interfere with the subsequent analysis. By using a simple dilution technique to normalize the ethanol content of samples to 20% prior to analysis, ethyl carbamate could be rapidly (40 min) and reliably measured to low concentrations ($4.2\,\mu g\,L^{-1}$) with a method performance that was comparable to GC/MS. De Melo Abreu *et al.* (75) validated the HPLC-FLD method of Herbert *et al.* (53) by collaborative trial. Despite a limited number of participants (six), the results obtained by laboratories using the HPLC-FLD method (two) were comparable to those using GC/MS.

Although HPLC-FLD may be suitable for the analysis of ethyl carbamate in alcoholic beverages, there might be limitations for other matrices. Park *et al.* (63) used propyl carbamate as internal standard and liquid–liquid extraction (ethyl acetate) to isolate ethyl carbamate from soy sauce prior to derivatization with 9-xanthydrol. Both ethyl carbamate and propyl carbamate were well separated and free from interfering peaks in the chromatograms, but the LOD in soy sauce was about five times higher ($20\,\mu g\,L^{-1}$) than that obtained by Herbert *et al.* for alcoholic beverages. In the same publication, Park *et al.* (63) also described a method for the analysis of soy sauce using HPLC/MS/MS. Samples were prepared by SPE using dichloromethane for elution, and propyl carbamate as internal standard. After removal of the solvent, residues were dissolved in water, filtered, and injected directly into an HPLC/MS/MS system that was fitted with an electrospray interface (ESI). The SRM precursor to product ion transitions was m/z 90 → 62 for ethyl carbamate and m/z 104 → 62 for propyl carbamate. The LOD for ethyl carbamate in soy sauce using this technique was $0.05\,\mu g\,L^{-1}$ for a sample-to-final-extract ratio of $5\,g\,mL^{-1}$. It is likely that with further development, for example, from the use of ENV+ SPE columns, and reduced costs of ownership, the LC/MS/MS procedure could become a primary screening tool for ethyl carbamate in a wide range of matrices.

3.1.4 OCCURRENCE AND EXPOSURE

3.1.4.1 Alcoholic Beverages

Published data taken on ethyl carbamate concentrations in alcoholic beverages since 2000 are given in Table 3.1.3. The data taken from surveys and investigations carried out in Brazil, Germany, Korea, and the United Kingdom show that the range of ethyl carbamate values in alcoholic beverages varies widely. The stone-fruit spirits and other alcoholic beverages distilled from crops such as sugar cane and cassava account for the widest range and highest mean concentrations. The low mean amounts of ethyl carbamate found in wine and whisky are consistent with the successful action taken by these industries to reduce concentrations in recent years.

TABLE 3.1.3 Published data on ethyl carbamate concentrations in alcoholic beverages (since 2000).

Alcoholic beverage	Number of samples	Mean, μg kg^{-1}	Range, μg kg^{-1}	Reference no.
Fermented beverages				
Fortified wines	15	32	14–60	78, 79
Sake	2	123	81–164	78, 79
Wine, red	5	13.6[b]	<10–18	78, 79
Wine, white	7	9.4[b]	<10–24	78, 79
Distilled beverages				
Cachaças	126	770	13–5,700	80
Grappa	6	25[b]	5–68	80
Korean distilled spirits	10	3.4	ND[a]–15.4	81
Tiquira (Manioc brandy)	37	2,353	190–10,000	81
Soju	—[a]	—[a]	0.6–15.7	82
Soju	7	3.0	0.8–10.1	83
Spirits and liqueurs (brandy, calvados, rum)	9	42[b]	<10–170	78, 79
Stone-fruit spirits (brandies)	7	898[b]	<10–6,130	78, 79
Stone-fruit spirits	631	1,400	10–18,000	84
Stone-fruit spirits	70	1,210	70–7,700	67
Whiskey (America)	6	161	64–400	80
Whisky (Scottish)	13	127	65–340	80
Whisky (United Kingdom)	205	29[b]	<10–239	85

[a]Data not available.

[b]Derived value from reported data (one-half of the reported LOD value used to calculate the mean)

The occurrence of ethyl carbamate in alcoholic beverages was reviewed by expert committees in 2005 and again in 2007. The 6004 results reported to the Joint FAO/WHO Expert Committee on Food Additives (JECFA) (86) by national authorities were mainly from wine. The highest mean concentration of ethyl carbamate of 122 $\mu g\,kg^{-1}$ was found in sake, while the highest amounts were found in brandies, although few data were reported. Because of the sparse number of results for important categories such as stone-fruit brandies, the European Food Safety Authority (EFSA) issued a call for additional data on ethyl carbamate occurrence in alcoholic beverages. Monitoring results were received from seven EU member states covering the period 1998–2006, together with data from North America on products originating from the EU over the period 2002–2006. The summary of the EFSA data (87) given in Table 3.1.4 has been ordered by decreasing upper-bound mean ethyl carbamate concentrations. Positive results (88%) and mean concentrations reported by EU member states were much higher than those reported by North America (56% positive) for products originating from the EU. Although the latter sample set comprised 84% wine samples, which had relatively low amounts of ethyl carbamate, the range of lower mean concentrations may reflect progress in level reduction by manufacturers for these more recent samples, i.e., over the period 2002–2006. As expected, ethyl carbamate concentrations were particularly high in the stone-fruit brandies and spirits, although the highest mean quantity was found in tequila. The amount of ethyl carbamate found in the latter product (1233 $\mu g\,kg^{-1}$) was comparable to concentrations measured in some distillates (upper-bound mean 1435 $\mu g\,kg^{-1}$), i.e., tested at an interim step during production (87).

3.1.4.2 Foodstuffs

In contrast to data on alcoholic beverages, data on ethyl carbamate concentrations in foods are sparse. Table 3.1.5 gives a summary of foodstuffs found to contain quantifiable amounts of ethyl carbamate. The data, taken from surveys and investigations carried out since 2000 in Korea and the United Kingdom show that, as expected, ethyl carbamate concentrations are very much lower than those in alcoholic beverages. The highest amounts of ethyl carbamate were found predominantly in toasted breads (1.4–14.7 $\mu g\,kg^{-1}$) and soy sauces (not detected—129 $\mu g\,kg^{-1}$). A UK seasonal product that is fortified with ethanol, i.e., Christmas pudding, was found to contain 20 $\mu g\,kg^{-1}$ ethyl carbamate. Very few data were reported to JECFA and EFSA and with the exception of soy sauces, these generally contained low or non-detectable quantities (87).

3.1.4.3 Exposure Estimates

Published information on national estimates of ethyl carbamate intake is scarce. Noh *et al.* (83) estimated the average daily intake of ethyl carbamate

TABLE 3.1.4 Concentrations of ethyl carbamate in alcoholic beverages from EU member states as reported to EFSA.[a]

Alcoholic beverage	Incidence/ number	Ethyl carbamate, μg kg⁻¹			
		Mean[b]	Median[b]	P95[c]	Range
EU MONITORING DATA[d]					
Tequila	84/84	1,233	800	5,397	70–6,730
Stone-fruit brandy	2,912/3,244	848–851	330	3,399	ND[e]–22,000
Other fruit brandy	281/328	663–667	215	4,187	ND[e]–7,920
Miscellaneous spirits	64/86	590	290	1,745	ND[e]–6,000
Gin	1/1	580	—	—	—
Vodka	57/60	386–387	365	846	ND[e]–2,140
Rum	10/11	325–328	280	755	ND[e]–1,020
Cachaça	19/19	229	110	478	40–730
Brandy	19/42	123–129	0–30	395	ND[e]–2,100
Sake	2/2	123	—	—	81–164
Liqueur	2/4	45–47	6–7	146	ND[e]–170
Whisky	196/210	41	22	78	ND[e]–1,000
Fortified wine	15/15	32	29	49	14–60
Wine	11/17	10–11	11	21	ND[e]–24
Beer	1/13	—	—	—	ND[e]–1
Cider	0/1	—	—	—	ND[e]
NORTH AMERICA DATA[f]					
Armagnac	69/71	246	219	503	ND[g]–630
Fruit brandy	168/186	100	27	284	ND[g]–3,133
Brandy	135/137	78	45	345	ND[g]–642
Whisky	1,076/1,122	40	30	106	ND[g]–509
Fortified wine	965/1,000	39	26	113	ND[g]–404
Grappa	242/270	32	24	87	ND[g]–192
Cognac	247/256	30	24	67	ND[g]–191
Liqueur	252/356	21–22	9	74	ND[g]–405
Miscellaneous spirits	370/632	17–19	7	58	ND[g]–1,060
Rum	14/19	16–17	12	45	ND[g]–57
Gin	30/53	9–11	6	28	ND[g]–60
Vodka	33/101	4–8	ND–5	17	ND[g]–49
Cooler	14/93	3–7	ND–5	13	ND[g]–68
Wine	12,001/23,278	5–7	5	78	ND[g]–180
Fruit wine	8/44	2–6	ND–5	10	ND[g]–17
Beer	88/1,208	0.6–5	ND–5	6	ND[g]–33
Cider	3/26	0.9–5	ND–5	8	ND[g]–9

[a]Reference 87.
[b]Range indicates lower- and upper-bound values, i.e., with results set to zero or the LOD, respectively, for samples with no detectable concentrations.
[c]95th percentile.
[d]Data from EU member state analyses over 1998–2006.
[e]Value at or below LOD (0.1–400 μg kg⁻¹).
[f]Product originating from EU member states over 2002–2006.
[g]Value at or below LOD of 5 μg kg⁻¹.

TABLE 3.1.5 Summary of ethyl carbamate levels in foodstuffs (since 2000).

Foodstuff	Number of samples	Mean, $\mu g\,kg^{-1}$	Range, $\mu g\,kg^{-1}$	Reference no.
Cereal products				
Bread products (UK)	25[a]	<1	<1–5.2	66
Toasted (white)	3	7.2	5.8–8.1	66
Bread, domestic[b] (UK white + wholemeal)	3	5.1	2.1–6.7	66
Toasted	5	10.9	6.1–14.7	66
Bread (UK white + wholemeal)	7	1.3[c]	0.6–2.3[c]	59
Toasted	8	3.2[c]	1.4–5.0[c]	59
Bread, domestic[b] (UK white + wholemeal)	36	5.8[c]	3.1–17.5[c]	59
Toasted	6	8.9[c]	4.8–12.1[c]	59
Bread, domestic[b] (UK speciality)	3	9.4[c]	6.2–12.2[c]	59
Christmas pudding (UK)	1	—	20	78, 79
Fermented dairy products	—[d]	—[d]	ND[d]–1.9	83
Kimchi	20	3.5	ND[d]–16.2	81
Chinese cabbage	14	1.4	0.1–4.3	83
Sauerkraut (in white wine)	1	—	29	78, 79
Soybean paste	7	2.3	ND[d]–7.9	81
Soybean paste	12	1.1	ND[d]–3.8	83
Soy sauce				
UK retail	10	5.5[e]	5–10	78, 79
Korean and Japanese-style	136	10.7	ND[d]–128.9	88
Japanese	7	19.4	3.6–58.9	83
Korean traditional	4	16.7	1.4–49.2	83
Korean (regular)	5	14.6	ND[d]–19.5	81
Korean (traditional)	15	17.1	ND[d]–73	81
Korean (fermented)	7	27.2[e]	0.5–99.9	63
Korean (mixed fermented + acid hydrolyzed)	6	1.3[e]	0.5–4.0	63
Vinegar	10	7.8[e]	5–33	78, 79
Vinegar	5	1.2	0.3–2.5	81
Yeast extract	1	—	41	78, 79

[a]Composite bread group samples, e.g., white, wholemeal, pita prepared from 158 individual samples.

[b]Produced using a domestic bread machine.

[c]Dry weight basis.

[d]Data not available.

[e]Derived value from reported data (one-half of the reported LOD value used to calculate the mean).

from Korean fermented foods to be 6.0 ng kg^{-1} bw per day (equivalent to 0.3 μg per person per day for the average Korean aged 3–64 years); alcoholic beverages, kimchi (fermented Chinese cabbage), and soy sauce were significant contributors of ethyl carbamate intake (83–88). JECFA (86) prepared international estimates of ethyl carbamate intake using data supplied by Australia (1.4 μg per person per day), New Zealand (1.4 μg per person per day), and South Korea (0.6 μg per person per day). Using the five regional diets of the WHO Global Environment Monitoring System (GEMS/Food) database (90), JECFA (86) estimated the mean intake of ethyl carbamate from food, excluding alcoholic beverages, to be approximately 1 μg per person per day (equivalent to about 15 ng kg^{-1} bw per day). When the intake of ethyl carbamate for a high-percentile consumer of wine was modeled, this resulted in a total estimated intake of ethyl carbamate from foods and beverages of up to 5 μg per person per day (equivalent to 80 ng kg^{-1} bw per day). JECFA also noted that the consumption of stone-fruit brandies, which contain higher amounts of ethyl carbamate, could lead to higher intakes of ethyl carbamate.

Following the ethyl carbamate evaluation, JECFA requested an opinion from EFSA on the risks to human health related to the presence of ethyl carbamate in food and alcoholic beverages, in particular stone-fruit brandies. In excess of 33,000, testing results from analyses covering the period 1998–2000, together with FAO and World Drink Trends data, were used to derive exposure estimates for ethyl carbamate from the consumption of alcoholic beverages (87). In contrast, only very few food results for ethyl carbamate were obtained and the intake value of 1 μg per person per day used by JECFA was used for risk assessments. From these data, a dietary exposure of 17 ng kg^{-1} bw per day was estimated from food for an average 60-kg person who does not consume alcohol, rising to 65 ng kg^{-1} bw per day for consumers of a variety of different alcoholic beverages. The highest exposure to ethyl carbamate was estimated for exclusive consumers of fruit brandy with exposure at a 95th percentile consumption level of 558 ng kg^{-1} bw per day.

3.1.4.3.1 Risk Assessment The most recent evaluation of cancer risk from the consumption of food and beverages was carried out by JECFA (86) and EFSA (87) using the margin of exposure (MOE) approach for risk characterization. Using data from the NTP long-term study of carcinogenicity in mice (31), JECFA determined that for cancer, the dose that caused the lowest but measurable response (tumor incidence), i.e., the so-called bench mark dose (BMD) lower confidence limit (BMDL), was 0.3 mg kg^{-1} bw per day. Although the increased incidence of alveolar and bronchiolar adenoma or carcinoma was considered to be the critical response for dose–response analysis, data for mice with Harderian gland tumors were also analyzed. The range of BMD and BMDL (the one-sided lower 95% confidence limit of the BMD) values obtained for each end point used is given in Table 3.1.6.

The MOE and ethyl carbamate intake estimates from the JECFA and EFSA evaluations are given in Table 3.1.7. The MOE is obtained by dividing

TABLE 3.1.6 Ranges of BMD and BMDL values for tumors associated with administration of ethyl carbamate.

Tumor type	Range of BMD values[a], $mg\,kg^{-1}$ bw per day	Range of BMDL values[a], $mg\,kg^{-1}$ bw per day
Lung adenoma or carcinoma	0.50–50.63	0.26–20.51
Harderian gland adenoma or carcinoma	0.47–40.76	0.28–20.61

[a]Data from JECFA (86).

TABLE 3.1.7 MOE estimates for different exposure scenarios to ethyl carbamate in subjects aged 15 years and older.

	Ethyl carbamate intake, $ng\,kg^{-1}$ bw per day[a]		MOE	
Exposure scenario	LB[b]	UB[c]	LB[b]	UB[c]
Food excluding alcoholic beverages	17	17 (15)	18,000	18,000 (20,000)
Whole population (food and alcoholic beverages)	33	55 (80)	9,000	5,460 (3,800)
Consumers of alcoholic beverages	37	65	8,180	4,620
Consumers of alcoholic beverages at the 95th percentile				
Beer	17	100	18,000	3,000
Wine	52		5,810	
Spirits	63		4,740	
Fruit brandy	558		538	

[a]Data from EFSA (87) and (in parentheses) JECFA (86).
[b]Lower-bound mean.
[c]Upper-bound mean.

the BMDL value ($0.3\,mg\,kg^{-1}$ bw per day) by the estimated intake, and a MOE greater than 10,000 is usually considered to be of low health concern (91, 92). While both JECFA and EFSA concluded that the exposure to ethyl carbamate from foods excluding alcoholic beverages was a low health concern, the MOE for all intakes, i.e., food and alcoholic beverages combined, was of concern. For consumers of particular brands of stone-fruit brandy or tequila, with higher-than-average amounts of ethyl carbamate, these MOEs could be even lower. The expert committees have recommended that mitigation measures should be taken to reduce the quantities of ethyl carbamate in certain alcoholic beverages.

3.1.5 FORMATION ROUTES

3.1.5.1 General Overview

Considerable literature concerning formation routes has been accumulated since the discovery of ethyl carbamate in foodstuffs in the early 1970s (for earlier reviews, see Reference 19). While most studies relate to alcoholic beverages, the mechanisms identified may well apply to foods in which the precursors of ethyl carbamate are also known to occur.

Figure 3.1.5 summarizes the current knowledge with respect to key precursors, their sources, and the main formation routes for ethyl carbamate. Cyanide (from cyanogenic glycosides), carbamyl phosphate, citrulline, and urea (from arginine catabolism) have emerged as the major precursors of ethyl carbamate in alcoholic beverages and foods. Cyanate, formed from the oxidation of cyanide—and the possible dissociation of carbamyl phosphate, citrulline, and urea (19)—is believed to be the ultimate precursor, and its subsequent reaction with ethanol to give ethyl carbamate is well known (19). While cyanate and carbamyl phosphate react spontaneously with ethanol at low temperatures, the formation of EC from citrulline and urea appears to require higher temperatures.

3.1.5.2 From Hydrogen Cyanide

Cyanide ion has been identified as one of the most important precursors of ethyl carbamate. It is formed by enzymatic action and thermal cleavage of cyanogenic glycosides, which occur in a wide range of important human foodstuffs. Some examples include amygdalin in stone fruits (93), epiheterodendrin in barley (94), linamarin and lotaustralin in cassava, and a hitherto unidentified cyanide source in sugar cane (93), all of which are used in the production of spirit drinks.

3.1.5.2.1 Formation in Spirit Drinks The formation of ethyl carbamate in spirit drinks can occur before, during, and after the distillation process. Until recently, formation before distillation, i.e., in the fermentation mash, was believed to occur via urea (see Section 3.1.5.3). However, the occurrence of ethyl carbamate in cherry mash fermentations using yeast that had been genetically modified to suppress urea formation (via arginine degradation) (95) indicates that other as-yet-unknown mechanisms may operate. In most cases, the transfer of ethyl carbamate formed during fermentation to the final spirit is likely to be small, principally because ethyl carbamate has a relatively high boiling point (185 °C). The majority of ethyl carbamate found in spirit drinks is formed post distillation, especially on exposure to light in the wavelength range 350–475 nm (19, 84, 96). Besides light, other factors influencing ethyl carbamate formation from cyanide are pH, ethanol concentration, temperature, presence of vicinal dicarbonyl compounds (e.g., diacetyl,

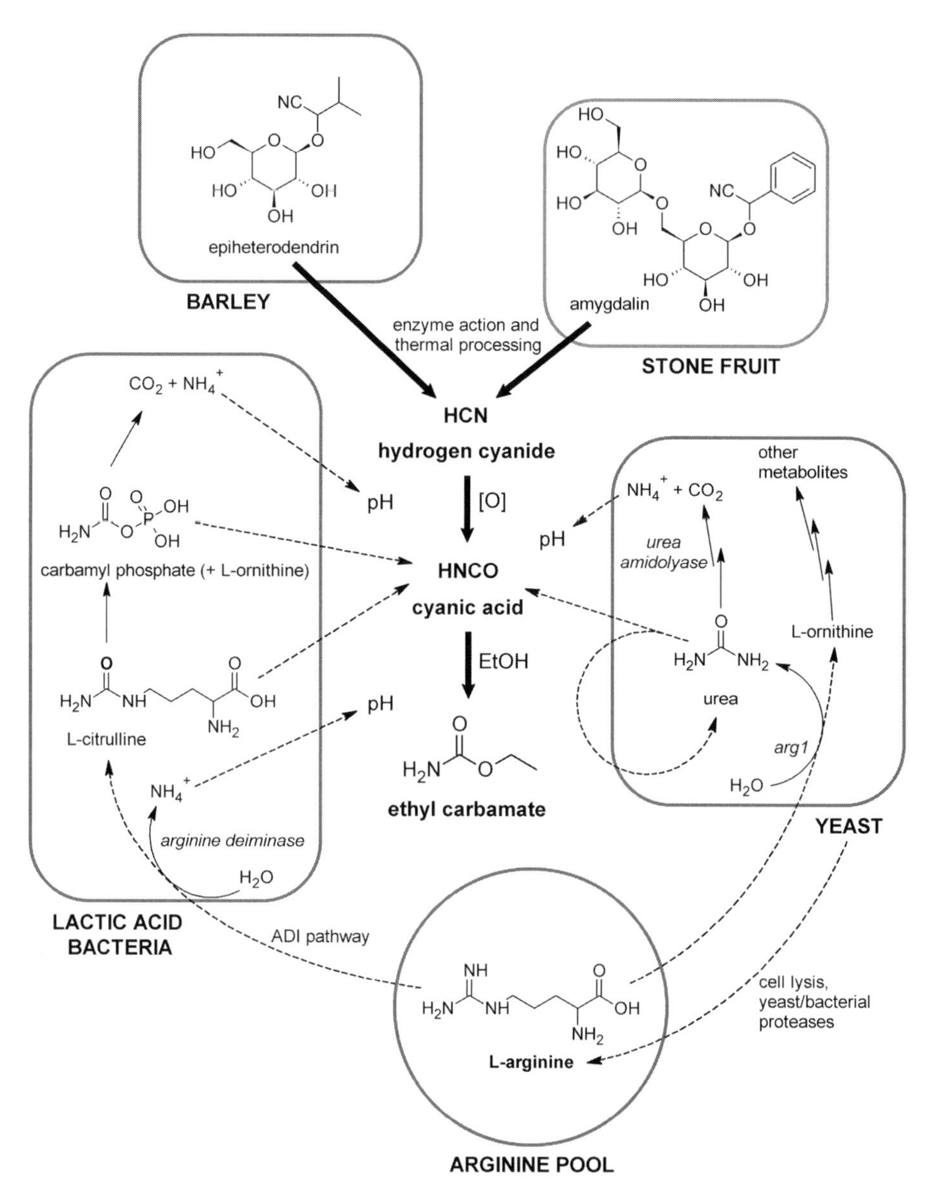

Figure 3.1.5 Summary of key precursors, interrelationships, and reactions leading to ethyl carbamate in alcoholic beverages (*arg1* = arginase).

2,3-pentanedione, and methyl glyoxal), and the concentrations of copper or iron ions (18, 97, 98).

It is generally accepted that the formation of ethyl carbamate proceeds via the oxidation of cyanide to cyanate. Copper ions, present in the distillation apparatus used to manufacture spirit drinks, have been implicated in both the oxidation of cyanide to cyanate and the subsequent formation of ethyl carbamate. Aresta *et al.* (98) showed that the formation of ethyl carbamate from cyanide increased with an increasing concentration of Cu^{II} ions and the process was found to be independent of dissolved oxygen. These data supported a coordination chemistry mechanism according to Equations 1 and 2, rather than a free radical process via, for example, oxygen-initiated hydroperoxides (19). Although oxygen was not directly implicated in the oxidation of cyanide, it is likely that it is involved in the oxidation of Cu^{I} to Cu^{II}. Furthermore, the coordination of Cu^{II} to cyanate makes the carbon more susceptible to nucleophilic attack, first yielding a carbamate moiety (Eq. 3), which by ethanolysis is converted into ethyl carbamate (Eq. 4). A competing reaction with water could account for the observed evolution of carbon dioxide in model solutions (Eq. 5) (98).

$$CN^- + 2OH^- + 2Cu^{II} \rightarrow CNO^- + 2Cu^{I} + H_2O \quad (1)$$

$$2CNO^- + Cu^{II} \rightarrow Cu(NCO)_2 \quad (2)$$

$$Cu(NCO)_2 + 2H_2O \rightarrow Cu(OOCNH_2)_2 \quad (3)$$

$$Cu(OOCNH_2)_2 + 2EtOH \rightarrow 2EtOHCONH_2 + Cu(OH)_2 \quad (4)$$

$$Cu(OOCNH_2)_2 + 2H_2O \rightarrow Cu(OH)_2 + 2CO_2 + 2NH_3 \quad (5)$$

The dependence of ethyl carbamate on metal ions has also been observed with both Fe^{II} and Fe^{III} ions, the latter species being present in stainless steel distillation equipment (98).

The mechanism of light-induced ethyl carbamate formation from cyanide, on the other hand, still remains obscure. Although an interrelationship has previously been established between ethanol, vicinal dicarbonyl compounds such as diacetyl, and light (16), it should be noted that cyanide complexes of copper and iron are also photochemically active.

3.1.5.3 From *N*-Carbamyl Precursors

The *N*-carbamyl compounds, urea, citrulline, and carbamyl phosphate are mainly derived from the yeast and bacterial catabolism of arginine (see Fig. 3.1.5), an abundant amino acid in many foods, especially grapes.

In aqueous solutions, urea and ammonium cyanate comprise an equilibrium pair (see Fig. 3.1.6) and at urea concentrations of 5–100 μg mL^{-1}, i.e., typical of wine, the degree of dissociation has been estimated to be in the range 50–90% at 100 °C (19). Ethyl carbamate is then formed from the well-known alcoholysis reaction of isocyanates (99), which is believed to occur via a concerted

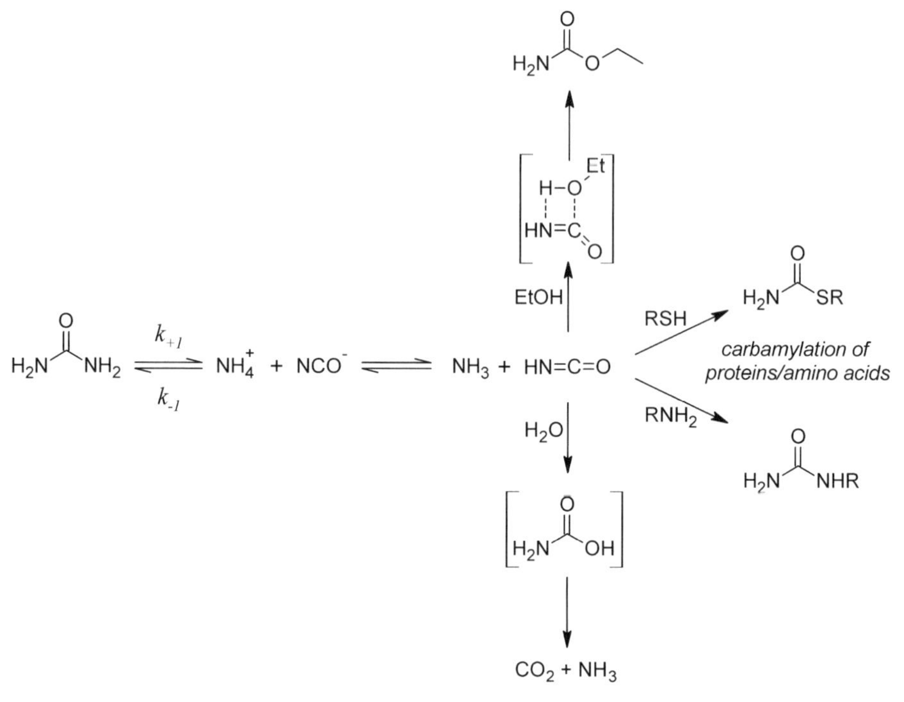

Figure 3.1.6 Formation of ethyl carbamate from urea in aqueous ethanol and competing reactions in foodstuffs.

nucleophilic addition of the alcohol to the N=C bond of the isocyanate (100). In addition to hydrolysis of the isocyanate, competing reactions with other nucleophiles present in foodstuffs, such as the amino and thiol groups of proteins and amino acids, may also occur (see Fig. 3.1.6) (101, 102). The addition of Cu^{II} ions has been shown to promote the formation of ethyl carbamate from urea and ethanol (98), presumably via the mechanisms given in Equations 1–4. It is very probable that citrulline reacts with aqueous ethanolic solutions in a similar way, either via urea or by decomposing directly to cyanic acid.

In the case of carbamyl phosphate, the aqueous decomposition reactions have been studied by various authors (103, 104). The half-life for the release of phosphate, which occurs predominantly between pH 2–4 and pH 6–8, is about 45 min at 37 °C for each pH range. Below pH 4, phosphate, carbon dioxide, and ammonia are released in a sequence involving carbamic acid (or possibly cyanic acid), whereas at higher pH values, decomposition proceeds by a monomolecular elimination of cyanic acid. It is possible that the spontaneous reaction of carbamyl phosphate with aqueous ethanol in the cold, as described by Ough *et al.* (105), may occur via the cyanate mechanism discussed earlier.

3.1.5.3.1 Formation in Wine Most of the ethyl carbamate content of wine is formed post fermentation where the concentrations of ethanol, urea, and citrulline as well as temperature and time of storage are the important parameters (54). The amounts of carbamyl phosphate produced by yeast in wine are very low and it is not considered a significant source of ethyl carbamate. Daylight seems not to have any significant impact on ethyl carbamate formation (106).

Urea is a major source of ethyl carbamate in wines (107) and its excretion and utilization during alcoholic fermentation is governed mainly by the yeast strain used and environmental conditions. Urea is produced in yeast from the catabolism of arginine to ornithine by the enzyme arginase. Although wine yeast strains contain an enzyme (amydolyase) that can degrade urea into ammonia and carbon dioxide, this pathway is usually suppressed during alcoholic fermentation when a good nitrogen source (e.g., arginine) is available (108) (see Fig. 3.1.5). In addition to the strain of yeast and arginine concentrations, other factors affecting the utilization of urea by yeast during alcoholic fermentation include aeration and the concentrations of other amino acids (e.g., glutamine), ammonia, ethanol, and urea (109, 110).

In the production of most red and some white wines, a secondary fermentation using lactic acid bacteria (LAB), the so-called malolactic fermentation (MLF), is usually carried out after the alcoholic yeast fermentation. Wine LAB of the genus *Lactobacillus, Pediiococcus*, and *Oenococcus* are typically used to deacidify wine by the bioconversion of dicarboxylic L-malic acid to monocarboxylic lactic acid, with the subsequent improvement in wine sensory properties. These LAB metabolize arginine via the arginine deiminase (ADI) pathway (111) and can influence ethyl carbamate concentrations in wine by the excretion and subsequent utilization of the ethyl carbamate precursor citrulline and, to a lesser extent, carbamyl phosphate (112). This potential is strain-dependent and influenced by pH and ethanol concentrations in wine (113–116). The accumulation of citrulline has also been associated with the adventitious bacterial contamination of fortified wines (117). The main sources of arginine in wine post alcoholic fermentation occur via yeast cell lysis, proteolytic activity of residual yeast enzymes, and also by bacterial exoproteases (114).

3.1.5.3.2 Formation in Other Foodstuffs An accumulation of citrulline from the catabolism of arginine by soy *Pediiococci* and subsequent ethanolysis is believed to be responsible for the formation of ethyl carbamate in soy sauces during pasteurization (118).

Azodicarbonamide ($H_2NCON_2CONH_2$; CAS No. 123-77-3), used as a blowing agent in closures for, for example, beer bottles and as a food additive, has also been suggested as an ethyl carbamate precursor (119–121). Azodicarbonamide is also permitted as a flour improver for baking in some countries, and in dough it is converted into the potential ethyl carbamate precursor biurea (122). However, the contribution of azodicarbonamide to the amount of ethyl carbamate formed in bread was very low (119–121).

Temperature and extended fermentation appear to be the determinants of ethyl carbamate formation in bread. In a study of UK breads, concentrations of ethyl carbamate in commercial products were virtually not detected, whereas bread prepared using domestic bread machines contained higher amounts (66, 123). The two main factors that contributed to higher ethyl carbamate concentrations were shown to be reduced evaporative losses and probably higher precursor quantities of yeast-derived ethanol and urea from the extended fermentation. In experiments with bread dough that had been fortified with ethyl carbamate, it was shown that losses of ethyl carbamate due to diffusivity and volatility were significantly higher in bread baked in a convection oven, compared with bread baked within the enclosed chamber of a domestic bread machine (66, 123). Irrespective of bread type, toasting led to increases of between two- and threefold in mean ethyl carbamate concentrations (dry weight basis) (59, 123).

3.1.5.4 Conclusions for Other Foodstuffs

Because cyanide and *N*-carbamyl precursors are widespread, there is no reason to believe that ethyl carbamate might not be formed from the reaction of ethanol with these compounds in many foodstuffs. In products where ethanol and heat have been used during processing, e.g., Christmas puddings, higher amounts of ethyl carbamate might be expected (see Table 3.1.5).

3.1.6 MITIGATION

Because of the prevalence of ethyl carbamate in alcoholic beverages, considerable efforts have been made by industry and researchers alike to reduce concentrations of ethyl carbamate in these products. Principal among these has been a series of preventative actions drawn from scientific research and produced jointly by U.C. Davis, the Wine Institute, and the US Food and Drug Administration (124). These actions covering best practices in viticulture, juice nutrient status, yeast strains and LAB, urease application, sur lie aging, distillation, and shipment and storage have helped US wine growers and distillers to control the formation of ethyl carbamate in their products. Many other preventative measures to avoid ethyl carbamate in alcoholic beverages have been proposed including the use of the enzyme rhodanase to remove cyanide (125), the addition of copper and silver salts to precipitate cyanide in the must (126), double distillation (127), elimination of Cu^{II} ions with cationic-exchange resins (127), and storage of bottles at low temperature in the dark (106). However, many of the measures are not economically viable or in the case of copper salts, can cause environmental problems. Some of the more recent mitigation measures that have been evaluated are summarized in Table 3.1.8.

TABLE 3.1.8 Published measures (since 1999) to reduce or control ethyl carbamate in alcoholic beverages.

Application	Process	Control measure	Reference no.
Distilled spirits	Raw materials selection	Fruit quality (damage, microbiology)	84
		Molecular markers or colorimetric tests to identify barley cultivars that are low in cyanogenic glycosides	128, 129
	Mashing	Stone removal prior to mashing to remove the source of cyanide	84
	Fermentation	Genetically engineered *Saccharomyces cerevisiae* with reduced arginase activity, thus blocking the pathway to urea production	95
	Distillation	Distillation using a catalytic converter with a high copper surface area	98, 80, 130
		Separation of tailing at an alcoholic strength of greater than 50% vol	130
		Optimum design and operation of distillation apparatus	98, 131
Winemaking	Alcoholic fermentation	Genetically engineered *S. cerevisiae* capable of metabolizing urea	108
		Monitoring of urea/ammonia in wine and must	132, 133
		Predictive modeling of ethyl carbamate formation during storage	54
		Use of urease	134
	Malolactic fermentation	Removal of residual yeast lees	114
		Use of pure cultures of *Oenococcus oeni*	113, 114
		Control of pH	113, 114

3.1.7 LEGISLATION

Canada was the first country to introduce maximum limits for ethyl carbamate in a range of alcoholic beverages. In the United States, voluntary targets have been agreed with wine producers, and exporters to that country have been notified that they must develop programs to meet these target concentrations. Some European countries have also recommended maximum amounts for

TABLE 3.1.9 International maximum limits/specifications for ethyl carbamate in alcoholic beverages.

	Ethyl carbamate by commodity, $\mu g\,L^{-1}$ or $\mu g\,kg^{-1}$						
Country	Wine	Fortified wine	Distilled spirits	Sake	Fruit brandies	Other	Reference no.
Brazil	—	—	150[a]	—	—	—	135
Canada	30	100	150	200	400	—	136
Czech Republic	30	100[b]	150	200	400[c]	—	87
France	—	—	150	—	1000	—	87
Germany	—	—	—	—	800	—	87
Switzerland	—	—	—	—	—	1000[d]	137
United States[e]	15	60	—	—	—	—	62

[a]Limit to be attained by 2010.
[b]Fruity wines and liqueurs.
[c]Fruity distillates and fruity, mixed, and other spirits.
[d]Spirit drinks.
[e]Voluntary agreement between wine producers and the US Food and Drug Administration.

ethyl carbamate in alcoholic beverages, although there are currently no harmonized regulatory limits in the European Union (see Table 3.1.9). There are no maximum concentrations for ethyl carbamate in foods.

Both JECFA and EFSA have advised that exposure to ethyl carbamate from the consumption of alcoholic beverages is of concern, and that measures to reduce concentrations should be taken. EFSA has also recommended that such measures should focus on hydrogen cyanide and other precursors of ethyl carbamate to prevent formation during storage.

3.1.8 FUTURE PROSPECTS AND CONCLUSIONS

There is a vast literature on the toxicology of ethyl carbamate from which it must be concluded that dietary exposure to ethyl carbamate represents a potential risk to human health.

A wide range of analytical methods is available for the analysis of ethyl carbamate, many of which have been validated by collaborative trial. The availability of lower cost analytical instrumentation has led to the increased use of mass spectrometry and reductions in sample preparation time. In this respect, the recent use of LC/MS/MS for the rapid identification of ethyl carbamate in complex matrices looks promising.

The main precursors of ethyl carbamate in wine and distilled spirit drinks have been identified and factors affecting formation have been established. However, detail of the reaction mechanisms, particularly that of light-induced ethyl carbamate formation in distilled spirits, is still lacking. Much less is known about the formation routes in fermented foods, although the amounts of ethyl carbamate formed appear to be relatively low. Since the precursors of ethyl carbamate in alcoholic beverages may also be widespread in foods, the mechanisms and risk factors identified for alcoholic beverages are likely to be applicable.

The main dietary exposure to ethyl carbamate is from the consumption of alcoholic beverages, in particular the stone-fruit brandies. Consequently, a wide range of measures to reduce and control ethyl carbamate has been identified and successfully implemented. However, it seems that problems could exist with, for example, the smaller producers of distilled spirits, who may not be able to afford the technologies established primarily by the large manufacturers. These smaller enterprises require simpler and economically viable methods to reduce ethyl carbamate in their products.

REFERENCES

1. Field, K.J., Lang, C.M. (1988). Hazards of urethane (ethyl carbamate): a review of the literature. *Laboratory Animals*, *22*, 255–262.
2. Benson, R.W., Beland, F.A. (1997). Modulation of urethane (ethyl carbamate) carcinogenicity by ethyl alcohol: a review. *International Journal of Toxicology*, *16*, 521–544.
3. Law, L.W. (1947). Urethane (ethyl carbamate) therapy in spontaneous leukemias in mice. *Proceedings of the National Academy of Sciences of the United States of America*, *33*, 204–210.
4. International Agency for Research on Cancer (1974). *Monographs on the Carcinogenic Risk of Chemicals to Man: Some Antithyroid and Related Substances, Nitrofurans and Industrial Chemicals*, International Agency for Research on Cancer, Lyon, pp. 111–140.
5. Schmeltz, I., Chiong, K.G., Hoffmann, D. (1978). Formation and determination of ethyl carbamate in tobacco and tobacco smoke. *Journal of Analytical Toxicology*, *2*, 265–268.
6. Nettleship, A.H., Henshaw, P.S., Meyer, H.L. (1943). Induction of pulmonary tumours in mice with ethyl carbamate. *Journal of the National Cancer Institute*, *4*, 309–319.
7. Di Paolo, J.A. (1948). Studies on chemical mutagenesis utilizing nucleic acid components, urethane, and hydrogen peroxide. *The American Naturalist*, *86*, 49–56.
8. Mirvish, S.S. (1968). The carcinogenic action and metabolism of urethane and N-hydroxyurethane. *Advances in Cancer Research*, *11*, 42.
9. Joint FAO/WHO Expert Committee on Food Additives (1972). *WHO Food Additives Series, 1972, No. 4. Evaluation of Mercury, lead, cadmium and the food additives amaranth, diethylpyrocarbonate, and octyl gallate*, http://www.inchem.org/documents/jecfa/jecmono/v004je06.htm (accessed October 2007).
10. Solymosy, F., Antoni, F., Fedorcsak, I. (1978). On the amounts of urethane formed in diethyl pyrocarbonate treated beverages. *Journal of Agricultural and Food Chemistry*, *26*, 500–503.

11. Ough, C.S. (1976). Ethylcarbamate in fermented beverages and foods. II. Possible formation of ethylcarbamate from diethyl dicarbonate addition to wine. *Journal of Agricultural and Food Chemistry*, *24*, 328–331.
12. Löfroth, G., Gejvall, T. (1971). Diethyl pyrocarbonate; formation of urethan in treated beverages. *Science*, *174*, 1248–1250.
13. Ough, C.S. (1976). Ethylcarbamate in fermented beverages and foods. I. Naturally occurring ethylcarbamate. *Journal of Agricultural and Food Chemistry*, *24*, 323–8.
14. Conacher, H.B., Page, B.D. (1986). Ethyl carbamate in alcoholic beverages: a Canadian case history. Proceedings of Euro Food Tox II, Interdisciplinary Conference on Natural Toxicants in Food, Zurich, 1986, pp. 237–242.
15. Adams, P., Baron, F.A. (1965). Esters of carbamic acid. *Chemical Reviews*, *65*, 567–602.
16. Baumann, U., Zimmerli, B. (1987). Zur bildungsweise von ethylcarbamat (urethan) insteinobstdestillaten. *Mitteilungen aus dem Gebiete der Lebensmitteluntersuchung und Hygiene*, *78*, 317–314.
17. Baumann, U., Zimmerli, B. (1986). Origination of urethane (ethyl carbamate) in alcoholic beverages. *Schweizerische Zeitschrift fur Obst- und Weinbau*, *122*, 602–607.
18. Battaglia, R., Conacher, H.B.S., Page, B.D. (1990). Ethyl carbamate (urethane) in alcoholic beverages and foods: a review. *Food Additives and Contaminants*, *7*, 477–496.
19. Zimmerli, B., Schlatter, J. (1991). Ethyl carbamate: analytical methodology, occurrence, formation, biological activity and risk assessment. *Mutation Research—Genetic Toxicology Testing and Biomonitoring of Environmental or Occupational Exposure*, *259*, 325–350.
20. Schlatter, J., Lutz, W.K. (1990). The carcinogenic potential of ethyl carbamate (urethane): risk assessment at human dietary exposure levels. *Food and Chemical Toxicology*, *28*, 205–211.
21. Lewis, R.J. (1996). *Sax's Dangerous Properties of Industrial Materials*, 9th edn, Van Nostrand Reinhold, New York.
22. Hoffler, U., El-Masri, H.A., Ghanayem, B.I. (2003). Cytochrome P450 2E1 (CYP2E1) is the principal enzyme responsible for urethane metabolism: comparative studies using CYP2E1-null and wild-type mice. *Journal of Pharmacology and Experimental Therapeutics*, *305*, 557–564.
23. Guengerich, F.P., Kim, D.H. (1991). Enzymatic oxidation of ethyl carbamate to vinyl carbamate and its role as an intermediate in the formation of 1,N6-ethenoadenosine. *Chemical Research in Toxicology*, *4*, 413–421.
24. Guengerich, F.P., Kim, D.H., Iwasaki, M. (1991). Role of human cytochrome P-450 IIE1 in the oxidation of many low molecular weight cancer suspects. *Chemical Research in Toxicology*, *4*, 168–179.
25. Park, K.K., Liem, A., Stewart, B.C., Miller, J.A. (1993). Vinyl carbamate epoxide, a major strong electrophilic, mutagenic and carcinogenic metabolite of vinyl carbamate and ethyl carbamate (urethane). *Carcinogenesis*, *14*, 441–450.
26. International Agency for Research on Cancer (2007). *Volume 96: Alcoholic Beverage Consumption and Ethyl Carbamate (Urethane) 6–13 February 2007*,

http://monographs.iarc.fr/ENG/Meetings/vol96-summary.pdf (accessed October 2007).

27. Waddell, W.J., Marlowe, C., Pierce, J. (1987). Inhibition of the localization of urethane in mouse tissues by ethanol. *Food and Chemical Toxicology*, *25*, 527–531.
28. Yamamoto, T., Pierce, J., Hurst, H.E., Chen, D., Waddell, W.J. (1988). Inhibition of the metabolism of urethane by ethanol. *Drug Metabolism and Disposition*, *16*, 355–358.
29. Kurata, N., Hurst, H.E., Kemper, R.A., Waddell, W.J. (1991). Studies on induction of metabolism of ethyl carbamate in mice by ethanol. *Drug Metabolism and Disposition*, *19*, 239–240.
30. Beland, F.A., Benson, R.W., Mellick, P.W. *et al.* (2005). Effect of ethanol on the tumourigenicity of urethane (ethyl carbamate) in B6C3F1 mice. *Food and Chemical Toxicology*, *43*, 1–9.
31. National Toxicology Program (NTP) (2004). *Toxicology and Carcinogenesis Studies of Urethane, Ethanol, and Urethane/Ethanol in B6C3F1 Mice (Drinking Water Studies)*, NIH Publication No 04-4444. Department of Health and Human Services, Public Health Service, National Institutes of Health, Research Triangle Park, NC.
32. Sotomayor, R.E., Collins, T.F.X. (1990). Mutagenicity, metabolism, and DNA interactions of urethane. *Toxicology and Industrial Health*, *6*, 71–108.
33. Cheng, M., Conner, M.K., Alarie, Y. (1981). Multicellular in vivo sister-chromatid exchanges induced by urethane. *Mutation Research*, *88*, 223–231.
34. Cheng, M., Conner, M.K., Alarie, Y. (1981). Potency of some carbamates as multiple tissue sister chromatid exchange inducers and comparison with known carcinogenic activities. *Cancer Research*, *41*, 4489–4492.
35. Choy, W.N., Mandakas, G., Paradisin, W. (1996). Co-administration of ethanol transiently inhibits urethane genotoxicity as detected by a kinetic study of micronuclei induction in mice. *Mutation Research—Genetic Toxicology and Environmental Mutagenesis*, *367*, 237–244.
36. Hernandez, L.G., Forkert, P.G. (2007). Inhibition of vinyl carbamate-induced mutagenicity and clastogenicity by the garlic constituent diallyl sulfone in F1 (Big Blue® × A/J) transgenic mice. *Carcinogenesis*, *28*, 1824–1830.
37. Hernandez, L.G., Forkert, P.G. (2007). In vivo mutagenicity of vinyl carbamate and ethyl carbamate in lung and small intestine of F1 (Big Blue® × A/J) transgenic mice. *International Journal of Cancer*, *120*, 1426–1433.
38. Hubner, P., Groux, P.M., Weibel, B. *et al.* (1997). Genotoxicity of ethyl carbamate (urethane) in Salmonella, yeast and human lymphoblastoid cells. *Mutation Research—Genetic Toxicology and Environmental Mutagenesis*, *390*, 11–19.
39. Hoffler, U., Dixon, D., Peddada, S., Ghanayem, B.I. (2005). Inhibition of urethane-induced genotoxicity and cell proliferation in CYP2E1-null mice. *Mutation Research—Fundamental and Molecular Mechanisms of Mutagenesis*, *572*, 58–72.
40. Sakano, K., Oikawa, S., Hiraku, Y., Kawanishi, S. (2002). Metabolism of carcinogenic urethane to nitric oxide is involved in oxidative DNA damage. *Free Radical Biology and Medicine*, *33*, 703–714.
41. Tomisawa, M., Suemizu, H., Ohnishi, Y. *et al.* (2003). Mutation analysis of vinyl carbamate or urethane induced lung tumours in rasH2 transgenic mice. *Toxicology Letters*, *142*, 111–117.

42. Nomura, T. (1975). Transmission of tumours and malformations to the next generation of mice subsequent to urethan treatment. *Cancer Research*, *35*, 264–266.
43. US National Library of Medicine (2008). *Hazardous Substances Data Bank*, http://toxnet.nlm.nih.gov/cgi-bin/sis/htmlgen?HSDB (accessed 18 September 2008).
44. Lide, D.R. (2000). *CRC Handbook of Chemistry and Physics*, 81st edn, CRC Press LLC, Boca Raton, FL.
45. O'Neil, M.J. (2001). *The Merck Index—An Encyclopedia of Chemicals, Drugs, and Biologicals*, 13th edn, Merck & Co., Inc.
46. Seidell, A. (1941). *Solubilities of Organic Compounds*, D. Van. Nostrand Co., Inc., New York.
47. Perry, R.H., Green, D. (1984). *Perry's Chemical Handbook. Physical and Chemical Data*, 6th edn, McGraw-Hill, New York.
48. Lewis, R.J. (1999). *Sax's Dangerous Properties of Industrial Materials*, 10th edn, John Wiley & Sons, Inc., New York.
49. Hansch, C., Leao, A., Hoekman, D. (1995). *Exploring QSAR—Hydrophobic, Electronic, and Steric Constants*. American Chemical Society, Washington, D.C.
50. Walker, G., Winterlin, W., Fouda, H., Seiber, J. (1974). Gas chromatographic analysis of urethane (ethyl carbamate) in wine. *Journal of Agricultural and Food Chemistry*, *22*, 944–947.
51. Bailey, R., North, D., Myatt, D. (1986). Determination of ethyl carbamate in alcoholic beverages by methylation and gas chromatography with nitrogen-phosphorus thermionic detection. *Journal of Chromatography*, *369*, 199–202.
52. Giachetti, C., Assandri, A., Zanolo, G. (1991). Gas chromatographic-mass spectrometric determination of ethyl carbamate as the xanthylamide derivative in Italian aqua vitae (grappa) samples. *Journal of Chromatography*, *585*, 111–115.
53. Herbert, P., Santos, L., Bastos, M., Barros, P., Alves, A. (2002). New HPLC method to determine ethyl carbamate in alcoholic beverages using fluorescence detection. *Journal of Food Science*, *67*, 1616–1620.
54. Hasnip, S., Caputi, A., Crews, C., Brereton, P. (2004). Effects of storage time and temperature on the concentration of ethyl carbamate and its precursors in wine. *Food Additives and Contaminants*, *21*, 1155–1161.
55. Funch, F., Lisbjerg, S. (1988). Analysis of ethyl carbamate in alcoholic beverages. *Zeitschrift fur Lebensmittel-Untersuchung und-Forschung*, *186*, 29–32.
56. Canas, B.J., Havery, D.C., Robinson, L.R., Sullivan, M.P., Joe, J., Diachenko, G.W. (1989). Ethyl carbamate levels in selected fermented foods and beverages. *Journal of the Association of Official Analytical Chemists*, *72*, 873–876.
57. Canas, B.J., Joe, J., Diachenko, G.W., Burns, G. (1994). Determination of ethyl carbamate in alcoholic beverages and soy sauce by gas chromatography with mass selective detection: collaborative study. *Journal of AOAC International*, *77*, 1530–1536.
58. Jagerdeo, E., Dugar, S., Foster, G.D., Schenck, H. (2002). Analysis of ethyl carbamate in wines using solid-phase extraction and multidimensional gas chromatography/mass spectrometry. *Journal of Agricultural and Food Chemistry*, *50*, 5797–5802.
59. Hamlet, C.G., Jayaratne, S.M., Morrison, C. (2005). Application of positive ion chemical ionisation and tandem mass spectrometry combined with gas chroma-

tography to the trace level analysis of ethyl carbamate in bread. *Rapid Communications in Mass Spectrometry*, *19*, 2235–2243.

60. Lau, B.P., Weber, D., Page, B.D. (1987). Gas chromatographic-mass spectrometric determination of ethyl carbamate in alcoholic beverages. *Journal of Chromatography, A*, *402*, 233–241.
61. Kim, Y.K.L., Koh, E., Chung, H.J., Kwon, H. (2000). Determination of ethyl carbamate in some fermented Korean foods and beverages. *Food Additives and Contaminants*, *17*, 469–475.
62. Whiton, R.S., Zoecklein, B.W. (2002). Determination of ethyl carbamate in wine by solid-phase microextraction and gas chromatography/mass spectrometry. *American Journal of Enology and Viticulture*, *53*, 60–63.
63. Park, S.K., Kim, C.T., Lee, J.W. *et al.* (2007). Analysis of ethyl carbamate in Korean soy sauce using high-performance liquid chromatography with fluorescence detection or tandem mass spectrometry and gas chromatography with mass spectrometry. *Food Control*, *18*, 975–982.
64. NIST Chemistry WebBook (2005). *NIST Standard Reference Database Number 69, June 2005 Release*, http://webbook.nist.gov/chemistry/ (accessed September 2007).
65. Lachenmeier, D.W. (2005). Rapid screening for ethyl carbamate in stone-fruit spirits using FTIR spectroscopy and chemometrics. *Analytical and Bioanalytical Chemistry*, *382*, 1407–1412.
66. Hamlet, C.G., Jayaratne, S.M. (2003). *Ethyl Carbamate in Bread, Toast and Similar Products. A Report Prepared for the UK Food Standards Agency*, RHM Technology Ltd, High Wycombe, UK.
67. Lachenmeier, D.W., Frank, W., Kuballa, T. (2005). Application of tandem mass spectrometry combined with gas chromatography to the routine analysis of ethyl carbamate in stone-fruit spirits. *Rapid Communications in Mass Spectrometry*, *19*, 108–112.
68. Lachenmeier, D.W., Nerlich, U., Kuballa, T. (2006). Automated determination of ethyl carbamate in stone-fruit spirits using headspace solid-phase microextraction and gas chromatography-tandem mass spectrometry. *Journal of Chromatography, A*, *1108*, 116–120.
69. Woo, I.S., Kim, I.H., Yun, U.J., Chung, S.K., Choi, S.W., Park, H.D. (2001). An improved method for determination of ethyl carbamate in Korean traditional rice wine. *Journal of Industrial Microbiology and Biotechnology*, *26*, 363–368.
70. Mirzoian, A., Mabud, M. (2006). Comparison of methods for extraction of ethyl carbamate from alcoholic beverages in gas chromatography/mass spectrometry analysis. *Journal of AOAC International*, *89*, 1048–1051.
71. Tareke, E., Rydberg, P., Karlsson, P., Eriksson, S., Tornqvist, M. (2002). Analysis of acrylamide, a carcinogen formed in heated foodstuffs. *Journal of Agricultural and Food Chemistry*, *50*, 4998–5006.
72. Eerola, S., Hollebekkers, K., Hallikainen, A., Peltonen, K. (2007). Acrylamide levels in Finnish foodstuffs analysed with liquid chromatography tandem mass spectrometry. *Molecular Nutrition and Food Research*, *51*, 239–247.
73. Kim, C.T., Hwang, E.S., Lee, H.J. (2007). An improved LC-MS/MS method for the quantitation of acrylamide in processed foods. *Food Chemistry*, *101*, 401–409.

74. Commission Regulation (EC) No. 761/1999 of 12 April 1999, Oj No. L99/4, 14.4.1999.
75. De Melo Abreu, S., Alves, A., Oliveira, B., Herbert, P. (2005). Determination of ethyl carbamate in alcoholic beverages: an interlaboratory study to compare HPLC-FLD with GC-MS methods. *Analytical and Bioanalytical Chemistry*, *382*, 498–503.
76. Arthur, C.L., Pawliszyn, J. (1990). Solid phase microextraction with thermal desorption using fused silica optical fibers. *Analytical Chemistry*, *62*, 2145–2148.
77. Kataoka, H., Lord, H.L., Pawliszyn, J. (2000). Applications of solid-phase microextraction in food analysis. *Journal of Chromatography, A*, *880*, 35–62.
78. Food Standards Agency (2005). *Survey of ethyl carbamate in foods and beverages, Food Survey Information Sheet 78/05*, http://www.food.gov.uk/science/surveillance/fsis2005/fsis7805 (accessed September 2007).
79. Hasnip, S., Crews, C., Potter, N., Christy, J., Chan, D., Bondu, T., Matthews, W., Walters, B., Patel, K. (2007). Survey of ethyl carbamate in fermented foods sold in the United Kingdom in 2004. *Journal of Agricultural and Food Chemistry*, *55*, 2755–2759.
80. Andrade-Sobrinho, L.G., Boscolo, M., Lima-Neto, B.S., Franco, D.W. (2002). Carbamato de etila em bebidas alcoólicas (Cachaça, Tiquira, Uísque e Grapa). *Quimica Nova*, *25*, 1074–1077.
81. Kim, Y.K.L., Koh, E., Chung, H.J., Kwon, H. (2000). Determination of ethyl carbamate in some fermented Korean foods and beverages. *Food Additives and Contaminants*, *17*, 469–475.
82. Ha, M.S., Kwon, K.S., Kim, M. *et al.* (2006). Exposure assessment of ethyl carbamate in alcoholic beverages. *Journal of Microbiology and Biotechnology*, *16*, 480–483.
83. Noh, I.W., Ha, M.S., Han, E.M. *et al.* (2006). Assessment of the human risk by an intake of ethyl carbamate present in major Korean fermented foods. *Journal of Microbiology and Biotechnology*, *16*, 1961–1967.
84. Lachenmeier, D.W., Schehl, B., Kuballa, T., Frank, W., Senn, T. (2005). Retrospective trends and current status of ethyl carbamate in German stone-fruit spirits. *Food Additives and Contaminants*, *22*, 397–405.
85. Food Standards Agency (2000). *Survey for ethyl carbamate in whisky, food surveillance information sheet 2/00, May 2000*, http://archive.food.gov.uk/fsainfsheet/2000/no2/2whisky.htm (accessed September 2007).
86. Joint FAO/WHO Expert Committee on Food Additives (JECFA) (2005). *Summary and conclusions from the Sixty-fourth meeting, Rome, 8–17 February 2005, JECFA/64/SC.* http://www.who.int/ipcs/food/jecfa/summaries/summary_report_64_final.pdf (accessed November 2007).
87. European Food Safety Authority (2007). Opinion of the Scientific Panel on contaminants in the food chain on a request from the European Commission on ethyl carbamate and hydrocyanic acid in food and beverages. *The EFSA Journal*, *551*, 1–44.
88. Koh, E., Kwon, H. (2007). Quantification of ethyl carbamate in soy sauce consumed in Korea and estimated daily intakes by age. *Journal of the Science of Food and Agriculture*, *87*, 98–102.

89. Ha, M.S., Hu, S.J., Park, H.R. *et al.* (2006). Estimation of Korean adult's daily intake of ethyl carbamate through Korean commercial alcoholic beverages based on the monitoring. *Food Science and Biotechnology*, *15*, 112–116.
90. World Health Organization (1998). *GEMS/Food Regional Diets. Regional Per Capita Consumption of Raw and Semi-processed Agricultural Commodities.* WHO, Geneva.
91. Larsen, J.C. (2006). Risk assessment of chemicals in European traditional foods. *Trends in Food Science and Technology*, *17*, 471–481.
92. Barlow, S., Renwick, A.G., Kleiner, J. *et al.* (2006). Risk assessment of substances that are both genotoxic and carcinogenic. Report of an International Conference organized by EFSA and WHO with support of ILSI Europe. *Food and Chemical Toxicology*, *44*, 1636–1650.
93. Jones, D.A. (1998). Why are so many food plants cyanogenic? *Phytochemistry*, *47*, 155–162.
94. Erb, N., Zinsmeister, H.D., Lehmann, G., Nahrstedt, A. (1979). A new cyanogenic glycoside from *Hordeum vulgare*. *Phytochemistry*, *18*, 1515–1517.
95. Schehl, B., Senn, T., Lachenmeier, D.W., Rodicio, R., Heinisch, J.J. (2007). Contribution of the fermenting yeast strain to ethyl carbamate generation in stone fruit spirits. *Applied Microbiology and Biotechnology*, *74*, 843–850.
96. Suzuki, K., Kamimura, H., Ibe, A., Tabata, S., Yasuda, K., Nishijima, M. (2001). Formation of ethyl carbamate in umeshu (plum liqueur). *Journal of the Food Hygienic Society of Japan*, *42*, 354–358.
97. Baumann, U., Zimmerli, B. (1988). Accelerated ethyl carbamate formation in spirits. *Mitteilungen aus dem Gebiete der Lebensmitteluntersuchung und Hygiene*, *79*, 175–185.
98. Aresta, M., Boscolo, M., Franco, D.W. (2001). Copper(II) catalysis in cyanide conversion into ethyl carbamate in spirits and relevant reactions. *Journal of Agricultural and Food Chemistry*, *49*, 2819–2824.
99. Satchell, D.P.N., Satchell, R.S. (1975). Acylation by ketens and isocyanates. A mechanistic comparison. *Chemical Society Reviews*, *4*, 231–250.
100. Raspoet, G., Nguyen, M.T., McGarraghy, M., Hegarty, A.F. (1998). The alcoholysis reaction of isocyanates giving urethanes: evidence for a multimolecular mechanism. *Journal of Organic Chemistry*, *63*, 6878–6885.
101. Stark, G.R., Stein, W.H., Moore, S. (1960). Reactions of the cyanate present in aqueous urea with amino acids and proteins. *The Journal of Biological Chemistry*, *235*, 3177–3181.
102. Smyth, D.G. (1967). Carbamylation of amino and tyrosine hydroxyl groups. *The Journal of Biological Chemistry*, *242*, 1579–1591.
103. Halmann, M., Lapidot, A., Samuel, D. (1962). Kinetic and tracer studies of the reactions of carbamoyl phosphate in aqueous solution. *Journal of the Chemical Society (Resumed)*, 1944–1957.
104. Allen, C.M., Jones, M.E. (1964). Decomposition of carbamylphosphate in aqueous solutions. *Biochemistry*, *3*, 1238–1247.
105. Ough, C.S., Crowell, E.A., Gutlove, B.R. (1988). Carbamyl compound reactions with ethanol. *American Journal of Enology and Viticulture*, *39*, 239–242.

106. Tegmo-Larsson, I.-M., Spittler, T.D. (1990). Temperature and light effects on ethyl carbamate formation in wine during storage. *Journal of Food Science*, *55*, 1166–1167.

107. Ough, C.S., Crowell, E.A., Mooney, L.A. (1988). Formation of ethyl carbamate precursors during grape juice (chardonnay) fermentation. I. addition of amino acids, urea, and ammonia: effects of fortification on intracellular and extracellular precursors. *American Journal of Enology and Viticulture*, *39*, 243–249.

108. Coulon, J., Husnik, J.I., Inglis, D.L. *et al.* (2006). Metabolic engineering of *Saccharomyces cerevisiae* to minimize the production of ethyl carbamate in wine. *American Journal of Enology and Viticulture*, *57*, 113–124.

109. An, D., Ough, C.S. (1993). Urea excretion and uptake by wine yeasts as affected by various factors. *American Journal of Enology and Viticulture*, *44*, 35–40.

110. Monteiro, F.F., Bisson, L.F. (1991). Amino acid utilization and urea formation during vinification fermentations. *American Journal of Enology and Viticulture*, *42*, 199–208.

111. Liu, S.Q., Pritchard, G.G., Hardman, M.J., Pilone, G.J. (1996). Arginine catabolism in wine lactic acid bacteria: is it via the arginine deiminase pathway or the arginase-urease pathway? *Journal of Applied Bacteriology*, *81*, 486–492.

112. Mira, D.O., Liu, S.Q., Patchett, M.L., Pilone, G.J. (2000). Ethyl carbamate precursor citrulline formation from arginine degradation by malolactic wine lactic acid bacteria. *FEMS Microbiology Letters*, *183*, 31–35.

113. Arena, M.E., Manca De Nadra, M.C. (2005). Influence of ethanol and low pH on arginine and citrulline metabolism in lactic acid bacteria from wine. *Research in Microbiology*, *156*, 858–864.

114. Terrade, N., Mira, D.O. (2006). Impact of winemaking practices on arginine and citrulline metabolism during and after malolactic fermentation. *Journal of Applied Microbiology*, *101*, 406–411.

115. Liu, S.Q., Pilone, G.J. (1998). A review: arginine metabolism in wine lactic acid bacteria and its practical significance. *Journal of Applied Microbiology*, *84*, 315–327.

116. Mira De Orduna, R., Patchett, M.L., Liu, S.Q., Pilone, G.J. (2001). Growth and arginine metabolism of the wine lactic acid bacteria *Lactobacillus buchneri* and *Oenococcus oeni* at different pH values and arginine concentrations. *Applied and Environmental Microbiology*, *67*, 1657–1662.

117. Azevedo, Z., Couto, J.A., Hogg, T. (2002). Citrulline as the main precursor of ethyl carbamate in model fortified wines inoculated with *Lactobacillus hilgardii*: a marker of the levels in a spoiled fortified wine. *Letters in Applied Microbiology*, *34*, 32–36.

118. Matsudo, T., Aoki, T., Abe, K. *et al.* (1993). Determination of ethyl carbamate in soy sauce and its possible precursor. *Journal of Agricultural and Food Chemistry*, *41*, 352–356.

119. Canas, B.J., Diachenko, G.W., Nyman, P.J. (1997). Ethyl carbamate levels resulting from azodicarbonamide use in bread. *Food Additives and Contaminants*, *14*, 89–94.

120. Dennis, M.J., Massey, R.C., Ginn, R., Willette, P., Crews, C., Parker, I. (1997). The contribution of azodicarbonamide to ethyl carbamate formation in bread and beer. *Food Additives and Contaminants*, *14*, 101–108.

121. Dennis, M.J., Massey, R.C., Ginn, R. *et al.* (1997). The effect of azodicarbonamide concentrations on ethyl carbamate concentrations in bread and toast. *Food Additives and Contaminants*, *14*, 95–100.

122. Carey, R., Dobson, S., Ball, E. (1999). Azodicarbonamide. *IPCS Concise International Chemical Assessment Document*, *16*, 1–23.
123. Hamlet, C.G., Jayaratne, S.M., Morrison, C. (2005). *Processing Contaminants in Bread from Bread Making Machines. A Report Prepared for the UK Food Standards Agency*, RHM Technology Ltd, High Wycombe, UK.
124. Butzke, C.E., Bisson, L.F. (1997). *Ethyl Carbamate Preventative Action Manual*, US Food and Drug Administration, Washington, D.C. http://www.cfsan.fda.gov/~frf/ecaction.html (accessed September 2007).
125. Taki, N., Imamura, L., Takebe, S., Kobashi, K. (1992). Cyanate as a precursor of ethyl carbamate in alcoholic beverages. *Japanese Journal of Toxicology and Environmental Health*, *38*, 498–505.
126. Laugel, P., Bindler, F. (1992). *Le carbamate d'ethyle dans les eaux-de-vie de fruits à noyaux. Elaboration et Connaissance des Spiritaux*. Tec&Doc, Paris, France.
127. Riffkin, H.L., Wilson, R., Howie, D., Müller, S. (1989). Ethyl carbamate formation in the production of pot still whisky. *Journal of the Institute of Brewing*, *95*, 115–119.
128. Swanston, J.S., Thomas, W.T.B., Powell, W., Young, G.R., Lawrence, P.E., Ramsay, L., Waugh, R. (1999). Using molecular markers to determine barleys most suitable for malt whisky distilling. *Molecular Breeding*, *5*, 103–109.
129. Swanston, J.S. (1999). Quantifying cyanogenic glycoside production in the acrospires of germinating barley grains. *Journal of the Science of Food and Agriculture*, *79*, 745–749.
130. Weltring, A., Rupp, M., Arzberger, U. *et al.* (2006). Ethyl carbamate: analysis of questionnaires about production methods of stone-fruit spirits at German small distilleries. *Deutsche Lebensmittel-Rundschau*, *102*, 97–101.
131. Bruno, S.N.F., Vaitsman, D.S., Kunigami, C.N., Brasil, M.G. (2007). Influence of the distillation processes from Rio de Janeiro in the ethyl carbamate formation in Brazilian sugar cane spirits. *Food Chemistry*, *104*, 1345–1352.
132. González-Rodríguez, J., Pérez-Juan, P., Luque de Castro, M.D. (2002). Method for monitoring urea and ammonia in wine and must by flow injection-pervaporation. *Analytica Chimica Acta*, *471*, 105–111.
133. Francis, P.S. (2006). The determination of urea in wine—a review. *Australian Journal of Grape and Wine Research*, *12*, 97–106.
134. Miyagawa, K., Sumida, M., Nakao, M. *et al.* (1999). Purification, characterization, and application of an acid urease from *Arthrobacter mobilis*. *Journal of Biotechnology*, *68*, 227–236.
135. Ministério da Agricultura (2005). *Diário Oficial da União, No. 124, Instrução Normativa No. 13 de 29 de junho de 2005. que aprova o Regulamento Técnico para Fixação dos Padroes de Identidade e Qualidade para Aguardente de Cana e para Cachaça*, Ministério da Agricultura, Sao Palo, Brasil.
136. Health Canada (2007). *Canadian Standards ("Maximum Limits") for Various Chemical Contaminants in Foods*, http://hc-sc.gc.ca/fn-an/securit/chem-chim/contaminants-guidelines-directives_e.html (accessed September 2007).
137. World Trade Organization (WTO) (2002). *Committee on Sanitary and Phytosanitary Measures—Notification—Switzerland—Alcoholic beverages, G/SPS/N/CHE/29, 2 April 2002*, http://www.ipfsaph.org/En/default.jsp (accessed September 2002).

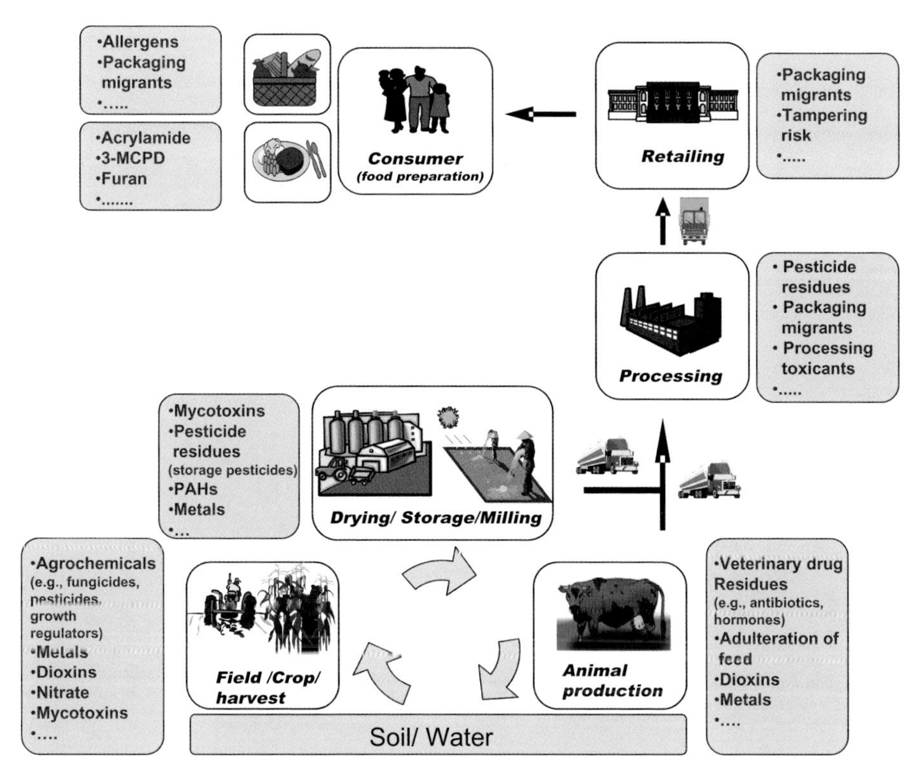

Figure 1.1 Overview of possible routes of chemical contamination of food (simplified).

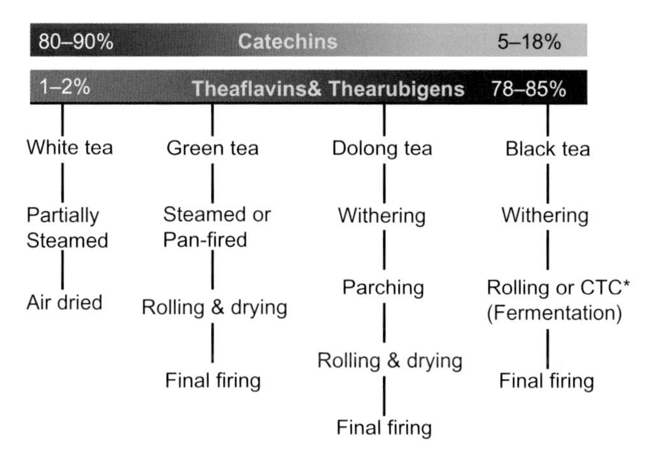

Figure 9.6 Overview of traditional tea manufacture and its impact on polyphenol content (*CTC = crushing/tearing/curling).

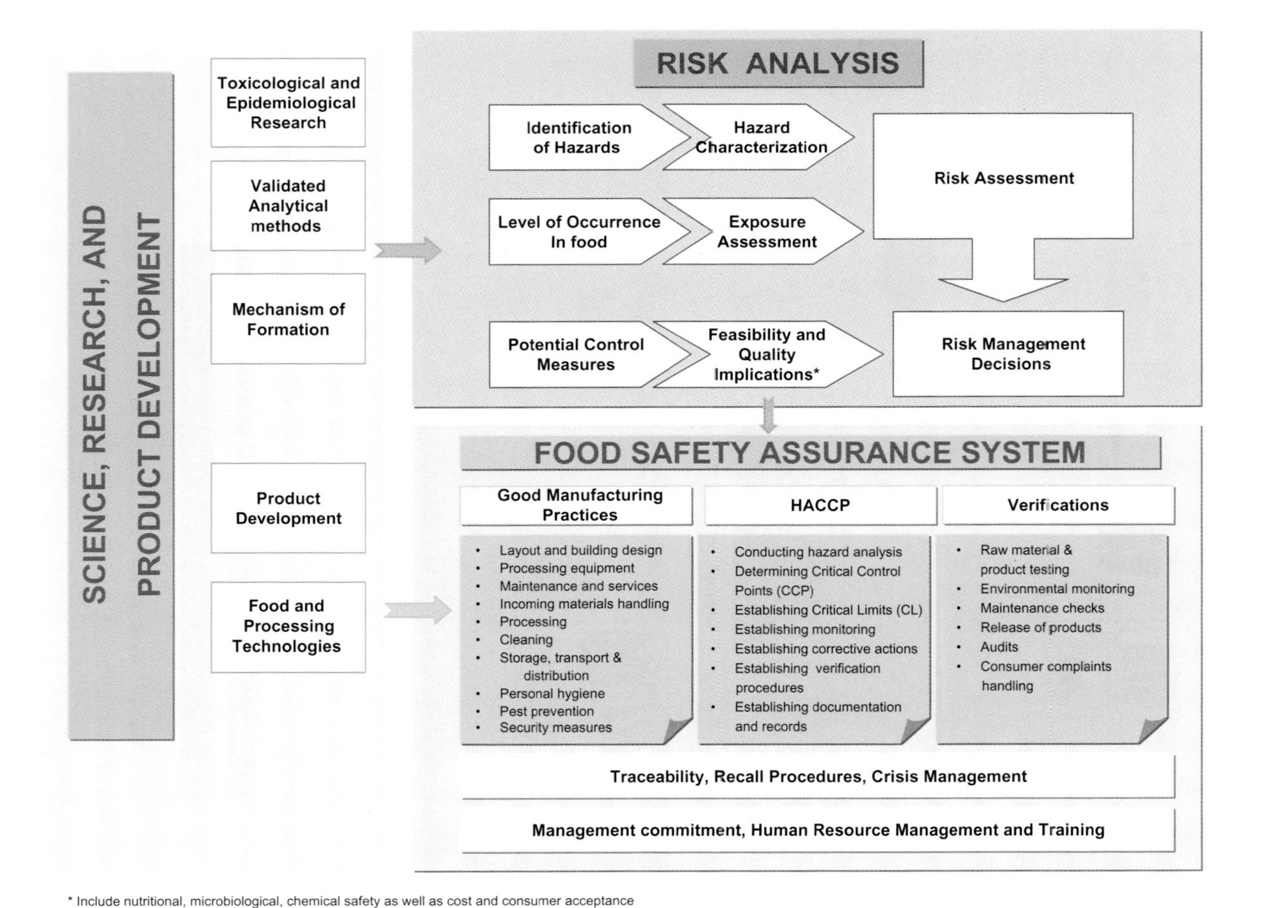

* Include nutritional, microbiological, chemical safety as well as cost and consumer acceptance

Figure 7.3 Illustration of the management of processing contaminants from fundamental research to the risk management decision and integration in the industry food safety assurance programs (adapted from Reference 42).

3.2

BIOGENIC AMINES

LIVIA SIMON SARKADI
Budapest University of Technology and Economics, 1111 Budapest, Müegyetem rkp 3, Hungary

3.2.1 INTRODUCTION

3.2.1.1 Brief History of Biogenic Amines

The history of biogenic amine research began in 1678 with the discovery of the crystallization of spermine phosphate from human semen by Anthony van Leeuwenhoek (1), the inventor of the microscope. The name spermine was given to the base by A. Ladenburg and J. Abel in 1888 (2). The correct structure of spermine was established by H.W. Dudley *et al.* in 1926 (3).

In 1885, two other related bases were isolated from decomposing animal material by L. Brieger (4), and in 1886, Ladenburg confirmed their structures by synthesis (5). They were named putrescine and cadaverine because of their origin and foul smell of putrefaction.

Agmatine was first identified in 1910 by A. Kossel in herring sperm (6) and was known as an intermediate in the polyamine metabolism of various bacteria, fungi, parasites, and marine fauna. In 1994, it was discovered that agmatine was expressed and stored also in many mammalian organs (7).

Histamine was first discovered in 1910 by Sir H.H. Dale and G. Barger (8) as a contaminant of ergot, a plant fungus that infects cereals and other grasses. In the same year, Dale and Laidlaw (9) isolated histamine from animal tissues. Histamine had previously been synthesized by A. Windaus and W. Vogt in 1907 (10), but it was not known to occur naturally in the animal body or elsewhere. The word "histamine" comes from *histos*, which means tissue.

Process-Induced Food Toxicants: Occurrence, Formation, Mitigation, and Health Risks, Edited by Richard H. Stadler and David R. Lineback

Biogenic Amines (e.g., Histamine): At a Glance	
Historical	Biogenic amine research began some 120 years ago with the identification of spermine, putrescine, and cadaverine. In the early 20th century histamine was isolated from animal tissues. Biogenic amines are of interest due to their potential risk for human health but also in their role as chemical markers of food spoilage and poor processing and storage conditions.
Analysis	HPLC is the most common technique used coupled to either UV-absorbing or fluorescence detection. Reversed phase HPLC is usually coupled with either pre-or post-column derivatization of the analytes.
Occurrence in Food	Found in many different foods, especially fermented foods, dairy products; the biogenic amine patterns in vegetables and meat differ. The main compounds encountered are histamine, tyramine, cadaverine, putrescine, spermidine, and spermine.
Main Formation Pathways	Formation from their amino acid precursors by enzyme-catalyzed decarboxylation under conditions that favor bacterial growth, i.e., histidine → histamine, tyrosine → tyramine, etc.
Mitigation in Food	Appropriate handling and processing of fish and meat under sanitary conditions and controlled temperatures. Bacterial activity is increased at raised temperatures, so poor storage conditions exacerbate the problem. Cooking can destroy the bacteria but not the toxic agent (histamine). In commercial fermented food applications, the utilization of amine negative bacterial starter cultures or mixed starter cultures has been suggested.
Health Risks	Histamine intoxication, often called scombrotoxicosis, is a common seafood-borne disease associated with the consumption of spoiled scombroid fish such as tuna, mackerel, and sardines. The typical symptoms like flushing, urticaria, and palpitations mimic those of allergy so histamine fish poisoning can easily be misdiagnosed. However, food poisoning may also occur in conjunction with potentiating factors such as drugs, alcohol, gastrointestinal diseases. Putrescine, spermine, spermidine and cadaverine have no adverse health effects, but they may react with nitrite to form carcinogenic nitrosamines especially in meat products that contain nitrite and nitrate as curing agents.
Regulatory Information or Industry Standards	The EU has established a legislative limit for histamine in fishery products particularly in those fish species that are associated with a high amount of histidine. Some countries have regulated the amounts of histamine in different foods at the national level. Generally, upper limits of 100 mg/kg histamine in foods and 2 mg/L in beverages are suggested.
Knowledge Gaps	The safety assessment of biogenic amines is difficult because knowledge concerning long-term toxic effects and dose-response are often incomplete. A database on biogenic amines in different foods at the national and regional level will enable better exposure assessments and subsequently lead to dietary recommendations.
Useful Information (web) & Dietary Recommendations	No specific recommendations from a dietary perspective, as the risks are associated with food spoilage.

Tyramine was first isolated from the posterior salivary glands of *Octopus macropus* by M. Henze in 1913 (11).

3.2.1.2 General Considerations on Biogenic Amines

Biogenic amines are aliphatic, aromatic, or heterocyclic low-molecular-mass biomolecules. They are formed and degraded during normal cellular metabolism. Biogenic amines play a variety of physiological roles, such as regulation of the digestion, the central and peripheral nervous systems, and blood pressure. The polyamines, spermidine and spermine, are essential for cell growth and proliferation. They are involved in nucleic acids (DNA, RNA) and protein biosynthesis and are also mediators of hormones (12, 13).

Biogenic amines in food are mainly formed as a result of microbial decarboxylation of amino acids. The most important biogenic amines in food are histamine (Him), derived from histidine; tyramine (Tym), derived from tyrosine; and cadaverine (Cad; 1,5-diaminopentane), which is derived from lysine. Putrescine (Put; 1,4-diaminobutane) is formed either from ornithine or from arginine via agmatine (Agm; 1-amino-4-guanidinobutane). Polyamines spermidine (Spd; [*N*-(3-aminopropyl)-1,4-diaminobutane]) and spermine (Spm; [*N*,*N*′-bis(3-aminopropyl) 1,4 diaminobutane]) arise from putrescine (14). Structures of the main biogenic amines are shown in Fig. 3.2.1.

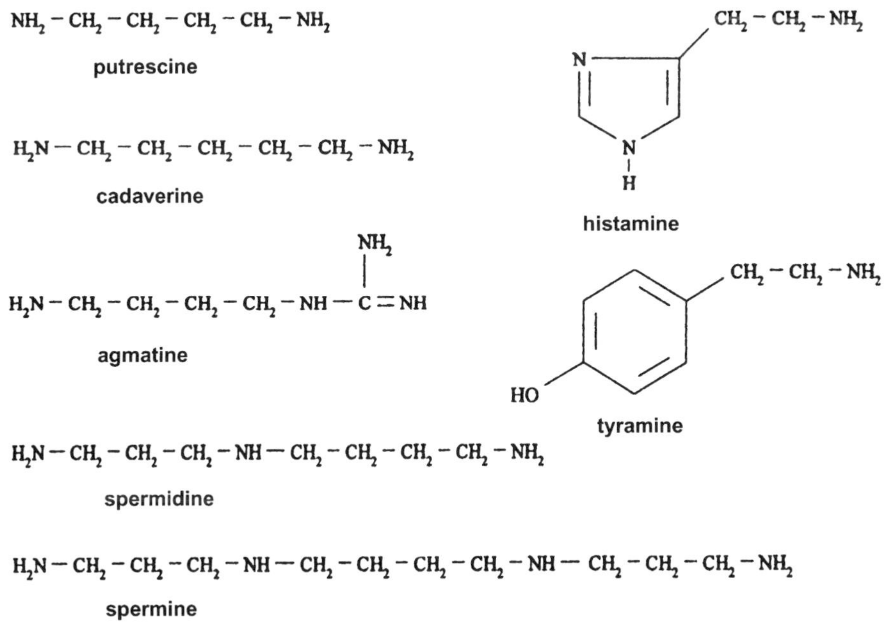

$NH_2-CH_2-CH_2-CH_2-CH_2-NH_2$

putrescine

$H_2N-CH_2-CH_2-CH_2-CH_2-CH_2-NH_2$

cadaverine

$H_2N-CH_2-CH_2-CH_2-CH_2-NH-C(NH_2)=NH$

agmatine

$H_2N-CH_2-CH_2-CH_2-NH-CH_2-CH_2-CH_2-CH_2-NH_2$

spermidine

$H_2N-CH_2-CH_2-CH_2-NH-CH_2-CH_2-CH_2-CH_2-NH-CH_2-CH_2-CH_2-NH_2$

spermine

$CH_2-CH_2-NH_2$ (imidazole ring: N, N–H)

histamine

$CH_2-CH_2-NH_2$ (benzene ring, HO)

tyramine

Figure 3.2.1 Structures of the main biogenic amines.

High amounts of exogenous biogenic amines, especially histamine and tyramine, in the human diet may contribute to a wide variety of toxic effects. The allergy-like symptoms may include sneezing and congestion of the nose, headache, breathing difficulty, bronchial asthma, gastric disorders, diarrhea, hypotension, cardiac palpitations, and urticarial exanthema (14).

Biogenic amines in food are of great interest not only for their potential risk to human health but also because they could have a role as chemical indicators of unwanted microbial contamination and processing conditions. Numerous bacteria have been reported to possess amino acid decarboxylase activity, such as *Clostridium, Enterobacter, Escherichia, Lactobacillus, Pediococcus, Proteus, Pseudomonas*, and *Salmonella* (15).

There are significant differences in the biogenic amine composition of the two major types of food, of plant and animal origin. Vegetable-type foods contain high amount of putrescine, spermine, and spermidine but significantly lower amount of histamine than do animal-derived foods. Generally, the vegetable-type foods may be considered low-risk products with regard to the presence of biogenic amines, while the products of microbial fermentation (cheese, sausage, fish, wine, beer, sauerkraut) may contain relatively high amounts of the biogenic amines (16).

However, while biogenic amines can cause several problems for susceptible consumers, there is a general absence of specific legislation setting limits on biogenic amines in food. The European Union (17) established legislative limit values only for histamine in fish, since histamine has been implicated in causing the most frequent foodborne intoxications. Some countries have regulated the maximum amounts of histamine in different foods at a national level. Generally, upper limits of 100 mg histamine/kg in food and 2 mg/L in beverages have been suggested. There are recommendations for tyramine (100–800 mg tyramine/kg) and for 2-phenylethylamine (30 mg/kg) in food (18).

Knowledge of biogenic amines in fermented foods is necessary to make an assessment of the health hazards arising from the consumption of these products, and also it can provide information to improve food quality with respect to biogenic amine content.

Several studies have monitored the biogenic amine formation and occurrence in food. The first general monograph on biogenic amines was published by Guggenheim in 1920 (19). Recent information on the topic is given in some valuable reviews (16, 20–26).

3.2.2 OCCURRENCE IN FOOD

Biogenic amines are commonly found in many foods, especially in fermented foods and beverages. Fermentation is an ancient preservation method to increase the shelf life of various foods such as cheese, sausage, wine, beer, and sauerkraut. Bacteria of the genera *Acetobacter, Bifidobacterium, Brevibacterium, Lactobacillus, Micrococcus, Propionibacterium*, and *Streptococcus* are used as starter cultures for the large-scale production of fermented foods and beverages in food technology. The types of food, the applied technology, and hygienic conditions during manufacture and/or storage are the main influencing factors in the qualitative and quantitative occurrence of biogenic amines in food.

3.2.2.1 Dairy Products

Cheese is one of the oldest human foods. All cheese results from a lactic acid fermentation of milk. Proteolysis of casein during cheese ripening leads to an increase in free amino acids. Due to the action of the raw milk flora or the starter strains, the formation of biogenic amines, especially histamine and tyramine, very often occurs in cheese. High amounts of tyramine may cause an increase in blood pressure; this symptom is known as "cheese reaction."

Cheese represents an ideal environment for biogenic amine production because of the great availability of amino acids and the presence of bacteria. The production of biogenic amines in cheese has been mainly attributed to the activity of nonstarter microorganisms, but the role of starter *Lactobacillus* cannot be excluded. Numerous bacteria used in cheese production have been reported to possess the amino acid decarboxylase activity (Table 3.2.1). The data indicate that starter cultures should be carefully checked for their potential to form biogenic amines during cheese processing conditions.

Several factors may contribute to biogenic amine formation in cheese, such as the microbiological quality and the type of raw milk, the use of starter cultures, and the conditions and time of the ripening process. Higher ripening temperature and pH, and low salt concentration may contribute to the ability of the microbes to produce biogenic amines.

The content of biogenic amines in several kinds of cheese has been studied and their amounts varied greatly even within the same variety (20, 27–39). Generally, the main amine found in cheese was tyramine, followed by histamine, putrescine, and cadaverine (Table 3.2.2).

In particular, the microbial population of raw milk can influence biogenic amine formation in cheese, even when thermal treatments are applied. The effect of milk quality (unpasteurized, pasteurized) and the type of milk (cow, ewe, goat) on biogenic amine content of cheese have been extensively studied (41–46). Cheese produced from ewes or goat milk showed lower amounts of biogenic amines in comparison with cows' milk cheese.

Other milk products such as yogurt and kefir have little or no detectable amounts of tyramine if made from pasteurized milk (47).

3.2.2.2 Meat and Meat Products

Besides the fermentation of dairy products, a variety of meats can also be fermented. Dry sausage and salami are valuable meat products prepared by specific processing. Fermented dry sausage is defined as a mixture of comminuted fat and lean meat, salt, nitrate and/or nitrite, sugar, and different spices,

TABLE 3.2.1 Microorganisms involved in cheese fermentation and their ability to form biogenic amines.

Cheese	Microorganism	
Cottage	*Lactococcus lactis* (Tym, Trpm)	*Leuconostoc cremoris* (Tym)
Cream	*L. cremoris*, *L. diacetilus* (Him), *S. thermophilus*, *L. bulgaricus* (Tym)	
Mozzarella	*S. thermophilus*, *L. bulgaricus*	
Brie	*Lactococcus lactis*, *L. cremoriz* (Him, Tym)	*Penicillium camemberti*, *P. candidum*, *Brevibacterium linens*
Camembert	*Lactococcus lactis*, *L. cremoris*	*Penicillium camemberti*, *Brevibacterium linens*
Blue	*Lactococcus lactis*, *L. cremoris*	*Penicillium roqueforti*
Brick	*Lactococcus lactis*, *L. cremoris*	*Brevibacterium linens*
Limburger	*Lactococcus lactis*, *L. cremoris*	*Brevibacterium linens*
Muenster	*Lactococcus lactis*, *L. cremoris*	*Brevibacterium linens*
Roquefort	*Lactococcus lactis*, *L. cremoris*	*Penicillium roqueforti*
Cheddar	*Lactococcus lactis*, *L. cremoris*, *E. durans* (Tym)	*Lactobacillus casei*, *L. plantarum* (Tym, Put, Him)
Colby	*Lactococcus lactis*, *L. cremoris*, *E. durans*	*Lactobacillus casei* (Him, Put, Cad, Tym)
Edam	*Lactococcus lactis*, *L. cremoris*	
Gouda	*Lactococcus lactis*, *L. cremoris*, *L. diacetilus*	
Swiss	*Lactococcus lactis*, *L. helveticus*, *S. thermophilus*	*Propionibacterium shermanii*, *P. freudenreichii*
Parmesan	*Lactococcus lactis*, *L. cremoris*, *S. thermophilus*	*L. bulgaricus*

Data from Reference 27.
Tym, tyramine; Trpm, tryptamine; Him, histamine; Put, putrescine; Cad, cadaverine.

TABLE 3.2.2 Ranges of biogenic amine contents in major types of cheese (mg/kg).

Cheese	Histamine	Tyramine	Putrescine	Cadaverine
Brie	10–600		0–5	0–2
Camembert	0–1000	0–4000	0–250	0–500
Blue	0–1900	0–2500	0–80	0–110
Cheddar	0–2100	0–1500	5–300	2–350
Edam	1–500	0–900	1–190	4–190
Emmental	5–2500	0–700	0–300	0–180
Gouda	10–900	10–900	0–400	1–300
Swiss	4–2500	0–700	—	—
Parmesan	10–581	0–840	1–90	0–250

Data from References 14 and 40.

which are stuffed into casings, subjected to fermentation, and then allowed to dry. The spontaneous fermentation of dry sausages involves the participation of lactic acid bacteria (LAB), coagulase negative cocci (CNC; mostly *Staphylococcus* and *Kocuria* species), and yeasts and molds. Most of the commercially available meat starter cultures contain mixtures of LAB and CNC (48).

High amounts of proteins and the proteolytic activity during ripening provide free amino acids as the precursors for biogenic amines. Biogenic amines are formed in meat and meat products, as in various other protein-rich foods, as a consequence of spoilage or as an undesired by-product of a principally desired microbial activity during fermentation of raw sausage.

The quality of raw materials influences the composition and the concentration of biogenic amines produced during the ripening of sausages (49, 50). The main amines in fresh meat used for fermented sausage production are spermidine and spermine, and, to a lesser extent, putrescine (51). High amounts of putrescine and the presence of other amines have been attributed to microbial growth and depend on meat freshness.

Biogenic amines have been proposed as a quality index for fresh meat and processed meat (52). The amounts of putrescine, cadaverine, and tyramine seem to be good indicators of the quality of different meats. Based on the determination of biogenic amine content in red (adult bovine) and white (chicken) meat, it was concluded that cadaverine concentration could be used to monitor spoilage in both kinds of meat. Also, tyramine contents appeared to be useful to control red meat storage (53). An index based on the ratio of spermidine/spermine concentrations was considered appropriate for the evaluation of stored chicken meat quality (54).

Several reports showed that tyramine and putrescine are the most abundant biogenic amines found in dry sausages (22, 55–59). Tyramine is mainly related to the activity of fermentative LAB while putrescine and cadaverine are usually the result of the action of non-fermentative strains. For health concerns, significant histamine was also surveyed in various meat and meat products (sausages, beef, and poultry meat) (60). Table 3.2.3 indicates the ranges of some biogenic amine content in different meats and meat products.

Starter cultures are frequently used in sausage manufacturing in order to shorten the ripening time, ensure color development, enhance flavor, and improve product safety. The primary genera of bacteria utilized as starter cultures are *Lactobacillus* sp., *Pediococcus* sp., and *Micrococcus/Staphylococcus* sp. (48, 62). Sausages fermented with starter culture had lower amounts of tyramine and histamine than naturally fermented sausages (63–65). The control of fermentation by introducing competitive lactic acid bacterial starter strains is an important method proposed to influence the formation of biogenic amines by preventing the growth of amine-producing bacteria in meat products, which leads to health-related benefits (48).

Moreover, it is known that many other factors such as sausage diameter, pH, water activity, and NaCl may influence the formation of biogenic amines in dry fermented sausage. A larger diameter might lead to more favorable

TABLE 3.2.3 Ranges of biogenic amine contents in meat and meat products (mg/kg).

Type of meat	Histamine	Tyramine	Putrescine	Cadaverine	Spermine	Spermidine
Beef	nd–8	nd–35	nd–40	nd–68	nd–16	nd–3
Pork	nd–15	1–14	nd–38	nd–66	nd–35	nd–2
Chicken	nd–3	1–30	nd–4	1–140	1–210	1–8
Ham	1–10	2–20	2–22	2–4	1–28	2–10
Fermented sausage	20–260	50–450	60–550	4–120	12–36	12–20
Salami	1–350	1–280	1–520	1–930	nd	1–8

Data from Reference 61.
nd, not detected.

environment (lower NaCl concentration and drying level, higher water activity) for the growth of microorganisms and for the development of biogenic amines. Bover-Cid *et al.* (66) observed increasing putrescine and tyramine concentration with increasing diameter of sausage. Similar results were reported by Komprda *et al.* (67). Trevino *et al.* (68) found lower biogenic amine content at the edge of sausages in comparison with the central part.

3.2.2.3 Fish

Scombroid fish poisoning is a foodborne intoxication caused by certain spoiled *scombroid* fish such as mackerel (*Scomber* spp.), tuna (*Thunnus* spp.), saury (*Cololabis* saira), and bonito (*Sarda* spp.). Recently, it has been reported that non-scombroid fish have also been implicated to cause identical symptoms (69). Non-scombroid fish include mahimahi (*Coryphaena* spp), sardines (*Sardinella* spp.), pilchards (*Sardina pilchardus*), marlin (*Makaira* spp.), bluefish (*Pomatomus* spp.), sockeye salmon (*Oncorhynchus nerka*), yellowtail (*Seriola lalandii*), and Australian salmon (*Arripis trutta*). These fish species have significant amounts of histidine in their muscle tissues that serve as a substrate for bacterial HDC.

Although histamine is the main compound responsible for intoxication after fish consumption, the toxicity may be increased by other amines, such as putrescine and cadaverine. Enterobacteriaceae species (*Morganella morganii, Klebsiella pneumoniae, Proteus vulgaris*, and *Hafnia alvei*) are the most important biogenic amine-forming bacteria in fish.

Many studies on fish have reported that fresh fish contain only low amounts of biogenic amines but considerably increased amounts have been observed in fish handled under poor hygiene or stored under inappropriate conditions. Upon investigation of the effect of storage temperature on biogenic amine formation, it was established that the elevated accumulation of histamine and other biogenic amines occurs at higher temperatures (70–74). However, several other studies have also demonstrated that histamine and other biogenic amines can accumulate in fish stored at low temperatures (75, 76).

Mietz and Karmas (77) were the first who proposed a quality index (biogenic amine index [BAI]) based on the increases in putrescine, cadaverine, and histamine, and decreases in spermine and spermidine during fish storage. The index is calculated by the content of biogenic amines in mg/kg:

$$\text{BAI} = (\text{histamine} + \text{putrescine} + \text{cadaverine})/(1 + \text{spermidine} + \text{spermine})$$

Based on the correlation between amines in fish and organoleptic properties, the BAI may be adjusted to the organoleptic quality classes as follows: fish with a BAI value below 1 is considered to be of first quality, between 1 and 10 is borderline quality, whereas BAI values above 10 indicate a very poor microbial quality (decomposed fish) (78).

Veciana-Nogues *et al.* (79) suggested that the amounts of tyramine should also be included in the Mietz and Karmas index and proposed a BAI, based on the sum of histamine, putrescine, cadaverine, and tyramine, to describe the freshness of tuna.

Fermented fish products (pickled fish, sauces, pastes) are traditionally produced in Mediterranean countries. The amount of biogenic amines in fishery products depends on the specific technology of curing, drying, salting, smoking, and marinating in combination with the temperature, time, and storage conditions. Higher values of biogenic amines have been found in many of these products (80).

The European Union (17) regulated maximum concentrations of histamine in fish products. The ranges of histamine contents in different fish and fish products are shown in Table 3.2.4.

TABLE 3.2.4 Ranges of histamine content in different fish and fish products (mg/kg).

Type of fish product	Histamine
Frozen mackerel	1–20
Smoked mackerel	1–1788
Canned mackerel	nd–210
Frozen herring	1–4
Salted herring	5–121
Canned herring	1–479
Smoked sardine	42–99
Salted sardine	14–150
Canned sardine	3–2000
Canned anchovy	1–54
Canned bonito	10–36
Canned tuna	1–402

Data from Reference 81.
nd, not detected.

3.2.2.4 Wine

Alcoholic beverages constitute another category of fermented products that sometimes bear substantial quantities of biogenic amines. Wine is known to contain many biologically active compounds. The amounts and compositions of these compounds depend on the type of grapes and their degree of ripeness, climate, and soil of the viticultural area, as well as vinification techniques. Amino acids represent the main source of nitrogen for both yeast and malolactic bacteria during wine fermentation, and also serve as substrate for volatile aroma compounds for biogenic amine production in wine. The amines form primarily during and after the spontaneous malolactic fermentation process by decarboxylation of the precursor free amino acids. Alcohol may potentiate the biological effect of biogenic amines present in wine by inhibiting the catabolism of amines (21).

In general, white wines have rarely been implicated, while red wines have often provoked physiological distress because red wines contain higher amount of histamine than white wines do. These differences may be due to the different fermentation processes. Red wine is produced from whole grapes whereas white wine is produced from grape juice without the skins. It means that red wine is liable to be contaminated by amine-producing microorganisms.

Some countries have regulations for the maximum content of histamine allowed in wine. The recommended upper limit for histamine in wine has been reported to be 10 mg/L in Austria, 5–6 mg/L in Belgium, 8 mg/L in France, 2 mg/L in Germany, 10 mg/L in Hungary, 3.5 mg/L in the Netherlands, and 10 mg/L in Switzerland (21).

The presence of biogenic amines in wines is well documented in the literature (82–92). Predominant biogenic amines in wine are histamine, tyramine, putrescine, and agmatine. The production of histamine, tyramine, and putrescine by LAB isolated from wine has been studied by different authors (93–95). Some LAB strains are responsible for the histamine, tyramine, and phenylethylamine concentrations, but not for putrescine concentrations in wine. Putrescine mainly originates from the grape must since it is, besides agmatine and spermidine, the most abundant amine in grapes (95).

Surveys made on wines showed that winemaking technology had greater effect on biogenic amine formation in wines than geographical origin, grape variety, and year of vintage (86–91). Ratios of putrescine to tyramine were successfully used to differentiate between white and red wines (88). Table 3.2.5 shows the ranges of biogenic amine contents in red and white wines.

TABLE 3.2.5 Ranges of biogenic amine contents in wines (mg/L).

Wine	Histamine	Tyramine	Putrescine	Cadaverine	Spermine	Spermidine	Agmatine
Red wine	nd–14.5	nd–11.9	1–14	0.2–3	nd–2	0.1–3	nd–53
White wine	nd–0.2	nd–12	0.5–10	0.1–1.5	0.1–2.5	0.1–7	nd–32

Data from Reference 61.
nd, not detected.

3.2.2.5 Beer

Beer is defined as an alcoholic beverage from starch-containing raw materials serving as a source for maltose and glucose, which are fermented by brewers yeast. Although barley malt is the most important cereal, wheat, wheat malt, corn, rice, and millet are also used as starch-containing adjuncts or extenders and sources for fermentable sugars.

In beer production, alcoholic fermentation takes place by the action of selected strains of the yeast. Beers are classified into two groups: top- and bottom-fermented based on whether yeast floats or sinks by the end of fermentation. Besides *Saccharomyces cerevisiae* (top fermenting) and *Saccharomyces carlsbergensis* (bottom fermenting), various wild yeasts, together with LAB, are involved in the brewing process of special local beers.

The total biogenic amine content of beer is influenced by the barley variety used in the brewing process, malting technology, wort processing, and the conditions during fermentation (96, 97). Higher amounts of histamine and tyramine in some European beers indicate microbial contamination during brewing (96–107).

Considering the bacterial origin of biogenic amines, Loret *et al.* established a beer biogenic amine index (Beer BAI) (107) similarly to what has been done for fish by Mietz and Kamas (77). The Beer BAI calculation formula consists of the ratio of biogenic amines of bacterial origin (Him, Put, Cad, Tym, phenylethylamine; Phem, tryptamine; Trpm) to the natural biogenic amine found in the malt (Agm). Each biogenic amine concentration is expressed in mg/L:

$$\mathrm{BAI} = (\mathrm{Him} + \mathrm{Put} + \mathrm{Cad} + \mathrm{Tym} + \mathrm{Phem} + \mathrm{Trpm})/(1 + \mathrm{Agm})$$

Beer BAI value reflects the microbiological quality of the fermentation process. If the BAI value is lower than 1.0, it means the beer has been produced by a non-contaminated fermentation process (high microbiologic quality). If the BAI value is between 1.0 and 10.0, it means moderate contamination by decarboxylating bacteria (intermediate level of microbiological quality), and higher than 10.0 BAI value means the beer is highly contaminated by amine-producing bacteria (poor microbiological quality).

The nonalcoholic beers do not have significantly lower amounts of biogenic amines than the majority of regular beers, indicating that methods used to produce them do not result in the decrease of amines (104).

No official maximum or limits have been set for histamine or tyramine concentrations in beers. According to a recommended guideline, 6-mg tyramine ingestion within a 4-h period is considered a safe amount for beers (99). The ranges of histamine and tyramine contents detected in different types of beer are shown in Table 3.2.6.

3.2.2.6 Sauerkraut

Sauerkraut has been very popular in many European countries due to its sensorial properties and favorable nutritional value. Sauerkraut is produced

TABLE 3.2.6 Ranges of biogenic amine contents in different types of beer (mg/L).

Type of beer	Histamine	Tyramine	Putrescine	Cadaverine	Agmatine
Top-fermented					
Ale	0.5–2.0	1.9–17.4	2.6–9.7	nd–4.2	1.1–15.7
Stout and porter	nd–3.2	1.1–9.1	2.4–8.2	nd–1.9	5.2–10.8
Weissbier	nd–2.4	1.3–33.6	2.4–6.7	0.4–17.7	3–10.7
Kriek	1.6–14.0	7.6–36.4	3.5–5.1	1.9–15.2	1.1–3.4
Trappsite	nd–1.6	1.8–4.5	3.2–9.1	nd–1.9	3.2–19.4
Bottom-fermented					
Lager	nd–2.6	1.6–20.9	1.5–9.7	nd–6.2	0.9–27.2
Pils	nd–17.0	0.5–46.8	2.6–8.8	nd–31.4	5.1–40.9
Dortmünder	nd–1.3	1.4–7.6	3–5.6	nd–0.5	6.1–13.4
Bock	nd–2.9	2.1–10.2	2.1–12.4	nd–1.5	0.4–21.5
Nonalcoholic	nd–3.2	2.1–31.5	1.6–4.7	nd–5.3	2.6–16.8

Data from Reference 61.
nd, not detected.

from shredded white cabbage. Fermentation process can be carried out using either spontaneous fermentation (which relies on the LAB occurring naturally on vegetables) or controlled fermentation (using a starter culture of *Lactobacillus* species). Among microorganisms contributing to sauerkraut production, *Leuconostoc mesenteroides* is of special importance in initiating the lactic acid fermentation. The next phase is characterized by the activity of homofermentative, no-gas-producing LAB with higher occurrence of *Lactobacillus plantarum*. The last phase of fermentation is dominated by heterofermentative lactic acid strains such as *Lactobacillus brevis, Pediococcus*, and *Enterococccus* (108). Table 3.2.7 shows the microorganisms involved in sauerkraut fermentation and their ability to form biogenic amines.

Biogenic amine content of sauerkraut is highly influenced by cabbage variety, technology (temperature, pH value change, oxygen access, or sodium chloride content), and bacterial contamination or microbial starters used for

TABLE 3.2.7 Microorganisms involved in sauerkraut fermentation and their ability to form biogenic amines.

Species	Biogenic amines
Leuconostoc mesenteroides	Tyramine
Lactobacillus plantarum	Tyramine, putrescine, phenylethylamine
Lactobacillus brevis	Tyramine, putrescine, tryptamine
Pediococcus	Tyramine
Enterococcus	Tyramine, cadaverine, phenylethylamine

Data from Reference 27.

mercaptopropionic acid, FMOC, 6-aminoquinolyl-*N*-hydroxysuccinimidyl carbamate (AQC), 2-naphthyloxycarbonyl chloride (NOC), and fluorescamine (4-phenylspiro[furan-2-(3H),1′-phthalan]-3,3′-dione). Dansyl chloride is a widely used fluorogenic agent in chromatographic procedures for amine analysis. However, it is a nonspecific reagent that will react also with alkaloids, amino acids, phenols, aliphatic alcohols, purine bases, and sugars. After chromatography, the dansyl derivatives are visualized under a UV light source (360 nm). Using this reagent, a concentration as low as 1 nmol (10^{-9} mol) can be detected. Dabsyl chloride is also suitable for modifying polyamines for HPLC. The detection is in the visible light range (436 nm). The detection limit is less than 2 pmol. The principles for benzoylation are the same as for dansylation. HPLC of benzoyl chloride derivatives is not as sensitive, but sample preparation is simple, and detection is achieved with a UV detector. The limit of detection is between 0.01 and 1 nmol. Oxycarbonylchlorides such as FMOC or NOC react quantitatively at ambient temperatures within a few minutes with primary and secondary amino groups forming stable carbamates except for histamine. The sensitivity of detection is at the picomole concentration.

HPLC is the most frequently used technique for biogenic amine determination. Depending on the type of interaction between stationary phase, mobile phase, and samples, several separation mechanisms can be used in HPLC, e.g., normal phase, reversed-phase chromatography or reversed-phase chromatography with ion pairing (50, 127–129). The traditional HPLC of biogenic amines has involved the use of either UV-absorbing or fluorescent detection (53, 116–134). RP-HPLC of biogenic amines involves either pre-column or post-column derivatization procedures with dansyl chloride (135–137), benzoyl chloride (69, 138), FMOC (139), AQC (140), NOC (141), and *para*-nitrobenzoyloxycarbonyl chloride (PNZ-Cl) (142), with detection limits at the fmol concentrations. HPLC method with mass spectrometry detection (HPLC-MS/MS) has become a popular and powerful technique for the determination of underivatized biogenic amines (118, 124).

OPLC is theoretically and practically a planar layer version of HPLC. OPLC involves a microchamber in which a membrane under external pressure covers the adsorbent layer and solvent is introduced by means of a pump. OPLC combines the advantages of classical TLC, HPTLC, and HPLC in the analysis of biogenic amines. Simon Sarkadi *et al.* (143) developed an OPLC method for determination of biogenic amines as dansyl derivatives in food by use of a Personal OPLC BS50 Chromatograph (OPLC-NIT Ltd., Budapest, Hungary). Kovács *et al.* (144) developed a step-wise gradient elution system for determination of dansylated biogenic amines in some vegetables by OPLC. This latter method was used for different food and beverage analyses (125, 145).

CE and related techniques are relatively new tools in biogenic amine analysis. Biogenic amines have been determined in food by CE with pulsed amperometric detection (PAD) (146) and indirect UV detection (147–149). There are other methods using capillary zone electrophoresis (CZE) with lamp-induced fluorescence detection (150), conductometric detection (151), and laser-induced fluorescence (LIF) detection (152). Micellar electrokinetic chromatography (MECC) methods were developed for the separation of biogenic amines in foods labeled with *N*-(4-aminobutyl)-*N*-ethylisoluminol (153), fluorescein thiocarbamate derivatives (154), *N*-substituted benzamides (73, 155), AQC (156), or benzoyl chloride (157). A new methodology was developed to separate and determine amines by cyclodextrin-modified CE with UV or LIF detection (158).

GC is considerably less applied for the quantification of biogenic amines. This is probably due to the extensive pre-purification methods that are necessary to obtain derivatizable, amine-containing extracts from biological materials. However, there are some useful GC methods developed for the determination of biogenic amines in food (159–161).

TLC is a simple and rapid procedure, considering that several samples can be run in parallel on the same plate. Depending on the sorbent layer adsorption, partition or ion-exchange methods can be used for separating biogenic amines by TLC (162–164).

Ionic compounds such as biogenic amines are often better separated by IEC following post-column colorimetry with ninhydrin, using the equipment for amino acid analysis (68, 165–167) or conductivity detection using HPLC equipped with an ion-exchange column (117).

Micellar liquid chromatography (MLC) followed by PAD was also successfully applied for biogenic amine analysis (168).

3.2.3.3 Rapid Methods Applied for Biogenic Amine Determination

Besides the modern food analytical techniques (chromatography, electrophoresis), biochemical methods such as different types of biosensors, enzyme-based sensors, and immunosensors have been developed, with the aim to provide a rapid means of detecting biogenic amines in food.

The control of food quality and freshness is of growing interest for both consumer and food industry. The conventionally used techniques are expensive, slow, need well-trained operators, and in most cases required time-consuming sample pretreatment. Food industry needs simple, nondestructive screening methods that correlate information available on the product with the stage of freshness.

Biosensors consist of a molecular recognition element (such as enzyme, antibody, receptor, or microorganisms) and a chemical or physical transducer (electrochemical, mass, optical, and thermal) (169). Several biosensors have been described so far to offer rapid screening methods for industrial food quality testing (169–187). Table 3.2.8 shows the most frequently used biosensor methods for detecting biogenic amines in food.

The other rapid technique for the quantitative measurement of histamine in food is the solid-phase enzyme-linked immunosorbent assay (ELISA). Most

TABLE 3.2.8 Biosensors frequently used for detecting biogenic amines in food.

Food	Biocomponent	Transducer	Detection range
Fish	DAO	Amp.	<6 mM
Anchovy	DAO	Amp.	1×10^{-6} to 5×10^{-5} mol/L
Fruit, vegetables	DAO	Amp.	2×10^{-6} to 2×10^{-3} mol/L
Fruit, vegetables	PAO	Amp.	2×10^{-6} to 1×10^{-3} mol/L
Meat	XOD	O_2 electrode	<4 mM
Prawn	Orn-carbamyl transferase Nucleoside phosphorylase XOD	Amp.	40 mM
Fish	Hypoxanthine oxidase XOD	Amp.	1×10^{-7} to 1×10^{-5} mol/L
Fish	Amine oxidase, peroxidase	Amp.	1100 mM
Fish	DAO	Amp.	1100 mM
Seafoods	Histamine oxidase	Amp.	$<9.5 \times 10^{-7}$ M
Fish	Amine oxidase, peroxidase	Amp.	1020 mM

Data from Reference 169.
Amp., amperometric detection; XOD, xanthine oxidase.

immunochemical methods (enzyme immunoassays) for the detection of histamine are based on antibodies against *N*-amino derivatives of histamine synthesized by reaction with, for example, *p*-benzoquinone or propionic acid esters. Because the antibodies used in these tests are reactive only with the histamine adduct, chemical derivatization of histamine is necessary before analysis (188–191).

The first polyclonal antihistamine antibodies recognizing intact histamine were prepared by Schneider *et al.* (192) and incorporated in a commercial ELISA test kit. Recently, several commercial tests have been available. Commercial competitive direct ELISA immunoassay method was successfully used for the detection of histamine in cheese (193) and wines (194, 195). Simon Sarkadi *et al.* (196) applied competitive indirect ELISA method for detection of histamine in different foods (sauerkraut, cheese, fish, milk, and wine). The enzyme immunoassay seems to be a reliable technique for simple and rapid determination of histamine in food.

3.2.4 FORMATION

Biogenic amines are formed in food from their amino acid precursors by decarboxylation. The reaction requires the availability of free amino acids, the presence of decarboxylase-positive microorganisms, and conditions allowing bacterial growth and decarboxylase activity (16, 197). Free amino acids either occur as such in foods or may be liberated by proteolysis during processing or storage. Decarboxylase-producing microorganisms may be part of the associ-

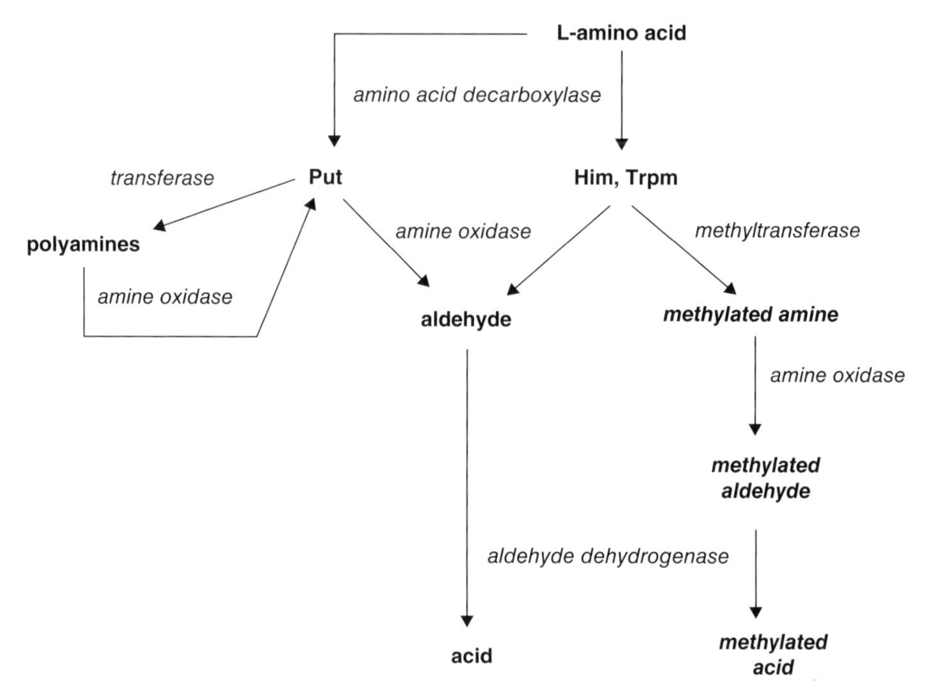

Figure 3.2.2 The metabolic pathways for formation and degradation of biogenic amines.

ated flora of a particular food or may be introduced by contamination before, during, or after processing the food. In the case of fermented foods and beverages, the applied starter cultures may also affect the production of biogenic amines.

Figure 3.2.2 shows the metabolic pathways for formation and degradation of biogenic amines. In the formation of all biogenic amines, there is a specific key enzyme with amino acid decarboxylase activity: aromatic amino acid decarboxylase for tryptamine, histidine decarboxylase (HDC) for histamine, arginine decarboxylase (ADC) for agmatine, and ornithine decarboxylase (ODC) for putrescine. In the degradation of these amines, there are some oxidoreduction reactions catalyzed by amine oxidases (MAO, monoamine oxidase; DAO, diamine oxidase) and aldehyde dehydrogenases. In the degradation of monoamines, but not in that of polyamines, there are methylation reactions using *S*-adenosylmethionine as the high-energy methyl donor (198).

There are two ways of histamine metabolism in the human body. The major process is when nitrogen in the imidazole cycle is methylated by histamine *N*-methyltransferase (HNMT) at the formation of *N*-methylhistamine, which is further oxidized by monoamino oxidase to *N*-methylimidazolylacetic acid.

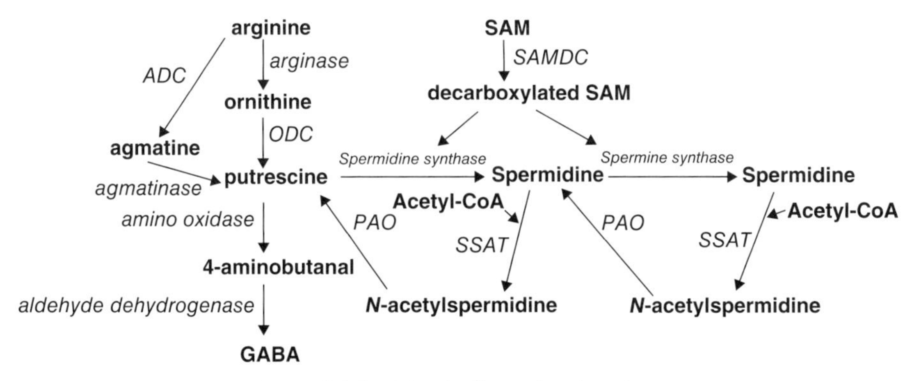

Figure 3.2.3 Metabolism of polyamines.

This enzyme is very selective for histamine detoxification and involves *S*-adenosylmethionine as donor of methyl group. The other possibility is the oxidation of histamine by DAO to imidazolylacetic acid (198).

A very specific feature of polyamine metabolism is their interconversion pathway, which enables the cells to transform one polyamine into another (Fig. 3.2.3).

Putrescine is a precursor for spermidine and spermine. The biosynthesis of putrescine involves the decarboxylation of either ornithine by ODC or arginine by ADC, and subsequent conversion of the agmatine into putrescine by agmatinase. Putrescine is then converted into spermidine and spermidine by sequential transfer of aminopropyl groups donated by decarboxylated *S*-adenosylmethionine, in reactions catalyzed by spermidine and spermine synthase, respectively. Decarboxylated *S*-adenosylmethionine is produced from *S*-adenosylmethionine (SAM) by the enzyme *S*-adenosylmethionine decarboxylase (SAMDC). In the return direction of the interconversion pathway, spermine and spermidine can be acetylated by spermidine/spermine N^1-acetyltransferase (SSAT) to produce compounds suitable for oxidation by polyamine oxidase (PAO), finally yielding spermidine from spermine and putrescine from spermidine (198).

GABA (γ-aminobutyric acid) is an intermediate of putrescine catabolism. Regulation of growth, differentiation, or cell maturation may be related to the biochemical equilibrium between putrescine and GABA.

3.2.5 MITIGATION

Numerous efforts have been made in food science and in the food industry to reduce or to prevent formation of biogenic amines in food. There are some trials to reduce biogenic amines in foods: handling and processing under sanitary conditions, utilizing of some amine-negative starter cultures, adding of some probiotic bacterial strains alone or in combination with the starter culture, high-pressure processing (HPP), or low-dose gamma irradiation.

Improved manufacturing practices have led to better hygiene and to substitution of cultures less likely to produce decarboxylation. Using high-quality fresh meat and GMPs greatly reduced the risk of biogenic amines formation in processed meat products (49, 50, 52, 199).

The presence of a selected starter culture or using mixed microbial strains is one of the ways to reduce the development of biogenic amines in fermented food (63, 112, 200, 201). Mixed starter cultures (*Lactobacillus sakei, Staphylococcus carnosus*, and *Staphylococcus xylosus*) greatly reduced (about 90%) the presence of putrescine, cadaverine, and tyramine in fermented Spanish sausages (202). A 50% decrease in biogenic amine content was observed also in sausages fermented by *Lactobacillus curvatus* CTC371 in association with a proteolytic strain of *S. xylosus* (203). In Turkish sausages, the addition of mixed starter cultures (*L. sakei, Pediococcus pentosaceus, S. xylosus*, and *S. carnosus*) avoided the formation of putrescine, but not of tyramine (64). Similar decreases in tyramine, cadaverine, and histamine concentration in other sausages were observed using amine-negative mixed (staphylococci plus lactobacilli) starter cultures (204).

High-pressure homogenization (HPH) is one of the most encouraging alternatives to traditional thermal treatment for food preservation. Its effectiveness in the deactivation of pathogenic and spoilage microorganisms in model and real systems is well documented (205–208).

Low dose of gamma irradiation has also been considered a useful method of preservation to extend the shelf life of chilled, stored fish and to reduce microbial population as well as biogenic amine formation in fish and fish products (209–212).

Vacuum or modified atmospheres seem to be successful treatment to reduce the biogenic amine concentration in fish and meat products (213–216).

Bodmer *et al.* (217) developed two methods for wine production that include low-histamine technology: Methode Schlumberger for sparkling wines and Método Estévez for sherry wines, named after two wineries in Austria and Spain, respectively.

3.2.6 EXPOSURE

The relationship between biogenic amines and human pathologies has been known for many years. Biogenic amines are involved in carcinogenesis and tumor invasion (ornithine-derived polyamines and histamine), allergy and immune response in general (histamine), and neurological disorders such as Parkinson's disease, Alzheimer's disease, and depression (serotonin, dopamine, histamine). Pathological disorders mainly related to histamine are cancer, oxidative stress, and allergy (218–222). Tyramine is involved in migraine, hypertension, Parkinson's disease, and depression (223). Polyamines are involved in neural injury, muscular dystrophy, Alzheimer's disease, and cancer (224–227).

The biogenic amine content of various foods has been widely studied because of their potential toxicity. Food poisoning may occur especially in conjunction with potentiating factors such as monoamine oxidase inhibiting (MAOI) drugs, alcohol, gastrointestinal diseases, and other amines. Only histamine has legal limits established for food but some countries have regulated the maximum amounts of other biogenic amines in different foods.

Tyramine and 2-phenylethylamine have been proposed as the initiators of hypertensive crisis in certain patients and of dietary-induced migraine. Aged cheeses are the most frequently reported food and most serious of the case reports. The phenomenon is the so-called cheese reaction caused by high amounts of tyramine in cheese. McCabe *et al.* recently published a critical review on tyramine in foods and MAOI drugs (227).

Table 3.2.9 summarizes the percentage of published values of foods by categories that have been found to contain clinically significant amounts of tyramine.

The presence of 6-mg tyramine in one or two usual servings is thought to be sufficient to cause a mild adverse event, while 10–25 mg will produce a severe adverse event in those using MAOI drugs (228, 229). For unmedicated adults, 200–800 mg of dietary tyramine is needed to induce a mild rise (30 mm Hg) in blood pressure (230).

Histamine intoxication, also called scombrotoxicosis, is still one of the most common seafood intoxications. Histamine poisoning is usually related to the consumption of scombroid fish such as tuna, mackerel, and sardines (231–235).

However, reliable statistics about its incidence do not exist because the poisoning incidents are often unreported because of the mild nature of the

TABLE 3.2.9 Percentage of food analyzed for tyramine that had clinically significant (>6 mg) quantities.

Food	Total number of published values	Number with clinical significant levels	
		Number	%
Cheese	71	18	25
Asian dishes (fermented)	37	12	30
Meat/meat products	74	20	27
Sauces (fish, shrimp, soy)	21	5	24
Tap beer	34	4	12

Data from Reference 227.

TABLE 3.2.10 Outbreaks of HFP in the United States, the United Kingdom, and Australia.

Country	Period	Number of outbreaks	Percentage of all seafood outbreaks	Total ill
United States	1990–2000	103	43	680
United Kingdom	1992–1999	47	—	—
Australia	1990–2000	10	31	28

Data from Reference 242.

illness, lack of adequate systems for reporting foodborne diseases, or ignorance by medical personnel who misdiagnose histamine poisoning as a food allergy (236, 237). Japan, the United States (238), and the United Kingdom (239) are the countries with the highest number of reported incidents, although this possibly reflects better reporting systems. Frequent incidents have been reported all over the world (240, 241). Table 3.2.10 shows the outbreaks of histamine fish poisoning (HFP) in some countries.

Putrescine, spermine, spermidine, and cadaverine have no adverse health effects, but they may react with nitrite to form carcinogenic nitrosoamines especially in meat products that contain nitrite and nitrate salts as curing agents (51, 243).

Determination of the exact toxicity threshold of biogenic amines in individuals is extremely difficult since the toxic dose strongly depends on the efficiency of the detoxification mechanisms of each individual.

3.2.7 HEALTH RISKS

The biogenic amine content of various foods has been widely studied because of their potential toxicity. Biogenic amines in foods are not necessarily hazardous. In healthy individuals, low amounts of biogenic amines are easily metabolized in the gut to physiologically less active degradation products during the food intake process. This intestinal detoxifying system includes specific enzymes such as MAO, DAO, and HNMT. However, upon intake of high amounts of biogenic amines in foods, the detoxification system is unable to eliminate these amines sufficiently or when it is genetically deficient (244). Moreover, the amine metabolism could be inhibited by simultaneous ingestion of alcoholic beverages, certain drugs, or MAO inhibitors.

The biogenic amines are categorized primarily as either vasoactive or psychoactive. The physiological effects of vasoactive amines (mainly tyramine) include peripheral vasoconstriction, increased cardiac output, increased respiration, elevated blood glucose, and release of norepinephrine. Histamine poisoning symptoms include headache, nausea, vomiting, diarrhea, itching, oral burning sensation, red rash, and hypotension (14). Polyamines, such as putrescine, cadaverine, spermidine, and spermine, although not exerting a direct toxic effect, can potentiate the toxic effects of tyramine and histamine by competing for the detoxifying enzymes (16), and act as precursors of carcinogenic nitrosoamines (243). Furthermore, some authors (13, 245) have reported that diets low in polyamines can have beneficial effects in reducing tumor growth.

3.2.8 RISK MANAGEMENT

The primary goal of risk management associated with food is to protect public health by controlling all risks as effectively as possible through the selection and implementation of appropriate measures. Risk management decisions should take into account the whole food chain from primary production to consumption. In the food processing chain, managing risks should be based on scientific knowledge of the microbiological hazards and an understanding of the primary production, processing and manufacturing technologies, handling during food preparation, storage, transport, retail, and catering.

Hazard identification is the determination of whether or not a specific constituent in food is causally linked to particular health effects. Hazard is usually determined experimentally in controlled toxicology studies with known doses or exposures to the toxic agent under study. Statistical considerations have resulted in the use of a maximum tolerated dose (MTD), the highest practical dose that can be administered, in most studies carried out in laboratory animals.

In the specific context of a food safety assessment, the World Health Organization (2000) has defined hazard as a biological, chemical, or physical agent in, or condition of, food with the potential to cause an adverse health effect.

The main European regulations, which include different aspects of the management of microbial risks as well as strategies for control and prevention of risks for consumer protection, are as follows (246):

—Regulation (EC) No. 852/2004 of the European Parliament (EUP) and of the Council on the hygiene of foodstuffs;

—Regulation (EC) No. 853/2004 of the EUP and of the Council laying down specific hygiene rules for food of animal origin;

—Regulation (EC) No. 854/2004 laying down specific rules for the organization of official controls on products of animal origin intended for human consumption;

—Regulation (EC) No. 178/2002 laying down the general principles and requirements of food law, establishing the European Food Safety Authority and laying down procedures in matters of food safety;

—Regulation (EC) No. 882/2004 on official controls performed to ensure the verification of compliance with feeding and food law, animal health, and animal welfare;

—Commission Regulation (EC) No. 2073/2005 on microbiological criteria for foodstuffs.

Risk assessment on biogenic amines is difficult because knowledge concerning the acute and especially long-term toxic effects of amines and dose–response is incomplete. However, it is well known that some types of foods may pose a risk on sensitive persons and for patients using MAO- and DAO-inhibiting drugs.

Shalaby (240) reviewed the oral toxicity of histamine and other biogenic amines in foods for humans. He considered that histamine-induced poisoning is, in general, slight at ≤40 mg, moderate at >40 mg and severe at >100 mg. Based on an analysis of recent poisoning episodes, the following guidelines for histamine content of fish have been suggested:

safe for consumption	<5 mg/100 g
potentially toxic	5–20 mg/100 g
probably toxic	20–100 mg/100 g
toxic	>100 mg/100 g

In the United Kingdom, guidelines for histamine content in fish are as follows (239):

safe for consumption	<10 mg/100 g
potentially toxic	10–50 mg/100 g
probably toxic	50–100 mg/100 g
toxic	>100 mg/100 g

The US FDA guidelines established for tuna, mahimahi, and related fish, specify 50 mg/100 g as the toxicity concentration and 5 mg/100 g as the defect action concentration because histamine is not uniformly distributed in fish that has undergone temperature abuse. Therefore, if 5 mg/100 g is found in one section, there is a possibility that other units may exceed 50 mg/100 g (247). FDA requires the use of the AOAC fluorometric method (248).

In Australia and New Zealand, the concentration of histamine in a composite sample of fish or fish products, other than crustaceans and mollusks, must not exceed 20 mg/10 g. A composite sample is a "sample taken from each lot, comprising five portions of equal mass from five representative samples."

The European Union requires that nine samples are to be taken from each batch of fish species of the following families: Scombridae, Clupeidae, Engraulidae, and Coryphaenidae. These samples must fulfill the following requirements: mean value of all samples must not exceed <10 mg/100 g; two samples may be >10 mg/100 but <20 mg/100; and no sample may exceed 20 mg/100.

Examinations must be carried out in accordance with reliable, scientifically recognized methods such as HPLC (17).

Risk management is very important also in winemaking. Reviewing current practices in grape and wine processing regarding risk assessment, one cannot find a well-documented approach in this area. The majority of wine production involves three main production concepts. These are conventional (current), sustainable, and modern (use of genetically modified organisms) applications (249). The most extensive production technique is the conventional technique, which is the classic method with some modifications and improvements including yeast usage (250), bacterial culture usage for malolactic fermentation (251), enzyme usage (252), and fermentation in controlled conditions (253). Organic and biodynamic viticulture are rare where chemicals such as fertilizers, pesticides, sulfur dioxide, and sorbic acid are reduced or excluded in the production process.

Plahuta and Raspor (254) compared the hazards to human health and the environment from six wines that have been produced by different viticulture and winemaking technology such as conventional, organic, integrated production of grape and wine (IPGW), biodynamic, and two methods using genetically modified organisms (GFLV; grapevine fan leaf virus-resistant and polysaccharide degrading genetically modified wine yeast). They established medium risk regarding biogenic amines in the case of organic and biodynamic wines, and low risk in the case of the other investigated wines.

3.2.9 CONCLUSION

Fermented foods and beverages usually contain significant amounts of biogenic amines that are present as a result of food processing or storage. Since some of these components, especially histamine and tyramine, have deleterious effects on humans, it is important to minimize exposure to these amines in food. The evidence of numerous undesirable reactions after intake of histamine and/or tyramine-containing foods represents a challenge for the food industry to produce foods with histamine and tyramine concentrations as low as possible.

For the assessment of the risk of orally ingested biogenic amines, one should take a number of factors into consideration; besides the use of alcohol and specific medicine also the concentration of histamine and/or tyramine in a specific food, the amounts of that food consumed, the presence of other amines in that food, and the amine content of other dietary components. Alcohol and medicine may amplify the biological effect of histamine and/or tyramine by inhibiting the catabolism of amines.

The reviewed data support the view that it is important to set up a database on biogenic amine content of different foods at the European level in order to make recommendations for their concentrations in different diets. Furthermore, products with a risk of high histamine content should be labeled to prevent increased adverse reactions in persons with suspected histamine intolerance.

REFERENCES

1. van Leeuwenhoek, A. (1678). Observationes de natis e semine genitali animalculis. *Philosophical Transactions of the Royal Society London*, *12*, 1040–1043.
2. Ladenburg, A., Abel, J. (1888). Ueber das Aethylenimin (Spermin?). *Berichte der Deutschen Chemischen Gesellschaft*, *21*, 758–766.
3. Dudley, H.W., Rosenheim, O., Starling, W.W. (1926). The chemical constitution of spermine. III. Structure and synthesis. *Biochemical Journal*, *20* (5), 1082–1094.
4. Brieger, L. (1885). *Ueber Ptomaine*, August Hirschwald, Berlin.
5. Ladenburg, A. (1886). Über die Identität des Cadaverin mit dem Pentamethyldiamin. *Berichte der Deutschen Chemischen Gesellschaft*, *19*, 2585–2586.
6. Kossel, A. (1910). Ueber das Agmatin. *Zeitschrift fuer Physiologische Chemie*, *66*, 257–261.
7. Li, G., Regunathan, S., Barrow, C.J., Eshraghi, J., Cooper, R., Reis, D.J. (1994). Agmatine: an endogenous clonidine-displacing substance in the brain. *Science*, *263*, 966–969.
8. Barger, G., Dale, H.H. (1910). 4-β-Aminoethylglyoxaline (β-iminazolylethylamine) and the other active principles of ergot. *Journal of the Chemical Society*, *97*, 2592–2595.
9. Dale, H.H., Laidlaw, P.P. (1910). The physiological action of β-imidazolethylamine. *Journal of Physiology*, *41*, 318–344.
10. Windaus, A., Vogt, W. (1907). Synthese des imidazolaethylamins. *Chemische Berichte*, *40*, 3691–3694.
11. Henze, M. (1913). *p*-Oxyphenylaethylamin, das speicheldruesengift der cephalopoden. *Zeitschrift fuer Physiologische Chemie*, *87*, 51.
12. Bardocz, S., Grant, G., Brown, D.S., Ralph, A., Pusztai, A. (1993). Polyamines in food-implications for growth and health. *Journal of Nutritional Biochemistry*, *4*, 66–71.
13. Bardocz, S. (1995). Polyamines in food and their consequences for food quality and human health. *Trends in Food Science and Technology*, *6*, 341–346.
14. Beutling, D.M. (1996). *Biogene Amine in der Ernaehrung*, Springer Verlag, Berlin, Heidelberg, New York.
15. Bover-Cid, S., Holzapfel, W. (2000). Biogenic amine production by bacteria, in *COST 917 Biogenically Active Amines in Food*, Vol. 4 (eds D.M.L. Morgan, A. White, F. Sánchez-Jiménez, S. Bardócz), EC Publication, Luxembourg, pp. 20–29.
16. Halasz, A., Barath, A., Simon Sarkadi, L., Holzapfel, W.H. (1994). Biogenic amines and their production by micro-organisms in food. *Trends in Food Science and Technology*, *5*, 42–49.
17. European Council Directive (1991). *91/493/EEC The health conditions for the production and the placing on the market of fishery products.*
18. Brink, B. ten, Damink, C., Joosten, H.M.L.J. Huis, in't Veld, J.H.J. (1990). Occurrence and formation of biologically active amines in foods. *International Journal of Food Microbiology*, *11*, 73–84.

19. Guggenheim, M. (1920). *Die Biogenen Amine und ihre Bedeutung für die Physiologie und Pathologie des pflanzlichen und tierischen Stoffwechsels*, Springer, Berlin.
20. Stratton, J.E., Hutkins, R.W., Taylor, S.L. (1991). Biogenic amines in cheese and other fermented foods: a review. *Journal of Food Protection*, *54*, 460–470.
21. Lehtonen, P. (1996). Determination of amines and amino acids in wine—A review. *American Journal of Enology and Viticulture*, *47*, 127–133.
22. Suzzi, G., Gardini, F. (2003). Biogenic amines in dry fermented sausages: a review. *International Journal of Food Microbiology*, *88*, 41–54.
23. Falus, A., Grosman, N., Darvas, Zs. (eds) (2004). *Histamine Biology*, SpringMed Publishing Ltd., Hungary; S. Karger AG, Switzerland.
24. Kalac, P., Krausova, P. (2005). A review of dietary polyamines: formation, implications for growth and health and occurrence in foods. *Food Chemistry*, *90*, 219–230.
25. Kalac, P. (2006). Biologically active polyamines in beef, pork and meat products: a review. *Meat Science*, *73*, 1–11.
26. Onal, A. (2007). A review: current analytical methods for the determination of biogenic amines in foods. *Food Chemistry*, *103*, 1475–1486.
27. Simon Sarkadi, L. (2005). Biogenic amines in fermented vegetables and dairy products, in *COST 917 Biogenically Active Amines in Food*, Vol. *7* (eds D.M.L. Morgan, F. Bauer, A. White), EC Publication, Luxembourg, pp. 210–215.
28. Clasadonte, M.T., Zerbo, A., Cuccia, T. (1995). Biogenic amines variations in Sicilian pecorino during the ripening process. *Industrie Alimentari*, *34*, 599–603.
29. El Sayed, M.M. (1996). Biogenic amines in processed cheese available in Egypt. *International Dairy Journal*, *6*, 1079–1086.
30. Petridis, K.D., Steinhart, H. (1996). Biogenic amines in hard cheese production: 1. Factors influencing the biogenic amine content of the finished product using Emmental cheese as an example. *Deutsche Lebensmittel-Rundschau*, *92* (4), 114–120.
31. Vale, S., Gloria, B.A. (1998). Biogenic amines in Brazilian cheeses. *Food Chemistry*, *63*, 343–348.
32. Simon Sarkadi, L., Hodosi, E. (1998). Formation of biogenic amine during cheese processing, in *COST 917 Biogenically Active Amines in Food*, Vol. *2* (eds S. Bardócz, A. White, Gy. Hajós), EC Publication, Luxembourg, pp. 8–10.
33. Valsamaki, K., Michaelidou, A., Polychroniadou, A. (2000). Biogenic amine production in Feta cheese. *Food Chemistry*, *71*, 259–266.
34. Fernandez Garcia, E., Tomillo, J., Nunez, M. (2000). Formation of biogenic amines in raw milk Hispanico cheese manufactured with proteinases and different levels of starter culture. *Journal of Food Protection*, *11*, 1551–1555.
35. Innocente, N., D'Agostini, P. (2002). Formation of biogenic amines in a typical semi hard Italian cheese. *Journal of Food Protection*, *65*, 1498–1501.
36. Durlu Ozkaya, F. (2002). Biogenic amine content of some Turkish cheeses. *Journal of Food Processing and Preservation*, *26*, 259–265.
37. Novella Rodriguez, S., Veciana-Nogues, M.T., Izquierdo Pulido, M., Vidal Carou, M.C. (2003). Distribution of biogenic amines and polyamines in cheese. *Journal of Food Science*, *3*, 750–755.

38. Martuscelli, M., Gardini, F., Torriani, S., Mastrocola, D., Serio, A., Chaves-Lopez, C., Schirone, M., Suzzi, G. (2005). Production of biogenic amines during the ripening of Pecorino Abruzzese cheese. *International Dairy Journal*, *15*, 571–578.
39. Komprda, T., Smela, D., Novicka, K., Kalhotka, L., Sustova, K., Pechova, P. (2007). Content and distribution of biogenic amines in Dutch-type hard cheese. *Food Chemistry*, *102*, 129–137.
40. Jarish, R. (2000). Histamine intolerance (HIT)—an overlooked disease, in *COST 917 Biogenically Active Amines in Food*, Vol. 4 (eds D.M.L. Morgan, A. White, F. Sánchez-Jiménez, S. Bardócz), EC Publication, Luxembourg, pp. 30–34.
41. Schneller, R., Good, P., Jenny, M. (1997). Influence of pasteurized milk, raw milk and different ripening cultures on biogenic amine concentrations in semi-soft cheeses during ripening. *Zeitschrift für Lebensmittel Untersuchung und Forschung A*, *204*, 265–272.
42. Ordonez, A.I., Ibanez, F.C., Torre, P., Barcina, Y. (1997). Formation of biogenic amines in Idiazabal ewe's milk cheese, effect of ripening, pasteurisation and starter. *Journal of Food Science*, *60*, 1371–1375.
43. Novella Rodriguez, S., Veciana-Nogues, M.T., Trujillo Mesa, A.J., Vidal Carou, M.C. (2002). Profile of biogenic amines in goat cheese made from pasteurized and pressurized milks. *Journal of Food Science*, *67*, 2940–2944.
44. Novella Rodriguez, S., Veciana-Nogues, M.T., Roig Sagues, A.X., Trujillo Mesa, A.J., Vidal Carou, M.C. (2004). Comparison of biogenic amine profile in cheeses manufactured from fresh and stored (4 °C, 48 hours) raw goat's milk. *Journal of Food Protection*, *67*, 110–116.
45. Pinho, O., Pintado, A.I.E., Gomes, A.M.P., Pintado, M.M.E., Malcata, F.X., Ferreira, I.M.P.L.V.O. (2004). Interrelationships among microbiological physicochemical, and biochemical properties of Terincho cheese, with emphasis on biogenic amines. *Journal of Food Protection*, *67*, 2779–2785.
46. Simon Sarkadi, L., Kiss, K. (2006). Comparative analysis of the biogenic amines in cheeses. *Élelmezési Ipar*, *60* (6, 7), 168–172.
47. Guzel Seydim, Z.B., Seydim, A.C., Greene, A.K. (2003). Comparison of amino acid profiles of milk, yoghurt, and Turkish kefir. *Milchwissenschaft, Milk Science International*, *58*, 155–160.
48. Ammor, M.S., Mayo, B. (2007). Selection criteria for lactic acid bacteria to be used as functional starter cultures in dry sausage production: an update. *Meat Science*, *76*, 138–146.
49. Bover-Cid, S., Izquierdo Pulido, M., Vidal Carou, M.C. (2000). Influence of hygienic quality of raw materials on biogenic amine production during ripening and storage of dry fermented sausages. *Journal of Food Protection*, *63* (11), 1544–1550.
50. Bover-Cid, S., Miguelez Arrizado, M.S., Latorre Moratalla, L.L., Vidal Carou, M. C. (2006). Freezing of meat raw materials affects tyramine and diamine accumulation in spontaneously fermented sausages. *Meat Science*, *72*, 62–68.
51. Hernandez Jover, T., Izquierdo Pulido, M., Veciana-Nogues, M.T., Marine Font, A., Vidal Carou, M.C. (1997). Biogenic amine and polyamine contents in meat and meat products. *Journal of Agricultural and Food Chemistry*, *45*, 2098–2102.
52. Bauer, F. (2006). Assessment of process quality by examination of the final product. 1. Assessment of the raw material. *Fleischwirtschaft*, *86* (7), 106–107.

53. Vinci, G., Antonelli, M.L. (2002). Biogenic amines: quality index of freshness in red and white meat. *Food Control*, *13*, 519–524.

54. Silva, C.M.G., Gloria, M.B.A. (2002). Bioactive amines in chicken breast and thigh after slaughter and during storage at 4 °C and in chicken based meat products. *Food Chemistry*, *78*, 241–248.

55. Hernandez Jover, T., Izquierdo Pulido, M., Veciana-Nogues, M.T., Vidal Carou, M.C. (1996). Biogenic amine sources in cooked cured shoulder pork. *Journal of Agricultural and Food Chemistry*, *44*, 3097–3101.

56. Roig Sagues, A.X., Hernandez Herrero, M., Lopez Sabater, E.I., Rodriguez Jerez, J.J., Mora Ventura, M.T. (1999). Microbiological events during the elaboration of "fuet", a Spanish ripened sausages. Relationships between the development of histidine- and tyrosine-decarboxylase containing bacteria and pH and water activity. *European Food Research and Technology*, *209*, 108–112.

57. Paulsen, P., Bauer, F. (1999). The formation of biogenic amines during maturation of Austrian fermented sausage. *Ernaehrung*, *23* (2), 61–63.

58. Senoz, B., Isikli, N., Coksoyler, N. (2000). Biogenic amines in Turkish sausages (Sucuk). *Journal of Food Science*, *65*, 764–767.

59. Parente, E., Martuscelli, M., Gardini, F., Grieco, S., Crudele, M.A., Suzzi, G. (2001). Evolution of microbial populations and biogenic amine production in dry sausages produced in Southern Italy. *Journal of Applied Microbiology*, *90*, 882–891.

60. Cvrtila, Z., Kozacinski, L. (2003). The histamine in meat and meat products. *Meso*, *5* (5), 42–45.

61. Morgan, D.M.L., Bauer, F., White, A. (eds) (2005). *COST 917 Biogenically Active Amines in Food*, Vol. 7, EC Publication, Luxembourg, pp. 247–320.

62. Eerola, S., Maijala, R., Roig Sagues, A.X., Salminen, M., Hirvi, T. (1996). Biogenic amines in dry sausages as affected by starter culture and contaminant amine-positive lactobacillus. *Journal of Food Science*, *61*, 1243–1246.

63. Bover-Cid, S., Izquierdo Pulido, M., Vidal Carou, M.C. (1999). Effect of proteolytic starter cultures of Staphylococcus spp. on biogenic amine formation during the ripening of dry fermented sausages. *International Journal of Food Microbiology*, *46* (2), 95–104.

64. Ayhan, K., Kolsarici, N., Ozkan, G.A. (1999). The effects of a starter culture on the formation of biogenic amines in Turkish soudjoucks. *International Journal of Food Microbiology*, *53*, 183–188.

65. Gardini, F., Martuscelli, M., Crudele, M.A., Paparella, A., Suzzi, G. (2002). Use of *Staphylococcus xylosus* as a starter culture in dried sausages: effect on the biogenic amine content. *Meat Science*, *61*, 275–283.

66. Bover-Cid, S., Schoppen, S., Izquierdo Pulido, M., Vidal Carou, M.C. (1999). Relationship between biogenic amine contents and the size of dry fermented sausages. *Meat Science*, *51*, 305–311.

67. Komprda, T., Smela, D., Pechova, P., Kalhotka, L., Stencl, J., Klejdus, B. (2004). Effect of starter culture, spice mix and storage time and temperature on biogenic amine content of dry fermented sausages. *Meat Science*, *67*, 607–616.

68. Trevino, E., Beil, D., Steinhart, H. (1997). Formation of bioorganic amines during the maturity process of raw meat products, for example of cervelat sausages. *Food Chemistry*, *60*, 521–526.

69. Tsai, Y.H., Kung, H.F., Lee, T.M., Chen, H.C., Chou, S.S., Wei, C.I., Hwang, D.F. (2005). Determination of histamine in canned mackerel implicated in a food borne poisoning. *Food Control*, *16*, 579–585.
70. Ababouch, L., Afilal, M.E., Benabdeljelil, H., Busta, F.F. (1991). Quantitative changes in bacteria, amino acids and biogenic amines in sardine (*Sardina pilchardus*) stored at ambient temperature (25–28 °C) and in ice. *International Journal of Food Science and Technology*, *26*, 297–306.
71. Feldhusen, F., Josefowski, B., Helle, N. (1999). Biogenic amines in fish and fish products with regard to technological aspects, in *COST 917 Biogenically Active Amines in Food*, Vol. *3* (eds S. Bardócz, J. Koninkx, M. Grillo, A. White), EC Publication, Luxembourg, pp. 45–49.
72. Krizek, M., Pavlicek, T., Vacha, F. (2002). Formation of selected biogenic amines in carp meat. *Journal of the Science of Food and Agriculture*, *82*, 1088–1093.
73. Krizek, M., Vacha, F., Vorlova, L., Lukasova, J., Cupakova, S. (2004). Biogenic amines in vacuum-packed and non-vacuum-packed flesh of carp (*Cyprinus carpio*) stored at different temperatures. *Food Chemistry*, *88* (2), 185–191.
74. Lehane, L., Olley, J. (2000). Histamine fish poisoning revisited. *International Journal of Food Microbiology*, *58*, 1–37.
75. Jorgensen, L.V., Huss, H.H., Dalgaard, P. (2000). The effect of biogenic amine production by single bacterial cultures and metabiosis on cold-smoked salmon. *Journal of Applied Microbiology*, *89*, 920–934.
76. Hernandez-Herrero, M.M.; Duflos, G.; Malle, P.; Bouquelet, S. (2002). Amino acid decarboxylase activity and other chemical characteristics as related to freshness loss in iced cod (*Gadus morhua*). *Journal of Food Protection*, *65* (7), 1152–1157.
77. Mietz, J.L., Karmas, E. (1977). Chemical quality index of canned tuna as determined by high-pressure liquid chromatography. *Journal of Food Science*, *42*, 155–158.
78. Karmas, E. (1981). Biogenic amines as indicator of seafood freshness. *Lebensmittel Wissenschaft und Technologie*, *14*, 273–275.
79. Veciana-Nogues, M.T., Marine Font, A., Vidal Carou, M.C. (1997). Biogenic amines as hygienic quality indicators of tuna. Relationships with microbial counts, ATP-related compounds, volatile amines, and organoleptic changes. *Journal of Agricultural and Food Chemistry*, *45*, 2036–2041.
80. Brinker, B., Helle, N., Feldhusen, F. (2005). Biogenic amines in fishery products, in *COST 917 Biogenically Active Amines in Food*, Vol. 7 (eds D.M.L. Morgan, F. Bauer, A. White), EC Publication, Luxembourg, pp. 176–180.
81. Nordic Council of Ministers (2002). *Present Status of Biogenic Amines in Foods in Nordic Countries*, Nordic Council of Ministers, ISBN 92-893-0773-0.
82. Soulferos, E., Marie Lyse, B., Bertrand, A. (1998). Correlation between the content of biogenic amines and other wine compounds. *American Journal of Enology and Viticulture*, *49*, 266–277.
83. Hajos, G., Sass Kiss, A., Szerdahelyi, E., Bardocz, S. (2000). Changes in biogenic amine content of Tokaj grapes, wines, and aszu-wines. *Journal of Food Science*, *65*, 1142–1144.
84. Leitao, M.C., Teixeira, H.C., Crespo, M.T.B., San Romao, M.V. (2000). Biogenic amines occurrence in wine amino acid decarboxylase and proteolytic activities

expression by *Oenococcus oeni*. *Journal of Agricultural and Food Chemistry*, *48*, 2780–2784.

85. Caruso, M., Fiore, C., Contursi, M., Salzano, G., Paparella, A., Romano, P. (2002). Formation of biogenic amines as criteria for the selection of wine yeasts. *World Journal of Microbiology and Biotechnology*, *18*, 159–163.
86. Csomos, E., Heberger, K., Simon Sarkadi, L. (2002). Principal component analysis of biogenic amines and polyphenols in Hungarian wines. *Journal of Agricultural and Food Chemistry*, *50* (13), 3768–3774.
87. Arozarena, I., Casp, A., Martín, R., Navarro, M. (2000). Multivariate differentiation of Spanish red wines according to region and variety. *Journal of the Science of Food and Agriculture*, *80*, 1909–1917.
88. Simon Sarkadi, L., Csomos, E. (2002). Free amino acid and biogenic amine contents of Hungarian wines. *Polish Journal of Food and Nutrition Sciences*, *11/52*, (SI 2), 106–110.
89. Torrea, D., Ancin, C. (2002). Content of biogenic amines in a Chardonnay wine obtained through spontaneous and inoculated fermentations. *Journal of Agricultural and Food Chemistry*, *50*, 4895–4899.
90. Heberger, K., Csomos, E., Simon Sarkadi, L. (2003). Principal component and linear discriminant analyses of free amino acids and biogenic amines in Hungarian wines. *Journal of Agricultural and Food Chemistry*, *51* (27), 8055–8060.
91. Landete, J.M., Ferrer, S., Polo, L., Pardo, I. (2005). Biogenic amines in wines from three Spanish regions. *Journal of Agriculture and Food Chemistry*, *53*, 119–1124.
92. Bover-Cid, S., Izquierdo Pulido, M., Marine Font, A., Vidal Carou, M.C. (2006). Biogenic mono-, di- and polyamine contents in Spanish wines and influence of a limited irrigation. *Food Chemistry*, *96*, 43–47.
93. Le Jeune, C., Lonvaud Funel, A., ten Brink, B., Hofstra, H., van der Vossen, J.M.B.M. (1995). Development of a detection system for histidine decarboxylating lactic acid bacteria on DNA probes, PCR and activity test. *Journal of Applied Bacteriology*, *78*, 316–326.
94. Moreno Arribas, M.V., Polo, M.C., Jorganes, F., Munoz, R. (2003). Screening of biogenic amine production by lactic acid bacteria isolated from grape must and wine. *International Journal of Food Microbiology*, *84*, 117–123.
95. Landete, J.M., Ferrer, S., Pardo, I. (2005). Which are the lactic acid bacteria responsible of histamine production in wine? *Journal of Applied Microbiology*, *99*, 580–586.
96. Kalac, P., Krizeck, M. (2003). A review of biogenic amines and polyamines in beer. *Journal of the Institute of Brewing*, *109*, 123–128.
97. Romero, R., Bagur, M.G., Sanchez Vinas, M., Gazquez, D. (2003). The influence of the brewing process on the formation of biogenic amines in beers. *Analytical and Bioanalytical Chemistry*, *376* (2), 162–167.
98. Izquierdo Pulido, M., Hernandez Jover, T., Marine Font, A., Vidal Carou, M.C. (1996). Biogenic amines in European beers. *Journal of Agricultural and Food Chemistry*, *44*, 3159–3163.
99. Izquierdo Pulido, M., Albala Hurtado, S., Marine Font, A., Vidal Carou, M.C. (1996). Biogenic amines in Spanish beers: differences among breweries. *Zeitschrift für Lebensmittel Untersuchung und Forschung*, *203*, 507–511.

100. Kalac, P., Hlavata, V., Krizeck, M. (1997). Concentrations of five biogenic amines in Czech beers and factors affecting their formation. *Food Chemistry*, *58*, 209–214.

101. Gloria, M.B.A., Izquierdo Pulido, M. (1999). Levels and significance of biogenic amines in Brazilian beers. *Journal of Food Composition and Analysis*, *12*, 129–136.

102. Halasz, A., Barath, A., Holzapfel, W.H. (1999). The influence of starter culture selection on sauerkraut fermentation. *Zeitschrift für Lebensmittel Untersuchung und Forschung*, *208*, 434–438.

103. Simon Sarkadi, L., Kovács, A., Hodosi, E., Holzapfel, W.H. (1999). Biogenic amine content of foods and its relation to microbial activity and food quality, in *COST 917 Biogenically Active Amines in Food*, Vol. *3* (eds S. Bardócz, J. Koninkx, M. Grillo, A. White), EC Publication, Luxembourg, pp. 14–19.

104. Izquierdo Pulido, M., Marine Font, A., Vidal Carou, M.C. (2000). Effect of tyrosine and tyramine formation during beer fermentation. *Food Chemistry*, *70*, 329–332.

105. Slomkowska, A., Ambroziak, W. (2002). Biogenic amine profile of the most popular Polish beers. *European Food Research and Technology*, *215*, 380–383.

106. Kalac, P., Savel, J., Krizek, M., Pelikanova, T., Prokopova, M. (2002). Biogenic amine formation in bottled beer. *Food Chemistry*, *79* (4), 431–434.

107. Loret, S., Deloyer, P., Dandrifosse, G. (2005). Levels of biogenic amines as a measure of the quality of the beer fermentation process: data from Belgian samples. *Food Chemistry*, *89*, 519–525.

108. Kalac, P., Spicka, J., Krizek, M., Steidlova, S., Pelikanova, T. (1999). Concentrations of seven biogenic amines in sauerkraut. *Food Chemistry*, *67*, 275–280.

109. Halasz, A., Barath, A., Holzapfel, W.H. (1999). The influence of starter culture selection on sauerkraut fermentation. *European Food Research and Technology*, *208*, 434–438.

110. Kalac, P., Spicka, J., Krizek, M., Pelikanova, T. (2000). Changes in biogenic amines concentrations during sauerkraut storage. *Food Chemistry*, *69*, 309–314.

111. Kalac, P., Spicka, J., Krizek, M., Pelikanova, T. (2000). The effects of lactic acid bacteria inoculants on biogenic amines formation in sauerkraut. *Food Chemistry*, *70*, 355–359.

112. Spicka, J., Kalac, P., Bover-Cid, S., Krizek, M. (2002). Application of lactic acid bacteria starter cultures for decreasing the biogenic amine levels in sauerkraut. *European Food Research and Technology*, *215*, 509–514.

113. Ansorena, D., Montel, M.C., Rokka, M., Talon, R., Eorola, S., Rizzo, A., Raemackers, M., Demeyer, D. (2002). Analysis of biogenic amines in northern and southern European sausages and role of flora amine production. *Meat Science*, *61*, 141–147.

114. Nout, M.J.R. (1994). Fermented foods and food safety. *Food Research International*, *27*, 291–294.

115. Künsch, U., Scharer, H., Temperli, A. (1989). *Biogene Amine als Qualitaetsindikatoren von Sauerkraut*. In XXIV Vortragstagung der Deutschen Gesellschaft fur Qualitaetsforschung, Qualitaetsaspekte von Obst und Gemuse, Ahrensburg, Germany. Vortragstagung Deutsch Gesellschaft Qualitatsforsch, Qualitatsaspekte Obst Gemuse, Ahrendburg, Germany.

116. Eerola, S., Hinkkanen, R., Lindfors, E., Hirvi, T. (1993). Liquid chromatographic determination of biogenic amines in dry sausages. *Journal of AOAC International*, *76* (3), 575–577.
117. Cinquina, A.L., Cali, A., Longo, F., De Santis, L., Severoni, A., Abballe, F. (2004). Determination of biogenic amines in fish tissues by ion-exchange chromatography with conductivity detection. *Journal of Chromatography, A*, *1032*, 73–77.
118. Gosetti, F., Mazzucco, E., Gianotti, V., Polati, S., Gennaro, M.C. (2007). High performance liquid chromatography/tandem mass spectrometry determination of biogenic amines in typical Piedmont cheeses. *Journal of Chromatography, A*, *1149*, 151–157.
119. Bonneau, L., Carre, M., Martin-Tanguy, J. (1994). Polyamine metabolism during seedling development in rice. *Plant Growth Regulation*, *15*, 83–92.
120. Moret, S., Bortolomeazzi, R., Lercker, G. (1992). Improvement of extraction procedure for biogenic amines in foods and their high-performance liquid chromatographic determination. *Journal of Chromatography*, *591*, 175–180.
121. Pemberton, I.J., Smith, G.R., Forbes, T.D., Hensarling, C.M. (1993). Technical note: an improved method for extraction and quantification of toxic phenethylamines from *Acacia berlandieri*. *Journal of Animal Science*, *71* (2), 467–470.
122. Busto, O., Mestres, M., Guasch, J., Borrull, F. (1995). Determination of biogenic amines in wine after clean-up by solid-phase extraction. *Chromatographia*, *40*, 404–410.
123. Busto, O., Minacle, M., Guasch, J., Borrull, F. (1997). Solid phase extraction of biogenic amines from wine before chromatographic analysis of their AQC derivatives. *Journal of Liquid Chromatography and Related Technologies*, *20*, 743–755.
124. Calbiani, F., Careri, M., Elviri, L., Mangia, A., Pistara, L., Zagnoni, I. (2005). Rapid assay for analyzing biogenic amines in cheese: matrix solid-phase dispersion followed by liquid chromatography-electrospray-tandem mass spectrometry. *Journal of Agricultural and Food Chemistry*, *53*, 3779–3783.
125. Simon Sarkadi, L., Kovács, A. (2002). Biogenic amine determination in food by different techniques. *G.I.T. Laboratory Journal*, *6*, 11–13.
126. Slocum, R.D., Flores, H.E. (1991) *Biochemistry and Physiology of Polyamines in Plants*, CRC Press, Inc., Boca Raton, FL.
127. Hernandez Jover, T., Izquierdo Pulido, M., Veciana-Nogues, M.T., Vidal Carou, M.C. (1996). Ion pair liquid chromatographic determination of biogenic amines in meat and meat products. *Journal of Agricultural and Food Chemistry*, *44*, 2710–2715.
128. Hlabangana, L., Hernandez Cassou, S., Saurina, J. (2006). Determination of biogenic amines in wines by ion-pair liquid chromatography and post-column derivatization with 1,2-naphthoquinone-4-sulphonate. *Journal of Chromatography, A*, *1130*, 130–136.
129. Lavizzari, T., Veciana-Nogues, M.T., Bover-Cid, S., Marine Font, A., Vidal Carou, M.C. (2006). Improved method for the determination of biogenic amines and polyamines in vegetable products by ion-pair high-performance liquid chromatography. *Journal of Chromatography, A*, *1129*, 67–72.
130. Busto, O., Miracle, M., Guasch, J., Borrull, F. (1997). Determination of biogenic amines in wines by high-performance liquid chromatography with on-column fluorescence derivatization. *Journal of Chromatography, A*, *757*, 311–318.

131. Sass Kiss, A., Szerdahelyi, E., Hajos, G. (2000). Study of biologically active amines in grapes and wines by HPLC. *Chromatographia*, *51*, S316.

132. Loukou, Z., Zotou, A. (2003). Determination of biogenic amines as dansyl derivatives in alcoholic beverages by high-performance liquid chromatography with fluorimetric detection and characterization of the dansylated amines by liquid chromatography–atmospheric pressure chemical ionization mass spectrometry. *Journal of Chromatography, A*, *996*, 103–113.

133. Mafra, I., Herbert, P., Santos, L., Barros, P., Alves, A. (1999). Evaluation of biogenic amines in some Portuguese quality wines by HPLC fluorescence detection of OPA derivates. *American Journal of Enology and Viticulture*, *50*(1), 128–132.

134. Smela, D., Pechova, P., Komprda, T., Klejdus, B., Kuban, V. (2003). Liquid chromatographic determination of biogenic amines in a meat product during fermentation and long-term storage. *Czech Journal of Food Science*, *21*, 167–175.

135. Romero, R., Gazquez, D., Bagur, M.G., Sanchez-Vinas, M. (2000). Optimization of chromatographic parameters for the determination of biogenic amines in wines by reversed-phase high-performance liquid chromatography. *Journal of Chromatography, A*, *871*, 75–83.

136. Loukou, Z., Zotou, A. (2003). A comparative survey of the simultaneous ultraviolet and fluorescence detection in the RP-HPLC determination of dansylated biogenic amines in alcoholic beverages. *Chromatographia*, *58*, 579–585.

137. Soufleros, E.H., Bouloumpasi, E., Zotou, A., Loukou, Z. (2007). Determination of biogenic amines in Greek wines by HPLC and ultraviolet detection after dansylation and examination of factors affecting their presence and concentration. *Food Chemistry*, *101*, 704–716.

138. Kirschbaum, J., Rebscher, K., Bruckner, H. (2000). Liquid chromatographic determination of biogenic amines in fermented foods after derivatization with 3,5-dinitrobenzoyl chloride. *Journal of Chromatography, A*, *881*, 517–530.

139. Kirschbaum, J., Luckas, B., Beinert, W.D. (1994). Precolumn derivatization of biogenic amines and amino acids with 9-pluorenylmethyl chloroformate and heptylamine. *Journal of Chromatography, A*, *661*, 193–199.

140. Busto, O., Guasch, J., Borrull, F. (1996). Determination of biogenic amines in wine after precolumn derivatization with 6-aminoquinolyl-*N*-hydroxysuccinimidyl carbamate. *Journal of Chromatography, A*, *737*, 205–213.

141. Kirschbaum, J., Busch, I., Brückner, H. (1997). Determination of biogenic amines in food by automated pre-column derivatization with 2-naphthyloxycarbonyl chloride (NOC-Cl). *Chromatographia*, *45S*, 263–268.

142. Kirschbaum, J., Meier, A., Brückner, H. (1999). Determination of biogenic amines in fermented beverages and vinegars by pre-column derivatization with *para*-nitrobenzyloxycarbonyl chloride (PNZ-Cl) and reversed-phase LC. *Chromatographia*, *49*, 117–124.

143. Simon Sarkadi, L., Kovács, A., Mincsovics, E. (1997). Determination of biogenic amines by personal OPLC. *Journal of Planar Chromatography*, *10*, 59–60.

144. Kovács, A., Simon Sarkadi, L., Mincsovics, E. (1998). Step-wise gradient separation and quantification of dansylated biogenic amines in vegetables using personal OPLC. *Journal of Planar Chromatography*, *11* (1), 43–46.

145. Csomos, E., Simon Sarkadi, L., Katay, G., Kiraly Veghely, Z., Diofasi, L., Tyihak, E. (2000). Determination of biologically active compounds in wines using overpressured-layer chromatography. *Czech Journal of Food Sciences*, *18*, 184–185.
146. Sun, X., Yang, X., Wang, E. (2003). Determination of biogenic amines by capillary electrophoresis with pulsed amperometric detection. *Journal of Chromatography, A*, *1005*, 189–195.
147. Arce, L., Rios, A., Valcarcel, M. (1998). Direct determination of biogenic amines in wine by integrating continuous flow clean-up and capillary electrophoresis with indirect UV detection. *Journal of Chromatography, A*, *803*, 249–260.
148. Lange, J., Thomas, K., Wittmann, C. (2002). Comparison of a capillary electrophoresis method with high-performance liquid chromatography for the determination of biogenic amines in various food samples. *Journal of Chromatography, B: Analytical Technology Biomedical Life Science*, *779*, 229–239.
149. Ruiz Jimenez, J., Luque de Castro, M.D. (2006). Pervaporation as interface between solid samples and capillary electrophoresis: determination of biogenic amines in food. *Journal of Chromatography, A*, *1110*, 245–253.
150. Zhang, L.Y., Sun, M.X. (2004). Determination of histamine and histidine by capillary zone electrophoresis with pre-column naphthalene-2,3-dicarboxaldehyde derivatization and fluorescence detection. *Journal of Chromatography, A*, *1040*, 133–140.
151. Kvasnicka, F., Voldrich, M. (2006). Determination of biogenic amines by capillary zone electrophoresis with conductometric detection. *Journal of Chromatography, A*, *1103*, 145–149.
152. Cortacero Ramirez, S., Arraez Roman, D., Segura Carretero, A., Fernandez Gutierrez, A. (2007). Determination of biogenic amines in beers and brewing-process samples by capillary electrophoresis coupled to laser-induced fluorescence detection. *Food Chemistry*, *100*, 383–389.
153. Liu, Y.M., Cheng, J.K. (2003). Separation of biogenic amines by micellar electrokinetic chromatography with on-line chemiluminescence detection. *Journal of Chromatography, A*, *1003*, 211–216.
154. Nouadje, G., Simeon, N., Dedieu, F., Nertz, M., Puig, P., Couderc, F. (1997). Determination of twenty eight biogenic amines and amino acids during wine aging by micellar electrokinetic chromatography and laser-induced fluorescence detection. *Journal of Chromatography, A*, *765* (2), 337–343.
155. Krizek, M., Pelikanova, T. (1998). Determination of seven biogenic amines in foods by micellar electrokinetic capillary chromatography. *Journal of Chromatography, A*, *815*, 243–250.
156. Kovács, A., Simon Sarkadi, L., Ganzler, K. (1999). Determination of biogenic amines by capillary electrophoresis. *Journal of Chromatography, A*, *836*, 305–313.
157. Su, S.C., Chou, S.S., Chang, P.C., Hwang, D.F. (2000). Determination of biogenic amines in fish implicated in food poisoning by micellar electrokinetic capillary chromatography. *Journal of Chromatography, B: Biomedical Science Applications*, *749*, 163–169.
158. Male, K.B., Luong, J.H.T. (2001). Derivatization, stabilization and detection of biogenic amines by cyclodextrin-modified capillary electrophoresis–laser-induced fluorescence detection. *Journal of Chromatography, A*, *926*, 309–317.

159. Fernandes, J.O., Ferreira, M.A. (2000). Combined ion-pair extraction and gas chromatography–mass spectrometry for the simultaneous determination of diamines, polyamines and aromatic amines in Port wine and grape juice. *Journal of Chromatography, A*, *886*, 183–195.
160. Fernandes, J.O., Judas, I.C., Oliveira, M.B., Ferreira, I.M.P.L.V.O., Ferreira, M.A. (2001). A GC-MS method for quantitation of histamine and other biogenic amines in beer. *Chromatographia*, *53-S*, S327–S331.
161. Hwang, B.S., Wang, J.T., Choong, Y.M. (2003). A rapid gas chromatographic method for the determination of histamine in fish and fish products. *Food Chemistry*, *82*, 329–334.
162. Shalaby, A.R. (1999). Simple, rapid and valid thin layer chromatographic method for determining biogenic amines in foods. *Food Chemistry*, *65*, 117–121.
163. Shakila, R.J., Vasundhara, T.S., Kumudavally, K.V. (2001). A comparison of the TLC-densitometry and HPLC method for the determination of biogenic amines in fish and in fishery products. *Food Chemistry*, *75*, 255–259.
164. Lapa Guimaraes, J., Pickova, J. (2004). New solvent systems for thin-layer chromatographic determination of nine biogenic amines in fish and squid. *Journal of Chromatography, A*, *1045*, 223–232.
165. Simon Sarkadi, L., Holzapfel, W.H. (1994). Determination of biogenic amines in leafy vegetables by amino acid analyser. *Zeitschrift für Lebensmittel Untersuchung und Forschung*, *198*, 230–233.
166. Standara, S., Vesela, M., Drdak, M. (2000). Determination of biogenic amines in cheese by ion exchange chromatography. *Nahrung*, *44*, 28–31.
167. Csomos, E., Simon Sarkadi, L. (2002). Characterisation of Tokaj wines based on free amino acids and biogenic amines using ion-exchange chromatography. *Chromatographia*, *56-S*, S185–S188.
168. Gil Agusti, M., Carda Broch, S., Monferrer Pons, L., Esteve Romero, J. (2007). Simultaneous determination of tyramine and tryptamine and their precursor amino acids by micellar liquid chromatography and pulsed amperometric detection in wines. *Journal of Chromatography, A*, *1156*, 288–295.
169. Mello, L.D., Kubota, L.T. (2002). Review of the use of biosensors as analytical tools in the food and drink industry. *Food Chemistry*, *77*, 237–256.
170. Scheller, F., Schubert, F. (1992) *Biosensors*, Elsevier, Amsterdam.
171. Ohashi, M., Nomura, F., Suzuki, M., Otuka, M., Adachi, O., Arakawa, N. (1994). Oxygen-sensor-based simple assay of histamine in fish using purified amine oxidase. *Journal of Food Science*, *59*, 519–522.
172. Wittmann, C., Riedel, K., Schmid, R.D. (1997). Microbial and enzyme sensors for environmental monitoring, in *Handbook of Biosensors and Electronic Noses: Medicine, Food and the Environment* (ed. E. Kress-Rogers), CRC Press, Inc., Boca Raton, FL, pp. 299–332.
173. Bouvrette, P., Male, K.B., Luong, J.H.T., Gibbs, B.F. (1997). Amperometric biosensor for diamine using diamine oxidase purified from porcine kidney. *Enzyme and Microbial Technology*, *20* (1), 32–38.
174. Esti, M., Volpe, G., Massignan, L., Compagnone, D., La Notte, E., Palleschi, G. (1998). Determination of amines in fresh and modified atmosphere packaged

fruits using electrochemical biosensors. *Journal of Agricultural and Food Chemistry*, *46*, 4233–4237.

175. Draisci, R., Volpe, G., Lucentini, L., Cecilia, A., Federico, R., Palleschi, G. (1998). Determination of biogenic amines with an electrochemical biosensor and its application to salted anchovies. *Food Chemistry*, *62*, 225–232.
176. Shin, S.J., Yamanaka, H., Endo, H., Watanabe, E. (1998). Development of an octopine biosensor and its application to the estimation of scallop freshness. *Enzyme and Microbial Technology*, *23*, 10–13.
177. Qiong, C., Tuzhi, P., Liju, Y. (1998). Silk fibroin/cellulose acetate membrane electrodes incorporating xanthine oxidase for the determination of fish freshness. *Analytica Chimica Acta*, *369* (3), 245–251.
178. Tombelli, S., Mascini, M. (1998). Electrochemical biosensors for biogenic amines: a comparison between different approaches. *Analytica Chimica Acta*, *358*, 277–284.
179. Carsol, M.A., Mascini, M. (1999). Diamine oxidase and putrescine oxidase immobilized reactors in flow injection analysis. *Talanta*, *50*, 141–148.
180. Niculescu, M., Frebort, I., Pec, P., Galuszka, P., Mattiasson, B., Csoregi, E. (2000). Amine oxidase based biosensors for histamine detection. *Electroanalysis*, *12*, 369–375.
181. Niculescu, M., Nistor, C., Frebort, I., Pec, P., Mattiasson, B., Csoregi, E. (2000). Redox hydrogel based amperometric bienzyme electrodes for fish freshness monitoring. *Analytical Chemistry*, *72*, 1591–1597.
182. Hibi, T., Senda, M. (2000). Enzymatic assay of histamine by amperometric detection of H2O2 with a peroxidase-based sensor. *Bioscience Biotechnology and Biochemistry*, *64*, 1963–1966.
183. Sotzing, G.A., Phend, J.N., Grubbs, R.H., Lewis, N.S. (2000). Highly sensitive detection and discrimination of biogenic amines utilizing arrays of polyaniline/carbon black composite vapour detectors. *Chemistry of Materials*, *12*, 593–595.
184. Zeng, K., Tachikawa, H., Zhu, Z., Davidson, V.L. (2000). Amperometric detection of histamine with a methylamine dehydrogenase polypyrrole-based sensor. *Analytical Chemistry*, *72*, 2211–2215.
185. Niculescu, M., Nistor, C., Ruzgas, T., Frebort, I., Kebela, M., Peb, P., Csoregi, E.I. (2001). Detection of histamine and other biogenic amines using biosensors based on amine oxidase. *Inflammatory Research* Supplement, 2, S146–S148.
186. Lange, J., Wittmann, C. (2002). Enzyme sensor array for the determination of biogenic amines in food samples. *Analytical and Bioanalytical Chemistry*, *372*, 276–283.
187. Wang, Y., Sotzing, G.A., Weiss, R.A. (2003). Conductive polymer foams as sensors for volatile amines. *Chemistry of Materials*, *15*, 375–377.
188. Hammar, E., Berglund, A., Hedin, A., Rustas, K., Ytterstrom, U., Kerblom, E. (1990). An immunoassay for histamine based on monoclonal antibodies. *Journal of Immunological Methods*, *128*, 51–58.
189. Rauch, P., Rychestsky, P., Hochel, I., Bilek, R., Guesdon, J.L. (1992). Enzyme immunoassay of histamine in foods. *Food and Agricultural Immunology*, *4*, 67–72.
190. Kruger, C., Sewing, U., Stengel, G., Kema, I., Westermann, J., Manz, B. (1995). ELISA for the determination of histamine in fish. *Archiv für Lebensmittelhygiene*, *46*, 115–119.

191. Serrar, D., Breband, R., Bruneau, S., Denoyel, G.A. (1995). The development of a monoclonal antibody-based ELISA for the determination of histamine in food: application to fishery products and comparison with the HPLC assay. *Food Chemistry*, *54*, 85–91.
192. Schneider, E., Usleber, E., Martlbauer, E. (1996). Production and characterization of antibodies against histamine, in *Immunoassays for Residue Analysis: Food Safety* (eds R.C. Beier, L.H. Stanker), ACS Symposium Series 621, American Chemical Society, Washington, DC, pp. 413–421.
193. Aygun, O., Schneider, E., Scheuer, R., Usleber, E., Gareis, M., Martlbauer, E. (1999). Comparison of ELISA and HPLC for the determination of histamine in cheese. *Journal of Agricultural and Food Chemistry*, *47*, 1961–1964.
194. Marcobal, A., Polo, M.C., Martin-Alvarez, P.J., Moreno-Arribas, M.V. (2005). Biogenic amine content of red Spanish wines: comparison of a direct ELISA and an HPLC method for the determination of histamine in wines. *Food Research International*, *38* (4), 387–394.
195. Rupasinghe, H.P.V., Clegg, S. (2007). Total antioxidant capacity, total phenolic content, mineral elements, and histamine concentrations in wines of different fruit sources. *Journal of Food Composition and Analysis*, *20*, 133–137.
196. Simon Sarkadi, L., Gelencser, E., Vida, A. (2003). Immunoassay method for detection of histamine in foods. *Acta Alimentaria Hungary*, *32* (1), 89–93.
197. Silla Santos, M.H. (1996). Biogenic amines: their importance in foods. *International Journal Food Microbiology*, *29*, 213–231.
198. Medina, M.Á., Urdiales, J.L., Rodríguez-Caso, C., Javier Ramírez, F., Sánchez-Jiménez, F. (2003). Biogenic amines and polyamines: similar biochemistry for different physiological missions and biomedical applications. *Critical Reviews in Biochemistry and Molecular Biology*, *38* (1), 23–59.
199. Blixt, Y., Borch, E. (2002). Comparison of shelf-life of vacuum-packaged pork and beef. *Meat Sciences*, *60*, 370–376.
200. Leuschner, R.G., Hammes, W.P. (1998). Degradation of histamine and tyramine by *Brevibacterium linens* during surface ripening of Munster cheese. *Journal of Food Protection*, *61*, 874–878.
201. Bover-Cid, S., Izquierdo Pulido, M., Vidal Carou, M.C. (2001). Effectiveness of a *Lactobacillus sakei* starter culture in the reduction of biogenic amine accumulation as a function of the raw material quality. *Journal of Food Protection*, *64*, 367–373.
202. Bover-Cid, S., Izquierdo Pulido, M., Vidal Carou, M.C. (2000). Mixed starter cultures to control biogenic amine production in dry fermented sausages. *Journal of Food Protection*, *63*, 1556–1562.
203. Bover-Cid, S., Izquierdo Pulido, M., Vidal Carou, M.C. (2001). Effect of the interaction between a low tyramine-producing *Lactobacillus* and proteolytic staphylococci on biogenic amine production during ripening and storage of dry sausages. *International Journal of Food Microbiology*, *65*, 113–123.
204. Maijala, R., Eerola, S., Lievonen, S., Hill, P., Hirvi, T. (1995). Formation of biogenic amines during ripening of dry sausages as affect by starter cultures and thawing time of raw materials. *Journal of Food Science*, *69*, 1187–1190.
205. Lanciotti, R., Gardini, F., Sinigaglia, M., Guerzoni, M.E. (1996). Effects of growth conditions on the resistance of some pathogenic and spoilage species to high pressure homogenization. *Letters in Applied Microbiology*, *22*, 165–168.

206. Wuytack, E.Y., Diels, A.M.J., Michiels, C.W. (2002). Bacterial inactivation by high-pressure homogenisation and high hydrostatic pressure. *International Journal of Food Microbiology*, *77*, 205–212.
207. Kheadr, E.E., Vachon, J.F., Paquin, P., Fliss, I. (2002). Effect of dynamic high pressure on microbiological, rheological and microstructural quality of Cheddar cheese. *International Dairy Journal*, *12*, 435–446.
208. Lanciotti, R., Patrignani, F., Iucci, L., Guerzoni, M.E., Suzzi, G., Belletti, N., Gardini, F. (2007). Effects of milk high pressure homogenization on biogenic amine accumulation during ripening of ovine and bovine Italian cheeses. *Food Chemistry*, *104*, 693–701.
209. Abu Tarboush, H.M., Al Kahtani, H.A., Atia, M., Abou Arab, A.A., Bajaber, A.S., El Mojaddidi, M.A. (1996). Irradiation and postirradiation storage at $2 \pm 2\,°C$ of tilapia (*Tilapia nilotica* × *T. aurea*) and Spanish mackerel (*Scomberomorus commerson*): sensory and microbial assessment. *Journal of Food Protection*, *59* (10), 1041–1048.
210. Ray, B. (1996). *Fundamental food Microbiology*, CRC Press, Inc., Boca Raton, FL.
211. Mendes, R., Silva, H.A., Nunes, M.L., Empis, J.M.A. (2005). Effect of low-dose irradiation and refrigeration on the microflora, sensory characteristics and biogenic amines of Atlantic horse mackerel (trachurus trachurus). *European Food Research and Technology*, *221*, 329–335.
212. Kim, J.H., Ahn, H.J., Lee, J.W., Park, H.J., Ryu, G.H., Kang, I.J., Byun, M.W. (2005). Effects of gamma irradiation on the biogenic amines in pepperoni with different packaging conditions. *Food Chemistry*, *89*, 199–205.
213. Nadon, C.A., Ismond, M.A.H., Holley, R. (2001). Biogenic amines in vacuum-packaged and carbon dioxide-controlled atmosphere-packaged fresh pork stored at 1.5 °C. *Journal of Food Protection*, *64*, 220–227.
214. Sivertsvik, M., Jekstud, W.K., Rosnes, J.T. (2002). A review of modified atmosphere packaging of fish and fishery products: significance of microbial growth, activities, and safety. *Journal of International Food Science Technology*, *37*, 107–127.
215. Ozogul, F., Taylor, K.D.A., Quantick, P., Ozogul, Y. (2002). Biogenic amines formation in Atlantic herring (*Clupea harengus*) stored under modified atmosphere packaging using a rapid HPLC method. *International Journal of Food Science and Technology*, *37*, 515–522.
216. Ruiz Capillas, C., Moral, A. (2002). Effect of controlled and modified atmospheres on the production of biogenic amines and free amino acids during storage of hake. *European Food Research and Technology*, *214* (6), 476–481.
217. Bodmer, S., Imark, C., Kneubuhl, M. (2000). Low-histamine technology, the genuine way to avoid histamine-intolerance, in *COST 917 Biogenically Active Amines in Food*, Vol. 4 (eds D.M.L. Morgan, A. White, F. Sánchez-Jiménez, S. Bardócz), EC Publication, Luxembourg, pp. 35–40.
218. Heleniak, E., O'Desky, I. (1999). Histamine and prostaglandins in schizophrenia: revisited. *Medical Hypotheses*, *52*, 37–42.
219. Medina, M.A., Quesada, A.R., Nunez de Castro, I., Sanchez-Jimenez, F. (1999). Histamine, polyamines, and cancer. *Biochemical Pharmacology*, *57*, 1341–1344.
220. Hellstrand, K., Brune, M., Naredi, P., Mellqvist, U.H., Hansson, M., Gehlsen, K.R., Hermodsson, S. (2000). Histamine: a novel approach to cancer immunotherapy. *Cancer Investigation*, *18*, 347–355.

221. Hart, P.H., Grimbaldeston, M.A., Finlay-Jones, J.J. (2001). Sunlight, immunosuppression and skin cancer: role of histamine and mast cells. *Clinical and Experimental Pharmacology and Physiology*, *28*, 1–8.

222. Hellstrand, K. (2002). Histamine in cancer immunotherapy: a preclinical background. *Seminars in Oncology*, *29*, 35–40.

223. Premont, R.T., Gainetdinov, R.R., Caron, M.G. (2001). Following the trace of elusive amines. *Proceedings of the National Academy of Sciences of the United States of America*, *98*, 9474–9475.

224. Seiler, N., Douaund, F. (1998). Polyamine metabolism and transport in the mammalian organism, in *COST 917 Biogenically Active Amines in Food*, Vol. 2 (eds S. Bardocz, A. White, Gy. Hajos), EC Publication, Luxembourg, pp. 19–38.

225. Bernstein, H.G., Muller, M. (1999). The cellular localization of the L-ornithine decarboxylase/polyamine system in normal and diseased central nervous systems. *Progress in Neurobiology*, *57*, 485–505.

226. Davidson, N.E., Hahm, H.A., McCloskey, D.E., Woster, P.M., Casero, R.A., Jr. (1999). Clinical aspects of cell death in breast cancer: the polyamine pathway as a new target for treatment. *Endocrine-Related Cancer*, *6*, 69–73.

227. McCabe Sellers, B.J., Staggs, C.G., Bogle, M.L. (2006). Critical review. Tyramine in foods and monoamine oxidase inhibitor drugs: a crossroad where medicine, nutrition, pharmacy, and food industry converge. *Journal of Food Composition and Analysis*, *19*, S58–S65.

228. McCabe, B.J., Frankel, E.H., Wolfe, J.J. (2003). *Handbook of Food and Drug Interactions*, CRC Press, Inc., Boca Raton, FL, pp. 457–479.

229. Bieck, P.R., Antonin, K.H. (1988). Oral tyramine pressor tests the safety of monoamine oxidase inhibitor drugs: comparison of brofaromine and tranylcypromine in healthy subjects. *Journal of Clinical Psychopharmacology*, *8*, 237–245.

230. Da Prada, M., Zurcher, G. (1992). Tyramine content of preserved and fermented foods or condiments of Far Eastern cuisine. *Psychopharmacology*, *106*, 532–534.

231. Veciana-Nogue, M.T., Marine Font, A., Vidal Carou, M.C. (1997). Biogenic amines as hygienic quality indicators of tuna. Relationships with microbial counts, ATP-related compounds, volatile amines and organoleptic changes. *Journal of Agricultural and Food Chemistry*, *45*, 2041–2041.

232. Wu, M.L., Yang, C.C., Yang, G.Y., Ger, J., Deng, J.F. (1997). Scombroid fish poisoning: an overlooked marine food poisoning. *Veterinary and Human Toxicology*, *39*, 236–241.

233. Center for Disease Control and Prevention (2000). *Morbidity and Mortality Weekly Report*, *49*, 398–400.

234. Becker, K., Southwick, K., Reardon, J., Berg, R., MacCormack, J.N. (2001). Histamine poisoning associated with eating tuna burgers. *JAMA—Journal of the American Medical Association*, *285*, 1327–1330.

235. Sumner, J., Ross, T. (2002). A semi-quantitative seafood safety risk assessment. *International Journal of Food Microbiology*, *77*, 55–59.

236. Taylor, S.L. (1986). Histamine food poisoning: toxicology and clinical aspects. *Critical Reviews in Toxicology*, *17* (2), 91–128.

237. Lehane, L., Olley, J. (1999). *Histamine (Scombroid) Fish Poisoning: a Review in a Risk-Assessment Framework*, National Office of Animal and Plant Health, Canberra.

238. Sours, H.E., Smith, D.G. (1980). Outbreaks of foodborne disease in the United States, 1972–1978. *Journal of Infectious Diseases*, *142*, 122–125.
239. Scoging, A. (1998). Scombrotoxic (histamine) fish poisoning in the United Kingdom: 1987 to 1996. *Communicable Disease and Public Health*, *1*, 204–205.
240. Shalaby, A.R. (1996). Significance of biogenic amines to food safety and human health. *Food Research International*, *29*, 675–690.
241. Fletcher, G.C., Summers, G., van Veghel, P.W.C. (1998). Levels of histamine and histamine-producing bacteria in smoked fish from New Zealand markets. *Journal of Food Protection*, *61* (8), 1064–1070.
242. Sumner, J., Ross, T., Ababouch, L. (2004). Application of risk assessment in the fish industry. No. 442, *FAO Fisheries Technical Paper*, Rome, FAO. 2004. p. 78.
243. Oliveira, C.P., Gloria, M.B.A., Barbour, J.F., Scanlan, R.A. (1995). Nitrate, nitrite and volatile nitrosoamines in whey-containing food products. *Journal of Agriculture and Food Chemistry*, *43*, 967–969.
244. Lehane, L., Olley, J. (2000). Review—Histamine fish poisoning revisited. *International Journal of Food Microbiology*, *58*, 1–37.
245. Eliassen, K.A., Reistad, R., Risoen, U., Ronning, H.F. (2002). Dietary polyamine. *Food Chemistry*, *78*, 273–280.
246. Reilly, A. (2006). Managing Microbiological Risk in the Fish Processing Chain. Presentation at the FAO EUROFISH Workshop on Seafood Safety, 13–15 December 2006, Copenhagen.
247. FDA (2001). *Fish and Fishery Products Hazards and Controls Guide*, 3rd edn, Office of Seafood, Washington, DC, p. 326.
248. Rogers, P., Staruszkiewicz, W. (1997). Gas chromatographic method for putrescine and cadaverine in canned tuna and mahimahi and fluorometric methods for histamine (minor modification of AOAC Official Method 977.13): collaborative study. *Journal of AOAC International*, *80* (3), 591–602.
249. Fisher, R., Nolke, G., Schillberg, S., Twyman, R.M. (2004). Improvement of grapevine using current biotechnology, in O.A. de Sequeira, J.C. Sequeira (eds), 1st IS on Grapevine, *Acta Horticulturae*, *652*, 3833–3890.
250. Deak, T. (2004). Progress and prospects in the ecology of wine yeasts, in *Food Micro, the 19th International ICFMH Symposium*, Portoroz, Slovenia, pp. 245.
251. Henick Kling, T. (1992). Malolactic fermentation, in *Wine Microbiology and Biotechnology* (ed. G. Fleet), Harwood Academic Publishers, Chur, pp. 289–326.
252. Canal Llauberes, R.M. (1992). Enzymes in winemaking, in *Wine Microbiology and Biotechnology* (ed. G. Fleet), Harwood Academic Publishers, Chur, pp. 477–506.
253. Cantarelli, C. (1989). Factors affecting the behaviour of yeast in wine fermentation, in *Biotechnology Application in Beverage Production* (eds C. Cantarelli, C. Lanzarini), Elsevier Applied Science, New York, pp. 127–151.
254. Plahuta, P., Raspor, P. (2007). Comparison of hazards: current vs. GMO wine. *Food Control*, *18* (5), 492–502.

4

PRESERVATION

4.1

N-NITROSAMINES, INCLUDING *N*-NITROSOAMINOACIDS AND POTENTIAL FURTHER NONVOLATILES

MICHAEL HABERMEYER AND GERHARD EISENBRAND

Department of Chemistry, Division of Food Chemistry and Toxicology, University of Kaiserslautern, Erwin-Schroedinger-Str. 52, Kaiserslautern 67663, Germany

4.1.1 INTRODUCTION

Today's knowledge about the human exposure to carcinogenic *N*-nitroso compounds is based on studies from more than 40 years of research. One of the first publications that dealt with *N*-nitroso compounds in food was by Ender *et al.* (1). The authors described the occurrence of a hepatotoxic factor in herring meal produced from sodium nitrite-preserved herring that was used as feed for sheep. Although appropriate analytical methods were not available yet, the group identified *N*-nitrosodimethylamine (NDMA), obviously present at rather high concentrations in the feed, as the most probable causative agent. The treatment of food with nitrite was suspected to be the causative factor for the formation of *N*-nitrosamines. As a result of these findings, a large series of analytical studies to uncover the presence of *N*-nitroso compounds in food was triggered. The challenging and difficult analytical detection of *N*-nitroso compounds was greatly improved with the advent of the highly specific and ultrasensitive gas chromatographic detector by Fine and Rounbehler (2), called thermal energy analyzer (TEA). It relies on thermal cleavage of the

Process-Induced Food Toxicants: Occurrence, Formation, Mitigation, and Health Risks,
Edited by Richard H. Stadler and David R. Lineback

***N*-Nitrosamines: At a Glance**	
Historical	First publications on the toxicity of *N*-nitroso compounds (NOCs) in the mid-1950s, identifying *N*-nitrosodimethylamine (NDMA) as a potent liver toxine and carcinogen.
Analysis	GC coupled to a thermal energy analyzer (TEA) or GC/HLPC with high-resolution mass spectrometry.
Occurrence in Food	In cured and smoked meats and meat products. Processing by direct drying techniques (increased NOx exposure) usually equates to higher amounts of NOCs, and initial concerns were raised in the case of beer (malt). NOCs have also been found at trace levels in other foods such as cocoa, cheese, powdered milks and eggs, coffee, instant soups, and spices.
Main Formation Pathways	Classical reaction of an amine or *N*-nitrosatable amino acid with a nitrosating agent (e.g., nitrous acid, formed from nitrite under acidic conditions, or NOx). The pKa value of the amino group is decisive for the nitrosation rate. NOC formation at basic conditions is of major importance for industrial/workplace situations.
Mitigation in Food	Reduction of nitrite in curing salt, use of inhibitors such as ascorbic acid. Modifications of the firing techniques (e.g., lower operating temperatures, indirect heating) reduces NOx and consequently NOCs formation. The amount of nitrite in cured meats has decreased significantly over the past years.
Health Risks	IARC has classified a number of nitrosamines as "probably" (Group 2A) or "possibly" (Group 2B) carcinogenic to humans. Most NOCs are pre-carcinogens and upon metabolic activation afford electrophilic intermediates that react with cellular components such as protein, RNA, and DNA. Thus, a major concern is their carcinogenic and genotoxic potential. EFSA has calculated a MoE for NDMA based on average consumption at >10,000, and concluded a low concern for this compound.
Regulatory Information or Industry Standards	In the USA, limits (NDMA or total *N*-nitrosamines) for ham, bacon, malt beverages, and barley malt have been established. Since 1998 there are also notification levels for several NOC in drinking water. In Switzerland the limit for beer (total *N*-nitrosamines) is 0.5 µg/kg. EU Directive 93/11/EEC defines 10 µg/kg total *N*-nitrosamines for rubber nipples and pacifiers. The EU Cosmetic Directive 76/768/ECC includes a negative list of NOC for cosmetics and also regulations for cosmetic ingredients to avoid NOC formation. Levels of NOC in industry workplaces are regulated in Germany by the TRGS 552 (technical guidelines for hazardous substances).
Knowledge Gaps	Sources of exposure such as cosmetics, occupational settings, and cigarette smoke warrant continuous monitoring and development/refinement of mitigation measures also from these non-food sources. Endogenous formation of carcinogenic NOCs has not been adequately evaluated in terms of relevance to human cancers. Further, the development of appropriate and valid biomarkers of exposure to assess the relevance of *in vivo* formation of NOCs is a priority.
Useful Information (web) & Dietary Recommendations	http://ec.europa.eu/health/ph_risk/committees/04_sccp/docs/sccp_o_121.pdf http://www.epa.gov/iris/subst/0045.htm http://www.inchem.org/documents/jecfa/jecmono/v35je13.htm Avoid the over-heating of cured and smoked foods; cook/prepare foods to the appropriate "doneness."

N–NO bond and detection of the liberated NO radical through the chemiluminescence signal generated by its reaction with ozone (3). *N*-nitroso compounds were subsequently detected in many areas of human environment, although mostly in low concentrations.

N-nitroso compounds are a class of potent human carcinogens. The strong carcinogenic potential of NDMA was described in 1956 by Barnes and Magee (4). In the following years, over 300 compounds were investigated and approximately 90% were carcinogenic in animal experiments (5, 6).

4.1.2 CHEMISTRY, FORMATION, AND OCCURRENCE

N-nitroso compounds are divided into two groups: *N*-nitrosamines and *N*-nitrosamides (Fig. 4.1.1).

N-nitrosamines are relatively stable compounds found as contaminants in the environment. In contrast, *N*-nitrosourea, -carbamate, and -guanidine derivatives, all belonging to the class of *N*-nitrosamides, are directly reactive chemicals that are rather unstable.

In virtually any situation where nitrosating agents encounter *N*-nitrosatable amino compounds, *N*-nitroso compounds can be formed. The classical situation reflects the reaction of an amine with nitrous acid, generated from nitrite at acidic conditions. Under proton catalysis, nitrite and nitrous acid generate the actual nitrosating species, either dinitrogen trioxide (N_2O_3) or tetroxide (N_2O_4) or nitrous acidium ion $NO^{+}{\bullet}H_2O$. The mechanism of nitrosation is exemplified for a secondary amine in Fig. 4.1.2.

In aqueous acidic medium, it is only the unprotonated amine nitrogen that undergoes nitrosation. At a given pH value, therefore the pK_a value of a given amine is decisive for the nitrosation rate. Since the nitrosating species (N_2O_3) under these conditions is formed from two molecules of HNO_2 by proton catalysis, nitrosation of strongly basic dialkylamines ($pK_a > 9.5$) has a rate optimum at around pH 3.5, close to the pK_a value of HNO_2 (7). In contrast to strongly basic amines, weakly basic amines, such as morpholine ($pK_a = 8.7$) are much more rapidly nitrosated than strongly basic ones. Moreover, in the presence of formaldehyde, nitrosation is strongly catalyzed and proceeds even at neutral or basic pH values, because a methylene immonium intermediate is formed that is undergoing fast nitrosation by nitrite anion (Fig. 4.1.3), a reac-

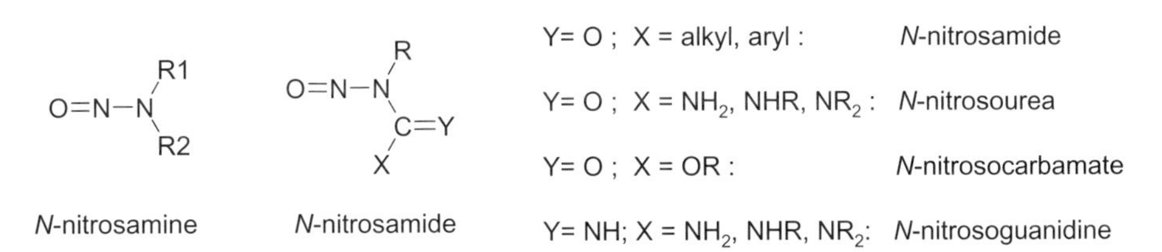

Figure 4.1.1 Basic chemical structures of *N*-nitroso compunds.

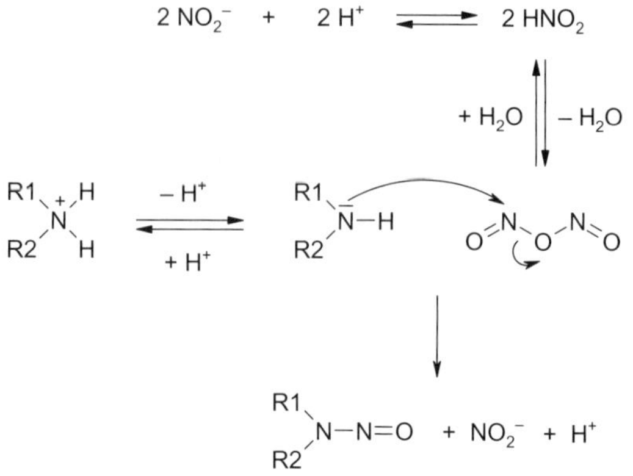

$$2\,NO_2^- + 2\,H^+ \rightleftharpoons 2\,HNO_2$$

$$2\,HNO_2 \underset{+H_2O}{\overset{-H_2O}{\rightleftharpoons}} O{=}N{-}O{-}N{=}O$$

$$R1R2NH_2^+ \underset{+H^+}{\overset{-H^+}{\rightleftharpoons}} R1R2N{-}H$$

$$R1R2N{-}H + O{=}N{-}O{-}N{=}O \longrightarrow R1R2N{-}N{=}O + NO_2^- + H^+$$

Figure 4.1.2 Mechanism of secondary amine nitrosation.

$$R1R2N{-}H + HCHO \underset{}{\overset{-OH^-}{\rightleftharpoons}} R1R2\overset{+}{N}{=}CH_2 \xrightarrow{NO_2^-} R1R2N{-}CH_2{-}O{-}N{=}O \longrightarrow R1R2N{-}N{=}O + HCHO$$

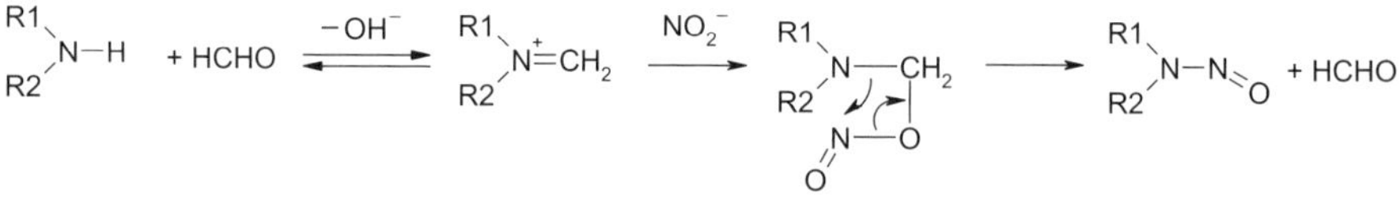

Figure 4.1.3 Formaldehyde-catalyzed nitrosation of secondary amines in neutral to alkaline media.

tion of great importance at specific working places, for example, those where exposure to metal cutting fluids might occur. The mechanism has been elucidated by Keefer and Roller (8). Because acid-catalyzed nitrosation is inappreciable at $pH > 5$, *N*-nitroso compounds in food, consumer products, or in specific working place situations appear to be preferentially formed from exposure to atmospheric nitrogen oxides (NOx).

Acid-catalyzed nitrosation has been suspected, first by Druckrey and coworkers (9), as a possible way of *in vivo* formation of *N*-nitroso compounds in the acid environment of the stomach. Due to the fact that the free and not the protonated amine is nitrosated, *in vivo* formation of *N*-nitrosamines in the stomach is governed by the basicity of the amine nitrogen and the pH of the gastric medium.

Tertiary amines can also generate nitrosamines by a reaction first described by Smith and Loeppky, as dealkylative nitrosation (10).

With few exceptions, the nitrosation rate of a tertiary amine in general is considerably lower as compared with that of a secondary amine. Nitrosation of primary amines in aqueous solution results in unstable diazonium intermediates that react with water to alcohols.

The reaction of α-amino acids, dietary aldehydes, and nitrite has been found to lead to the formation of a largely unexplored group of heterocyclic *N*-nitrosamines, 3-nitroso-oxazolidinones (NOZ) (11). The reaction of certain

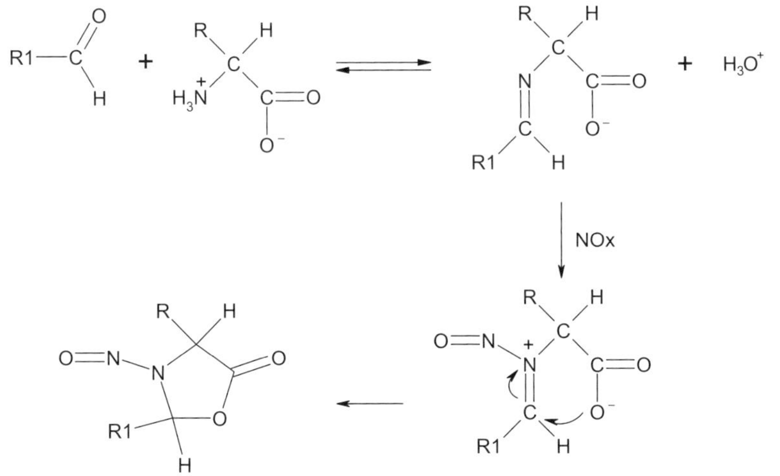

3-nitroso-oxazolidin-5-one

Figure 4.1.4 Formation of 3-nitroso-oxazolidin-5-one.

carbonyl compounds, such as aldehydes, with amino acids leads to the formation of imines or immonium species that might undergo *N*-nitrosation, resulting in the formation of NOZs, as exemplified in Fig. 4.1.4. It is not known as yet, however, whether such compounds indeed are formed in food or *in vivo*.

The nitrosating reaction can also be inhibited, for example, in the presence of ascorbic acid, primary amines, tannins, or other phenolic compounds (12). Total exposure to *N*-nitroso compounds (NOC) results from exogenous exposure with preformed *N*-nitrosamines from endogenous exposure as a result of intake of precursor compounds that might be subject to *in vivo* formation of NOC. Scheme 4.1.1 shows potential contributing factors to total human exposure.

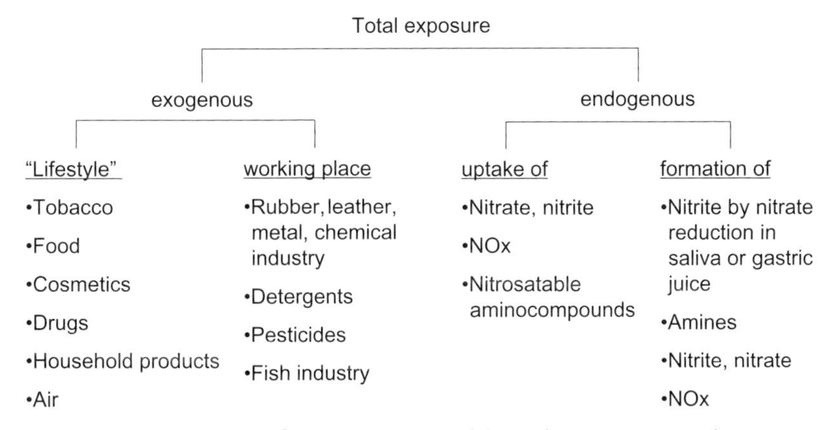

Scheme 4.1.1 Human exposure with *N*-nitroso compounds.

4.1.2.1 Food

Contamination of food with nitrosamines is primarily caused by their formation through various processes during manufacture, storage, and/or cooking, in rare cases also through migration from packaging materials (13, 14). In addition, nitrite might also be produced by microorganisms from nitrate. Factors of major significance for nitrosamine formation in the food are the presence of *N*-nitrosatable amines and of nitrosating agents (15). Nitrogen oxides primarily may result from addition of nitrate and/or nitrite. Additionally, a variety of technical processes to which food or food constituents are exposed, such as drying, kilning, and smoking are prone to bring food constituents in close contact with NOx. Nitrosamines most frequently found are NDMA, *N*-nitrosopyrrolidine (NPYR), *N*-nitrosopiperidine (NPIP), and *N*-nitrosothiazolidine (NTHZ), whereas *N*-nitrosodiethylamine (NDEA), *N*-nitrosodibutylamine (NDBA), and *N*-nitrosomorpholine (NMOR) have been found only sporadically. The group of the so-called nonvolatile *N*-nitroso compounds consists mainly of *N*-nitrosated amino acids, including the *N*-nitroso products of sarcosine (NSAR), 3-hydroxyproline and proline (NPRO), thiazolidine-4-carboxylic acid (NTCA), oxazolidine-4-carboxylic acid (NOCA), and *N*-nitroso-2-methyl-thiazolidine-4-carboxylic acid (NMTCA) as well as its oxazolidine analog *N*-nitroso-2-methyl-oxazolidine-4-carboxylic acid (NMOCA) (16). Figure 4.1.5 provides an overview of the respective compounds. NPRO and NTCA are most frequently found in foods; the other compounds have been reported only sporadically. Of note, with the exception of NSAR, which is a relatively weak carcinogen, *N*-nitrosated amino acids are not mutagenic and not carcinogenic. The occurrence of the compounds in food is discussed considering different food categories.

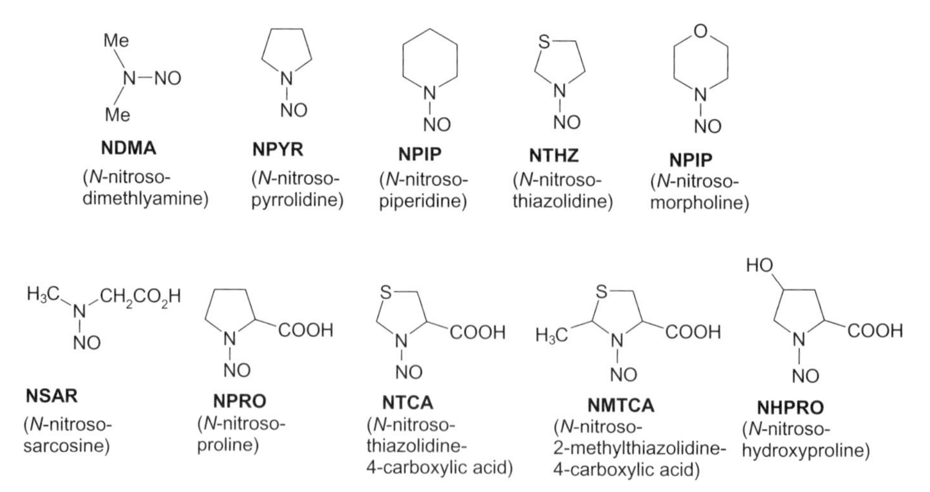

Figure 4.1.5 Structures of volatile and nonvolatile *N*-nitroso compounds in food.

4.1.2.1.1 Meat and Meat Products The formation of *N*-nitrosamines in meat and meat products can occur as a result of various processing techniques such as smoking, salting, and/or curing. Salting and curing of meat are used since ancient times for preservation purposes. In general, curing is achieved by treatment of meat with curing salt mixtures consisting of sodium chloride, sodium/potassium nitrite, and sodium/potassium nitrate. Often so-called curing aids are also used, including ascorbic and erythorbic acid, gluconic-5-acid lactone, acidulans, and glucose. The characteristic cured meat color that develops is attributed to the formation of heat-stable nitrogen monoxide (NO) complexes with muscle myoglobine. Curing, in general, is a dry or wet process and involves the development of a curing microflora that is responsible for the generation of NO from nitrite and nitrate. The microflora also is important for flavor development.

A German study from the early eighties showed that volatile *N*-nitrosamines were found in 32% of almost 400 meat and sausage samples (17, 18). Similar results were found in the United States at that time (19). The reduction of nitrite content in curing salt, together with decreased addition of nitrate and the regular use of nitrosation inhibitors, such as ascorbic acid and/or tocopherols, resulted in significantly reduced contents of volatile *N*-nitrosamines in meat products as shown in studies from the late eighties (see Table 4.1.1).

Studies in model systems showed that the content of NPYR in cured meat products can increase as a consequence of thermal treatment or processing, whereas the amount of NDMA appeared to remain stable (24). NPYR is mainly formed by heat-induced decarboxylation of nitrosoproline. In addition, meat products may contain up to 20 mg/kg pyrrolidine that might be subjected to direct nitrosation in the presence of nitrite in food (15). In addition to the generation of NPYR by thermal decarboxylation of NPRO, *N*-terminal NPRO-containing peptides might also act as progenitors (24–26). Relative high amounts of peptide-bound NPRO can be released by enzymatic degradation or chemical hydrolysis (27). In addition to nitrite and NOx, further nitrosating agents contributing to nitrosamine formation, especially in the lipid phase of meat products, have to be taken into consideration. For instance, nitrogen oxides were found to react with unsaturated fatty acids and cholesterol to form nitrite esters, potential nitrosating agents (28).

TABLE 4.1.1 Volatile N-nitrosamines in heat-treated meat products.

Country	Concentration, μg/kg	*N*-nitrosamine	Year (Reference)
Sweden	nd–8	NDMA, NPYR	1988 (20)
Italy	nd–5	NDMA, NPYR, NDBA	1988 (21)
Germany	nd–2.5	NDMA, NPIP, NPYR	1991 (22)
China	nd–14	NDMA, NDEA, NPYR	1988 (23)

nd, not detectable.

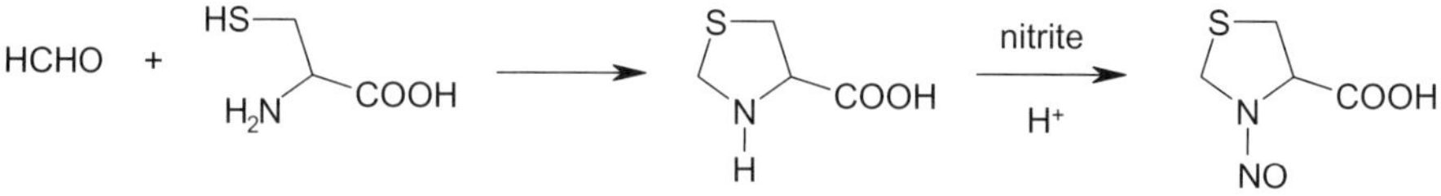

$$\mathrm{HCHO} + \mathrm{HS{-}CH_2{-}CH(NH_2){-}COOH} \longrightarrow \text{thiazolidine-4-COOH (N-H)} \xrightarrow[H^+]{\text{nitrite}} \text{thiazolidine-4-COOH (N-NO)}$$

Figure 4.1.6 NTCA formation in food from cysteine and formaldehyde.

In addition to volatile *N*-nitrosamines, nonvolatile *N*-nitroso compounds have been found in nitrite-treated or smoked meat products in much higher concentrations, e.g., NPRO in concentrations up to 360 μg/kg in cured meat (24, 25, 29). Thermal decarboxylation of nitrosated amino acids generates the respective volatile *N*-nitrosamines (NPRO → NPYR; NSAR → NDMA). Decarboxylation of NHPRO leads to NHPYR, a nonvolatile nitrosamine. NTCA was found in different smoke-treated meat products in amounts up to 1 mg/kg (30, 31), in uncooked bacon up to 1.4 mg/kg (32), and up to 13.7 mg/kg in fried bacon (31). *N*-nitrosothiazolidine (NTHZ), a noncarcinogenic nitrosamine, has been found in smoke-treated products in concentrations up to 10 μg/kg (30, 33, 34). Higher concentrations (up to 1010 μg/kg) were found in fried bacon (35). NOCA was identified in smoked mutton (40–70 μg/kg) together with MeNOCA, the 5-methyl derivative, in concentrations of 30–120 μg/kg (16).

Formation of NTCA in food during the heating process can be rationalized by reaction of the amino acid cystein with formaldehyde to thiazolidine-4-carboxylic acid (TCA) that is subsequently nitrosated to NTCA (see Fig. 4.1.6) (31). Analogous reactions of serine or threonine with formaldehyde lead to the formation of NOCA and MeNOCA, respectively (16).

4.1.2.1.2 Fish and Fish Products In fish and fish products, NDMA is the most commonly detected volatile *N*-nitrosamine. NDMA contents of products in Western countries have mostly been found below 1 μg/kg. Tricker *et al.* (22) identified NDMA in 17 different fish products (cooked or smoked fish samples and conserved fish) in concentrations from 0.5 to 8.0 μg/kg. Contents appeared to depend on the way of preparation. Fish heated over an open gas flame (high NOx-content) showed a 30-fold increase of NDMA, whereas cooking with an electric hot plate showed no such effect (36). NDMA contents in differently processed fish products from different countries are exemplified in Table 4.1.2. Asian fish products showed higher NDMA concentrations, probably as a consequence of specific preparation or conservation methods (curing, smoking) (14). In a study from Finland on smoked fish, NDMA of up to 28.5 μg/kg was found (40).

In single cases, NPYR was identified in samples of fish or mussels in concentrations <1 μg/kg (20, 37). Presence of NPRO was reported in 8 of 63 fish samples with a mean concentration of 6 μg/kg (41). Additionally, NTCA was found in this study in six samples (27 to 344 μg/kg). Canadian fish samples showed NTCA concentrations at a mean of 67 μg/kg, with one sample of

TABLE 4.1.2 Examples of volatile *N*-nitrosamines concentrations in fish products.

Fish	Country	Number	NDMA, μg/kg		Other *N*-nitrosamines	Reference no.
			Mean	Range		
Fresh, cooked	England	94	0.2	5–10	NPYR	37
Smoked	Japan	22		0.5–13	—	36
Cooked	Japan	26	4.3	nd–30	NDEA	38
Fish products	Japan	42	5.9	nd–39	NDEA, NDBA	38
Cooked, smoked	The Netherlands	53	0.4	nd–2.1	—	39
Fish products	China	63		nd–131	—	23
Smoked	Sweden	61	1.3	nd–12	NPYR	20
Fish products	Sweden	20	0.9	nd–4.4	—	20
Fresh	Germany	8	3.0	0.5–8.0	—	22
Smoked	Germany	4	1.4	0.6–2.6	—	22
Fish products	Germany	5	2.1	0.7–5.3	—	22

nd, not detectable.

TABLE 4.1.3 Concentrations of NDMA in beer from the German market.

	NDMA, µg/kg		
Number of samples	Mean	Range	Year (Reference)
158	2.7	<0.5–68	1977/78 (42)
401	0.28	<0.5–9.2	1980 (45)
454	0.44	<0.5–7	1981 (44)
72	<0.1	<0.2–1.3	1983 (46)
175	<0.1	<0.2–0.8	1984 (46)
87	0.1	<0.2–1.0	1985 (46)
71	0.1	<0.2–0.6	1986 (46)
580	0.1	<0.2–1.7	1987 (46)
13	0.15	<0.2–0.6	1989/90 (22)

smoked herring at 1600µg/kg NTCA (31). In samples of smoked oyster, concentrations of 167µg/kg NTCA and 109µg/kg NTHZ were reported (30).

4.1.2.1.3 Beer and Other Alcoholic Beverages In a publication by Spiegelhalder *et al.* (42), contamination of beer with *N*-nitrosamines was reported for the first time. In 70% of 158 different beer samples tested, NDMA was identified at a mean concentration of 2.7µg/kg. The highest concentration found was 68µg/kg. Other *N*-nitrosamines like NDEA and NPYR were only found very rarely (17). Studies investigating the origin of this contamination unraveled that the contamination with *N*-nitrosamines arose during thermal treatment of the malt (kilning). Precursors of NDMA in malt are hordenin, gramine, and dimethylamine, the latter in concentrations of 5.5–12mg/kg (15). Hordenin and gramine are naturally occurring alkaloids with a tertiary amine structure and are produced in barley during germination processes. Their nitrosation leads to the formation of NDMA. Gramine is converted more rapidly than hordenin; however, hordenin concentrations are higher (43). It has been found that the heating technique is of major importance. Heating the air in the kilning process by direct firing resulted in substantial higher amounts of NDMA in the malt (up to 1080µg/kg) compared with indirect heating using heat exchanger technology (44). In particular, gas burner flames operating at high temperatures (>1500°C) generate substantial amounts of nitrogen oxides from atmospheric nitrogen in the drying air. NOx react with the precursor amines in the malt and this entails markedly increased NDMA formation in the kiln. Modifications in the firing technique (e.g., lower operating temperatures and indirect heating) reduced NOx generation and therefore NDMA formation (16, 22). Table 4.1.3 shows the trend of NDMA concentrations in beer in Germany from 1977 to 1990. A similar decrease of NDMA concentrations in beer was subsequently observed also in other countries (see Table 4.1.4).

In the United States and Canada, a reduction of NDMA contaminations down to 1–5% of earlier levels has been reported by Scanlan and Barbour

TABLE 4.1.4 NDMA concentrations in beer in different countries.

	NDMA, μg/kg	
Country	Mean	Year (Reference)
United States/Canada	0.074	1991 (47)
France	0.23	1991 (48)
Sweden	0.1	1994 (49)
Spain	0.11	1996 (50)

(47). Although NDMA amounts in beer have been markedly reduced, consumption of beer still is a major factor with respect to the total dietary exposure with *N*-nitrosamines. Approximately 30% of the exposure via food results from the consumption of beer, contributing about 0.1 μg of NDMA per person and day (20, 22). The daily uptake of NDMA resulting from the consumption of beer in France and Spain was estimated to be 0.02–00.03 μg per person (48, 50). Other alcoholic beverages were also studied for contamination with volatile *N*-nitrosamines. Goff and Fine (51) examined wine, sherry, liqueur, gin, brandy, vodka, rum, and whiskey. Only the latter showed some NDMA contamination with concentrations between 0.3 and 3 μg/kg. In a study by Walker *et al.* (52), the occurrence of volatile *N*-nitrosamines such as NDMA, NDEA, and *N*-nitrosodi-*N*-propylamine (NDPA) at low concentrations (<1 μg/kg) in brandy and cider from France has been reported. In 11 out of 15 different whiskey samples examined, NDMA was found at concentrations up to 1.2 μg/kg, whereas in wine, volatile *N*-nitrosamines were not detected (20, 22).

4.1.2.1.4 Milk Products and Cheese In general, only rather low concentrations of NDMA were found in cheese and powdered milk.

In two of eight cheese samples, NDMA was detected at concentrations of 0.8–1.1 μg/kg (22). This confirmed results from an earlier study with 200 cheese samples where a low contamination was reported (17). A single study reports the occurrence of NDBA (up to 1.7 μg/kg) along with NDMA (up to 0.84 μg/kg) in cheese from Italy. However, these findings were not confirmed by other studies (53).

In powdered milk, NDMA might be formed when direct drying processes are used (52). Studies by Sen and Seaman (54) reported a mean NDMA concentration of 0.4 μg/kg in Canadian samples, whereas in two studies in Europe, only trace amounts of NDMA were found in milk powder (mean: 0.07 μg/kg NDMA) (20, 22). In the United States, mean NDMA concentrations in milk powder products of 0.7–1.05 μg/kg were reported (55).

4.1.2.1.5 Other Foods *N*-nitrosamines were also found in other foods processed by direct drying techniques, encompassing increased NOx exposure. Low amounts of *N*-nitrosamines (<1 μg/kg, primarily NDMA) have been reported for instant coffee, infant formula, cocoa, powdered egg, and instant soup (20, 22, 54). Spices showed contaminations with NDMA (mean: 0.3 µg/kg, max: 1.4 µg/kg), NPYR (mean: 1.75 µg/kg, max: 29 µg/kg), and NPIP (mean: 1.8 µg/kg, max: 23 µg/kg) (22). Peak contaminations with NPIP and NPYR were found in ground pepper, also containing high amounts of respective precursor amines such as pyrrolidine (41–91 mg/kg) and piperidine (535–642 mg/kg) (15). NDMA (max: 16 µg/kg) and NPYR (max: 6.1 µg/kg) were also found in dried chilies and in dried chili powder (56).

In single cases of fermented vegetables (Chinese cabbage and radish), NDMA (up to 13 µg/kg), NPIP (up to 14 µg/kg), and NPYR (up to 96 µg/kg) have been reported (57). In a study investigating *Brassica oleracea* from India, NDMA (15–16 µg/kg), NPYR (19–25 µg/kg), NPRO (2.1–3.5 µg/kg), and NTCA (0.7 µg/kg) were found (58).

4.1.2.1.6 Human Exposure to* N*-Nitrosamines by Uptake of Food Results from a series of studies estimating the daily exposure per person with volatile *N*-nitrosamines from food in different countries are summarized in Table 4.1.5. NDMA is the most common *N*-nitrosamine found in food and thus contributes primarily to the total human exposure with volatile *N*-nitrosamines. Table 4.1.5 also shows that the mean exposure to volatile *N*-nitrosamines calculated for Germany in 1980 (1.1 µg/person/day) had been reduced to 0.2–0.3 µg/kg in 1990, obviously as a result of mitigation measures such as, in particular, the improved kilning technology for barley malt production (14). Other mitigation measures also contributed, e.g., reduction of nitrite content in curing salt or addition of nitrosation inhibitors such as ascorbic acid or tocopherols.

For most industrialized countries, the estimated average present day intake of volatile *N*-nitrosamines is approximately 0.2–0.3 µg/person, respectively 3.3–5 ng/kg body weight (14, 20, 65). Beer, meat products, and fish are considered the main sources of exposure.

However, other nonvolatile *N*-nitroso compounds have not been identified yet. By using a summary approach detecting the "apparent total *N*-nitroso compounds" (ATNC), the daily exposure to nonvolatile *N*-nitroso compounds was estimated to be 10–100 µg/person (16, 65, 66) with a substantial contribution to this exposure attributed to beer and bacon consumption (66, 67). It has to be underlined that most nonvolatile *N*-nitroso compounds identified in food are biologically not active. It is assumed that the fraction of characterized nonvolatile *N*-nitroso compounds is approximately 10% of the total ATNC content (16).

4.1.3 TOXICOLOGY, TOXICOKINETICS

Preformed and endogenously formed *N*-nitrosamines are well absorbed from the gastrointestinal tract. The rate of absorption varies for different *N*-nitrosamines and for different sections of the gastrointestinal tract. *N*-nitrosamines are distributed by the bloodstream and are rapidly metabolized in the liver, particularly at low concentrations (68). Most of the *N*-nitrosamines

TABLE 4.1.5 Development of estimated daily exposure to preformed nonvolatile *N*-nitrosamines with food in different countries (μg/person/day), reflecting the effect of introducing mitigation measures from the seventies to the nineties.

Country	NDMA	NPYR + NPIP	Comment	Year (Reference)
Finland	0.18 (m); 0.06 (w)	ns	58% from beer[a]	1967–1972 (59)
			22% from cured and smoked fish[a]	
			20% from meat[a]	
	0.1	ns	75% from beer[a]	1989 (40)
			25% from smoked fish[a]	
The Netherlands	0.4	<0.1	71% from beer[a]	1976–1978 (39)
	<0.1	<0.1	[b]	1984/1985 (60)
Japan	1.8	ns	94% from fish[c] (91%[a])	1976–1980 (61)
	0.5	ns	88% from fish[c]	1982 (38)
England	0.1[d]	0.4	35% from meat[a]	1978 (37)
	0.6	0.1		1987 (62)
Sweden	0.1	0.2	6[illegible]% from meat[a]	1980–1986 (20)
			32% from beer	
France	0.25	ns		1980[e] (63)
	0.19	ns	33.5% from alcoholic beverages[a]	1987–1992 (64)
			22% from vegetables[a]	
			12.5% from meat[a]	
Germany	1.0 (m), 0.6 (w)	0.2 (m), 0.1 (w)	64% from beer (m)[a]	1979–1980 (44)
	0.5 (m), 0.4 (w)	0.2 (m), 0.1 (w)	36% from beer (m)[a]	1981 (44)
	0.3 (m), 0.2 (w)	0.03 (m), 0.03 (w)	36% (m), 44% (w) from meat[a]	1989–1990 (22)
			30% (m), 16% (w) from beer[a]	
			15% (m), 19% (w) from fish[a]	

[a]Referring to NDMA.
[b]Results from a 24-h duplicate portion study.
[c]Including other volatile *N*-nitrosamines.
[d]Without beer.
[e]Estimated 1992 with data from 1980.
m, males; w, women; ns, not specified.

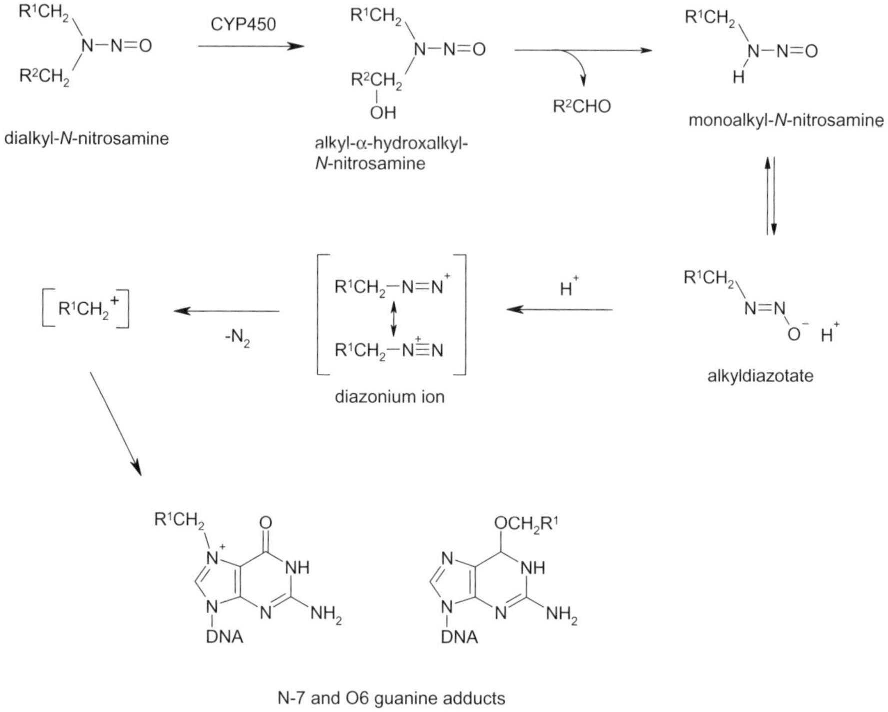

Figure 4.1.7 Metabolic activation of dialkyl-*N*-nitrosamines.

are precarcinogens and subject to metabolic activation. With few exceptions, the elimination of intact parent compounds is negligible due to the effective metabolization of *N*-nitrosamines. The metabolic activation leads to the formation of electrophilic intermediates that react with nucleophilic centers of cellular components such as proteins, RNA, and DNA (Fig. 4.1.7).

The first step in the metabolic activation is the hydroxylation at the α-C-position and is mediated by cytochrome P450 monooxygenases (CYP450). It results in the formation of an α-hydroxy-*N*-nitrosamine, termed a proximal carcinogen. This unstable compound dissociates rapidly into an aldehyde and monoalkylnitrosamine, the latter rearranging into the corresponding diazonium intermediate. The diazonium electrophile reacts with cellular macromolecules such as DNA, RNA, or protein, forming covalent adducts with appropriate nucleophilic centers (69).

The metabolic activation of *N*-nitroso compounds by CYP450 depends on the molecular structure of the compounds, such as length or branching of substituents or presence of other functional groups. It also varies for different CYP450 isoenzymes. Of major importance for metabolic activation of *N*-nitrosamines in humans are CYP450 2E1, 2A3, 2A6, and 3A4 (70, 71). Due to the varying expression of CYP450 isoenzymes in different organs, *N*-nitroso compounds might also be metabolically activated in extrahepatic tissues, e.g., lung, gastrointestinal tract, or other organs.

N-nitrosamines with long side chains might also be hydroxylated at positions other than the α-C atom. The resulting non-α-hydroxylated *N*-nitrosamines are chemically much more stable and might be directly excreted or further oxidized to the respective acids or carbonyl compounds. Such metabolites might also act as substrates for metabolism by phase II enzymes, undergoing conjugation reactions, followed by excretion of the respective conjugates via bile or urine.

The DNA-damaging effect is generally accepted to be the causative factor for the carcinogenicity of *N*-nitrosamines. Metabolically generated reactive electrophilic compounds lead to alkylation of DNA bases, mainly at N-7, O6, N-3 of guanine (see Fig. 4.1.7), N-1, N-3, N-7 of adenine, and the O2- and O4-position of thymine. Although N-7 alkylguanine, in general, is the major DNA base adduct of *N*-nitrosamines, it is assumed that O6-alkylguanine, O4-alkylthymidine, and O2-alkylthymidine, mutations that result in DNA mismatches and miscoding, are more relevant mutagenic lesions entailing carcinogenesis.

N-nitroso-oxazolidinones, such as NOZ-5 and NOZ-2 do not require CYP450-mediated activation because they represent stabilized α-hydroxy nitrosamines, liberating electrophilic diazonium intermediates by ester hydrolysis (Fig. 4.1.8) (72, 73). These compounds have been shown to be strong mutagens and DNA-damaging agents (74–76).

As a consequence of the previously described mitigation measures for foods and cosmetic products, a consistent reduction of contamination with *N*-nitrosamines, and thus of human exposure, down to about 0.2–0.3 μg/person/day (see Table 4.1.5), respectively 3.3–5.0 ng/kg body weight/day (based on a body weight of 60 kg) has been achieved. Based on these data, a margin of exposure (MOE) can be calculated. An MOE describes the ratio between human exposure and a dose level from long-term carcinogenicity experiments

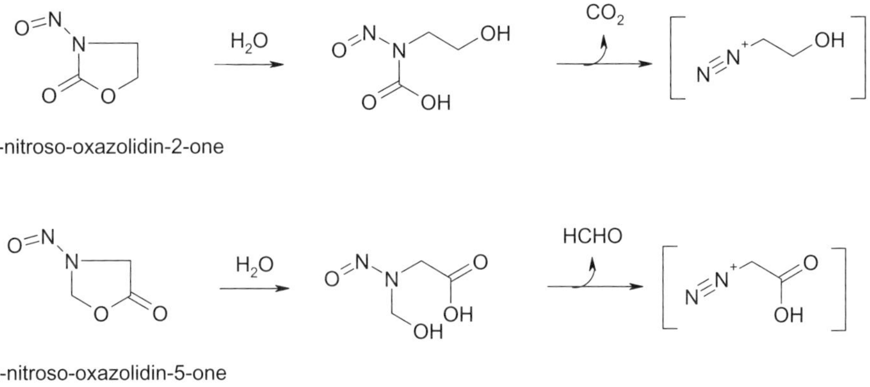

Figure 4.1.8 Formation of electrophilic diazonium intermediates by hydrolyzation of NOZ.

TABLE 4.1.6 MOE calculation for NDMA.

Carcinogen	BMDL10, mg/kg bw/d	Estimated human exposure, ng/kg bw/d	MOE
NDMA	0.06[a]	3.3–5.0	12,000–18,200

[a]Data reviewed in O'Brien *et al.* (78).

inducing a certain tumor response. Either the T25 is chosen, reflecting the dose rate in mg/kg/day which will give 25% tumors at a specific site, or the benchmark dose lower limit (BMDL10), obtained by modeling dose–response data to arrive at the 95% confidence interval of a dose resulting in a 10% tumor response. The EFSA Scientific Committee expresses the view that an MOE of 10,000 or higher would be of low concern (77). The MOE calculation for the most common *N*-nitrosamine in food, NDMA, is shown in Table 4.1.6.

The MOE range currently indicated for dietary exposure might allow utilizing an MOE of 10,000 as reference when evaluating exposure from other sources, such as cosmetics, various consumer products, or working place exposure to prioritize mitigation measures.

4.1.4 CONCLUDING REMARKS

N-nitroso compounds are present in the human environment in foods, cosmetics, and other nonfood products and are taken up by the consumer, giving rise to systemic exposure. The main proportion of this consumer exposure is provided by food. Based on average consumption figures, sustained development and application of mitigation measures has brought about an MOE of >10,000 for NDMA, the main NOC found in food. It is proposed that this might serve as a reference value when other sources of exposure are to be evaluated, including for instance, cases where the risk of *in vivo* nitrosation of a given drug or food additive or constituent is to be assessed or exposure to NOC from nonfood products is to be evaluated. Moreover, exposure at specific working places needs continuous monitoring and further development/refinement of preventive measures to achieve reduction in a way to approach the MOE from food consumption.

Importantly, humans are not only exposed to preformed NOC from environmental media and/or the working place. One important further source of exposure that outweighs by far those discussed here is by smoking and intake of tobacco-specific nitroamines. Another important point to consider is the endogenous formation of NOC. Most nitrosated amino acids excreted in the urine and utilized as biomarkers for endogenous nitrosation are noncarcinogenic, with the exception of *N*-nitrososarcosine, which is rather weak carcinogen. There have been many attempts to assess endogenous formation of NOC, predominantly by NPRO measurement in urine, stool, and other biological media. Although it has been clearly demonstrated that carcinogenic NOC might easily be formed when the appropriate precursors are being taken up, endogenous formation of carcinogenic NOC has not really been evaluated adequately as a process of relevance to human cancer. It is well known that there is endogenous formation of nitrosating agents and that certain disorders, such as inflammatory diseases, bacterial, viral, or parasite infections and the like, can substantially increase endogenous formation of nitrosating agents and thus enhance the risk of forming carcinogenic NOC *in vivo*. It is still not clearly established yet, whether NPRO is a valid biomarker for endogenous formation of NOC other than those arising from nitrosation of amino acids or peptides. It is thus very important to develop in the future appropriate biomarkers that might allow obtaining a realistic estimate of *in vivo* formation of carcinogenic NOC, to better approach the answer of their relevance to human health.

REFERENCES

1. Ender, F., Havre, G., Helgebostad, A., Koppang, N., Madsen, R., Ceh, L. (1964). Isolation and identification of a hepatotoxic factor in herring meal produced from sodium nitrite preserved herring. *Naturwissen*, *51*, 637–638.
2. Fine, D.H., Rounbehler, D.P. (1975). Trace analysis of volatile N-nitroso compounds by combined gas chromatography and thermal energy analysis (TEA). *J. Chromatogr.*, *109*, 271–279.
3. Fine, D.H., Lieb, D., Rufeh, F. (1975). Principle of operation of the thermal energy analyzer for the trace analysis of volatile and non-volatile N-nitroso compounds. *J. Chromatogr.*, *107*, 351–357.
4. Barnes, J.M., Magee, P.N. (1954). Some toxic properties of dimethylnitrosamine. *Br. J. Ind. Med.*, *11*, 167–174.
5. Preussmann, R. (1990). Carcinogenicity and structure-activity relationships of N-nitroso compounds: a review, in *The Significance of N-Nitrosation of Drugs* (ed. H.G. Eisenbrand *et al.*), Gustav Fischer Verlag, pp. 3–17.
6. Reed, P.I. (1996). N-Nitroso compounds, their relevance to human cancer and further prospects for prevention. *Eur. J. Cancer Prev.*, *5*, 137–147.
7. Mirvish, S.S. (1975). Formation of N-nitroso compounds. Chemistry, kinetics and in vivo occurrence. *Toxicol. Appl. Pharmacol.*, *31*, 325–351.
8. Keefer, L.K., Roller, P.P. (1973). N-Nitrosation by nitrite ion in neutral and basic medium. *Science*, *181*, 1245–1246.
9. Druckrey, H., Preussmann, R., Schmähl, D., Müller, M. (1961). Erzeugung von Magenkrebs durch Nitrosamine an Ratten. *Naturwissen*, *48*, 165.
10. Smith, P.A.S., Loeppky, R.N. (1967). Nitrosative cleavage of tertiary amines. *J. Am. Chem. Soc.*, *89*, 1147–1157.
11. Loeppky, R.N., Yu, H., Tang, E. (2002). Endogenous nitrosation: potentially toxic nitrosaminolactones from dietary amino acids and aldehydes. *Chem. Res. Toxicol.*, *15*, 1675.
12. Loeppky, R.N., Bao, Y.T., Bae, J.Y., Yu, L., Shevlin, G. (1994). Blocking nitrosamine formation: understanding the chemistry of nitrosamine formation, in

Nitrosamines and Related N-Nitroso Compounds: Chemistry and Biochemistry (eds R.N. Loeppky, C.J. Michejda), American Chemical Society, Washington, DC, pp. 52–65.
13. Sen, N.P. (1988). Migration and formation of N-nitrosamines from food contact materials, in *Food and Packaging Interacts*, (ed. J.H. Hotchkiss), American Chemical Society, Washington, DC, ACS Symposium Series 365, pp. 146–158.
14. Bartsch, H., Spiegelhalder, B. (1996). Environmental exposure to N-nitroso compounds (NNOC) and precursors: an overview. *Eur. J. Cancer Prev.*, *1*, 11–17.
15. Pfundstein, B., Tricker, A.R., Theobald, E., Spiegelhalder, B., Preussmann, R. (1991). Mean daily intake of primary and secondary amines from foods and beverages in West Germany in 1989–1990. *Food Chem. Toxicol.*, *29*, 733–739.
16. Tricker, A.R., Kubacki, S.J. (1992). Review of the occurrence and formation of non-volatile N-nitroso compounds in foods. *Food Addit. Contam.*, *9*, 39–69.
17. Spiegelhalder, B., Eisenbrand, G., Preussmann, R. (1980). Occurrence of volatile nitrosamines in food: a survey of the West German market. *IARC Sci. Publ.*, *31*, 467–479.
18. Spiegelhalder, B., Eisenbrand, G., Preussmann, R. (1980). Volatile nitrosamines in food. *Oncology*, *37*, 211–216.
19. Sen, N.P., Seaman, S., Miles, W.F. (1976). Dimethylnitrosamine and nitrosopyrrolidine in fumes produced during the frying of bacon. *Food Cosmet. Toxicol.*, *14*, 167–170.
20. Osterdahl, B.G. (1988). Volatile nitrosamines in foods on the Swedish market and estimation of their daily intake. *Food Addit. Contam.*, *5*, 587–595.
21. Gavinelli, M., Fanelli, R., Bonfanti, M., Davoli, E., Airoldi, L. (1988). Volatile nitrosamines in foods and beverages: preliminary survey of the Italian market. *Bull. Environ. Contam. Toxicol.*, *40*, 41–46.
22. Tricker, A.R., Pfundstein, B., Theobald, E., Preussmann, R., Spiegelhalder, B. (1991). Mean daily intake of volatile N-nitrosamines from foods and beverages in West Germany in 1989–1990. *Food Chem. Toxicol.*, *29*, 729–732.
23. Song, P.J., Hu, J.F. (1988). N-Nitrosamines in Chinese foods. *Food Chem. Toxicol.*, *26*, 205–208.
24. Janzowski, C., Eisenbrand, G., Preussmann, R. (1978). Occurrence of N-nitrosamino acids in cured meat products and their effect on formation of N-nitrosamines during heating. *Food Cosmet. Toxicol.*, *16*, 343–348.
25. Tricker, A.R., Perkins, M.J., Massey, R.C., Bishop, C., Key, P.E., McWeeny, D.J. (1984). Incidence of some non-volatile N-nitroso compounds in cured meats. *Food Addit. Contam.*, *1*, 245–252.
26. Tricker, A.R., Perkins, M.J., Massey, R.C., McWeeny, D.J. (1985). N-Nitrosopyrrolidine formation in bacon. *Food Addit. Contam.*, *2*, 247–252.
27. Dunn, B.P., Stich, H.F. (1984). Determination of free and protein-bound N-nitrosoproline in nitrite-cured meat products. *Food Chem. Toxicol.*, *22*, 609–613.
28. Ross, H.D., Henion, J., Babish, J.G., Hotchkiss, J.H. (1987). Nitrosating agents from the reaction between methyl oleate and dinitrogen trioxide: identification and mutagenicity. *Food Chem.*, *23*, 207–222.
29. Tricker, A.R., Perkins, M.J., Massey, R.C., McWeeny, D.J. (1985b). Some nitrosoamino acids in bacon adipose tissue and their contribution to the total N-nitroso compound concentration. *Z. Lebensm.-Unters. Forsch.*, *180*, 379–383.

30. Helgason, R., Ewen, S.W.B., Jaffray, B., Stowers, J.M., Outram, J.R., Pollock, J.R.A. (1984). N-Nitrosamines in smoked meats and their relation to diabetes, in *N-Nitroso Compounds: Occurrence, Biological Effects and Relevance to Human Cancer*, Vol. 57 (eds I.K. O'Neill, R.C. von Borstel, C.T. Miller, J. Long, H. Bartsch), IARC Sci Publ., Lyon, France, pp. 911–920.
31. Sen, N.P., Baddoo, P.A., Seaman, S.W. (1986). N-Nitrosothiazolidine and N-nitrosothiazolidine-4-carboxylic acid in smoked meats and fish. *J. Food Sci.*, *51*, 821–825.
32. Pensabene, J.W., Fiddler, W. (1985). Effect of N-nitrosothiazolidine-4-carboxylic acid on formation of N-nitrosothiazolidine in uncooked bacon. *J. Assoc. Off. Anal. Chem.*, *68*, 1077–1080.
33. Lijinsky, W., Kovatch, R.M., Keefer, L.K., Saavedra, J.E., Hansen, T.J., Miller, A.J., Fiddler, W. (1988). Carcinogenesis in rats by cyclic N-nitrosamines containing sulphur. *Food Chem. Toxicol.*, *26*, 3–7.
34. Ellen, G., Egmond, E., Sahertian, E.T. (1986). N-nitrosamines and residual nitrite in cured meats from the Dutch market. *Z. Lebensm.-Unters. Forsch.*, *182*, 14–18.
35. Massey, R.C., Key, P.E., Jones, R.A., Logan, G.L. (1991). Volatile, non-volatile and total N-nitroso compounds in bacon. *Food Addit. Contam.*, *8*, 585–598.
36. Maki, T., Tamura, Y., Shimamura, Y., Naoi, Y. (1991). Estimate of the volatile nitrosamine content of Japanese food. *Bull. Environ. Contam. Toxicol.*, *25*, 257–261.
37. Gough, T.A., Webb, K.S., Coleman, R.F. (1988). Estimate of the volatile nitrosamine content of UK food. *Nature*, *272*, 161–163.
38. Yamamoto, M., Iwata, R., Ishiwata, H., Yamada, T., Tanimura, A. (1984). Determination of volatile nitrosamine levels in food and estimation of their daily intake in Japan. *Food Chem. Toxicol.*, *22*, 61–64.
39. Stephany, R.W., Schuller, P.L. (1980). Daily dietary intakes of nitrate, nitrite and volative N-nitrosamines in the Netherlands using the duplicate portion sampling technique. *Oncology*, *37*, 203–210.
40. Penttilä, P.L., Räsänen, L., Kimppa, S. (1990). Nitrate, nitrite, and N-nitroso compounds in Finnish foods and the estimation of the dietary intakes. *Z. Lebensm. -Unters. Forsch.*, *190*, 336–340.
41. Sen, N.P., Tessier, L., Seaman, S.W., Baddo, A.P. (1985). Volatile and nonvolatile nitrosamines in fish and the effect of deliberate nitrosation under simulated gastric conditions. *J. Agric. Food Chem.*, *33*, 264–268.
42. Spiegelhalder, B., Eisenbrand, G., Preussmann, R. (1979). Contamination of beer with trace quantities of N-nitrosodimethylamine. *Food Cosmet. Toxicol.*, *17*, 29–31.
43. Mangino, M.M., Scanlan, R.A. (1985). Nitrosation of the alkaloids hordenine and gramine, potential precursors of N-nitrosodimethylamine in barley malt. *J. Agric. Food Chem.*, *33*, 699–705.
44. Spiegelhalder, B. (1983). Vorkommen von Nitrosaminen in der Umwelt, in *Das Nitrosaminproblem* (ed. R. Preussmann), Verlag Chemie, Weinheim, Germany, pp. 235–344.
45. Frommberger, R. (1985). Nitrat, Nitrit, Nitrosamine in Lebensmitteln pflanzlicher Herkunft. *Ernährungsumschau*, *33*, 47–50.

46. Ministerium für Ernährung und Ländlichen Raum Baden-Württemberg (ed.). *Aus der Arbeit der Chemischen Landesuntersuchungsanstalt Stuttgart, Jahresberichte 1983, 1984, 1985, 1986, 1987*, Württembergische Landesbibliothek, Signature ZCa 437.

47. Scanlan, R.A., Barbour, J.F. (1991). N-Nitrosodimethylamine content of US and Canadian beers, in *N-Nitroso Compounds: Occurrence, Biological Effects and Relevance to Human Cancer*, Vol. 57 (eds I.K. O'Neill, R.C. von Borstel, C.T. Miller, J. Long, H. Bartsch), IARC Sci Publ., Lyon, France, pp. 242–243.

48. Mavelle, T., Bouchikhi, B., Debry, G. (1991). Contamination des bières par la N-nitrosodimethylamine. *Sci. Aliments*, *11*, 163–170.

49. Österdahl, B.G., Cuibe, A., Brädemark, P. (1994). Mindre nitrosodimethylamin i starköl. *Vår Föda*, *46*, 269–272.

50. Izquierdo-Pulido, M., Barbour, J.F., Scanlan, R.A. (1996). N-nitrosodimethylamine in Spanish beers. *Food Chem. Toxicol.*, *34*, 297–299.

51. Goff, E.U., Fine, D.H. (1979). Analysis of volatile N-nitrosamines in alcoholic beverages. *Food Cosmet. Toxicol.*, *17*, 569–573.

52. Walker, E.A., Castegnaro, M., Garren, L., Toussaint, G., Kowalski, B. (1979). Intake of volatile nitrosamines from consumption of alcohols. *J. Natl. Cancer Inst.*, *63*, 947–951.

53. Dellisanti, A., Cerutti, G., Airoldi, L. (1996). Volatile N-nitrosamines in selected Italian cheeses. *Bull. Environ. Contam. Toxicol.*, *57*, 16–21.

54. Sen, N.P., Seaman, S. (1981). Volatile N-nitrosamines in dried foods. *J. Assoc. Off. Anal. Chem.*, *64*, 1238–1242.

55. Scanlan, R.A., Barbour, J.F., Bodyfelt, F.W., Libbey, L.M. (1994). N-Nitrosodimethylamine in nonfat dry milk, in *Nitrosamines and Related N-Nitroso Compounds: Chemistry and Biochemistry* (eds R.N. Loeppky, C.J. Michejda), American Chemical Society, Washington, DC, pp. 34–41.

56. Tricker, A.R., Siddiqi, M., Preussmann, R. (1988). Occurrence of volatile N-nitrosamines in dried chillies. *Cancer Lett.*, *38*, 271–273.

57. Poirier, S., Ohshima, H., de-Thé, G., Hubert, A., Bourgade, M.C., Bartsch, H. (1987). Volatile nitrosamine levels in common foods from Tunisia, south China and Greenland, high-risk areas for nasopharyngeal carcinoma (NPC). *Int. J. Cancer*, *39*, 293–296.

58. Kumar, R., Mende, P., Tricker, A.R., Siddiqi, M., Preussmann, R. (1990). N-nitroso compounds and their precursors in *Brassica oleracea*. *Cancer Lett.*, *54*, 61–65.

59. Dich, J., Järvinen, R., Knekt, P., Penttilä, P.L. (1996). Dietary intakes of nitrate, nitrite and NDMA in the Finnish Mobile Clinic Health Examination Survey. *Food Addit. Contam.*, *13*, 541–552.

60. Ellen, G., Egmond, E., Van Loon, J.W., Sahertian, E.T., Tolsma, K. (1990). Dietary intakes of some essential and non-essential trace elements, nitrate, nitrite and N-nitrosamines, by Dutch adults: estimated via a 24-hour duplicate portion study. *Food Addit. Contam.*, *7*, 207–221.

61. Maki, T., Tamura, Y., Shimamura, Y., Naoi, Y. (1980). Estimate of the volatile nitrosamine content in Japanese food. *Bull. Environ. Contam. Toxicol.*, *25*, 257–261.

62. MAFF Ministry of Agriculture, Fisheries and Food (1987). *Nitrate, Nitrite and N-nitroso Compounds in Foods*. Food Surveillance Paper 20, HMSO, London.
63. Cornee, J., Lairon, D., Velema, J., Guyader, M., Berthezene, P. (1992). An estimate of nitrate, nitrite, and N-nitrosodi-methylamine concentrations in French food products or food groups. *Sci. Aliments*, *12*, 155–197.
64. Biaudet, H., Mavelle, T., Debry, G. (1992). Mean daily intake of N-nitrosodimethylamine from foods and beverages in France in 1987–1992. *Food Chem. Toxicol.*, *32*, 417–421.
65. van Maanen, J.M., Dallinga, J.W., Kleinjans, J.C. (1996). Environmental exposure to N-nitroso compounds and their precursors. *Eur. J. Cancer Prev.*, *5*, 29–31.
66. Challis, B.C. (1996). Environmental exposures to N-nitroso compounds and precursors: general review of methods and current status. *Eur. J. Cancer Prev.*, *5*, 19–26.
67. MAFF Ministry of Agriculture, Fisheries and Food (1992). *Nitrate, Nitrite and N-nitroso Compounds in Food: Second Report*. Food Surveillance Paper No. 32, HMSO, London.
68. Mirvish, S.S. (1995). Role of N-nitroso compounds (NOC) and N-nitrosation in etiology of gastric, esophageal, nasopharyngeal and bladder cancer and contribution to cancer of known exposures to NOC. *Cancer Lett.*, *93*, 17–48.
69. Shu, L., Hollenberg, P.F. (1997). Alkylation of cellular macromolecules and target specificity of carcinogenic nitrosodialkylamines: metabolic activation by cytochromes P450 2B1 and 2E1. *Carcinogenesis*, *18*, 801–810.
70. Shu, L., Hollenberg, P.F. (1996). Identification of the cytochrome P450 isozymes involved in the metabolism of N-nitrosodipropyl-,N-nitrosodibutyl- and N-nitroso-n-butyl-n-propylamine. *Carcinogenesis*, *17*, 839–848.
71. Yang, C.S., Smith, T.J., Hong, J.-Y., Zhon, S. (1994). *Kinetics and Enzymes Involved in the Metabolism of Nitrosamines*, American Chemical Society, Washington, DC, ACS Symposium Series 553, pp. 176–178.
72. Singer, S.S. (1985). Decomposition reactions of (hydroxyalkyl) nitrosoureas and related compounds: possible relationship to carcinogenicity. *J. Med. Chem.*, *28*, 1088–1093.
73. Dennehy, M.K., Loeppky, R.N. (2005). Mass spectrometric methodology for the determination of glyoxaldeoxyguanosine and O6-hydroxyethyldeoxyguanosine DNA adducts produced by nitrosamine bident carcinogens. *Chem. Res. Toxicol.*, *18*, 556–565.
74. Lijinsky, W., Reuber, M.D. (1983). Carcinogenicity of hydroxylated alkylnitrosoureas and of nitrosooxazolidones by mouse skin painting and by gavage in rats. *Cancer Res.*, *43*, 214–221.
75. Mirvish, S.S., Markin, R.S., Lawson, T.A., Nickols, J.G. (1988). Induction of hyperplastic liver nodules in Wistar and MRC-Wistar rats by phenobarbital and the liver carcinogens acetoxime, 1-nitroso-5,6-dihydrouracil and 3-nitroso-2-oxazolidinone. *Cancer Lett.*, *41*, 211–216.
76. Lijinsky, W., Thomas, B.J., Kovatch, R.M. (1992). Systemic and local carcinogenesis by directly acting N-nitroso compounds given to rats by intravesicular administration. *Carcinogenesis*, *13*, 1101–1105.

77. EFSA (2005). Opinion of the scientific committee on a request from EFSA related to a harmonised approach for risk assessment of substances which are both genotoxic and carcinogenic. *EFSA J.*, *282*, 1–31.

78. O'Brien, J., Renwick, A.G., Constable, A., Dybing, E., Müller, D.J., Schlatter, J., Slob, W., Tueting, W., van Benthem, J., Williams, G.M., Wolfreys, A. (2006). Approaches to the risk assessment of genotoxic carcinogens in food: a critical appraisal. *Food Chem. Toxicol.*, *44*, 1613–1635.

4.2

FOOD IRRADIATION

EILEEN M. STEWART
Agriculture, Food and Environmental Science Division, Agri-Food and Biosciences Institute (AFBI), Newforge Lane, Belfast BT9 5PX, UK

4.2.1 INTRODUCTION

Of all the food processing technologies ever developed, there have been few more thoroughly researched than that of food irradiation. Use of the technology dates back to 1896 when the first documented proposal to use ionizing radiation to preserve food by destroying spoilage microorganisms was published in Germany. However, despite many years of scientific research and numerous publications on the safety and benefits of using ionizing radiation to preserve food, it is a largely underused technology, being greeted warily by consumers and processors alike.

4.2.2 THE IRRADIATION PROCESS

The process of food irradiation involves the exposure of food to ionizing radiation in a controlled manner. The ionizing radiation can be generated by machine-generated electron beams (maximum energy 10 MeV), X-rays (5 MeV), or by gamma rays produced from the radioisotopes cobalt-60 (1.17 and 1.33 MeV) and cesium 137 (0.662 MeV).

Irradiation, like ultraviolet treatment of drinking water or pasteurization of food, is a process whereby energy is imparted to a product to bring about a desired effect. Food irradiation is unique because the ionizing radiation is

Process-Induced Food Toxicants: Occurrence, Formation, Mitigation, and Health Risks,
Edited by Richard H. Stadler and David R. Lineback

penetrating and transfers energy without a significant rise in temperature, thus it is termed a "cold process." The amount of energy transferred to the foodstuff is known as the "dose" and generally a higher dose means that more energy is absorbed by the product and the greater the induced effect (1).

The amount of irradiation energy that a product absorbs is measured as a specially derived unit known as the "gray" (symbolized Gy). The Gy is the amount of energy that one kilogram of product receives from ionizing radiation, with one Gy being equivalent to one joule (a unit of energy; symbolized by the letter J) per kilogram (1 J/kg). There is an acceptable dose range for a given product and technologically it can be stated that there is a dose window, about 10 Gy to 100 kGy, representing all useful applications of food irradiation outside which no application is currently known (2).

The actual dose of ionizing radiation employed to treat foodstuffs represents a balance between the amount needed to produce a desired effect and the dose that can be tolerated by the product without undergoing an undesired effect. For example, irradiation can cause organoleptic changes, such as off-flavors in foods of animal origin or dairy products, and can potentially cause softening in fresh fruits and vegetables, as well as increasing the permeability of tissue. However, as ionizing radiation slows down the rate of ripening of fresh fruits and vegetables, shelf life can be extended when products are properly stored and/or packaged (1). Factors that influence the successful use of the technology include the quality of the raw materials, the dose applied, the temperature of irradiation, the type of packaging, if used, and storage conditions pre- and postirradiation (3).

In regard to foodstuffs, irradiation can offer a number of applications for improving overall safety and quality (4, 5).

4.2.3 APPLICATIONS OF FOOD IRRADIATION

Ionizing radiation has been used for many years for the sterilization of medical supplies (4) such as orthopedic and cardiovascular devices as well as disposable products and those for wound management. Radiation treatment is also a recognized and well-established process for bringing about a wide range of improvements of materials, including mechanical strength, electrical properties, heat resistance, chemical resistance, color changes, and surface alterations. The technology is routinely used to treat cosmetics and toiletries, pharmaceutical raw materials, veterinary products, laboratory disposables, dyes and colorants, horticultural products, and food packaging.

The benefits of using ionizing radiation to treat foods include the following:

- improvement of safety and shelf-life extension by reduction of numbers of vegetative pathogens and spoilage organisms;

- extension of shelf life by delaying mold growth;
- reduction in the bacterial load on herbs, spices, and seasonings to make them safe to consume and incorporate into foods;
- disinfestation of grains, grain products, and tropical fruits to prevent damage to the commodity and meet quarantine regulations;
- inhibition of sprouting in stored tuber and bulb crops such as garlic, onions, and potatoes;
- sterilization of foods so they can be stored at room temperature for extended periods of time (3–5).

Due to the fact that irradiation causes practically no increase in the temperature of the foodstuff being processed, it can kill microorganisms without thawing frozen food. It can also be used to treat hermetically sealed products without the risk of recontamination or reinfestation of properly packaged foods. Some products may have to be irradiated under particular conditions such as at low temperature or in an oxygen-free environment, while others may have to undergo multiple or hurdle processing such as using a combination of radiation and heat (1, 6).

Food irradiation should, however, not be used by food producers, processors, and consumers as a substitute for good manufacturing practices or proper food handling practices but as a complementary technology, and although the process can offer the aforementioned benefits, it is not a panacea for all problems relating to food safety and quality. For example, if toxins, such as those produced by the spores of *Clostridium botulinum*, were already present in a foodstuff prior to irradiation, these would not be inactivated. Viruses have been shown to be particularly resistant to ionizing radiation, mainly due to their low moisture content and the small size of their genetic material; thus, it is generally accepted that irradiation is not a good method for their elimination. Furthermore, even if foodstuffs were to be irradiated, they would be equally susceptible to post-processing contamination as their nonirradiated counterparts.

Only a small proportion of the world's foodstuffs is irradiated and in 2000 it was estimated that ~100,000 tons per annum were treated worldwide. Dried herbs, spices, and seasonings comprise the largest proportion of food that is irradiated. Many herbs and spices are grown and processed under poor hygienic conditions and are vulnerable to contamination with microorganisms. Although the low water content of dried herbs and spices will inhibit microbial growth, any microorganisms present will generally survive and grow when added to food unless cooked, and furthermore, may grow if the dried product comes into contact with moisture. The irradiation treatment of dried herbs, spices, and seasonings is appropriate to ensure hygienic quality, especially for food consumed cold or without sufficient heating. In addition, because little or no heat is produced during radiation processing, as noted earlier, the microbial load is reduced without causing any adverse effects to the volatile oil component, an important quality characteristic of many of these ingredients (1, 3).

Like other food processing technologies, irradiation causes some chemical changes in food with the majority of chemical substances formed not being unique to irradiation, but similar to those either occurring naturally in some foods or formed as a result of conventional processing methods such as cooking.

4.2.4 RADIATION CHEMISTRY

The changes induced in food by ionizing radiation can either be the result of "direct" or "indirect" action. In direct action, a sensitive target, such as the DNA of a living organism, can be damaged directly by a particle or ray. In indirect action, changes to food are mainly caused as the result of the effect of ionizing radiation on water. As water is a major component of most foodstuffs, water radiolysis is of particular importance in food irradiation.

When pure water is irradiated, a number of highly reactive entities are formed as shown in the following equation (7):

$$H_2O \rightarrow \bullet OH(2.7) + e^-_{aq}(2.7) + \bullet H(0.55) + H_2(0.55) + H_2O_2(0.71) + H_3O^+(2.7)$$

where: $\bullet OH$ = hydroxyl radical
e^-_{aq} = aqueous (or solvated or hydrated) electron
$\bullet H$ = hydrogen atom
H_2 = hydrogen
H_2O_2 = hydrogen peroxide
H_3O^+ = solvated (or hydrated) proton

These are transient in nature and disappear in fractions of a second by reacting with each other and/or with other food components. Hydrogen and hydrogen peroxide are the only stable end products of water radiolysis and are largely lost prior to consumption of the foodstuff. Hydrogen peroxide is an oxidizing agent but of significantly less importance than the highly reactive short-lived radical and hydrated electron intermediates. The hydroxyl radical is a powerful oxidizing agent; the hydrated electron is a strong reducing agent while hydrogen atoms are slightly weaker reducing agents. Overall, both oxidation and reduction reactions take place when food containing water is irradiated (7, 8).

Radiation effects have been found in all the major components of food, i.e., carbohydrates, proteins, and lipids. However, even at the high doses used for sterilization, the changes are so small that they are difficult to detect and are similar to those produced by other food processing technologies such as canning.

4.2.4.1 Carbohydrates

The radiation chemistry of carbohydrates is complex and numerous radiolyic products are possible (7, 9). In foods with a high water content, the carbohydrates react mainly with hydroxyl radicals to form ketones, aldehydes, or acids. It is worthy of note that at least 34 radiolytic products are formed when glucose is irradiated, and it has been shown that the average degree of polymerization of starch is significantly reduced upon irradiation, leading to a reduction in viscosity (10). Generally, the radiolysis products of carbohydrates are formed in proportion to the dose applied.

4.2.4.2 Proteins

As for carbohydrates, the radiation chemistry of proteins is complex, which is not difficult to understand since proteins are comprised of about 20 amino acids, each with its own unique structure and composition (7, 11). The nature of the radiolytic products is dependent on the amino acid. Although irradiation can alter proteins, this does not create a significant nutritional problem as the amino acids generally survive the process due to the fact that they are protected within the complex structure of the protein. It has been demonstrated that doses as high as 50 kGy do not significantly alter protein quality, although the degradation and aggregation of proteins upon irradiation may alter the viscosity of the proteins, such as egg white, where irradiation can impair the whipping quality. Enzymes present in foods, however, are quite stable upon irradiation. Thus, even when foods are sterilized, although they do not spoil microbiologically, they will spoil enzymatically. As a consequence, foods intended for long-term storage must also be heat treated to prevent enzymatic spoilage.

4.2.4.3 Lipids

The lipid, or fat, portion of food consists primarily of triglycerides, which are comprised of fatty acids esterified with glycerol. Changes in lipids due to ionizing radiation can be brought about in two ways:

- by catalyzing their reaction with molecular oxygen, i.e., autoxidation;
- by the action of the high-energy radiation (direct or indirect) on the lipid molecules (7, 12, 13).

As for proteins, an understanding of the basic mechanisms involved in the chemical changes occurring in lipids upon irradiation has been obtained using model systems such as pure triglycerides (7, 13). The chemical reactions resulting from the irradiation of lipids are affected by parameters such as:

- the composition of the lipid (saturated or unsaturated);
- the presence of other substances (antioxidants);

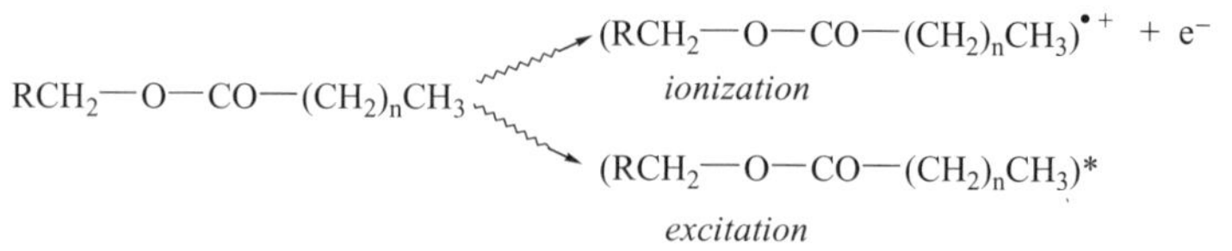

$$RCH_2{-}O{-}CO{-}(CH_2)_nCH_3 \rightsquigarrow (RCH_2{-}O{-}CO{-}(CH_2)_nCH_3)^{\bullet +} + e^- \quad \textit{ionization}$$

$$RCH_2{-}O{-}CO{-}(CH_2)_nCH_3 \rightsquigarrow (RCH_2{-}O{-}CO{-}(CH_2)_nCH_3)^* \quad \textit{excitation}$$

Figure 4.2.1 Irradiation of lipids: primary effect of incident electrons.

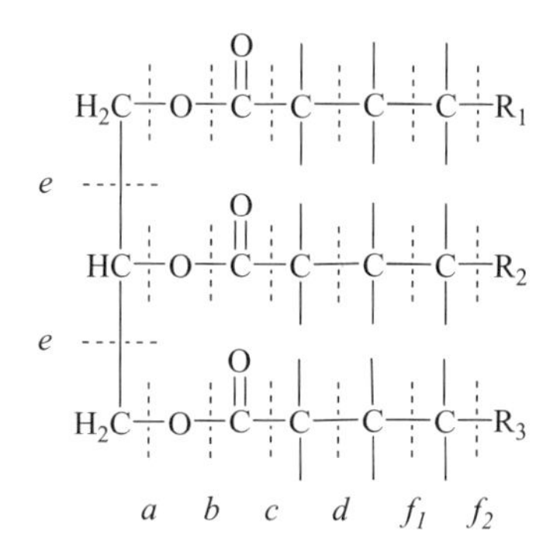

Figure 4.2.2 Possible triglyceride cleavages and products.

- whether the lipid is in the liquid or solid form;
- the irradiation conditions employed and treatment of the food postirradiation, i.e., the storage atmosphere and temperature, which is also of particular importance with lipids (14).

When lipids are irradiated, the primary effect of incident electrons leads to cation radicals and excited molecules as shown in Fig. 4.2.1 (7, 14). According to Nawar (13), the radiolysis of lipids is thought to involve primary ionization, followed by migration of the positive charge either toward the carboxyl group or double bonds. It is thought that cleavage occurs preferentially at positions near the carbonyl group (a, b, c, d, e) but can also occur at other locations (f_1, f_2) (Fig. 4.2.2) (7, 14). Approximately, 16 different free radicals have been proposed to be preferentially produced by cleavage of bonds in the vicinity of the carbonyl group.

For example, cleavage between carbons 1 and 2 (c) of a fatty acid results in the production of a free radical, which may either accept or lose a hydrogen atom to yield the C_{n-1} saturated (alkane) or unsaturated (1-alkene) compounds, respectively (15). If the fatty acid is cleaved between carbons 2 and 3 (d), this may result in a free radical that will further decompose to yield the C_{n-2} alkane and 1-alkene hydrocarbons. As well as hydrocarbons, each triglyceride produces a normal aldehyde and a 2-alkylcyclobutanone when irradi-

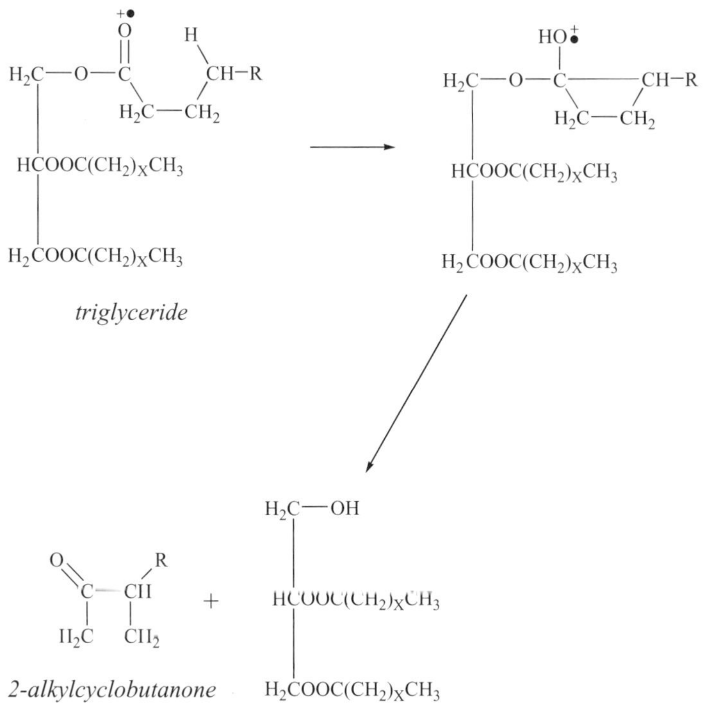

Figure 4.2.3 Proposed pathway for the formation of 2-alkylcyclobutanones.

ated. Both contain the same number of carbons as the parent fatty acid from which they are formed, as well as the methyl and ethyl esters of the parent fatty acids.

It was proposed that the formation of the 2-alkylcyclobutanone results from cleavage at the acyl–oxygen bond in triglycerides with the pathway involving a six-membered ring intermediate (Fig. 4.2.3) (7, 13, 16). The cyclobutanones so formed contain the same number of carbon atoms as the parent fatty acid, and the alkyl group is located in ring position 2.

To date, the cyclobutanones are the only cyclic compounds reported in the radiolytic products of saturated triglycerides. If the fatty acid composition of a lipid is known, then the products formed upon irradiation can be predicted to a certain degree, a fact that has been used for the development of a number of detection methods for irradiated food containing lipid. Thus, for example, if the fatty acids palmitic, stearic, oleic, and linoleic acid are exposed to irradiation, then the respective 2-dodecyl-, 2-tetradecyl-, 2-tetradecenyl-, and 2-tetradecadienyl- cyclobutanones will be formed (17). Currently, two standardized methods are available for the detection of the major radiolytic hydrocarbons and 2-alkylcyclobutanones in fat-containing foods (18).

4.2.5 DETECTION METHODS FOR IRRADIATED FOODS

Irradiated food in international trade should conform to the provisions of a Codex Alimentarius Commission (CAC) General Standard for Irradiated Food and a recommended Code of Practice (19). Within the European Union (EU), framework Directive 1999/2/EC sets out general and technical aspects for carrying out food irradiation, labeling of irradiated foods, and the conditions for authorizing the process, while the implementing Directive 1999/3/EC establishes an initial "positive list" specifying food categories that may be irradiated and freely traded within the EU (20). Currently the list includes only dried aromatic herbs, spices, and vegetable seasonings, and until this list is complete, EU member states may continue to apply their own existing national authorizations of irradiated foodstuffs not included in the initial positive list.

Under Directive 1999/2/EC (20), irradiated foods or food ingredients have to be labeled accordingly using the words "irradiated" or "treated with ionizing radiation" and there is no minimum quantity below which the irradiation would not need to be declared. The indication of treatment should in all cases be given on the documentation accompanying or referring to irradiated foodstuffs. On the subject of additional measures concerning control of foodstuffs, Directive 1999/2/EC lays down quality standards for laboratories and requires the use of validated methods of analysis where available.

Prior to the mid-1980s, little progress had been made in the development of detection methods for irradiated foods. Lack of progress was partly due to the fact that the changes known to occur in food upon irradiation are minimal, as indicated previously, and often similar to those that take place during other food processes, thus making the development of reliable detection methods difficult. However, between 1985 and 1995, significant research took place resulting in a wide range of detection methods being developed, 10 of which have been adopted as European Standards by the European Committee for Standardization (CEN) and these are listed in Table 4.2.1 (18). Although the table lists the foods for which the various methods have been validated by inter-laboratory blind trials, it should be noted that the methods can be applied to other foodstuffs. As well as these methods being adopted by national standardization institutes such as the British Standards Institute (BSi), they have also been adopted by the CAC as General Methods and are referred to in the Codex General Standard for Irradiated Foods in Section 6.4 on "Postirradiation verification" (18). The methodologies are based on physical, chemical, biological, or microbiological changes induced in food upon irradiation. A number of methods are outlined here but more detail can be obtained in reviews by Delincée (21, 22), Stewart (23), Marchioni (24), and in the more detailed publication by McMurray *et al.* (25).

The technique of electron spin resonance (ESR) spectroscopy is used to detect free radicals produced in irradiated food. In foodstuffs with a high moisture content, such as meat and vegetables, the radicals induced upon

TABLE 4.2.1 European standards for the detection of irradiated foodstuffs.[a]

Standard	Description
EN1784:2003	Foodstuffs—detection of irradiated food containing fat—gas chromatographic analysis of hydrocarbons *Validated with raw meat, Camembert, fresh avocado, papaya, mango*
EN1785:2003	Foodstuffs—detection of irradiated food containing fat—gas chromatographic/mass spectrometric analysis of 2-alkylcyclobutanones *Validated with raw chicken, pork, liquid whole egg, salmon, Camembert*
EN1786:1996	Foodstuffs—detection of irradiated food containing bone—method by ESR spectroscopy *Validated with beef bones, trout bones, chicken bones—expected that method can be applied to all meat and fish species containing bone*
EN1787:2000	Foodstuffs—detection of irradiated food containing cellulose, method by ESR spectroscopy *Validated with pistachio nut shells, paprika powder, fresh strawberries*
EN1788:2001	Foodstuffs—detection of irradiated food from which silicate minerals can be isolated, method by thermoluminescence *Validated with herbs and spices as well as their mixtures, shellfish including shrimps and prawns, both fresh and dehydrated fruits and vegetables, potatoes*
EN13708:2001	Foodstuffs—detection of irradiated food containing crystalline sugar by ESR spectroscopy *Validated with dried figs, dried mangoes, dried papayas, raisins*
EN13751:2002	Detection of irradiated food using photostimulated luminescence *Validated with shellfish, herbs, spices, seasonings*
EN13783:2001	Detection of irradiated food using Direct Epifluorescent Filter Technique/Aerobic Plate Count (DEFT/APC)—Screening method *Validated with herbs and spices*
EN13784:2001	DNA comet assay for the detection of irradiated foodstuffs—Screening method *Validated with chicken bone marrow, chicken muscle, pork muscle, almonds, figs, lentils, linseed, rosé pepper, sesame seeds, soya beans, sunflower seeds*
EN14569:2004	Microbiological screening for irradiated foodstuffs—Screening method *Validated for chilled or frozen chicken fillets (boneless) with or without skin*

[a]Reference 18.

irradiation disappear rapidly. However, if the food contains a component with a relatively high dry matter, the radicals may be trapped and remain sufficiently stable to be detected by ESR spectroscopy. Examples would include meat containing bone, crustaceans such as prawns, which have a hard shell, fruits with seeds attached such as strawberries, nuts with shells on, some dry food ingredients such as spices, and dried fruits containing crystalline sugars (18, 23). Characteristic ESR signals obtained from irradiated products can be used to differentiate them from nonirradiated products, although it should be noted that absence of such a characteristic signal does not constitute proof that the food has not been irradiated (24). Currently, there are three CEN and Codex Standards for the identification of irradiated foods based on the use of ESR spectroscopy (Table 4.2.1).

As noted previously, a number of analytical methods are based on the detection of radiolytic products from the lipids contained within foodstuffs. In 1970, Nawar and Balboni (26) reported the feasibility of detecting irradiated pork meat by analysis of six "key hydrocarbons," namely tetradecene ($C_{14:1}$), pentadecane ($C_{15:0}$), hexadecene ($C_{16:1}$), heptadecane ($C_{17:0}$), hexadecadiene ($C_{16:2}$), and heptadecene ($C_{17:1}$), which are typically produced from the three major fatty acids of pork fat, i.e., palmitic, stearic, and oleic acid. Experimental work by Nawar and coworkers (27) demonstrated tetradecene, hexadecadiene, and heptadecene to be the most promising hydrocarbons produced in irradiated chicken meat since they were found in the highest concentrations and were absent or present at low amounts in nonirradiated samples. Following development of a routine method for the detection of these key hydrocarbons, research demonstrated that the method can be widely applied to irradiated fat-containing foods including meat and meat products, frog legs, fish, shrimp, Brazilian beans, Camembert cheese, and sponge cake prepared with irradiated liquid egg (18, 23, 24).

As mentioned before, a routine method based on the detection of 2-alklycyclobutanones was developed for fat-containing foodstuffs, with 2-dodecylcyclobutanone (2-DCB) and 2-tetradecylcyclobutanone (2-TCB) being the two markers most commonly used for identification purposes. The cyclobutanones have been identified in irradiated foods treated with irradiation doses as low as 0.1 kGy and, to date, have not been detected in nonirradiated foods or microbiologically spoiled produce. The specificity of the compounds as irradiation markers has been demonstrated in extensive experimental work, which showed that they are not produced by cooking, by packaging in air, vacuum, or carbon dioxide, or during storage (28). As most foods contain some fat, the method is applicable to a wide range of products such as chicken meat, ground beef, pork, liquid whole egg, Camembert cheese, mango, and papaya (23, 24, 28–30). Detection of irradiated ingredients such as irradiated liquid whole egg in cakes is also possible (23).

Thermoluminescence (TL) and photostimulated luminescence (PSL) are the most sensitive and commonly used methods for the identification of irradiated herbs, spices, and seasonings (23, 24, 31). The methods are based on the detection of luminescence signals originating from mineral grains adhering to the product even though they usually account for <1% of sample weight. The methods can be applied to any foodstuff from which silicate minerals can be extracted, including herbs, spices, their mixtures, fresh fruits and vegetables (e.g., strawberries, mushrooms, papayas, and potatoes), dehydrated fruits and vegetables (e.g., sliced apples, carrots, leeks, and onions) as well as shellfish including shrimps and prawns (18).

In the United Kingdom, the Food Standards Agency (FSA), as part of its authenticity program, conducts surveys to detect unlabeled irradiated foods. In a 2001 survey, a total of 543 different products on sale in the United Kingdom were sampled, including herbs, spices, dietary supplements, prawns, and shrimp (32). Results revealed that of the 138 dietary supplements analyzed, 44 had been wholly irradiated and a further 14 samples contained irradiated ingredients. Five out of 202 prawn and shrimp samples and 1 out of 203 herb and spice samples were also found to be irradiated. None of the foods sampled were appropriately labeled as "irradiated" or "treated with ionizing radiation." In 2005, 17 EU member states reported checks on foods available in the market place (33). A total of 7011 food samples were checked for irradiation treatment of which about 4% were irradiated and/or not labeled. The products found to be irradiated and not correctly labeled included herbs, spices, food supplements, tea, noodles, and dried mushrooms.

The results of these surveys provide good evidence of the reliability of the detection methods in place and can be used to reassure consumers that unlabeled irradiated foods on sale in the market place can be identified and the labeling regulations enforced.

4.2.6 TOXICOLOGY STUDIES

As alluded to earlier, despite the fact that there is an abundance of evidence to support the safety, wholesomeness, and potential benefits of using food irradiation, consumers are still wary of the process. Their concerns include a fear that radioactivity will be induced in the food. However, this is unfounded as the background radioactivity already present in food far exceeds that induced by the sources of ionizing radiation used for treatment at the recommended energy levels (7, 34).

Prior to food irradiation being introduced as a food processing method, clear evidence and assurance had to be obtained that it would not have any adverse toxicological, nutritional, and microbiological effects on the foods being treated while producing the desired results. Extensive studies conducted worldwide since the early 1950s have indicated that the compounds so formed in irradiated foodstuffs are generally the same as those produced during other food processing technologies such as cooking, canning, and pasteurization, and that any differences are not at risk to the consumer (35). To quote from Molins (36), "If cooked foods had not been eaten by humans since the dawn of time, the chemical changes induced in food by cooking would have given scientists material for centuries of research before food control authorities would approve—if ever—the cooking process." Thus, there is a long history of safety studies on irradiated food and it can surely be said that the safety of no other food processing technology has been as thoroughly tested (34).

One of the routes by which the issue of safety was addressed included the International Project in the Field of Food Irradiation (IFIP), or the "Karlsruhe Project," which commenced in 1970 and was in existence until 1982. During this time period, a large number of animal feeding studies were carried out and over 70 reports generated. The data generated were subsequently reviewed at a series of international meetings organized by the World Health Organization (WHO), which were often jointly held with the Food and Agriculture Organization of the United Nations (FAO) and the International Atomic Energy Agency (IAEA). In 1980, a Joint FAO/IAEA/WHO Expert Committee on the Wholesomeness of Irradiated Food (JECFI) met in Geneva and their landmark report was published in 1981 (37). In this report the Committee concluded that the "irradiation of any food commodity up to an overall average dose of 10 kGy presents no toxicological hazard; hence, toxicological testing of foods so treated is no longer required." The Committee also found that irradiation up to 10 kGy "introduces no special nutritional or microbiological problems." Overall, JECFI examined 100 compounds from irradiated beef, pork, ham, and chicken.

As a result of the JECFI report (37), in 1983 the CAC adopted the Codex General Standard for Irradiated Foods and the Recommended Code of Practice for the Operation of Radiation Facilities Used for the Treatment of Foods as referred to earlier in the chapter (19). Under the General Standard, the upper dose limit was set at 10 kGy. The CAC, US Food and Drug Administration (FDA), and the UK Advisory Committee on Irradiated and Novel Foods also adopted the 10-MeV limit for electron radiation and 5-MeV limit for X-rays, as did the governments of most countries, which permitted the irradiation of certain foodstuffs.

Among the most extensive toxicological studies carried out on irradiated food were those undertaken by Raltech Scientific Services, a well-known testing laboratory in the United States. A report published by Thayer *et al.* in 1987 (38) succinctly summarized the comprehensive nutritional, genetic, and toxicological studies of shelf-stable chicken sterilized by ionizing radiation undertaken by Raltech. These investigations were initiated by the US Army in 1976, with supervision being transferred to the US Department of Agriculture (USDA) in October 1980. The reports were reviewed and accepted in their final form in 1984.

During the Raltech studies, a total of 230,000 broiler chickens were processed producing the 134 metric tons of chicken meat needed for this work. The broiler meat (consisting of 18% skin and 82% lean meat with added NaCl and sodium tripolyphosphate, stuffed with cellulose casings, and heated to an internal temperature of 73–80 °C to inactivate enzymes) was used to produce the following four diets: (i) frozen control (FC), canned *in vacuo* and frozen; (ii) thermally processed (TP) control, canned *in vacuo* and thermally processed at 115.6 °C; (iii) GAM containing enzyme-inactivated chicken meat, canned *in vacuo* and sterilized by exposure to gamma radiation at −20 ± 15 °C from a cobalt-60 source, giving a minimum absorbed dose of 46 kGy and a maximum dose of 68 kGy; and (iv) ELE containing the enzyme-inactivated chicken meat, vacuum-packed in 26-mm-thick slices in laminated foil packages and sterilized by exposure to 10-MeV electrons at −25 ± 15 °C, giving an average dose of 58 kGy. Another diet, known as CLD, was used as the negative or husbandry control diet serving as a carrier for the chicken meat in the other four diets (34, 38). The diets produced underwent (i) nutritional studies, (ii) teratology studies, (iii) chronic toxicity, oncogenicity, and multigeneration reproductive studies, and (iv) genetic toxicity studies, and the following is a brief summary of the experimental work and results reported in the review by Thayer *et al.* (38).

The nutritional studies examined the protein efficiency ratios (PER) for rats and mice at the same time evaluating possible antivitamin effects of irradiated meat. ANRC casein was added to the reference standard diet (CLD) for evaluation of PER. Results showed that all the diets containing the chicken meat had higher PER values than the casein standard and were not significantly affected by any of the ways the chicken had been processed.

Genetic toxicology studies of the diets were undertaken using a number of methods. A modified protocol of the Ames test (Salmonella-microsomal mutagenicity assay) was used employing *Salmonella typhimurium* TA1535, which detects mutagens causing base-pair substitutions, and *S. typhimurium* TA1537, TA1538, TA100, and TA98, which detect various kinds of frameshift mutagens. Results from the studies led to the conclusion that the manner in which the chicken was processed, either irradiated or nonirradiated, did not affect the response of the Salmonella-microsomal mutagenicity test system to known mutagens. In addition, no positive results were observed for any of the chicken diets in the absence of the known mutagens. Sex-linked recessive lethal mutations were also tested in a series of studies using Canton-S *Drosophila melanogaster*. None of the four chicken meat diets produced evidence of sex-linked recessive lethal mutations, although it was observed that there was a significant reduction in the egg hatchability of cultures of *D. melanogaster* reared on the gamma-irradiated chicken meat diet (GAM). Despite additional testing carried out to confirm these results, it was concluded that although the irradiated chicken meat was not mutagenic in the test system used, the number of offspring from *D. melanogaster*-fed diets containing chicken was consistently reduced, particularly those containing irradiated chicken. The cause of the observed reduction in the offspring of *Drosophila* or its biological significance as related to humans is not known.

A series of teratology studies were undertaken with mice, hamsters, rats, and rabbits with pregnant females being exposed to the test meats as well as positive control substances, these being all-trans retinoic acid for mice, hamsters and rats, and thalidomide for rabbits. Using *in vivo* studies, it was found that the chicken diets, either irradiated or nonirradiated, induced a teratogenic response when consumed by the pregnant animals. However, when the positive controls were administered to the animals, there were significant incidences of resorbed embryos and congenital malformations in both soft and skeletal body tissues. Consumption of any of the four processed chicken meat diets did not induce significant incidences of resorbed embryos or congenital malformations.

In these studies undertaken at Raltech, chronic feeding studies were conducted in mice and beagle dogs with the five test diets being provided to the animals *ad libitum*. The findings of these studies showed that all five diets supported the growth of the beagles to maturity, although it was noted that the actual body weight and consumption in dogs fed on the negative or husbandry control diet (CLD) were significantly lower than for the beagles consuming the diets containing chicken meat. Moreover, there were no obvious signs of diet-related toxicity in any of the animals, although it was found that the dogs fed GAM diets had lower body weights throughout adulthood than males fed the FC diet. Many of the FC-fed dogs became obese thereby indicating that the difference in body weight between the FC- and GAM-fed animals was not evidence of toxicity. Additionally, there was no evidence of any oncogenic effect from any of the diets. The ability to breed was found to be greater in the female beagles, which consumed the GAM diet, than in dogs fed the other diets and there was no evidence of reproductive toxicity. In the experimental work carried out on mice, the only impaired reproduction noted was for comparatively decreased fertility in mice fed the TP diet as there were no significant differences observed in fertility between mice fed the GAM, ELE, or FC diets. No significant differences in frequency of stillbirths, numbers of viable offspring born, and survival to weaning were observed when groups of mice fed the irradiated chicken meat were compared with mice consuming the FC diet. Mice that consumed the CLD diet were found to have lower mean body weights throughout life than those that ate the diets containing the chicken meat with many mice becoming obese when the FC, TP, GAM, and ELE diets were consumed. Although the mean body weights of the female mice consuming the chicken diets did not differ significantly, it was observed that male mice fed GAM had lower body weights than those fed the other meat diets. The latter observation was attributed to the decreased survival among heavier weight animals in the GAM group, although overall survival for the male mice was not significantly different among the four meat-containing diets. The differences in survival between the CLD group of mice and those fed the meat diets were probably caused by non-neoplastic disease processes such as myocardial degeneration and cardiomyopathy, which were common in the mice consuming the meat diets, the incidence being highest in the GAM group and lowest in the TP group for both sexes. Immune complex glomerulonephropathy was the most common renal lesion, with the incidence being lowest in the

CLD and TP groups, and approximately equal in those fed the FC, GAM, and ELE diets.

Overall, the incidence of tumor development was highest in both males and females fed the FC diet. The female mice that consumed the ELE diet had the lowest incidence of tumors being significantly lower than the group fed the FC diet. For male mice, the lowest incidence of tumors was among those animals fed the GAM diet, although there were no significant differences among the groups consuming the chicken diets.

Thus, overall the Raltech studies consistently produced negative results in all of the tests undertaken. The whole study took 7 years to complete, cost $8 million, and was undoubtedly the most comprehensive safety evaluation ever undertaken on irradiated food (34, 38). Initially, results from the experimental work caused concern. Preliminary evaluation of the data indicated that mice fed the radiation-sterilized chicken showed an increase in kidney damage, and their survival rate was decreased. And, as noted previously, male dogs fed the irradiated chicken appeared to have lower body weights than dogs given the TC diet (34). However, upon re-examination of some of the data, it was concluded that neither glomerulonephropathy or survival in mice nor body weight in dogs was affected by irradiation of chicken meat in the diet (34, 39, 40).

In 1992, the WHO had an expert committee evaluate literature and data available since 1980, taking into account over 500 studies evaluating the safety of irradiated food (7). The critical review undertaken by this committee indicated that "food irradiation is a thoroughly tested food technology" and that "safety studies have so far shown no deleterious effects." For further information, Table 1 in Chapter 6 (Radiological and Toxicological Safety of Irradiated Foods) of the publication by Diehl (34) is based on, with some modifications, the data presented in the WHO report of 1994 (8). It summarizes the long-term or chronic feeding studies undertaken on rats and demonstrates the wide spectrum of food items and dose levels studied, as well as showing the international nature of the work and the variety of sponsors and institutions involved over the 25-year time span during which the studies were undertaken (34).

As noted by Molins (36), two of the most extensive reviews on the assessment of the wholesomeness of irradiated foods are those written by Thayer (41) and Diehl and Josephson (42). The paper by Thayer (41) reviews many of the studies undertaken on chicken as well as looking at the nutritional adequacy of irradiated foodstuffs, while Diehl and Josephson (42) cover the radiological safety, microbiological safety, nutritional adequacy, and toxicological safety of irradiated foods. However, prior to these publications, the paper by Brynjolfsson in 1985 (40) reviewed the major findings in the wholesomeness studies on irradiated foods with the author concluding that the process of food irradiation was ready for industrial applications and "could be effectively regulated for the benefit of the consumer."

In 1997, a further Joint WHO/FAO/IEA Study Group was convened to review data relating to irradiated foods irradiated to doses above 10 kGy and to consider whether a maximum irradiation dose needed to be specified. The report published in 1999 concluded that "food irradiated to any dose appropriate to achieve the intended technologically objective is both safe to consume and nutritionally adequate" (43). In assessing risk, the Study Group also concluded that "irradiation to high doses is essentially analogous to conventional thermal processing, such as the canning of low-acid foods, in that it eliminates biological hazards (i.e. pathogenic and spoilage micro-organisms) from food materials intended for human consumption, but does not result in the formation of physical or chemical entities that could constitute a risk."

Mainly as a result of the findings of the report of the Study Group on High-Dose Irradiation, and after significant debate, in March 2003, the 35th Session of the Codex Committee on Food Additives and Contaminants agreed to amend the General Standard to make a general removal of the upper dose limit. Consequently, the Revised Codex General Standard for Irradiated Foods (Codex Stan 106-1983, Rev.1-2003) was adopted in July 2003 (44), recognizing the use of doses above 10 kGy when necessary for achieving a legitimate technological purpose and states that "the maximum absorbed dose delivered to a food should not exceed 10 kGy, except when necessary to achieve a legitimate technological purpose" (45).

Around the same time as the WHO was reviewing the upper dose limit, the European Commission's Scientific Committee on Food (SCF) was requested to advise the Commission on the appropriateness of specifying a maximum dose for treatment of certain products. Consequently, in April 2003, the SCF expressed a revision of its 1986 opinion on the irradiation of food (46). The Committee concluded that since only very limited toxicological studies have been undertaken for foods treated with doses greater than 10 kGy and none provided for deep frozen convenience foods irradiated above 10 kGy, it could not accept the suggested removal of the upper limit of 10 kGy. The SCF was still of the opinion that it is appropriate to specify a maximum dose for the treatment of certain food products by ionizing radiation and that irradiated foods should continue to be evaluated individually, taking into account the technological need and their safety. The review by the SCF included the findings of relatively new work being undertaken at that time on the safety of the 2-alkylcyclobutanones which, as noted earlier, are used as specific markers for irradiated foodstuffs.

Studies on the toxicological potential of the cyclobutanones commenced mainly in the late 1990s. Initial work reported by Delincée and Pool-Zoebel (47) investigated the effect of a sample of 2-DCB on rat colon cells (*in vitro* and *in vivo*) and on colon cells from human biopsy samples using the comet assay, a single-cell microgel electrophoresis technique. Results from this work indicated that 2-DCB, derived from palmitic acid, had a slight genotoxic potential in rat and in human colon cells. However, the actual identity and purity of the 2-DCB was not verified prior to the experimental work being undertaken, and later characterization of the compound was not possible. The equivocal findings of the aforementioned work on the toxicity of the cyclobutanones were discussed by the authors along with the limitations of the test systems used, and they cautioned against misinterpretation of the results, including interpretation of the data to infer that the cyclobutanones are carcinogens. It was concluded that a possible risk from 2-DCB must be at a very low level and that in order to assess and quantify this minimal risk from the intake of 2-DCB with irradiated food, more experimental work would be required.

A further series of toxicity studies was subsequently undertaken using well-characterized cyclobutanones of high purity. Delincée *et al.* (48, 49) tested 2-TCB, produced from stearic acid, for its cytotoxic and genotoxic potential. This group of workers studied the effect of 2-TCB on established human colon tumor cell lines, which generally function as models for *in vitro* experiments. In this instance, HT29 stem cells and HT29 clone 19A cells were used with cytotoxicity being measured by tetrazolium salt reduction assays (MTT and WST-1) and genotoxicity by measuring DNA damage using the comet assay. Experimental results showed that neither cytotoxic nor genotoxic effects were induced by 2-TCB at an incubation time of 30 min at 37 °C. However, after incubation times of 1–2 days at concentrations of 2-TCB of >50 μM, cytotoxicity did appear although the authors did note that the concentrations tested were very high compared with assumed human intake.

Further work from Raul's group (49, 50) studied whether the cyclobutanones could modulate carcinogenesis in an experimental animal model of colon carcinogenesis. Rats received a daily solution of 2-TCB or 2-tetradecenylcyclobutanone (2-tDeCB), produced from oleic acid, at a concentration of 0.005% in 1% ethanol as drinking fluid with the average consumption of the cyclobutanones being approximately 1.6 mg per rat per day. Control animals received 1% ethanol. After a 2-week period, all rats were injected with the chemical carcinogen azoxymethane (AOM), once a week for 2 weeks. Data reported from this study suggested that cyclobutanones may not initiate colon carcinogenesis *per se* but in the long-term may be promoters of intestinal tumor formation. These workers found that 6 months after injection with AOM, the total number of tumors in the colon was threefold higher in the cyclobutanone-treated rats than in the rats treated with AOM only. As for the other work reported previously (47, 48), the daily amount of pure cyclobutanones given to the rats corresponded to a pharmacological dose of 3.2 mg/kg body weight, which was not comparable to the amount ingested by humans consuming irradiated foods. The latter has been estimated to be ≤5–10 μg/kg body weight. According to Raul *et al.* (50), the amount of cyclobutanones ingested by consuming 200 g of chicken irradiated at 3 kGy can be estimated at 80 μg. In addition, such food products may also contain several components that may reduce the bioavailability of the cyclobutanones.

Horvatovich *et al.* (51) investigated whether the presence of 2-TCB and 2-tDeCB could be detected in the adipose tissues of rats that had consumed these compounds. In this work, laboratory rats received freshly prepared drinking fluid containing 0.005% 2-TCB and 2-tDeCB daily over a 4-month period. Both compounds were recovered in the adipose tissues of the rats that had consumed them, with less than 1% of the cyclobutanones ingested daily being excreted in the feces. Results also demonstrated that both compounds crossed the intestinal barrier, entered the bloodstream and were stored in the adipose tissue of an animal. The amounts detected in both the adipose tissue and feces of the rats were very low compared with the high amounts of cyclobutanones ingested.

In July 2002 (52), the EU SCF assessed the results of these toxicological studies on the 2-alkylcyclobutanones as summarized in the report by Burnouf *et al.* (49). The Committee came to the conclusion that as the adverse effects observed referred almost entirely to *in vitro* studies, it was not appropriate on the basis of these results "to make a risk assessment for human health associated with the consumption of 2-alkylcyclobutanones present in irradiated fat-containing food." Furthermore, they went on to state that the genotoxicity of these compounds had not been established by the standard genotoxicity assays nor were there adequate animal feeding studies in existence to determine no observed adverse effect levels (NOAELs) for various cyclobutanones (52). The fact that the cyclobutanones were present in the radiation-sterilized chicken meat tested in the Raltech studies (38, 53), which strongly supported the safety of irradiated food, should also be used as evidence as to the safety of these radiolytic products.

Further work reported by Sommers (54) evaluated the capacity of 2-DCB to induce mutations using the *Escherichia coli* tryptophan reverse mutation (Trp) assay. *E. coli* tester strains WP2 and WP2 *uvr*A, with or without exogenous metabolic activation, were exposed to 0, 0.05, 0.1, 0.5, and 1 mg per well 2-DCB using the Miniscreen version of the assay. Results showed that 2-DCB did not induce mutations in the *E. coli* Trp assay, which was in agreement with the negative results obtained in historical data but in contrast to the potential genotoxicity reported by Delincée and coworkers (48, 49).

Sommers and Schiestl (55) used the *Salmonella* mutagenicity test and the yeast deletion (DEL) assay to evaluate the genotoxic potential of 2-DCB. The results obtained were in agreement with those previously reported by Sommers (54) demonstrating the absence of genotoxicity of purified 2-DCB. Sommers and Mackay (56) also studied the ability of 2-DCB to increase the expression of DNA damage-inducible genes in *E. coli* that contained stress-inducible promoters fused to β-galactosidase reporter genes, and to induce the formation of 5-fluorouracil (5-FU)-resistant mutants in *E. coli*. Their findings showed that 2-DCB did not increase expression of DNA damage-inducible genes in *E. coli* or the formation of 5-FU-resistant mutants.

In their paper published in 2004, Marchioni *et al.* (57) presented the results of a collaborative study on the toxicology of 2-alkylcyclobutanones. It was concluded that although the results of the study pointed toward the toxic, genotoxic, and even tumor-promoting activity of certain highly purified cyclobutanones, it should be emphasized that the experimental data are inadequate to characterize a possible risk associated with the consumption of irradiated fat-containing foods. The authors noted that other food components may influence the reactions of the cyclobutanones not evident from their experiments on the purified compounds and that further knowledge is needed about the kinetics and metabolism of these compounds in the living organism.

Knoll *et al.* (58) explored the relative sensitivities of human colon cells, representing different stages of tumor development and healthy colon tissues, to 2-DCB. HT29clone19A cells, LY97 adenoma cells, and primary human epithelial cells were exposed to 2-DCB. It was observed that the cyclobutanone was cytotoxic in a time- and dose-dependent manner in LT97 adenoma cells and in freshly isolated primary cells but not in the human colon tumor cell line. An associated induction of DNA damage by 2-DCB was noted in the LT97 cells and in freshly isolated colonocytes, while no strand breaks were detectable in HT29clone19A cells. When LT97 adenoma cells were incubated on a long-term basis with lower concentrations of 2-DCB, cytogenic effects were observed. Knoll and coworkers thus concluded that 2-DCB was genotoxic in healthy human colon epithelial cells and in cells representing preneoplastic colon adenoma. They could not, however, speculate if the cell-specific effects of the compound were as a result of differences in cellular uptake, metabolism, DNA repair, or other pathways. It was also reported that 2-DCB induces chromosomal aberrations in a human colon adenoma, which are associated with human cancer. However, the authors note that the doses needed to cause the genetic alterations probably by far exceed the normal exposure situation. They went on to state that the amount of 2-DCB found in a 100-g beef burger equates to ~3.3 μg or ~14 nmol (59), is some 50,000-fold lower than effective concentrations *in vitro*, and thus possibly too low to have a significant impact on human health (58).

Gadgil and Smith (60) determined the mutagenic potential of 2-DCB using the Ames assay and compared the acute toxicity of 2-DCB with the food additive cylohexanone and *t*-2-nonenal using the Microtox assay. Cyclohexanone and 2-nonenal are both carbonyl compounds like 2-DCB. The results of this experimental work suggested that 2-DCB is not mutagenic to the Salmonella strains tested and that its acute toxicity is between that of cyclohexanone and 2-nonenal. The work indicated that 2-DCB is similar in toxicity to cyclohexanone, which has generally recognized as safe (GRAS) status in the United States, and was 10 times less toxic than *t*-2-nonenal, a normal food constituent of cooked ground beef, and an approved food additive (GRAS status flavorant). In their 2004 Scientific Status Summary on "Irradiation and Food Safety" for the Institute of Food Technologists (IFT), Smith and Pillai (61) concluded that such results indicate that 2-DCB has very low toxicity thereby not warranting concern.

Further studies by Gadgil and Smith (62) followed on from the work by Horvatovich *et al.* (51), as mentioned previously, examined the fate of 2-DCB in rats after consumption by looking at its recovery from the feces and adipose tissue of rats and if any urinary metabolites could be identified. From the findings of this work, it was observed that between 3% and 11% of the total amount of 2-DCB given to the rats was recovered from the feces while approximately 0.33% was recovered from adipose tissue. These results indicated that most of the 2-DCB is metabolized and excreted or stored in tissues other than adipose.

The most recent study carried out on the cyto- and genotoxicity potential of 2-alkylcyclobutanones of varying chain length in different cell lines was conducted by Hartwig *et al.* (63). Studies undertaken included the (i) impact of the cyclobutanones on the growth of *S. typhimurium* TA97 strain; (ii) the mutagenic potential of the cyclobutanones using the Ames test; (iii) *in vitro* experiments to study the exposure of the human colon tumor cells HT 29 stem, HT 29 clone 19A to cyclobutanones; (iv) use of different test systems to investigate cytotoxicity (trypan blue exclusion test to assess membrane integrity; tetrazolium salt reduction assays applying both MTT and WST-1; assessment of cytotoxicity in logarithmically growing cell by colony forming ability); (v) genotoxicity of cyclobutanones in human colon tumor cells as determined by the comet assay in the absence or presence of Fpg protein; and (vi) measurement of DNA strand breaks and Fpg-sensitive sites in human cells measured by alkaline unwinding. Results from this extensive work indicated that the cyclobutanones have cytotoxic properties both in bacteria and human cells. However, the authors pointed out that the cytotoxic effects vary with the nature of the actual compound and the cells investigated. As a common trend, which was more pronounced in bacteria, they observed that the shorter the chain of the cyclobutanone the higher the toxic effect. No mutagenic potential was detected for any cyclobutanone using the Ames test, which is in agreement with previously published results for 2-DCB (54, 61). For 2-DCB, both the comet assay and alkaline unwinding revealed an increase in Fpg sensitivities after a 24-h incubation period but not after 30 min as determined by the comet assay. The result for short-term incubation differs to that found previously where a lack of genotoxicity was found (47). The data derived from the alkaline unwinding experiments demonstrated a genotoxic potential of all compounds investigated with the intensity of DNA damage being dependent on the length of the fatty acid side chain, the degree of unsaturation, and the cell line applied. Hartwig *et al.* (63) concluded that, taken together, the results indicate a genotoxic potential of purified cyclobutanones in mammalian cells and seem to contradict the outcome of the Raltech studies (38) where, as noted earlier, no genotoxicity was observed with radiation-sterilized chicken meat. However, as for other work carried out previously, the authors state that the effects of the cyclobutanones should be elucidated in more detail with complementary studies being needed to clarify mechanisms of action and an adequate risk assessment for human exposure undertaken.

The European Commission's SCF reviewed the safety of irradiated food in 1986, 1992, 1998, and 2003, and on all occasions agreed that the irradiation of food is safe up to an overall average dose of 10 kGy. In its 2003 report (46), it referred to the toxicity studies of the cyclobutanones by Burnouf *et al.* (49) and stated that the genotoxicity of these compounds could not be considered as having been established and that the cytotoxicity was observed at concentrations of some 0.30–31.25 mg/mL medium, which were about three orders of magnitude greater than the levels of 17 μg/g reported in the lipids of chicken irradiated at 59 kGy (38). SCF also noted that in contrast to the work of Burnouf and coworkers (49), no mutagenic activity was detected in studies with *D. melanogaster* and mice fed chicken irradiated at 55.8 and 59 kGy reported in the Raltech studies.

Lastly, another group of compounds that have generated some concern with regard to irradiated food is benzene and its derivatives. The results of studies carried out by the Federation of American Societies for Experimental Biology, as reported by Chinn (64), reached the conclusion that the small amounts of benzene generated in irradiated beef, i.e., 15 ppb in beef irradiated with a dose of 56 kGy compared with the 3 ppb measured in nonirradiated beef, did not constitute a significant risk. Experimental work undertaken by Health Canada (65) measured approximately 3 ppb benzene in beef treated with an irradiation dose of 1.5–4.5 kGy. It was noted that this amount of benzene was significantly lower than the naturally occurring concentrations of 200 ppb found in haddock and 62 ppb in eggs as reported by McNeal *et al.* (66). Thus, the risk of benzene exposure from irradiated foods is considered negligible (61).

4.2.7 CONCLUSIONS

Food can be classified as one of life's risks although food-related risks are incredibly low in context of many other daily living risks (67). Risks most commonly associated with the consumption of food include those related to (i) food-poisoning microorganisms; (ii) viruses such as the norovirus infection, which has been known to occur on cruise ships; (iii) human-related transmissible spongiform encephalopathies (TSEs) such as Creutzfeldt–Jakob disease (CJD); (iv) natural toxins in foods such as mycotoxins or aflatoxins; (v) agrochemical residues including pesticides, veterinary residues, or fertilizers—the list could go on. As pointed out by Sommers *et al.* (68) the carcinogens in foods, such as acrylamide, benzene, formaldehyde, furan, and nitrosamines, are naturally occurring, or formed as a result of thermal processing. It should, however, be borne in mind that the benefits of consuming food are great and as well as providing sustenance, essential micronutrients, and so on, certain foods also contain other compounds such as antioxidants, which help fight against illnesses such as heart disease and cancer. Thus, the risks and benefits must be weighed.

The same philosophy can be applied to food irradiation as there are pros and cons to treating food with ionizing radiation (69, 70). Given the significant amount of research undertaken on the application and safety of the technology, along with the fact that it is considered safe by national and international bodies such as the WHO, FAO, and others including the American Medical Association and the Institute of Food Technologists, it can be concluded that consuming irradiated foods as part of a healthy balanced diet would be difficult to conceive as a risk (68, 71).

REFERENCES

1. Farkas, J. (2001). Food irradiation. A technique for preserving and improving the safety of food, in *Food Microbiology: Fundamentals and Frontiers*, 2nd edn (eds M. O. Doyle, L.R. Beuchat, T.J. Montville), ASM Press, Washington, DC, pp. 567–592.
2. Ehlermann, E. (2001). Process control and dosimetry in food irradiation, in *Food Irradiation—Principles and Applications* (ed. R. Molins), John Wiley & Sons, Inc., Ch. 15, pp. 387–414.
3. Stevenson, M.H. (1990). The practicalities of food irradiation. *Food Technology International Europe*, *2*, 73–77.
4. Molins, R. (ed.) (2001). *Food Irradiation—Principles and Applications*, John Wiley & Sons, Inc.
5. Sommers, C.H., Fan, X. (eds) (2006). *Food Irradiation and Technology*, IFT Press, Blackwell Publishing.
6. Patterson, M. (2001). Combination treatments involving food irradiation, in *Food Irradiation—Principles and Applications* (ed. R. Molins), John Wiley & Sons, Inc., Ch. 12, pp. 313–328.
7. Stewart, E.M. (2001). Food irradiation chemistry, in *Food Irradiation—Principles and Applications* (ed. R. Molins), John Wiley & Sons, Inc., Ch. 14, pp. 37–76.
8. WHO (1994). *Safety and Nutritional Adequacy of Irradiated Food*, World Health Organization, Geneva.
9. Dauphin, J-F., Saint-Lèbe, L.R. (1977). Radiation chemistry of carbohydrates, in *Radiation Chemistry of Major Food Components* (eds P.S. Elias, A.J. Cohen), Elsevier Scientific Publishing Company, Amsterdam, Oxford, New York, Ch. 5, pp. 131–220.
10. Farkas, J., Sharif, M.M., Koncz, A. (1990). Detection of some irradiated spices on the basis of radiation induced damage of starch. *Radiation Physics and Chemistry*, *36*, 621–627.
11. Delincée, H. (1983). Recent advances in radiation chemistry of proteins, in *Recent Advances in Food Irradiation* (eds P.S. Elias, A.J. Cohen), Elsevier Biomedical Press, Amsterdam, the Netherlands, pp. 129–147.
12. Nawar, W.W. (1986). Volatiles from food irradiation. *Food Reviews International*, *2*, 45–78.
13. Nawar, W.W. (1978). Reaction mechanisms in the radiolysis of fats: a review. *Journal of Agricultural and Food Chemistry*, *26*, 21–25.
14. Delincée, H. (1983). Recent advances in radiation chemistry of lipids, in *Recent Advances in Food Irradiation* (eds P.S. Elias, A.J. Cohen), Elsevier Biomedical Press, Amsterdam, the Netherlands, pp. 89–114.
15. Dubravic, M.F., Nawar, W.W. (1968). Radiolysis of lipids: mode of cleavage of simple triglycerides. *Journal of the American Oil Chemists Society*, *45*, 656–660.
16. LeTellier, P.R., Nawar, W.W. (1972). 2-Alkylcyclobutanones from radiolysis of triglycerides. *Lipids*, *7*, 75–76.

17. Elliott, C.T., Hamilton, L., Stevenson, M.H., McCaughey, W.J., Boyd, D.R. (1995). Detection of irradiated chicken meat by analysis of lipid extracts for 2-substituted cyclobutanones using an enzyme linked immunosorbent assay. *Analyst*, *120*, 2337–2341.
18. EC (2008). Food irradiation—analytical methods, Information on analytical methods for the detection of irradiated foods standardised by the European Committee for Standardisation (CEN). http://ec.europa.eu/food/food/biosafety/irradiation/anal_methods_en.htm (accessed 19 September 2008).
19. CAC (1984). Codex "General Standards for Irradiated Foods and Recommended International Code of Practice for the Operation of Radiation Facilities used for the Treatment of Foods", Codex Alimentarius Commission, Vol. XV, 1st edn, Rome, Italy.
20. EC (1999). Directive 1999/2/EC of the European Parliament and of the Council of 22 February 1999. *Official Journal of the European Communities*, *L 066*, 16–23.
21. Delincée, H. (1998). Detection of food treated with ionizing radiation. *Food Science and Technology*, *9*, 73–82.
22. Delincée, H. (2002). Analytical methods to identify irradiated food—a review. *Radiation Physics and Chemistry*, *63*, 455–458.
23. Stewart, E.M. (2001). Detection methods for irradiated foods, in *Food Irradiation—Principles and Applications* (ed. R. Molins), John Wiley & Sons, Inc., Ch. 3, pp. 347–386.
24. Marchioni, E. (2006). Detection of irradiated foods, in *Food Irradiation Research and Technology* (eds C.H. Sommers, X. Fan), IFT Press, Blackwell Publishing, Ch. 6, pp. 85–104.
25. McMurray, C.H., Stewart, E.M., Gray, R., Pearce, J. (eds) (1996). *Detection Methods for Irradiated Foods—Current Status*, Royal Society of Chemistry, Special Publication No. 171, Cambridge.
26. Nawar, W.W., Balboni, J.J. (1970). Detection of irradiation treatment in foods. *Journal of the Association of Official Analytical Chemists*, *53*, 726–729.
27. Nawar, W.W., Zhu, R., Yoo, Y.J. (1990). Radiolytic products of lipids as markers for the detection of irradiated meats, in *Food Irradiation and the Chemist* (eds D.E. Johnston, M.H. Stevenson), Royal Society of Chemistry, Special Publication No. 86, Cambridge, pp. 13–24.
28. Stevenson, M.H. (1994). Identification of irradiated foods. *Food Technology*, *48*, 141–144.
29. Stewart, E.M., Graham, W.D., Moore, S.K., McRoberts, W.C., Hamilton, J.T.G. (2000). 2-Alkylcyclobutanones as markers for the detection of irradiated mango, papaya, Camembert cheese and salmon meat. *Journal of the Science of Food and Agriculture*, *80*, 121–130.
30. Stewart, E.M., McRoberts, W.C., Hamilton, J.T.G., Graham, W.D. (2001). Isolation of lipid and 2-alkylcyclobutanones from irradiated foods by supercritical fluid extraction. *Journal of AOAC International*, *84*, 976–986.
31. Sanderson, D.C.W., Carmichael, L.A., Naylor, J.D. (1995). Photostimulated luminescence and thermoluminescence techniques for the detection of irradiated food. *Food Science and Technology Today*, *9*, 150–154.

32. FSA (2002). Survey for Irradiated Foods—Herbs, Spices, Dietary Supplements and Prawns and Shrimps. Food Survey Information Sheet Number 25/02, Food Standards Agency, June 2002.

33. EU (2007). Report from the Commission on food irradiation for the year 2005. *Official Journal of the European Union, 2007*, C122/03–C122/21.

34. Diehl, J.F. (1995). Chemical effects of radiation, in *Safety of Irradiated Foods*, 2nd edn, Marcel Dekker, Inc., New York, Ch. 3, pp. 43–88.

35. GAO (2000). Food irradiation. Available research indicates that benefits outweigh risks. United States General Accounting Office, Report to Congressional Requesters, GAO/RCED-00-217, August 2000.

36. Molins, R.A. (2001). Introduction, in *Food Irradiation—Principles and Applications* (ed. R. Molins), John Wiley & Sons, Inc., Ch. 1, pp. 1–22.

37. WHO (1981). Wholesomeness of Irradiated Food: a Report of a Joint FAO/IAEA/WHO Expert Committee on Food Irradiation. WHO Technical Report Series, 659, World Health Organization, Geneva.

38. Thayer, D.W., Christopher, J.P., Campbell, L.A., Ronning, D.C., Dahlgren, R.R., Thomson, G.M., Wierbicki, E. (1987). Toxicology studies of irradiation-sterilized chicken. *Journal of Food Protection*, *50*, 278–288.

39. FDA (1986). Irradiation in the production, processing, and handling of food. *Federal Register*, *51*, 13376–13399.

40. Brynjolfsson, A. (1985). Wholesomeness of irradiated foods: a review. *Journal of Food Safety*, *7*, 107–126.

41. Thayer, D.W. (1994). Wholesomeness of irradiated foods. *Food Technology*, *48*, 132–135.

42. Diehl, J.F., Josephson, E.S. (1994). Assessment of wholesomeness of irradiated foods: a review. *Acta Alimentaria*, *23*, 195–214.

43. WHO (1999). High-Dose Irradiation: Wholesomeness of Food Irradiated with Doses above 10 kGy. Report of a Joint FAO/IAEA/WHO Study Group, WHO Technical Report Series 890, World Health Organization, Geneva.

44. FAO (2003). Codex Alimentarius Commission Adopts More than 50 New Food Standards. FAO Newsroom, 9 July 2003. http://www.fao.org/english/newsroom/news/2003/20363-en.html (accessed 19 September 2008).

45. CAC (2003). Revised Codex General Standard for Irradiated Foods. Codex Alimentarius Commission. CODEX STAN 106-1983, REV. 1-2003. Joint FAO/WHO Food Standards Programme, FAO, Rome.

46. SCF (2003). Revision of the Opinion of the Scientific Committee on Food on the Irradiation of Food (expressed on 4 April 2003), SCF/CS/MF/IRR/24 Final, 24 April 2003. http://ec.europa.eu/food/fs/sc/scf/out193_en.pdf (accessed 19 September 2008).

47. Delincée, H., Pool-Zobel, B-L. (1998). Genotoxic properties of 2-dodecylcyclobutanone, a compound formed on irradiation of food containing fat. *Radiation Physics and Chemistry*, *52*, 39–42.

48. Delincée, H., Soika, C., Horvatovich, P., Rechkemmer, G., Marchioni, E. (2002). Genotoxicity of 2-alkylcyclobutanones, markers for an irradiation treatment of fat-containing food—Part I: cyto- and genotoxic potential of 2-tetradecylcyclobutanone. *Radiation Physics and Chemistry*, *63*, 431–435.

49. Burnouf, D., Delincée, H., Hartwig, A., Marchioni, E., Miesch, M., Werner, D. (2002). *Toxicological Study to Assess the Risk Associated with the Consumption of Irradiated Fat-Containing Food*, Bundesforschungsanstalt für Ernährung (BFE), Karlsruhe, Germany (in German and French but with English Summary and Conclusions).
50. Raul, F., Gossé, F., Delincée, H., Hartwig, A., Marchioni, E., Miesch, M., Werner, D., Barnouf, D. (2002). Food-borne radiolytic compounds (2-alkylcyclobutanones) may promote experimental colon carcinogenesis. *Nutrition and Cancer*, *44*, 188–191.
51. Horvatovich, P., Raul, F., Miesch, M., Burnouf, D., Delincée, H., Hartwig, A., Werner, D., Marchioni, E. (2002). Detection of 2-alkylcyclobutanones, markers for irradiated foods, in adipose tissues of animals fed with these substances. *Journal of Food Protection*, *65*, 1610–1613.
52. SCF (2002). Statement of the Scientific Committee on Food on a Report on 2-Alkylcyclobutanones (expressed on 3 July 2002), SCF/CS/NF/IRR/26 ADD 3 Final, 3 July 2002. http://ec.europa.eu/food/fs/sc/scf/out135_en.pdf (accessed 19 September 2008).
53. Crone, A.V.J., Hamilton, J.T.G., Stevenson, M.H. (2002). The detection of 2-dodecylcyclobutanone in radiation-sterilised meat stored for several years. *International Journal of Food Science and Technology*, *27*, 691–696.
54. Sommers, C.H. (2003). 2-Dodecylcyclobutanone does not induce mutations in the *Escherichia coli* tryptophan reverse mutation assay. *Journal of Agricultural and Food Chemistry*, *51*, 6367–6370.
55. Sommers, C.H., Schiestl, R.H. (2002). 2-Dodecylcyclobutanone does not induce mutations in the *Salmonella* mutagenicity test or intrachromosomal recombination in *Saccharomyces cerevisiae*. *Journal of Food Protection*, *67*, 1293–1298.
56. Sommers, C.H., Mackay, W.J. (2005). DNA damage-inducible gene expression and formation of 5-fluorouracil-resistant mutants in *Escherichia coli* exposed to 2-dodecylcyclobutanone. *Food Chemistry and Toxicology*, *70*, C254–C257.
57. Marchioni, E., Raul, F., Burnouf, D., Miesch, M., Delincée, H., Hartwig, A., Werner, D. (2004). Toxicological study on 2-alkylcyclobutanones—results of a collaborative study. *Radiation Physics and Chemistry*, *71*, 145–148.
58. Knoll, N., Weise, A., Claussen, U., Sendt, W., Marian, B., Glei, M., Pool-Zobel, B.L. (2006). 2-Dodecylcyclobutanone, a radiolytic product of palmitic acid, is genotoxic in primary human colon cells and in cells from preneoplastic lesions. *Mutation Research*, *594*, 10–19.
59. Gadgil, P., Hachmeister, K.A., Smith, J.S., Kropf, D.H. (2002). 2-Alkylcyclobutanones as irradiation indicators in irradiated ground beef patties. *Journal of Agricultural and Food Chemistry*, *50*, 5746–5750.
60. Gadgil, P., Smith, J.S. (2004). Mutagenicity and acute toxicity evaluation of 2-dodecylcyclobutanone. *Food Chemistry and Toxicology*, *69*, C713–C716.
61. Smith, J.S., Pillai, S. (2004). Irradiation and food safety. *Food Technology*, *58*, 48–55.
62. Gadgil, P., Smith, J.S. (2006). Metabolism of 2-dodecylcyclobutanone, a radiolytic compound present in irradiated beef. *Journal of Agricultural and Food Chemistry*, *54*, 4896–4900.
63. Hartwig, A., Pelzer, A., Burnouf, D., Titéca, H., Delincée, H., Briviba, K., Soika, C., Hodapp, C., Raul, F., Miesch, M., Werner, D., Horvatovich, P., Marchioni, E. (2007).

Toxicological potential in 2-alkylcyclobutanones—specific radiolytic products in irradiated fat-containing food—bacteria and human cell lines. *Food and Chemical Toxicology*, *45*, 2581–2591.

64. Chinn, H.I. (1979). Further toxicological considerations of volatile products, in *Evaluation of the Health Aspects of Certain Compounds Found in Irradiated Beef*, Life Sciences Research Office, Federation of American Societies for Experimental Biology, Bethesda, MD, Ch. 1, pp. 1–29.
65. Health Canada (2002). Irradiation of ground beef: summary of submission process. October 29. Food Directorate, Food Products and Health Branch, Ottawa. http://www.hc-sc.gc.ca/fn-an/alt_formats/hpfb-dgpsa/pdf/securit/gbeef_submission-soumission_viande_hachee-eng.pdf (accessed 19 September 2008).
66. McNeal, T.P., Nyman, P.J., Diachenko, G.W., Hollifield, H.C. (1993). Survey of benzene in foods using headspace concentration techniques and capillary gas chromatography. *Journal of AOAC International*, *76*, 1213–1219.
67. Shaw, I. (2005). *Is It Safe to Eat?* Springer, Berlin, Heidelberg, New York.
68. Sommers, C.H., Delincée, H., Scott Smith, J., Marchioni, E. (2006). Toxicological safety of irradiated foods, in *Food Irradiation Research and Technology* (eds C.H. Sommers, X. Fan), IFT Press, Blackwell Publishing, Ch. 4, pp. 43–62.
69. Stewart, E.M. (2004). Food irradiation: more pros than cons? *Biologist*, *51*, 91–96.
70. Stewart, E.M. (2004). Food irradiation: more pros than cons? Part 2. *Biologist*, *51*, 141–144.
71. Satin, M. (1996). *Food Irradiation: A Guidebook*, 2nd edn, CRC Press.

4.3

BENZENE

ADAM BECALSKI[1] AND PATRICIA NYMAN[2]

[1] *Food Research Division, Bureau of Chemical Safety, Health Products and Food Branch, Health Canada, Address Locator 2203D, 251 Sir F. Banting driveway, Ottawa, Ontario K1A 0L2, Canada*

[2] *Center for Food Safety and Applied Nutrition, U.S. Food and Drug Administration, College Park, MD 20740, USA*

4.3.1 INTRODUCTION

Benzene is a volatile organic compound (VOC) that has caused cancer in workers exposed to high concentrations in workplace air (1). It is used in the manufacture of other chemicals, detergents, and plastics, and is released into the environment from gasoline vapors and the burning of fossil fuels. Several reviews briefly mention the occurrence of benzene and other VOCs in foods (2–6). Benzene can occur in food as a result of processed-induced changes resulting from high-temperature chemical transformations, ionizing radiation, and from the reaction of added or naturally occurring precursors. In the absence of good manufacturing practices, benzene also can be introduced into food as an environmental contaminant or as a contaminant of food additives or flavors. One or more of these sources potentially could be associated with benzene found in food.

Roasting foods at high temperatures can cause the pyrolysis of organic matter, resulting in benzene formation. This occurs either from the recombination of intermediates or by the degradation of compounds containing a benzene moiety such as phenylalanine. Benzene also can be introduced into food as a result of smoking or grilling over an open flame, especially with wood or

Process Induced Food Toxicants: Occurrence, Formation, Mitigation, and Health Risks, Edited by Richard H. Stadler and David R. Lineback

Benzene: At a Glance	
Background	Benzene is used in the manufacture of chemicals, detergents, and plastics and released into the environment from gasoline vapors and the burning of fossil fuels.
Analysis	Several analytical techniques have been developed to measure benzene and other volatile organic compounds (VOCs) in food, the most predominant being static and dynamic purge and trap headspace coupled to GC/MS or GC/flame ionization detection (FID).
Occurrence in Food	Occurs in very low amounts (low μg/kg) in a wide range of foods such as meat products, poultry, fish, fruit, roasted coffee, and nuts. Surveys in foods and beverages have been conducted in a few countries including the USA, UK, Italy, South Korea, Australia, and Canada.
Main Formation Pathways	Three main pathways can lead to benzene formation in food under certain conditions, i.e., (i) thermal decomposition and rearrangement of precursors, (ii) as a byproduct of irradiation, and (iii) from the combination of benzoate and ascorbic acid in the presence of trace metal ions and oxygen.
Mitigation in Food	Benzene concentrations in food are generally very low. Mitigation measures by industry have focussed on beverages, and guidelines have been published that address key factors, viz. the combination of benzoate and ascorbic acid, the use of chelating agents and nutritive sweeteners, storage conditions, and shelf life. Benzene formation from the decomposition of phenylalanine can be minimized by grilling and roasting foods at lower temperatures and/or minimizing the dose of ionizing radiation. The uptake of benzene in grilled fatty foods and concentration from charcoal and wood emissions can be reduced depending on the choice of fuel.
Health Risks	Benzene is a carcinogen that causes tumors in rodents at multiple sites and leukemia in humans. Inhalation of contaminated air is the primary route of non-occupational benzene exposure for non-smokers, and mainstream cigarette smoke is the primary route for smokers. Dietary exposure to benzene is not considered a significant pathway (approx. 1.5% of the total exposure for non-smokers).
Regulatory Information or Industry Standards	No regulatory limits have been established for benzene in food and beverages Guidance levels for drinking water vary from country to country, but most are within the same order of magnitude (1–10 ppb). Many countries consider their guidance level for benzene in drinking water as an appropriate standard for beverages and ask manufacturers to reformulate or withdraw products that exceed these amounts.
Knowledge Gaps	Factors governing the amount of benzene formed in complex matrices containing benzoate and ascorbic acid.
Useful Information (web) & Dietary Recommendations	(i) Guidance Document to Mitigate the Potential for Benzene Formation in Beverages. The International Council of Beverages Associations (ICBA) 2006, http://www.icba-net.org (ii) Health Risk Assessment: Benzene in Beverages. Chemical Health Hazard Assessment Division, Food Directorate, Health Canada 2006, 1-14, http://www.hc-sc.gc.ca/fn-an/securit/chem-chim/food-aliment/benzene/benzene_hra-ers-eng.php (iii) Benzene in Drinking Water. World Health Organization 2003, WHO/SDE/WSH/03.04/24, http://www.who.int/water_sanitation_health/dwq/benzene.pdf No specific dietary recommendations.

charcoal. Benzene concentrations greater than 10 mg/m^3 were detected in flue gases from glowing charcoal (7).

Food irradiation processes are currently allowed in about 60 countries to inactivate pathogenic bacteria and parasites and for phytosanitary purposes. Negligible levels of benzene formation can occur in some irradiated foods under certain conditions and currently is believed to form from oxidative and radiolytic cleavage of phenylalanine (8). Some studies also report low-level benzene formation from irradiation of model solutions of benzoate and experimentally prepared foods preserved with benzoate (9).

In the early 1990s, benzene was found in beverages containing sodium or potassium benzoate and ascorbic acid. Benzoates may be either naturally occurring or added to beverages as an antimicrobial agent. Ascorbic acid also may be naturally occurring or added as a preservative or nutrient. Several studies showed that under certain conditions, benzene can be formed in beverages containing benzoate and ascorbic acid (10–12). Benzene forms through a series of reactions in which a hydroxyl radical is thought to decarboxylate benzoic acid to form benzene (13).

Other sources of benzene contamination include its migration into food from plastic cookware (14–16) and sorption into food from the emissions of automobiles, burning of fossil fuels, and fires. In some cases, agricultural products were thought to have become contaminated during farming/harvesting due to the use of motorized equipment (17). Finally, substances intentionally added to food may contain benzene due to the manner in which they were produced. Liquid smoke flavoring obtained through the incomplete combustion of wood is an example (18).

4.3.2 ANALYSIS

A number of analytical techniques have been developed for the quantitative determination of benzene and other VOCs in food and for qualitative analysis of flavor and freshness (19, 20). Most of these methods were developed for multi-residue analysis and include benzene as a target analyte. This section will focus primarily on quantitative methods developed for the determination of benzene and VOCs in general. Table 4.3.1 specifically identifies methods with reported performance standards for benzene in food and beverages, i.e., limits of detection (LOD) and quantitation (LOQ), recoveries, linearity, and relative standard deviation for repeatability (RSD_r) and/or reproducibility (RSD_R).

Static and dynamic purge and trap (P&T) headspace (HS) sampling followed by gas chromatography/mass spectrometry (GC/MS) and GC/flame ionization detection (FID) are the predominant techniques used to determine benzene and other VOCs in food (20, 31). In the 1990s, vacuum distillation (VD) and solid-phase microextraction (SPME) were introduced as methods to extract VOCs from various matrices (28, 32). Although VD and SPME have

TABLE 4.3.1 Quantitative methods for the determination of benzene in food and beverages.

Method	Matrix	GC column, P&T trap	Conditions	Method performance	Reference no.
P&T GC/MS	Table-ready foods	Column: DB-624 (30 m, 0.32 mm, 1.8 µm) Trap: Vocarb 3000	P&T: 10-min purge at 40 mL/min; desorb at 335 °C; cryofocus at −150 °C GC: 40 °C for 10 min; 8 °C/min to 140 °C; 5 °C/min to 180 °C and hold 3 min MS: full scan	LOQ_{avg} = 9.3 ng/g Avg. Rec. (RSDr)%: 104 (30.1) at 10 ng/g 103 (11.6) at 100 ng/g In nine table-ready foods RSD_R = 14.5 (50 ng/g)	21
P&T GC/FID	Various foods	Column: HP5 (50 m, 0.32 mm, 1.0 µm) Trap: Tenax	P&T: 9 min at 30 mL/min; desorb at 180 °C GC: 70 °C for 4 min; 10 °C/min to 200 °C and hold 8 min	LOQ = 1 ng/g in water; Standard additions at 0.5, 1, and 2 times the suspected concentration	11
Blender P&T GC/MS	Various foods	Column: Rtx-5 (30 m, 0.25 mm, 1.0 µm) Trap: OV-1/Tenax/Silica Gel	P&T: 11 min @ 40 mL/min; desorb @ 350 °C; with cryofocusing GC/MS: (i) 30 °C, 3 °C/min to 60 °C, 5 °C/min to 120 °C, 10 °C/min to 250 °C and hold 4 min (full scan) (ii) 30 °C, 3 °C/min to 60 °C, 20 °C/min to 250 °C (SIM)	LOD ≤ 100 pg/mL R^2 = 0.995 in water (0–12.5 ng/mL)	22
Static HS GC/PID	Eggs	Column: 5% methyl silicone (30 m, 0.53 mm, 1 µm)	HS: 2-h equilibration at 50 °C GC: 40 °C with N_2 gas flow at 15 mL/min PID: 10 eV lamp at 180 °C	LOD = 0.002 µg/mL in egg whites Rec.: 97% whites; 26% whole eggs; 24% yolks (0.05 µg/mL)	23
Static HS GC/MS	Various foods and food contact materials	Column: DB-624 (30 m, 0.33 mm, 1.8 µm)	Automated HS: 30 min in 100 °C HS oven GC: 50 °C isothermal, 50:1 split ratio MS: SIM	LOD = 2 ng/g in foods	24

Static HS GC/MS	Beverages	Column: HP-PLOT-Q (30 m, 0.32 mm, 20 µm)	Automated HS: 15 min in 60 °C HS oven GC: 100 °C, 10 °C/min to 225 °C and hold 12.5 min MS: SIM	LOD = 0.04 ng/g in beverages RSD_R <28%	25
Static HS GC/MS	Fruit juices, fruit drinks, and soft drinks	Column: DB-624 (30 m, 0.32 mm, 1.8 µm)	HS: 1-h equilibration at room temperature (RT) GC: –20 °C for 1 min, 10 °C/min to 200 °C and hold 1 min MS: SIM and full scan	LOD = 0.02 ng/g R^2 = 0.993 in water (0.05–2.5 ng/g)	10
Static HS GC/MS or GC/FID	Bottled water	Column: (i) DB-5 (30 m, 0.32 mm, 1.0 µm); (ii) Ultra 2 (30 m, 0.32mm, 0.52 µm)	HS: 30-min equilibration at RT or 35 °C GC: –20 °C for 1 min, 10 °C/min to 200 °C MS: full scan	Interlaboratory study with three laboratories LODs = 0.2–1 ppb RSD_R = 14% (4.1 ng/g)	26
Static HS GC/MS	Beverages	Column: DB-624 (30 m, 0.25 mm, 1.4 µm)	Automated HS: 30-min equilibration at RT GC: 35 °C for 1 min, 10 °C/min to 100 °C, 35 °C/min to 250 °C and hold 5 min MS: SIM	LOD = 0.2 ng/mL R^2 > 0.999 (0.5–50 g/mL) Rec. (RSDr) % in water at 5 ng/mL = 85 (5)	27
VD/GC/MS	Milk	Column: Rtx-VMS (60 m, 0.25 mm, 1.4 µm)	VD (28, 29) GC: –25 °C for 6 min, 50 °C/min to 220 °C and hold 7.15 min MS: Full scan	LOD = 0.1 ng/mL Rec. in milk at 20 ng/mL = 97 ± 11.5%	30

not been as widely reported for the quantitative determination of benzene in food, these techniques are at the forefront of analytical method development and merit mention and future investigation.

4.3.2.1 Sample Preparation and Matrix Effects

Prior to HS analysis, solid and viscous foods may need to be homogenized by blending with water. To avoid benzene losses, samples should be chilled (0–4 °C) prior to homogenization and then homogenized as quickly as possible. The samples should remain chilled during homogenization and while test portions are being transferred. Liquids can be chilled and analyzed undiluted. Matrix effects during HS analysis can vary significantly depending on the food. Benzene in water and beverages usually is quantified by isotope dilution with a d_6-benzene internal standard. For complex foods, quantification of benzene based solely on isotope dilution may not adequately compensate for matrix effects, in which case the method of standard additions may be necessary.

4.3.2.2 Static HS Analysis

In static HS, test portions of samples are thermally equilibrated in a sealed vial until the volatiles reach equilibrium between the samples and vapor phases. A sample of the gas phase is then injected onto the chromatographic column either with a gastight syringe or by an automated HS sampler. Cryogenic trapping and oven cooling are often used to focus the VOCs at the head of the GC column. Cryogenic trapping increases sensitivity and improves resolution by reducing bandwidth broadening. Automated HS sampling is preferable for quantitative analysis, because the thermal equilibration time, vial pressurization, and injection volume can be precisely controlled, thus reducing the variability of analytical results.

Stein and Narang used static HS GC equipped with a photoionization detector to determine benzene and other VOCs in egg whites, whole eggs, and egg yolks. Benzene was quantified on the basis of a fortified egg standard or a series of egg standards (23). Page *et al.* conducted an interlaboratory study in which three laboratories analyzed bottled drinking water for benzene and other VOCs by using static HS with either GC/MS or GC/FID; 182 samples were analyzed. GCs were equipped with either an on-column or split-splitless injector and a cryogenically cooled oven. Benzene found in the samples was quantified by using external standards. The overall RSD_R for benzene was determined to be 14% (26). Jickells *et al.* used static HS GC/MS to determine the amount of benzene that migrates into foods from nonstick cookware and microwave susceptors during cooking. Calibration curves were constructed with blank foods fortified with benzene and d_6-benzene. The LOD was reported to be 2 ng/g in various foods, but no other performance characteristics were provided (24).

Several studies reported using static HS GC/MS to conduct surveys of various fruit juices and other beverages containing added benzoate and ascorbic acid. In 1992, static HS GC/MS was used by Page *et al.* to conduct a survey of fruit juices and beverages. The GC was equipped with an on-column injector and a cryogenically cooled oven. A gastight syringe was used to inject a portion of the HS on-column (10).

Health Canada used HS GC/MS to conduct a survey of soft drinks and other beverages containing benzoate and ascorbic acid. A GC autosampler equipped with a 100-μL syringe and a split-splitless injector was used for HS sampling. In order to avoid benzene formation that can occur from the precursor compounds at elevated temperatures, test portions were equilibrated at room temperature for 30 min and then a portion of HS was injected into the GC in the splitless mode. The syringe was flushed with air between injections. Benzene found in the samples was quantified by isotope dilution with d_6-benzene (27).

The US Food and Drug Administration (FDA) subsequently reported that short-term exposure (15 min) to temperatures as high as 105 °C did not result in benzene formation in beverages containing benzoate and ascorbic acid. The FDA methods subject test portions to a 15-min thermal equilibration in an automated 60 °C HS oven. The GCs were equipped with a ZB-624 and a PLOT-Q (porous layer open tubular) capillary GC columns. The PLOT-Q column eliminated the need to cryogenically focus benzene at the head of the GC column. This column is constructed with a polystyrene-divinylbenzene stationary phase that has a high adsorptivity for gases and apolar compounds such as benzene. Benzene found in the beverages was quantified by isotope dilution with d_6-benzene (25).

4.3.2.3 Dynamic P&T HS Analysis

The US Environmental Protection Agency (EPA) method 524.2 is a P&T HS GC/MS method adapted for the analysis of VOCs in foods (11, 21, 33, 34). Liquid, semisolid, and solid foods are extracted by a continuous flow of inert gas. The liberated volatiles are trapped on a column packed with an adsorbent material. Rapid thermal desorption is used to release the VOCs from the trap. During the desorption phase, the VOCs are back flushed onto the chromatographic column. Cryogenic trapping focuses the VOCs at the head of the GC column. Careful selection of the trap adsorbent is critical for achieving high recoveries, sharp peaks, and good resolution. Hydrophobic adsorbents are essential to prevent interference from water vapor. Adsorbent traps packed with 2,6-diphenylene-oxide polymer resin (Tenax) and graphite are often used for the determination of benzene by P&T HS analysis.

McNeal *et al.* used P&T sampling with GC/FID to determine benzene in over 50 foods. Standard additions were conducted at 0.5, 1, and 2 times the suspected benzene concentration in the food (11). Fabietti *et al.* used P&T sampling with GC/MS to determine benzene and toluene in 60 soft drink samples. Benzene was quantified by isotope dilution with d_6-benzene (35). Heikes *et al.* used P&T sampling with GC/MS to determine benzene and other VOCs in 234 table-ready foods. An intralaboratory study of the method was conducted with four table-ready foods fortified with 50 ng/g benzene. Average recoveries for duplicate analyses performed by two analysts were 110% and 103%, respectively, and the RSD_R was 14.5% (21). Barshick *et al.* developed a method for the determination of benzene that interfaced a blender to a P&T GC/MS (22). This system minimized the loss of benzene by allowing P&T sampling while homogenizing the samples. Several foods were analyzed using both the blender P&T GC/MS and a conventional P&T GC/MS. Improved sensitivity and reproducibility were achieved with the blender P&T GC/MS system. Comparable results were obtained for various foods analyzed by the conventional and blender P&T systems.

4.3.2.4 HS SPME Analysis

HS SPME also has been applied to the analysis of liquid, semisolid, and solid foods. Test portions are sealed in HS vials and a needle assembly is used to pierce the septum of the vial. A fused silica fiber internally coated with an adsorptive phase is extended from the needle into the HS above the sample. HS SPME extraction is complete when the volatiles have reached equilibrium between the sample, HS, and fiber coating. The equilibration time can be decreased by heating the sample. The reusable fibers are selected on the basis of the affinity of the analyte to the liquid polymer coating of the fiber. Fiber coatings such as polydimethylsiloxane (PDMS) and PDMS/carboxen are thermally stable up to 300 °C and are often used for the analysis of volatiles such as benzene. For analysis, the concentrated volatiles are thermally desorbed from the fiber in a splitless GC injection port. For optimum peak shape, a narrow bore GC liner should be used to ensure a high linear flow of the GC carrier gas and efficient transfer of the desorbed volatiles to the GC column. Analyte sensitivity can be optimized by focusing the desorbed volatiles to the head of the column either by a low initial GC temperature or by cryofocusing (36, 37).

In 2006, the UK's Food Standards Agency (FSA) conducted a survey of benzene in beverages (38). Some of the survey results were determined by a private laboratory using a proprietary HS SPME GC/MS method. A test portion was transferred to an HS vial and fortified with d_6-benzene as the internal standard. The HS was sampled by adsorption onto a 75-µm PDMS/carboxen fiber. No other analytical conditions were provided. It can be reasonably assumed that the methodology was similar to HS SPME methods developed for the analysis of water (39–41).

4.3.2.5 VD Analysis

In VD analysis, volatiles in liquid, semisolid, and solid samples are trapped by placing a sample in a chamber and distilling under vacuum at room temperature. Most of the water vaporized from the sample is condensed with a condenser column cooled to 5 °C. The compounds that pass through the condenser are collected in a cryoloop that has been cooled with liquid nitrogen to −196 °C. At the completion of the VD, flow through the cryoloop is directed to the GC while the distillate is heated ballistically to 120 °C. Transfer of the distilled volatiles to the GC column is completed in about 2.5 min. For quantitative analysis, internal standard-based matrix correction is applied, thus obviating the need for standard additions. The recoveries of the VOCs are determined with respect to a suitable internal standard that reflects the effects of the matrix on the VOC as a function of boiling point and relative volatility (water to air partitioning) (28).

Hiatt (29) used VD/GC/MS with surrogate-based matrix correction to determine benzene, other VOCs, and semi-volatile compounds in fish tissue. In a similar study, Hiatt and Pia (30) used VD/GC/MS to determine VOCs in milk. The matrix correction for benzene was derived on the basis of boiling point and the water–air partitioning of d_6-benzene.

4.3.3 OCCURRENCE IN FOOD

The occurrence of benzene in foods has been investigated by way of targeted sampling and market basket sampling as part of total diet studies (TDS). TDS are designed to monitor chemical contaminants in table-ready foods and to estimate the average dietary intake from eating those foods. In general, TDS data and data from targeted sampling show that the benzene concentrations in food seldom exceed low ppb levels.

4.3.3.1 General Occurrence

The US Food and Drug Administration (FDA) conducted a 5-year TDS investigation of VOCs in table-ready foods. Benzene was found in 68 of the 70 foods analyzed at concentrations ranging from 1 to 190 ppb (ng/g or ng/mL). The highest amount found was in cooked ground beef (42). However, the FDA recently recommended that the TDS benzene data be used with caution. An FDA evaluation found that the TDS analytical method produced unreliable benzene results in some foods, particularly foods containing benzoates (see Section 4.3.4) (21, 43). The P&T method that was used to determine benzene subjected samples to a 30-min helium sparge at 100 °C. The FDA evaluation found that some of the TDS benzene data may represent artifacts that can occur under the P&T conditions.

In 1993, the United Kingdom conducted a TDS investigation of VOCs in 20 food groups collected from 10 UK locations. Benzene was found in samples of carcass meat, offal (meat waste products), meat products, poultry, fish, fruit products, and nuts. Benzene concentrations in these food groups ranged from less than 1 ppb in fruit products to 18 ppb in offal. All other food groups were found to contain 1 ppb benzene or less (44).

In a 1995 survey conducted in the Unites States, 22 raw and cooked foods were analyzed. Most of the foods were found to contain less than 1 ppb benzene with the exception of peanuts and canned olives, which were found to contain about 2 ppb benzene. Several protein-rich foods were analyzed including raw and cooked fish, chicken, and pork. There was no significant difference in the amount of benzene found in the raw and cooked samples, and the concentrations remained below 1 ppb (22). In another 1995 US survey, benzene was not found in 22 of the 37 foods and beverages analyzed. Fifteen samples were found to contain benzene at levels ranging from 0.5 to 9 ppb with the highest concentration found in a flavored noncarbonated beverage (34). In two other surveys conducted in the United States, benzene amounts ranged from 0.03 to 0.12 ppb in whole milk (19 samples) and from less than 5 to 30 ppb in whole eggs (five samples) (23, 30).

In a 1996 survey of 24 fruits and vegetables, benzene was found at concentrations ranging from 27 to 56 ppb (dry weight) in the peels of apples, kiwifruit, and oranges. No benzene was found in the pulp of those fruits or the remaining fruits and vegetables (45). In a 2005 Brazilian study of fruits and the corresponding commercial juices, benzene was found in mango and guava juices; the concentrations were not reported (46). In a 1983 study of two varieties of mango fruit, benzene was found in both varieties at an estimated concentration of 30 and 40 ppb (47).

4.3.3.2 Food Handling and Processing

Benzene potentially can be introduced into food from smoking, roasting, or exposure to ionizing irradiation. In one study, benzene concentrations in an alder wood smoke chamber ranged from 1.9 to 2.2 mg/m^3 (48). Despite these concentrations, benzene was reportedly much less absorbed by meat products than other smoke components. However, there is a lack of data in the literature on benzene contamination of meat resulting from smoke curing with wood products (49). During the manufacture of liquid smoke flavoring, a water-based phase is separated from an organic phase, which likely contains some benzene. Benzene is partially soluble in water (0.1%), and some cross-transfer likely occurs. There is not much data on benzene in liquid smoke flavoring except for the analysis of two products, which were found to contain 21 and 121 ppb benzene (11).

Roasting of food can potentially generate benzene from pyrolytic decomposition of food ingredients. Most of the data on VOCs released during roasting are reported for roasted coffee beans. In a study reported in 1966, benzene was identified in the HS of roasted beans by GC/MS (50). In more recent studies, benzene concentrations in roasted coffee beans were estimated to be about 0.1 to 0.15 ppm (51) as compared with 0.1 ppb in brewed coffee (22). Some studies failed to detect benzene in roasted coffee beans, perhaps as a result of the analytical method used for the analyses (52, 53). Benzene also was identified in some samples of green coffee (54) but not others (55, 56), which may suggest possible environmental contamination during harvesting, transportation, and/or storage of the beans.

Some foods treated with high doses of ionizing radiation can form low amounts of benzene. In the United States and many other countries, the amount of ionizing radiation (dose) approved for use depends on the type of food and the desired technical effect (57, 58). Low doses (1 kGy or less) are used to control insects and parasites in fresh fruits and vegetables and to delay ripening and sprouting. Medium doses (1 to 10 kGy) are used to reduce pathogenic microorganisms and extend the shelf life of foods such as meat and poultry products. High doses (greater than 10 kGy) are used to disinfect or sterilize food. For example, doses greater than 10 kGy are used to sterilize meat products for the NASA space flight program. In the United States, the dosage of ionizing radiation used to treat food typically ranges from 0.15 kGy for the treatment of fresh fruits and vegetables to 7 kGy for frozen meat and up to 30 kGy for spices and seasoning (57, 59). Chemical analysis of irradiated foods has shown that most of the radiolytic by-products are identical to by-products identified in foods treated by traditional food processing (58). Studies have shown that meat products treated with very high doses of ionizing radiation can form low amounts of benzene from the degradation of phenylalanine. In an experiment, 19 ppb benzene formed in a beef sample irradiated with 56 kGy (8, 60), which is about 10 times higher than permitted under current regulations. A linear extrapolation to a dose approved for frozen meat might be expected to form approximately 3 ppb, which is equivalent to the amount of benzene found in the cooked control samples (2 to 3 ppb). No benzene was found in the nonirradiated control. It is important to point out that the dose of ionizing radiation used in this experiment (56 kGy) likely produced undesirable changes in appearance, flavor, and aroma that would be unacceptable to consumers (61).

Solvent extraction of vegetable oils with hexane or other organic solvents can introduce benzene into the oil. In India, relatively high concentrations of benzene were found in refined and unrefined soy oil, i.e., 3.1 and 32 ppm, respectively (62). Trace amounts of benzene and toluene reported in Italian Parma ham were probably of environmental origin as the ratio of these contaminants was similar to that found in gasoline, i.e., three to one (63).

4.3.3.3 Foods Containing Benzoate and Ascorbic Acid

In the early 1990s, it was found that benzene could form in certain beverages containing potassium or sodium benzoate and ascorbic acid (vitamin C). Ascorbic acid might be either naturally occurring in fruit juice or added as a preservative or nutrient. Benzoate is used as an antimicrobial agent and is particularly effective at the low pH levels found in many beverages. Limited data on erythorbic acid (the isomer of vitamin C) in beverages also show that it can form benzene (43). Studies conducted with aqueous solutions of benzoate and ascorbic acid showed that benzene formed at approximately

300 ppb when the test solutions were exposed to exaggerated conditions of heat and UV light in the absence of competing reactions (see Section 4.3.4) (11). The FDA and the beverage industry also found benzene in a few commercial beverages containing benzoate and ascorbic acid. The amount found in these beverages exceeded the EPA maximum contaminant level (MCL) of 5 ppb benzene in drinking water. In response to these findings, the beverage industry reformulated affected products to eliminate or minimize benzene formation.

Benzoic acid occurs naturally in many fruits (46, 64). In cranberries and lingonberries, concentrations of 480 ppm (65) and 600–1300 ppm (66) have been reported. In cranberries, most of the acid is in the form of esters or glycosides; while in lingonberries, the acid is unbound. Benzoic acid also was found in other foods (67), most notably in cinnamon at concentrations ranging from 131 to 461 ppm (68). In addition to beverages, benzoate is used as an antimicrobial in other foods such as jams and jellies (69). The Canadian maximum permitted concentration for benzoate in beverages is 1000 ppm (70). The European Union (EU) established a 150 ppm maximum concentration for benzoate in beverages (71). In the United States, benzoate may be used in food in amounts not to exceed current good manufacturing practice (1000 ppm) (72). In a 2003 survey of benzoic acid in Korean beverages, the highest concentration found was 470 ppm (73).

In the 1990s, several surveys were conducted to investigate the amount of benzene found in beverages and other foods containing added or naturally occurring benzoate. In a survey conducted in Canada, 74 samples of fresh expressed fruit juices, fruit drinks, and carbonated beverages were analyzed (10). None of the samples were found to contain benzene above the Canadian maximum acceptable concentration (MAC) of 5 ppb in drinking water. The highest amount of benzene found was 3.8 ppb in a carbonated beverage with declared benzoates. Benzene concentrations in carbonated beverages with declared benzoate (six samples) and without declared benzoate (20 samples) ranged from 0.01 to 3.8 ppb and 0.03 to 0.12 ppb, respectively. Retail fruit juices with declared benzoate (10 samples) and without declared benzoate (13 samples) were found to contain benzene concentrations that ranged from 0.14 to 1.5 ppb and 0.02 to 0.24 ppb, respectively. Fruit drinks with declared benzoate (seven samples) were found to contain benzene concentrations that ranged from 0.05 to 0.28 ppb. Three cranberry juice samples without declared benzoates were found to contain benzene in amounts that ranged from 0.46 to 1.8 ppb. The cranberry juices likely contained naturally occurring benzoic acid.

In a survey conducted in the United States, more than 50 foods and beverages were analyzed (11). The foods and beverages were selected on the basis of previous reports of naturally occurring benzene or because the products contained naturally occurring or added benzoate and ascorbic acid. Most of the 26 beverages analyzed were found to contain less than 1 ppb benzene. Benzene concentrations as high as 3 ppb were found in two diet carbonated beverages and a cocktail mix with added benzoate. The highest benzene concentration found was 121 ppb in liquid smoke. Little or no benzene was found in foods containing naturally occurring benzoates and ascorbic acid. However, most foods and beverages containing added benzoate and ascorbate were found to contain benzene amounts ranging from less than 1 to 38 ppb.

In 2001, an Italian survey was conducted to determine benzene and other VOCs in 60 beverages (35). Benzene was detected at low concentrations in all the samples. Regular and diet soft drinks, orange juice, and carbonated orange juice were analyzed and found to contain benzene amounts that ranged from 0.7 to 2.9 ppb.

In 2005, a few US beverages containing benzoate and ascorbic acid were once again found to contain benzene above the US and Canadian 5 ppb maximum level for benzene in drinking water (74, 75). The reoccurrence of this problem prompted global investigations of benzene in beverages. In general, the surveys were targeted at products that contained added benzoate and ascorbic acid, and most samples were collected from retail markets.

In a survey conducted in the United Kingdom, 107 of the 150 soft drinks did not contain benzene greater than 1 ppb (38). The majority of these beverages contained benzoate and ascorbic acid. Thirty-eight samples were found to contain benzene that ranged from 1 to 10 ppb. Four products had benzene concentrations of more than 10 ppb. The highest benzene concentration found was 28 ppb in a diet soft drink.

In a survey conducted in the United States, 199 soft drinks and other beverages were analyzed (43, 76). More than 90% of the samples analyzed did not contain benzene greater than 5 ppb and most samples contained less than 1 ppb. Nine beverage products were found to contain benzene above the US EPA MCL of 5 ppb for drinking water; all of the products contained added ascorbic acid and either added benzoate (seven products) or an unknown amount of naturally occurring benzoic acid (two cranberry juice products). The concentration of benzene as high as 89 ppb was found in a black cherry beverage. The amount of benzene found in this product was likely the result of the product's handling in the retail market. The sample analyzed was purchased from a retail store several months beyond the manufacturer's sell-by date. Multiple lots of a diet soft drink with added benzoate and ascorbic acid also were found to contain benzene above the EPA MCL; the highest concentration found in this product was 79 ppb. Two lots of a diet or light cranberry juice with added ascorbic acid were found to contain 5.4 and 9.9 ppb benzene. In comparison, cranberry juices with added ascorbic acid and sugar were found to contain 2.3 ppb benzene or less. Subsequent analyses of the products reformulated by the manufacturers showed that benzene concentrations were significantly reduced compared with levels found in the original formulations, and all were well below 5 ppb.

In a survey conducted in South Korea, 36 of 37 beverages analyzed were found to contain benzene at concentrations ranging from 1.7 to 263 ppb (77). All of the beverages contained benzoate and ascorbic acid. Repeat sampling of 30 of those products was conducted in order to analyze the products soon after their date of manufacture. Twenty-seven of the products were found to contain benzene at concentrations ranging from 5.7 to 87.7 ppb.

In Australia, a Food Standards Australia New Zealand (FSANZ) survey included 68 flavored beverages (78). Among the samples analyzed, 29 were found to contain benzene of more than 1 ppb; four samples contained benzene greater than 5 ppb, and five samples contained benzene of more than 10 ppb. Benzene concentrations found in the beverages ranged from less than 1 to 40 ppb.

In Ireland, 76 samples of soft drinks, squashes, and flavored water were analyzed (79). Sixty-nine samples contained no benzene. Only two diet products were found to contain benzene greater than 10 ppb; the highest concentration found was 91 ppb in one product that was analyzed after the "best before" date. Five additional samples of that product were analyzed before the "best before" date, and the benzene concentrations found ranged from 9 to 17 ppb.

In Canada, a survey of 124 soft drinks and beverages was conducted (27). Ninety percent of the samples analyzed were found to contain 2.5 ppb or less benzene. Approximately 60% of the beverages analyzed were found to contain less than 1 ppb benzene. Six products were found to contain benzene above the Canadian MAC of 5 ppb for benzene in drinking water; two of those products were found to contain benzene above the World Health Organization (WHO) guideline of 10 ppb in drinking water. The highest benzene concentration found was 23 ppb in a low-calorie soft drink specifically marketed to children. Two cranberry cocktail products without declared benzoate and sweetened with sugar were found to contain 0.6 and 1.3 ppb benzene. Subsequent analyses of products reformulated by the manufacturer showed that benzene concentrations were significantly reduced compared with levels found in the original formulations.

In the UK, Canadian, and US surveys, wide variations in benzene concentration were observed for different lots of the same product. For example, a UK low-calorie lemon beverage was found to contain benzene at concentrations that ranged from 11 to 28 ppb; a Canadian cocktail mix was found to contain amounts of benzene that ranged from 2.5 to 13 ppb; and a US diet orange beverage was found to contain concentrations of benzene that ranged from none detected to 82 ppb. Product formulation and storage conditions likely contributed to the variability observed in different lots of the same product (see Section 4.3.5). Results from these surveys also suggest that diet or light beverages formulated with artificial sweeteners and low in sugar have the potential to form higher amounts of benzene.

4.3.4 FORMATION

Three pathways of benzene formation in food will be presented: (i) benzene formation from thermal decomposition and rearrangement of precursors,

(ii) benzene formation as a by-product of irradiation, and (iii) benzene formation from benzoate and ascorbic acid.

4.3.4.1 Thermal Decomposition

Most research on benzene formation during frying, roasting, and grilling has focused on the decomposition of phenylalanine. Compared with other amino acids, phenylalanine is fairly unstable; more than 99.5% will decompose when heated for 20 min in a 400 °C oven (80). Toluene was the main pyrolytic by-product of phenylalanine via benzylic cleavage at 700 and 800 °C; benzene was a minor by-product (81, 82). In a more detailed study, which proposed a mechanism of phenylalanine decomposition, the yield of benzene from phenylalanine increased from 2.1% to 7.6% when the temperature was increased from 650 to 850 °C. The yield of toluene at both temperatures remained stable at about 21% (83). A recent study failed to detect both benzene and toluene during pyrolysis of phenylalanine at 700 °C, perhaps because the data acquisition was begun too late (84). In another study of amino acid decomposition, only toluene was detected when phenylalanine was heated for 30 min at 300 °C, perhaps due to analytical limitations (85). The temperatures described here are likely higher than those typically used for food processing. Relatively low concentrations of benzene in roasted coffee might be related to the low abundance of phenylalanine in green coffee (86) and the fact that the roasting temperatures seldom exceed 250 °C (87).

4.3.4.2 Irradiation

The formation of benzene by ionizing radiation is also thought to proceed as a result of the decomposition of phenylalanine (8). In addition, ionizing radiation will cause benzoic acid to break down to form benzene. Compared with a control sample, higher amounts of benzene formed when an experimental turkey meat prepared with 0.1% added benzoate was irradiated at 2 kGy (no quantitative data reported) (9). The irradiation of other foods rich in benzoates could potentially produce benzene. For example, irradiation of an aqueous 8-mM solution of benzoic acid with gamma rays showed that hydroxyl radicals ($OH^{\cdot}$) generated by the radiolysis of water were responsible for the decarboxylation of benzoic acid to form benzene. It is thought that during this reaction, hydroxyl radicals generate an unstable benzoic acid radical (C_6H_5—$COO^{\cdot}$), which readily loses carbon dioxide (CO_2) to form a benzene radical. The benzene radical ultimately forms benzene from hydrogen abstraction from a suitable donor molecule (88, 89). The hydroxyl radical also reacted with benzoic acid to form *o*-, *m*-, and *p*-hydroxybenzoic acid with a combined yield that was 50% greater than benzene. The addition of 2.3-mM ethanol as a hydroxyl radical scavenger produced a reduction of benzene as measured indirectly by a decrease in CO_2 production. The CO_2 yield was the same in a pH range of 3 to 10.8 but decreased by a factor of 3 in deoxygenated water.

A postirradiation effect resulting in the decarboxylation of benzoic acid may have occurred as indicated by a 40% increase in CO_2 production.

4.3.4.3 Interaction between Benzoate and Ascorbic Acid

The determination of benzene in soft drinks and other beverages prompted investigations into the pathways of benzene formation from benzoate and ascorbic acid (11–13). Gardner and Lawrence (13) proposed that benzene could form in foods and beverages containing benzoate and ascorbic acid through a series of reactions (Eqs 4.3.1–4.3.4). Equation 4.3.1 was slightly modified from Gardner and Lawrence to show the ascorbic acid anion ($HAsc^-$) that occurs under acidic conditions. Under these conditions, a hydroxyl radical ($OH^{\cdot}$) forms by an ascorbic acid-assisted pathway of hydrogen peroxide (H_2O_2) formation mediated by catalytic amounts of iron or copper salts (Eqs 4.3.1–4.3.3). Iron (Fe^{2+}) or copper (Cu^+) ions react with H_2O_2 to form the hydroxyl radical (Eq. 4.3.4).

$$Cu^{2+} + HAsc^- \rightarrow HAsc^{\cdot} + Cu^+ \tag{4.3.1}$$

$$Cu^+ + O_2 \rightarrow Cu^{2+} + O_2^{\cdot -} \tag{4.3.2}$$

$$2O_2^{\cdot -} + 2H^+ \rightarrow H_2O_2 + O_2 \tag{4.3.3}$$

$$Cu^+ + H_2O_2 \rightarrow OH^{\cdot} + OH^- + Cu^{2+} \tag{4.3.4}$$

In theory, the hydroxyl radical generates an unstable benzoic acid radical that will readily lose CO_2 to form benzene (see Section 4.3.4.2). Equation 4.3.4 is similar to Fenton's reaction in which catalytic amounts of iron react with H_2O_2 to form the hydroxyl radical.

Gardner and Lawrence conducted a series of experiments on test solutions containing 8.0-mM ascorbic acid, 6.25-mM benzoic acid, and 10.5-mM hydrogen peroxide in a 50-mM sodium phosphate buffer at pH 3. The test solutions were incubated in a 25 °C water bath. Benzene formation was observed within 15 min, presumably due to trace amounts of iron or other trace metals present in the distilled water and other reagents used to prepare the test solutions. No benzene formation was observed when 0.1-mM desferioxamine was added as an iron-chelating agent. Gardner and Lawrence also investigated the influences of metal ions on benzene formation by adding copper and iron to the test solutions. Both cupric and ferrous sulfates were equally effective catalysts at 0.05 mM. However, the benzene yield decreased by half when the ferrous sulfate concentration was increased to 1 mM. The same negative correlation between the iron concentration and the benzene yield was observed by Chang and Ku (12), and in unpublished studies conducted by Health Canada. In contrast, the benzene yield increased when the copper concentration was increased from 0.05 to 1 mM. The pH of the test solutions also was an important factor influencing the benzene yield. Test solutions containing 0.25-mM cupric sulfate showed a near-linear dependence from pH 2 to 6. The highest amount of benzene was observed at a pH of 2. Test solutions with and without H_2O_2 were found to form similar amounts of benzene.

A simpler model was employed by McNeal *et al.* (11). In that study, test solutions prepared with 0.04% (2.8 mM) sodium or potassium benzoate and 0.025% (1.4 mM) ascorbic acid generated about 300 ppb benzene after 20 h at either 45 °C or exposure to strong UV light (wavelength and power were not reported). The benzoate and ascorbic acid concentrations were reportedly comparable to the concentrations found in commercial beverages. Only 4 ppb benzene was found when these test solutions were stored in the dark at room temperature. However, the benzene concentrations in these test solutions increased to 266 ppb after 8 days at room temperature. No additional benzene was formed when the heated or irradiated test solutions were stored at room temperature for 8 days. Benzene was not found in control solutions containing either benzoate or ascorbic acid and subjected to the conditions described. Interestingly, when benzaldehyde was substituted for benzoate, 74 ppb benzene was formed. This finding may be important since benzaldehyde is often used to simulate cherry flavor in food and beverages. The exact mechanism of benzene formation is unknown, but it is unlikely that benzene would form as a result of direct oxidation of benzaldehyde or by the Cannizzaro reaction, which forms the benzoic acid precursor. Potentially, a mechanism similar to the decomposition of benzoic acid might be involved, i.e., loss of carbon monoxide from a benzaldehyde radical.

In a study on the oxidation of ascorbate in the absence of catalytic metals, 50-mM phosphate buffer solutions were found to contain 0.3-μM iron and 0.13-μM copper (90). Similar amounts of metal ions may have shown catalytic activity in studies on benzene formation from benzoate and ascorbic acid. In the study conducted by McNeal *et al.* (11), unbuffered test solutions were prepared and no additional iron and copper salts were added. Chang and Ku (12) also prepared unbuffered test solutions of 0.04% sodium benzoate and 0.025% ascorbic acid; no additional iron or copper salts were added. In both studies, benzene formed in the test solutions after 8 days at room temperature. These results suggest that trace metal ions in the laboratory water or reagents may be sufficient to mediate hydroxyl radical formation. Chang and Ku also showed that the addition of chelating agents, ethylenediamine-tetraacetic acid (EDTA) or diethylenetriamine pentaacetic acid (DTPA) at 0.1 mM, prevented the formation of benzene in the benzoate/ascorbic acid test solutions. Ethanol at 100 mM was shown by Chang and Ku to reduce the yield of benzene by approximately 90%.

Many factors can affect the oxidation of ascorbic acid and the subsequent generation of hydroxyl radicals in aqueous solutions. These factors may include the type of metal ion, pH of the solution, exposure to heat and UV light, and the effects of different chelating agents. All of these factors interact in a complex manner. For example, the rate of oxidation of ascorbic acid by Cu^{2+}

is somewhat less rapid than the rate of oxidation by Fe^{3+} at a pH of 3.45. The addition of EDTA to sequester Cu^{2+} or Fe^{3+} will slow down the reaction approximately 100-fold (91). At a pH of 7, Cu^{2+} has a stronger catalytic activity than Fe^{3+}. However, the Fe^{3+}–EDTA complex remains a very efficient catalyst of ascorbic acid oxidation, while the Cu^{2+}–EDTA complex is completely inactive (92). Various iron chelates differ in their ability toward reduction and subsequent ability to react with H_2O_2. For example, the Fe^{3+}–EDTA complex is reduced quite rapidly by superoxide ($O_2^{\cdot -}$) to Fe^{2+}–EDTA. Subsequently, the Fe^{2+}–EDTA complex reacts readily with H_2O_2 to form hydroxyl radicals. Complexes of Fe^{3+} with DTPA or desferal are more resistant to the reduction. However, the Fe^{2+}–DTPA complex can still catalyze the formation of hydroxyl radicals (93). Another important consideration is possible synergistic effects of copper and iron salts on hydroxyl radical formation (94). Differences in the rates of oxidation of Cu^{2+} and Fe^{3+} and associated complexes make it difficult to compare results from various model systems. In summary, these studies suggest that the use of chelating agents to sequester metal ions may not necessarily reduce or eliminate hydroxyl radical formation.

Sodium and potassium benzoates are effective antimicrobial agents in beverages at a pH range of 2.5 to 4.0. Under these conditions, benzoate is converted to free benzoic acid, and benzene formation can proceed according to the series of reactions proposed by Gardner and Lawrence (13). Beverages high in sugar that contain benzoate and ascorbic acid may have less potential to form benzene. Sugars can react with and inactivate hydroxyl radicals. This effect is suggested by the data observed in surveys conducted in Canada and the United States (27, 76). For example in the Canadian survey, a regular calorie beverage with 10 times as much sugar was found to contain one-sixth the amount of benzene as the equivalent diet beverage.

The effects of temperature and light were not always obvious. For example in the Canadian survey, the highest benzene concentrations (19 and 23 ppb) were found in two low-sugar soft drinks packaged in thermally sealed transparent pouches. After a 10-h exposure to sunlight, no additional benzene formed in these low-sugar soft drinks (X.L. Cao and A. Becalski, unpublished, 2006). In comparison, McNeal *et al.* (11) observed about 300 ppb benzene formation when test solutions containing benzoate and ascorbic acid were exposed to UV light for 20 h. Unpublished studies conducted at the FDA showed that concentrations of benzene increased in some beverages containing benzoate and ascorbic acid after heating for 24 h in a 60 °C oven. Unpublished studies conducted by Health Canada showed that benzene increased in certain beverages after heating for 30 min at 100 °C (X.L. Cao and A. Becalski, unpublished, 2006). These discrepancies could be attributed to differences in the light source, the use of UV stabilizers in the packaging materials, and the composition of beverages and test solutions, temperature, and the exposure time.

The effectiveness of EDTA as a chelating agent was not always obvious for some beverages formulated with EDTA and calcium. The International

Council of Beverages Associations and the American Beverage Association state in their guidelines that the effectiveness of EDTA may be lessened in products fortified with calcium and other minerals (95, 96). In such cases, calcium could compete with iron or copper for EDTA. Metal ions remaining in solution will be free to initiate formation of the hydroxyl radical, which potentially can react with benzoic acid to form benzene.

Caution should be exercised in the interpretation of benzene results reported for beverages and other foods containing benzoate and ascorbic acid. The mere presence of benzoate and ascorbic acid in a product does not mean that benzene will form. Several factors may be interacting to increase or decrease the formation of benzene, including the presence of $OH^{\cdot}$ scavengers, exposure to elevated temperatures and light, the concentrations of benzoate and ascorbic acid, the presence of trace metal ions in the product ingredients, the presence of chelating agents, the amount of oxygen present in the products, changes in pH, the order of added ingredients, and so on. In addition, environmental and/or process-induced contamination may be another potential source of benzene contamination. For example, contamination of a mineral water occurred as a result of an improperly maintained charcoal filter (97). All of these factors could make it difficult to differentiate between the exact source and associated amount of benzene reported for beverages and other foods.

4.3.5 MITIGATION

With the exception of beverages, efforts to mitigate processed-induced benzene contamination in food have not been necessary. This may be due to the low benzene concentrations generally found in food. Benzene formation from the decomposition of phenylalanine can be minimized by grilling and roasting foods at lower temperatures. Similarly, benzene formation from the decomposition of phenylalanine during food irradiation is controlled by minimizing the dose of ionizing radiation or irradiating meat in a frozen state. The uptake of benzene in grilled fatty foods and the benzene concentration from charcoal and wood emissions can be reduced depending on the choice of fuel. A recent study showed that benzene emissions from flue gases could be reduced by an order of magnitude depending on the composition of the charcoal used during grilling (7).

The mitigation of benzene formation in beverages containing benzoate and ascorbic acid has received much attention recently. Both the International Council of Beverages Associations and the American Beverage Association have published guidelines that describe mitigation strategies that can be used by the beverage industry to reformulate affected products (95, 96). According to Equations 4.3.1 through 4.3.4 (see Section 4.3.4), benzene formation in beverages requires four reactants—benzoate, ascorbic acid, trace metal ions, and oxygen. In principle, the exclusion of any one of these reactants would be an option that could be used to mitigate benzene formation.

Due to the ubiquitous nature of oxygen and trace metal ions, a practical approach to benzene mitigation in food and beverages would be the elimination or reduction of benzoate or ascorbic acid. The EU currently permits up to 150 ppm benzoate in beverages, and Canada and the United States currently permit up to 1000 ppm (see Section 4.3.3.3). In the recent survey conducted in Canada, a beverage reformulated without benzoate was analyzed along with two samples of the original formulation that contained benzoate. Concentrations of benzene as high as 19 ppb were found in samples of the original formulation. No benzene was found in the reformulated sample (27). In the 2006 survey conducted in the United States, similar results were reported for beverages reformulated by eliminating or reducing benzoate (43). Alternative antimicrobial methods or agents will need to be considered for products reformulated by eliminating or reducing benzoate.

The elimination or reduction of ascorbic acid also can be an effective mitigation strategy. However, ascorbic acid may be naturally occurring in some fruit juices and is often added as a preservative or nutrient. In the 2006 US survey, most of the products with declared benzoate and no added ascorbic acid were found to contain less than 1 ppb benzene. However, the Canadian survey found benzene in amounts as high as 5.6 ppb in some products with declared benzoate and no ascorbic acid. Beverages containing fruit juices with declared benzoate but without declared ascorbic acid could still form benzene from reactions associated with benzoate and naturally occurring ascorbic acid. Another option to control benzene formation may be to avoid the addition of ascorbic acid to foods with naturally occurring benzoate such as cranberries and lingonberries (see Section 4.3.3.3). In addition, diet or light beverages are low in sugars, and sugars are known hydroxyl radical scavengers. Chelating agents such as EDTA may be added to beverages containing benzoate and/or ascorbic acid to reduce the potential for benzene formation; however, EDTA may be less effective in products fortified with calcium (see Section 4.3.4.3).

Another mitigation strategy might be the use of vacuum-sealed packaging to reduce the amount of oxygen in the finished product. Model studies conducted *in vitro* showed that under anaerobic conditions, benzene formation did not occur as a result of the decarboxylation of benzoate (98). However, there is a lack of data on the activity of oxygen in beverages. Other factors that can reduce exposure to benzene in beverages are minimizing exposure of products to high temperatures and/or UV light during the manufacture, storage, distribution, and shelf life of beverages.

The International Council of Beverages Associations and American Beverage Association have developed accelerated testing procedures that manufacturers can use to evaluate beverage formulations under extreme conditions over the product's shelf life (95, 96). The American Beverage Association guidelines recommended heating samples from 40 to 60 °C for 24 h or as long as 14 days.

4.3.6 EXPOSURE

Several studies concluded that inhalation of contaminated air was the primary route of nonoccupational benzene exposure for nonsmokers, and mainstream cigarette smoke was the primary route for smokers. The average inhalation exposures were estimated to be approximately 3.3 μg/kg bw/day (0.2 mg/day, 60-kg adult) for nonsmokers and 33.3 mg/kg bw/day (2 mg/day, 60-kg adult) for smokers. Dietary exposure to benzene was not considered a significant pathway (99–101).

Table 4.3.2 summarizes estimated dietary exposure to benzene determined by several organizations including the US National Research Council (NRC), Health Canada, the US FDA, the WHO, and the European Commission (EC) Joint Research Center. In general, estimated exposures are expressed in nanograms per kilogram body weight per day (ng/kg bw/day) and are derived from intake values using the amount of benzene found in particular foods and beverages and the average adult body weight (circa 60–64 kg).

Differences in the estimates can be attributed to the survey data and methodologies used to derive the dietary exposures. For example, concentrations of benzene in eggs have been reported to range from 500 to 1900 ppb benzene (101). In contrast, McNeal *et al.* found less than 2 ppb benzene in eggs and other foods and concluded that benzene found in some foods may have resulted from laboratory contamination (11). The FDA recently stated that the TDS analytical method produced unreliable results for benzene in some foods. FDA scientists recommended that the benzene data be viewed with caution while the Agency considers removing TDS benzene data from the FDA web site (43).

Some of the higher estimates of dietary exposure may have been derived using the US TDS data and the higher benzene concentrations reported for

TABLE 4.3.2 Estimated dietary exposures determined by various international organizations.

Source	Estimated dietary exposure	Reference no.
US NRC, 1980	4200 ng/kg bw/day (252 μg/day, 60-kg adult; food and drinking water)	102
Health Canada, 1992	120–325 ng/kg bw/day (food)	103
Canadian Environmental Protection Act, 1993	40–130 ng/kg bw/day (food and drinking water)	3
US FDA, 1997	5000 ng/kg bw/day (food)	104
WHO, 2003	3000 ng/kg bw/day (180 μg/day, 60-kg adult; food)	6
EC, 2005	3–50 ng/kg bw/day (food; eggs with no benzene or 2 ppb benzene)	101
	200–800 ng/kg bw/day (food; eggs with 500–1900 ppb benzene)	

eggs. The EC data were determined on the basis of the average intake of food and beverages in different regions in the EU. Three scenarios were used that included intake from fruit juices, carbonated soft drinks, fish products, and either (1) eggs containing 500–1900 ppb benzene, (2) no eggs, or (3) eggs containing 2 ppb benzene. By using scenario one, the estimated dietary exposure was determined to be 200 to 800 ng/kg bw/day; by using scenarios two and three, the estimated dietary exposures were determined to be 3 to 50 ng/kg bw/day. This may be a more realistic value because it excludes questionable data that could exaggerate dietary benzene exposure. By using 3 to 50 ng/kg bw/day, the estimated dietary exposure to benzene represents approximately 1.5% or less of the total exposure for nonsmokers and 0.2% or less for smokers. It is important to note that the total inhalation and dietary exposures may be more representative of exposures in developed countries. In addition, the EC exposure does not reflect the 2006 survey data on benzene in beverages conducted by various international food safety organizations. The survey results likely had short-term impact on dietary exposure for some individuals consuming large volumes of products containing elevated benzene levels. Mitigation strategies implemented by the beverage industry in 2006 should reduce the dietary exposure for those individuals.

4.3.7 HEALTH RISK—A RECENT ASSESSMENT OF BEVERAGES

Benzene is a carcinogen that causes tumors in rodents at multiple sites and leukemia in humans (1). Several international organizations have conducted risk assessments based on exposure to benzene using various models. These models use a number of methods to estimate an exposure level associated with low levels of risk on the basis of cancer and noncancer end points. A risk level for cancer end points may be defined as a unit risk value over a lifetime; noncancer end points may be defined as a tolerable daily intake (TDI) or reference dose (Rfd). The uncertainty associated with the dietary exposure and an estimated low dose effect level has resulted in some uncertainty in the assessed health risk from dietary exposure to benzene.

In 2006, Health Canada and the US FDA each assessed the risk associated with short-term exposure to beverages containing benzene at trace to low concentrations. In order to conduct the risk assessment, Health Canada used both a TDI value of 0.36 μg/kg bw/day benzene that was based on experimental animal data available during the previous 1991 assessment as well as a range of oral slope factors calculated by the US EPA derived on the basis of a human cancer end point (leukemia). The slope factors were used to calculate the oral exposure (0.018 to 0.066 μg/kg bw/day) associated with a cancer risk level of 10^{-6}.

A worst case probable daily intake (PDI) for children (5 to 11 years of age) was determined on the basis of daily consumption of a cherry-flavored beverage with 18 ng/mL benzene. This product contained the highest benzene concentration found in the preliminary Health Canada survey. The PDI was determined to be 180 ng/kg bw/day which was equivalent to the dose associated with a lifetime risk of 10^{-5} and was approximately three times higher than the US EPA-derived dose associated with a cancer risk of 10^{-6}. However, the product was reformulated and the old formulation depleted from the retail market by mid-June 2006. As a result, the potential long-term exposure to benzene from these beverages was much reduced. On the basis of this limited exposure, the health risk was considered negligible over the short-term that these products were available on the market and of relatively low concern to human health over the long term. The FDA also concluded that the concentrations of benzene found in soft drinks and other beverages did not pose a safety concern for consumers (43, 103).

4.3.8 RISK MANAGEMENT

Regulatory limits for benzene in food and beverages have not been established, except for bottled water where the FDA has adopted the US EPA MCL of 5 ppb as a quality standard (105). In general, the low concentrations found have not generated the level of concern that would warrant the establishment of formal risk management procedures or regulatory limits. Recent reports of elevated levels of benzene in soft drinks and beverages initiated a series of actions by food safety organizations and the beverage industry to verify these findings, identify the source of contamination, evaluate the health risk, and reduce the concentrations found to meet acceptable quality standards.

Guidelines for benzene in drinking water vary from country to country. The WHO guideline for benzene in drinking water is 10 ppb. The United States and Canada established an MCL of 5 ppb (74, 75). Australia and the EU established a reference level of 1 ppb (106, 107). These guidance levels formed the basis for risk management decisions made by food safety organizations and the beverage industry. Several countries conducted surveys to determine the amount of benzene in soft drinks and beverages. Survey results and consumer advisories were published on agency web sites. In addition, guidance documents were developed by the International Council of Beverages Associations and American Beverage Association to educate the beverage industry on strategies for mitigating benzene formation in beverages (95, 96). The guidance documents advised manufacturers that the main factors affecting benzene formation in beverages were the combination of benzoic acid sources and ascorbic acid, heat, and time. Formulation control strategies were recommended to minimize benzene formation.

With the exception of South Korea, surveys conducted in several countries found that only a few products contained benzene above established guidance levels for drinking water (38, 43, 77–79, 108). The following summarizes some of the risk management strategies announced on agency web sites:

- In South Korea, beverages with greater than 10 ppb benzene were withdrawn from the market and manufacturers were instructed to reformulate their products to minimize benzene formation (77).
- An "ALARA" (as low as reasonably achievable) approach has been applied by Health Canada, with consideration given to the lifetime cancer risk, in the case of genotoxic carcinogens that have not been directly added to the food but are present for some other reason (e.g., environmental or processing contaminants) (103). The industry trade association "Refreshments Canada" and manufacturers of drinks with excessive concentrations of benzene were notified about the findings and Health Canada's ALARA strategy. Beverage manufacturers responded by reformulating the affected products to prevent benzene formation and have stopped further shipments of the old formulations to retail markets. A follow-up survey conducted in 2007 found that benzene concentrations in beverages remained safe for consumers (109).
- The FDA contacted beverage manufacturers whose products were found to contain greater than 5 ppb benzene. Those manufacturers reformulated the affected products to minimize or eliminate benzene formation. The FDA "plans to continue its testing program for benzene in soft drinks and other beverages and will inform the public and manufacturers as new data become available (43)."
- The UK's FSA "has drawn on the WHO guidelines for safe levels in drinking water as an appropriate comparator. The FSA has asked the soft drinks industry to ensure that amounts of benzene are kept as low as practicable." Furthermore, the FSA asked the relevant manufacturers to remove from sale products with greater than 10 ppb benzene (38).

The general consensus of regulatory authorities was that the risk posed by benzene in beverages was negligible in comparison with the overall exposure from environmental sources. Nevertheless, the presence of benzene in beverages was considered avoidable and warranted mitigation to eliminate or minimize benzene concentrations while maintaining microbial safety.

The reoccurrence of this problem in 2006 is another important consideration. Regulatory authorities believed that the issue had been corrected in the 1990s when the conditions leading to benzene formation in beverages were first elucidated. The recent worldwide findings of a small number of beverages containing ppb levels of benzene suggest that new beverage products could enter the retail market without due consideration of product formulation. For this reason, ongoing testing programs are essential to ensure the safety of food and beverages.

4.3.9 CONCLUSION

The occurrence of benzene in food and beverages has been evaluated in targeted surveys and market basket studies. In general, benzene concentrations reported are low and seldom exceed low ppb levels. The available data suggest that dietary exposure to benzene is negligible in comparison with the overall exposure from environmental sources. Improper handling and process-induced and environmental contamination can lead to increased benzene concentrations in food and beverages. Mitigation strategies have been effective at lowering amounts of benzene in affected products. Ongoing testing programs for beverages containing benzoate and ascorbic acid are essential to ensure that benzene levels are minimized.

REFERENCES

1. Agency for Toxic Substances and Disease Registry (2005). Draft Toxicological Profile for Benzene, Division of Toxicology and Environmental Medicine, Agency for Toxic Substances, Public Health Services, US Department of Health and Human Services, Atlanta, GA.
2. Moffat, C.F., Whittle, K.J. (1999). Polycyclic aromatic hydrocarbons, petroleum and other hydrocarbon contaminants, in *Environmental Contaminants of Food* (eds C.F. Moffat, K.J. Whittle), Sheffield Academic Press, Sheffield, pp 364–429.
3. Environment Canada, Health and Welfare Canada (1993). Benzene. Priority substances list assessment report. http://www.hc-sc.gc.ca/ewh-semt/alt_formats/hecs-sesc/pdf/pubs/contaminants/psl1-lsp1/benzene/benzene_e.pdf (accessed 17 November 2006).
4. Claydon, M., Evans, M., Gennart, G.-P., Roythorne, C., Simpson, B. (1999). Environmental exposure to benzene; 2/99; Concawe, Brussels.
5. The National Toxicology Program (2005). Report on Carcinogens, 11th edn, US Department of Health and Human Services. http://ntp.niehs.nih.gov/index.cfm?objectid=72016262-BDB7-CEBA-FA60E922B18C2540 (accessed 12 September 2008).
6. World Health Organization (2003). Benzene in drinking water. WHO/SDE/WSH/03.04/24. http://www.who.int/water_sanitation_health/dwq/benzene.pdf (accessed 19 January 2007).
7. Olsson, M., Petersson, G. (2003). Benzene emitted from glowing charcoal. *The Science of the Total Environment*, *303*, 215–220.
8. Sommers, S.H., Delincee, H., Smith, J.S., Marchioni, E. (2006). Toxicological safety of irradiated foods, in *Food Irradiation—Research and Technology* (eds C.H. Sommers, X. Fan), Blackwell Publishing, Ames, pp. 43–61.
9. Zhu, M.J., Mendonca, A., Min, B., Lee, E.J., Nam, K.C., Park, K., Du, M., Ismail, H.A., Ahn, D.U. (2004). Effects of electron beam irradiation and antimicrobials on the volatiles, colour, and texture of ready-to-eat turkey breast roll. *Journal of Food Science*, *69*, C382–C387.
10. Page, B.D., Conacher, H.B.S., Weber, D., Lacroix, G. (1992). A survey of benzene in fruits and retail fruit juices, fruit drinks, and soft drinks. *Journal of AOAC International*, *75*, 334–340.
11. McNeal, T.P., Nyman, P.J., Diachenko, G.W., Hollifield, H.C. (1993). Survey of benzene in foods by using headspace concentration techniques and capillary gas chromatography. *Journal of AOAC International*, *76*, 1213–1219.

12. Chang, C., Ku, K. (1993). Studies on benzene formation in beverages. *Journal of Food and Drug Analysis*, *1*, 385–393.
13. Gardner, L.K., Lawrence, G.D. (1993). Benzene production from decarboxylation of benzoic acid in the presence of ascorbic acid and a transition-metal catalyst. *Journal of Agricultural and Food Chemistry*, *41*, 693–695.
14. Grob, K., Frauenfelder, C., Artho, A. (1990). Uptake by foods of tetrachloroethylene, trichloroethylene, toluene and benzene from air. *Zeitschrift fur Lebensmittel-Untersuchung und-Forschung*, *191*, 435–441.
15. Jickells, S.M., Crews, C., Castle, L., Gilbert, J. (1990). Headspace analysis of benzene in food contact materials and its migration into foods from plastics cookware. *Food Additives and Contaminants*, *7*, 197–205.
16. Varner, S.L., Hollifield, H.C., Anrzejewski, D. (1991). Determination of benzene in polypropylene food-packaging materials and food-contact paraffin waxes. *Journal of the Association of Official Analytical Chemists*, *74*, 367–374.
17. Kokot-Helbling, K., Schmid, P., Schlatter, C. (1995). Vergleich der aufnahme von flüchtigen organischen verbindungen (benzol, toluol, xylol und tetrachloethen) aus lebensmitteln mit der aufnahme aus der luft. *Mittenlungen aus dem Gebiete der Lebensmittelunterschung und Hygiene*, *86*, 556–565.
18. Wittkowski, R., Baltes, W., Jennings, W.G. (1990). Analysis of liquid smoke and smoked meat volatiles by headspace gas chromatography. *Food Chemistry*, *37*, 135–144.
19. Wilkes, J.G., Conte, E.D., Kim, Y., Holcomb, M., Sutherland, J.B., Miller, D.W. (2000). Sample preparation for the analysis of flavours and off-flavours in foods. *Journal of Chromatography, A*, *880*, 3–33.
20. Kopp, B., Gilbert, J. (1984). Analysis of food contaminants by headspace gas chromatography, in *Analysis of Food Contaminants*, Elsevier, Amsterdam, pp. 117–130.
21. Heikes, D.L., Jensen, S.R., Fleming-Jones, M.E. (1995). Purge and trap extraction with GC-MS determination of volatile organic compounds in table-ready foods. *Journal of Agricultural and Food Chemistry*, *43*, 2869–2875.
22. Barshick, S.-A., Smith, S.M., Buchanan, M.V., Guerin, M.R. (1995). Determination of benzene content in food using a novel blender purge and trap GC/MS method. *Journal of Food Composition and Analysis*, *8*, 244–257.
23. Stein, V.B., Narang, R.S. (1990). A simplified method for the determination of volatiles in eggs using headspace analysis with photoionization detector. *Archives of Environmental Contamination and Toxicology*, *19*, 593–596.
24. Jickells, S.M. (1993). Gas chromatographic/mass spectrometric determination of benzene in nonstick cookware and microwave susceptors and its migration into foods on cooking. *Journal of AOAC International*, *76* (4), 760–764.
25. US Food and Drug Administration (2006). Determination of benzene in soft drinks and other beverages. http://www.cfsan.fda.gov/~dms/benzmeth.html (accessed 19 May 2006).
26. Page, B.D., Conacher, H., Salminen, J. (1993). Survey of bottled drinking water sold in Canada. Part 2. Selected volatile organic compounds. *Journal of AOAC International*, *76* (1), 26–31.
27. Cao, X.-L., Casey, V., Seaman, S., Tague, B., Becalski, A. (2007). Analysis of benzene in soft drinks and other beverages by isotope dilution headspace gas chromatography and mass spectrometry. *Journal of AOAC International*, *90*, 479–484.

28. Hiatt, M.H. (1995). Vacuum distillation coupled with gas chromatography/mass spectrometry for the analysis of environmental samples. *Analytical Chemistry*, *67*, 4044–4052.
29. Hiatt, M.H. (1997). Analyses of fish tissue by vacuum distillation/gas chromatography/mass spectrometry. *Analytical Chemistry*, *69* (6), 1127–1134.
30. Hiatt, M.H., Pia, J.H. (2004). Screening processed milk for volatile organic compounds using vacuum distillation/gas chromatography/mass spectrometry. *Archives of Environmental Contamination and Toxicology*, *46* (2), 189–196.
31. Kolb, B., Ettre, L.S. (2006). *Static Headspace-Gas Chromatography, Theory and Practice*, 2nd edn, John Wiley & Sons, Inc.., Hoboken, NJ.
32. Zhang, Z., Pawliszyn, J. (1993). Headspace solid-phase microextraction. *Analytical Chemistry*, *65*, 1843–1852.
33. Eichelberger, J.W., Budde, W.L. (1989). *Method 524.2, Measurement of Purgeable Organic Compounds in Water by Capillary Column Gas Chromatography/mass Spectrometry, Revision 3.0*, US Environmental Protection Agency, Cincinnati, OH.
34. McNeal, T.P., Hollifield, H.C., Diachenko, G.W. (1995). Survey of trichloromethanes and other volatile chemical contaminants in processed foods by purge-and-trap capillary gas chromatography with mass selective detection. *Journal of AOAC International*, *78*, 391–397.
35. Fabietti, F., Delise, M., Piccioli Bocca, A. (2001). Investigation into benzene and toluene content of soft drinks. *Food Control*, *12*, 505–509.
36. Kataoka, H., Lord, H.L., Pawliszyn, J. (2000). Applications of solid-phase microextraction in food analysis. *Journal of Chromatography, A*, *800*, 35–62.
37. Pawliszyn, J. (1997). *Solid Phase Microextraction: Theory and Practices*, Wiley-VCH.
38. UK Food Standards Agency. (2006) Benzene in soft drinks. Food Survey Information Sheet 06. UK Food Standards Agency. http://www.food.gov.uk/science/surveillance/fsisbranch2006/fsis0606 (accessed 1 November 2006).
39. Nilsson, T., Pelusio, F., Montanarella, L., Larsen, B., Facchetti, S., Madsen, J. (1995). An evaluation of solid-phase microextraction for analysis of volatile organic compounds in drinking water. *Journal of High Resolution Chromatography*, *18*, 617–624.
40. Menéndez, J.C., Sánchez, M.L., Uría, J.E., Martínez, E., Sanz-Medel, A. (2000). Static headspace, solid-phase microextraction and headspace solid-phase microextraction for BTEX determination in aqueous samples by gas chromatography. *Analytica Chimica Acta*, *415*, 9–20.
41. Potter, D., Pawliszyn, J. (1992). Detection of substituted benzenes in water at pg/ml level using solid-phase microextraction and gas chromatography-ion trap mass spectrometry. *Journal of Chromatography*, *625*, 247–255.
42. Fleming-Jones, M.E., Smith, R.E. (2003). Volatile organic compounds in foods: a five year study. *Journal of Agricultural and Food Chemistry*, *51*, 8120–8127.
43. US Food and Drug Administration. (2006) Data on Benzene in Soft Drinks and Other Beverages. http://www.cfsan.fda.gov/~dms/benzdata.html (accessed 19 May 2006) and http://www.cfsan.fda.gov/~dms/benzqa.html (accessed 12 July 2007).
44. Food Standards Agency. (1995) Benzene and Other Aromatic Hydrocarbons in Food, Average UK Dietary Intakes, Food Surveillance Information Sheet, Number

58. Ministry of Agriculture, Fisheries and Food UK, Food Standards Agency. http://archive.food.gov.uk/maff/archive/food/infsheet/1995/no58/58benz.htm (accessed 2 April 2007).
45. Gorna-Binkul, A., Keymeulen, R., Van Langenhove, H., Buszewski, B. (1996). Determination of monocyclic aromatic hydrocarbons in fruit and vegetables by gas chromatography-mass spectrometry. *Journal of Chromatography, A*, *734*, 297–302.
46. Lourdes Cardeal, Z., Guimaraes, E.M., Parreira, F.V. (2005). Analysis of volatile compounds in some typical Brazilian fruits and juices by SPME-GC method. *Food Additives and Contaminants*, *22*, 508–513.
47. Engel, K.-H., Tressl, R. (1983). Studies on the volatile components of two mango varieties. *Journal of Agricultural and Food Chemistry*, *31*, 796–801.
48. Kjallstrand, J., Petersson, G. (2001). Phenolic antioxidants in alder smoke during industrial meat curing. *Food Chemistry*, *74*, 85–89.
49. Rozum, J. (1998). Smoke flavourings in processed meats, in *Flavour of Meat, Meat Products and Seafoods* (ed. F. Shahidi), Blackie Academic & Professional, London, pp. 342–354.
50. Heins, J.T., Maarse, H., Noever de Brauw, M.C., Weurman, C. (1966). Direct food vapour analysis and component identification by a coupled capillary GLC-MS arrangement. *Journal of Gas Chromatography*, *4*, 395–397.
51. Silwar, R., Kamperschrorer, H., Tressl, R. (1987). Gaschromatographisch-massenspektrometrische untersuchungen des rostkaffeearomas. Quantitative bestimmung wasserdampfflüchtige aromastoffe. *Chemie, Mikrobiologie, Technologie der Lebensmittel*, *10*, 176–187.
52. Sanz, C., Maeztu, L., Zapelena, M.J., Bello, J., Cid, C. (2002). Profiles of volatile compounds and sensory analysis of three blends of coffee: influence of different proportions of Arabica and Robusta and influence of roasting coffee with sugar. *Journal of the Science of Food Agriculture*, *82*, 840–847.
53. Mondello, L., Casilli, A., Tranchida, P.Q., Dugo, P., Costa, R., Festa, S., Dugo, G. (2004). Comprehensive multidimensional GC for the characterization of roasted coffee beans. *Journal of Separation Science*, *27*, 442–450.
54. Spadone, J.-C., Takeoka, G., Liardon, R. (1990). Analytical investigation of Rio off-flavour in green coffee. *Journal of Agricultural and Food Chemistry*, *38*, 226–233.
55. Mathieu, F., Malosse, C., Frerot, B. (1998). Identification of the volatile components released by fresh coffee berries at different stages of ripeness. *Journal of Agricultural and Food Chemistry*, *46*, 1106–1110.
56. Cantergiani, E., Brevard, H., Krebs, Y., Feria-Morales, A., Amado, R., Yeretzian, C. (2001). Characterisation of the aroma of green Mexico coffee and identification of mouldy/earthy defect. *European Food Research and Technology*, *212*, 648–657.
57. Morehouse, K.M. (1999) Food Irradiation: The treatment of foods with ionizing radiation. US Food and Drug Administration. http://www.cfsan.fda.gov/~dms/opa-fdir.html (accessed 3 April 2007).
58. Diehl, J.F. (1995). *Safety of Irradiated Foods*, 2nd edn, Marcel Dekker, Inc., New York.
59. Phytosanitary, I. (2002). Treatment of imported fruits and vegetables: final rule. *Federal Register*, 65016–65029.

60. Merritt, C., Jr, Angelini, P., Graham, R.A. (1978). Effects of radiation parameters on the formation of radiolysis products in meat and meat substances. *Journal of Agricultural and Food Chemistry*, *26*, 29–35.
61. Ahn, D.U., Lee, E.J. (2004). Mechanisms and prevention of off-odour production and colour changes in irradiated meat, in *Irradiation of Food and Packaging; Recent Developments* (eds V. Komolprasert, K.M. Morehouse), American Chemical Society, Washington, DC, pp. 43–74.
62. Masohan, A., Parsad, G., Khanna, M.K., Chopra, S.K., Rawat, B.S., Garg, M.O. (2000). Estimation of trace amounts of benzene in solvent-extracted vegetable oils and oil seed cakes. *Analyst*, *125*, 1687–1689.
63. Bolzoni, L., Barbieri, G., Virgili, R. (1996). Change in volatile compounds of Parma ham during maturation. *Meat Science*, *43*, 301–310.
64. Nagayama, T., Nishijima, M., Yasuda, K., Saito, K., Kamimura, H., Ibe, A., Ushiyama, H., Nagayama, M., Naoi, Y. (1983). Benzoic acid in fruits and fruit products. *Journal of the Food Hygienic Society of Japan*, *24*, 416–422.
65. Zuo, Y., Wang, C., Zhan, J. (2002). Separation, characterization, and quantitation of benzoic and phenolic antioxidants in American cranberry fruit by GC-MS. *Journal of Agricultural and Food Chemistry*, *50*, 3789–3794.
66. Visti, A., Viljakainen, S., Laakso, S. (2003). Preparation of fermentable lingonberry juice through removal of benzoic acid by *Saccharomyces cerevisiae* yeast. *Food Research International*, *36*, 597–602.
67. Sieber, R., Butikofer, U., Bosset, J.O., Ruegg, M. (1989). Benzoic acid as a natural component of foods—a review. *Mittenlungen aus dem Gebiete der Lebensmittel-unterschung und Hygiene*, *80*, 345–362.
68. Nagayama, T., Nishijima, M., Yasuda, K., Saito, K., Kamimura, H., Ibe, A., Ushiyama, H., Naoi, Y., Nishima, T. (1986). Benzoic acid in agricultural food products and processed foods. *Journal of the Food Hygienic Society of Japan*, *27*, 316–325.
69. Chipley, J.R. (2005). Sodium benzoate and benzoic acid, in *Antimicrobials in Food*, 3rd edn (eds P.M. Davidson, J.N. Sofos, A.L. Branen), CRC Press, Boca Raton, FL, pp. 11–48.
70. Health Canada. (2007). Food and Drugs Act, Food and Drug Regulations (C.R.C., c. 870). http://laws.justice.gc.ca/en/ShowFullDoc/cr/C.R.C.-c.870///en (accessed 12 September 2008).
71. European Commission. (1995). Food additives other than colours and sweeteners. *Council Directive No 95/2/EC.* http://ec.europa.eu/food/fs/sfp/addit_flavor/flav11_en.pdf (accessed 12 September 2008).
72. Direct Food Substances as Generally Recognized as Safe. US Code of Federal Regulations, Pt 184 (ss 184. 1021), Title 21 (2007).
73. Yoon, H.J., Cho, Y.H., Park, J., Lee, C.H., Park, S.K., Cho, Y.J., Han, K.W., Lee, J.O., Lee, C.W. (2003). Assessment of estimated daily intakes of benzoates for average and high consumers in Korea. *Food Additives and Contaminants*, *20*, 127–135.
74. Health Canada. (1987). Benzene—Supporting Document for the Guidelines for Canadian Drinking Water Quality. http://www.hc-sc.gc.ca/ewh-semt/pubs/water-eau/benzene/index-eng.php (accessed 22 September 2008).
75. Maximum Contaminant Levels for Organic Contaminants. US Code of Federal Regulations, Pt 141 (ss 141. 61), Title 40, (2007).

76. Nyman, P.J., Diachenko, G.W., Perfetti, G.A., McNeal, T.P., Hiatt, M.H., Morehouse, K.M. (2008). Survey results of benzene in soft drinks and other beverages by headspace gas chromatography/mass spectrometry. *Journal of Agricultural and Food Chemistry*, *56*, 571–576.
77. Korean Food and Drug Administration. (2006). Benzene in beverages. Bulletin # 938. http://www.kfda.go.kr (accessed 24 April 2007).
78. Food Standards Australia New Zealand. (2006). Food surveillance, Australia New Zealand. *Food Standards Australia New Zealand*. http://www.foodstandards.gov.au/_srcfiles/Autumn_Winter_2006.pdf (accessed 13 April 2007).
79. Food Safety Authority of Ireland. (2006). FSAI issues reassurance on levels of benzene in soft drinks. Food Safety Authority of Ireland. http://www.fsai.ie/news/press/pr_06/pr20060612.asp (accessed 8 February 2007).
80. Douda, J., Basiuk, V.A. (2000). Pyrolysis of amino acids: recovery of starting materials and yields of condensation products. *Journal of Analytical and Applied Pyrolysis*, *56*, 113–121.
81. Vollmin, J., Kriemler, P., Omura, I., Seibl, J., Simon, W. (1966). Structural elucidation with a thermal fragmentation-gas chromatography-mass spectrometry combination. *Microchemical Journal*, *11*, 73–86.
82. Giacobbo, H., Simon, W. (1964). Methodik zur pyrolyse und anschliesenden gas-chromatographischen analyse von probemengen unter einem mikrogram. *Pharmaceutica Acta Helvetiae*, *39*, 162–167.
83. Patterson, J.M., Haidar, N.F., Papadopoulos, E.P., Smith, W.T. (1973). Pyrolysis of phenylalanine, 3,6-dibenzyl-2,5-piperazinedione, and phenethylamine. *Journal of Organic Chemistry*, *38*, 663–666.
84. Wang, S., Liu, B.Z., Su, Q.D. (2004). Pyrolysis-gas chromatography/mass spectrometry as a useful technique to evaluate the pyrolysis pathways of phenylalanine. *Journal of Analytical and Applied Pyrolysis*, *71*, 393–403.
85. Kato, S., Kurata, T., Fujimaki, M. (1971). Thermal degradation of aromatic amino acids. *Agricultural and Biological Chemistry*, *35*, 2106–2112.
86. De Maria, C.A.B., Trugo, L.C., Aquino Neto, F.R., Moreira, R.F.A., Alviano, C.S. (1996). Composition of green coffee water-soluble fractions and identification of volatiles formed during roasting. *Food Chemistry*, *55*, 203–207.
87. Flament, I. (2002). *Coffee Flavour Chemistry*, John Wiley & Sons, Ltd, Chichester, England.
88. Matthews, R.W., Sangster, D.F. (1965). Measurement by benzoate radiolytic decarboxylation of relative rate constants for hydroxyl radical reactions. *Journal of Physical Chemistry*, *69*, 1938–1946.
89. Kochi, J.K. (1973). Oxidation-reduction reactions of free radicals and metal complexes, in *Free Radicals*, Vol. 1 (ed. J.K. Kochi), John Wiley & Sons, Inc., New York, pp. 591–683.
90. Buettner, G.R. (1988). In the absence of catalytic metals, ascorbate does not autoxidize at pH 7: Ascorbate as a test for catalytic metals. *Journal of Biochemical and Biophysical Methods*, *16*, 20–40.
91. Martell, A.E. (1982). Chelates of ascorbic acids: formation and catalytic properties, in *Ascorbic Acid: Chemistry, Metabolism and Uses* (eds P.A. Seib, B.M. Tolbert), ACS, Washington, DC, pp. 153–178.

92. Buettner, G.R. (1986). Ascorbate autoxidation in the presence of iron and copper chelates. *Free Radical Research Communications*, *1*, 349–353.
93. Buettner, G.R. (1987). Activation of oxygen by metal complexes and its relevance to autoxidative processes in living systems. *Bioelectrochemistry and Bioenergetics*, *18*, 29–36.
94. Buettner, G.R., Jurkiewicz, B.A. (1996). Catalytic metals, ascorbate and free radicals: combinations to avoid. *Radiation Research*, *145*, 532–541.
95. American Beverage Association. (2006). American Beverage Association Guidance Document to Mitigate the Potential for Benzene Formation in Beverages, American Beverage Association, 1101 Sixteenth Street, NW, Washington, DC, 20036–24877.
96. International Council of Beverages Association. (2006). Guidance Document to Mitigate the Potential for Benzene Formation in Beverages, The International Council of Beverages Associations (ICBA). http://www.icba-net.org (accessed 3 April 2007).
97. James, G. (1990). Perrier recalls its water in US after benzene is found in bottles. *The New York Times*, 10 February, p. 1.
98. Winston, G.W., Cederbaum, A.I. (1982). Oxidative decarboxylation of benzoate to carbon dioxide by rat liver microsomes: a probe for oxygen radical production during microsomal electron transfer. *Biochemistry*, *21*, 4265–4270.
99. Wallace, L.A. (1989). Major sources of benzene exposure. *Environmental Health Perspectives*, *82*, 165–169.
100. Wallace, L. (1996). Environmental exposure to benzene: an update. *Environmental Health Perspectives*, *104* (Suppl. 6), 1129–1136.
101. Bruinen de Bruin, Y., Kotzias, D., Kephalopoulos, S. (2005). *2005 HEXPOC Human Exposure Characterization of Chemical Substances; Quantification of Exposure Routes*, Institute for Health and Consumer Protection, European Commission Joint Research Centre, Italy.
102. National Research Council. (1980). *Drinking Water and Health*, Vol. 3, US National Research Council, National Academy Press, Washington, DC.
103. Health Canada (2006). *Health Risk Assessment: Benzene in Beverages*, Chemical Health Hazard Assessment Division, Food Directorate, pp. 1–14. http://www.hc-sc.gc.ca/fn-an/securit/chem.-chim/food-aliment/benzene/benzene_hra-ers_e.html (accessed 22 September 2008).
104. Diachenko, G. (1997). FDA Update on Packaging and Contaminants Issues, 22nd Annual Winter Meeting of the Toxicology Forum, Washington, DC.
105. Beverages. US Code of Federal Regulations, Part 165 (ss165. 110), Title 21, (2007).
106. Council of the European Community. (1998). Quality of water intended for human consumption. The Drinking Water Directive (DWD), Council Directive 98/83/EC. http://ec.europa.eu/environment/water/water-drink/index_en.html (accessed 22 September 2008).
107. Australian Water and Wastewater Association. (1996). Australian Drinking Water Guidelines. The National Health & Medical Research Council. http://www.nhmrc.gov.au/publications/synopses/_files/eh19.pdf (accessed 22 September 2008).

108. Health Canada (2006). Survey of benzene in soft drinks and other beverage products. http://www.hc-sc.gc.ca/fn-an/surveill/other-autre/benzene_survey_enquete_e.html (accessed 22 September 2008).
109. Cao, X.-L.; Casey, V. (2008). An improved method for determination of benzene in soft drinks at sub-ppb levels. *Food Additives and Contaminants*, *25(4)*, 401–405.

5

HIGH-PRESSURE PROCESSING

Alexander Mathys and Dietrich Knorr
Berlin University of Technology, Department of Food Biotechnology and Food Process Engineering, Koenigin-Luise-Str. 22, Berlin D-14195, Germany

5.1 INTRODUCTION

The initial intention for the re-emergence of high hydrostatic pressure treatment of foods about 20 years ago, after the initial work by Hite (1) and subsequent key studies by Clouston and Wills (2) as well as Gould and Sale (3), was to find gentle, low-intensity, low-energy-consuming processes leading to safe, high-quality products with extended shelf life as an alternative to conventional thermal processing. Consequently, most of the efforts of high-pressure research concentrated on microbial inactivation, on product modification, and on process development. Relatively little attention has been given to the impact of high pressure on nutrient, toxin, or allergen inactivation kinetics or on the understanding of mechanisms of such inactivations or potential generation of undesirable food constituents, toxin, and allergens due to high-pressure treatment (4, 5).

The evolution of high-pressure processing (HPP) started with the development of cannon for the military. A summary of the early cannon development that aimed to contain higher and higher pressure is given by Crossland (6). Based on this research, vessel designs for laboratory experiments became available, which followed in first measurements of the compressibility of water (7) and other fluids (8). First use of high pressure for biological studies was presented by Regnard (9) and Certes (10). Regnard studied the effects of

Process-Induced Food Toxicants: Occurrence, Formation, Mitigation, and Health Risks,
Edited by Richard H. Stadler and David R. Lineback

pressures ranging up to 100 MPa on a wide variety of aquatic organisms. In 1899, high-pressure experiments with microorganisms in a food sample were performed by Hite (1). Investigations of bacterial spores under pressure followed in 1903 by Chlopin and Tammann (11), which found that bacterial spores were resistant to hydrostatic pressure. This was also reported by Hite *et al.* (12) in 1914 and confirmed in more detail by experiments with different spore strains from about 300 to 1200 MPa (13). Larson *et al.* (13) were the first to show the differences in the inactivation of vegetative and sporulated cells of *Bacillus subtilis* at the highest applied pressure up to 1200 MPa.

Currently, most of the industrial applications of high pressure, which were introduced in Japan in 1990 and in Europe and the United States in 1996, are used for pasteurization purposes with some applications also geared toward product modification such as gelatinization of proteins and starch (4, 14–16).

5.2 HPP

5.2.1 Basic Principles

The use of high-pressure technology in food processing has increased steadily during the past 10 years (Fig. 5.1a) and in 2007, 110 industrial installations existed worldwide with volumes from 35 to 420 L and an annual production volume of more than 120,000 tons (Tonello Samson, C., 2007, NC Hyperbaric, Spain, personal communication). Most of the vessel volume is used for meat and vegetable products (Fig. 5.1b).

Two different concepts of HPP have been developed for different kinds of foods (17) (Fig. 5.2). By using the internal intensifier (Fig. 5.2b), the maximum size of the particulates is limited by the rating of valves and pumps.

Both concepts include an intensifier, where the simplest practical system is a single-acting hydraulically driven pump. The two main parts of an intensifier are the low-pressure and the high-pressure cylinder. A double-acting arrangement enables a continuous, uniform flow, while one of the double-acting pistons is delivering, the other cylinder is being charged during its intake stroke (Fig. 5.3).

Pressure-assisted heating as an emerging technology can heat up and cool down products, and it allows accurate control of the treatment intensity required for pasteurization or sterilization (18). Although it is widely accepted that pressure-assisted thermal sterilization is environmentally friendly and can retain the fresh-like characteristics of foods better than heat treatment (19), it has not yet been successfully introduced into the food industry due to the limited knowledge on inactivation mechanisms of bacterial spores.

As an example of the reduction in spore resistance to heat Heinz and Knorr (18) compared the inactivation data at 800 MPa of Rovere *et al.* (20) with the generally accepted botulinum cook at ambient pressure (21) in Fig. 5.4a. By using the F-value concept (Eq. 5.1), it is possible to compare the thermal effect

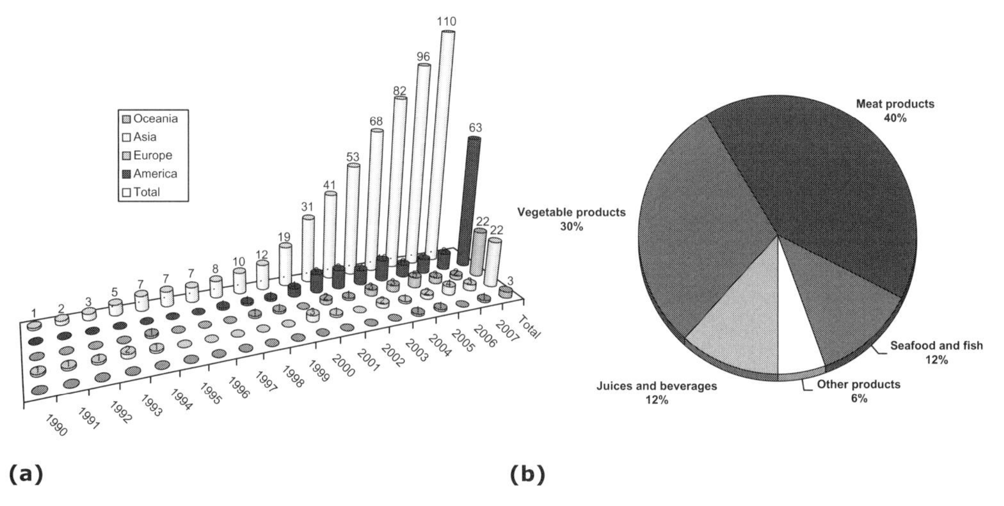

Figure 5.1 (a) HPP equipment in the world and (b) total vessel volume versus food industries (Tonello Samson, C., 2007, NC Hyperbaric, Spain, personal communication).

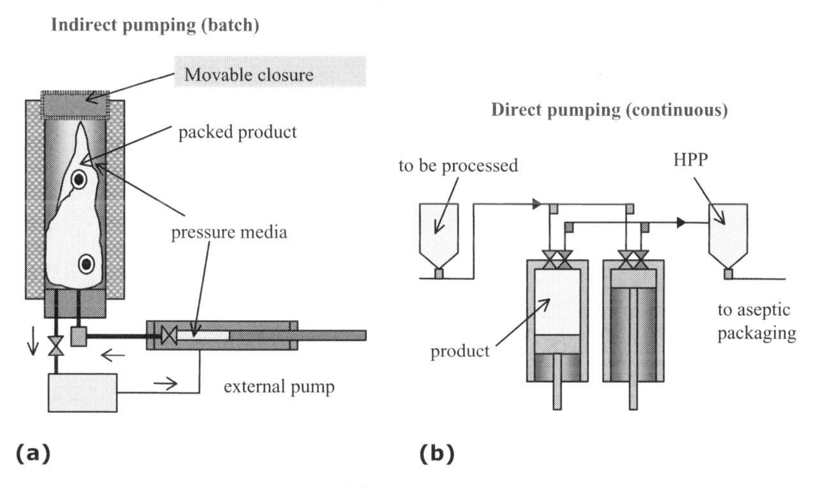

Figure 5.2 Two concepts of HPP: (a) indirect (batch) and (b) direct (continuous) pumping (17). Reprinted with kind permission of Springer Science and Business Media.

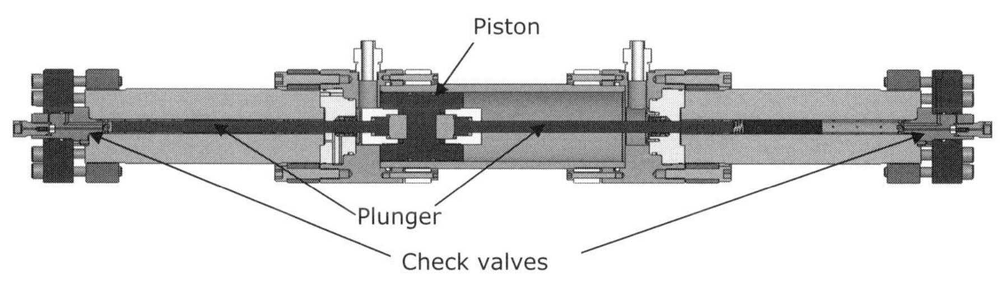

Figure 5.3 Double-acting intensifier (Hernando, A., 2007, NC Hyperbaric, Spain, personal communication).

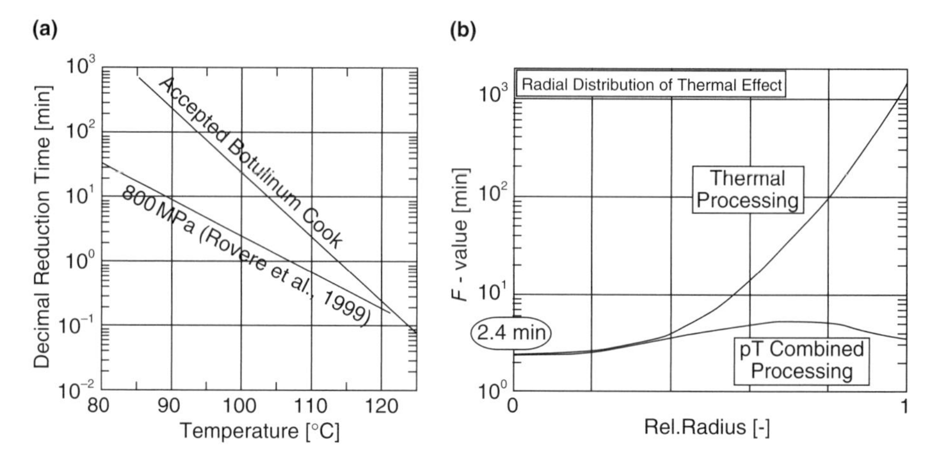

Figure 5.4 Comparison between (a) conventional and (b) *p*-*T* combined sterilization; rel. Radius 0 = center, 1 = periphery of container (18). Reprinted with kind permission of Springer Science and Business Media.

of treatments where the temperatures are a function of time $T(t)$, N_0 is the initial population, and N the survival count.

$$F \equiv \int_0^t 10^{\frac{T(t)-T_{\text{ref}}}{z}} \, dt = D_{\text{ref}} \log_{10}\left(\frac{N}{N_0}\right) \tag{5.1}$$

In Fig. 5.4a, both decimal reduction times (D-value) converge at 121.1 °C (250 °F) and this value has been chosen as the reference (ref) temperature T_{ref} for calculation of the F-value in Fig. 5.4b. The z-value represents the temperature increase for a decimal reduction of the D-value. In the food canning industry, it is common to use 121.1 °C as T_{ref} and a z-value of 10 °C, which is derived from the slope in Fig. 5.4a. For the inactivation data at 800 MPa of Rovere *et al.* (20), z yields 18 °C. An F-value of 2.4 min is equivalent to a 12 log-cycle reduction of *Clostridium botulinum* with $D_{\text{ref}} = 0.2$ min ($F = 36$ min for *Geobacillus stearothermophilus* with $D_{\text{ref}} = 3$ min) and is regarded as the minimum thermal load that must be applied to every part of the product.

5.2.2 Thermodynamics of High Isostatic Pressure

The fundamental behavior of a thermodynamic system is summarized in the four laws of thermodynamics.

The zeroth law states that if two systems are in equilibrium with a third, they are in equilibrium with each other. It essentially states that the equilibrium relationship is an equivalence relation.

The first law is the law of conservation of energy, where dU is the increase in internal energy of the system, dq is the amount of heat energy added to the system and $\mathrm{d}w$ is the amount of volumetric work done on the system (e.g., pressurization):

$$\mathrm{d}U = \mathrm{d}w + \mathrm{d}q + \sum_{i=1}^{j} \mu_i \mathrm{d}N_i \tag{5.2}$$

with the chemical potential μ and the number of particles N of type i in the system. The last term can be removed, if the closed system has just one single component ($j = 1$).

The second law summarizes the tendency of intensive thermodynamic properties, such as pressure, temperature, etc., to equalize as time goes by, or $\mathrm{d}S \geq 0$, where S is the entropy of the system. The heat term is generally related to the entropy by:

$$\mathrm{d}q = T\mathrm{d}S \tag{5.3}$$

Combining the first and the second law of thermodynamics (Eqs 5.2 and 5.3), where the term $-p\mathrm{d}V$ represents the change of volumetric work $\mathrm{d}w$, one achieves an equation which is convenient for situations involving variations in internal energy, with changes in volume V and entropy:

$$\mathrm{d}U = -p\mathrm{d}V + T\mathrm{d}S \tag{5.4}$$

The third law of thermodynamics states that at the absolute zero temperature, the entropy is at a minimum and all thermodynamic processes cease.

During pressure buildup, all these fundamental relationships have to be considered in their functional relationship with temperature and pressure.

By using the thermodynamic potentials,

Internal energy	$U(S, V)$
Enthalpy	$H(S, p)$
Helmholtz free energy	$A(T, V)$
Gibbs function of free energy	$G(T, p)$

the fundamental equations are expressed as:

$$\mathrm{d}U(S, V) = -p\mathrm{d}V + T\mathrm{d}S \tag{5.5}$$

$$\mathrm{d}H(S, p) = V\mathrm{d}p + T\mathrm{d}S \tag{5.6}$$

$$\mathrm{d}A(T, V) = -S\mathrm{d}T - p\mathrm{d}V \tag{5.7}$$

$$\mathrm{d}G(T, p) = V\mathrm{d}p - S\mathrm{d}T \tag{5.8}$$

The Gibbs function of free energy d$G(T, p)$ (Eq. 5.8) is the fundamental equation for a system where pressure and temperature are the independent variables.

For the case of a single component system, there are three standard material properties from which all others may be derived.

According to the first fundamental theorem of thermodynamics and for a constant pressure, heat capacity *cp* is usually defined as:

$$c_p \equiv \left(\frac{\partial H}{\partial T}\right)_p = T\left(\frac{\partial S}{\partial T}\right)_p \tag{5.9}$$

The isobaric coefficient of thermal expansion can be defined as (22)

$$\alpha_p \equiv \frac{1}{V}\left(\frac{\partial V}{\partial T}\right)_p = \frac{1}{\rho}\left(\frac{\partial \rho}{\partial T}\right)_p \tag{5.10}$$

with the density ρ.

Isothermal compressibility as an intrinsic physical property of the material is defined by Equation 5.11 (22) and exhibits a high variability in gases, liquids, and solids:

$$\beta_T \equiv \frac{1}{\nu}\left(\frac{\partial \nu}{\partial p}\right)_T = \frac{1}{\rho}\left(\frac{\partial \rho}{\partial p}\right)_T \tag{5.11}$$

with the specific volume v.

These properties are seen to be the three possible second derivative of the Gibbs free energy with respect to temperature and pressure.

On the basis of the first law of thermodynamics (Eq. 5.2) and incorporation of the isothermal compressibility (Eq. 5.11), the volumetric work dw from A to B can be expressed as:

$$\int_A^B \mathrm{d}w = \int_A^B pv\beta_T \mathrm{d}p \tag{5.12}$$

By using the second law of thermodynamics (Eq. 5.3), which derives a relation for enthalpy, entropy, and temperature,

$$\left(\frac{\partial H}{\partial T}\right)_p = T\left(\frac{\partial S}{\partial T}\right)_p \tag{5.13}$$

the adiabatic-isentropic heating of a system can be obtained by combining Equation 5.13 with the basic equations for the compressibility of a system (Eq. 5.11) and specific heat capacity at constant pressure (Eq. 5.9):

$$\left(\frac{\partial T}{\partial p}\right)_s = \frac{\beta_T T}{c_p \rho} \tag{5.14}$$

A temperature rise is accompanied by a dissipation of heat within and through the pressure vessel, which is dependent on the vessel size, rate of compression, heat transfer parameters as well as initial and boundary conditions.

The phase transition between two states of matter can be characterized by the Clausius–Clapeyron relation (Eq. 5.15), which can give a relation of the temperature dependence of the melting pressure.

$$\frac{dp}{dT} = \frac{\Delta H}{T\Delta V} \quad (5.15)$$

Chemical reactions under pressure and temperature are dependent on both parameters. An adequate equation on the temperature dependence of the rate constant k was first published by Arrhenius (23) in Equation 5.16:

$$\left(\frac{\partial \ln k}{\partial T}\right)_p = -\frac{E_a}{R_m T^2} \quad (5.16)$$

where E_a represents the activation energy ($kJ\,mol^{-1}$) and R_m the molar gas constant $8.3145\,cm^3\,MPa\,K^{-1}\,mol^{-1}$.

Eyring (24, 25) derived a similar expression (Eq. 5.17) for the pressure dependence of k:

$$\left(\frac{\partial \ln k}{\partial p}\right)_T = -\frac{\Delta V^{\#}}{R_m T} \quad (5.17)$$

with the activation volume of the reaction $\Delta V^{\#}$ [$cm^3\,mol^{-1}$].

Limitations of Equation 5.17 are the pressure dependence of $\Delta V^{\#}$ and the order of reaction, which might vary at different pressure levels.

5.2.3 Adiabatic Heating

Toepfl *et al.* (19) summarized the relevant knowledge of adiabatic heating during HPP. Considering that during compression all compressible materials change their temperature, an adiabatic heating will occur (Eq. 5.14). This temperature rise resulted due to inner friction that occurs when fluids are compressed adiabatically to extreme pressure. When this behavior is known, calculation of the thermal profile during the compression phase is possible. Different pressure transmitting media result in a various adiabatic heating (26), which could be obtained for some fluid food systems with the help of equations for estimating physical properties of mixture of pure substances. An example is the calculation of different water and sucrose solutions as a model system for orange juice (27). Some adiabatic heat of compression profiles are shown in Fig. 5.5, where *n*-hexane and bis-(2-ethylhexyl) sebacate (trivial name sebacate) are pressure-transmitting media. The lack of thermodynamic data on real foods under pressure has been limited to the study and calculation of compression temperature increase (19).

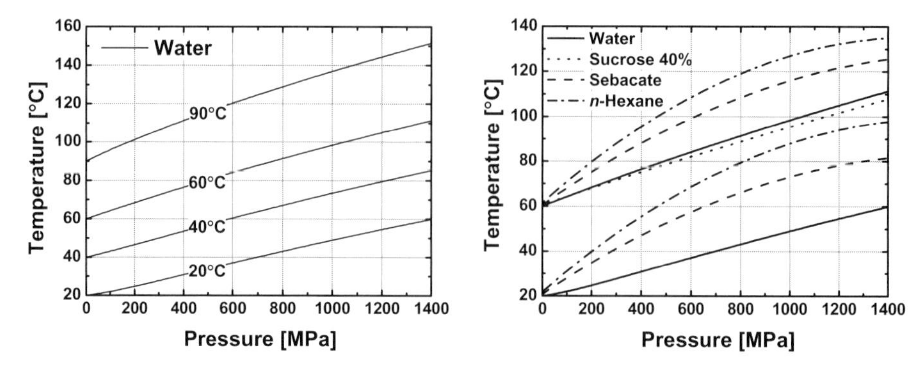

Figure 5.5 Adiabatic heat of compression in water, sucrose solution with 40% solid content, bis-(2-ethylhexyl) sebacate (trivial name sebacate) and *n*-hexane (26).

TABLE 5.1 Adiabatic heat of compression in different food systems.[a]

Substances at 25 °C	Temperature increase per 100 MPa [°C]
Water	~3.0
Mashed potato	~3.0
Tomato salsa	~3.0
2%-fat milk	~3.0
Salmon	~3.2
Chicken fat	~4.5
Beef fat	~6.3
Olive oil	From 8.7 to <6.3[b]
Soy oil	From 9.1 to <6.2[b]

[a]According to reference 28.
[b]Substances exhibited decreasing T as pressure increased.

Mainly practical measurements can demonstrate the differences of the adiabatic heating in real food systems (Table 5.1).

The main ingredient in most food is water and thus the thermodynamic properties of water can be utilized to estimate the temperature increase upon compression of high-moisture foods. The compression heating in fat-containing foods could be up to three times higher than for water (28). In a situation in which organic solvents or oils are used as pressure-transmitting medium and the food matrix has high water content, a difference in compression temperature increase between the food and the medium would occur. The transfer of heat from the pressure-transmitting medium into the product could be utilized to increase the temperature of the food system during and after the adiabatic heating (19).

5.3 CHEMICAL AND MATRIX EFFECTS

The DFG-Senate Commission for Food Safety (SKLM) considered the high-pressure treatment of foodstuffs already in 1998, and published a further opinion in 2004 with an extended product range (29).

Commonly, chemical reactions with a negative reaction and activation volume are dominant under high pressure, because of the principle of Le Chatelier and Braun, which states that when a chemical system at equilibrium experiences a change, the system will shift in order to minimize that change. Examples are dissociation reactions in water or buffer systems as well as formations of covalent bonds by cycloadditions. These cycloadditions of appropriate reaction partners were not observed under realistic production conditions (30, 31). Radical formation is inhibited by pressure and the partial volumes and the reactivity are influenced by interactions between the dissolved species and the solvent (32–34).

5.3.1 Vitamins

Published literature on the effects of HPP on nutrient composition indicates that the effects are minimal, and compare favorably with classic food processing treatments such as pasteurization or sterilization. Most of the detected changes were observed in model or buffer systems. The food matrix has a protective effect and has to be considered in this ease. Water-soluble vitamins, e.g. vitamin C, the vitamins B1, B2, B6, and folic acid, were not or were less affected by high-pressure processing. No significant effects by high pressure on fat-soluble vitamins, like vitamin A, vitamin E, vitamin K, and provitamin A (35, 36), as well as chlorophyll at low temperatures, were observed (37–39).

5.3.2 Lipids

The review of available published literature suggests that very high-pressure treatment could possibly result in lipid oxidation. Increases in peroxidized lipids or bioactive peptides could pose health risks, and may cause quality deterioration of food products. In one study, increased lipid oxidation in pressurized meat was reported at 800 MPa heated at 19 °C for 20 min. Pressures of around 600 MPa per 3-min treatment are not expected to impact the nutritional safety or quality of ready-to-eat meats. Thus, contradictory statements exist regarding the oxidation of fats in foodstuffs through high-pressure treatment. The main problem is that the reported changes are often not clearly distinguished from changes occurring during storage (40, 41). A lot of parameters such as pH value, water content, fatty acid spectrum, degree of oxidation before pressure treatment, pro- and antioxidants as well as residual enzymatic activities have an impact on the pressure-induced changes and the progress of oxidation during storage. However, proposals for treatment of foods at higher pressure should be considered on a case-by-case basis.

5.3.3 Carbohydrates

Most carbohydrates are not affected by high pressure (29). As mentioned, reactions with negative reaction and activation volumes are accelerated under pressure. As an example methylglycosides may be hydrolyzed under pressure into aglycone and MeOH. Because of the slightly positive activation volume, disaccharides are stable at process conditions. At very high pressure over 1000 MPa solvolysis reactions of the glycosidic bonds can occur (32). Concerning the effects on water-binding and gel-forming properties, polysaccharides can be affected by high isostatic pressure (42). These examples relate to the functional properties and do not involve structural changes.

5.3.4 Proteins

High pressure-induced reversible or irreversible changes of the protein native structure (43, 44) are analogous to the changes occurring with heat and chemicals, but the residual molecular structure can vary significantly. Knorr *et al.* (45) have published a review regarding the changes of the protein structure under pressure and temperature.

For analyzing the pressure and temperature landscape of proteins, a two-state transition is assumed, which is at equilibrium at the phase transition line (Fig. 5.6, e.g., between native and denatured). These phenomena can be modeled by an empirical model with a thermodynamic background by using the three-dimensional free energy landscape in response to pressure and temperature (Fig. 5.6).

In Gibbs function of free energy (Eq. 5.8), pressure and temperature are the independent variables. Integration of Equation 5.8 using a Taylor series expansion up to second-order terms (46) yields Equation 5.18:

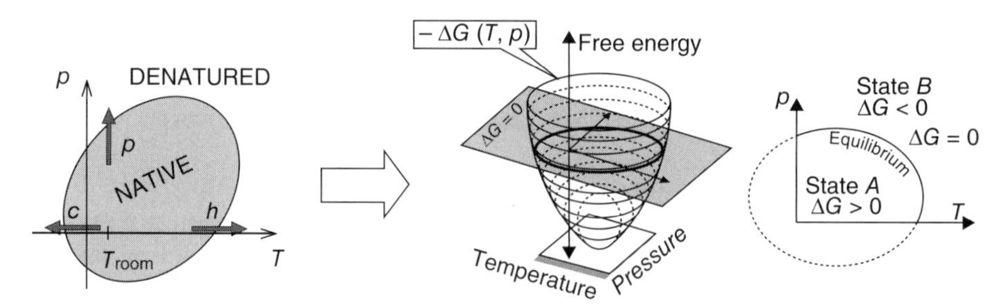

Figure 5.6 Relation between cold (*c*), pressure (*p*) and heat (*h*) denaturation of proteins and three-dimensional free energy landscape in response to pressure and temperature (18). Reprinted with kind permission of Springer Science and Business Media.

$$\Delta G = \Delta G_0 + \Delta V_0(p - p_0) - \Delta S_0(T - T_0) + (\Delta\beta/2)(p - p_0)^2 - (\Delta c_p/2T_0)(T - T_0)^2 + \Delta\alpha(p - p_0)(T - T_0) \tag{5.18}$$

where Δ denotes the change of the corresponding parameter during unfolding.

This quadratic two-variable approximation of the difference in Gibbs free energy yields to an ellipsoidal phase transition line in the pressure and temperature landscape for equilibrium condition ($\Delta G = 0$ in Fig. 5.6) (47). This model can be treated as a general approach, where state *A* and state *B* could denote molecular or physicochemical states. In the case of spore inactivation, state *A* could represent the recoverable and *B* the not recoverable spores, respectively. The transition line would run along the equilibrium $\Delta G = 0$ as elliptical shape at various pressures and temperatures (18).

The mode of action of the high hydrostatic pressure in the treated foods is not fully understood but it is theorized that this treatment inactivates the microbial flora by inactivating overall enzyme activity in the living cells, thus interrupting all cellular functions during the high-pressure phase. In contrast to heat, HPP does not denature covalent bonds, which in turn leaves primary protein structure largely unaffected. According to the document of the DFG-Senate Commission for Food Safety (SKLM) (29) high pressure influences the quaternary structure of the protein through hydrophobic interactions, the tertiary structure through reversible unfolding, and the secondary structure through irreversible unfolding. There are different rheological properties of pressure-induced gels in comparison to heat-induced gels. The higher protease sensitivity of pressure-modified proteins is probably related to a higher water-binding capacity.

Another interesting point is that the resistance to proteolysis of hamster and cattle prion proteins could be reduced by high-pressure processing at higher temperatures over 80 °C (48, 49). The activity and substrate specificity of enzymes can be also affected by high pressure. There might be a risk of the formation of undesirable substances by reactivation of enzymes during storage. Also an increase of the enzyme activity under pressure could be observed. The substrate specificity of enzymes in the field of foodstuffs is not well understood. As an example peroxidases are inactivated at low pressures in the presence of some substrates but not others (50, 51). Currently, no formations of toxic compounds caused by changed substrate specificity under pressure were observed (29).

Butz *et al.* (52) showed that the antioxidative and antimutagenic potential of fruit and vegetable juices remains intact after high-pressure treatment.

The main issue in terms of safety aspects is to prove the potential formations of biological active peptides. Peptides with pyroglutamate (2-oxyprolin) at the *N*-terminal are more resistant to breakdown by peptidases and such substances are occasionally biologically active (29). It was observed that the conversion of glutamine into pyroglutamate (2-oxyprolin) is favored under elevated pressure and temperature conditions (53–55).

5.3.5 Water

Water is essential to all known forms of life, represents the major component of most food systems itself, and is typically used as the pressure-transmitting liquid. Extensive data and formulations of the main thermodynamic properties of water are available from the International Association for the Properties of Water and Steam (IAPWS) and in the database from the National Institute of Standard and Technology (NIST). Most of the time, data and formulations are valid up to 1000 MPa. In this work, data were extrapolated up to 1400 MPa.

After compression of 1 kg water up to 1400 MPa, a maximum volumetric work of 112 kJ kg^{-1} is performed on the system according to Equation 5.12. In Fig. 5.7, the specific volumetric work w in pure water as function of different pressure levels at 20 °C is presented.

Bridgman (22, 56) was the first who determined the phase diagram of water as a function of temperature and pressure. The phase transition lines of water and its different ice modifications according to Bridgman are shown in Fig. 5.8a. At this time, 12 different crystal structures plus two amorphous states are known. At the transition from the liquid to the solid state, Ice I represents a specialty since only this ice modification shows a positive volume change ΔV.

In Fig. 5.8b, the density ρ (NIST) of pure water with adiabatic lines due to compression (Eq. 5.14) in the p-T landscape is shown. At high pressures, the compressibility becomes less and there is a volume contraction of 10%, 17%, or 23% at 400, 800, or 1400 MPa, respectively. Lower starting temperatures

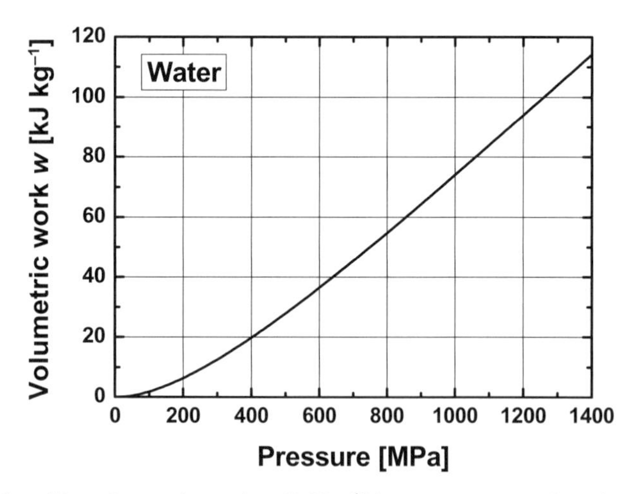

Figure 5.7 Specific volumetric work w [kJ kg^{-1}] in pure water as function of different pressure level at 20 °C according to Equation 5.12. The isothermal lines of w would be interfered with each other. At high pressure level, the function has linear behavior because compressibility becomes less.

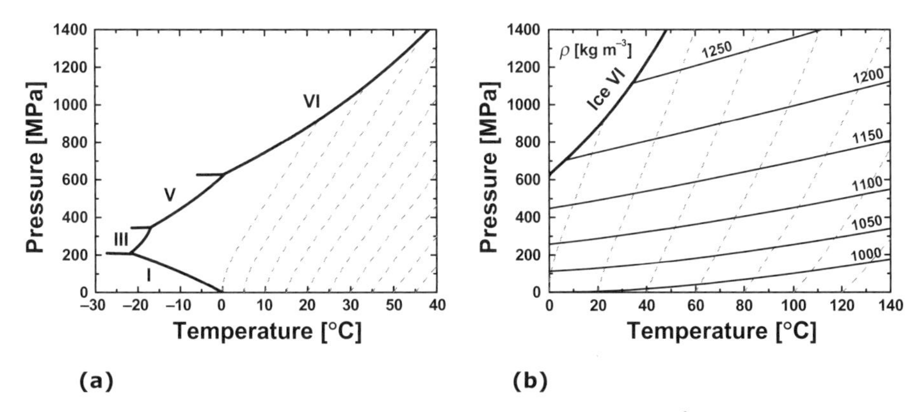

Figure 5.8 (a) Phase diagram (23, 57) and (b) density ρ [kg m^{-3}] with adiabatic lines due to compression (--) of water in the *p*-*T* landscape (NIST). The volume contraction of water at 400, 800, or 1400 MPa with 50 °C is 10%, 17%, or 23%, respectively.

have significantly lower (approx. 2 K per 100 MPa) temperature increases due to compression than higher starting temperatures (up to 5 K per 100 MPa over 80 °C).

5.3.6 Dissociation Reactions

A combination of pH value, temperature, and pressure can act synergistically (57), leading to increased microbial inactivation. During the first phase of basic inactivation studies, buffer systems to obtain constant pH values and medium properties are in common use. However, the dissociation equilibrium in water and buffer solutions varies with pressure and temperature (58–64). Mathys (65) discussed this problem in detail.

This change of the pK_a value plays a major role in different sensitive reaction, but its behavior has rarely been investigated. Distèche (58) developed one of the first pH electrodes up to 150 MPa. At present, the dissociation equilibrium shift cannot be measured directly in solid food, but few optical measurement methods during pressure treatment up to 250 MPa (63, 66) and 450 MPa (67, 68) have been developed for *in situ* pH measurement during pressure treatment of liquids. However, these experimental methods are limited and not suitable for the investigation of combined thermal and pressure dependencies. Due to the shift of dissociation equilibria during heating and/or compression, the results of inactivation experiments are prone to error if not designed correctly.

During shift of dissociation equilibria, the pH value would be changed with all reaction partners, but no change of the concentration difference on one site occurs during the reaction. For example in water at 0.1 MPa and 20 °C with nearly equal hydroxonium and hydroxide ($[OH^-]$) concentrations, there would be neutral conditions ($[H^+] = [OH^-]$). Under pressure, there are increased hydroxonium and hydroxide concentrations, but still "neutral" conditions. Consequently, the pH shift alone cannot exactly describe the dissociation equilibrium shift and thus the pK_a shift is used.

By using the regression equation from Marshall and Frank (60), the negative logarithm of the ion product K_w ($[mol\,kg^{-1}]^2$) of water substance (Eq. 5.19) can be calculated with a deviation of ±0.01 at <200 °C, ±0.02 at ≤374 °C, ±0.3 at >374 °C, and ±0.05–0.3 at high pressures.

$$-\log_{10}(K_w) = -\log_{10}([H_3O^+]-[OH^-]) \tag{5.19}$$

The modeled dissociation equilibrium shift in pure water showed large variations with pressure and temperature (Fig. 5.9). Especially the high-temperature area is a matter of particular interest for the application of supercritical water in food technology (Fig. 5.9a). After extrapolation up to 1400 MPa (Fig. 5.9b), an estimation about the K_w characteristics at the current technical limit for spore inactivation studies could be given.

For any equilibrium reaction, the pressure and temperature dependence of the equilibrium constant is described by Planck's equation (69) (Eq. 5.20):

$$\left(\frac{d\ln(K)}{dp}\right)_T = \frac{\Delta V}{R \cdot T} \tag{5.20}$$

where p is the pressure (MPa), T the absolute temperature (K), ΔV the reaction volume at atmospheric pressure ($cm^3\,mol^{-1}$), and R the gas constant $8.3145\,cm^3\,MPa\,K^{-1}\,mol^{-1}$.

The reaction volume ΔV is equal to the difference of the partial volumes of products and reactants. It includes the volume change because of alteration

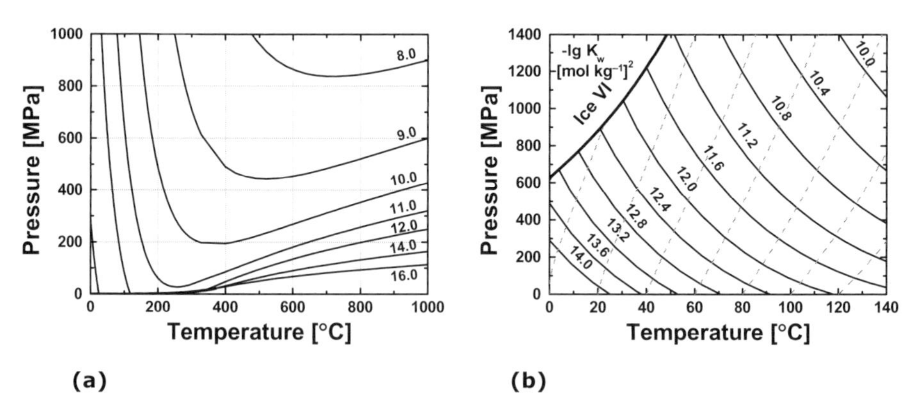

Figure 5.9 Dissociation equilibrium shift (negative logarithm of the ion product K_w $[mol\,kg^{-1}]^2$) in pure water under different p-T conditions up to 1000 MPa and 1000 °C (a) and with adiabatic lines (--) up to 140 °C extrapolated up to 1400 MPa (b) (60, 65).

in binding length and angle $\Delta V_{\text{intrinsic}}$ as well as solvational properties $\Delta V_{\text{solvational}}$ (Eq. 5.21):

$$\Delta V = \Delta V_{\text{intrinsic}} + \Delta V_{\text{solvational}} \tag{5.21}$$

For industrial applications, such property changes of food matrices are important. Figure 5.10 shows *in situ* pH shifts in food systems (milk, pea soup, baby mashed carrots with maize, herring with tomato sauce; for the detailed ingredients, see Annex) and *N*-(2-Acetamido)-2-aminoethanesulfonic acid ACES buffer in dependence of temperature. To remove all external effects, all data were divided by the pH-values of the temperature stable phosphate buffer. Hence, in Fig. 5.10, pH differences between food and phosphate buffer are shown. ACES buffer had the highest pH shift and all other food systems had lower temperature dependencies.

To investigate the temperature and pressure dependence of the dissociation equilibrium shift, different buffer systems were modeled. The created *p-T* diagrams of the calculated pK_a value have shown different changes (Fig. 5.11).

Different initial temperatures and the adiabatic heating (dashed line) due to the pressure buildup process required the implementation of the specific temperature dependences, respectively. These temperature dependences can be very high and could compensate the advantages of the low isothermal pressure dependence.

The diversity of iso-pK_a lines in buffer solutions resulted mainly from the different dissociation reactions and consequently different standard molar enthalpies ($\Delta_r H^0$) or reaction volumes (ΔV).

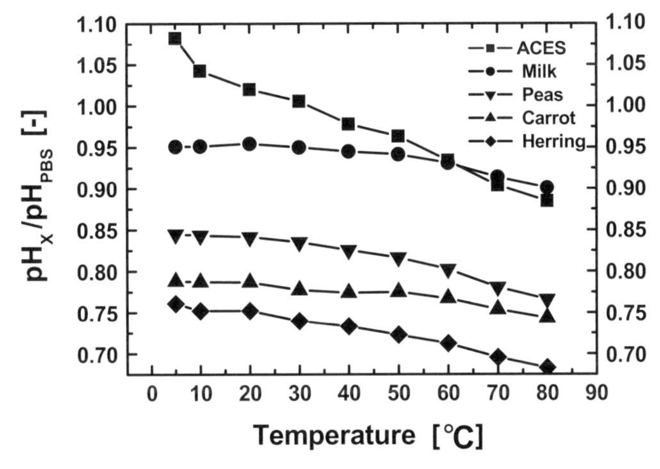

Figure 5.10 Real pH shift in food systems (milk, pea soup, baby mashed carrots with maize, herring with tomato sauce) and 0.01 M ACES buffer in dependence of temperature (65).

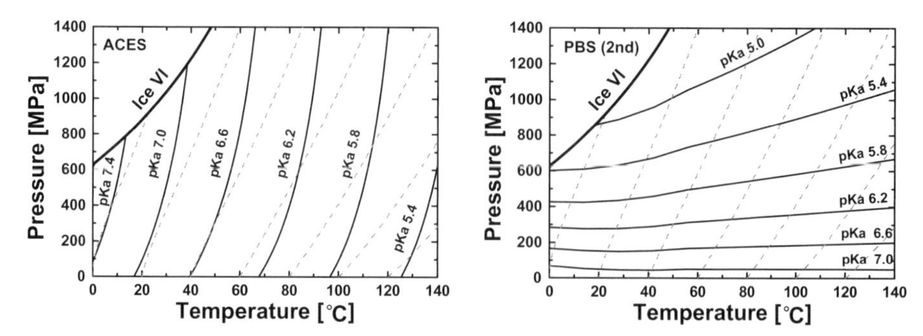

Figure 5.11 Modeling of the iso-pK$_a$ lines in different buffer systems under different *p-T* conditions with adiabatic lines (--) and phase transition line from pure water (65).

However, to verify and to evaluate the modeled data, *in situ* measurements for the dissociation equilibrium shift have to be developed. Up to now, such *in situ* method is not available and no appropriate online measurement device exists.

Applying the these models, sensitive reactions such as cell inactivation or enzyme reactions can be better anticipated in planning experimental designs.

For skim milk, Schraml (70) measured the pH shift up to 88 °C (compare Fig. 5.10). The dissociation of water and elimination of phosphate ions by crystallization of calcium phosphate decreased the pH value in this case (71). The crystallization led to a dissociation equilibrium shift of phosphoric acid, whereby rearrangement of the equilibrium, phosphoric acid dissociated and more hydroxonium ions (H_3O^+) were generated. Small changes of the pH value could affect different reactions or inactivations in food matrices. Dannenberg (72) found, that the so-called "fouling" of skim milk on hot surfaces is drastically increased, when the pH value was decreased from 6.6 to 6.5–6.4. A possible explanation could be increased protein aggregation, because of decreased electrostatic repulsion, when proteins approach the isoelectric point (71). The isoelectric point is the pH value at which a particular molecule or surface carries no net electrical charge. The complex matrix relations in skim milk during thermal processing provide ideas about how many relevant mechanisms might occur under high-pressure, high-temperature processing, but limited data exist.

5.4 MICROBIAL EFFECTS

5.4.1 Vegetative Cells

The German Research Foundation (DFG) Senate Commission for Food Safety (SKLM) (29) suggests that vegetative cells of bacteria relevant for foodstuffs are destroyed by hydrostatic pressures ranging from 150 to 800 MPa. They derive evidence for this postulation from numerous investigations including pathogenic microorganisms. The kinetics of inactivation show a steady decrease, but at higher inactions a leveling-off at longer treatment times ("tailing") was observed. There are different theories in the literature to explain this tailing phenomenon. In any case the pressure-induced inaction of vegetative cells strongly depends on the food matrix (73–78).

The main lethal effect of pressure treatment on cells is the pressure-induced breakdown of membrane permeability. Hence, the difference between the internal and external pH value could collapse or approach zero, resulting in loss of cell viability (79). The internal pH is critical for the control of many cellular processes, such as ATP synthesis, RNA and protein synthesis, DNA replication, and cell growth, and it plays an important role in secondary transport of several compounds (80). It has a particularly inhibitory effect on membrane ATPase, a very important enzyme in the acid–base physiology of cells (81).

5.4.2 Bacterial Spores

Bacterial endospores, as compared with vegetative cells, display a considerably higher resistance to high pressure, as described by the DFG-SKLM (29). Highly resistant spores of *Clostridium* and *Bacillus* species are key bacteria for the safety or the spoilage of low-acid (heat-treated) preserved food. The neurotoxin A from *Clostridium botulinum* has a lethal dose LD_{50} for humans of $1\,ng\,kg^{-1}$. Strains of this species can tolerate extremely high pressures over 1 GPa at room temperature. Using high-pressure thermal sterilization as combined technique it is possible to inactivate such food-relevant bacterial endospores. The required inactivation temperature and/or time can be lowered by combination with high pressure (65, 74, 82–85). Under certain combinations of pressure and temperature these highly resistant spores can be protected or stabilized (26, 65, 86). The main important issue is the pressure-induced spore germination (2, 3, 18, 26, 65).

Spore germination and subsequent growth can cause food spoilage and potential toxin formation, which may ultimately lead to foodborne disease. The neurotoxin A from *C. botulinum* has a lethal dose LD_{50} for humans of $1\,ng\,kg^{-1}$ (87). Infant botulism is the most common type of botulism and may occur in children in the first year of life, when *C. botulinum* spores populate the intestines, germinate, and produce toxins. The toxin can lead to the feared paralysis of respiratory muscles, which ultimately leads to death (88). Hence, probably stabilized dormant *C. botulinum* spores are also a risk factor. Another worse scenario is pulmonary anthrax, because of the germination of *Bacillus anthracis* in lung macrophage (89).

The loss of resistance properties is the crucial step during spore germination, which is of highest importance for sterilization techniques and research. The germination of spores has been studied and reviewed extensively recently

(18, 90–92). The main interesting germination process for this discussion is pressure-induced germination. Pressures between 100 and 200 MPa activate the germinant receptors, which lead to pathway similar to nutrient germination (91, 93). The whole reaction and/or kinetic depends strongly on temperature (94). Heinz and Knorr (94) assumed that pressure and temperature can trigger the de-immobilization of cortex lytic enzyme (CLE) activities and cause inactivation of the same enzymes by structural unfolding, which probably completely inhibits the CLE activity. Very high pressures of 500–600 MPa open the Ca^{2+}-DPA (dipicolinic acid) channels (95), which may result in an incomplete germination process (85, 95).

For the food industry, the hydrophobicity and agglomeration behavior is of high importance as possible reasons for adhesion to surfaces (96), microbial "fouling" and non-$\log_{10}$-linear shoulder as well as tailing formation (97).

The importance of spore agglomerations becomes evident by the following. Agglomerates always produce one colony per each plate. Consequently, agglomerates of unknown cell numbers are always counted as one spore until all spores in the agglomerate are inactivated. Beyond this agglomeration and disintegration can change the colony-forming units per milliliter.

Thus, it appears that agglomerations in spore suspensions need to be considered by modeling of the thermal (98) and probably pressure inactivation.

5.4.3 Viruses

According to the DFG-SKLM (29), the multiplicity of virus types and their structures are too large to formulate a general statement at the present time. An increased risk compared with untreated foodstuffs is presently not recognizable.

5.5 ALLERGENIC POTENTIAL

The assessment of the influence of high-pressure treatment of the allergenicity of foods should be performed in comparison with traditional food technological processes, in particular heat treatment, as suggested by the DFG-SKLM (29). Furthermore it was summarized that allergenicity can be altered after technological processing, for example, by formation of new allergens or epitopes.

Most technological processes, in particular thermal processing, showed a partial inactivation of the allergenic potential (99, 100). The same tendency is shown with high-pressure generated data. There is very little evidence for an increase in allergenicity from food processing (101–105), but according to the DFG-SKLM (29) an increase of the allergenic potential through high-pressure treatment of foods thus seems to be unlikely. Unfortunately only a few studies were performed. Jankiewicz *et al.* found reduced IgE reactivity of an extract from celery tuber after high-pressure processing at 600 MPa (100). The results showed that the allergenic potential of the pressure-treated vegetable was graded between that of raw and of cooked celery. Kato *et al.* (106) reported that after high-pressure treatment at 500 MPa the major allergens of rice are released from the grains in a liquid medium. Grimm *et al.* (107) used circular dichroism (CD) spectroscopy to show that the recombinant major allergen from apple revealed changes in the secondary structure. The authors detected a decrease of the α-helical regions and an increase of the β-sheet structures. It was also mentioned that subsequent to high-pressure treatment, apples were tolerated in challenge tests without symptoms by five individuals allergic to apples.

In any case more research needs to be done to evaluate these first experiments. High-pressure experiments have to be performed in the real food matrix and special attention should be paid to the "big eight" allergies to milk, egg, peanut, tree nut, seafood, shellfish, soy, and wheat.

5.6 CONCLUSION

In accordance with documents put forward by the DFG-SKLM (29), high-pressure treatment can cause **chemical changes** in foodstuffs. This involves preferentially those reactions and conformational changes that are associated with a reduction in volume. So far, vitamins, colors, and flavors examined seem to be largely unaffected if compared with conventional thermal processes. However, attention is demanded concerning some reactions of food ingredients that could lead to chemical changes. For example, in the following reaction types (29): I = dissociation of organic acids and amines, the reversibility and reactivity of the dissociated species; II = cyclization reactions, for example, reactions of quinones with dienes (Diels–Alder) as well as 2 + 2 cycloadditions; III = formation of ammonium, sulfonium, and phosphonium salts, reversibility, and reactivity of ions formed under pressure; and IV = hydrolysis reactions of ethers, esters, acetals, and ketals.

The possible formation of bioactive peptides in protein-rich food requires explicit investigation. Detailed investigation on the effect of high pressure on the conformation of proteins in appropriate systems is required. In particular, protein fibrils from β-sheets associated through wrongly folded protein aggregates should be investigated in more detail, because they appear in diseases such as the TSE (transmissible spongiform encephalopathy) diseases.

The contrary results on the high-pressure-induced oxidation of fats require more clarification.

It has been proposed that the **microbiological** safety of high-pressure-treated food has to be proven by a case-by-case evaluation using realistic concentrations of the relevant bacterial species. It is essential to characterize the hygiene-relevant target organisms (29). An in-depth understanding of the mechanisms of the pressure-induced inactivation of bacteria is required for the intentional utillzation of synergistic or antagonistic effects of high-pressure treatment with the food matrix.

The mechanistic background of the inactivation of **bacterial endospores** is still a topic of debate. Attention should be focused on different germination reactions induced by different pressure levels. The observed stabilization needs more clarification. Data on the inactivation of produced bacterial toxins have to be determined. For investigations of the detailed microbial inactivation mechanism(s) and dependence on pressure and temperature, a homogeneous microbial suspension in a very simple matrix with the lowest possible pressure- and temperature-dependent changes is required.

According to the safety assessment carried out by the DFG-SKLM (29), investigations of high-pressure-treated food carried out so far show no evidence of an increased adverse **toxicological** potential compared with unprocessed or thermally preserved food.

From the **allergenic** point of view, relatively minor changes are induced in food by high pressure, as compared with those observed during thermal processing. There is no significant evidence that the allergenic potential of foodstuffs might be increased by high-pressure treatment (29).

Case-by-case examination of the suitability of the **packaging** material must be carried out to ensure that the quality of the foodstuff is maintained during the storage period. As concerns the effects of pressure on components of packaging materials, it should be examined whether the physicochemical properties of polymers change so much under pressure as to result in an accelerated diffusion of plasticizers, such as phthalates (29).

In conclusion, evidence of any microbial, toxicological, or allergenic risks as a consequence of the high-pressure treatment has not been observed in investigations on high-pressure-treated foodstuffs. However, these results are based on only a few studies and have to be proven by case-by-case examination (29).

In addition, protocols have been developed (108) on how to perform and report high-pressure experiments and tests. Work on modeling pressure and temperature inhomogeneities during pressure treatments and subsequent scale-up issues of high-pressure equipment is being carried out (109, 110) and attempts for identifying time/temperature/pressure integrators for process control are in progress (111–113).

APPENDIX

pH shift in real food systems (T)- Ingredients lists

Milk: Milsani UHT (Sachsenmilch AG, Germany), homogenized, 3.5% fat, 3.7% protein, 4.8% carbohydrates

Pea soup: Maggi pea soup with bacon (Maggie GmbH, Germany), 4.5% fat, 4.8% protein, 8.0% carbohydrates

Baby mashed carrots with maize: (Hipp GmbH & Co Vertriebs KG, Germany), 2.1% fat, 1.3% protein, 6.7% carbohydrates

Herring with tomato sauce: (Produced for Kaiser Tengelmann AG, Germany), 60% herring filet

REFERENCES

1. Hite, B.H. (1899). The effect of pressure in the preservation of milk—a preliminary report. *West Virginia Agricultural and Forestry Experiment Station Bulletin*, *58*, 15–35.
2. Clouston, J.G., Wills, P.A. (1969). Initiation of germination and inactivation of *Bacillus pumilus* spores by hydrostatic pressure. *Journal of Bacteriology*, *97*, 684–690.
3. Gould, G.W., Sale, A.J.H. (1970). Initiation of germination of bacterial spores by hydrostatic pressure. *Journal of General Microbiology*, *60*, 335.
4. Hendrickx, M., Knorr, D. (2002). *Ultra High Pressure Treatment of Foods*, Food Engineering Series (ed. G.V. Barbosa Canovas), Kluwer Academic/Plenum Publisher, New York.
5. Rastogi, N.K., Raghavarao, K.S.M., Balasubramaniam, V.M., Niranjan, K., Knorr, D. (2007). Opportunities and challenges in high pressure processing of foods. *Critical Reviews in Food Science and Nutrition*, *47*, 1–44.
6. Crossland, B. (1995). The development of high pressure equipment, in *High Pressure Processing of Foods* (eds D.A. Ledward et al.), Nottingham University Press, Nottingham, pp. 7–26.
7. Perkins, J. (1820). On the compressability of water. *Philosophical Transactions of the Royal Society*, *110*, 324–329.
8. Perkins, J. (1826). On the progressive compression of water by a high degree of force, with trials on the effect of other fluids. *Philosophical Transactions of the Royal Society*, *116*, 541–547.
9. Regnard, P. (1884). Effet des hautes pressions sur les animaux marins. *C.R. Séances Soc. Biol.*, *36*, 394–395.
10. Certes, A. (1884). Sur la culture, à l'abri des germes atmosphériques, des eaux et des sediments rapportés par les expeditions du "Travailleur" du "Talisman"; 1882–1883. *Comptes Rendus de l'Academie des Sciences*, *98*, 690–693.
11. Chlopin, G.W., Tammann, G. (1903). Über den Einfluß hoher Drücke auf Mikroorganismen. *Zeitschrift für Hygiene und Infektionskrankheiten*, *45*, 171–204.
12. Hite, B.H., Giddings, N.J., Weakley, C.E. (1914). The effect of pressure on certain micro-organisms encountered in the preservation of fruits and vegetables. *West Virginia Agricultural and Forestry Experiment Station Bulletin*, *146*, 2–67.
13. Larson, W.P., Hartzell, T.B., Diehl, H.S. (1918). The effect of high pressures on bacteria. *Journal of Infectious Diseases*, *22*, 271–279.
14. Palou, E., López-Malo, A., Barbosa-Cánovas, G.V., Swanson, B.G. (1999). High-pressure treatment in food preservation, in *Handbook of Food Preservation* (ed. M.S. Rahman), Marcel Dekker, Inc., New York, pp. 533–576.

15. Cheftel, J.C. (1995). Review: high pressure, microbial inactivation and food preservation. *Food Science and Technology International*, *1*, 75–90.
16. Tauscher, B. (1995). Pasteurization of food by hydrostatic high pressure: chemical aspects. *Zeitschrift für Lebensmittel-Untersuchung und-Forschung*, *200*, 3–13.
17. Rovere, P. (2002), Industrial-scale high pressure processing of foods, in *Ultra High Pressure Treatments of Foods* (eds M.E.G. Hendrickx and D. Knorr), Kluwer Academic/Plenum Publishers, New York, pp. 251–268.
18. Heinz, V., Knorr, D. (2002). Effects of high pressure on spores, in *Ultra High Pressure Treatments of Foods* (eds M.E.G. Hendrickx and D. Knorr), Kluwer Academic/ Plenum Publishers, New York, pp. 77–114.
19. Toepfl, S., Mathys, A., Heinz, V., Knorr, D. (2006). Review: potential of emerging technologies for energy efficient and environmentally friendly food processing. *Food Reviews International*, *22*, 405–423.
20. Rovere, P., Lonnerborg, N.G., Gola, S., Miglioli, L., Scaramuzza, N., Squarcina, N. (1999). Advances in bacterial spores inactivation in thermal treatments under pressure, in *Advances in High Pressure Bioscience and Biotechnology* (ed. H. Ludwig), Springer-Verlag, Berlin, pp. 114–120.
21. Stumbo, C.R. (1948). Bacteriological considerations relating to process evaluation. *Food Technology*, *2*, 115–132.
22. Bridgman, P.W. (1912), Water in the liquid and five solid forms under pressure. *Proceedings of American Academy of Arts and Sciences*, *47*, 439–558.
23. Arrhenius, S.A. (1889). Über die Reaktionsgeschwindigkeit bei der Inversion von Rohrzucker durch Säuren. *Zeitschrift für physikalische Chemie*, *4*, 226–248.
24. Eyring, H. (1935). The activated complex in chemical reactions. *Journal of Chemical Physics*, *3*, 107–115.
25. Eyring, H. (1935). The activated complex and the absolute rate of chemical reactions. *Chemical Reviews*, *17*, 65–77.
26. Ardia, A. (2004). Process considerations on the application of high pressure treatment at elevated temperature levels for food preservation. PhD Thesis, Berlin, Berlin University of Technology, 94.
27. Ardia, A., Knorr, D., Heinz, V. (2004). Adiabatic heat modelling for pressure build-up during high-pressure treatment in liquid-food processing. *Food and Bioproducts Processing*, *82*, 89–95.
28. Ting, E., Balasubramaniam, V.M., Raghubeer, E. (2002). Determining thermal effects in high pressure processing. *Journal of Food Technology*, *56*, 31–35.
29. Eisenbrand, G. (2005), Safety assessment of high pressure treated foods—Opinion of the Senate Commission on Food Safety (SKLM) of the German Research Foundation (DFG) (shortened version). *Molecular Nutrition & Food Research*, *49*(*12*), 1168–1174.
30. Gruppe, C., Marx, H., Kübel, J., Ludwig, H., Tauscher, B. (1998). Cyclization reactions of food components to hydrostatic high pressure, in *High Pressure Research in the Bioscience and Biotechnology* (ed. K. Heremans), Leuven University Press, Leuven, Belgium, pp. 339–342.
31. Kübel, J., Ludwig, H., Tauscher, B. (1998). Diels-Alder reactions of food relevant compounds under high pressure: 2,3-dimethoxy-5-methyl-p-benzoquinone and

myrcene, in *High Pressure Food Science, Bioscience and Chemistry* (ed. N.S. Isaacs), The Royal Society of Chemistry, Cambridge, UK, pp. 271–276.

32. Tauscher, B. (1995). Pasteurization of food by hydrostatic high pressure: chemical aspects. *Zeitschrift für Lebensmitteluntersuchung und-Forschung A*, *200*/Nr.1, 3–13.
33. Butz, P., Tauscher, B. (1998). Food chemistry under high hydrostatic pressure, in *High Pressure Food Science, Bioscience and Chemistry* (ed. N.S. Isaacs), The Royal Society of Chemistry, Cambridge, UK, pp. 133–144.
34. Butz, P., Tauscher, B. (2002). Emerging technologies: chemical aspects. *Food Research International*, *35*, 279–284.
35. Fernandez Garcia, A., Butz, P., Bognar, A., Tauscher, B. (2001). Antioxidative capacity, nutrient content and sensory quality of orange juice and an orange-lemon-carrot juice product after high pressure treatment and storage in different packaging. *European Food Research and Technology*, *213*, 290–296.
36. Sanchez-Moreno, C., Plaza, L., de Ancos, B., Cano, M.P. (2003). Vitamin C, provitamin A carotenoids, and other carotenoids in high-pressurized orange juice during refrigerated storage. *Journal of Agricultural and Food Chemistry*, *51*, 647–653.
37. Tauscher, B. (1998), Effect of high pressure treatment to nutritive substances and natural pigments, in *Fresh Novel Foods by High Pressure*, VTT Symposium 186 (ed. K. Autio), Technical Research Center of Finland, pp. 83–95.
38. May, T., Tauscher, B. (1998). Influence of pressure and temperature on chlorophyll a in alcoholic and aqueous solutions, in *Process Optimization and Minimal Processing of Foods* (eds J.C. Olivera, F.A.R. Olivera), Copernicus Programme Proceedings of the Third Main Meeting, Vol. 4: High Pressure, pp. 57–59.
39. Van Loey, A., Ooms, V., Weemaes, C., Van den Broeck, I., Ludikhuyze, L., Indrawati, Denys, S., Hendrickx, M. (1998). Thermal and pressure-temperature degradation of chlorophyll in broccoli (*Brassica oleracea* L. *italica*) juice: a kinetic study. *Journal of Agricultural and Food Chemistry*, *46*, 5289–5294.
40. Angsupanich, K., Ledward, D.A. (1998). Effects of high pressure on lipid oxidation in fish, in *High Pressure Food Science, Bio- Science and Chemistry* (ed. N.S. Isaacs), The Royal Society of Chemistry, Cambridge, pp. 284–288.
41. Cheah, P.B., Ledward, D.A. (1995). High-pressure effects on lipid oxidation. *Journal of the American Oil Chemists Society*, *72*, 1059–1063.
42. Pfister, M.K.-H., Butz, P., Heinz, V., Dehne, L.I., Knorr, D., Tauscher, B. (2000). Bundesinstitut für gesundheitlichen Verbraucherschutz und Veterinärmedizin, in *Der Einfluss der Hochdruckbehandlung auf chemische Veränderungen in Lebensmitteln*, Eine Literaturstudie, Berlin (BgVV-Hefte), pp. 17–22.
43. Heremans, K. (1982). High pressure effects on proteins and other biomolecules. *Annual Review of Biophysics and Bioengineering*, *11*, 1–21.
44. Cheftel, J.C. (1992). Effects of high hydrostatic pressure on food constituents: an overview, in *High Pressure and Biotechnology* (eds C. Balny *et al.*), John Libbey Eurotext, Montrouge, pp. 195–209.
45. Knorr, D., Heinz, V., Buckow, R. (2006). High pressure application for food biopolymers. *Biochimica et Biophysica Acta*, *1764*, 619–631.

46. Smeller, L. (2002). Pressure-temperature phase diagram of biomolecules. *Biochimica et Biophysica Acta, 1595*, 11–29.
47. Zhang, J., Peng, X., Jonas, A., Jonas, J. (1995). NMR study of the cold, heat, and pressure unfolding of ribonuclease A. *Biochemistry, 34*, 8631–8641.
48. Garcia, A.F., Heindl, P., Voigt, H., Buttner, M., Wienhold, D., Butz, P., Starke, J., Tauscher, B., Pfaff, E. (2004). Reduced proteinase K resistance and infectivity of prions after pressure treatment at 60 degrees C. *Journal of General Virology, 85*, 261–264.
49. Heinz, V., Kortschack, F. (2002). Method for modifying the protein structure of PrPsc in a targeted manner. Germany, patent, (WO 02/49460).
50. Garcia, A.F., Butz, P., Tauscher, B. (2002). Mechanism-based irreversible inactivation of horseradish peroxidase at 500 MPa. *Biotechnology Progress, 18*, 1076–1081.
51. Fernandez Garcia, A., Butz, P., Lindauer, R., Tauscher, B. (2002). Enzyme-substrate specific interactions: in situ assessments under high pressure, in *First International Conference on High Pressure Bioscience and Biotechnology. 2002*, Elsevier Science Ltd, pp. 189–192.
52. Butz, P., FernandezGarcia, A., Lindauer, R., Dieterich, S., Bognar, A., Tauscher, B. (2003). Influence of ultra high pressure processing on fruit and vegetable products. *Journal of Food Engineering, 56*, 233–236.
53. Butz, P., Fernandez, A., Schneider, T., Starke, J., Tauscher, B., Trierweiler, B. (2002). The influence of high pressure on the formation of diketopiperazine and pyroglutamate rings. *High Pressure Research, 22*, 697–700.
54. Schneider, T., Butz, P., Ludwig, H., Tauscher, B. (2003). Pressure-induced formation of pyroglutamic acid from glutamine in neutral and alkaline solutions. *Lebensmittel-Wissenschaft und-Technologie, 36*, 365–367.
55. Fernandez Garcia, A., Butz, P., Trierweiler, B., Zoller, H., Starke, J., Pfaff, E., Tauscher, B. (2003). Pressure/temperature combined treatments of precursors yield hormone-like peptides with pyroglutamate at the N terminus. *Journal of Agricultural and Food Chemistry, 51*, 8093–8097.
56. Bridgman, P.W. (1911). Water in the liquid and five solid forms, under pressure. *Proceedings of American Academy of Arts and Sciences, 47*, 441–558.
57. Zipp, A., Kauzmann, W. (1973). Pressure denaturation of metmyoglobin. *Biochemistry, 12*, 4217–4228.
58. Distèche, A. (1959). pH measurements with a glass electrode withstanding 1500 kg/cm^2 hydrostatic pressure. *Review of Scientific Instruments, 30*, 474–478.
59. North, N.A. (1973). Dependence of equilibrium constants in aqueous solutions. *The Journal of Physical Chemistry, 77*, 931–934.
60. Marshall, W.L., Franck, E.U. (1981). Ion product of water substance, 0–1000 °C, 1–10 000 bars new international formulation and its background. *Journal of Physical and Chemical Reference Data, 10*, 295–304.
61. Hamann, S.D. (1982). The influence of pressure on ionization equilibria in aqueous solutions. *Journal of Solution Chemistry, 11*, 63–68.
62. Kitamura, Y., Itoh, T. (1987). Reaction volume of protonic ionization for buffering agents. Prediction of pressure dependence of pH and pOH. *Journal of Solution Chemistry, 16*, 715–725.

63. Quinlan, R.J., Reinhart, G.D. (2005). Baroresistant buffer mixtures for biochemical analyses. *Analytical Biochemistry*, *341*, 69–76.
64. Bruins, M.E., Matser, A.M., Janssen, A.E.M., Boom, R.M. (2007). Buffer selection for HP treatment of biomaterials and its consequences for enzyme inactivation studies. *High Pressure Research*, *27*, 101–107.
65. Mathys, A. (2008). Inactivation mechanisms of *Geobacillus* and *Bacillus* spores during high pressure thermal sterilization. PhD Thesis. Berlin, Berlin University of Technology, 161.
66. Hayert, M., Perrier-Cornet, J.-M., Gervais, P. (1999). A simple method for measuring the pH of acid solutions under high pressure. *The Journal of Physical Chemistry A*, *103*, 1785–1789.
67. Stippl, V.M., Delgado, A., Becker, T.M. (2002). Optical method for the *in-situ* measurement of the pH-value during high pressure treatment of foods. *High Pressure Research*, *22*, 757–761.
68. Stippl, V.M., Delgado, A., Becker, T.M. (2004). Development of a method for the optical in-situ determination of pH value during high-pressure treatment of fluid food. *Innovative Food Science & Emerging Technologies*, *5*, 285–292.
69. Planck, M. (1887). Über das Prinzip der Vermehrung der Entropie. *Annalen der Physik*, *32(2)*, 426–503.
70. Schraml, J. (1993). Zum Verhalten konzentrierter Produkte bei der Belagbildung an heißen Oberflächen. PhD Thesis, München, TU München-Weihenstephan.
71. Kessler, H.G. (1996). *Lebensmittel-und Bioverfahrenstechnik—Molkereitechnologie*, Verlag A. Kessler, Freising.
72. Dannenberg, F. (1986). Zur Reaktionskinetik der Molkenproteindenaturierung und deren technologischer Bedeutung. PhD Thesis, München, TU München-Weihenstephan.
73. Garcia-Graells, C., Hauben, K., Michiels, C.W. (1998). High-pressure inactivation and sublethal injury of pressure-resistant *Escherichia coli* mutants in fruit juices. *Applied and Environmental Microbiology*, *64*, 1566–1568.
74. San Martin, M.F., Barbosa-Canovas, G.V., Swanson, B.G. (2002). Food processing by high hydrostatic pressure. *Critical Reviews in Food Science and Nutrition*, *42*, 627–645.
75. Smelt, J.P.P.M., Hellemons, J.C., Wouters, P.C., van Gerwen, S.J.C. (2002). Physiological and mathematical aspects in setting criteria for decontamination of foods by physical means. *International Journal of Food Microbiology*, *78*, 57–77.
76. Ulmer, H.M., Ganzle, M.G., Vogel, R.F. (2000). Effects of high pressure on survival and metabolic activity of *Lactobacillus plantarum* TMW1.460. *Applied and Environmental Microbiology*, *66*, 3966–3973.
77. Ulmer, H.M., Herberhold, H., Fahsel, S., Gänzle, G., Winter, R., Vogel, R.F. (2002). Effects of pressure-induced membrane phase transitions on inactivation of HorA, an ATP-dependent multidrug resistance transporter, in *Lactobacillus plantarum*. *Applied and Environment Microbiology*, *68*, 1088–1095.
78. Karatzas, A.K., Bennik, M.H.J. (2002). Characterization of a *Listeria monocytogenes* Scott A isolate with high tolerance towards high hydrostatic pressure. *Applied and Environment Microbiology*, *8*, 3138–3189.

79. Nannen, N.L., Hutkins, R.W. (1991). Proton-translocating adenosine triphosphatase activity in lactic acid bacteria. *Journal of Dairy Science*, *74*, 747–751.
80. Belguendouz, T., Cachon, R., Divie's, C. (1997). pH homeostasis and citric acid utilization: differences between *Leuconostoc mesenteroides* and *Lactococcus lactis*. *Current Microbiology*, *35*, 233–236.
81. Hoover, D.G., Metrick, C., Papineau, A.M., Farkas, D.F., Knorr, D. (1989). Biological effects of high hydrostatic pressure on food microorganisms. *Food Technology*, *43*, 99–107.
82. Heinz, V., Knorr, D. (1996). High pressure inactivation kinetics of *Bacillus subtilis* cells by a three-state-model considering distribution resistance mechanisms. *Food Biotechnology*, *10*, 149–161.
83. Margosch, D., Ehrmann, M.A., Gaenzle, M.G., Vogel, R.F. (2003). Rolle der Dipicolinsäure bei der druckinduzierten Inaktivierung bakterieller Endosporen. 5. Fachsymposium Lebensmittelmikrobiologie der VAAM und DGHM. 2003. Seeon.
84. Reddy, N.R., Solomon, H.M., Fingerhut, G.A., Rhodehamel, E.J., Balasubramaniam, V.M., Palaniappan, S. (1999). Inactivation of *Clostridium botulinum* type E spores by high pressure processing. *Journal of Food Safety*, *19*, 277–288.
85. Wuytack, E.Y., Boven, S., Michiels, C.W. (1998). Comparative study of pressure-induced germination of *Bacillus subtilis* spores at low and high pressures. *Applied and Environmental Microbiology*, *64*, 3220–3224.
86. Margosch, D. (2005). Behaviour of bacterial endospores and toxins as safety determinants in low acid pressurized food. PhD Thesis, München, TU München.
87. Morin, R.S., Kozlovac, J.P. (2000). Biological Safety: principles and practices, in *Biological Safety: Principles and Practices* (eds D.O. Fleming, D.L. Hunt), ASM Press, Washington, DC, pp. 261–272.
88. BfR (2001). Rare but dangerous: food poisoning from *Clostridium botulinum*. *Federal Institute for Risk Assessment*, 25/2001, 10.09.2001.
89. Guidi-Rontani, C., Weber-Levy, M., Labruyere, E., Mock, M. (1999). Germination of *Bacillus anthracis* spores within alveolar macrophage. *Molecular Microbiology*, *31*, 9–17.
90. Moir, A., Corfe, B.M., Behravan, J. (2002). Spore germination. *Cellular and Molecular Life Sciences*, *59*, 403–409.
91. Setlow, P. (2003). Spore germination. *Current Opinion in Microbiology*, *6*, 550–556.
92. Moir, A. (2006). How do spores germinate? *Journal of Applied Microbiology*, *101*, 526–530.
93. Wuytack, E.Y., Soons, J., Poschet, F., Michiels, C.W. (2000). Comparative study of pressure- and nutrient-induced germination of *Bacillus subtilis* spores. *Applied and Environmental Microbiology*, *66*, 257–261.
94. Heinz, V., Knorr, D. (1998). High pressure germination and inactivation kinetics of bacterial spores, in *High Pressure Food Science, Bioscience and Chemistry* (ed. N.S. Isaacs), The Royal Society of Chemistry, Cambridge, pp. 435–441.
95. Paidhungat, M., Setlow, B., Daniels, W.B., Hoover, D., Papafragkou, E., Setlow, P. (2002). Mechanisms of induction of germination of *Bacillus subtilis* spores by high pressure. *Applied and Environmental Microbiology*, *68*, 3172–3175.

96. Wiencek, K.M., Klapes, N.A., Foegeding, P.M. (1990). Hydrophobicity of *Bacillus* and *Clostridium* spores. *Applied and Environment Microbiology*, *56*, 2600–2605.
97. Furukawa, S., Narisawa, N., Watanabe, T., Kawarai, T., Myozen, K., Okazaki, S., Ogihara, H., Yamasaki, M. (2005). Formation of the spore clumps during heat treatment increases the heat resistance of bacterial spores. *International Journal of Food Microbiology*, *102*, 107–111.
98. Mathys, A., Heinz, V., Schwartz, F.H., Knorr, D. (2007). Impact of agglomeration on the quantitative assessment of *Bacillus stearothermophilus* heat inactivation. *Journal of Food Engineering*, *81*, 380–387.
99. Besler, M., Steinhut, H., Paschke, A. (2001). Stability of food allergens and allergenicity of processed foods. *Journal of Chromatography B-Analytical Technologies in the Biomedical and Life Sciences*, *756*, 207–228.
100. Jankiewicz, A., Baltes, W., Bögl, K.W., Dehne, L.I., Jamin, A., Hoffmann, A., Haustein, D., Vieths, S. (1997). Influence of food processing on the immunochemical stability of celery allergens. *Journal of the Science of Food and Agriculture*, *75*, 359–370.
101. Malanin, K., Lundberg, M., Johansson, S.G.O. (1995). Anaphylactic reaction caused by neoallergens in heated pecan nut. *Allergy*, *50*, 988–991.
102. Maleki, S.J., Chung, S.-Y., Champagne, E.T., Raufman, J.-P. (2000). The effects of roasting on the allergenic properties of peanut proteins. *Journal of Allergy and Clinical Immunology*, *106*, 763–768.
103. Chung, S.-Y., Butts, C.L., Maleki, S.J., Champagne, E.T. (2003). Linking peanut allergenicity to the processes of maturation, curing, and roasting. *Journal of Agricultural and Food Chemistry*, *51*, 4273–4277.
104. Bleumink, E., Berrens, L. (1966). Synthetic approaches to the biological activity of ß-lactoglobulin in human allergy to cow's milk. *Nature*, *212*, 541–543.
105. Cariollo, T., de Castro, R., Cuevas, M., Caminero, J., Cabrera, P. (1991). Allergy to limpet. *Allergy*, *46*, 515–519.
106. Kato, T., Katayama, E., Matsubara, S., Omi, Y., Matsuda, T. (2000). Release of allergenic proteins from rice grains induced by high hydrostatic pressure. *Journal of Agricultural and Food Chemistry*, *48*, 3124–3129.
107. Grimm, V., Scheibenzuber, M., Rakoski, J., Behrendt, H., Blümelhuber, G., Meyer-Pittroff, R., Ring, J. (2002). Ultra-high pressure treatment of foods in the prevention of food allergy. *Allergy*, *57*, 102.
108. Balasubramaniam, V.M., Ting, E.Y., Stewart, C.M., Robbins, J.A. (2004). Recommended laboratory practices for conducting high-pressure microbial inactivation experiments. *Innovative Food Science & Emerging Technologies*, *5*, 299–306.
109. Hartmann, C., Delgado, A., Szymczyk, J. (2003). Convective and diffusive transport effects in a high pressure induced inactivation process of packed food. *Journal of Food Engineering*, *59*, 33–44.
110. Knoerzer, K., Pablo, J., Gladman, S., Versteeg, C., Fryer, P. (2007). A computational model for temperature and sterility distributions in a pilot-scale high-pressure high-temperature process. *American Institute of Chemical Engineers Journal*, *53*, 2996–3010.

111. Minerich, P.L., Labuza, T.P. (2003). Development of a pressure indicator for high hydrostatic pressure processing of foods. *Innovative Food Science & Emerging Technologies*, *4*, 235–243.

112. Rumpold, B.A. (2005). Impact of high hydrostatic pressure on wheat, tapioca, and potato starches. PhD Thesis, Berlin, Berlin Technical University, 120.

113. Van der Plancken, I., Grauwet, T., Oey, I., Van Loey, A., Hendrickx, M. (2008), Impact evaluation of high pressure treatment on foods: considerations on the development of pressure-temperature-time integrators (pTTIs). *Trends in Food Science & Technology*, *19(6)*, 337–348.

6

ALKALI AND/OR ACID TREATMENT

6.1

DIETARY SIGNIFICANCE OF PROCESSING-INDUCED LYSINOALANINE IN FOOD

MENDEL FRIEDMAN
Western Regional Research Center, Agricultural Research Service, United States Department of Agriculture, Albany, CA 94710, USA

6.1.1 INTRODUCTION

Exposure of food proteins to high pH and heat induces two major changes: racemization of all amino acids to D-isomers and concurrent formation of cross-linked amino acids such as lysinoalanine. The discovery of lysinoalanine (LAL) formed during the exposure of proteins to high pH in 1964 (1–3) would by itself probably have been of interest largely to protein chemists. Interest grew when only 3 years later it was found that the introduction of LAL into food proteins under conditions widely used in the processing of foods resulted in adverse effects on nutritional quality and in kidney damage in rodents. The further discovery that LAL is also produced *in vivo* by microbes, animals, and humans and that these amino acids may govern antimicrobial activities and be related to the aging of tissues further broadened interest in LAL and the related cross-linked amino acids.

Protein-containing foods and feeds are often treated with alkali in the course of preparing protein concentrates and isolates for dietary and other uses. For example, in preparing soy protein concentrates, the usual step is to extract soybeans with aqueous alkali and then precipitate the protein from the

Process-Induced Food Toxicants: Occurrence, Formation, Mitigation, and Health Risks,
Edited by Richard H. Stadler and David R. Lineback

Lysinoalanine (LAL): At a Glance	
Historical	Discovered in the mid-1960s upon alkaline treatment of proteins. Heat or alkali treatment of proteins during food processing leads to structural changes and the formation of cross-links, such as LAL, in the proteins.
Analysis	Several methods have been reported, the most common being GC/MS, HPLC, and ion chromatography.
Occurrence in Food	Present in many proteinaceous foods, including heated milk, protein concentrates, and isolates of dairy products such as acid casein, processed fish, meat, and eggs, cooked corn products, hydrolyzed vegetable protein, legumes, and cereals used in the feed and food industry.
Main Formation Pathways	High pH and thermal input (temperature / time) favor the formation of LAL. The reaction essentially involves alkali catalysed β-elimination to form a dehydroprotein that undergoes cross-linking with ε-NH_2-protein and subsequent hydrolysis to afford LAL.
Mitigation in Food	The use of additives prior to alkali treatment may minimize LAL formation.
Nutritional Consequences	Potential adverse effects on nutritional quality of foods, particularly in low lysine proteins (wheat gluten, corn protein), where lysine is a nutritionally limiting amino acid. Animal feeding studies indicate that alkali-treated protein isolates impact the *in vivo* digestibility and consequently the weight gain of the animals.
Health Risks	Free LAL induces partially reversible kidney lesions in rodents. No histological changes to the kidney have so far been observed in other animals. There is supportive evidence that some of the biological properties of LAL are linked to metal ion chelation. Affinity toward Cu(II) is pronounced and may be the underlying mechanism of kidney damage observed particularly in rodents.
Regulatory Information or Industry Standards	No regulatory limits have been set in foods.
Knowledge Gaps	The safety of LAL and related compounds for humans has not yet been fully resolved. There is currently a lack of understanding of fundamental mechanisms of action at the cellular level, location of LAL in proteins, and metabolism in animals and humans.
Useful Information (web) & Dietary Recommendations	Protein-bound LAL is a normal and ubiquitous constituent of the human diet and thus there are no specific recommendations from a dietary perspective.

resulting solution at the isoelectric point (4). Similar alkali treatments are used in recovering proteins from cereal grains and milling by-products, oilseeds such as cottonseed, flaxseeds, safflower seeds, and peanuts, and dairy proteins such as sodium caseinate. Alkali procedures are also used to induce fiber-forming properties for use in textured soybean foods (meat analogue vegetable soy protein), for preparing peeled fruits and vegetables, and for destroying microorganisms (5, 6). Such treatments, however, may also cause side reac-

TABLE 6.1.1 Lysinoalanine content of different foods.

Food	Lysinoalanine, μg/g
Cereal products	200–390
Cheese for babies	201
Chicken meat	370
Eggs	160–1,820
Gelatin	250
Infant formulas, dry	150–920
Infant formulas, liquid	160–2,120
Meat products	140–540
Milk, UHT-treated	186
Milk, sterilized	200–1,160
Milk powders	150–1,620
Noodles, Chinese	480
Skipjack, dry	410
Sodium caseinate	430–6,900
Soy protein isolate	370–1,300
Whipping agents	6,500–53,150

tions. These include formation of new amino acids, which may affect the digestibility, nutritional quality, and safety of the treated proteins. Foods containing LAL are widely consumed (Table 6.1.1).

This chapter is largely limited to LAL. It covers the following aspects: (i) formation and reduction in food proteins; (ii) distribution in food; and (iii) metabolism, nutrition, and toxicity. The tables and figures summarize experimental data largely based on our studies, which are the basis for the following overview on LAL.

6.1.2 LYSINOALANINE FORMATION IN FOOD PROTEINS

Figure 6.1.1 depicts a postulated mechanism of LAL formation (7). The mechanism involves an OH^- ion-catalyzed transformation of the ε-NH_2 group of lysine to a LAL side chain via elimination and cross-link formation. The rate-determining second-order elimination reaction thus depends directly on the concentration of both OH^- ions and susceptible lysine side chains. This elimination reaction generates a dehydroalanine residue whose conjugated carbon–carbon double bond then reacts with the ε-NH_2 of lysine to form an LAL cross-link. This nucleophilic addition reaction is governed by the number and location of available ε-NH_2 and reactive dehydroalanine groups in the protein chain. Elsewhere, we offer detailed mathematical analyses of the kinetics that govern the indicated and related transformations (8–10).

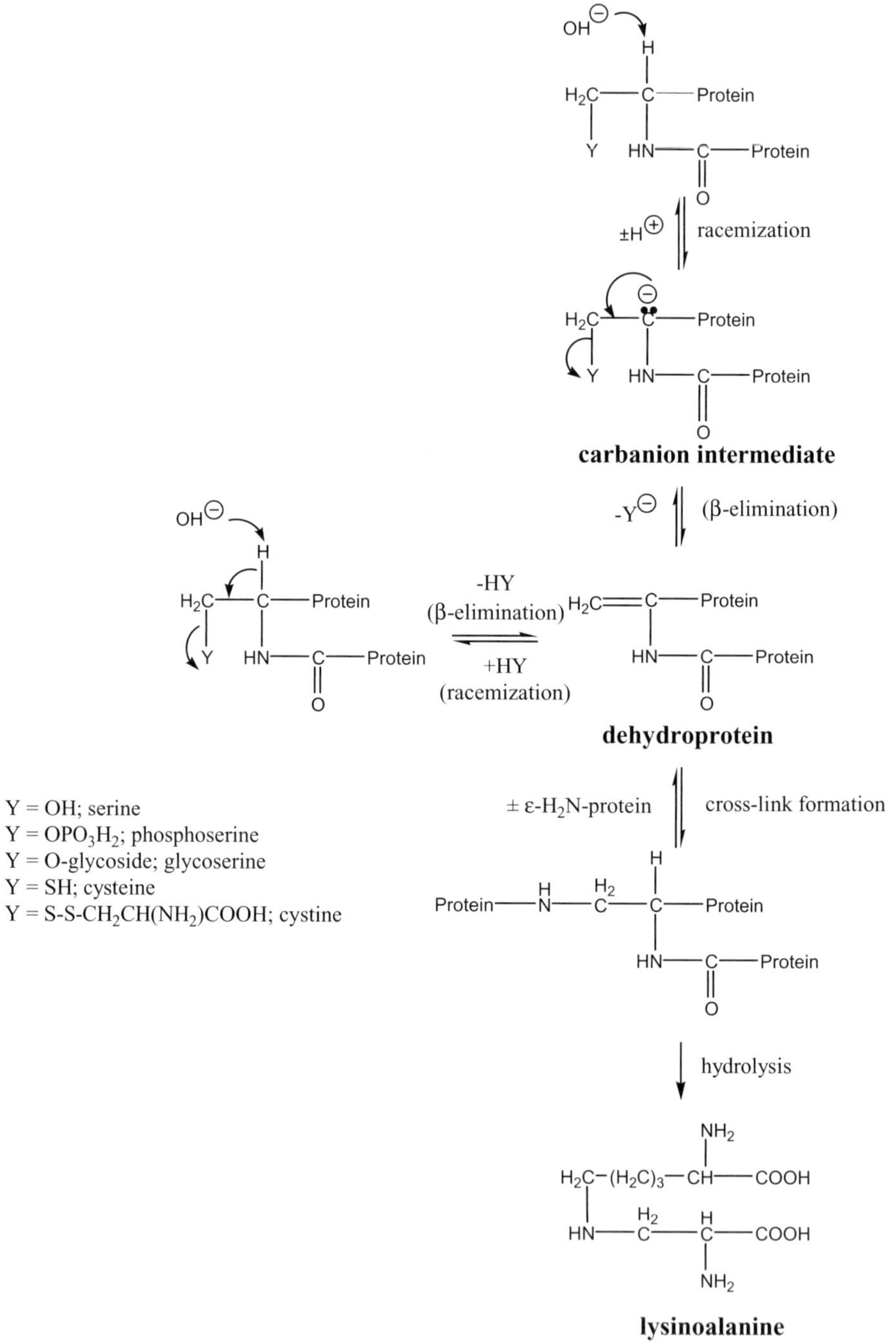

Y = OH; serine
Y = OPO_3H_2; phosphoserine
Y = O-glycoside; glycoserine
Y = SH; cysteine
Y = S-S-$CH_2CH(NH_2)COOH$; cystine

Figure 6.1.1 Base-catalyzed lysinoalanine formation.

Since nucleophilic addition may take place on either side of the carbon–carbon double bond of dehydroalanine, the resulting product is a mixture of L- and D-isomers (LL- and LD-lysinoalanines). Moreover, since some of the original L-lysine residues can undergo a hydroxide ion-catalyzed racemization to D-lysine before LAL formation, the D-lysine can, in principle, generate two additional isomers (DD- and DL-lysinoalanines) (11–13). These have different chemical and biological properties (14–17).

Two types of LAL protein cross-links are possible: intramolecular and intermolecular. The introduction of intramolecular cross-links leaves the molecular weight of the treated protein largely unchanged, whereas the molecular weight increases proportionately with the number of intermolecular cross-links. Nutritional and biological properties of LAL-containing proteins are probably strongly influenced by the relative numbers of the two types of cross-links.

6.1.3 RELATED PROCESSING-INDUCED AMINO ACIDS

Figure 6.1.2 shows the structures of several other novel amino acids that can be formed concurrently with LAL (7, 18). These include β-aminoalanine (19), dehydroalanine (20–23), histidinoalanine (24–26), histidinomethylalanine (27), lanthionine (7, 19, 28, 29), lysinomethylalanine (27), ornithinoalanine (30, 31), and phenylethylaminoalanine (32, 33).

6.1.4 ANALYSIS: LAL CONTENT OF PROCESSED PROTEINS AND FOOD PRODUCTS

Complex foods such as dairy, soybean, and meat products contain numerous compounds, which may interfere with the analysis both during hydrolysis (incomplete recovery of LAL) and quantitative determination (overlapping peaks). For example, we found that high concentrations of starch caused poor recovery of LAL during acid and alkaline hydrolysis used for proteins (34). Several different methods have been reported to measure the content of LAL in foods. These include ion-exchange chromatography (34–36), GC-MS (Fig. 6.1.3) (12), GC-FID (38), HPLC (39), and TLC (40). Detailed discussion of the reported methods is beyond the scope of this chapter. To our knowledge, there is still a need for a method that can determine all four possible LAL stereoisomers (LL, LD, DL, and DD) in a single analysis.

6.1.4.1 Factors Favoring Lysinoalanine Formation

Figure 6.1.4 depicts pH-induced degradation trends of wheat gluten and lactalbumin. The destruction of cystine, serine, threonine, and arginine residues is

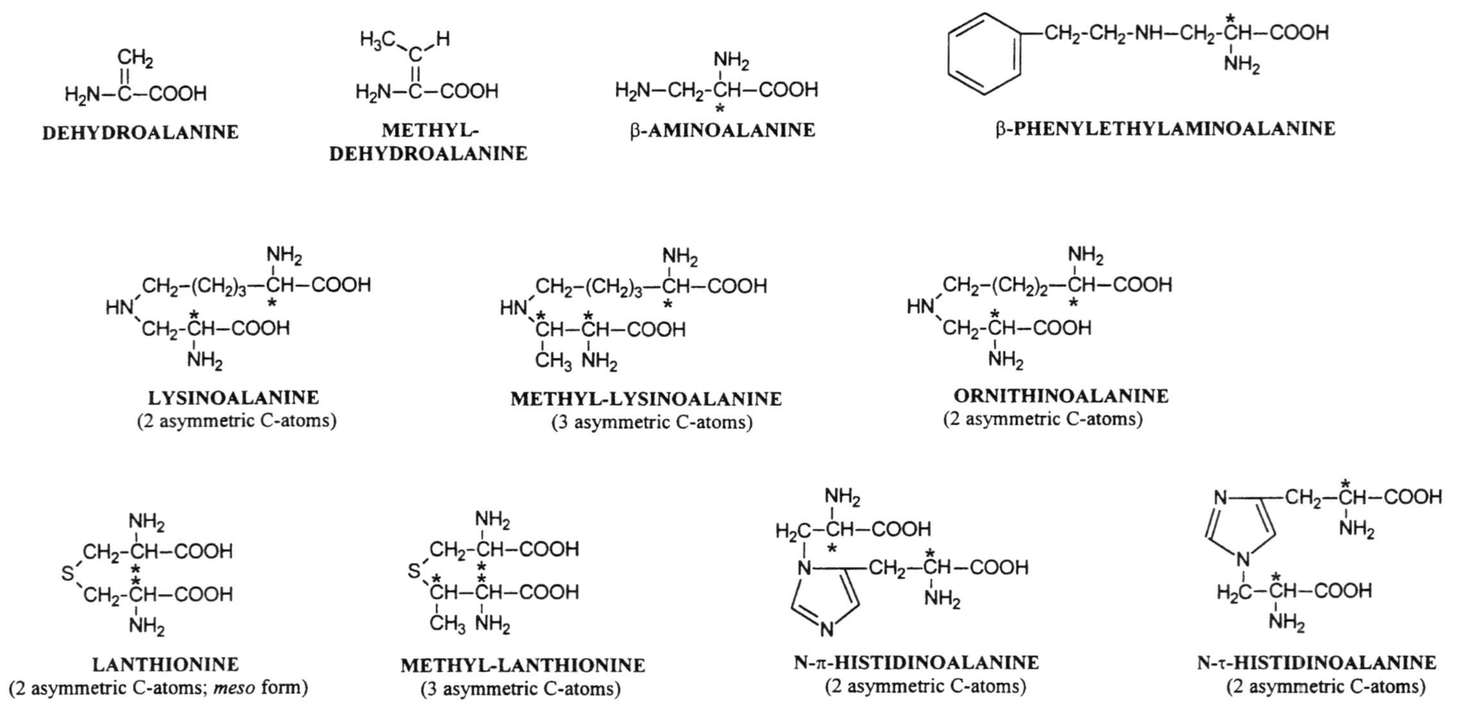

Figure 6.1.2 Structures of dehydro and cross-linked amino acids present in foods, body tissues, and lantibiotics.

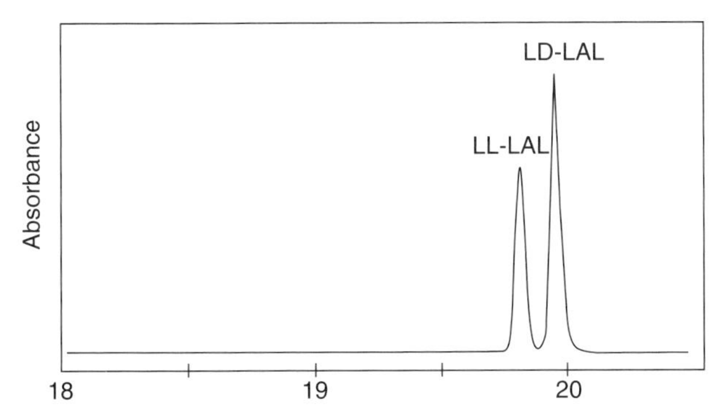

Figure 6.1.3 Separation of LL-LAL and LD-LAL isomers. Adapted from Reference 37.

accompanied by the formation of LAL illustrated in Figs 6.1.5 and 6.1.6 with soy and wheat gluten other proteins. Studies on the influence of pH on LAL content of wheat gluten and proteins showed that LAL began to appear at pH 9 and increased continually up to pH 12.5, and then decreased at pH 13.9 (34). At the very high pH, LAL is both formed and degraded. Arginine, cystine, lysine, serine, and threonine were also modified during the alkaline treatment of soy proteins. Thus, when a 1% solution of soy protein was heated in 0.1 N sodium hydroxide (pH 12.5) at 75 °C for various time periods, LAL formation progressively increased for about 3 h. Beyond that time, the concentration started to decrease. These results and those mentioned earlier for the pH dependence of LAL formation indicate that for each protein, conditions may exist where LAL is formed as fast as it is destroyed. The nature of the degradation products is not known. Disappearance of arginine and lysine started at about 35 °C and of serine and threonine at about 45 °C. The treatment also induced the loss of cystine residues (not shown) and the appearance of LAL residues. LAL residues started to appear at 25 °C (0.69 g/100 g) and continuously increased up to 4.13 g/100 g at 85 °C.

Table 6.1.2 shows that the LAL content of seven alkali-treated proteins ranges from 0.32 g/16 g N for zein to 8.52 g/16 g N for bovine serum albumin. This wide variation appears to be associated with the corresponding variations in the number of cystine and lysine residues along the protein chains that serve as LAL precursors.

6.1.4.2 Separating Lysinoalanine and D-Amino Acid Formation

Because LAL formation requires the participation of the ε-NH_2 group of lysine, acylation of the amino group is expected to prevent LAL formation but not racemization (42). Table 6.1.3 shows that this is indeed the case, since alkali treatment of casein resulted in formation of both D-aspartic acid and LAL, whereas the corresponding treatment of acylated casein produced the

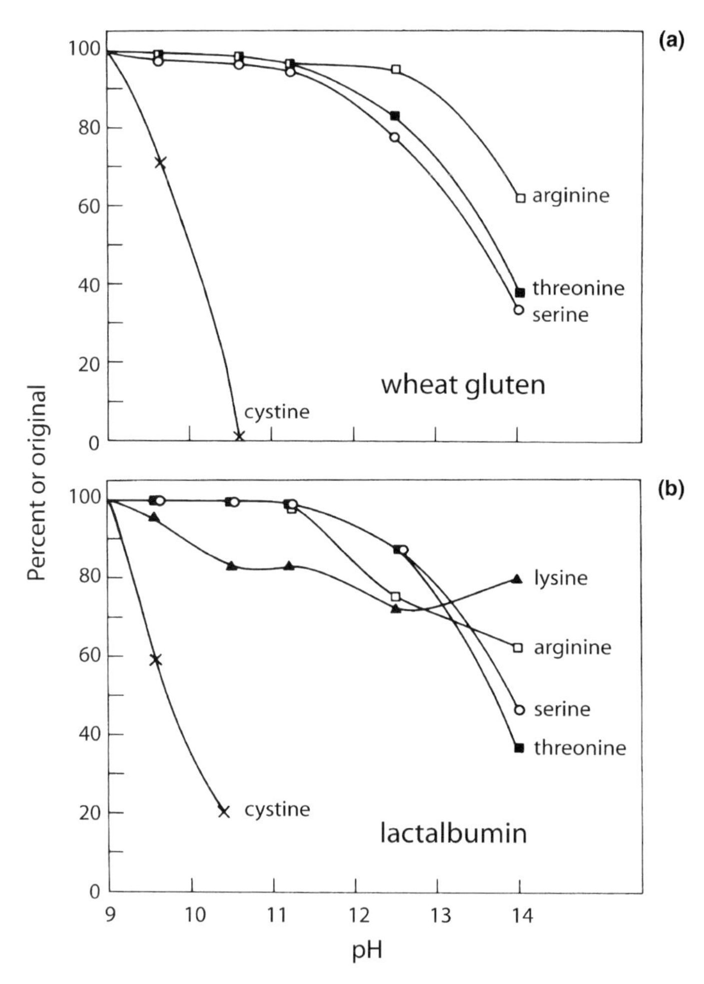

Figure 6.1.4 Effect of pH on the extent of alkali-induced degradation of susceptible amino acid residues in protein. Conditions: 1% protein; 65 °C; 3 h. Adapted from Reference 41.

same amount of D-aspartic acid but no LAL. This approach permits partitioning nutritional and other consequences of racemization from LAL production. Since the LAL but not the D-amino acid content of different proteins treated under the same conditions varies by a factor of about 30 (Table 6.1.2), it is possible to produce proteins with varying LAL to D-amino acid ratios, depending on whether the product is to be used in human or ruminant nutrition.

6.1.4.3 Preventing Lysinoalanine Formation

Several possible approaches can be used to prevent or minimize LAL formation. These include acetylation and succinylation of amino groups (7, 8, 20, 42, 44–47). These include the use of additives prior to alkali treatment, the use of

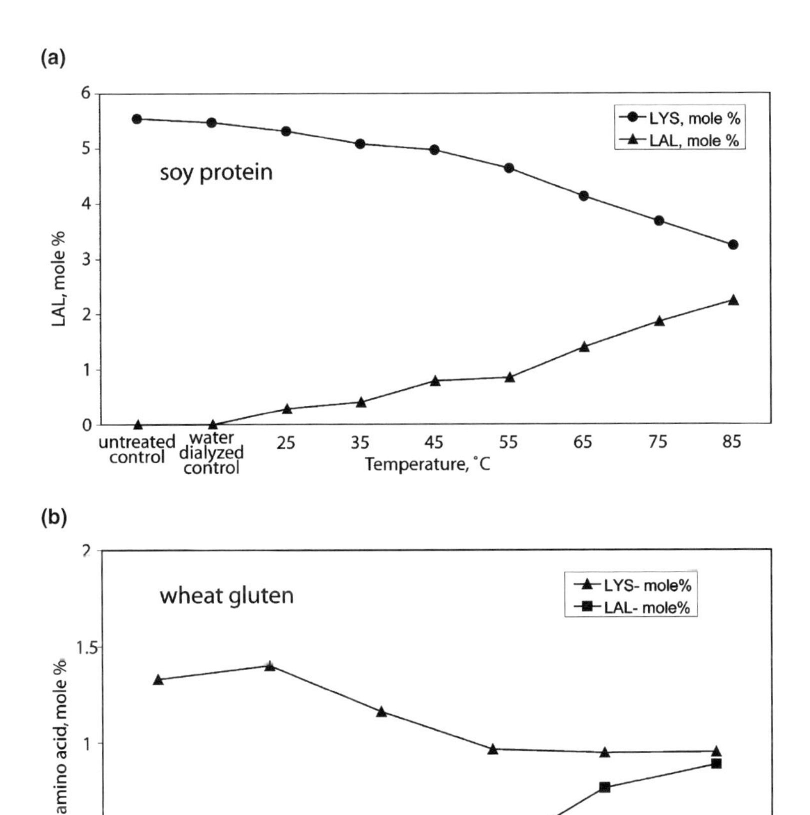

Figure 6.1.5 (a) Effect of temperature of treatment on lysinoalanine and lysine content of alkali-treated 1% soy protein, 0.1 N sodium hydroxide, 3 h. (b) Effect of pH treatment on lysine and lysinoalanine content of alkali-treated 1% wheat protein, 65 °C, 3 h. Adapted from Reference 41.

ascorbic acid, malic acid, cysteine, *N*-acetylcysteine, reduced glutathione (Table 6.1.4), sodium sulfite (7), glucose (48), ammonia (49), metal salts, and biogenic amines (33). Dephosphorylation (50) of casein and casein-derived plastein formation (51), and the use of a nitrogen atmosphere (52) also minimized LAL concentrations.

6.1.4.4 Dehydroalanine in Food Proteins

We developed a procedure for detecting dehydroalanine in alkali-treated proteins based on addition of the SH group of 2-mercaptethylpyridine to the

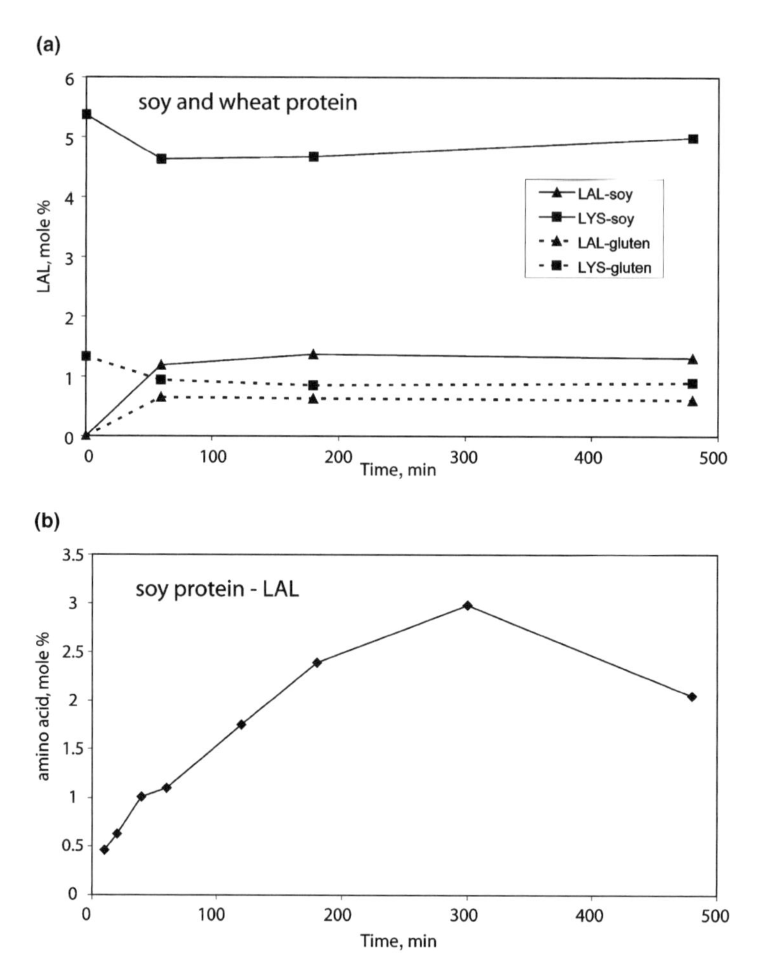

Figure 6.1.6 (a) Effect of time on lysinoalanine and lysine content of alkali-treated 1% soy or wheat protein, 1 N sodium hydroxide, 65 °C. (b) Effect of time on lysinoalanine content of alkali-treated 1% soy protein, 0.1 N sodium hydroxide, 75 °C. Adapted from Reference 41.

TABLE 6.1.2 Variation in LAL content of alkali-treated proteins.

Protein	LAL, g/16 g N	Isomeric ratio [LL-LAL]/[LL-LAL + [LD-LAL]
Zein	0.32	
Wheat gluten	0.95	0.50
Fish protein concentrate	2.75	0.40
Bovine hemoglobin	3.36	0.50
Casein	4.40	0.51
Lactalbumin	5.38	0.51
Bovine serum albumin	8.52	0.50

Conditions: 1% protein; 0.1 N sodium hydroxide; 75 °C; 3 h. Adapted from References 11 and 12.

TABLE 6.1.3 Aspartic acid racemization and LAL content of alkali-treated casein and acetylated casein.

Protein	D/L Asp	LAL (mol %)
Casein, untreated	0.023	0.0
Casein + alkali	0.387	2.35
Acetylated casein + alkali	0.336	0.0

Conditions: 1% protein; 0.1 N sodium hydroxide; 65 °C; 3 h. Adapted from Reference 43.

TABLE 6.1.4 Effect of thiols on lysinoalanine content (g/100 g protein) of alkali-treated soybean proteins.

Additive			
None	Cysteine	*N*-acetyl-cysteine	Reduced glutathione
4.04	1.69	1.72	1.05

Conditions: 1% protein; 1 mM thiol, pH 12.5; 65 °C; 3 h. Adapted from Reference 34.

TABLE 6.1.5 Dehydroalanine content of alkali-treated casein and acetylated casein.

Protein	Dehydroalanine, g/16 g N
Casein, untreated	0
Casein, acetylated	0
Casein, alkali-treated	0.33
Casein, acetylated, alkali-treated	1.39

Conditions: 1% protein; pH 12.5; 70 °C; 3 h. Adapted from References 20 and 45.

double bond of dehydroalanine to form S-β-(2-pyridylethyl) cysteine (2-PEC) (21). The cysteine derivative can be assayed by amino acid analysis (46, 47, 53). The method revealed the presence of a significant amount of dehydroalanine in alkali-treated casein and acetylated casein (Table 6.1.5). Dehydroalanine residues, which are present in alkali-treated proteins and which occur naturally in peptide antibiotics can, in principle, act as alkylating agents *in vivo*, analogous to that observed with acrylamide (54).

6.1.4.5 Physicochemical Properties of Lysinoalanine-Containing Food Proteins

To manufacture fibrous proteins resembling meat fibers, the alkaline protein solution or dope is injected through a punctured plate into an acid bath, where the pH is reduced to the isoelectric point of the protein (55, 56). The precipitated coagulated fibers are then washed and used to manufacture vegetable meat analogues. The critical pH for preparing dope solutions from proteins

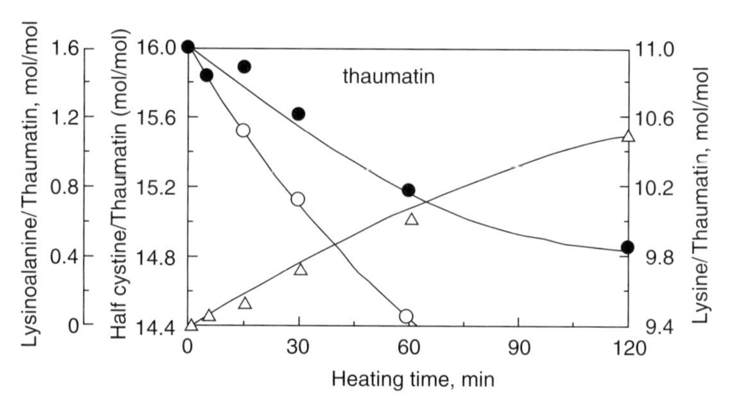

Figure 6.1.7 Changes in the number of half-cystine (○), lysine (●), and lysinoalanine residues (Δ) per mole of thaumatin on heating. Adapted from Reference 63.

was higher than 11. After 2 h at pH 12.5 and 60 °C, about half of the cysteine/cystine residues had reacted to form LAL. After 90 s at pH 12.5 and 65 °C, soy proteins contained 0.2 g/100 g of LAL and whey protein, 0.5 g/100 g. For casein, the amount of LAL formed increased with time (57). Cysteine inhibited LAL formation and NaCl enhanced it. LAL (85 mg/100 g protein) was also present in extruded fibers (pH 12, 40–50 °C) from suspensions of chicken and carp muscle proteins (58). Alkali treatment of soy protein increased emulsifying activity index from 47% to 99.5% and reduced protein digestibility (59).

The formation of heat-induced LAL cross-links contributed to the enhanced strength of collagen fibers, the main connective tissue component and most abundant animal protein (60). Related changes include those of the gel strength of yogurt due to protein cross-linking (61), aggregations and consequent changes in physicochemical properties of casein (62), heat-induced cross-linking in the sweet protein thaumatin (63) (Fig. 6.1.7), and changes in physicochemical properties in egg white lipid (64) and wheat gluten films (65).

6.1.5 LYSINOALANINE CONTENT OF FOODS

6.1.5.1 Cereals

Exposure of cereal protein to alkali and heat has a number of benefits. These include (i) prevention of microbial attack on moist grain, (ii) enhancement of storage properties of grain (66), (iii) enhancement of solubility of gluten proteins (67), and (iv) improvement in the quality of corn tortillas.

Wheat gluten contains high concentrations of the LAL precursor cystine and about 2% lysine (68). The factors that influence LAL formation in cereal proteins have been extensively studied by Friedman *et al.* (7, 41, 42, 69), Fujimaki *et al.* (70), and Takeuchi *et al.* (71). Conditions that favor LAL formation include high pH, temperature, and time of exposure. Since protein concentration did not affect LAL formation, intra- rather than intermolecular cross-links are formed in gluten proteins. Hydrolysis of gluten by pronase followed by alkali exposure of the resulting peptides formed less LAL as compared with the undigested protein (72).

The maximum amount of LAL formed when wheat gluten was treated with 10% sodium carbonate (Na_2CO_3) by weight at pH 10.5 at 100 °C for 60 min was 22 μmol/kg (73). Addition of malic acid to the Chinese wheat noodles treated with 5% Na_2CO_3 before drying resulted in the reduction of LAL content to 14.4 mg/kg. Treatment of instant noodles with malic acid completely prevented LAL formation, presumably because the acid protonates the reactive ε-NH_2 groups of lysine side chains to the nonreactive ε-NH_3^+ form. The LAL content of Finnish whole grain cereal powder ranged up to 265 mg/kg; apple porridge powder, up to 241 mg/kg; and rice porridge powder, up to 19 mg/kg (74).

Exposure of barley to sodium hydroxide induced the formation of 1.16 g of LAL/kg of the grain (75). Since barley contains about 10% protein whose lysine content is between 2% and 3% (76), about one-half of the lysine seems to have participated in LAL formation.

Liming of corn meal during the preparation of tortillas involves exposure to calcium hydroxide and heat. Studies by Sanderson *et al.* (77) showed that (i) up to 1339 μg of LAL/g of protein was present in corn flour exposed to high concentrations of sodium hydroxide; and (ii) the LAL content of tortillas was 810 mg/kg protein and of commercial masa, 200 mg/kg. In related studies, we found that the LAL content of white dent corn tortillas before baking was 0.16 g/16 g N and after baking, 0.36 g/16 g N (78). This result shows that the baking process induces the formation of LAL. Additional studies showed that alkali-treated high-lysine corn protein isolate treated with sodium hydroxide contained 1.42 g LAL/16 g N. The corresponding value for calcium hydroxide treated samples was 1.51 g LAL/16 g N. These observations suggest that consumers of tortillas prepared from more nutritious high-lysine corn may ingest high amounts of LAL. Our studies also showed that cystine content of the high-lysine corn protein was nearly lost as a result of both treatments.

6.1.5.2 Legumes

Exposing soy proteins to alkaline conditions (pH 8–14) for various time periods (10–480 min) and temperatures (25–95 °C at 10 °C intervals) destroyed all of the cystine and part of the arginine, lysine, serine, and threonine residues at the higher pHs and temperatures (34, 45). These losses were accompanied by the appearance of LAL and unknown ninhydrin-positive compounds. LAL formation was suppressed by protein acylation of amino groups with acetic and succinic anhydrides and by addition of SH-containing compounds, copper salts, and glucose. Free and protein-bound LAL was stable to acid but not to basic conditions used for protein hydrolysis.

LAL concentrations of less than 500 μg/g in mildly treated rapeseed products were similar to those present in casein, soy, and other food products (79). The results demonstrate the need for careful control of reaction conditions during protein extraction at high pH in order to minimize LAL formation.

6.1.5.3 Dairy Products

All milk proteins have a high lysine content, which can react with the dehydroalanine derived from phosphoserine side chains to form LAL. Since casein isolation on an industrial scale involves isoelectric precipitation followed by neutralization at alkaline pH (80, 81), and since heat is widely used to pasteurize and process milk (82), the effect of alkali and heat on the formation of LAL and other unusual amino acids has been widely studied in order to define its significance for food quality and nutrition. The LAL content of raw and pasteurized milk (in mg/kg protein) ranged up to 15; of whey protein concentrate, up to 145; of ultra-high-temperature (UHT)-heated milk, up to 400; of autoclaved milk, up to 880; of sodium caseinate, up to 1530; and of calcium caseinate, up to 1560 (10, 83–92). The LAL content of 54 Finnish milk protein concentrates and isolates ranged from 0 to 143 mg/kg of protein (74). The corresponding values for milk-based infant formulas, baby foods, and formula diets exceed 300 mg/kg in some cases. Commercial milk products contain significant amounts of LAL.

The LAL content of processed cheeses with added caseinates ranged from 50 to 1070 mg/kg protein (74, 93, 94). The mean value for imitation mozzarella cheeses of 54 mg/kg was about 50 times greater than the corresponding value for the natural cheeses. The LAL content of pasteurized milk (in mg/kg protein) was 0.44; of natural mozzarella cheeses, 0.4 to 4; and of processed and imitation mozzarella cheeses, 15 to 421. These observations are the basis for the suggestion that the LAL content appears to be a good indicator to distinguish natural from imitation mozzarella cheeses (39).

Because peptide bonds in cheese proteins slowly hydrolyze during storage due to the action of proteolytic enzymes (94a), and since cleavage of peptide bonds affects both LAL formation and racemization (37, 95, 96), comparisons of LAL content should be made with cheeses of the same storage history.

6.1.5.4 Infant Formulas

The LAL content of commercial powdered infant formulas ranges up to 920 ppm and of liquid formulas up to 2120 ppm (Table 6.1.1). The amino acid content of 13 liquid and powdered milk-based infant formulas produced in Italy showed a marked difference in LAL content between liquid and powdered samples (97). The powders contained low concentrations of LAL, whereas the LAL content of the liquid samples ranged up to 1032 mg/kg protein. LAL content was a sensitive index of heat damage, and correlated with other indices such as hydroxymethylfurfural content and reduction in

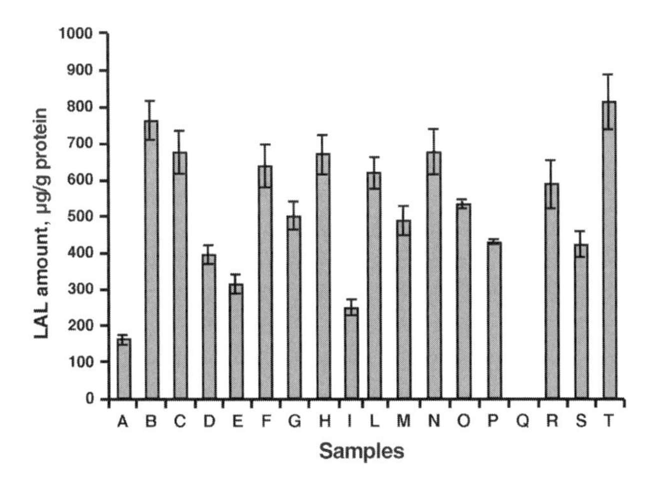

Figure 6.1.8 Fourfold variation in LAL contents of commercial formulas for enteral nutrition. Adapted from Reference 98.

protein quality. Figure 6.1.8 illustrates the wide variations in LAL content of different samples of infant formulas (98).

6.1.5.5 Eggs

LAL was present in the egg white and yolk of hen and quail eggs sold in Japan (99). The values ranged up to 93 mM/kg protein. Heating egg white in boiling water for 30 min induced LAL formation at the natural storage pH of 8 used for eggs. LAL formation increased with pH, time, and temperature of treatments and was accompanied by degradation of cystine. The LAL content of boiled eggs increased with storage time of the raw eggs (100).

6.1.5.6 Fish

High pH solutions are used to solubilize fish muscle proteins (101). LAL was detected in commercial fish meals, fish solubles, and bone meal at concentrations of 6, 10, and 29 µmol/kg, respectively (102). These authors also found that (i) the LAL content of the dry fish meals was not changed after exposure to irradiation by γ-rays; (ii) dry LAL was not, whereas solutions of free and protein-bound LAL were susceptible to degradation by γ-radiation; and (iii) γ-radiation did not induce LAL formation in bovine serum albumin, lysozyme, and ovalbumin.

6.1.5.7 Meats

Alkaline treatment of the protein–polysaccharide complex of cartilage used to separate the protein part from the polysaccharide chain induces the formation of LAL, which was accompanied by loss of cystine, serine, and lysine residues (103). Alkaline extraction of deboning residues from poultry and red meats may be used for the recovery of additional food-grade proteins (104). Such treatments induce the formation of varying amounts of LAL (105). The presence of high concentrations of LAL and of histidinoalanine (70 and 250 nmol/mg protein, respectively), in lime-processed gelatin used in photography is attributed to the non-gelatin protein components which are co-extracted with gelatin from beef bones and hides (106). The formation of lanthionine and LAL residues in insulin occurred during its isolation and purification by preparative HPLC (107).

Inactivation of biological properties of animal proteins by heat and alkali was assessed by measuring the selective destruction of cystine and accompanying formation of lanthionine and LAL residues (108). Related studies showed that cooking methods did not affect the LAL content of hamburgers (109), and that the LAL content of the protein fraction may be a useful indicator for protein quality of processed meats (110).

6.1.5.8 Wool and Other Protein Fibers

Wool, animal and human hair, feathers, silk, and skin are keratin proteins composed of amino acids linked by peptide bonds. The peptide units are cross-linked by a large number of intra- and intermolecular disulfide bonds (111–115). Extensive efforts have been made to improve the textile properties of wool by altering the structure of keratin fibers by heating, steaming, and mild alkali treatment. These efforts included an assessment of the roles of LAL and lanthionine formed as a result of such treatments.

Alkali treatment and steaming of wool induce the formation of LAL (3). LAL residues in wool behaved as interchain cross-links (29). Studies with insoluble wool are much easier to conduct than with soluble proteins. Processed feather meal contains lanthionine (116). LAL formation in wool induced by phosphate salts depended on pH of the salt solution and structure of the phosphate anion (117). Silk worms synthesize an enzyme that catalyzes lanthionine formation *in vivo* (118). We do not know whether exposure of human hair to high-pH shampoo and other treatments induces the formation of LAL (114).

6.1.5.9 Yeast Proteins

Alkaline conditions used to disrupt yeast cells to facilitate extractions of yeast proteins induced formation of significant amounts of LAL (119). The amount (3.6 g/16 g N) formed during the isolation of the proteins by a high-alkali, low-temperature process (pH 12.5, 65 °C, 2 h) was seven times greater than the value (0.49 g/16 g N) observed when the protein was isolated by a low-alkali, high-temperature process (pH 10.5, 85 °C, 4 h).

6.1.6 NUTRITION AND SAFETY

6.1.6.1 Digestibility and Nutritional Quality

The following examples illustrate the adverse and beneficial effects of alkali treatment on protein nutrition. (89, 95, 120–126). Figure 6.1.9 shows that impairment of *in vitro* digestibility of casein is inversely proportional to the amounts of LAL and D-amino acids formed. Table 6.1.6 and Fig. 6.1.10 show that alkali treatment of soy protein isolates adversely affected *in vivo* digestibility in rats and body weight gain in baboons. In contrast, the lower digestibility of LAL-containing soybean proteins reduced the rate of degradation of the modified proteins in the rumen of cattle by bacterial enzymes (127–129). Such reduction is beneficial for ruminant nutrition since it improves N retention and the nutritional value of the proteins consumed by cattle and sheep.

Possible causes for the reduction in digestibility and nutritional quality include destruction of proteolytic enzyme substrates such as arginine and lysine, isomerization of L-amino acids to less digestible D-forms, formation of inter- and intramolecular cross-links, which hinder access of proteolytic enzymes, inhibition by LAL of proteolytic metalloenzymes such as carboxypeptidase (130–134), and alkali-induced formation of trypsin- and chymotrypsin-inhibiting peptides (126, 135).

Generally, the extent of nutritional damage of alkali treatment associated with loss of lysine may depend on the original lysine content of a protein. Decrease in lysine due to LAL formation in a high-lysine protein such as casein, high-lysine corn protein, or soy protein isolate may have a less adverse effect than in a low-lysine protein such as wheat gluten or corn protein, where lysine is a nutritionally limiting amino acid (136, 137).

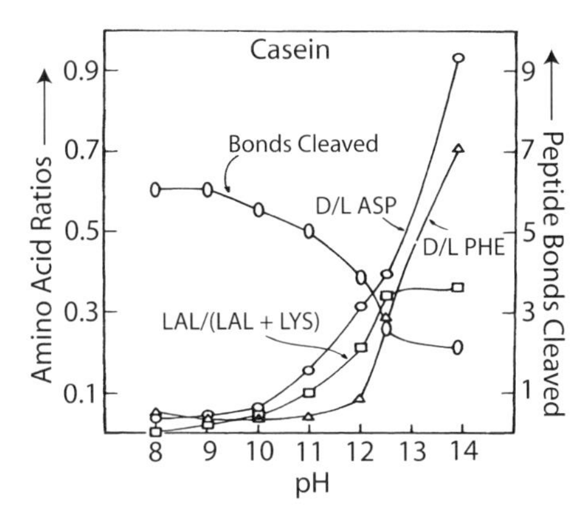

Figure 6.1.9 Inverse relationship between LAL and D-amino acid content of casein and extent of digestibility of peptide bonds by trypsin. Adapted from Reference 45. It is not possible to estimate the respective contributions of concurrently formed LAL and D-amino acids to the inhibition of proteolysis.

TABLE 6.1.6 Digestibilities and net protein utilization (NPU) of toasted and alkali-treated soy proteins in rats.

Product	Digestibility	Assimilability	NPU
Casein	98.3 ± 0.49	82.0 ± 2.85	80.6
Toasted soy	97.0 ± 20.59	64.3 ± 21.93	62.4
Alkali-treated soy	83.2 ± 20.06	34.0 ± 21.48	28.3

Listed values are in %. Adapted from Reference 6.

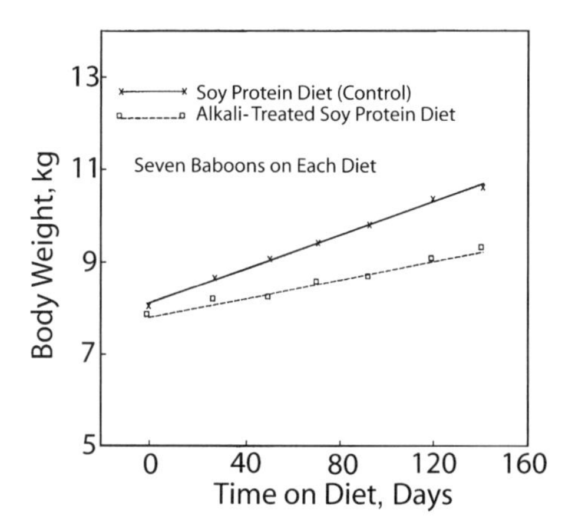

Figure 6.1.10 Effect on weight gain of long-term feeding baboons a 20% toasted and alkali-treated, LAL-containing soy protein diets. Adapted from Reference 6. Weight gain was significantly less for the LAL-containing diet.

6.1.6.2 Lysinoalanine as a Source of Lysine

The protein efficiency ratio (PER) of wheat gluten in rats increased from 0.65 to 1.62 on addition of 0.3% lysine and to 1.19 with an equimolar amount of LAL (138). In contrast, although the PER of 0.62 of gluten in mice also increased (to 0.85) with lysine addition, it decreased to 0.31 with LAL. The growth-depressing effect of LAL in mice may be due to their inability to transform some of the LAL to lysine compared with rats. Robbins *et al.* (139) showed that LAL is completely unavailable to the rat as source of lysine.

We determined the biological utilization of LAL as a source of lysine in a growth assay in weanling male mice in which all lysine in a synthetic amino acid diet was replaced by a molar equivalent of LAL (127). The replacement produced an amount of weight gain equivalent to that expected from a diet containing 0.05% L-lysine. On a molecular equivalent basis, LAL was 3.8% as potent as lysine in supporting weight gain in mice.

6.1.6.3 Metabolism of Lysinoalanine

Metabolic studies with radioactive-labeled LAL and alkali-treated lactalbumin, and fish and soy proteins showed that LAL is partly released by digestive enzymes and then absorbed by the intestine (140). The non-absorbed part was partly degraded by the intestinal microflora to CO_2. The absorbed part was eliminated in the urine of rats, mice, and hamster largely as free LAL, although some of the LAL was also excreted as acetylated derivatives. In contrast to rodents, quail excreted little free LAL. In rats, those derivatives that were excreted slowly were concentrated in the cortex of the kidney susceptible to nephrocytomegaly. Similar observations were made by Abe *et al.* (73) and by Struthers *et al.* (141, 142).

The stereochemistry of synthetic LAL did not change after consumption, absorption, and excretion in the urine (14). However, free LAL in urine of rats fed protein-bound LAL consisted mostly of the LL-isomer. This result could be due to the reduced ability of intestinal proteases to hydrolyze peptide bonds involving LD-LAL, resulting in a lower absorption rate of this isomer. The lower absorption is also consistent with the reported 20 to 100-fold greater effectiveness of free LAL to induce nephrocytomegaly as compared with bound LAL.

There seems to be a species dependence in the LAL-metabolizing activity of crude extracts of kidneys (143). Relative degradation rates (in nmol/h/g wet kidney tissue) were as follows: human, 70; pig, 100; cow, 110; mouse, 145; chicken, 163; rat, 185; rabbit, 264; Japanese quail, 1551. The low degrading activity in human kidneys indicates that humans may be more sensitive to the biological effects of LAL than the other animals.

6.1.6.4 Nephrocytomegaly—Kidney Toxicity

Feeding alkali-treated proteins to rats induces changes in kidney cells. These changes are characterized by enlargement of the nucleus and cytoplasm and disturbances in DNA synthesis and mitosis. These lesions, which have been attributed to LAL (15, 144–152), are designated as nephrocytomegaly (karyomegaly). The affected cells are epithelial cells of the straight portion (*pars recta*) of the proximal renal tubules (Fig. 6.1.11). Enlarged nuclei tend to have more than the diploid complement of DNA, unusual chromatin patterns, and proteinaceous inclusions. Increases in total nonchromosomal protein parallel increases in nuclear volume. These events suggest disruption of normal regulatory function of the *pars recta* cells.

The renal tubular epithelial kidney cells of all animals increased in both size and in DNA content. Necrosis of the cells was characterized by cytoplasmic edema and vacuolization, loss of microvilli, and increased lysosomal and cytoplasmic inclusions. β-Aminoalanine, ornithinoalanine, and β-phenylethylaminoalanine induced similar rat kidney lesions at higher doses

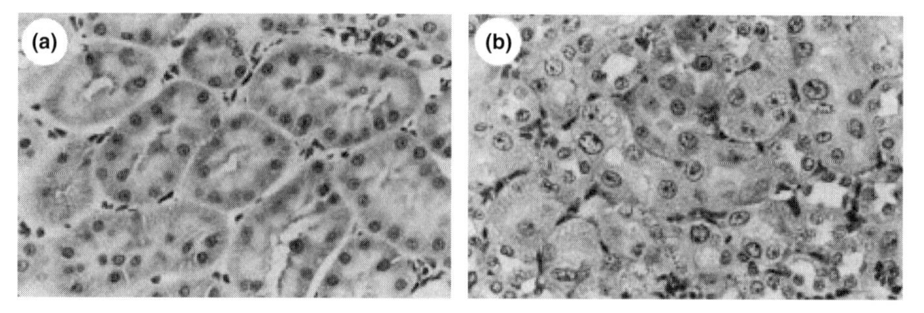

Figure 6.1.11 Photomicrographs of outer medullary stripe of kidneys from rats fed 20% soy protein diets for 8 weeks: (a) control diet; note uniformity of *pars recta* cells; (b) alkali-treated protein (2630 ppm of dietary LAL); note cytoplasmic and nuclear enlargement of the *pars recta* cells. Adapted from Reference 144.

than did LAL (15, 32). The amount of LAL required for induction of cytomegaly in rats was similar to that present in some commercial foods (Table 6.1.1). The cytomegaly was partly reversed following discontinuance of the alkali-treated soy protein diets.

A difficulty in formulating a simple relationship between LAL and nephrocytomegaly is that proteins of equal LAL content produce different biological responses. Thus, O'Donovan (153) reported that feeding rats alkali-treated soy protein led to severe nephrocytomegaly, while a different protein with the same LAL content did not produce lesions. Alkali-treated soy protein (supplying 1400–2600 ppm LAL) induced nephrocytomegaly, whereas 2500 ppm LAL derived from alkali-treated lactalbumin did not (146). Generally, free LAL is a much more potent inducer of kidney damage than is the same concentration of protein-bound LAL.

The divergent observations about relative potencies of various alkali-treated proteins in inducing kidney lesions could arise from dietary factors and from the combined effects of other kidney-damaging compounds present in the diet. Because D-Ser is formed concurrently with LAL, and since it also induced kidney lesions (154, 155), serine may potentiate the action of LAL. This aspect is examined in more detail in Chapter 6.2 on D-amino acids.

The mechanism of the observed cellular action of LAL is not well understood. Based on the observed inhibition of metalloenzymes by LAL and the observed high affinity of copper ions for LAL and metalloenzymes, Pearce and Friedman (17) suggested that the damage observed in the proximal tubules probably arises from interaction of LAL with copper(II) of metallothioneins within epithelial cells. Generally, LAL may interfere with the mechanism by which the kidney conserves copper by displacing histidine as the major low-molecular-weight carrier of copper *in vivo*.

The observed high specificity of the LAL effect for the rat kidney is probably due to the fact that nephrotoxicity in the rat is related to the high content of L-amino acid oxidase activity (which presumably catalyzes the formation of LAL metabolites responsible for toxicity) compared with other species evaluated (156). In addition to the rat (minimum-nephrotoxic-effect concentration [MNEL] = 100 ppm LAL) and the mouse (MNEL = 1000 ppm LAL), nephrocytomegaly was not observed in the kidneys of hamsters, dogs, Rhesus monkeys, rabbits, and Japanese quail fed up to 10,000 ppm LAL for 4 to 9 weeks (15, 121, 157, 158).

LAL competitively inhibited lysyl-tRNA-synthetase of prokaryotic and eukaryotic cells. It was incorporated into proteins and inhibited incorporation of lysine by a cell-free eukaryotic protein-synthesizing system (159). Whether these actions at the cellular concentration are relevant to the induction of nephrocytomegalia is not known.

6.1.6.5 Chelation of Metal Ions

Because the structure of LAL contains two amino, one imino, and two carboxyl groups, which can participate in acid-based equilibria (Fig. 6.1.12) and serve as potential metal ion-chelating sites (Fig. 6.1.13), our research suggests, as mentioned earlier, that some of the biological properties of the molecule may be due to metal ion chelation (7, 20, 160). This prediction was later confirmed by several investigators (130, 133, 161–164). *In vivo* studies confirmed expectations from the *in vitro* results. LAL and Maillard products complexed with essential trace elements and enhanced renal resorption and excretion of copper in rats. The resorption-excretion was less pronounced with iron and zinc.

To further demonstrate this possibility, we have examined LAL (a mixture of the LD- and LL-isomers as well as the individual isomers) for its affinity toward a series of metal ions, of which copper(II) was chelated the most strongly (Fig. 6.1.13). On this basis, we have suggested a possible mechanism for kidney damage in the rat involving LAL interaction with copper within the epithelial cells of the kidneys.

As we described in detail elsewhere (16, 17), it is possible to determine pK values of LAL amino, imino, and carboxyl groups (Fig. 6.1.12), metal ion-binding constants of LAL isomers, and to predict the *in vivo* equilibria between histidine, the major low-molecular-weight copper carrier in plasma, and competing chelating agents such as LAL. A mathematical analysis was used to predict LAL plasma concentrations needed to displace histidine as the major copper carrier *in vivo*. The calculated values were 27 μM for LD-LAL, 100 μM for LL-LAL, and 49 μM for the mixture of the two. LD-LAL would be a better competitor for copper(II) *in vivo* than the LL-isomer, that is it will take about one-fourth as much LD-LAL as LL-LAL to displace the same amount of histidine from copper histidine. This difference could explain the greater observed toxicity of LD-LAL (15). The apparent direct relationship between the observed affinities of the two LAL isomers for Cu^{+2} ions *in vitro* and their relative toxicities in the rat kidney is consistent with our hypothesis that LAL

(a)

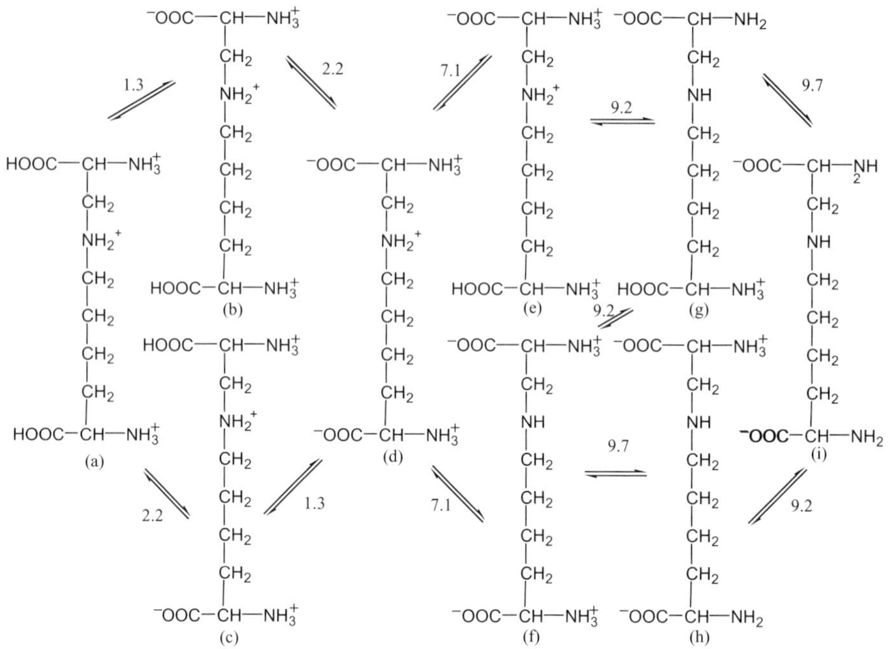

(b)

$K_1 = [H_4LAL][H]/[H_5LAL] = ([b] + [c])[H]/[a] = [b][H]/[a] + [c][H]/[a] = k_1 + k_4 =$
$10^{-1.3} + 10^{-2.2} = 10^{-1.25}$ (data from above figure)

$pK_1 = 1.25$

$1/K_2 = [H_4LAL]/[H][H_3LAL] = ([b] + [c])/[H][d] = [b]/[H][d] + [c]/[H][d] =$
$1/k_{14} + 1/k_{41} = 10^{+1.3} + 10^{+2.2} = 10^{+2.25}$

$pK_2 = 2.25$

$K_3 = [H_2LAL][H]/[H_3LAL] = ([e] + [f])[H]/[d] =$
$[e][H]/[d] + [f][H]/[d] = k_{142} + k_{143} = 10^{-7.1} + 10^{-7.1} = 10^{-6.8}$

$pK_3 = 6.8$

$K_4 = [HLAL][H]/[H_2LAL] = ([g] + [h]) \text{ X } [H]/([e] + [f]) =$
$([g][H] + [h][H])/([e] + [f]) = ([f]k_{1342} + [f]k_{1345})/([g][H]/k_{1243} + [g][H]/k_{1342}) =$
$[f](k_{1342} + k_{1345})/[(1/k_{1243} + 1/k_{1342})[f]k_{1342}] = (k_{1342} + k_{1345})/[(1/k_{1243} + 1/k_{1342})k_{1342}] =$
$(10^{-9.2} + 10^{-9.7})/[(10^{+9.2} + 10^{+9.2}) \text{ X } 10^{-9.2}] = 10^{-9.4}$

$pK_4 = 9.4$

$1/K_5 = [HLAL]/[H][LAL] = ([g] + [h])/[H][i] =$
$[g]/[H][i] + [h]/[H][i] = 1/k_{12345} + 1/k_{13452} = 10^{+9.7} + 10^{+9.2} = 10^{+9.8}$

$pK_5 = 9.8$

Figure 6.1.12 Acid–base ionization equilibria of lysinoalanine. Adapted from Reference 17. The determined pK values were used to determine LAL affinities to metal ions (16, 17).

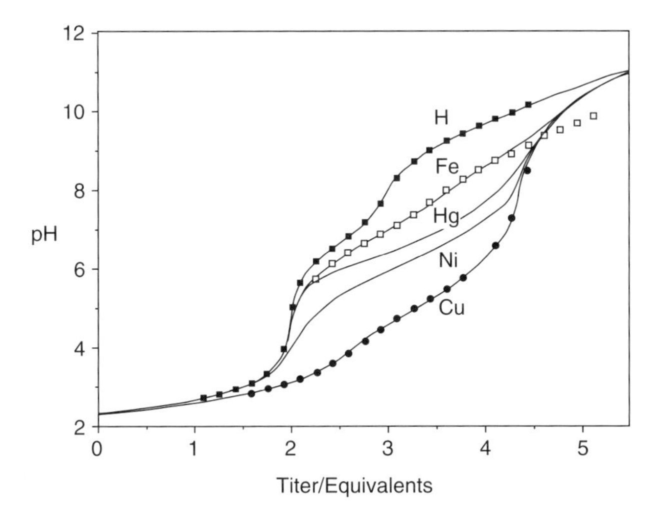

Figure 6.1.13 Potentiometric titration curves for LAL in the absence (H-curve) and presence of various divalent metal ions. The plots show the following relative affinities of metal ions for LAL: Cu > Ni > Hg > Fe. Adapted from Reference 17.

exerts some of its biological effects through chelation to copper and other metal ions *in vivo*.

6.1.6.6 Primate Studies

A short-term feeding study of alkali-treated soy protein diets to Rhesus monkeys revealed no apparent histological changes in kidney tissue of the test animals (15). To obtain additional information on this aspect, we evaluated nutritional and histopathological consequences of feeding toasted and alkali-treated soy flours to baboons (6). The untreated commercial toasted protein isolate contained 5.74 g lysine/16 g N. The corresponding value for LAL was 0.37. The alkali-treated proteins contained 4.96 g lysine/16 g N. The corresponding value for LAL was 1.61.

Figure 6.1.10 shows the growth curves for seven preadolescent male baboons each fed soy control and alkali-treated soy diets. There was no difference in growth rates over the 150-day period, since the slopes of the growth curves are not significantly different from each other. However, in absolute terms, weight gain of the baboons fed the treated soy diet (containing 370 mg of LAL/100 g air-dry diet = 3700 ppm) was about 20% lower than the gain observed with the control soy diet (containing 80 mg of LAL/100 g = 80 ppm of air-dry diet). The decreased weight gains may be due to the decreased digestibility and utilization of the treated soy protein compared with the control (Table 6.1.6).

Histological evaluation of the pancreas and kidney tissues from baboons fed toasted and alkali-treated soy diets indicated that as far as the difference due to soy processing treatment was concerned, this was essentially a negative study. The results of the baboon study show that although feeding alkali-treated soy protein containing a moderate amount of LAL to baboons for about 6 months adversely affected body mass, it apparently did not influence pancreatic and kidney histology.

Short-term feeding of LAL and Maillard product-containing formulas to healthy preterm babies did not appear to induce tubular kidney damage, as determined by urinary excretion of four kidney-derived enzymes (165). The authors concluded that the 10-day feeding study period may have been too short to cause significant changes in renal function.

6.1.7 CONCLUSIONS AND RESEARCH NEEDS

Although the safety of LAL and related compounds for humans remains unresolved, it is reassuring that the effect on kidneys in rodents was not apparent with primates. Resolution of the safety issue will depend on a better understanding of the fundamental mechanisms of action of these unusual amino acids at the cellular concentration. Mechanistic studies on the formation and biological action and fate of LAL *in vivo* could perhaps benefit from a proposed method for the synthesis of deuterium- and tritium-labeled free and protein-bound LAL (166). Such labeling should help ascertain the location of LAL in proteins and its metabolism in animals and humans.

Understanding food-processing conditions that govern LAL formation makes it possible to minimize or maximize the LAL content of foods and feeds depending on dietary needs. It may also facilitate the development of new LAL-containing, antimicrobial peptides effective against human pathogens as well as the inhibition of cross-linked amino acid (lysinoalanine, histidinoalanine, and lanthionine) formation *in vivo* to retard the aging process.

Because of its strong affinity for metal ions, free and protein-bound dietary LAL may be of value in the treatment of human diseases associated with retention of excess copper, iron, and mercury.

Because baboons gained less weight on an alkali-treated soy protein diet, it may be worthwhile to find out whether consumption by humans of alkali-treated cereal and legume proteins may help overcome obesity.

There is also the need to find out whether phenylethylamine, histamine, and tyramine in biogenic amine-rich foods react with dehydroalanine moieties during food processing and storage to form phenylethylaminoalanine and histaminoalanine, respectively. Whether lysine or lysinoalanine amino groups react with acrylamide during processing of foods also merits study (167). Research is continuing in this important area of food chemistry, food and microbial safety, nutrition, and medicine (168–172).

ACKNOWLEDGMENT

I am most grateful to Carol E. Levin for constructive contributions.

REFERENCES

1. Bohak, Z. (1964). N^{ε}-(DL-2-amino-2-carboxyethyl)-l-lysine, a new amino acid formed on alkaline treatment of proteins. *J. Biol. Chem.*, *239*, 2878–2882.
2. Patchornik, A., Sokolovsky, M. (1964). Chemical interactions between lysine and dehydroalanine in modified bovine pancreatic ribonuclease. *J. Am. Chem. Soc.*, *86*, 1860–1861.
3. Ziegler, K.L. (1964). New cross-links in alkali-treated wool. *J. Biol. Chem.*, *239*, 2713–2714.
4. Liener, I.E. (1994). Implications of antinutritional components in soybean foods. *Crit. Rev. Food Sci. Nutr.*, *34*, 31–67.
5. Friedman, M. (1999). Chemistry, nutrition, and microbiology of D-amino acids. *J. Agric. Food Chem.*, *47*, 3457–3479.
6. Friedman, M. (1999). Chemistry, biochemistry, nutrition, and microbiology of lysinoalanine, lanthionine, and histidinoalanine in food and other proteins. *J. Agric. Food Chem.*, *47*, 1295–1319.
7. Friedman, M. (1977). Crosslinking amino acids-stereochemistry and nomenclature, in *Protein Crosslinking, Nutritional and Medical Consequences* (ed. M. Friedman), Plenum, New York, pp. 1–27.
8. Friedman, M. (1982). Lysinoalanine formation in soybean proteins: kinetics and mechanism. *ACS Symp. Ser.*, *206*, 231–273.
9. Friedman, M., Williams, L.D. (1977). A mathematical analysis of kinetics of consecutive, competitive reactions of protein amino groups. *Adv. Exp. Med. Biol.*, *86B*, 299–319.
10. Hasegawa, K., Okamoto, N., Ozawa, H., Kitajima, S., Takado, Y. (1981). Limits and sites of lysinoalanine formation in lysozyme, α-lactalbumin and α_{s1}- and β-caseins by alkali treatment. *Agric. Biol. Chem.*, *45*, 1645–1651.
11. Friedman, M., Liardon, R. (1985). Racemization kinetics of amino acid residues in alkali-treated soybean proteins. *J. Agric. Food Chem.*, *33*, 666–672.
12. Liardon, R., Friedman, M., Philippossian, G. (1991). Racemization kinetics of free and protein-bound lysinoalanine (LAL) in strong acid media. Isomeric composition of bound LAL in processed proteins. *J. Agric. Food Chem.*, *39*, 531–537.
13. Tas, A.C., Kleipool, R.J.C. (1979). The stereoisomers of lysinoalanine. *Lebensm.-Wiss. Technol.*, *9*, 360–363.
14. De Weck-Gaudard, D., Liardon, R., Finot, P.-A. (1988). Stereomeric composition of urinary lysinoalanine after ingestion of free or protein-bound lysinoalanine in rats. *J. Agric. Food Chem.*, *36*, 717–721.
15. Feron, V.J., van Beek, L., Slump, P., Beems, R.B. (1978). Toxicological aspects of alkali-treated food proteins, in *Biological Aspects of New Protein Foods* (ed. J. Alder Nissen), Pergamon, Oxford, pp. 139–147.

16. Friedman, M., Pearce, K.N. (1989). Copper(II) and cobalt(II) affinities of LL- and LD-lysinoalanine diastereomers: implications for food safety and nutrition. *J. Agric. Food Chem.*, *37*, 123–127.
17. Pearce, K.N., Friedman, M. (1988). Binding of copper(II) and other metal ions by lysinoalanine and related compounds and its significance for food safety. *J. Agric. Food Chem.*, *36*, 707–717.
18. Finley, J.W., Friedman, M. (1977). New amino acid derivatives formed by alkaline treatment of proteins. *Adv. Exp. Med. Biol.*, *86* (B), 123–130.
19. Nashef, A.S., Osuga, D.T., Lee, H.S., Ahmed, A.I., Whitaker, J.R., Feeney, R.E. (1977). Effects of alkali on proteins. Disulfides and their products. *J. Agric. Food Chem.*, *25*, 245–251.
20. Friedman, M., Finley, J.W., Yeh, L.S. (1977). Reactions of proteins with dehydroalanines. *Adv. Exp. Med. Biol.*, *86* (B), 213–224.
21. Masri, M.S., Friedman, M. (1982). Transformation of dehydroalanine to S-β-(2-pyridylethyl)-L-cysteine side chains. *Biochem. Biophys. Res. Commun.*, *104*, 321–325.
22. Snow, J.T., Finley, J.W., Friedman, M. (1976). Relative reactivities of sulfhydryl groups with N-acetyl dehydroalanine and N-acetyl dehydroalanine methyl ester. *Int. J. Pept. Protein Res.*, *8*, 57–64.
23. Seebeck, F.P., Szostak, J.W. (2006). Ribosomal synthesis of dehydroalanine-containing peptides. *J. Am. Chem. Soc.*, *128*, 7150–7151.
24. Henle, T., Walter, A.W., Klostermeyer, H. (1993). Detection and identification of the cross-linking amino acids N^{τ}-and N^{π}-(2′-amino-2′-carboxy-ethyl)-l-histidine (“histidinoalanine”, HAL) in heated milk products. *Z. Lebensm.-Unters. Forsch.*, *197*, 114–117.
25. Walter, A.W., Henle, T., Klostermeyer, H. (1994). Histidinoalanin (HAL)—eine “neue” Crosslink-Aminosaeure in Milchproteinen. *Deutsche Milchwirtschaft*, *45*, 284, 286.
26. Taylor, C.M., Wang, W. (2007). Histidinoalanine: a crosslinking amino acid. *Tetrahedron*, *63*, 9033–9047.
27. Walter, A.W., Henle, T., Haessner, R., Klostermeyer, H. (1994). Studies on the formation of lysinomethylalanine and histidinomethylalanine in milk products. *Z. Lebensm.-Unters. Forsch.*, *199*, 1994.
28. Friedman, M., Noma, A.T. (1975). Methods and problems in chromatographic analysis of sulfur amino acids, in *Protein Nutritional Quality of Foods and Feeds. Part I. Assay Methods—Biological, Biochemical, and Chemical* (ed. M. Friedman), Marcel Dekker, New York, pp. 521–548.
29. Kearns, J.E., MacLaren, J.A. (1979). Lanthionine cross-links and their effects in solubility tests on wool. *J. Text. Inst.*, *12*, 534–536.
30. Asquith, R.S., Otterburn, M.S. (1977). Cystine/alkali reactions in relation to protein crosslinking. *Adv. Exp. Med. Biol.*, *86B*, 93–121.
31. Boschin, G., D’Agostina, A., Arnoldi, A. (2002). A convenient synthesis of some cross-linked amino acids and their diastereoisomeric characterization by nuclear magnetic resonance. *Food Chem.*, *78*, 325–331.
32. Jones, G.P., Hooper, P.T., Rivett, D.E., Tucker, D.J., Lambert, G., Billett, A. (1987). Nephrotoxic activity in rats fed diets containing DL-3-(N-phenylethylamino)-alanine. *Aust. J. Biol. Sci.*, *40*, 115–123.

33. Friedman, M., Noma, A.T. (1986). Formation and analysis of (phenylethyl)amino]alanine in food proteins. *J. Agric. Food Chem.*, *34*, 497–502.
34. Friedman, M., Levin, C.E., Noma, A.T. (1984). Factors governing lysinoalanine formation in soy proteins. *J. Food Sci.*, *49*, 1282–1288.
35. Wilkinson, J., Hewavitharana, A.K. (1997). Lysinoalanine determination in sodium caseinate using the LKB Alpha Plus Amino Acid Analyser. *Milchwissenschaft*, *52*, 423–427.
36. Friedman, M. (2004). Application of the ninhydrin reaction for analysis of amino acids, peptides, and proteins to agricultural and biomedical sciences. *J. Agric. Food Chem.*, *52*, 385–406.
37. Liardon, R., Friedman, M. (1987). Effect of peptide bond cleavage on the racemization of amino acid residues in proteins. *J. Agric. Food Chem.*, *35*, 661–667.
38. Montilla, A., Gomez-Ruiz, J.A., Olano, A., Castillo, M.D.D. (2007). A GC-FID method for analysis of lysinoalanine. *Mol. Nutr. Food Res.*, *51*, 415–422.
39. Pellegrino, L., Resmini, P., Noni, I.D., Masotti, F. (1996). Sensitive determination of lysinoalanine for distinguishing natural from imitation Mozzarella cheese. *J. Dairy Sci.*, *79*, 725–734.
40. Freimuth, U., Nötzold, H., Krause, W. (1980). On alkali treatment of proteins. 5. Formation of lysinoalanine in β casein and β-lactoglobulin. *Nahrung*, *24*, 351–237.
41. Friedman, M. (1979). Alkali-induced lysinoalanine formation in structurally different proteins. *ACS Symp. Ser.*, 225–235.
42. Friedman, M. (1978). Inhibition of lysinoalanine synthesis by protein acylation. *Adv. Exp. Med. Biol.*, *105*, 613–648.
43. Masters, P.M., Friedman, M. (1980). Amino acid racemization in alkali-treated food proteins—chemistry, toxicology, and nutritional consequences, in *Chemical Deterioration of Proteins* (eds J.R. Whitaker, M. Fujimaki), American Chemical Society, Washington, D.C, pp. 165–194.
44. Finley, J.W., Snow, J.T., Johnston, P.H., Friedman, M. (1977). Inhibitory effect of mercaptoamino acids on lysino-alanine formation during alkali treatment of proteins. *Adv. Exp. Med. Biol.*, *86* (B), 85–92.
45. Friedman, M., Gumbmann, M.R., Masters, P.M. (1984). Protein-alkali reactions: chemistry, toxicology, and nutritional consequences. *Adv. Exp. Med. Biol.*, *177*, 367–412.
46. Friedman, M. (1994). Improvement in the safety of foods by SH-containing amino acids and peptides. A review. *J. Agric. Food Chem.*, *42*, 3–20.
47. Friedman, M. (2001). Application of the S-pyridylethylation reaction to the elucidation of the structures and functions of proteins. *J. Protein Chem.*, *20*, 431–453.
48. Dworschák, E.F., Orsi, A., Zsigmond, A., Trezl, L., Rusznak, I. (1981). Factors influencing the formation of lysinoalanine in alkali-treated proteins. *Nahrung*, *25*, 44–46.
49. Mukai, K., Shimizu, Y., Murata, R., Matoba, T., Hasegawa, K. (1986). Effect of ammonia on lysinoalanine formation in proteins by alkali treatment. *Nihon Nogei Kagakkaishi*, *60*, 1009–1015.
50. Meyer, M., Klostermeyer, H., Kleyn, D.H. (1981). Reduced formation of lysinoalanine in enzymatically dephosphorylated casein. *Z. Lebensm.-Unters. Forsch.*, *172*, 446–448.

51. Nötzold, H., Winkler, H., Wiedemann, B., Ludwig, E. (1984). The effect of enzymatic modification on lysinoalanine formation in field-bean protein isolate and beta-casein. *Nahrung*, *28*, 299–308.
52. Steining, J., Montag, A. (1982). Studies on alternation of the lysine of food proteins. II formation of lysinoalanine. *Z. Lebensm.-Unters. Forsch.*, *175*, 8–12.
53. Friedman, M. (1973). *The Chemistry and Biochemistry of the Sulfhydryl Group in Amino Acids. Peptides and Proteins*, Pergamon Press, Oxford, p. 485.
54. Friedman, M. (2003). Chemistry, biochemistry, and safety of acrylamide. A review. *J. Agric. Food Chem.*, *51*, 4504–4526.
55. de Rham, O., van de Rovaart, P., Bujard, E., Mottu, F., Hidalgo, J. (1977). Fortification of soy protein with cheese whey protein and the effect of alkaline pH. *Cereal Chem.*, *54*, 238–245.
56. Hayakawa, I., Katsuta, K. (1981). A study of fibre spun from soy protein. *Nippon Shokuhin Kogyo Gakkaishi*, *28*, 347–354.
57. Dosako, S. (1979). Changes in rheological and chemical properties of soy protein dope solution through increase in aging time. *Agric. Biol. Chem.*, *43*, 803, 807, 809–814.
58. Hayasho, Y., Takahashi, H. (1990). Fibre formation from animal protein dispersion. *Nippon Shokuhin Kogyo Gakkaishi*, *37*, 935–939.
59. Wu, W., Hettiarachchy, N.S., Kalapathy, U., Williams, W.P. (1999). Functional properties and nutritional quality of alkali- and heat-treated soy protein isolate. *J. Food Qual.*, *22*, 119–133.
60. Gorham, S.D. (1992). Effect of chemical modifications on the susceptibility of collagen to proteolysis. II. Dehydrothermal cross-linking. *Int. J. Biol. Macromol.*, *14*, 129–138.
61. Lauber, S., Klostermeyer, H., Henle, T. (2001). On the influence of non-enzymatic crosslinking of caseins on the gel strength of yoghurt. *Nahrung*, *45*, 215–217.
62. Pellegrino, L., Boekel, M.A.J.S., Gruppen, H., Resmini, P., Pagani, M.A. (1999). Heat-induced aggregation and covalent linkages in beta-casein model systems. *Int. Dairy J.*, *9*, 255–260.
63. Kaneko, R., Kitabatake, N. (1999). Heat-induced formation of intermolecular disulfide linkages between thaumatin molecules that do not contain cysteine residues. *J. Agric. Food Chem.*, *47*, 4950–4955.
64. Handa, A., Gennadios, A., Hanna, M.A., Weller, C.L., Kuroda, N. (1999). Physical and molecular properties of egg-white lipid films. *J. Food Sci.*, *64*, 860–864.
65. Kayserilioglu, B.S., Stevels, W.M., Mulder, W.J., Akkas, N. (2001). Mechanical and biochemical characterisation of wheat gluten films as a function of pH and co-solvent. *Staerke*, *53*, 381–386.
66. Henning, P.H., Steyn, D.G. (1984). The response of different portions of the maize plant to NaOH treatment. *S. Arf. J. Anim. Sci.*, *14*, 142–143.
67. Batey, I.L., Gras, P.W. (1984). Effects of using defatted gluten as a substrate for solubilization with sodium hydroxide. *J. Food Technol.*, *19*, 109–114.
68. Friedman, M., Finot, P.A. (1990). Nutritional improvement of bread with lysine and γ-glutamyl-L-lysine. *J. Agric. Food Chem.*, *38*, 2011–2020.
69. Friedman, M. (1978). Wheat gluten-alkali reactions, Proceedings 10th Conference on Wheat Utilization Research, U. S. Department of Agriculture, Washington, DC, pp. 613–648.

70. Fujimaki, M., Haraguchi, T., Abe, K., Homma, S., Arai, S. (1980). Specific conditions that maximize formation of lysinoalanine in wheat gluten and fish protein concentrate. *Agric. Biol. Chem.*, *44*, 1911–1916.
71. Takeuchi, H., Kato, H., Fujimaki, M. (1978). Effects of temperature and alkali on formation of lysinoalanine in wheat gluten and soybean powder. *J. Food Hyg. Soc. Jpn.*, *19*, 44–49.
72. Haraguchi, T., Abe, K., Arai, S., Homma, S., Fujimaki, M. (1980). Lysinolalanine formation in wheat gluten: a conformational effect. *Agric. Biol. Chem.*, *44*, 1951–1952.
73. Abe, K., Ozawa, M., Homma, S., Fujimaki, M. (1981). Lysinoalanine formation in the processing of quick-served noodles. *Kaseigaku Zasshi*, *32*, 367–371.
74. Antila, P. (1987). The formation and determination lysinoalanine in foods containing milk proteins. *Meijeritiet. Aikak*, *45*, 1–12.
75. Davidson, J., McIntosh, A.D., Milne, E. (1982). Note on the lysinoalanine content of alkali-treated barley. *Anim. Food Sci. Technol.*, *7*, 217–220.
76. Friedman, M., Atsmon, D. (1988). Comparison of grain composition and nutritional quality in wild barley (*Hordeum spontaneum*) and in a standard cultivar. *J. Agric. Food Chem.*, *36*, 1167–1172.
77. Sanderson, J., Wall, J.S., Donaldson, G.L., Cavins, J.F. (1978). Effect of alkaline processing of corn on its amino acids. *Cereal Chem.*, *55*, 204–213.
78. Friedman, M. (1999). Lysinoalanine in food and in antimicrobial proteins. *Adv. Exp. Med. Biol.*, *459*, 145–159.
79. Deng, Q.Y., Barefoot, R.R., Diosady, L.L., Rubin, L.J., Tzeng, Y.M. (1990). Lysinoalanine concentrations in rapeseed protein meals and isolates. *Can. Inst. Food Sci. Technol. J.*, *23*, 140–142.
80. Jelen, P., Schmidt, T. (1976). Alkaline solubilization of heat-precipitated cheese whey protein. *J. Inst. Can. Sci. Technol. Aliment.*, *9*, 61–65.
81. Ward, L.S., Bastian, E.D. (1998). Isolation and identification of beta-casein A(1)-4P and beta-casein A(2)-4P in commercial caseinates. *J. Agric. Food Chem.*, *46*, 77–83.
82. Dehn-Muler, B., Muller, B., Erbersdobler, H.F. (1991). Studies on protein damage in UHT milk and milk products. *Milchwissenschaft*, *46*, 431–434.
83. Amarowicz, R., Olender, B., Smoczynski, S. (1988). Lysinoalanine levels in powdered milk. *Rocz. Panstw. Zakl. Hig.*, *39*, 31–34.
84. Amarowicz, R. (1991). Lysinoalanine. *Bromatol. Chem. Toksykol.*, *24*, 89–100.
85. Annan, W.D., Manson, W. (1981). The production of lysinoalanine and related substances during processing of proteins. *Food Chem.*, *6*, 255–261.
86. Creamer, L.K., Matheson, A.R. (1977). Action of alkali on caseins. *N. A. J. Dairy Sci. Technol.*, *12*, 253–259.
87. de Koning, P.J., van Rooijen, P.J. (1982). Aspects of formation of lysinoalanine in milk and milk products. *J. Dairy Res.*, *49*, 725–736.
88. Hasegawa, K., Kitajima, S., Takado, Y. (1981). An examination of intermolecular lysinoalanine formation using dephosphorylated casein and succinylated proteins. *Agric. Biol. Chem.*, *45*, 2133–2134.
89. Friedman, M., Zahnley, J.C., Masters, P.M. (1981). Relationship between in vitro digestibility of casein and its content of lysinoalanine and D -amino acids. *J. Food Sci.*, *46*, 127–131.

90. Fritsch, R.J., Hoffmann, H., Klostermeyer, H. (1983). Formation of lysinoalanine during heat treatment of milk. *Z. Lebensm.-Unters. Forsch.*, *176*, 341–345.
91. Isohata, T., Abe, K., Homma, S., Fujimaki, M., Arai, S. (1983). Involvement of unmasked serine residues in formation of lysinoalanine: verification by a study with dephosphorylated α_{s1}-casein and poly-L-serine. *Agric. Biol. Chem.*, *47*, 2633–2635.
92. Manson, W., Carolan, T. (1980). Formation of lysinoalanine from individual bovine caseins. *J. Dairy Res.*, *47*, 193–198.
93. Fritsch, R.J., Klostermeyer, H. (1981). Improved method for determination of lysinoalanine in food. *Z. Lebensm.-Unters. Forsch.*, *172*, 435–439.
94. Fritsch, R.J., Klostermeyer, H. (1981). Study on the occurrence of lysinoalanine in food which contain milk proteins. *Z. Lebensm.-Unters. Forsch.*, *172*, 440–445.
94a. Pearce, K.N., Karahalios, D., Friedman, M. (1988). Ninhydrin assay for proteolysis of ripening cheese. *J. Food Sci.*, *53*, 432–435, 438.
95. Hayashi, R., Kameda, I. (1980). Decreased proteolysis of alkali-treated protein: consequences of racemization in food processing. *J. Food Sci.*, *45*, 1430–1431.
96. Hayashi, R., Kameda, I. (1980). Conditions for lysinoalanine formation during exposure of protein to alkali. *Agric. Biol. Chem.*, *44*, 175–181.
97. Pompei, C., Rossi, M., Mare, F. (1987). Protein quality in commercial milk-based infant formulas. *J. Food Qual.*, *10*, 375–391.
98. Boschin, G., D'Agostina, A., Rinaldi, A., Arnoldi, A. (2003). Lysinoalanine content of formulas for enteral nutrition. *J. Dairy Sci.*, *86*, 2283–2287.
99. Murase, M., Goto, F. (1980). Effect of glucose on lysinoalanine formation. *Nippon Nogei Kagaku Kaishi*, *54*, 13–19.
100. Mukai, K., Matoba, T., Hasegawa, K. (1987). Influence of freshness of egg on lysinoalanine formation in boiled egg. *Kaseigaku Kenkyu*, *34*, 86–88.
101. Mohan, M., Ramachandran, D., Sanker, T.V., Anandan, R. (2007). Influence of pH on the solubility and conformational characteristics of muscle proteins from mullet (*Mugil cephalus*). *Process Biochem.*, *42*, 1056–1062.
102. Kume, T., Takehisa, M. (1984). Effect of gamma-irradiation on lysinoalanine in various feedstuffs and model systems. *J. Agric. Food Chem.*, *32*, 656–658.
103. Whiting, A.H. (1971). Isolation of lysinoalanine from the protein-polysaccharide complex of cartilage after alkali treatment. *Biochim. Biophys. Acta*, *243*, 332–336.
104. Ozimek, G., Jelen, P., Ozimek, L., Sauer, W., McCurdy, S.M. (1986). A comparison of mechanically separated and alkali extracted chicken protein for functional and nutritional properties. *J. Food Sci.*, *51*, 748–753.
105. Lawrence, R.A., Jelen, P. (1982). Formation of lysinoalanine in alkaline extracts of chicken protein. *J. Food Prot.*, *45*, 923–924.
106. Fuji, T., Kuboki, Y. (1985). Nongelatin components from lime-processed gelatins: presence of lysinoalanine and histidinoalanine. *J. Soc. Photogr. Sci. Jpn*, *48*, 359–366.
107. Schartmann, B., Gattner, H.G., Danho, W., Zahn, H. (1983). Increased yields of insulin by recycling of non-combined insulin chains. *Hoppe Seylers Z. Physiol. Chem.*, *364*, 179–186.
108. Chang, J.-Y., Knecht, R. (1991). Direct analysis of the disulfide content of proteins: methods for monitoring the stability and refolding process of cystine-containing proteins. *Anal. Biochem.*, *197*, 52–58.

109. Rodriguez-Estrada, M.T., Penazzi, G., Caboni, M.F., Bertacco, G., Lercker, G. (1997). Effect of different cooking methods on some lipid and protein components of hamburgers. *Meat Sci.*, *45*, 365–375.
110. Piva, G., Moschini, M., Fiorentini, L., Masoero, F. (2001). Effect of temperature, pressure and alkaline treatments on meat meal quality. *Anim. Feed Sci. Tech.*, *89*, 59–68.
111. Corfield, M.C., Wood, C., Robson, A., Williams, M.J., Woodhouse, J.M. (1967). The formation of lysinoalanine during the treatment of wool with alkali. *Biochem. J.*, *103*, 15C–16C.
112. Friedman, M., Whitfield, R.E., Tillin, S. (1973). Enhancement of the natural flame resistance of wool. *Text. Res. J.*, *43*, 212–217.
113. Friedman, M., Noma, A.T. (1970). Cystine content of wool. *Text. Res. J.*, *40*, 1073–1078.
114. Friedman, M., Orraca-Tetteh, R. (1978). Hair as an index of protein malnutrition, in *Nutritional Improvement of Food and Feed Proteins* (ed. M. Friedman), Plenum Press, New York, pp. 131–154.
115. Zahn, H., Gattner, H.G. (1997). Hair sulfur amino acid analysis. *Exp.* Suppl., 78, 239–258.
116. Han, Y., Parsons, C.M. (1991). Protein and amino acid quality of feather meal. *Poult. Sci.*, *70*, 812–822.
117. Touloupis, C., Vassiliadis, A. (1977). Lysinoalanine formation in wool after treatments with some phosphate salts. *Adv. Exp. Med. Biol.*, *86 B*, 187–195.
118. Shinbo, H. (1998). An enzyme that catalyzes the synthesis of lanthionine in the silkworm, *Bombyx mori*. *Comp. Biochem. Physiol.*, *91*, 301–308.
119. Shetty, J.K., Kinsella, J.E. (1980). Lysinoalanine formation in yeast proteins isolated by alkaline methods. *J. Agric. Food Chem.*, *28*, 798–800.
120. Chung, S.-Y., Swaisgood, H.E., Catignani, G.L. (1986). Effects of alkali treatment and heat treatment in the presence of fructose on digestibility of food proteins as determined by an immobilized digestive enzyme assay (IDEA). *J. Agric. Food Chem.*, *34*, 579–584.
121. De Groot, A.P., Slump, P., Feron, V.J., Van Beek, L. (1976). Effects of alkali treated proteins: feeding studies with free and protein bound lysinoalanine in rats and other animals. *J. Nutr.*, *106*, 1527–1538.
122. Slump, P. (1978). Lysinoalanine in alkali-treated proteins and factors influencing its biological activity. *Ann. Nutr. Aliment.*, *32*, 271–279.
123. Savoie, L., Parent, G., Galibois, I. (1991). Effects of alkali treatment on the in-vitro digestibility of proteins and the release of amino acids. *J. Sci. Food Agric.*, *56*, 363–372.
124. Savoie, L. (1984). Effect of protein treatment on the enzymatic hydrolysis of lysinoalanine and other amino acids. *Adv. Exp. Med. Biol.*, *177*, 413–422.
125. Swaisgood, H.E., Catignani, G.L. (1985). Digestibility of modified milk proteins: nutritional implications. *J. Dairy Sci.*, *68*, 2782–2790.
126. Possompes, B., Berger, J. (1986). Effect of severely alkali-treated casein on gastrointestinal transit and selected intestinal enzyme activities. *Adv. Exp. Med. Biol.*, *199*, 517–530.
127. Friedman, M., Gumbmann, M.R., Savoie, L. (1982). The nutritional value of lysinoalanine as a source of lysine for mice. *Nutr. Rep. Int.*, *26*, 937–943.

128. Nishino, N., Masaki, Y., Uchida, S. (1996). Changes in nitrogenous compounds and rates of in situ N loss of soybean meal treated with sodium hydroxide or heat. *J. Agric. Food Chem.*, *44*, 2667–2671.
129. Nishino, N., Uchida, S., Ohshima, M. (1995). Formation of lysinoalanine following alkaline processing of soya bean meal in relation to the degradability of protein in the rumen. *J. Sci. Food Agric.*, *68*, 59–64.
130. Hayashi, R. (1982). Lysinoalanine as a metal chelator. An implication for toxicity. *J. Biol. Chem.*, *257*, 13896–13898.
131. Friedman, M., Grosjean, O.-K., Zahnley, J.C. (1985). Carboxypeptidase inhibition by alkali-treated food proteins. *J. Agric. Food Chem.*, *33*, 208–213.
132. Friedman, M., Grosjean, O.-K.K., Zahnley, J.C. (1985). Metalloenzyme inhibition by lysinoalanine, phenylethylaminoalanine, and alkali-treated food proteins. *Fed. Proc.*, *44*, 6351.
133. Friedman, M., Grosjean, O.K., Zahnley, J.C. (1986). Inactivation of metalloenzymes by food constituents. *Food Chem. Toxicol.*, *24*, 897–902.
134. Friedman, M., Grosjean, O.K., Zahnley, J.C. (1986). Inactivation of metalloenzymes by lysinoalanine, phenylethylaminoalanine, alkali-treated food proteins, and sulfur amino acids. *Adv. Exp. Med. Biol.*, *199*, 531–560.
135. Possompes, B., Berger, J. (1991). Effect of severely alkali treated casein on gastrointestinal transit and selected intestinal enzyme activities, in *Nutritional and Toxicological Consequences of Food Processing* (ed. M. Friedman), Plenum, New York, pp. 517–530.
136. Friedman, M., Finot, P.A. (1991). Improvement in the nutritional quality of bread. *Adv. Exp. Med. Biol.*, *289*, 415–445.
137. Friedman, M. (1996). Food browning and its prevention: an overview. *J. Agric. Food Chem.*, *44*, 631–653.
138. Sternberg, M., Kim, C.Y. (1979). Growth response of mice and *Tetrahymena pyriformis* to lysinoalanine-supplemented wheat gluten. *J. Agric. Food Chem.*, *27*, 1130–1132.
139. Robbins, K.R., Baker, D.H., Finley, J.W. (1980). Studies on the utilization of lysinoalanine and lanthionine. *J. Nutr.*, *110*, 907–915.
140. Finot, P.A., Bujard, E., Arnaud, M. (1977). Metabolic transit of lysinoalanine (LAL) bound to protein and of free radioactive 14C]-lysinoalanine. *Adv. Exp. Med. Biol.*, *86 B*, 51–71.
141. Struthers, B.J., Dahlgren, R.R., Hopkins, D.T. (1977). Biological effects of feeding graded levels of alkali treated soybean protein containing lysinoalanine (N(ε) 2 carboxyethyl] L lysine) in Sprague Dawley and Wistar rats. *J. Nutr.*, *107*, 1190–1199.
142. Struthers, B.J., Brielmaier, J.R., Raymond, M.L. (1980). Excretion and tissue distribution of radioactive lysinoalanine, Nε-DL-(2-amino-2-carboxyethyl)-U-^{14}C-l-lysine (LAL) in Sprague-Dawley rats. *J. Nutr.*, *110*, 2065–2077.
143. Kawamura, Y., Hayashi, R. (1988). Lysinoalanine degrading enzymes of various animal kidneys. *Agric. Biol. Chem.*, *51*, 2289–2290.
144. Gould, D.H., MacGregor, J.T. (1977). Biological effects of alkali-treated protein and lysinoalanine: an overview. *Adv. Exp. Med. Biol.*, *86 B*, 29–48.

145. Karayiannis, N.I., Panopoulos, N.J., Bjeldanes, L.F., Macgregor, J.T. (1979). Lysinoalanine utilization by *Erwinia chrysanthemi* and *Escherichia coli*. *Food Cosmet. Toxicol.*, *17*, 319–320.
146. Karayiannis, N.I., MacGregor, J.T., Bjeldanes, L.F. (1979). Lysinoalanine formation in alkali-treated proteins and model peptides. *Food Cosmet. Toxicol.*, *17*, 585–590.
147. Slump, P., van Beek, L., Janssen, W.M., Terpstra, K., Lenis, N.P., Smits, B. (1977). A comparative study with pigs, poultry and rats of the amino acid digestibility of diets containing crude protein with diverging digestibilities. *Z. Tierphysiol. Tierernahr. Futtermittelkd.*, *39*, 257–272.
148. Slump, P. (1977). Determination of lysinoalanine with an automatic amino-acid analyzer. *J. Chromatogr.*, *135*, 502–507.
149. Sternberg, M., Kim, C.Y. (1977). Lysinoalanine formation in protein food ingredients. *Adv. Exp. Med. Biol.*, *86 B*, 73–84.
150. Woodward, J.C., Short, D.D. (1977). Renal toxicity of N(ε) (DL 2 amino 2 carboxyethyl) L lysine (Lysinoalanine) in rats. *Food Cosmet. Toxicol.*, *15*, 117–119.
151. Woodard, J.C. (1975). Renal toxicity of N(ε) (DL 2 amino 2 carboxyethyl) L lysine, lysinoalanine. *Vet. Pathol.*, *12*, 65–66.
152. Kolonkaya, D. (1986). Renal toxicity of soybean protein containing lysinoalanine. *Doga: Turk. Biyol. Derg.*, 394–402.
153. O'Donovan, C.J. (1976). Recent studies of lysinoalanine in alkali treated proteins. *Food Cosmet. Toxicol.*, *14*, 483–489.
154. Friedman, M. (1991). Formation, nutritional value, and safety of D-amino acids. *Adv. Exp. Med. Biol.*, *289*, 447–481.
155. Young, G.A., Kendall, S., Brownjohn, A.M. (1994). D-Amino acids in chronic renal failure and the effects of dialysis and urinary losses. *Amino Acids*, *6*, 283–293.
156. Leegwater, D.C. (1978). The nephrotoxicity of lysinoalanine in the rat. *Food Cosmet. Toxicol.*, *16*, 405.
157. Jonker, D., Woutersen, R.A., Van Bladeren, P.J., Til, H.P., Feron, V.J. (1993). Subacute (4-wk) oral toxicity of a combination of four nephrotoxins in rats: comparison with the toxicity of the individual compounds. *Food Chem. Toxicol.*, *31*, 125–136.
158. Jonker, D., Woutersen, R.A., Feron, V.J. (1996). Toxicity of mixtures of nephrotoxicants with similar or dissimilar mode of action. *Food Chem. Toxicol.*, *34*, 1075–1082.
159. Lifsey B.J.Jr, , Farkas, W.R., Reyniers, J.P. (1988). Interaction of lysinoalanine with the protein synthesizing apparatus. *Chem. Biol. Interact.*, *68*, 241–257.
160. Friedman, M. (ed.) (1974). *Protein-Metal Interactions*, Plenum Press, New York, p. 692.
161. Rehner, G., Walter, T. (1991). Effect of Maillard products and lysinoalanine on bioavailability of iron, copper and zinc. *Z. Ernahrungswiss.*, *30*, 50–55.
162. Friedman, M., Grosjean, O.K., Zahnley, J.C. (1985). Inactivation of metalloenzymes by lysinoalanine, phenylethylaminoalanine, alkali-treated food proteins, and sulfur amino acids, in *Nutritional and Toxicological Significance of Enzyme Inhibitors in Foods* (ed. M. Friedman), Plenum Press, New York, pp. 531–560.
163. Furniss, D.E., Vuichoud, J., Finot, P.A., Hurrell, R.F. (1989). The effect of Maillard reaction products on zinc metabolism in the rat. *Br. J. Nutr.*, *62*, 739–749.

164. Sarwar, G., L'Abbe, M.R., Trick, K., Botting, H.G., Ma, C.Y. (1999). Influence of feeding alkaline/heat processed proteins on growth and protein and mineral status of rats. *Adv. Exp. Med. Biol.*, *459*, 161–177.

165. Langhendries, J.P., Hurrell, R.F., Furniss, D.E., Hischenhuber, C., Finot, P.A., Bernard, A., Battisti, O., Bertrand, J.M., Senterre, J. (1992). Maillard reaction products and lysinoalanine: urinary excretion and the effects on kidney function of preterm infants fed heat-processed milk formula. *J. Pediatr. Gastroenterol. Nutr.*, *14*, 62–70.

166. Friedman, M., Boyd, W.A. (1977). A nuclear magnetic double resonance study of N-beta-bis-(beta'-chloroethyl) phosphonylethyl-DL-phenylalanine. *Adv. Exp. Med. Biol.*, *86 A*, 727–743.

167. Friedman, M., Levin, C.E. (2008). Review of methods for the reduction of dietary content and toxicity of acrylamide. *J. Agric. Food Chem.*, *56*, 6113–6140.

168. Cattaneo, S., Masotti, F., Pellegrino, L. (2008). Effects of overprocessing on heat damage of UHT milk. *Eur. Food Res. Technol.*, *226*, 1099–1106.

169. Corpet, D.E., Taché, S., Archer, M.C., Bruce, W.R. (2008). Dehydroalanine and lysinoalanine in thermolyzed casein do not promote colon cancer in the rat. *Food Chem. Toxicol.*, *46*, 3037–3042.

170. Danalev, D., Koleva, M., Ivanova, D., Vezenkov, L., Vassilev, N. (2008). Synthesis of two peptide mimetics as markers for chemical changes of wool's keratin during skin unhairing process. *Protein and Peptide Letters*, *15*, 353–355.

171. Bosch, L., Sanz, M.L., Montilla, A., Alegría, A., Farré, R., del Castillo, M.D. (2007). Simultaneous analysis of lysine, N^{ε}-carboxymethyllysine and lysinoalanine from proteins. *Journal of Chromatography B: Analytical Technologies in the Biomedical and Life Sciences*, *860*, 69–77.

172. Somoza, V., Wenzel, E., Weiß, C., Clawin-Rädecker, I., Grübel, N., Erbersdobler, H.F. (2006). Dose-dependent utilisation of casein-linked lysinoalanine, N(epsilon)-fructoselysine and N(epsilon)-carboxymethyllysine in rats. *Mol. Nutr. Food Res.*, *50*, 833–841.

6.2

DIETARY SIGNIFICANCE OF PROCESSING-INDUCED D-AMINO ACIDS

Mendel Friedman

Western Regional Research Center, Agricultural Research Service, United States Department of Agriculture, Albany, CA 94710, USA

6.2.1 INTRODUCTION

During food processing, naturally occurring L-amino acids may be transformed to their mirror image configuration D-isomers. Microorganism-synthesized D-amino acids are estimated to constitute about one-third of the human D-amino acid burden (1). A better understanding of the dietary significance of D-amino acids requires knowledge about the factors that induce racemization of L- to D-amino acids in food proteins during food processing. Processing-induced amino acid racemization includes those formed during exposure of food proteins to high pH, heat, and acids. The nutritional effectiveness of protein-bound essential D-amino acids depends on the amino acid composition, digestibility, and physiological utilization of released amino acids. Since an amino acid must be liberated by digestion before nutritional utilization can occur, the decreased susceptibility to digestion by proteolytic enzymes of D-D, D-L, and L-D peptide bonds in D-amino acid-containing proteins is a major factor adversely affecting the bioavailability of protein-bound D-amino acids (2–29). In addition, some D-amino acids may be toxic. For example, although D-Phe is nutritionally available as a source of L-Phe, high levels of D-Tyr

Process-Induced Food Toxicants: Occurrence, Formation, Mitigation, and Health Risks,
Edited by Richard H. Stadler and David R. Lineback

D-Amino Acids: At a Glance	
Historical	The racemization of amino acids in alkali-treated foods was first reported in the late 1970s.
Analysis	Several analytical methods are available to measure the content of D-amino acids in foods, including chiral phase GC, HPLC, and capillary electrophoresis.
Occurrence in Food	Widespread occurrence in many different foods, e.g., in (i) cows milk, originating from D-amino acids present as constituents of bacterial proteins in cows rumen, (ii) sourdough, (iii) in food products that contain monosodium glutamate as flavor enhancer, and (iv) other processed foods such as fruit juices, soy sauce, coffee, beer, and wine.
Main Formation Pathways	Formed during food processing, especially upon exposure of proteins to high heat, acid, or alkali conditions. It is estimated that one third of the human D-amino acid burden originates from micro-organisms.
Mitigation in Food	Less severe acid / alkali or heat treatments of food constituents or foods for the non-microbial sources.
Nutritional Consequences	Racemized food proteins appear to be less digestible. In mice, the relative nutritional value of several D-amino acids vs. the L-form is <20% based on growth response studies. Similar observations are made for peptides that harbor D-amino acids.
Health Risks	Rat studies and *in vitro* models have reported various toxic effects for certain D-amino acids including oxidative stress and nephrotoxicity. However, several beneficial effects have also been observed, including enhancement of detoxification mechanisms and inhibition of cell tumor growth.
Regulatory Information or Industry Standards	No regulatory limits have been set in foods.
Knowledge Gaps	Understanding the biological utilization and effects of D-amino acids in the free and protein-bound forms, their impact on the microflora of the intestine, and mechanisms of potential positive action.
Useful Information (web) & Dietary Recommendations	No specific recommendations from a dietary perspective.

and D-Cys inhibit the growth of mice. D-Lys is not utilized as source of the L-isomer and D-Ser is toxic. Some D-amino acids have been shown to impart beneficial effects *in vivo*. It is not known whether the biological effects of D-amino acids vary, depending on whether they are consumed in the free state or as part of protein that contains processing-induced D-amino acids. To cross-fertilize information among several disciplines and to stimulate further needed studies, this review attempts to integrate and correlate the widely scattered literature on D-amino acid nutrition and safety.

6.2.1.1 Analysis of D-Amino Acids

Because all of the amino acid residues in a protein (illustrated in Fig. 6.2.1) undergo racemization simultaneously, but at differing rates, assessment of the extent of racemization in a protein requires quantitative measurement of ~40 L- and D-optical isomers. Reported methods include the use of chiral phase gas chromatography (24), HPLC (30), HPLC with enzyme reactors (31), LC/MS (32, 33), capillary electrophoresis (34, 35), electrokinetic chromatography with detection by laser-induced fluorescence (36), and biosensors (37). Detailed discussion of these and related methods is beyond the scope of this chapter (21, 37–42).

6.2.1.2 Racemization Mechanisms

Three different mechanisms have been postulated to govern racemizations of amino acids during food processing. Figure 6.2.2 shows a mathematical derivation of the first-order kinetic equations that govern reversible *in vitro* base-catalyzed amino acid racemizations that proceed by the mechanism depicted in Fig. 6.2.3a. Figure 6.2.3b depicts a postulated mechanism for acid catalyzed racemization and Fig. 6.2.3c, a new mechanism for heat-induced racemization via an Amadori compound (fructose-L-phenylalanine) intermediate proposed by Pätzold and Brückner (32). Further studies are needed to demonstrate

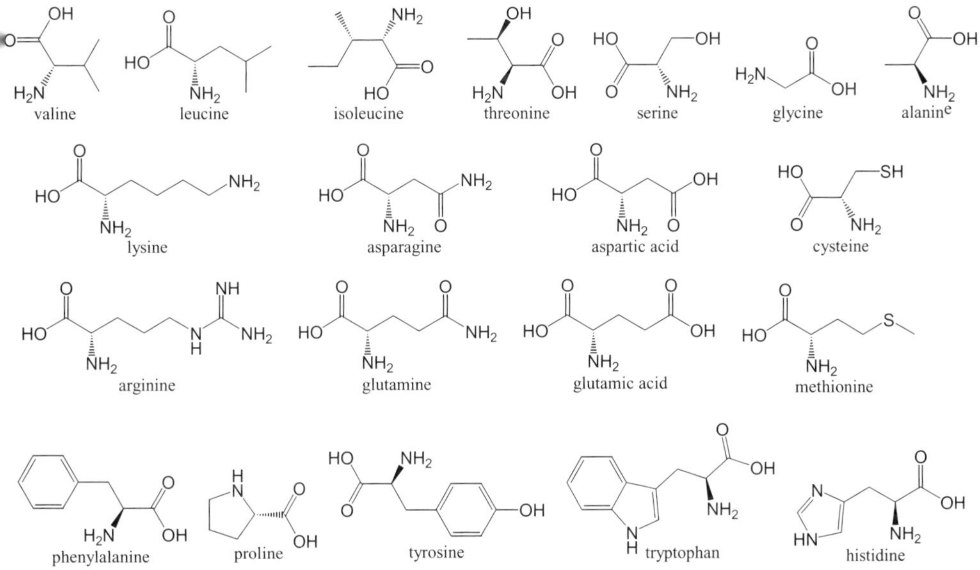

Figure 6.2.1 Structures of genetically coded amino acids subject to racemization during food processing.

$$\text{L-aminoacids} \underset{k'}{\overset{k}{\rightleftharpoons}} \text{D-aminoacids}$$

$$-dL/dt = kL - k'D$$

if $D_{t=0} \ll L_{t=0}$ then $L_{t=0} = L + D$ and $L_{t=0} - L = D$

$$-dL/dt = kL - k'(L_{t=0} - L) = kL - k'L_{t=0} + k'L$$

$$= (k + k')L - k'L_{t=0}$$

$$dL/dt = -(k + k')L + k'L_{t=0}$$

$$dL/dt + (k + k')L = k'L_{t=0}$$

$$dL/dt\, e^{(k+k')t} + (k + k')L\, e^{(k+k')t} = k'L_{t=0}\, e^{(k+k')t}$$

$$d/dt(Le^{(k+k')t}) = k'L_{t=0}\, e^{(k+k')t}$$

$$\int d(Le)(Le^{(k+k')t}) = \int (k'L_{t=0}\, e^{(k+k')t}) \cdot dt$$

$$Le^{(k+k')t} = [k/(k + k')]L_{t=0}\, e^{(k+k')t} + \text{constant}$$

$$L = [k'/(k + k')]L_{t=0} + \text{constant} \cdot e^{-(k+k')t}$$

At $t = 0$, $L = L_{t=0}$ and $e^{-(k+k')t} = 1$

$$L_{t=0} = [k'/(k + k')]L_{t=0} + \text{constant}$$

$$L_{t=0} - [k'/(k + k')]L_{t=0} = \text{constant}$$

$$L_{t=0}[1 - k'/(k + k')] = \text{constant}$$

$$L = [k'/(k + k')]L_{t=0} + L_{t=0}\, k/(k + k')e^{-(k+k')t}$$

if $L_{t=0} = L + D$

$$L = [k'/(k + k')](L + D) + (L + D)\, k/(k + k')e^{-(k+k')t}$$

$$L/(L + D) = k'/(k + k') + [k/(k + k')]e^{-(k-k')t}$$

$$\frac{L(k + k') - k'(L + D)}{(L + D)(k + k')} = [k/(k + k')]e^{-(k+k')t}$$

$$= \frac{Lk - Dk'}{(L + D)(k + k')}$$

$$\left[\frac{L - D(k'/k)}{L + D}\right] = e^{-(k+k')t}$$

$$= \left[\frac{L[(1 - (D/L)(k'/k)]}{L[(1 + (D/L)]}\right]$$

$$= \left[\frac{1 - (D/L)(k'/k)}{1 + D/L}\right]$$

$$\ln\left[\frac{1 - (D/L)(k'/k)}{1 + (D/L)}\right] = -(k + k')t$$

$$\ln\left[\frac{1 + (D/L)}{1 - (D/L)(k'/k)}\right] = -(k + k')t$$

if $k'/k = K' = 1/K_{eq}$; $K_{eq} = D/L$ ratio at equilibrium $= 1$

$$\ln\left[\frac{1 + D/L}{1 - K'(D/L)}\right] = (1 + K') \cdot k \cdot t \qquad \text{(eq. 1)}$$

An alternate mathematical treatment of the first-order kinetics of racemization gives equation 2. Equation 1 is operationally more efficient than equation 2 because we can measure D/L ratios with a 1-3% error compared to a 10-15% error when concentrations of D only are measured.

$$\ln(D_e/D_e - D_t) = (k + k')t = k_{(obs)}t \qquad \text{(eq. 2)}$$

where D_e = equilibrium value of D and $D_t = D$ at time t

Figure 6.2.2 Derivations of operational first-order kinetic equations for reversible amino acid racemization. Adapted from References 2 and 3.

the generality of this mechanism with other Amadori compounds. Figure 6.2.4 depicts a postulated mechanism for the *in vivo* inversion of an L-amino acid to its D-isomer catalyzed by pyridoxal phosphate and Figs 6.2.5–6.2.8 illustrate trends in racemization in soy proteins and coffee.

Table 6.2.1 lists major food categories that have been reported to contain D-amino acids and Table 6.2.2 shows the D-amino acid content of eight alkali-treated proteins. The cited observations indicate that heat and high pH widely used in processing of food induce the racemization of all amino acid residues by the indicated pathways. Table 6.2.3 summarizes the biological utilization of selected individual D-amino acids and peptides in mice.

6.2.2 DISTRIBUTION OF D-AMINO ACIDS IN FOOD

Masters and Friedman (3) found that the percent of D-Asp of the total Asp was 9 for textured soy protein, 10 for an infant soy formula, 13 for simulated breakfast strip bacon, and 17 for an imitation coffee cream whitener. This apparent first observation on the presence of D-amino acids in commercial foods was followed by numerous studies, summarized in Table 6.2.2, on the content of D-amino acids in a variety of foods. The possible origin of these D-amino acids and their dietary significance will be examined in this section.

6.2.2.1 Alfalfa Seeds

Ono *et al.* (50) discovered the alanine racemase-catalyzed formation of D-Ala in alfalfa seeds (*Medicago sativa* L.). The possible role of D-Ala in the physiology of the plant is not known.

6.2.2.2 Bread

Use of lactic acid bacteria and yeast in the fermentation of sourdough before baking results in the introduction of D-Ala and D-Glu into the dough (51). Baking of the dough into bread induces a 44% decrease in the total free D-amino acid content.

6.2.2.3 Dairy Products

Raw milk from ruminants (cows, goats, and sheep), but not human milk, contains the following D-amino acids: D-Ala, D-Asp, D-Glu, D-Lys, and D-Ser (52). D-amino acids in cow's milk originate from the digestion (autolysis) of peptides and proteins containing D-amino acid originating from microbial cell walls (peptidoglycan) proteins in the rumen of the cows. The origin of the D-amino acids in the other milks may arise from a different, unknown physiological event (53). Relatively high concentrations of these D-amino acids were also present in widely consumed ripened cheeses (30). The D-amino acid content varied among several cheeses evaluated, and changed during cheese production (54).

The fact that the D-Ala content of raw milk increased during storage at 4°C implies that D-Ala could serve as an indicator of contamination of milk by D-Ala-producing psychotrophic bacteria. Because mastitis is an infection of the udder caused by bacteria, milk from infected cows should have a higher D-amino acid content originating from the microorganisms. Pasteurization of milk does not seem to increase its D-amino content.

6.2.2.4 Fruits and Vegetables

Fruits (apples, grapes, oranges) and vegetables (cabbage, carrots, garlic, tomatoes) as well as the corresponding juices contain variable but measurable

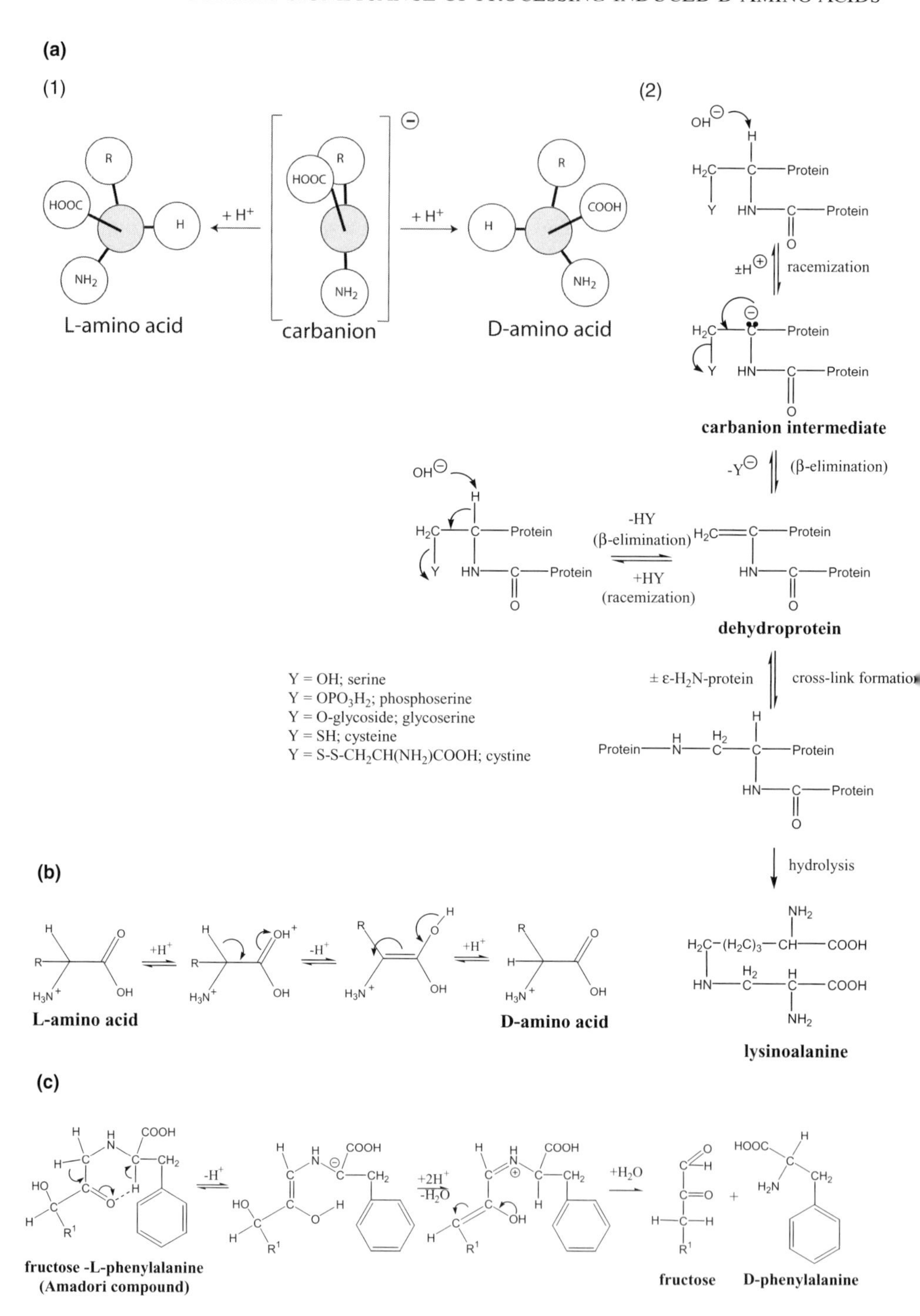
(a)
(1)
L-amino acid
carbanion
D-amino acid
(2)
racemization
carbanion intermediate
(β-elimination)
-HY
(β-elimination)
+HY
(racemization)
dehydroprotein
Y = OH; serine
Y = OPO3H2; phosphoserine
Y = O-glycoside; glycoserine
Y = SH; cysteine
Y = S-S-CH2CH(NH2)COOH; cystine
± ε-H2N-protein
cross-link formation
hydrolysis
lysinoalanine
(b)
L-amino acid
D-amino acid
(c)
fructose -L-phenylalanine
(Amadori compound)
fructose
D-phenylalanine

Figure 6.2.3 (a) Mechanisms of base-catalyzed racemization. (1) Proton abstraction-addition mechanism. An OH^- ion abstracts an H^+ from the RCH- of an amino acid [(RCH(NH_2)COOH] to form a negatively charged carbanion, which has lost its original asymmetry. The carbanion can then recombine with a proton from the solvent to regenerate the original amino acid, which is now (DL). Adapted from References 43 and 44. (2) Elimination-addition mechanism. The carbanion can also undergo an elimination reaction to form a dehydroalanine side chain. The resulting dehydroprotein can then react with H_2O to form a racemized amino acid side chain (e.g., DL-Ser). The concurrent formation of lysinoalanine is also shown. Adapted from Reference 45. (b) Acid-catalyzed racemization. Protonation of the carboxyl group of an L-amino acid facilitates removal of a proton to form dehydroalanine. Proton addition at two sides (*re* and *si* faces) of the double bond generates both D- and L-isomers. Adapted from Reference 46. (c) Heat-induced racemization of an Amadori compound. The indicated hydrogen bonding between the carbonyl group of the fructose moiety of the Amadori compound and the hydrogen atom of the asymmetric carbon atom of phenylalanine facilitates formation of a carbanion, which then undergoes reprotonation to form a D,L-mixture. Hydrolysis of the D,L-intermediate then takes place to regenerate fructose plus a mixture of L- and D-phenylalanine. Adapted from Reference 32.

amounts of D-amino acids including D-Ala, D-Arg, D-Asp, and D-Glu (30, 52). These amino acids could originate from plant sources, from soil and other microorganisms, and/or from heat treatments used to pasteurize juices. Specific D-amino acids could permit differentiating juices from biologically dissimilar fruits and could serve as an indicator for detecting bacterial activity and shelf life of fruit juices.

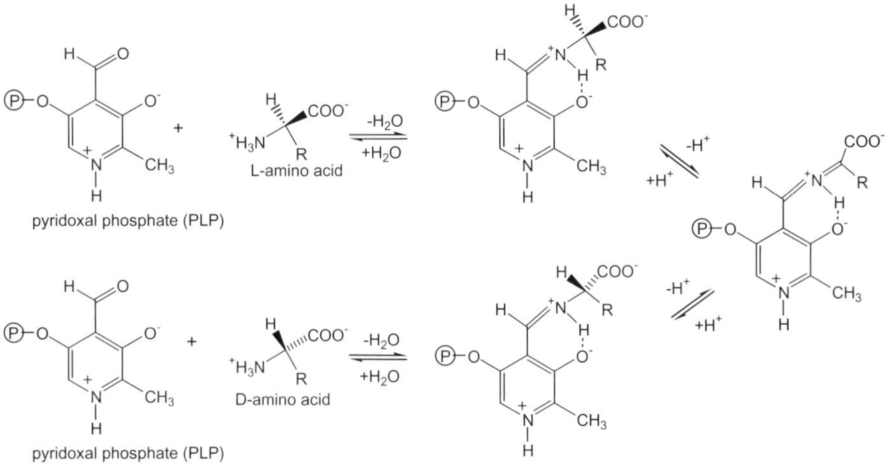

Figure 6.2.4 Catalysis of *in vivo* inversion of an L-amino acid by the pyridoxal phosphate (PLP) coenzyme of a microbial racemase involving stereoselective shifts of H atoms. Adapted from Reference 47.

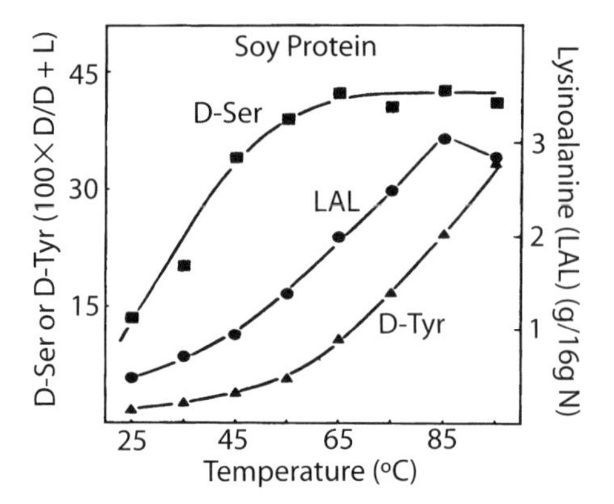

Figure 6.2.5 Effect of temperature on content of D-Ser, D-Tyr, and LAL of soybean protein. Adapted from References 18, 23, and 48.

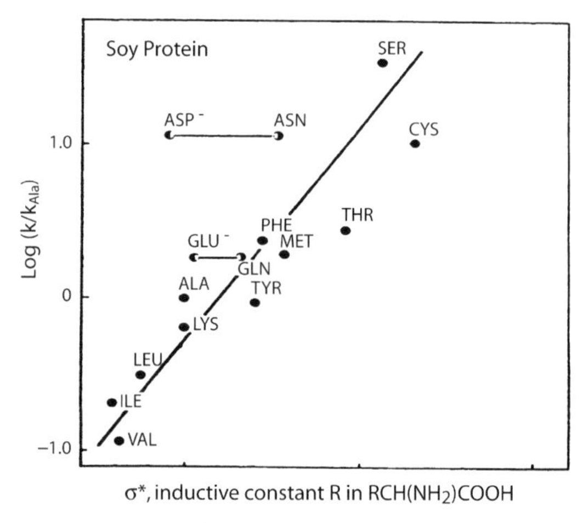

Figure 6.2.6 Relationship between racemization rates (k) relative to Ala and the inductive constant (σ*) of the amino acid side chain R in $RCH(NH_2)COOH$. Adapted from Reference 18. The indicated relative rates for soy protein are similar to those observed with other food proteins (46).

6.2.2.5 D-Glutamic Acid in Processed Foods

Monosodium glutamate (MSG), the sodium salt of the naturally occurring nonessential amino acid, is often added at levels of 0.2–0.9% to foods to improve flavor and palatability. A survey by Rundlett and Armstrong (55) showed that a variety of processed foods contain significant amounts of D-Glu (which does not possess flavor-enhancing properties). The %D ranged from 0.25 for pure MSG to 0.8 for soups, 1.6 for tomato products, 2.9 for crackers, 6.2 for milk products, 7.9 for sauces, 18 for vinegars, and 36 for sauerkraut juices.

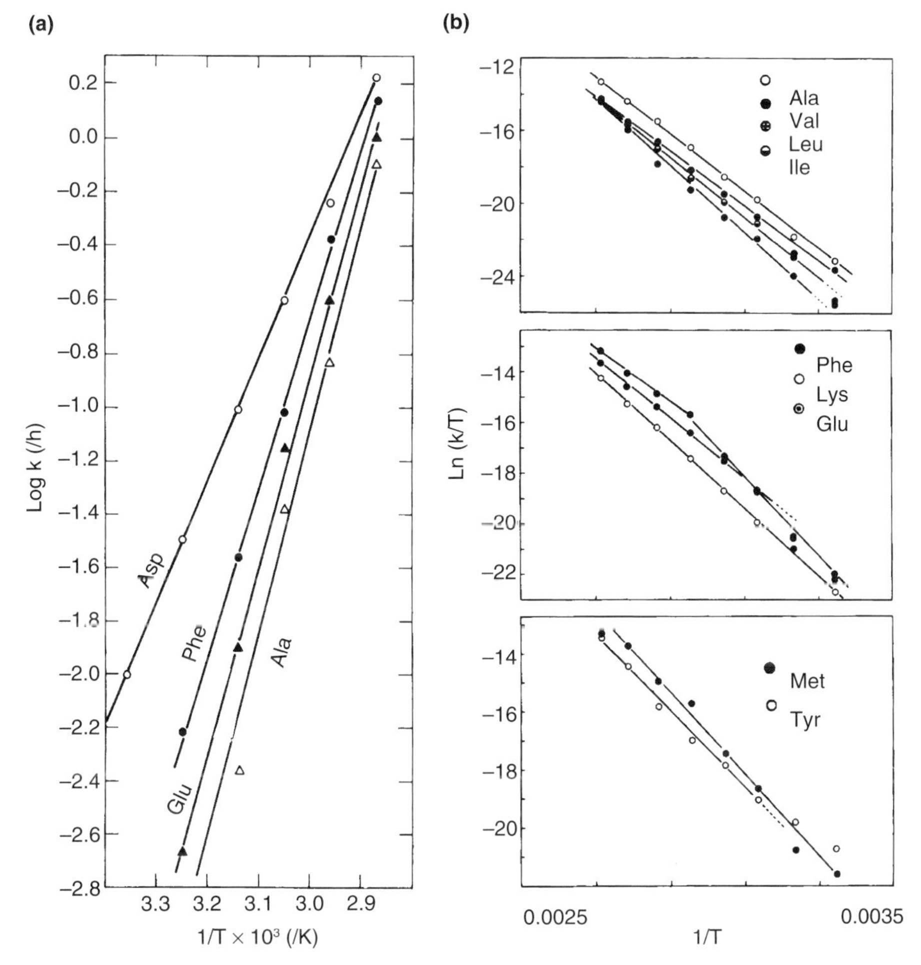

Figure 6.2.7 Arrhenius plots showing (a) time and (b) temperature dependence of the racemization of protein amino acid residues in soy proteins. Adapted from References 18 and 19.

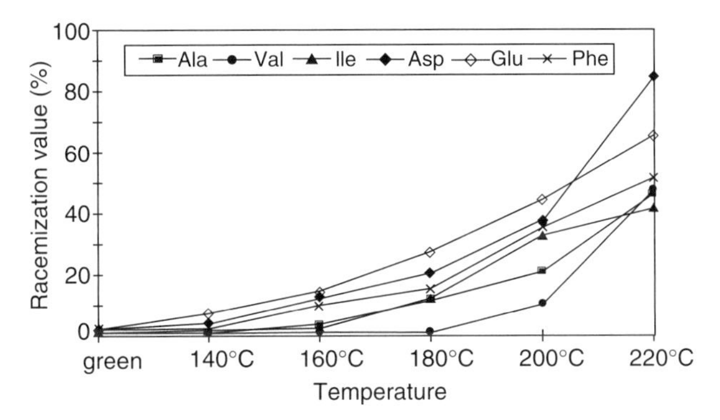

Figure 6.2.8 Heat-induced amino acid racemization in coffee. Adapted from Reference 49.

TABLE 6.2.1 Selected foods containing D-amino acids.

Alcoholic beverages: beer, wines, sake, vinegars
Alfalfa seeds
Bacterial starter cultures
Baked products: bread, cereals, crackers, dough, whcat flour
Bean products (legumes); soy flour (fermented)
Cacao powder; coffee; cream (sodium caseinate)
Corn (maize); corn meal: tacos and tortillas
Dairy products; milks; cheeses; infant formulas
Fish and fish products
Fruits and vegetables; fruit and vegetable juices
Plant saps and syrups
Meat and meat products
Sauces: mustard, pepper, soy
Soups; spices; flavor enhancers

The D-Glu could originate from microbial sources as well as from food-processing-induced racemization of free and protein-bound L-Glu. Table 6.2.2 shows that protein-bound L-Glu is one of the fastest racemizing amino acids. The possible impact of the D-isomer of MSG on the "Chinese Restaurant Syndrome" apparently has not been evaluated.

6.2.2.6 Eggs

Alkali pickling of duck eggs in a 4.2% NaOH/5% NaCl solution for 20 days at room temperature used to prepare the traditional Chinese pidan resulted in extensive racemization of amino acid residues and concurrent formation of lysinoalanine (56). The relative racemization rates appear to be similar to those observed with egg albumin and other pure proteins.

6.2.2.7 Honey

The D/L ratios of Leu, Phe, and Pro could serve as indicators of age, processing, and storage histories of honeys (57).

6.2.2.8 Fish Meal

Heating of laboratory-made herring meals at 125°C induced time-dependent formation of D-Asp (58). The D-Asp acid content of 12 fish meals from various sources ranged from 0.7% to 3.7%. The D-Asp content may indicate the severity of the thermal treatment during cooking or drying of fish meals (59).

6.2.2.9 Maize (Corn Grain)

Significant differences were observed in the D-amino acid content of some transgenic maize (corn) cultivars compared with standard varieties (60). The

TABLE 6.2.2 D-Amino acid content [(D/D + L) × 100) = %D] of eight alkali-treated proteins.[a]

Amino acid	Casein	Lactalbumin	Wheat gluten	Corn protein	Fish protein	Soybean protein	Bovine albumin	Hemoglobin
Ala	15.2	14.4	18.6	22.2	19.3	15.8	22.1	17.1
Val	2.6	2.7	4.0	4.9	3.1	2.5	3.5	4.0
Leu	7.4	5.0	7.2	7.8	6.8	6.3	8.2	6.6
Ile	3.3	3.1	4.0	5.5	3.6	3.9	5.7	5.0
Cys	—	32.1	32.0	43.7	22.8	21.0	23.0	30.0
Met	24.7	32.3	33.1	29.8	29.2	24.3	30.0	26.2
Phe	24.4	24.3	24.4	32.4	28.0	25.5	28.1	30.0
Lys	8.1	7.2	9.4	8.0	11.5	11.3	13.3	9.9
Asp	29.2	22.6	25.6	41.6	25.0	30.8	27.0	18.9
Glu	19.7	19.5	32.3	35.0	18.9	21.1	18.4	19.8
Ser	41.0	47.1	42.2	44.0	42.1	44.2	43.0	44.5
Thr	29.3	29.1	30.0	36.3	32.8	27.8	28.3	31.2
Tyr	15.0	18.9	19.5	35.5	16.3	13.7	15.3	22.6
LAL[b]	4.4	5.4	0.9	0.3	2.8	3.2	8.5	4.4

[a]Adapted from References 18, 19, 23, and 24.
[b]Mixture of (LD + LL) LAL isomers in g/16gN.

TABLE 6.2.3 Growth responses to D-amino acids and peptides in mice fed amino acid diets.

D-Amino acids and peptides	Relative nutritional value to L-form, %
D-Amino acids	
D-Methionine	79.5
D-Phenylalanine	51.6
D-Tryptophan	24.7
D-Leucine	12.4
D-Histidine	8.5
D-Valine	5.1
D-Threonine	3.1
D-Isoleucine	1.2
D-Lysine	–8.2
L,L + D,L-Lysinoalanine	3.8
Peptides and sulfoxides	
N-Acetyl-L-methionine	89.6
N-Acetyl-D-methionine	22.9
L-Methionyl-L-methionine	99.2
L-Methionyl-D-methionine	102.9
D-Methionyl-L-methionine	80.4
D-Methionyl-D-methionine	41.0
L-Methionine sulfoxide	87.5
D-Methionine sulfoxide	26.8
N-Acetyl-L-methionine sulfoxide	58.7
N-Acetyl-D-methionine sulfoxide	1.7
L-Methionine hydroxyl analog	55.4
D-Methionine hydroxyl analog	85.7

Adapted from References 8, 12–15, and 17.

authors suggest that this finding opens up new approaches that may permit comparing the composition of transgenic plant foods with their conventional counterparts in terms of their respective D-amino acid content.

6.2.2.10 Other Foods

Variable amounts of D-amino acids have been reported to be present in other fermented and processed foods including beer, coffee, vinegar, wine, bread toast, hamburger meat, ham, liquid spices, powdered milk, soy paste and soy sauce, plant saps, syrups, and fruit juice concentrates, alfalfa meals, collagen, and animal hair and skins (31–34, 46). Roasted coffee contained 10–40% of D-Asp, D-Glu, and D-Phe (61) (Fig. 6.2.8).

6.2.3 NUTRITION AND SAFETY

As part of a program to evaluate the chemistry and the nutritional and toxicological potential of novel amino acids, including D-amino acids, formed during food processing, we compared the weight gain in mice fed free amino acid diets in which the test D-amino acid was substituted for the L-isomer. The results obtained reflect the ability of mice to utilize D-amino acids in the complete absence of the L-form. In the case of essential amino acids, the mice must meet the entire metabolic demand from the D-isomeric forms.

Specifically, biological utilization by weanling male mice of D-amino acids was tested in a 14-day growth assay by using an all-free amino acid diet in which each L-amino acid was replaced by the D-isomer. The growth assay was standardized with six mice per group, with potency estimation being based of response of two to seven groups. Statistical evaluation of potencies of the D-amino acids was calculated as the slopes of growth curves (weight gain after 14 days per unit of dietary concentration) where linearity was approximated. Mean body weight gains were compared by Duncan's multiple range test using individual values. The results illustrated in Figs 6.2.9–6.2.11 show wide variation in the utilization of D-amino acids as a nutritional source of the corresponding L-isomers.

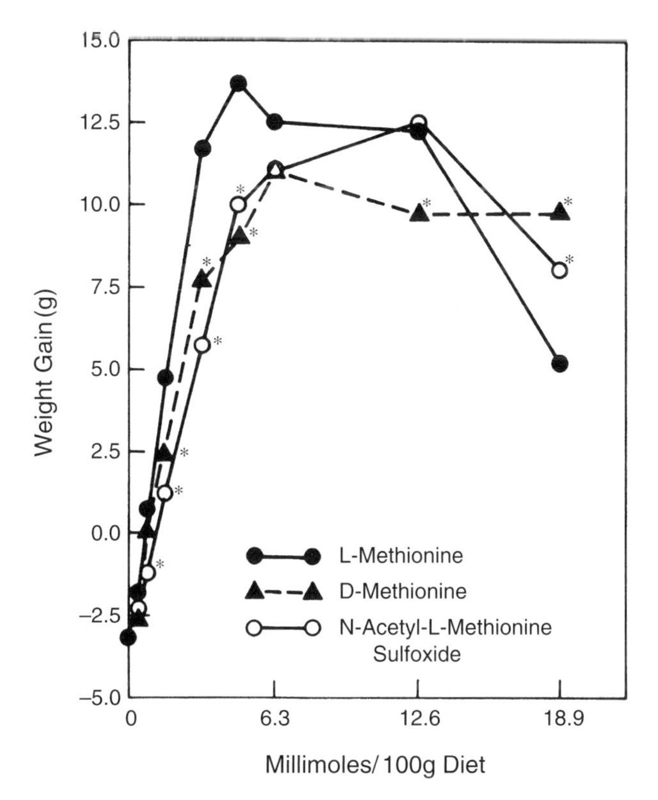

Figure 6.2.9 Weight gain in mice fed increasing dietary levels of L-Met, D-Met, and an isomeric methionine derivative for 14 days. Asterisk indicates significant difference from L-Met at the same dietary concentration. Adapted from References 14 and 15.

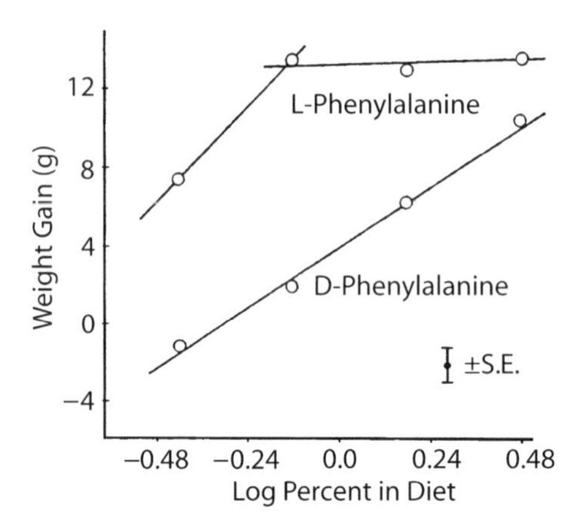

Figure 6.2.10 Relationships of weight gains to percent of L- and D-Phe isomers in amino acid diets fed to mice. The ranges for the S.E. value shown apply to both plots. Adapted from References 13, 62, and 63.

6.2.3.1 Biological Utilization and Safety of D-Amino Acids

6.2.3.1.1 Digestion of Racemized Proteins Friedman *et al.* (22) attempted to quantitatively define the susceptibility of alkali-treated casein to *in vitro* digestion by trypsin and chymotrypsin. Monitoring on a pH-stat, we observed an approximately inverse relationship between D-amino acid and lysinoalanine content on one hand and the extent of proteolysis measured on the other. The decreased susceptibility of alkali-treated casein to digestion could arise from loss of susceptible sites to enzyme cleavage. This aspect is examined in more detail in Chapter 6.1 on lysinoalanine.

6.2.3.2 Nutrition, Toxicity, and Medical Applications of Free D-Amino Acids

To facilitate understanding the complex nature of biological activities of D-amino acids, we integrate our results and results from selected published studies by other investigators into the following nutrition- and health-related aspects of D-amino acids listed alphabetically.

6.2.3.2.1 D-Alanine D-Ala induced cytotoxic oxidative stress in brain tumor cells (64). Because it is part of the structure of bacterial cell walls, D-Ala derived from microbial sources is present in many foods as well as in hydrolyzed protein fertilizers (34). The submandibular gland and oral epithelial cells, not ingested food or oral bacteria, appear to be the source of high levels of D-Ala and D-Asp in human saliva (65). D-Ala is a possible indicator of bacterial contamination (spoilage), heat treatment, and shelf life of fruit juices and other foods (31, 52). It can also be used in studies of molecular imaging (66).

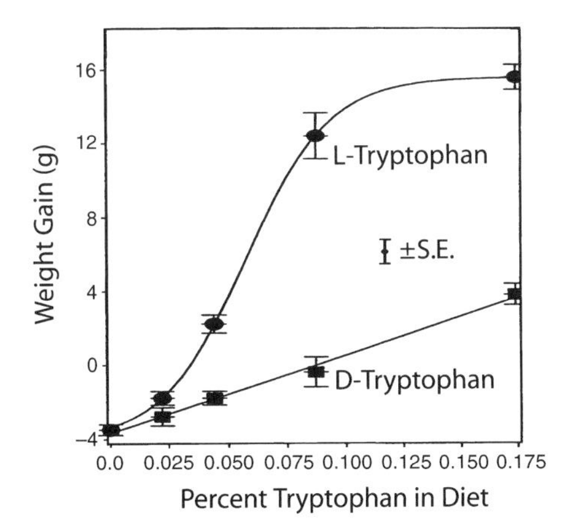

Figure 6.2.11 Relationships of weight gains to percent of L- and D- Trp isomers in amino acid diets fed to mice. Adapted from References 13, 62, and 63.

6.2.3.2.2 D-Arginine Both L- and D-Arg protected against oxygen-radical-induced injury of rat heart tissue (67) and against endotoxin shock in rabbits (68). D-Arg and other D-amino acids inhibited cell proliferation and tumor growth in rats (69), acted as a central nervous system stimulant, and exhibited anticonvulsant activity in humans (70). A D-Arg-containing peptide protected young chicks against memory loss (71). These observations suggest that dietary D-Arg may benefit human health.

6.2.3.2.3 D-Aspartic Acid D-Asp aggravated the pyelonephritis in rats induced by *Staphylococcus aureus* bacteria (72). Because D-Asp also prevented potassium and magnesium depletion in rats induced by cardiac drugs and diuretics (73), it may be of benefit to cardiac patients.

6.2.3.2.4 D-Cysteine Although L-CysSH had a sparing effect of L-Met consumed by mice, D-CysSH did not (14). In fact, D-CysSH imposed a metabolic burden as indicated by depressed growth when fed to mice with less than optimal levels of D-Met. The 24% decrease in weight gain of the D-CysSH plus L-Met amino-acid diet compared with L-Me alone implies that D-Cys is nutritionally antagonistic or toxic.

D-CysSH but not *N*-acetyl-D-cysteine lowered rat blood cyanide levels derived from acrylonitrile (21, 74, 75). D-CysSH is also reported to be involved in the detoxification and/or prevention of toxicities caused by other cyanides (76), the drug paracetamol (77), and other drugs (78–80). L-Cysteine-glutathione disulfide but not the D-cysteine analog protected mice against acetaminophen-induced liver damage (81).

6.2.3.2.5 D-Cystine Although L-Cys is not an essential amino acid for rodents, less L-Met is needed for growth if the diet contains L-Cys (14). Our results show that L-Cys is somewhat more efficient in sparing D-Met than in sparing L-Met in diets containing low levels (0.29%) of the two isomers. Supplementation of D-Met with an equal sulfur equivalent of L-Cys doubled growth. Thus, the overall response was equal to that produced by L-Met in the presence of L-Cys. In contrast, supplementation of suboptimal levels of L-Met with increasing concentrations of D-Cys reduced the growth rate of mice. These results imply that excess D-Cys in the diet is toxic.

6.2.3.2.6 D-Histidine D-His enhanced zinc accumulation, but reduced the fraction of zinc that was retained and absorbed by fish (82). Both D- and L-His enhanced the DNA degradation by hydrogen peroxide and ferric ions (83). D-His-induced cell injury is mediated by an iron-dependent formation of reactive oxygen species (84, 85). These observations suggest that D-His may be toxic to humans.

6.2.3.2.7 Lanthionine (LAN) Isomers LAN isomers and other D-amino acids are formed during the biosynthesis of microbial-derived peptide antibiotics (86–90) and during exposure of proteins to alkali and heat. Reaction of the SH group of CysSH and the double bond of dehydroalanine gives rise to one pair of optically active D- and L-isomers and one diastereomeric (*meso*) form of LAN (21). The mixture of DL + *meso*-LAN has a sparing effect on L-Met, as evidenced by a 27% greater weight gain when the two amino acids were fed together, than when fed suboptimal L-Met (14, 15).

6.2.3.2.8 D-Lysine In contrast to sulfur-containing amino acids, D-Lys is not utilized as a nutritional source of L-Lys by chicks, dogs, mice, rats, or humans, presumably because D-amino oxidase does not metabolize D-Lys.

Because of its low toxicity, D-Lys may be a better candidate to reduce radioactivity uptake by the kidneys during cancer therapy with radionuclides than is L-Lys (91–93).

6.2.3.2.9 Lysinoalanine (LAL) Isomers The amino acid LAL, formed concurrently with D-amino acids in alkali-treated proteins, has two asymmetric C-atoms, making possible four separate diastereoisomeric forms: DD, DL, LD, and LL. The nutritional value of the individual LAL isomers as a source of L-Lys is not known. Table 6.2.3 shows that a mixture of LL and LD isomers has a nutritional value for the mouse equivalent on a molar basis to 3.8% of L-Lys. For comparison, the table also lists the nutritional values of other D-amino acids we determined.

The four LAL isomers differ in their ability to chelate metal ions such as copper (94). The transformation of even a small fraction of L-Lys to D-Lys and to LAL adversely affects the nutritional quality of cereal proteins to a greater extent than would be the case for legume (soy) and animal (casein) proteins, whose L-Lys content is much higher (4, 95, 96).

6.2.3.2.10 D-Methionine The nutritional value of D-Met in mice approaches that of the L-Met (Table 6.2.2). D-Met appears to be poorly utilized by humans when consumed either orally or during total parenteral nutrition (TPN). One factor giving rise to inconsistencies in the utilization of D-Met is the dose dependency of the apparent potency of D-Met relative to its L-isomer, i.e., the dietary level of the D-form for any given growth response relative to that of the L-form, which would produce the same growth response. This dose dependency is a result of the nonlinear nature of the dose–response curves (Fig. 6.2.9). This complicates attempts to compare results from mice with those of other animal species. The latter studies often report data based on a single substitution of the D- for the L-isomer. The figures show that high levels of L-Met (but not of D-Met) are toxic since they inhibited growth of mice.

D-Met-containing solutions inhibited tumor cell growth *in vitro* (97) and protected against cisplatin ototoxicity (98, 99). A cascade of enzyme transformed D-Met to the L-form (100).

6.2.3.2.11 D-Phenylalanine The relative growth rate of mice fed D-Phe replacing the L-isomer in a free-amino-acid diet is concentration-dependent, ranging from 28.3% to 81.3% when compared with control diets containing the same amounts of L-Phe (Fig. 6.2.10) (13). The data suggest the absence of any antinutritional effects or toxicity from feeding either Phe isomer at twice the optimum dietary level.

6.2.3.2.12 D-Proline L-Pro is a nonessential amino acid. Oral feeding of an aqueous solution of D-Pro for 1 month to rats induced fibrosis and necrosis of kidney liver cells and elevation of serum enzymes (101). A subsequent study found no evidence that orally fed D-Pro and D-Asp induced acute toxicity in rats (102).

6.2.3.2.13 Selenomethionine Isomers Selenomethionine is a major source of selenium in the diets of both animals and humans. Comparison of the effects of oral consumption of seleno-L-methionine, seleno-DL-methionine, and selenized yeast on the reproduction of mallard ducklings revealed that although both seleno-methionine preparations were of similar toxicity, their potency was greater than that of selenium present in yeast (103). Heinz *et al.* (103) found that the survival of day-old ducklings consuming L-selenomethionine after 2 weeks was significantly lower (36%) than that of ducklings consuming the DL-isomer (100%). D-selenomethionine protected against adverse effects induced by space radiation (104).

6.2.3.2.14 D-Serine Protein-bound L-Ser racemizes to the D-isomer faster than any of the other amino acids (Table 6.2.2; Figs 6.2.5 and 6.2.6) (18). D-Ser has been reported to enlarge rat kidney cells (cytomegaly) similar to that observed with lysinoalanine. D-Ser may also be involved in neurotransmission in the mammalian brain (105, 106). Toxicity of D-Ser and D-Cys may arise from oxidative damage to cells induced by products of their metabolism such as H_2O_2 (107). Several approaches were used in an attempt to elucidate possible mechanisms of D-Ser renal toxicity (108–110). Sodium benzoate (108, 109), protein-deficient diets (111), and alpha-aminoisobutyric (112) acid attenuated D-Ser nephrotoxicity in rats.

D-Ser-induced kidney damage appears to be due to a lowering of the concentration of renal glutathione (GSH) that protects the kidneys against kidney-damaging reactive oxygen species (112). The decrease in glutathione concentration takes place during the metabolism of D-Ser by D-amino acid oxidase.

Because nephrotoxicity induced by D-Ser is similar to that caused by lysinoalanine, the question arises as to whether effects of these two amino acids on the rat kidney are competitive, additive, or synergistic. Because D-Ser acts as an agonist of *N*-methyl-D-aspartate (NMDA) receptor-mediated neurotransmission in the brain, it may be useful as "innovative pharmacologic strategy in schizophrenia," provided it is not neurotoxic (113).

The D-configuration of serine maintains the toxic conformation of the mushroom poison viroisin (114).

6.2.3.2.15 D-Threonine L-Thr is the second-limiting amino acid in maize (corn) proteins. Table 6.2.2 shows that at high pH, L-Thr racemizes rapidly to D-Thr. The utilization of D-Thr by the chick, rat, mouse, or human as a nutritional source of the L-isomer is insignificant (29).

6.2.3.2.16 D-Tryptophan L-Trp contributes to protein synthesis and regulates numerous physiological mechanisms (62, 115). These include serving as a precursor of the neurotransmitter serotonin and of the vitamin niacin. The relative nutritional potency of D-Trp compared with the L-Trp in mice is strongly dose-dependent, being inversely related to the dietary concentration and ranging from 29% to 64% (Fig. 6.2.11) (62). The maximum growth obtainable for L-Trp occurred at 0.174% in the diet. By increasing the dietary concentration of D-Trp up to 0.52%, growth passed through a maximum at 82% of that achieved with the L-isomer. This occurred at 0.44% of D-Trp.

Considerable species variation exists for the nutritive value of D-Trp (29). In chicks' diets, relative potency of the D- to the L-isomer has been reported to be 20%. D-Trp was well utilized by growing pigs (116). The value for humans is about 10%. Rats utilize D-Trp as efficiently as L-Trp. D-Trp can serve as a niacin precursor in rats to the same extent as does the L-form (117, 118). These studies also showed D-Trp has one-sixth of the activity of niacin. Both D-Phe and D-Trp taste sweet (62, 119–121).

6.2.3.2.17 D-Tyrosine Nutritionally, L-Tyr is classified as a semi-essential amino acid (122). Combinations of L-Tyr and L-Phe are complementary in supporting growth of mice (13). Thus, under conditions where L-Phe may be limiting, L-Tyr may supply half the requirement of L-Phe alone for chicks, mice, rats, and humans. Our feeding studies showed that with D-Tyr in an amino acid diet, growth inhibition was severe at a D-Tyr/L-Tyr ratio of 2:1, but was less so when the ratio was 1:1. Similar results were obtained with a casein diet supplemented with D-Tyr. The antimetabolic manifestation of D-Tyr may be ascribed to interference with the biosynthesis of vital neurotransmitters and proteins *in vivo* (13, 63).

Our findings demonstrated that D-Tyr, unlike L-Tyr, has no sparing effect for L-Phe in mice. In fact, growth inhibition may become evident when D-Tyr is present in the diet at level equal or greater than L-Phe. The growth-inhibiting effect of D-Tyr may be due to its toxicity. The potential for chronic toxicity following exposure to lower levels of D-Tyr remains unknown. Formation of D-tyrosyl-$tRNA^{Tyr}$ may be responsible for the toxicity of D-Tyr toward *Escherichia coli* (123).

6.2.3.2.18 D-Valine Administration of total parenteral nutrition (TPN) containing D-Leu, D-Met, D-Phe, and D-Val to hepatoma-bearing rats showed that D-Val inhibited tumor growth without negative effects on the host. D-Leu and D-Met also improved the nutritional status of the sick rats (124). These observations suggest that some D-amino acid diets may benefit cancer patients. It is also worth noting that D-amino acid-containing chemokines inhibited entry of human immunodeficiency virus type 1 (HIV-1) into cells (125). Possible antiviral activities of dietary D-amino acids merit further study.

In summary, mice provide a good animal model to study the biological utilization and biological effects of D-amino acids, both free and protein-bound. A major advantage of mouse bioassays is that they require about one-fifth of the test substance needed for rats and can be completed in 14 days (8, 12–15, 17).

6.2.4 RESEARCH NEEDS

There is a need to standardize methods designed to ascertain the role of D-amino acids in nutrition, food safety, microbiology, medicinal chemistry, and medicine. Because the organisms are forced to use the D-amino acid as the sole source of the L-form, the use of all-amino-acid diets in which the L-isomer is completely replaced with different levels of the corresponding D-amino acid may be preferable to supplementation of proteins with D-amino acids. D-amino acids along a peptide chain may be less utilized than the L-forms. The utilization of any D-amino acid may be affected by the presence of other D-amino acids in the diet. In addition to several research needs mentioned earlier, there is need for further research designed to find answers to the following questions:

(a) How do biological utilizations and effects of D-amino acids in the kidneys and brains of animals and humans vary, depending on whether they are consumed in the free state or as part of a food protein?

(b) Do D-amino acids and D-peptides alter the normal microflora of the intestine?

(c) Can poorly digestible racemized food proteins serve as dietary fiber in the digestive tracts of humans?

(d) Can poorly digestible racemized food proteins be used to treat obesity?

(e) Do metabolic interactions, antagonisms, or synergisms among free and protein-bound D-amino acids occur *in vivo*?

(f) Does racemization alter protein conformations and charge distributions at the isoelectric points, resulting in beneficial protective effects against protein-induced allergy by peanut and soy proteins, celiac disease by wheat gluten, and bacterial, plant, and venom toxic proteins that act by binding to specific cell receptor sites? Research is continuing in this important area of food chemistry, food and microbial safety, nutrition, and medicine (126–143).

REFERENCES

1. Leuchtenberger, W., Huthmacher, K., Drauz, K. (2005). Biotechnological production of amino acids and derivatives: current status and prospects. *Appl. Microbiol. Biotechnol.*, *69*, 1–8.
2. Masters, P.M., Friedman, M. (1979). Racemization of amino acids in alkali-treated food proteins. *J. Agric. Food Chem.*, *27*, 507–511.
3. Masters, P.M., Friedman, M. (1980). Amino acid racemization in alkali-treated food proteins—chemistry, toxicology, and nutritional consequences, in *Chemical Deterioration of Proteins* (eds J.R. Whitaker, M. Fujimaki), American Chemical Society, Washington, DC, pp. 165–194.
4. Friedman, M. (1991). Formation, nutritional value, and safety of D-amino acids. *Adv. Exp. Med. Biol.*, *289*, 447–481.
5. Friedman, M. (1992). Dietary impact of food processing. *Annu. Rev. Nutr.*, *12*, 119–137.
6. Friedman, M. (1996). Food browning and its prevention: an overview. *J. Agric. Food Chem.*, *44*, 631–653.
7. Friedman, M. (1999). Chemistry. biochemistry, nutrition, and microbiology of lysinoalanine, lanthionine, and histidinoalanine in food and other proteins. *J. Agric. Food Chem.*, *47*, 1295–1319.
8. Friedman, M., Gumbmann, M.R. (1979). Biological availability of ε-N-methyl-L-lysine, 1-N-methyl-L-histidine, and 3-N-methyl-L-histidine in mice. *Nutr. Rep. Int.*, *19*, 437–443.
9. Friedman, M. (2003). Nutritional consequences of food processing. *Forum Nutr.*, *56*, 350–352.
10. Friedman, M. (2004). Effects of food processing, in *Encyclopaedia of Grain Science* (eds H. Corke, C. Walker), Elsevier, Oxford, pp. 328–340.

11. Friedman, M. (2005). Biological effects of Maillard browning products that may affect acrylamide safety in food, in *Chemistry and Safety of Acrylamide in Food* (eds M. Friedman, D. Mottram), Springer, New York, pp. 135–155.
12. Friedman, M., Gumbmann, M.R. (1981). Bioavailability of some lysine derivatives in mice. *J. Nutr.*, *111*, 1362–1369.
13. Friedman, M., Gumbmann, M.R. (1984). The nutritive value and safety of D-phenylalanine and D-tyrosine in mice. *J. Nutr.*, *114*, 2089–2096.
14. Friedman, M., Gumbmann, M.R. (1984). The utilization and safety of isomeric sulfur-containing amino acids in mice. *J. Nutr.*, *114*, 2301–2310.
15. Friedman, M., Gumbmann, M.R. (1988). Nutritional value and safety of methionine derivatives, isomeric dipeptides and hydroxy analogs in mice. *J. Nutr.*, *118*, 388–397.
16. Friedman, M., Gumbmann, M.R., Masters, P.M. (1984). Protein-alkali reactions: chemistry, toxicology, and nutritional consequences. *Adv. Exp. Med. Biol.*, *177*, 367–412.
17. Friedman, M., Gumbmann, M.R., Savoie, L. (1982). The nutritional value of lysinoalanine as a source of lysine for mice. *Nutr. Rep. Int.*, *26*, 937–943.
18. Friedman, M., Liardon, R. (1985). Racemization kinetics of amino acid residues in alkali-treated soybean proteins. *J. Agric. Food Chem.*, *33*, 666–672.
19. Friedman, M., Masters, P.M. (1982). Kinetics of racemization of amino acid residues in casein. *J. Food Sci.*, *47*, 760–764.
20. Friedman, M., Noma, A.T. (1986). Formation and analysis of (phenylethyl)amino]alanine in food proteins. *J. Agric. Food Chem.*, *34*, 497–502.
21. Friedman, M., Noma, A.T., Wagner, J.R. (1979). Ion-exchange chromatography of sulfur amino acids on a single-column amino acid analyzer. *Anal. Biochem.*, *98*, 293–304.
22. Friedman, M., Zahnley, J.C., Masters, P.M. (1981). Relationship between in vitro digestibility of casein and its content of lysinoalanine and D-amino acids. *J. Food Sci.*, *46*, 127–131.
23. Liardon, R., Friedman, M. (1987). Effect of peptide bond cleavage on the racemization of amino acid residues in proteins. *J. Agric. Food Chem.*, *35*, 661–667.
24. Liardon, R., Friedman, M., Philippossian, G. (1991). Racemization kinetics of free and protein-bound lysinoalanine (LAL) in strong acid media. Isomeric composition of bound LAL in processed proteins. *J. Agric. Food Chem.*, *39*, 531–537.
25. Anonymous (1989). Mechanism of toxicity of lysinoalanine. *Nutr. Rev.*, *47*, 362–364.
26. Anonymous (1973). Nutritional value of alkali treated soy protein. *Nutr. Rev.*, *31*, 250–251.
27. Gumbmann, M.R., Friedman, M. (1987). Effect of sulfur amino acid supplementation of raw soy flour on the growth and pancreatic weights of rats. *J. Nutr.*, *117*, 1018–1023.
28. Gilani, G.S., Cockell, K.A., Sepehr, E. (2005). Effects of antinutritional factors on protein digestibility and amino acid availability in foods. *J. AOAC Int.*, *88*, 967–987.
29. Borg, B.S., Wahlstrom, R.C. (1989). Species and isomeric variation in the utilization of amino acids, in *Absorption and Utilization of Amino Acids*, Vol. 1 (ed. M. Friedman), CRC Press, Boca Raton, FL, pp. 155–171.

30. Brückner, H., Westhauser, T. (2003). Chromatographic determination of L- and D-amino acids in plants. *Amino Acids*, *24*, 43–55.
31. Voss, K., Galensa, R. (2000). Determination of L- and D-amino acids in foodstuffs by coupling of high-performance liquid chromatography with enzyme reactors. *Amino Acids*, *18*, 339–352.
32. Pätzold, R., Brüeckner, H. (2005). Mass spectrometric detection and formation of D-amino acids in processed plant saps, syrups, and fruit juice concentrates. *J. Agric. Food Chem.*, *53*, 9722–9729.
33. Pätzold, R., Brüeckner, H. (2006). Gas chromatographic detection of D-amino acids in natural and thermally treated bee honeys and studies on the mechanism of their formation as result of the Maillard reaction. *Eur. Food Res. Technol.*, *223*, 347–354.
34. Cavani, L., Ciavatta, C., Gessa, C. (2003). Determination of free L- and D-alanine in hydrolysed protein fertilisers by capillary electrophoresis. *J. Chromatogr., A*, *985*, 463–469.
35. Simó, C., Rizzi, A., Barbas, C., Cifuentes, A. (2005). Chiral capillary electrophoresis-mass spectrometry of amino acids in foods. *Electrophoresis*, *26*, 1432–1441.
36. Simó, C., Martin-Alvarez, P.J., Barbas, C., Cifuentes, A. (2004). Application of stepwise discriminant analysis to classify commercial orange juices using chiral micellar electrokinetic chromatography-laser induced fluorescence data of amino acids. *Electrophoresis*, *25*, 2885–2891.
37. Wcislo, M., Compagnone, D., Trojanowicz, M. (2007). Enantioselective screen-printed amperometric biosensor for the determination of d-amino acids. *Bioelectrochemistry*, *71*, 91–98.
38. Chen, S., Chen, N.-H. (2001). The HPLC analysis of the concentration and enantiomeric purity of selected amino acids in two highly fermented foods. *J. Chin. Chem. Soc.*, *48*, 757–762.
39. Jin, D., Miyahara, T., Oe, T., Toyo'oka, T. (1999). Determination of D-amino acids labelled with fluorescent chiral reagents, R(−)- and S(+)-4-(3-isothiocyanatopyrrolidin-1-yl)-7-(N, N-dimethylaminosulfonyl)-2,1,3-benzoxadiazoles, in biological and food samples by liquid chromatography. *Anal. Biochem.*, *269*, 124–132.
40. Sarkar, P., Tothill, I.E., Setford, S.J., Turner, A.P. (1999). Screen-printed amperometric biosensors for the rapid measurement of L- and D-amino acids. *Analyst*, *124*, 865–870.
41. Friedman, M., Orraca-Tetteh, R. (1978). Hair as an index of protein malnutrition, in *Nutritional Improvement of Food and Feed Proteins* (ed. Friedman, M.), Plenum Press, New York, pp. 131–154.
42. Friedman, M. (2004). Application of the ninhydrin reaction for analysis of amino acids, peptides, and proteins to agricultural and biomedical sciences. *J. Agric. Food Chem.*, *52*, 385–406.
43. Zagon, J., Dehne, L.I., Bogl, K.W. (1991). Mechanisms and occurrence of amino acid racemization in organisms and foods. Part I. *Ernaehrungs-Umschau*, *38*, 275–278.
44. Zagon, J., Dehne, L.I., Bogl, K.W. (1991). Isomerization of amino acids in foods. Part II. D-amino acids in foods and their physiological properties. *Ernaehrungs-Umschau*, *38*, 324–328.

45. Sawyer, T.K., Hruby, V.J., Hadley, M.E., Engel, M.H. (1983). α-Melanocyte stimulating hormone: chemical nature and mechanism of action. *Am. Zool.*, *23*, 529–540.
46. Friedman, M. (1999). Chemistry, nutrition, and microbiology of D-amino acids. *J. Agric. Food Chem.*, *47*, 3457–3479.
47. Soda, K. (1996). D-Amino acid metabolism and vitamin B6. enzymes. *Vitamin*, *70*, 103–113.
48. Friedman, M., Levin, C.E., Noma, A.T. (1984). Factors governing lysinoalanine formation in soy proteins. *J. Food Sci.*, *49*, 1282–1288.
49. Casal, S., Mendes, E., Oliveira, M.B.P.P., Ferreira, M.A. (2005). Roast effects on coffee amino acid enantiomers. *Food Chem.*, *89*, 333–340.
50. Ono, K., Yanagida, K., Oikawa, T., Ogawa, T., Soda, K. (2006). Alanine racemase of alfalfa seedlings (*Medicago sativa* L.): first evidence for the presence of an amino acid racemase in plants. *Phytochemistry*, *67*, 856–860.
51. Gobbetti, M., Simonetti, M.S., Rossi, J., Cossignani, L., Corsetti, A., Damiani, P. (1994). Free D- and L-amino acid evolution during sourdough fermentation and baking. *J. Food Sci.*, *59*, 881–884.
52. Gandolfi, I., Palla, G., Marchelli, R., Dossena, A., Puelli, S., Salvadori, C. (1994). D-Alanine in fruit juices: a molecular marker of bacterial activity, heat treatments and shelf life. *J. Food Sci.*, *59*, 152–154.
53. Pearce, K.N., Karahalios, D., Friedman, M. (1988). Ninhydrin assay for proteolysis of ripening cheese. *J. Food Sci.*, *53*, 432–435, 438.
54. Csapó, J., Varga-Visi, E., Lóki, K., Albert, C. (2006). The influence of manufacture on the free D-amino acid content of Cheddar cheese. *Amino Acids*, *32*, 39–43.
55. Rundlett, K.L., Armstrong, D.W. (1994). Evaluation of free D-glutamate in processed foods. *Chirality*, *6*, 277–282.
56. Chang, H.M., Tsai, C.F., Li, C.F. (1999). Changes of amino acid composition and lysinoalanine formation in alkali-pickled duck eggs. *J. Agric. Food Chem.*, *47*, 1495–1500.
57. Pawlowska, M., Armstrong, D.W. (1994). Evaluation of enantiomeric purity of selected amino acids in honey. *Chirality*, *6*, 270–276.
58. Luzzana, U., Mentasti, T., Morewtti, V.M., Albertini, A., Valfre, F. (1996). Aspartic acid racemization in fish meals as induced by thermal treatment. *Aquac. Nutr.*, *2*, 95–99.
59. Sarower, M.G., Matsui, T., Abe, H. (2003). Distribution and characteristics of D-amino acid and D-aspartate oxidases in fish tissues. *J. Exp. Zoolog. A Comp. Exp. Biol.*, *295*, 151–159.
60. Herrero, M., Ibanez, E., Martin-Alvarez, P.J., Cifuentes, A. (2007). Analysis of chiral amino acids in conventional and transgenic maize. *Anal Chem.*, *79*, 5071–5077.
61. Palla, G., Marchelli, R., Decennia, A., Casnati, G. (1989). Occurrence of D-amino acids in foods. *J. Chromatogr.*, *475*, 45–53.
62. Friedman, M., Cuq, J.L. (1988). Chemistry, analysis, nutritional value, and toxicology of tryptophan in food. A review. *J. Agric. Food Chem.*, *36*, 1079–1093.
63. Anon. (1985). Dietary D-tyrosine as an antimetabolite in mice. *Nutr. Rev.*, *43*, 156–158.

64. Stegman, L.D., Zheng, H., Neal, E.R., Ben-Yoseph, O., Pollegioni, L., Pilone, M.S., Ross, B.D. (1998). Induction of cytotoxic oxidative stress by D-alanine in brain tumour cells expressing *Rhodotorula gracilis* D-amino acid oxidase: a cancer gene therapy strategy. *Hum. Gene Ther.*, *9*, 185–193.
65. Nagata, Y., Higashi, M., Ishii, Y., Sano, H., Tanigawa, M., Nagata, K., Noguchi, K., Urade, M. (2006). The presence of high concentrations of free D-amino acids in human saliva. *Life Sci.*, *78*, 1677–1681.
66. Shikano, N., Nakajima, S., Kotani, T., Ogura, M., Sagara, J., Iwamura, Y., Yoshimoto, M., Kubota, N., Ishikawa, N., Kawai, K. (2007). Transport of d-1-(14)C]-amino acids into Chinese hamster ovary (CHO-K1) cells: implications for use of labelled d-amino acids as molecular imaging agents. *Nucl. Med. Biol.*, *34*, 659–665.
67. Suessenbacher, A., Lass, A., Mayer, B., Brunner, F. (2002). Antioxidative and myocardial protective effects of L-arginine in oxygen radical-induced injury of isolated perfused rat hearts. *Naunyn Schmiedebergs Arch. Pharmakol.*, *365*, 269–276.
68. Wiel, E., Pu, Q., Corseaux, D., Robin, E., Bordet, R., Lund, N., Jude, B., Vallet, B. (2000). Effect of L-arginine on endothelial injury and hemostasis in rabbit endotoxin shock. *J. Appl. Physiol.*, *89*, 1811–1818.
69. Szende, B. (1993). The effect of amino acids and amino acid derivatives on cell proliferation. *Acta Biomed Ateneo Parmense*, *64*, 139–145.
70. Navarro, E., Alonso, S.J., Martin, F.A., Castellano, M.A. (2005). Toxicological and pharmacological effects of D-arginine. *Basic Clin. Pharmacol. Toxicol.*, *97*, 149–154.
71. Mileusnic, R., Lancashire, C., Clark, J., Rose, S.P. (2007). Protection against Abeta-induced memory loss by tripeptide D-Arg-L-Glu-L-Arg. *Behav. Pharmacol.*, *18*, 231–238.
72. Koyuncuoglu, H., Gungor, M., Ang, O., Inanc, D., Ang-Kucuker, M., Sagduyu, H., Uysal, V. (1988). Aggravation by morphine and D-aspartic acid of pyelonephritis induced by i.v. inoculation of *Staphylococcus aureus* in rats. *Infection*, *16*, 42–45.
73. Iezhitsa, I.N., Spasov, A.A., Zhuravleva, N.V., Sinolitskii, M.K., Voronin, S.P. (2004). Comparative study of the efficacy of potassium magnesium L-, D- and DL-aspartate stereoisomers in overcoming digoxin- and furosemide-induced potassium and magnesium depletions. *Magnes. Res.*, *17*, 276–292.
74. Benz, F.W., Nerland, D.E., Pierce, W.M., Babiuk, C. (1990). Acute acrylonitrile toxicity: studies on the mechanism of the antidotal effect of D- and L-cysteine and their N-acetyl derivatives in the rat. *Toxicol. Appl. Pharmacol.*, *102*, 142–150.
75. Friedman, M., Cavins, J.F., Wall, J.S. (1965). Relative nucleophilic reactivities of amino groups and mercaptide ions in addition to reactions with α, β-unsaturated compounds. *J. Am. Chem. Soc.*, *87*, 3572–3582.
76. Huang, J., Niknahad, H., Khan, S., O'Brien, P.J. (1998). Hepatocyte-catalysed detoxification of cyanide by L- and D-cysteine. *Biochem. Pharmacol.*, *55*, 1983–1990.
77. McLean, A.E., Armstrong, G.R., Beales, D. (1989). Effect of D- or L-methionine and cysteine on the growth inhibitory effects of feeding 1% paracetamol to rats. *Biochem. Pharmacol.*, *38*, 347–352.
78. Takahashi, Y., Funakoshi, T., Shimada, H., Kojima, S. (1994). Comparative effects of chelating agents on distribution, excretion, and renal toxicity of gold sodium thiomalate in rats. *Toxicology*, *90*, 39–51.

79. Friedman, M. (1973). *Chemistry and Biochemistry of the Sulfhydryl Group in Amino Acids, Peptides, and Proteins*, Pergamon Press, Oxford, UK, p. 485.

80. Friedman, M. (1994). Improvement in the safety of foods by SH-containing amino acids and peptides. A review. *J. Agric. Food Chem.*, *42*, 3–20.

81. Berkeley, L.I., Cohen, J.F., Crankshaw, D.L., Shirota, F.N., Nagasawa, H.T. (2003). Hepatoprotection by L-cysteine-glutathione mixed disulfide, a sulfhydryl-modified prodrug of glutathione. *J. Biochem. Mol. Toxicol.*, *17*, 95–97.

82. Glover, C.N., Hogstrand, C. (2002). Amino acid modulation of in vivo intestinal zinc absorption in freshwater rainbow trout. *J. Exp. Biol.*, *205*, 151–158.

83. Tachon, P. (1990). DNA single strand breakage by H_2O_2 and ferric or cupric ions: its modulation by histidine. *Free Radic. Res. Commun.*, *9*, 39–47.

84. Rauen, U., Klempt, S., de Groot, H. (2007). Histidine-induced injury to cultured liver cells, effects of histidine derivatives and of iron chelators. *Cell. Mol. Life Sci.*, *64*, 192–205.

85. Yokel, R.A. (2006). Blood-brain barrier flux of aluminium, manganese, iron and other metals suspected to contribute to metal-induced neurodegeneration. *J. Alzheimers Dis.*, *10*, 223–253.

86. Friedman, M., Noma, A.T. (1975). Methods and problems in chromatographic analysis of sulfur amino acids, in *Protein Nutritional Quality of Foods and Feeds Part I. Assay Methods Biological, Biochemical, and Chemical* (ed. M. Friedman), Marcel Dekker, New York, pp. 521–548.

87. Mangoni, M.L., Papo, N., Saugar, J.M., Barra, D., Shai, Y., Simmaco, M., Rivas, L. (2006). Effect of natural L- to D-amino acid conversion on the organization, membrane binding, and biological function of the antimicrobial peptides bombinins H. *Biochemistry*, *45*, 4266–4276.

88. Sahl, H.G., Bierbaum, G. (1998). Lantibiotics: biosynthesis and biological activities of uniquely modified peptides from gram-positive bacteria. *Annu. Rev. Microbiol.*, *52*, 41–79.

89. McAuliffe, O., Ross, R.P., Hill, C. (2001). Lantibiotics: structure, biosynthesis and mode of action. *FEMS Microbiol. Rev.*, *25*, 285–308.

90. Dufour, A., Hindré, T., Haras, D., Le Pennec, J.P. (2007). The biology of lantibiotics from the lacticin 481 group is coming of age. *FEMS Microbiol. Rev.*, *31*, 134–167.

91. Bernard, B.F., Krenning, E.P., Breeman, W.A., Rolleman, E.J., Bakker, W.H., Visser, T.J., Macke, H., de Jong, M. (1997). D-lysine reduction of indium-111 octreotide and yttrium-90 octreotide renal uptake. *J. Nucl. Med.*, *38*, 1929–1933.

92. Boyd, B.J., Kaminskas, L.M., Karellas, P., Krippner, G., Lessene, R., Porter, C.J. (2006). Cationic poly-L-lysine dendrimers: pharmacokinetics, biodistribution, and evidence for metabolism and bioresorption after intravenous administration to rats. *Mol. Pharmacol.*, *3*, 614–627.

93. Lin, Y.C., Hung, G.U., Luo, T.Y., Tsai, S.C., Sun, S.S., Hsia, C.C., Chen, S.L., Lin, W.Y. (2007). Reducing renal uptake of ^{111}In-DOTATOC: a comparison among various basic amino acids. *Ann. Nucl. Med.*, *21*, 79–83.

94. Friedman, M., Pearce, K.N. (1989). Copper(II) and cobalt(II) affinities of LL- and LD-lysinoalanine diastereomers: implications for food safety and nutrition. *J. Agric. Food Chem.*, *37*, 123–127.

95. Friedman, M., Finot, P.A. (1990). Nutritional improvement of bread with lysine and γ-glutamyl-L-lysine. *J. Agric. Food Chem.*, *38*, 2011–2020.
96. Smith, G.A., Friedman, M. (1984). Effect of carbohydrates and heat on the amino acid composition and chemically available lysine content of casein. *J. Food Sci.*, *49*, 817–820, 843.
97. Sasamura, T., Matsuda, A., Kokuba, Y. (1999). Effects of D-methionine-containing solution on tumour cell growth in vitro. *Arzneimittelforschung*, *49*, 541–543.
98. Campbell, K.C., Meech, R.P., Rybak, L.P., Hughes, L.F. (2003). The effect of D-methionine on cochlear oxidative state with and without cisplatin administration: mechanisms of auto protection. *J. Am. Acad. Audiol.*, *14*, 144–156.
99. Wimmer, C., Mees, K., Stumpf, P., Welsch, U., Reichel, O., Sückfull, M. (2004). Round window application of D-methionine, sodium thiosulfate, brain-derived neurotrophic factor, and fibroblast growth factor-2 in cisplatin-induced ototoxicity. *Otol. Neurotol.*, *25*, 33–40.
100. Findrik, Z., Vasic-Racki, D. (2007). Biotransformation of D-methionine into L-methionine in the cascade of four enzymes. *Biotechnol. Bioeng.*, *98*, 956–967.
101. Kampel, D., Kupferschmidt, R., Lubec, G. (1990). Toxicity of D-proline, in *Amino Acid: Chemistry, Biology and Medicine* (eds G. Lubec, G.A. Rosenthal), Escom Science Publishers, Leiden, the Netherlands, pp. 1164–1171.
102. Schieber, A., Bruckner, H., Rupp-Classen, M., Specht, W., Nowitzki-Grimm, S., Classen, H.G. (1997). Evaluation of D-amino acid levels in rat by gas chromatography-selected ion monitoring mass spectrometry: no evidence for subacute toxicity of orally fed D-proline and D-aspartic acid. *J. Chromatogr., B Biomed. Sci. Appl.*, *691*, 1–12.
103. Heinz, G.H., Hoffman, D.J., LeCaptain, L.J. (1996). Toxicity of seleno-L-methionine, seleno-DL-methionine, high selenium wheat, and selenized yeast to mallard ducklings. *Arch. Environ. Contam. Toxicol.*, *30*, 93–99.
104. Kennedy, A.R., Ware, J.H., Guan, J., Donahue, J.J., Biaglow, J.E., Zhou, Z., Stewart, J., Vazquez, M., Wan, X.S. (2004). Selenomethionine protects against adverse biological effects induced by space radiation. *Free Radic. Biol. Med.*, *36*, 259–266.
105. Kappor, R., Kapoor, V. (1997). Distribution of D-amino acid oxidase (DAO) activity in the medulla and thoracic spinal cord of the rat: implications for a role for D-serine in autonomic function. *Brain Res.*, *771*, 351–355.
106. Gong, X.Q., Zabek, R.L., Bai, D. (2007). D-Serine inhibits AMPA receptor-mediated current in rat hippocampal neurons. *Can. J. Physiol. Pharmacol.*, *85*, 546–555.
107. Ercal, N., Luo, X., Matthews, R.H., Armstrong, D.W. (1996). In vitro study of the metabolic effects of D-amino acids. *Chirality*, *8*, 24–29.
108. Williams, R.E., Lock, E.A. (2004). D-serine-induced nephrotoxicity: possible interaction with tyrosine metabolism. *Toxicology*, *201*, 231–238.
109. Williams, R.E., Major, H., Lock, E.A., Lenz, E.M., Wilson, I.D. (2005). D-Serine-induced nephrotoxicity: a HPLC-TOF/MS-based metabonomics approach. *Toxicology*, *207*, 179–190.
110. Maekawa, M., Okamura, T., Kasai, N., Hori, Y., Summer, K.H., Konno, R. (2005). D-amino-acid oxidase is involved in D-serine-induced nephrotoxicity. *Chem. Res. Toxicol.*, *18*, 1678–1682.

111. Levine, S., Saltzman, A. (2003). Acute uremia produced in rats by nephrotoxic chemicals is alleviated by protein deficient diet. *Ren. Fail.*, *24*, 517–523.

112. Krug, A.W., Volker, K., Dantzler, W.H., Silbernagl, S. (2007). Why is D-serine nephrotoxic and alpha-aminoisobutyric acid protective? *Am. J. Physiol. Renal Physiol.*, *293*, F382–390.

113. Shoham, S., Javitt, D.C., Heresco-Levy, U. (2001). Chronic high-dose glycine nutrition: effects on rat brain cell morphology. *Biol. Psychiatry*, *49*, 876–885.

114. Zanotti, G., Kobayashi, N., Munekata, E., Zobeley, S., Faulstich, H. (1999). D-configuration of serine is crucial in maintaining the phalloidin-like conformation of viroisin. *Biochemistry*, *38*, 10723–10729.

115. Friedman, M., Levin, C.E., Noma, A.T., Montague, W.C., Jr, Zahnley, J.C. (1984). Comparison of tryptophan assays for food proteins, in *Progress in Tryptophan and Serotonin Research* (ed. H.G. Schlossberger), Walter de Gruyter, Berlin, pp. 119–123.

116. Arentson, B.E., Zimmerman, D.R. (1985). Nutritive value of D-tryptophan for the growing pig. *J. Anim. Sci.*, *60*, 474–479.

117. Shibata, K., Swabe, M., Fukuwatari, T., Sugimoto, E. (2000). Efficiency of D-tryptophan as niacin in rats. *Biosci. Biotechnol. Biochem.*, *64*, 206–209.

118. Carter, E.G.A., Carpenter, K.J., Friedman, M. (1982). The nutritional value of some niacin analogs for rats. *Nutr. Rep. Int.*, *25*, 389–397.

119. Finley, J.W., Friedman, M. (1973). New sweetening agents: N-formyl- and N′-acetylkynurenine. *J. Agric. Food Chem.*, *21*, 33–34.

120. Maehashi, K., Matano, M., Kondo, A., Yamamoto, Y., Udaka, S. (2007). Riboflavin-binding protein exhibits selective sweet suppression toward protein sweeteners. *Chem. Senses*, *32*, 183–190.

121. Manita, S., Bachmanov, A.A., Li, X., Beauchamp, G.K., Inoue, M. (2006). Is glycine "sweet" to mice? Mouse strain differences in perception of glycine taste. *Chem. Senses*, *31*, 785–793.

122. Mercer, L.P., Dodds, S.J., Smith, D.L. (1989). Dispensable, indispensable, and conditionally indispensable amino acids ratios in the diet, in *Absorption and Utilization of Amino Acids*, Vol. 1 (ed. M. Friedman), CRC Press, Boca Raton, FL, pp. 2–13.

123. Soutourina, O., Soutourina, J., Blanquet, S., Plateau, P. (2004). Formation of D-tyrosyl-tRNATyr accounts for the toxicity of D-tyrosine toward Escherichia coli. *J. Biol. Chem.*, *279*, 42560–42565.

124. Sasamura, T., Matsuda, A., Kokuba, Y. (1998). Nutritional effects of a D-methionine-containing solution on AH109A hepatoma-bearing rats. *Biosci. Biotechnol. Biochem.*, *62*, 2418–2420.

125. Liu, D., Madani, N., Li, Y., Cao, R., Choi, W.T., Kawatkar, S.P., Lim, M.Y., Kumar, S., Dong, C.Z., Wang, J., Russell, J.D., Lefebure, C.R., An, J., Wilson, S., Gao, Y.G., Pallansch, L.A., Sodroski, J.G., Huang, Z. (2007). Crystal structure and structural mechanism of a novel anti-human immunodeficiency virus and D-amino acid-containing chemokine. *J. Virol.*, *81*, 11489–11498.

126. Bosch, L., Sanz, M.L., Montilla, A., Alegría, A., Farré, R., del Castillo, M.D. (2007). Simultaneous analysis of lysine, N^{ε}-carboxymethyllysine and lysinoalanine from

proteins. *Journal of Chromatography B: Analytical Technologies in the Biomedical and Life Sciences*, *860*, 69–77.

127. Giuffrida, A., Tabera, L., Gonzalez, R., Cucinotta, V., Cifuentes, A. (2008). Chiral analysis of amino acids from conventional and transgenic yeasts. *Journal of Chromatography B: Analytical Technologies in the Biomedical and Life Sciences* [E-pub ahead of print June 11]

128. Wu, D., Kong, Y., Han, C., Chen, J., Hu, L., Jiang, H., Shen, X. (2008). D-Alanine: D-alanine ligase as a new target for the flavonoids quercetin and apigenin. *Int. J. Antimicrob.* Agents [E-pub ahead of print April 23].

129. Rosini, E., Molla, G., Rossetti, C., Pilone, M.S., Pollegioni, L., Sacchi, S. (2008). A biosensor for all D-amino acids using evolved D-amino acid oxidase. *J. Biotechnol.*, *135*, 377–384.

130. Duncker, S.C., Wang, L., Hols, P., Bienenstock, J. (2008). The D-alanine content of lipoteichoic acid is crucial for Lactobacillus plantarum-mediated protection from visceral pain perception in a rat colorectal distension model. *Neurogastroenterol. Motil.*, *20*, 843–850.

131. Morikawa, A., Hamase, K., Miyoshi, Y., Koyanagi, S., Ohdo, S., Zaitsu, K. (2008). Circadian changes of D-alanine and related compounds in rats and the effect of restricted feeding on their amounts, *J. Chromatogr. B. [E-pub aheadof print April 9).*

132. Schoenhusen, U., Voigt, J., Hennig, U., Kuhla, S., Zitnan, R., Souffrant, W.-B. (2008). Bacterial D-alanine concentrations as a marker of bacterial nitrogen in the gastrointestinal tract of pigs and cows. *Vet. Med.*, *53*, 184–192.

133. Grimble, G.K. (2007). Adverse gastrointestinal effects of arginine and related amino acids. *J. Nutr.*, *137*, 1693S–1701S.

134. Navarro, E., Alonso, S.J., Martin, F.A., Castellano, M.A. (2005). Toxicological and pharmacological effects of D-arginine. *Basic Clin. Pharmacol. Toxicol.*, *97*, 149–154.

135. Wu, G., Bazer, F.W., Cudd, T.A., Jobgen, W.S., Sung, W.K., Lassala, A., Li, P., Matis, J.H., Meininger, C.J., Spencer, T.E. (2007). Pharmacokinetics and safety of arginine supplementation in animals. J. *Nutr.*, *137*, 1673S–1680S.

136. Xin, L., Jie, L., Liu, C.-W., Zhao, S.-L. (2007). Determination of D-aspartic acid and D-glutamic acid in midbrain of Parkinson's disease mouse by reversed phase high performance liquid chromatography. Chin. *J. Anal. Chem.*, *35*, 1151–1154.

137. Dever, J.T., Elfarra, A.A. (2008). L-methionine toxicity in freshly isolated mouse hepatocytes is gender-dependent and mediated in part by transamination. *J. Pharmacol. Exp. Ther.*, *326*, 809–817.

138. Dong, F., Yang, X., Sreejayan, N., Ren, J. (2007). Chromium (D-phenylalanine)$_3$ improves obesity-induced cardiac contractile defect in ob/ob mice. *Obesity*, *15*, 2699–2711.

139. Ton, C., Parng, C. (2005). The use of zebrafish for assessing ototoxic and otoprotective agents. *Hear. Res.*, *208*, 79–88.

140. Soto, A., DelRaso, N.J., Schlager, J.J., Chan, V.T. (2008). D-Serine exposure resulted in gene expression changes indicative of activation of fibrogenic pathways and down-regulation of energy metabolism and oxidative stress response. *Toxicology*, *243*, 177–192.

141. Sasabe, J., Chiba, T., Yamada, M., Okamoto, K., Nishimoto, I., Matsuoka, M. Aiso, S. (2007). D-Serine is a key determinant of glutamate toxicity in amyotrophic lateral sclerosis. *EMBO J.*, *26*, 4149–4159.
142. Csapó, J., Varga-Visi, É., Lóki, K., Albert, C. (2006). Analysis of the racemization of tryptophan. *Chromatographia*, *63*, S101–S104.
143. Luzzana, U., Mentasti, T., Opstvedt, J., Nygård, E., Moretti, V. M., Valfrè, F. (1999). Racemization kinetics of aspartic acid in fish material under different conditions of moisture, pH, and oxygen pressure. *J. Agric. Food Chem.*, *47*, 2879–2884.

6.3

CHLOROPROPANOLS

JAN VELÍŠEK

Institute of Chemical Technology, Department of Food Chemistry and Analysis, Technická 1905, Prague 166 28, Czech Republic

6.3.1 INTRODUCTION

The collective term "chloropropanols" is used to define a group of chemical contaminants comprising three carbon alcohols and diols with one or two chlorine atoms that are hypothetically derived from glycerol (1,2,3-propanetriol). Six different compounds, (di)chloropropanols, chloropropanediols, and a closely related dichloropropane (Fig. 6.3.1), were identified as contaminants of the savory food ingredient acid-hydrolyzed vegetable protein (acid-HVP) in the 1970s and 1980s. Subsequent research in the 1990s revealed the presence of some of these contaminants in soy sauces and similar products manufactured using acid-HVP as an ingredient. Some chloropropanols were also found in a range of processed foods and food ingredients as a result of processing, manufacture, domestic cooking, or migration from packaging material during storage. These products are reviewed in Chapter 2.6.

The question over the use and safety of acid-HVP was first raised by Velíšek *et al.* at the Prague Institute of Chemical Technology in 1978 (1). Their work showed the presence of three chlorine-containing volatiles, namely 1,3-dichloropropan-2-ol (1,3-DCP), 2,3-dichloropropan-1-ol (2,3-DCP), and 3-chloropropan-1-ol (Fig. 6.3.1). Model studies with glycerol (2), triacylglycerols (2, 3), and natural lipids (4) suggested that glycerol monochlorohydrins, i.e., 3-chloropropane-1,2-diol (often called 3-monochloropropane-1,2-diol, 3-MCPD) and its positional isomer 2-chloropropane-1,3-diol (2-monochloropropane-

Process-Induced Food Toxicants: Occurrence, Formation, Mitigation, and Health Risks, Edited by Richard H. Stadler and David R. Lineback

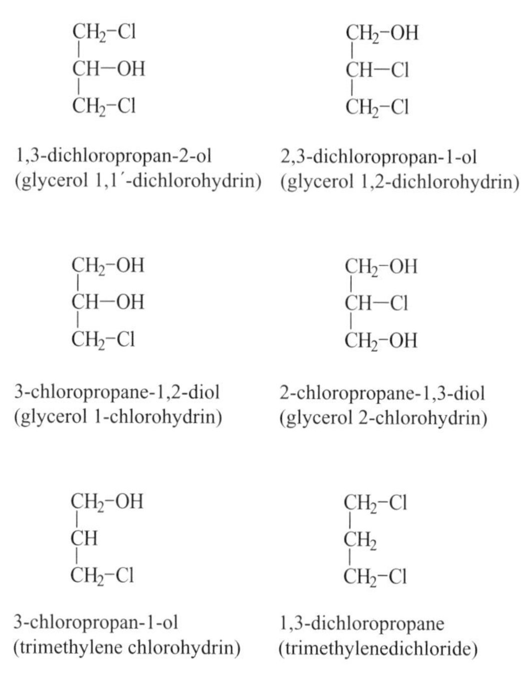

Figure 6.3.1 Structure, systematic, and trivial names of chloropropanols and chloroalkanes found in protein hydrolysates.

1,3-diol, 2-MCPD), may also become the constituents of commercial HVPs. 3-MCPD was later found as a component of HVPs by Davídek *et al.* in 1981 (5) and in 1987 its positional isomer 2-MCPD was identified (6). Investigations mainly done by industry until 1989 confirmed these findings and identified 1,3-dichloropropane as another minor constituent of acid-HVP (6). In the 1990s, MCPDs and DCPs have been often found in acid-HVPs, soy sauces, and related products (7–14).

Recently, it was found that the two chiral chloropropanols derived from prochiral L-glycerol (i.e., 3-MCPD and 2,3-DCP, respectively) occur in acid-HVPs as racemic mixtures of their (*R*)- and (*S*)-enantiomers (Fig. 6.3.2) (15) (J. Velíšek and M. Doležal, unpublished results).

Several other types of chlorine-containing organic compounds were identified as the minor constituents of HVPs, i.e., chlorinated sugar decomposition product 5-chloromethylfuran-2-carbaldehyde (16), chlorinated esters of levulinic (4-oxopentanoic) acid (e.g., 3-chloro-1-hydroxy-2-propyllevulinate, 3-chloro-2-hydroxy-1-propyllevulinate, and 3-chloropropyl-1,2-dihydroxy-1,2-dilevulinate) (17), and 3-chloro-Δ^5-ene analogs of 4-demethylsterols (sterols) including analogs of sitosterol, campesterol, stigmasterol, and cholesterol (18).

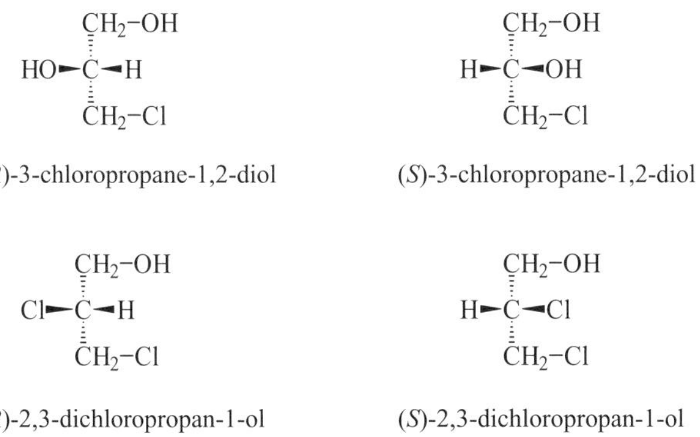

$$\begin{array}{c} CH_2{-}OH \\ HO \blacktriangleright C \blacktriangleleft H \\ CH_2{-}Cl \end{array}$$

(*R*)-3-chloropropane-1,2-diol

$$\begin{array}{c} CH_2{-}OH \\ H \blacktriangleright C \blacktriangleleft OH \\ CH_2{-}Cl \end{array}$$

(*S*)-3-chloropropane-1,2-diol

$$\begin{array}{c} CH_2{-}OH \\ Cl \blacktriangleright C \blacktriangleleft H \\ CH_2{-}Cl \end{array}$$

(*R*)-2,3-dichloropropan-1-ol

$$\begin{array}{c} CH_2{-}OH \\ H \blacktriangleright C \blacktriangleleft Cl \\ CH_2{-}Cl \end{array}$$

(*S*)-2,3-dichloropropan-1-ol

Figure 6.3.2 Structure of 3-chloropropane-1,2-diol and 2,3-dichloropropan-1-ol enantiomers.

6.3.2 OCCURRENCE IN PROTEIN HYDROLYSATES

6.3.2.1 Manufacture of Protein Hydrolysates

Protein hydrolysates generally can be divided throughout the world into two large principal groups, chemical hydrolysates of proteins that originated in Europe and enzymatic hydrolysates of proteins that originated in the Far East, respectively (19, 20).

6.3.2.1.1 Acid-Hydrolyzed Vegetable Proteins Chemical hydrolysates of proteins, also called hydrolyzed plant proteins (HPPs) or hydrolyzed vegetable proteins (HVPs), are commonly produced by hydrochloric acid (HCl) hydrolysis of various proteinaceous vegetable raw materials (acid-HVPs). These raw materials include defatted oilseed meals, the by-products from edible oil production (such as soybean meal, peanut meal, etc.), wheat gluten, maize and rice protein, and occasionally certain animal proteins (casein, keratin, etc.). The first commercial acid-HVPs were produced by Julius Michael Johannes Maggi (1846–1912), a Swiss entrepreneur, an inventor of Maggi spice and precooked soups, and the founder of the food company Maggi. In 1863, he developed a formula to bring added taste to meals (21). This type of material (acid-HVP) has been used constantly as a food condiment and ingredient since that time.

The manufacturing process for conventional acid-HVPs varies depending on the desired organoleptic properties of the end product. The source of the raw material, concentration of the acid, the temperature of the reaction, the time of the reaction, and other factors can all affect the organoleptic properties of the final product (22). Acid-HVPs are traditionally prepared by hydrolysis using approximately 20% (6 M) HCl. Acid hydrolysis is usually carried out at temperatures exceeding 100 °C either at the atmospheric pressure or under a given elevated pressure for up to about 8 h until the α-amino nitrogen content of the hydrolysate is in the range from approximately 35% to 55% of the total nitrogen content (19). The duration of the hydrolysis process is related to the type of protein being hydrolyzed and the pressure at which it is being hydrolyzed. After the period of hydrolysis and after cooling, the hydrolysate is neutralized by a suitable sodium alkali (sodium hydroxide or sodium carbonate), generally to a pH ranging from 4.5 to 7 at a temperature between 90 and 100 °C for 90–180 min, which results in the salt content of 35–50% (solid base) in the final neutralized hydrolysate. The pH to which the hydrolysate is adjusted depends on the nature of the proteinaceous material and the type of flavor required. The products in this stage of the process are raw acid-HVPs in liquid form. A filtration step is required to remove the insoluble residual solid material called humins that has not undergone hydrolysis. The filtrate can then be bleached or refined. Activated carbon treatment can be employed to remove both undesirable flavor and color components. Following further filtration the acid-HVP may, depending upon the application, be fortified with additional flavor-active compounds. The next step is a ripening process that takes several weeks and the ripe product is then available as a liquid acid-HVP at 30–40% dry matter (corresponding to 2–3% total nitrogen).

Alternatively, this final hydrolysate can be concentrated to a paste or dried to produce a powder (80–85% dry matter) depending on the need of the manufacturer. It can be also blended with other hydrolysates. The range of condiments based on acid-HVPs is large, and includes soy sauce, dark soy sauce, light soy sauce, mushroom soy sauce, oyster sauce, reduced soy sauce, seasoning sauce, shrimp-flavored soy sauce, thick soy sauce, and teriyaki sauce (22).

6.3.2.1.2 Soy Sauces Soy sauce or soya sauce is the best-known representative of the second group of protein hydrolysates. Soy sauce is made from soybeans (soya beans), roasted wheat, water, and salt. The sauce originated in ancient China in about 500s B.C. Its production in Japan is said to have been started by a priest Zen from China in about 1250 A.D. Soy sauces are commonly used in East and Southeast Asian cuisine and appear in some Western cuisine dishes (20).

The process of manufacture for naturally fermented soy sauce can be divided into four stages: preparation of *koji*, brine fermentation, filtration/pasteurization, and maturation. Koji is produced by cooking soybeans or defatted soybean meal and then mixing the resultant material with coarsely broken roasted wheat. The mixture is inoculated with a mold, either *Aspergillus oryzae* or *Aspergillus sojae*. The koji is then cultured for 2 to 3 days at 25–30 °C and then mixed with saltwater brine and yeast to give *moromi*. The moromi is fermented for 2 days and the resulting thick paste is pressed to obtain the raw soy sauce. The raw soy sauce is pasteurized and then filtered for clarity and stored for maturation over a period of months. The product is finally filtered, bottled, and distributed to the consumer. Short-term fermented soy sauce is produced in a similar manner except that the saltwater fermentation/ageing stage takes place at or above 40 °C and the process is completed within 90 days. *Shoyu* prepared by a total fermentative processes (naturally brewed) is differentiated by the label genuine fermented (20).

The protein treated with HCl can also be used as a raw material in a fermentation process that is faster than the traditional fermentation. Manufacture of some products involves mixing fermented soy sauces with acid-HVPs and ageing after mixing. Japanese soy sauces have been mixed, for example, with 40% chemical hydrolysate (on the nitrogen base) ranging from 100% chemical soy sauce to 100% genuine fermented soy sauce. All soy sauce varieties are sold in three different grades according to how they were produced. *Honjōzō hōshiki* contains 100% naturally fermented product, *shinshiki hōshiki* contains 30–50% naturally fermented product, and *aminosanekikongō hōshiki* contains 0% fermented product, respectively (20, 22).

6.3.2.2 Amounts in Protein Hydrolysates

6.3.2.2.1 Acid-HVPs Chloropropanols have most frequently been found in acid-HVPs that are widely used as seasonings and ingredients in a variety of processed savory food products, including many processed and pre-prepared foods, soups, gravy mixes, savory snacks, and bouillon cubes, with the typical amounts of use ranging from 0.1% to approximately 0.8% in these foodstuffs (23).

Generally, 3-MCPD is the most widely occurring chloropropanol in acid-HVPs. The other chloropropanols that can occur, albeit usually in smaller amounts, are 2-MCPD, 1,3-DCP, and 2,3-DCP. Investigations made in the 1980s showed that the relative proportions of the major chloropropanols (3-MCPD, 2-MCPD, 1,3-DCP, and 2,3-DCP) occurring in HVPs manufactured by the conventional technological procedures were approximately in the ratio of 1000:100:10:1 (24). These HVPs contained 100–800 mg/kg of 3-MCPD, 10–90 mg/kg of 2-MCPD, 0.1–6 mg/kg of 1,3-DCP, and 0.01–0.5 mg/kg of 2,3-DCP, respectively (25). Amounts of these chloropropanols in HVPs depended on the lipid content in the raw material, concentration and amount of HCl, temperature and pressure used, and duration of the hydrolysis process.

In 1994, following a review of the available data, the Scientific Committee for Food (SCF) (26) advised that residues of 3-MCPD in food products should be undetectable by the most sensitive analytical method and that all efforts should be undertaken to develop methods leading to products not containing chlorinated propanols. In May 1996, the Food Advisory Committee (FAC) recommended (27) that 3-MCPD should be reduced to the minimum detectable by the most sensitive and reliable analytical method available. However, since HVP manufacturers stated that they were unable to remove 3-MCPD entirely from their products, FAC further recommended that the food industry should be consulted on the timescale over which a switch could be made from acid-HVP to other products that do not contain 3-MCPD. As a consequence, manufacturers investigated ways of changing their production processes to further reduce or to eliminate 3-MCPD in their products. In October 1996, the FAC recommended (28) that within 18 months the industry should take all the steps necessary to ensure that 3-MCPD cannot be found in any foods or ingredients, regardless of the method of manufacture, when using a validated method of analysis capable of measuring down to 0.01 mg/kg, the lowest concentration of detection currently achievable. This meant that the use of acid-HVP could continue provided that it met the 0.01 mg/kg limit. In April 1998, the FAC confirmed its advice that industry should reduce the amount of 3-MCPD in foods and food ingredients to less than 0.01 mg/kg (29). Human health risk assessment of chloropropanols occurring in foods has recently been reviewed (30).

The presence of chloropropanols in food is of concern due to their toxicological properties (see Chapter 2.6). In view of 3-MCPD toxicity, the EC's Scientific Committee on Food has proposed a provisional tolerable daily intake (TDI) amount of 2 μg/kg body weight/day for the amount of 3-MCPD that can be consumed daily over a lifetime without appreciable harm to health (31). The TDI was adopted on 8 March 2001 and applies from 5 April 2002. Similarly, the Joint FAO/WHO Expert Committee on Food Additives (JECFA) set a provisional maximum tolerable daily intake (PMTDI) of 2 μg/kg body weight/day in 2001 (32). A regulatory limit of 0.02 mg/kg, based on a 40% dry matter content, has been adopted for 3-MCPD in acid-HVPs and soy sauces only and came into force in the European Union in 2002 (33). This corresponds to a maximum concentration of 0.05 mg/kg in the dry matter.

As a consequence, in the 1980s manufacturers started to investigate ways of changing their production processes to reduce or to eliminate 3-MCPD in their products (see Section 6.3.5). For example, acid-HVPs analyzed in 1992 contained 3-MCPD, 2-MCPD, 1,3-DCP, and 2,3-DCP at 19.4–549 mg/kg, 3.1–87.0 mg/kg, <0.01–4.03 mg/kg, and <0.01–0.45 mg/kg, respectively (14). Two surveys carried out by the UK government in 1990 and 1992 showed a decline in concentrations of 3-MCPD in acid-HVPs. Of the 39 samples tested in the 1990 survey, 33 exceeded the limit of detection for 3-MCPD of 1 mg/kg, with 23 of these being above the SCF's recommended limit of 10 mg/kg (34). In 1992, half of the 34 acid-HVP samples included in the survey contained 3-MCPD in excess of 1 mg/kg (35).

Four seasonings analyzed in 1998 had the concentrations of 3-MCPD between 0.428 and 0.0004 mg/kg and the concentrations of 2-MCPD between below the limit of detection and 0.125 mg/kg (8). Fifty samples of acid-HVP analyzed in 1999 revealed that 3-MCPD was undetectable in 21 samples (42%), with a further nine samples (18%) containing between 0.01 and 0.02 mg/kg. 3-MCPD was found to be present in eight samples (16%) between 0.02 and 0.05 mg/kg; in three samples (6%) between 0.05 and 0.1 mg/kg; and more than 0.1 mg/kg in nine samples (18%). The highest result was 2 mg/kg. Although the results show that amounts of 3-MCPD in acid-HVP have continued to decline, the results for a number of products were high relative to the limit (0.01 mg/kg) recommended by the FAC (28, 29).

TABLE 6.3.1 Concentrations of MCPD and DCP fatty acid esters in the neutralized raw acid-HVP and humins.

	Concentration in mg/kg	
Compound	Neutralized raw acid-HVP	Humins
3-Chloropropane-1,2-diol diesters	4	35
3-Chloropropane-1,2-diol monoesters	35	205
1,3-Dichloropropan-2-ol esters	8	65

Concerning 1,3-DCP, JECFA concluded in 1993 that this compound is genotoxic *in vivo* and consequently no amount is safe. Its content in food products should be reduced as much as possible. It means that it should be present in foods at the lowest technologically feasible concentration, which for acid-HVPs and soy sauces equates to not present at any concentration (36, 37). The maximum limit of 0.005 mg/kg for 1,3-DCP in soy sauces is the current limit of quantification and is considered to be as close as is possible to zero presence. For example, 1,3-DCP was found in only one sample in the 1990 survey (34), with none being detected in any of the 34 samples analyzed in the 1992 survey (35).

Besides MCPDs and DCPs, monoesters and diesters of 3-MCPD and esters of 1,3-DCP with higher fatty acids form during production of acid-HVPs (4, 15). Five major fatty acid (palmitic, stearic, oleic, linoleic, and linolenic) were the constituents of MCPD and DCP esters found in the raw neutralized hydrolysates. These esters were not identified in the final commercial acid-HVP used as a soup seasoning as they were effectively removed by filtration. The amount of the individual esters in the neutralized hydrolysate and humins is summarized in Table 6.3.1.

6.3.2.2.2 Soy Sauces In addition to the direct usage of acid-HVP as an ingredient for soy sauces and similar products, chloropropanols may also be formed in those soy sauces and related condiments where the manufacturing process of the sauce itself includes HCl treatment of soybean meal. Products made exclusively by means of fermentation generally do not contain chloropropanols, or, if present, they only occur in trace amounts.

In the 1990s, officials in the Joint Food Safety and Standards Group (JFSSG) have been notified that the authorities in Denmark, Germany, the Netherlands, and Sweden have found high concentrations of 3-MCPD (6–124 mg/kg) in several brands of soy sauce imported into and sold in EU member states. For example, soy sauces analyzed in 1989 and 1990 had their concentrations of 3-MCPD and 2-MCPD below the limit of detection and seasonings based on soy sauces contained 3-MCPD at the amount of 0.2–43 mg/kg and 2-MCPD at the amount of 0.3–7 mg/kg (7). Thirteen soy sauces analyzed in 1998 had 3-MCPD concentrations of between 0.0004 and 4.72 mg/kg, and 2-MCPD concentrations of between below the limit of detection and 0.256 mg/kg (8).

Analyses of soy sauces provided by the CVUA Stuttgart in Germany in 1997–2001 revealed that the mean concentration of 3-MCPD (14.23 mg/kg) found in 1997 decreased to 5.71 mg/kg in 1998, to 2.14 mg/kg in 1999, to 0.20 mg/kg in 2001, and to 0.04 mg/kg in 2001 (38). Analysis of 40 samples of soy, mushroom soy, oyster sauce, and other sauces purchased from retail outlets done by the Ministry of Agriculture Fisheries and Food (MAFF) in Great Britain in 1999 (9, 13) revealed that 3-MCPD was undetectable in 21 samples analyzed in the survey, with a further five samples (13%) containing very low amounts of between 0.01 and 0.02 mg/kg. However, nine samples (22%) contained 3-MCPD concentrations of more than 1 mg/kg, the highest being 30 mg/kg. In a survey carried out by the UK government in 2000 (11, 39), 17 out of 100 samples (of these samples 67 were soy sauces, the remaining samples were various other related sauces, i.e., mushroom soy sauce, oyster sauce, and teriyaki) contained quantifiable concentrations of 1,3-DCP, all of which contained concentrations of 3-MCPD greater than 0.02 mg/kg and also greater than the amount of 1,3-DCP observed. The highest amount of 1,3-DCP was 0.345 mg/kg, in a soy sauce. In a survey of 55 soy sauces and related products available in the United States in 2003 (10), 85% of the samples analyzed contained amounts higher than the detection limit of 0.005 mg/kg for 3-MCPD, 33% contained amounts higher than 1 mg/kg; the highest amount was 876 mg/kg. Thirty-nine of the samples analyzed for 3-MCPD also were analyzed for 1,3-DCP and 56% of these samples contained higher concentrations than the detection limit of 0.055 μg/kg; the highest amount was 9.8 mg/kg 1,3-DCP.

6.3.3 ANALYSIS

A variety of analytical methods have been used for the determination of chloropropanols in acid-HVP, soy sauces, and related products as well as in processed foods. The procedures nowadays used for the determination of MCPDs mainly rely on capillary gas chromatography/mass spectrometry of stable volatile derivatives and have been shown to be the methods of choice as 3-MCPD and its positional isomer 2-MCPD are polar diols having high boiling points, which complicates their direct analysis. Some of these methods used for the determination of MCPD are also suitable for the determination of DCPs. The sufficient volatility of 1,3-DCP and 2,3-DCP also enables their direct analysis by gas chromatography but, at the same time, makes the concentration of solvent extracts difficult due to some losses of these analytes. The overview of analytical methods currently used is given in Chapter 2.6.

6.3.4 FORMATION

The occurrence of chloropropanols in acid-HVP arises from their formation during the HCl-mediated hydrolysis step of the manufacturing process. It has

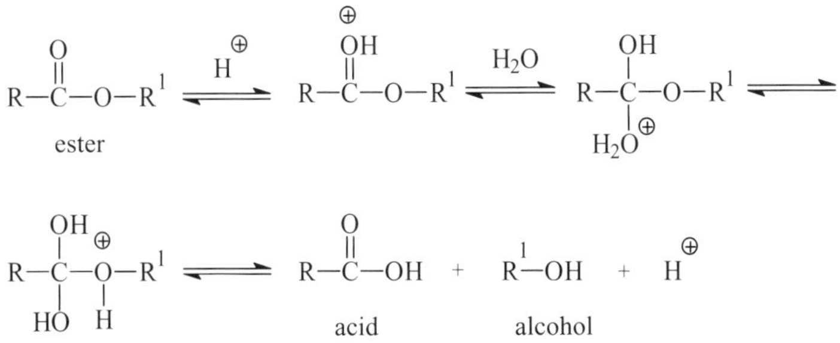

$$\underset{\text{ester}}{R{-}\overset{\overset{\large O}{\|}}{C}{-}O{-}R^1} \; \overset{H^{\oplus}}{\rightleftharpoons} \; R{-}\overset{\overset{\overset{\oplus}{OH}}{\|}}{C}{-}O{-}R^1 \; \overset{H_2O}{\rightleftharpoons} \; R{-}\underset{\underset{H_2O^{\oplus}}{|}}{\overset{\overset{OH}{|}}{C}}{-}O{-}R^1 \; \rightleftharpoons$$

$$R{-}\underset{\underset{HO}{|}}{\overset{\overset{OH}{|}}{C}}{-}\underset{\underset{H}{|}}{\overset{\oplus}{O}}{-}R^1 \; \rightleftharpoons \; \underset{\text{acid}}{R{-}\overset{\overset{O}{\|}}{C}{-}OH} \; + \; \underset{\text{alcohol}}{R^1{-}OH} \; + \; H^{\oplus}$$

Figure 6.3.3 Hydrolysis of esters in acid media.

been established (2–4, 40, 41) that during this hydrolytic stage, the acid reacts with residual lipids (i.e., triacylglycerols and phospholipids) present in the raw proteinaceous material as well as with glycerol formed by acid hydrolysis of glycerolipids.[1] For example, wheat gluten used for the production of HVPs contains about 0.5–3% of lipids, of which 30–36% constitutes neutral lipids, mainly triacylglycerols (42). Soybean meal (43) has similar amount of residual lipids (1.0–3.0%) principally composed from neutral lipids (about 30%) and phospholipids (60%). Experiments of Collier *et al.* (44) have shown that the major precursors of chloropropanols are triacylglycerols and, to a smaller extent, phospholipids and glycerol in decreasing order. DCPs evidently arise from HCl and MCPDs. It should be noted that the formation of chloropropanols cannot be avoided through the use of defatted protein sources.

6.3.4.1 Formation from Glycerol

HVPs contain relatively high amounts of glycerol that forms by hydrolysis of glycerolipids with HCl. Its content found in several types of commercial HVPs ranged from 750 to 3100 mg/kg (43). In acid solution (e.g., during HVP manufacture), the acid-catalyzed reaction (mechanistic designation $A_{AC}2$, where A denotes acid catalysis, $_{AC}$ indicates acyl–oxygen bond cleavage, digit 2 indicates bimolecular nature of the rate-determining step) is reversible (Fig. 6.3.3).

It has been known that the nucleophilic substitution of hydroxyl group in alcohols by chlorine anion derived from HCl leads to alkyl chlorides (45). HCl and alcohols form an equilibrium mixture and its composition depends on the structure of the alcohols, temperature, water content, and presence of other compounds (e.g., carboxylic acids) that may act as catalysts. Diols and triols

[1]The raw materials mainly contain triacylglycerols and glycerophospholipids. Triacylglycerols represent a storage form of energy for plant seeds. Triacylglycerols and partial fatty acid esters of glycerol (diacylglycerols and monoacylglycerols) belong to the group of lipids called homolipids, while glycerophospholipids are classified as heterolipids. Homolipids derived from glycerol, glycerophospholipids, and glyceroglycolipids together constitute one class of lipids known as glycerolipids.

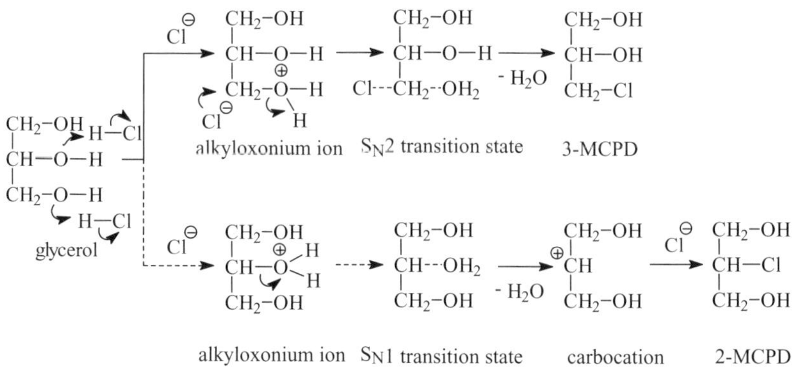

Figure 6.3.4 Reaction of glycerol with hydrochloric acid.

can form either monochloroderivatives or dichloroderivatives. Glycerol reacts with HCl with the formation of 3-MCPD and 2-MCPD. The distribution of both these isomers (approximately present in the ratio of 2:1) is the result of a nucleophilic substitution of the hydroxyl groups by chloride anion, in accord with statistical substitution of two equivalent primary hydroxyls and one secondary hydroxyl group (44). Hydroxyl groups of glycerol are first protonated by HCl to alkyloxonium ions (conjugated acids). With the primary hydroxyls, the next stage is an S_N2 reaction in which the chloride ion displaces a molecule of water from the alkyloxonium cation. This pathway is stereospecific and proceeds with inversion of configuration at the carbon that bears the leaving group (45). According to the theory, it yields a racemic mixture of both enantiomers of 3-MCPD. With the secondary hydroxyl group, this stage is an S_N1 reaction in which the alkyloxonium ion dissociates to a carbocation and water. Following its formation, the carbocation is captured by chloride ion under the formation of 2-MCPD (Fig. 6.3.4). The ratio of 3-chloro isomer to 2-chloro isomer is about 2.3.

The yield of chloropropanediols can be enhanced in the presence of carboxylic acids (such as acetic acid and to a much smaller extent by the addition of fatty acids or amino acids) that form with glycerol the corresponding esters (predominantly 1-acyl-*sn*-glycerol and to a smaller extent 2-acyl-*sn*-glycerol) in acid solutions (Fig. 6.3.5) (45). Monoacylglycerols readily eliminate hydroxyl groups yielding a cyclic acyloxonium ion intermediate. The reaction of the acyloxonium ion intermediate with chloride anion leads mostly to 2-ester of 3-MCPD. The corresponding 1-ester is a minor product. Both monoesters are hydrolyzed to 3-MCPD and 2-MCPD, respectively. For example, the ratio of MCPDs in the presence of acetic acid is higher than in the case of glycerol without acetic acid being about 6.4 as it is controlled by the steric and electronic effects arising from the terminal ester group, which directs substitution of chloride ion primarily to the CH_2 carbon atom. Analogous reactions can

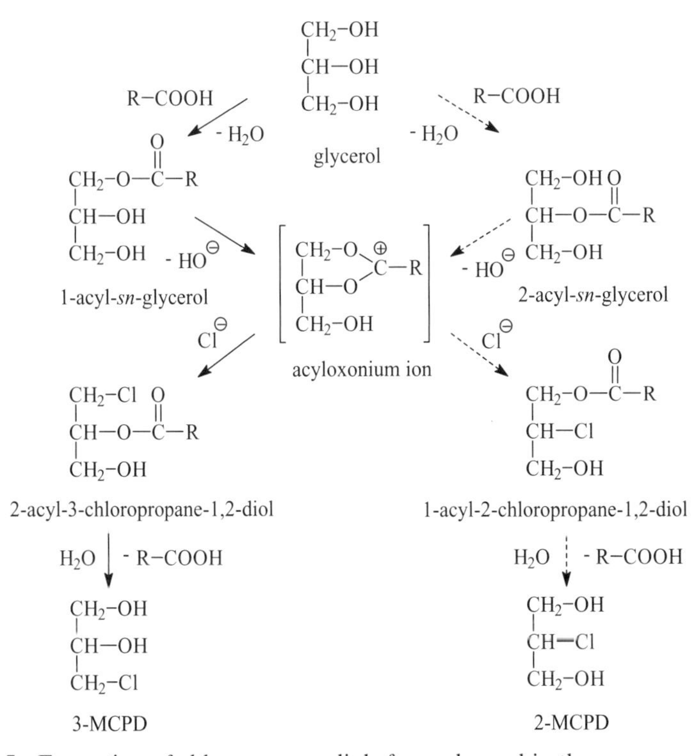

Figure 6.3.5 Formation of chloropropanediols from glycerol in the presence of acids.

proceed with partial acylglycerols (monoacylglycerols and diacylglycerols) derived from higher fatty acids. Another way in which chloropropanediols can arise is a direct substitution of the acyl groups or hydroxyl groups in monoacylglycerols by chloride anions.

Another possible mechanism for the formation of MCPDs from glycerol via intermediate glycidol can more likely proceed in low-moisture food (see Chapter 2.6).

MCPDs react with another molecule of HCl forming DCPs. Hydroxyl groups of MCPDs are eliminated as water and the intermediate carbocations combine with chloride anions yielding DCPs (45). The proposed simplified reaction mechanism starting from 3-MCPD and 2-MCPD is given in Fig. 6.3.6. Similarly to the formation of 2-MCPD from glycerol (Fig. 6.3.4), the chlorination of 3-MCPD at the secondary hydroxyl group can also yield 2,3-DCP.

6.3.4.2 Formation from Triacylglycerols

A simplified reaction scheme leading to the formation of 3-MCPD from triacylglycerols is given in Fig. 6.3.7 (reactions leading to 2-MCPD are not given). Triacylglycerols are first transformed to diacylglycerols by acid hydrolysis and

$CH_2{-}OH / CH{-}OH / CH_2{-}Cl$ (3-MCPD) $\xrightarrow{H^{\oplus}}$ $CH_2{-}\overset{\oplus}{O}H_2 / CH{-}OH / CH_2{-}Cl$ (alkyloxonium ion) $\xrightarrow[- H_2O]{}$ $\overset{\oplus}{C}H_2 / CH{-}OH / CH_2{-}Cl$ (carbocation) $\xrightarrow{Cl^{\ominus}}$ $CH_2{-}Cl / CH{-}OH / CH_2{-}Cl$ (1,3-DCP)

$CH_2{-}OH / CH{-}Cl / CH_2{-}OH$ (2-MCPD) $\xrightarrow{H^{\oplus}}$ $CH_2{-}\overset{\oplus}{O}H_2 / CH{-}Cl / CH_2{-}OH$ (alkyloxonium ion) $\xrightarrow[- H_2O]{}$ $\overset{\oplus}{C}H_2 / CH{-}Cl / CH_2{-}OH$ (carbocation) $\xrightarrow{Cl^{\ominus}}$ $CH_2{-}Cl / CH{-}Cl / CH_2{-}OH$ (2,3-DCP)

Figure 6.3.6 Formation of dichloropropanols from chloropropanediols.

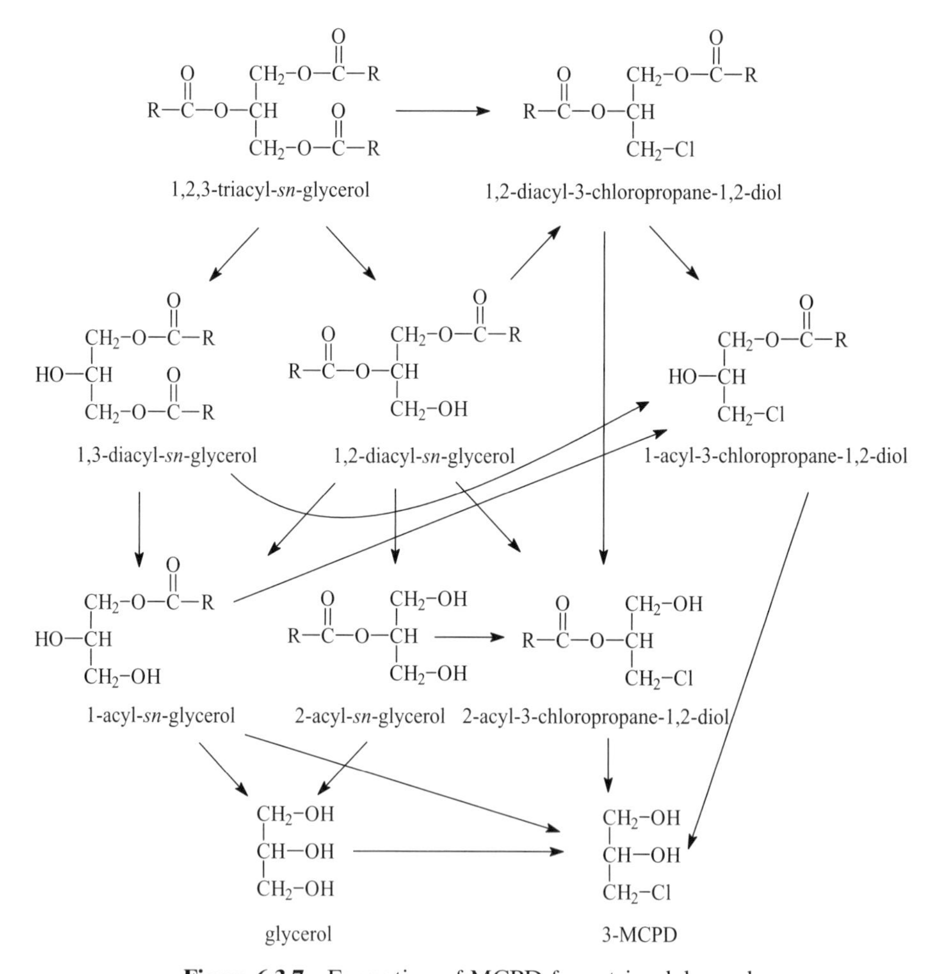

Figure 6.3.7 Formation of MCPD from triacylglycerols.

hydrolysis of diacylglycerols then leads to monoacylglycerols and finally to glycerol (Fig. 6.3.6). The principal reaction leading to 3-MCPD diesters is explained, in accord with the theory, as proceeding via the partial diacylglycerols with the ester group(s) providing anchimeric assistance through the formation of an acylated cyclic acyloxonium ion intermediate (44), analogously to the reaction of glycerol with HCl in the presence of carboxylic acids (Fig. 6.3.5). Diacylglycerols can be also hydrolyzed to monoacylglycerols that can again form the cyclic acyloxonium ions, which are opened by chloride anions to yield 3-MCPD monoesters. 3-MCPD diesters can be hydrolyzed to 3-MCPD monoesters and these to 3-MCPD. Another pathway that can take place in the frame of this complex reaction is a direct substitution of either the acyl group (acidolysis by HCl) or hydroxyl group by chloride anions in acylglycerols.

The ratio of chloropropanediols is again controlled by the steric and electronic effects arising from the terminal ester group, which directs substitution to the CH_2 carbon atom. Additionally, the regiospecificity is greater with 3-isomer to 2-isomer ratio of approximately 10:1 (44).

6.3.4.3 Formation from Phospholipids

Phospholipids are apparently hydrolyzed by HCl to totally deacylated derivatives, for example, (3-*sn*-phosphatidyl)choline (1,2-diacyl-*sn*-glycero-3-phosphocholine) yields *sn*-glycero-3-phosphocholine (Fig. 6.3.8). 1,2-Diacyl-*sn*-glycero-3-phosphatidylcholine can react with chloride ions yielding, for example, the

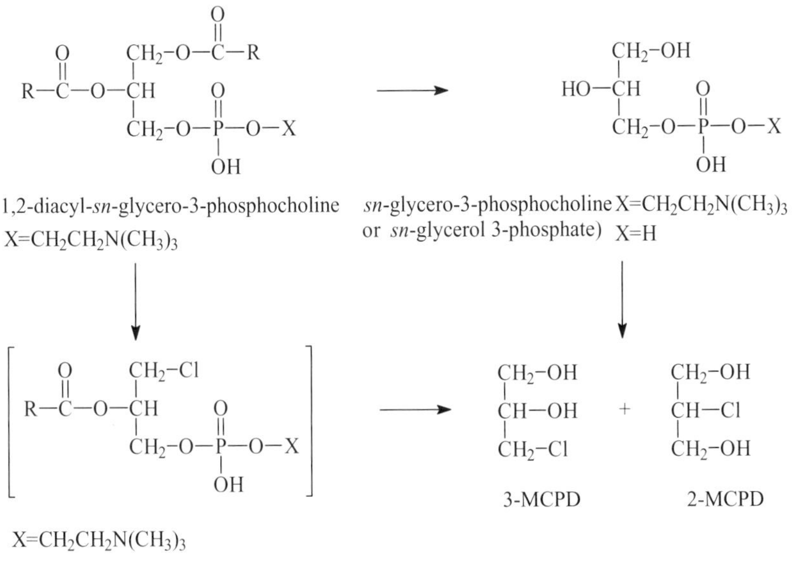

1,2-diacyl-*sn*-glycero-3-phosphocholine $X=CH_2CH_2N(CH_3)_3$

sn-glycero-3-phosphocholine $X=CH_2CH_2N(CH_3)_3$ or *sn*-glycerol 3-phosphate) $X=H$

$X=CH_2CH_2N(CH_3)_3$

3-MCPD

2-MCPD

Figure 6.3.8 Formation of MCPD from phospholipids.

corresponding 1-chloroderivative, which is hydrolyzed to 3-MCPD. The substituted phosphate group in the deacylated derivatives can be replaced by chloride ions, which leads to MCPDs. Similarly to glycerol, phospholipids show little regioselectivity due to facile intramolecular isomerization of *sn*-glycero-3-phosphocholine to *sn*-glycero-2-phosphocholine, and the ratio of the two isomers of MCPD was 2.8 (44).

6.3.5 MITIGATION

Measures were taken in the early 1980s by western European countries to reduce the concentrations of the only known dichloropropanols in conventionally produced acid-HVPs and related products, and the industry developed new production processes. As a results of these efforts, volatile dichloropropanols, namely 1,3-DCP and 2,3-DCP, have been virtually eliminated from HVPs by several removal systems implemented in 1982, by stripping the raw hydrolysate with overheated vapor at a reduced pressure (Maggi patent application, Buss apparatus and continuous operation) or by evaporation either operated in a continuous or batch apparatus. This step efficiently reduced the amount of volatile dichloropropanols[2] below 0.05 mg/kg.

The major acid-HVP contaminant, 3-MCPD, discovered in acid-HVPs in 1981, could not be removed by the stripping process as it does not distill with water vapor. It boils at the atmospheric pressure at 213 °C and its solubility in water is unlimited. Acid-HVP manufacturers have then implemented the necessary procedures during the late 1980s to minimize 3-MCPD formation and many manufacturers undertook reformulation of their products and/or they moved away from the use of acid-HVPs to enzyme-hydrolyzed and flavored vegetable proteins (46–51).

Actions taken to reduce 3-MCPD formation have an impact on organoleptic quality (taste, for example, free from raw flavor), yield, and chemical composition of acid-HVPs. The challenge for the manufacturers is to optimize the product so that the effects of changes in organoleptic properties experienced when using the improved methods of manufacture could be minimized. Parameters need to be optimized in order to balance the trade-off between low 3-MCPD concentrations and the organoleptic quality. It should be noted that different regional markets may require products with different organoleptic qualities to accommodate specific regional tastes. The individual approaches and combinations thereof to minimize amounts of 3-MCPD will have different effects on the organoleptic qualities of the final product and as such, manufacturers should take these effects into account when selecting a strategy to minimize 3-MCPD formation. Some manufacturers have stated

[2] 1,3-Dichloropropan-2-ol boils at 175 °C at the normal pressure and its solubility in water at 20 °C is 15.2 g per 100 g of water. It gives with water an azeotropic mixture boiling at 99 °C, which is composed of 76.8% of water and 23.2% of 1,3-dichloropropan-2-ol (43).

that, while it is technically possible to reduce 3-MCPD concentrations to less than 0.1 mg/kg, the organoleptic qualities of such products are adversely affected (52).

6.3.5.1 Practice for Removal of Chloropropanols

6.3.5.1.1 Acid-HVPs Three main approaches based on good manufacturing practice (GMP) have to be adopted to minimize the concentration of 3-MCPD in the final product. The first of these involves careful control of the acid hydrolysis step and the subsequent neutralization to minimize 3-MCPD formation. The alternative strategy entails destroying the 3-MCPD formed during the acid hydrolysis by employing a subsequent alkali treatment stage. Both of these processes require close control to minimize 3-MCPD formation and at the same time, prevent the occurrence of undesirable flavor components. Manufacturers mainly employ either or both of these strategies to minimize 3-MCPD in acid-HVPs in a production line. The third approach used a liquid extraction step to remove 3-MCPD from the final product.

With regard to the first strategy, the temperature and the heating time of the acid hydrolysis step must be simultaneously controlled and careful attention paid to the reaction conditions in the subsequent neutralization step. Typically, the hydrolysis reaction is initially carried out at a temperature between 60 and 95 °C. The temperature of the reaction is then increased at a rate of 0.01–0.3 °C until a temperature of 110 °C is attained. Once this maximum temperature is reached, it should be maintained for up to 2 h and then the resulting hydrolysate cooled, neutralized, and filtered.

Evidence suggests that the neutralization step can also be used as a method for reducing the amounts of chloropropanols in acid-HVP. If the pH of the hydrolysate is increased to 7.5 and heated to 100 °C for 1 h during the neutralization step, 3-MCPD concentrations of less than 1 mg/kg can be achieved in the final product (49).

Alternatively, the 3-MCPD that is formed during the acid hydrolysis step may be removed by subsequent alkaline treatment (48, 51). This alkaline treatment is in essence an extension of the neutralization process that follows acid hydrolyzation of the starting material and causes degradation of the chloropropanols present in the hydrolysate. The alkaline treatment can be performed before or after filtration of the hydrolysate. The hydrolyzed protein is treated with food-acceptable alkali such as potassium hydroxide, sodium hydroxide, ammonium hydroxide, or sodium carbonate to increase the pH to 8–9. This mixture is then heated between 90 and 100 °C for 90–180 min to obtain the product with the lowest intensity of off-flavor. Generally, alkaline treatments at higher pH and temperature will require shorter processing times and will lead to a product with a high intensity of off-flavor. Following alkaline treatment, the pH of the hydrolyzed protein is readjusted to a pH of 5.0–5.5 using a suitable acid (e.g., HCl) at a temperature of 10–50 °C. The hydrolysate may now be filtered to remove any insoluble residues and the final product obtained.

TABLE 6.3.2 Concentrations of 3-MCPD and 2-MCPD in acid-HVPs subjected to alkaline treatment.

Acid-HVP	Reaction conditions			Concentration in mg/kg	
No.	Temperature, °C	pH	Time, h	Residual 3-MCPD[a]	Residual 2-MCPD[a]
1	100	8.5	3	<0.01 (i)	0.025 (ii)
2	95	9.2	2	<0.01 (iii)	0.072 (iv)
3	95	9.2	2.5	<0.01 (iii)	<0.01 (iv)
4	93	8.6	3.25	<0.01 (v)	0.068 (vi)
5	93	8.3	4.0	<0.01 (vii)	0.01 (viii)
6	92	8.4	3.5	<0.01 (vii)	0.028 (viii)
7	90	8.5	3.5	<0.01 (iii)	0.283 (iv)

[a]Starting concentration (i) = 95.7 mg/kg, (ii) = 15.1 mg/kg, (iii) = 209 mg/kg, (iv) = 37.7 mg/kg, (v) = 150 mg/kg, (vi) = 22.5 mg/kg, (vii) = 207 mg/kg, (viii) = 32.7 mg/kg.

Use of an alkaline treatment when manufacturing acid-HVP has been shown to yield a final product with 3-MCPD concentrations of less than 0.01 mg/kg (51). As an example, Table 6.3.2 shows the amounts of 3-MCPD and 2-MCPD in acid-HVPs (mostly prepared from soybean meal) subjected to different alkaline treatments.

A third documented method for the reduction of chloropropanols in acid-HVP involves the use of a countercurrent liquid/liquid extraction step (47). Acid hydrolysis is performed and the resulting hydrolysate is passed though a pulsed chromatography column, operating with a throughput of between 50 and 1500 L hydrolysate per hour. Ethyl acetate, butan-1-ol, butan-2-ol, and isobutanol are all suitable solvents for this extraction step. Following extraction, 1–2% of the extraction solvent may remain in the hydrolysate and this must be removed by stream stripping. Because the steam stripping will introduce water to the hydrolysate, the water must be removed (readily achievable by vacuum evaporation). HCl is added to the resulting concentrated hydrolysate and the pH adjusted to 5.4. 3-MCPD concentrations of less than 0.5 mg/kg should be achievable in the final product.

6.3.5.1.2 Soy Sauces The presence of 3-MCPD and 1,3-DCP in soy sauces is avoidable. A number of different manufacturing processes are employed in the production of soy sauces, including traditional fermented products as well as cheaper lower grades that may involve the use of an acid treatment or even include acid-HVP as an ingredient (see Chapter 2.6). The process used will impact on whether the product contains 3-MCPD. Soy sauces that are produced solely by fermentation should be free from 3-MCPD. All other products may contain 3-MCPD and measures to prevent its occurrence are described earlier for acid-HVP.

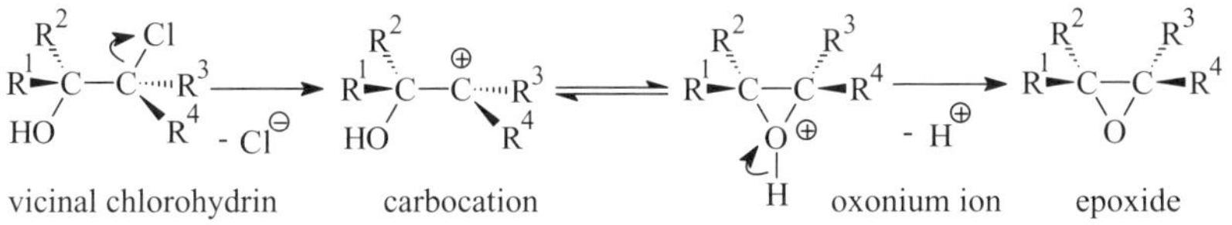

Figure 6.3.9 Mechanism of dehydrochlorination of vicinal chlorohydrins in neutral and acid media.

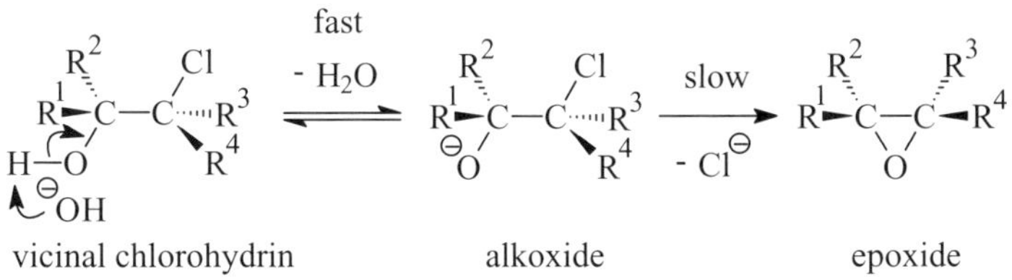

Figure 6.3.10 Mechanism of dehydrochlorination of vicinal chlorohydrins in alkaline media.

6.3.5.2 Principle of Alkali Treatment

6.3.5.2.1 Chloropropanediols The combination of a hydroxy group and a chlorine atom in neighboring carbon atoms is responsible for the most common reaction of vicinal chlorohydrins such as chloropropanediols and dichloropropanols, which is dehydrochlorination to form substituted oxiranes (1,2-epoxides) (45).

It was found that chloropropanediols have considerable stability (they decompose very slowly) in slightly acidic pH of the commercial acid-HVP (pH about 5.5) stored at room temperature (53–55). Generally, in slightly acidic and neutral media, the dehydrochlorination of vicinal chlorohydrins involves the elimination of the chloride anion resulting in the formation of an intermediate carbocation. The hydroxyl group of the chlorohydrin acts as a nucleophilic reagent, giving rise to a conjugated acid (a protonated epoxide) from which the epoxide arises by elimination of a proton (Fig. 6.3.9).

The decomposition of chloropropanediols in alkaline media is very rapid. The reaction with hydroxyl anions brings the alcohol function of the chlorohydrin to equilibrium with its corresponding alkoxide (alcoholate). The alkoxide oxygen attacks the carbon that bears the leaving chloride atom to give the epoxide. This step determines the reaction rate of dehydrochlorination. As in other nucleophilic substitution reactions, the nucleophile approaches the carbon from the side opposite the bond to the leaving chloride so that the intramolecular S_N reaction takes place with conversion of configuration at the carbon that bears the chloride leaving group. The reaction mechanism is outlined in Fig. 6.3.10. Both enantiomers of 3-MCPD decompose at the same rate in alkaline media (53). As a consequence, the residual 3-MCPD in acid-HVPs treated in alkaline media is a racemic mixture of both optical isomers (15).

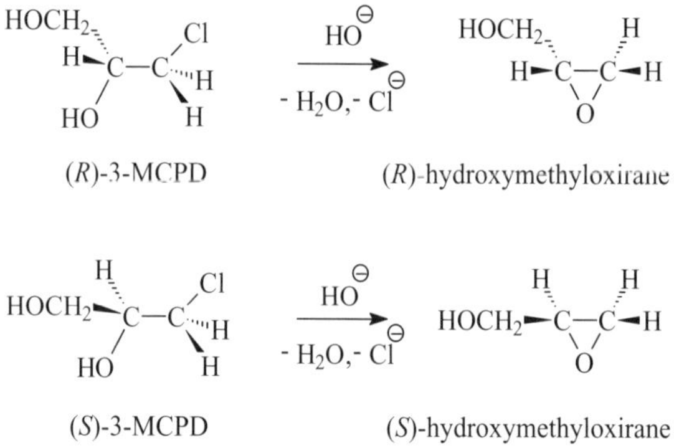

(*R*)-3-MCPD → (*R*)-hydroxymethyloxirane ($HO^{\ominus}$; $- H_2O, - Cl^{\ominus}$)

(*S*)-3-MCPD → (*S*)-hydroxymethyloxirane ($HO^{\ominus}$; $- H_2O, - Cl^{\ominus}$)

Figure 6.3.11 Dehydrochlorination of 3-chloropropane-1,2-diol to hydroxymethyloxirane.

X=OH, (*R*)-hydroxymethyloxirane X=OH, (*S*)-hydroxymethyloxirane

X=Cl, (*R*)-chloromethyloxirane X=Cl, (*S*)-chloromethyloxirane

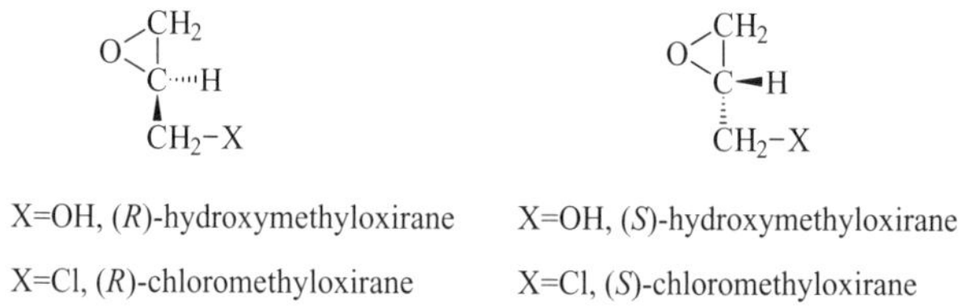

Figure 6.3.12 Structures of optically active hydroxymethyloxiranes in the Fischer projection.

Analogously to other vicinal chlorohydrins, (*R*)-3-MCPD ($R^1 = H$, $R^2 = CH_2OH$, $R^3 = R^4 = H$) gives (*R*)-hydroxymethyloxirane by base-promoted ring closure while (*S*)-3-MCPD ($R^1 = CH_2OH$, $R^2 = R^3 = R^4 = H$) yields enantiomeric (*S*)-hydroxymethyloxirane (Fig. 6.3.11). Thus, a racemic mixture of both isomers of hydroxymethyloxirane is obtained from the treatment of the acid-HVP with alkali. Structures of the chiral hydroxymethyloxiranes (also known as oxiranemethanols, oxiranylmethanols, or glycidols) in the Fischer projection are given in Fig. 6.3.12.

The hydroxymethyloxiranes produced are not stable but the rate of 3-MCPD decomposition is higher than the rate of the decomposition of hydroxymethyloxiranes. As a result, hydroxymethyloxiranes accumulate in some acid-HVPs treated with alkali (55). The concentration of hydroxymethyloxirane depends on the temperature and pH of the hydrolysate. For example, at 65 °C and pH 9 the maximum concentration of hydroxymethyloxirane, equal to about one-third of the quantity of the original 3-MCPD concentration, is reached within 5 min.

Epoxides are very reactive compounds in which the epoxide ring can open by a variety of nucleophiles (water, alcohols, thiols, amines, acids, etc.). Nucleophilic ring opening of epoxides has many of the features of an S_N2 reaction (inversion of configuration is observed at the carbon at which substitution occurs; Fig. 6.3.13). Asymmetrical epoxides are attacked at the less substituted,

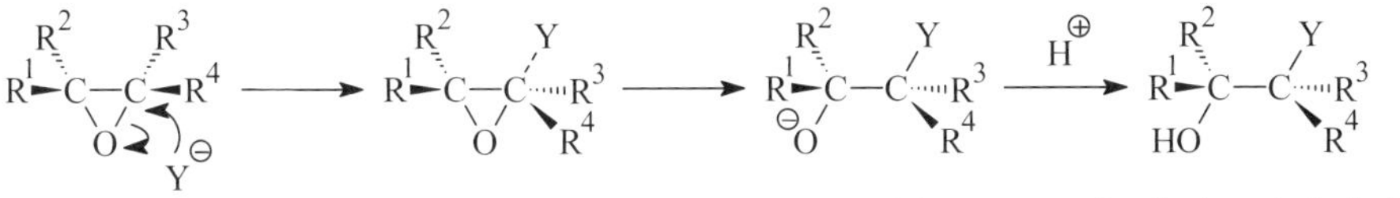

$$R^1\!-\!C(R^2)\!-\!C(R^3)R^4 \text{ (epoxide, O, } Y^{\ominus}) \longrightarrow R^1\!-\!C(R^2)\!-\!C(Y)R^3R^4 \text{ (transition state)} \longrightarrow R^1\!-\!C(R^2)(O^{\ominus})\!-\!C(Y)R^3R^4 \text{ (alkoxide)} \xrightarrow{H^{\oplus}} R^1\!-\!C(R^2)(HO)\!-\!C(Y)R^3R^4 \text{ (β-substituted alcohol)}$$

Figure 6.3.13 Nucleophilic ring opening of 1,2-epoxides.

less sterically hindered carbon of the ring from the side opposite to the carbon–oxygen bond (45).

In HVP, the oxirane rings of both optically active hydroxymethyloxiranes are opened mostly by the action of water with the formation of glycerol (55). Other nucleophiles present in HVP such as HCl, chlorides, ammonia, amino acids, and alcohols open the oxirane rings of hydroxymethyloxiranes to some extent. For example, the reaction of (*R*)-glycidol with HCl gives rise to (*R*)-chloropropane-1,2-diols. The major reaction product with ammonia was (*S*)-3-aminopropane-1,2-diol while (*R*)-3-aminopropane-1,2-diol formed from (*S*)-glycidol. Their racemic mixture has been found in the acid-HVP (56). Reactions of hydroxymethyloxiranes with amino group of amino acids in alkaline media lead to *N*-(2,3-dihydroxypropyl)amino acids (57), the reaction of hydroxymethyloxiranes with glycerol yields polyglycols (55).

Accordingly, symmetric cyclization of the prochiral 2-MCPD (elimination of proton occurs from either C-1 or C-3 hydroxyl group, followed by ring closure) always leads to a racemic mixture of both optically active hydroxymethyloxiranes. It is evident (Table 6.3.2) that 2-MCPD is more stable in alkaline media than 3-MCPD. In cases 3-MCPD is totally decomposed, 2-MCPD may still be found in the alkaline-treated HVPs and even becomes the major chloropropanol of these products (54).

6.3.5.2.2 Dichloropropanols Dehydrochlorination of 1,3-DCP involves symmetric cyclization (base-promoted ring closure), which leads to racemic intermediate chloromethyloxirane (epichlorohydrin) (58). (*S*)-Chloromethyloxirane is obtained by dehydrochlorination from the optically active (*R*)-2,3-dichloropropan-1-ol, while (*S*)-2,3-DCP forms (*R*)-chloromethyloxirane. The result of cyclization of racemic 2,3-DCP, which occurs in acid-HVPs, is racemic chloromethyloxirane (59). The epoxide ring of the reactive intermediate chloromethyloxirane then rapidly opens with water to give 3-MCPD. Racemic chloromethyloxirane converts to racemic 3-MCPD (60). The reactions of 3-MCPD in alkaline media are described previously.

6.3.6 CONCLUSION

Since 1989/1990, there have been a number of achievements including continual monitoring of the amount of 3-MCPD in acid-HVPs and imported soy sauces and issuing health hazard alerts where the concentration exceeded the guidelines. New technological approaches and new ingredients have been introduced to food manufacturers as alternatives to the classical procedures for acid-HVPs and they were adopted to minimize the concentration of 3-MCPD in the final product. In March 2001, the regulatory limit of 0.02 mg/kg, based on 40% dry matter content for 3-MCPD in acid-HVP and soy sauce, was adopted and came into force in April 2002. The results show a continuous decrease in frequency of contamination of acid-HVPs. The soy sauce contamination by 3-MCPD can be prevented by approaches based on GMP used for the production of acid-HVPs and by ensuring that 3-MCPD free acid-HVPs are used.

REFERENCES

1. Velíšek, J., Davídek, J., Hajšlová, J., Kubelka, V., Janíček, G., Mánková, B. (1978). Chlorohydrins in protein hydrolysates. *Zeitschrift Für Lebensmitteluntersuchung Und -Forschung*, *167*, 241–244.
2. Velíšek, J., Davídek, J., Kubelka, V., Bartošová, J., Tučková, A., Hajšlová, J., Janíček, G. (1979). Formation of volatile chlorohydrins from glycerol (triacetin, tributyrin) and hydrochloric acid. *Lebensmittel-Wissenschaft und-Technologie*, *12*, 234–236.
3. Davídek, J., Velíšek, J., Kubelka, V., Janíček, G., Šimicová, Z. (1980). Glycerol chlorohydrins and their esters as products of the hydrolysis of tripalmitin, tristearin and triolein with hydrochloric acid. *Zeitschrift für Lebensmitteluntersuchung und -Forschung*, *171*, 14–17.
4. Velíšek, J., Davídek, J., Kubelka, V., Janíček, G., Svobodová, Z., Šimicová, Z. (1980). New chlorine-containing organic compounds in protein hydrolysates. *Journal of Agricultural and Food Chemistry*, *28*, 1142–1144.
5. Davídek, J., Velíšek, J., Kubelka, V., Janíček, G. (1982). New chlorine containing organic compounds in protein hydrolysates, Proc. Euro Food Chem I, Vienna, Austria, 17–20 February, 1981, in *Recent Developments in Food Analysis* (eds. W. Baltes, P.B. Czedik-Eysenberg, W. Pfannhauser), Deerfield Beach, Florida, Weinheim, pp. 322–325.
6. Association Internationale de l'Industrie des Bouillons et Potages (AIIBP) (1992) IR/fcn/184, 16.01.92.
7. Wittmann, R. (1991). Bestimmung von Dichlorpropanolen und Monochlorpropandiolen in Würzen und würzehaltigen. *Lebensmitteln Zeitschrift für Lebensmitteluntersuchung und -Forschung*, *193*, 224–229.
8. Meierhans, D.C., Bruehlmann, S., Meili, J., Taeschler, C. (1998). Sensitive method for the determination of 3-chloro-propane-1,2-diol and 2-chloro-propane-1,3-diol by capillary gas chromatography with mass spectrometric detection. *Journal of Chromatography, A*, *802*, 325–333.
9. Macarthur, R., Crews, C., Davies, A., Brereton, P., Hough, P., Harvey, D. (2000). 3-Monochloropropane-1,2-diol (3-MCPD) in soy sauces and similar products available from retail outlets in the UK. *Food Additives and Contaminants*, *17*, 903–906.

10. Nyman, P.J., Diachenko, G.W., Perfetti, G.A. (2003). Survey of chloropropanols in soy sauces and related products. *Food Additives and Contaminants*, *20*, 909–915.
11. Crews, C., Hasnip, S., Chapman, S., Hough, P., Potter, N., Todd, J., Brereton, P., Matthews, W. (2003). Survey of chloropropanols in soy sauces and related products purchased in the UK in 2000 and 2002. *Food Additives and Contaminants*, *20*, 916–922.
12. Kwok, O.W., Yock, H.C., Leng, H.S. (2006). 3-Monochloropropane-1,2-diol (3-MCPD) in soy and oyster sauces: occurrence and dietary intake assessment. *Food Control*, *17*, 408–413.
13. MAFF (1999) Survey of 3-monochloropropane-1,2-diol (3-MCPD). Food Surveillance Information Sheet No. 187. http://archive.food.gov.uk/maff/archive/food/infsheet/1999/no187/187soy.htm (accessed 18 September 2008).
14. Van Bergen, C.A., Collier, P.D., Cromie, D.D.O., Lucas, R.A., Preston, H.D., Sissons, D.J. (1992). Determination of chloropropanols in protein hydrolysates. *Journal of Chromatography*, *589*, 109–119.
15. Velíšek, J., Doležal, M., Crews, C., Dvorák, T. (2002). Optical isomers of chloropropanediols: mechanisms of their formation and decomposition in protein hydrolysates. *Czech Journal of Food Sciences*, *20*, 161–170.
16. Velíšek, J., Ledahudcová, K., Pudil, F., Davídek, J., Kubelka, V. (1993). Chlorine-containing compounds derived from saccharides in protein hydrolysates I. Chloromethyl-2-furancarboxaldehyde. *Lebensmittel-Wissenschaft und-Technologie*, *26*, 38–41.
17. Velíšek, J., Ledahudcová, K., Kassahun, B., Doležal, M., Kubelka, V. (1993). Chlorine-containing compounds derived from saccharides in protein hydrolysates II. Levulinic acid esters in soybean meal hydrolysates. *Lebensmittel-Wissenschaft und-Technologie*, *26*, 430–433.
18. Velíšek, J., Davídek, J., Kubelka, V. (1986). Formation of Δ3,5-diene and 3-chloro-Δ5-ene analogues of sterols in protein hydrolysates. *Journal of Agricultural and Food Chemistry*, *34*, 660–662.
19. Manley, C.H., Fagerson, I.S. (1971). Aroma and taste characteristics of hydrolyzed vegetable protein. *Flavour Industry*, *2*, 686–690.
20. Yokotsuka, T. (1972). Recent technological problems related to the quality of Japanese shoyu. Proc. IV. International Food Symposium, Fermentation Technology Today, pp. 659–662.
21. Wedekind, F., Vincon, H. (1995). *Frank Wedekinds Maggi-Zeit*, Jürgen Häusser, Darmstadt, Germany.
22. Discussion paper for the 1st meeting of the codex committee on contaminants in food. Proposed draft code of practice for the reduction of chloropropanols in acid-hydrolysed vegetable proteins (acid-HVPs) and products that contain acid-HVPs (agenda item 14d).
23. MAFF (1999) Survey of 3-monochloropropane-1,2-diol (3-MCPD). Food Surveillance Information Sheet No. 181. http://archive.food.gov.uk/maff/archive/food/infsheet/1999/no181/181mcpd.htm (accessed 18 September 2008).
24. Velíšek, J., Davídek, J., Kubelka, V., Pánek, J. (1985). Doprovodné látky lipidů v bílkovinných hydrolyzátech (Accompanying compound of lipids in protein hydrolysates). *Sborník ÚVTIZ-Potravinárské Vědy*, *3*, 84–94.

25. Velíšek, J., Ledahudcová, K. (1993). Problematika organických sloučenin chloru v potravinárských hydrolyzátech bílkovin (Organic chlorine compounds in food protein hydrolysates. *Potravinárské Vědy*, *11*, 149–159.
26. EC (1997). Opinion on 3-Monochloropropanediol (3-MCPD), Expressed on 16 December 1994, in *Reports of the Scientific Committee for Food. Food Science and Techniques*, Thirty-sixth Series, Office for Official Publications of the European Community, Luxembourg, pp. 31–33.
27. Food Advisory Committee (1996). Press Release 6/96 Update on Chloropropanols in Hydrolysed Vegetable Protein.
28. Food Advisory Committee (1996). Press Release 13/96 Update on Chloropropanols in Hydrolysed Vegetable Proteins.
29. Food Advisory Committee (1998). Press Release 5/98 Update on 3-MCPD in Food and Food Ingredients.
30. Tritscher, A.M. (2004). Human health risk assessment of processing-related compounds in food. *Toxicology Letters*, *149*, 177–186.
31. SCF (2001). Opinion of the Scientific Committee on food on 3-monochloropropane-1,2-diol (3-MCPD) updating the SCF opinion of 1994. Adopted on 30 May 2001.
32. Joint FAO/WHO (2001). Expert Committee on Food Additives, Fifty-seventh meeting Rome 5–14, June.
33. EC (2001). Regulation No. 466/2001. Setting maximum amounts for certain contaminants in foodstuffs. Official Journal of the European Communities L77/1, 16 March, Luxembourg: Office for Official Publications of the European Communities.
34. MAFF (1991) Survey of hydrolysed vegetable proteins for chlorinated propanols. CSL Report FD 91/6.
35. MAFF (1993) Survey of chlorinated propanols in hydrolysed vegetable protein 1992. CSL Report FD 93/17.
36. COM (2001) Mutagenicity of 1,3-Dichloropropan-2-ol (1,3-DCP) and 2,3-Dichloropropan-1-ol (2,3-DCP), Statement-/01/S2-May 2001. http://www.advisorybodies.doh.gov.uk/com/statements.htm (accessed 18 September 2008).
37. COC (2001) Carcinogenicity of 1,3-Dichloropropan-2-ol (1,3 DCP) and 2,3-Dichloropropan-1-ol (2,3-DCP), Statement-COC/01/S1-January 2001. http://www.advisorybodies.doh.gov.uk/coc/cocdcp.htm (accessed 18 September 2008).
38. Chemisches und Veterinäruntersuchungsamt Stuttgart (2002). 3-MCPD in Lebensmitteln-Untersuchungen am CVUA Stuttgart im Jahr 2002. http://www.lebensmittel.org/lebensm/mcpd.htm (accessed 18 September 2008).
39. Food Standards Agency, Food Survey Information Sheet Number 15/01 (2001). Survey of 1,3-Dichloropropanol (1,3-DCP) in Soy Sauce and Related Products. http://www.food.gov.uk/science/surveillance/fsis2001/13dcpsoy (accessed 18 September 2008).
40. Velíšek, J., Davídek, J., Šimicová, Z., Svobodová, Z. (1982). Glycerol chlorohydrins and their esters-reaction products of lipids with hydrochloric acid. *Sborník VŠCHT v Praze E*, *53*, 55–65.

41. Velíšek, J., Davídek, J. (1985). Lipidy jako prekurzory organických sloučenin chloru v bílkovinných hydrolyzátech (Lipids as precursors of organic chlorine compounds in protein hydrolysates). *Potravinárské Vědy*, *3*, 11–18.
42. Macmurray, T.A., Morrison, W.R. (1970). Composition of wheat-flour lipids. *Journal of the Science of Food and Agriculture*, *21*, 520–528.
43. Velíšek, J. (1989). Organické sloučeniny chloru v potravinárských hydrolyzátech bílkovin (Organic Chlorine Compounds in Food Protein Hydrolysates). DSc Thesis, Institute of Chemical Technology, Prague, Czech Republic.
44. Collier, P.D., Cromie, D.D.O., Davies, A.P. (1991). Mechanism of formation of chloropropanols present in protein hydrolysates. *Journal of the American Oil Chemist's Society*, *68*, 785–790.
45. Carey, F.A. (2000). *Organic Chemistry*, 4th edn, The McGraw-Hill Companies, Inc., Boston.
46. Nestlé, S.A. (1990). Procédé de fabrication d'un condiment. European Patent No. 0226769.
47. Nestec, S.A. (1992). Process for elimination of chlorohydrins from protein hydrolysates. United States Patent No. 5079019.
48. Société des Produits Nestlé, S.A. (1992). Production of hydrolysed proteins. European Patent No. 0363771.
49. Unilever, N.V. (1992). Process for preparing improved hydrolysed protein. European Patent No. 0361596.
50. Unilever Patent Holding B.V. (1995). Process for improving improved hydrolyzed protein. United States Patent No. 5401527.
51. Société des Produits Nestlé, S.A. (1995). Process for reducing hydrolysed protein chlorohydrin content. European Patent No. 0505800.
52. Joint FAO/WHO Expert Committee on Food Additives (2003). Safety evaluation of certain food additives and contaminants. WHO Food Additives Series No. 48. http://www.inchem.org/documents/jecfa/jecmono/v48je01.htm (accessed 18 September 2008).
53. Doležal, M., Velíšek, J. (1992). Kinetics of 3-chloro-1,2-propanediol degradation in model systems. Proceedings of Chemical Reactions in Foods II, 24–26 September, Prague, Czech Republic, pp. 297–302.
54. Doležal, M., Velíšek, J. (1995). Kinetics of 2-chloro-1,3-propanediol degradation in model systems and in protein hydrolysates. *Potravavinárské Vědy*, *13*, 85–91.
55. Doležal, M. (1997). Dekontaminace potravinárských hydrolyzátů bílkovin (Decontamination of food protein hydrolysates), PhD Thesis, Faculty of Food and Biochemical Technology, Institute of Chemical Technology, Prague, Czech Republic.
56. Velíšek, J., Ledahudcová, K., Hajšlová, J., Pech, P., Kubelka, V., Víden, I. (1992). New 3-chloro-1,2-propanediol derived dihydroxypropylamines in hydrolyzed vegetable proteins. *Journal of Agricultural and Food Chemistry*, *40*, 1389–1392.
57. Velíšek, J., Davídek, T., Davídek, J., Kubelka, V., Víden, I. (1991). 3-Chloro-1,2-propanediol derived amino acids in protein hydrolysates. *Journal of Food Science*, *56*, 139–142.

58. Takeichi, T., Arihara, M., Ishimori, M., Tsuruta, T. (1980). Asymmetric cyclizations of some chlorohydrins catalyzed by optically active cobalt (salen) type complexes. *Tetrahedron*, *36*, 3391–3398.

59. Kasai, N., Sakaguchi, K. (1992). An efficient synthesis of (R)-carnitine. *Tetrahedron Letters*, *33*, 1211–1212.

60. Furrow, M.E., Schaus, S.E., Jacobsen, E.N. (1998). Practical access to highly enantioenriched C-3 building blocks via hydrolytic kinetic resolution. *Journal of Organic Chemistry*, *63*, 6776–6777.

PART II

GENERAL CONSIDERATIONS

7

APPLICATION OF THE HACCP APPROACH FOR THE MANAGEMENT OF PROCESSING CONTAMINANTS

YASMINE MOTARJEMI,[1] RICHARD H. STADLER,[2] ALFRED STUDER,[3] AND VALERIA DAMIANO[4]

[1]*Quality Management, Nestlé, 55 Avenue Nestlé, CH-1800 Vevey, Switzerland*
[2]*Nestlé Product Technology Centre Orbe, CH-1350 Orbe, Switzerland*
[3]*Nestlé Research Centre, Vers-chez-les-Blanc, CH-1000 Lausanne 26, Switzerland*
[4]*Nestlé Product Technology Centre Orbe, CH-1350 Orbe, Switzerland*

7.1 INTRODUCTION

Processing contaminants are essentially undesirable compounds that are formed during the treatment of food as a result of the interaction of their natural components or their ingredients (including food additives). In certain cases, however, the formation of such chemicals may be mediated by the metabolic activities of microorganisms. Typical examples are the formation of biogenic amines or ethyl carbamate during fermentation. However, the formation of other microbial toxins such as mycotoxins or bacterial toxins (e.g., *Staphylococcus aureus* or *Clostridium botulinum* toxins) is not within the scope of this book, although they may occur as the result of a failure at a processing step.

Humans have been exposed to a plethora of chemicals since the early days of the history of humankind. It is believed that *Homo erectus* living during the Paleolithic period already knew about fire and used it for the preparation of his food (1). It is most likely that since then, humankind has been exposed to some of these contaminants through the diet.

Process-Induced Food Toxicants: Occurrence, Formation, Mitigation, and Health Risks,
Edited by Richard H. Stadler and David R. Lineback

In recent years, there has been a growing awareness of and concern about the potential risks of processing contaminants, although often little is known about their true public health impact. Examples of these compounds are acrylamide, chloropropanols, furan, polycyclic aromatic hydrocarbons (PAHs) and toxic nitrogen compounds.

This chapter presents a short description of a selected number of these contaminants, an overview of their risk, and the possible measures to control them. It focuses on their management using the HACCP (Hazard Analysis and Critical Control Point) approach, and provides concrete examples of the application of HACCP to processing contaminants through two examples. For a more detailed coverage of each of the individual contaminants, the reader is referred to the specific topics in this book.

The main focus of this chapter is on the operations in the food and beverage industry, primarily as guidance on how to consider and manage processing contaminants in food production, although certain principles are also applicable to food preparation in and out of home, in catering and restaurant services. The chapter also serves as an insight for the regulatory and public health authorities on the needs and constraints of the food industry to meet regulatory requirements.

7.2 PROCESSING CONTAMINANTS: RISKS AND CONTROL MEASURES

Although by definition, processing contaminants are formed during a processing step, the control measure may not necessarily be at this step, but at different stages of the food chain, for example, where precursors are formed or are introduced into the food. For instance, to mitigate the formation of acrylamide, it may be possible to intervene (i) at the agronomical level, (ii) in designing the recipe, (iii) at the processing stage, or (iv) during final preparation.

7.2.1 Acrylamide

Acrylamide is carcinogenic to experimental animals and is classified by the International Agency for Research on Cancer (IARC) as probably carcinogenic to humans (Group 2A carcinogen) (2). Acrylamide itself is not genotoxic but glycidamide generated from acrylamide has been shown to have genotoxic effects at low amounts (3). Acrylamide is also a neurotoxicant; the "no observed effect level" (NOEL) in rats is found to be 0.2 mg/kg body weight/day.

Evidence of carcinogenicity in humans is contradictory. Human epidemiological studies in the Netherlands and Denmark have pointed to a possible association between acrylamide and an increased risk of ovarian and endometrial cancer in postmenopausal women as well as of renal cancer in the general population (4–6). However, other studies have failed to find an association between dietary exposure to acrylamide and cancer (7–12). Nevertheless, in its evaluation, the Joint FAO/WHO Expert Committee on Food Additives (JECFA) viewed acrylamide as a human health concern and recommended that the efforts to reduce acrylamide in food continue (13).

The formation of acrylamide is linked to the well-known Maillard reaction, a browning process that occurs during the heat treatment of food and which contributes to taste, aroma, and color. During this process, reducing sugars such as glucose and fructose react with amino acids to form acrylamide. Foods that are rich in carbohydrate and amino acids, particularly free asparagine, and are processed under conditions of high temperature and low moisture have the highest potential to form acrylamide. Examples are potato crisps, fried potatoes, breakfast cereals, biscuits/bakery wares, coffee, and coffee surrogates (beverages based on chicory and/or roasted cereal grains). Acrylamide is also present in staple food such as bread, but in low amounts. However, as bread is a common part of the daily diet, it may nevertheless contribute significantly to the intake of acrylamide.

Since the discovery of acrylamide in food in April 2002 (14), considerable research has been undertaken by industry and governments to understand the mechanism of formation and find ways to mitigate its formation. In some areas, significant progress has been made, for example, for French fries (15, 16).

However, for some other products such as coffee, the mitigation of acrylamide formation has proven to be more difficult since any change to the roasting process has significant impact on the organoleptic properties and acceptability of the product (17, 18). Recently, a commercial enzyme has been developed that may be used to reduce acrylamide formation in certain cereal-based products. The technique entails the addition of the enzyme asparaginase, which specifically converts asparagine to aspartic acid (19). Table 7.1 summarizes various control measures for acrylamide that can be applied in food manufacturing or in food preparation.

For further information, the reader is referred to the CIAA "Acrylamide Toolbox" that provides a consolidated overview on control measures for acrylamide in different food products (18). For more detailed information on acrylamide, refer to Chapter 2.1.

7.2.2 Chloropropanols

Chloropropanols, in particular 3-MCPD and 1,3-DCP, are compounds that are of concern due to their carcinogenicity. They have also shown to be genotoxic *in vitro*, although this has not been confirmed by *in vivo* studies as far as 3-MCPD is concerned. On the other hand, with regard to 1,3-DCP, JECFA concluded that this compound is genotoxic *in vivo* and consequently, no acceptable safe level can be established (20).

Chloropropanols have multiple sources and their formation follows different and complex pathways. Originally, when discovered in 1978, they were associated with acid-hydrolyzed vegetable proteins (acid-HVPs) used as flavoring in soups, gravy mixes, bouillon cubes and in particular, in soy sauce and

TABLE 7.1 Control measures for acrylamide in selected products.

Product category	Toolbox compartment			
	Agronomical	Recipe	Processing	Final preparation
French fries	Choose potato varieties with low sugar levels Storage conditions (>6 °C, maximum according to germination) Selection of potato with low amounts of reducing sugars (<0.4% fresh wt)	Not applicable	Blanch potato strips in hot water for a longer period of time to remove reducing sugars Cut thicker strips	Follow on-pack instructions Control the temperature and time of final cooking When cooking smaller amounts, reduce cooking time When frying, do not cook >175 °C Aim for a light golden color
Bread	For wheat grain, the importance of maintaining sulfur levels in the soil must be stressed to farmers	Avoid adding reducing sugars in the recipe The addition of calcium salts may reduce the formation of AA	Control the baking time and temperature to prevent excessive browning in the crust	When toasting bread, aim for a light golden color
Crispbread	For wheat grain, the importance of maintaining sulfur levels in the soil must be stressed to farmers	Not applicable	Non-fermented crispbread: control process temp. and oven speed Control the final moisture content	Not applicable
Biscuits/bakery wares	For wheat grain, the importance of maintaining sulfur levels in the soil must be stressed to farmers	Replacement of ammonium bicarbonate with other raising agents If possible avoid using fructose	Asparaginase is a tool for certain biscuit and cereal applications Do not overbake	Not applicable
Breakfast cereals	For wheat grain, the importance of maintaining sulfur levels in the soil must be stressed to farmers	Minimize reducing sugars in the cooking phase Consider the contribution of other inclusions, e.g., roasted nuts and dried fruits	Do not overbake or over-toast Manage the toasting to achieve a uniform color for the product	Not applicable

Adopted from the CIAA acrylamide (AA) Toolbox (18).

other liquid seasonings (21). During the production process of HVP at high temperature (100–130 °C, for 4–24 h), hydrochloric acid interacts with residual vegetable lipids (triacylglycerols and phospholipids) in the proteinaceous material to form chloropropanols (for more details, see Chapter 6.3).

3-MCPD can also be formed in other foods that contain chloride and fats/oils and that have been subjected to heat treatment, as typically encountered during baking, roasting, and grilling. Its occurrence in cereal-based products such as pastries, biscuits, crackers, or bread as well as meat products, e.g., burgers, has been reported by several research groups (22–25).

In some of these foods, 3-MCPD may actually be present in relatively high amounts. For instance, the UK Food Standard Agency reports up to 1 mg/kg of 3-MCPD in toasted bread, which is among the highest contents, excluding soy sauces and acid-HVPs (22, 26). The reaction is also enhanced in low-moisture systems (<15%) (27).

While new manufacturing techniques such as enzymatic hydrolysis have allowed the content of 3-MCPD and 1,3-DCP to be reduced in HVP and products thereof, more research is still needed to understand the mechanism of formation of these compounds in heat-processed foods and the ways to control them.

More recent studies have shown that in foods other than those containing acid-HVP, 3-MCPD may be present in a bound form, i.e., as 3-MCPD esters (see Chapter 2.6). In fact, it has been shown that the enzyme lipase can "release" 3-MCPD from the chloroester precursors in certain dry culinary products during storage (28). A gut model experiment conducted under different lipase activities suggests that a similar reaction may occur *in vivo*, i.e., the formation of 3-MCPD from chloroesters (29). Such a mechanism may explain the presence of 3-MCPD in foods such as salami, cold smoked fish, or cheese (feta, parmesan), which have not undergone heat treatment (25).

Depending on the product and the pathway that favors chloropropanol formation, potential control measures are:

- raising the pH in the production of acid-HVP, for instance, by using sodium hydroxide;
- product formulation:
  - selection of raw ingredients, e.g., minimize barley or wheat grains
  - addition of sodium bicarbonate and disodium carbonate
  - addition of amino compounds, e.g., glutathione and cysteine;
- reducing glycerol and lipids;
- reducing the amount of minerals (NaCl, KCl, $MgCl_2$);
- control of thermal treatment (time–temperature) and pH, i.e., avoiding extended baking times, or roasting at maximum 200 °C[1];

[1] 3-MCPD is formed at temperatures above 170 °C and is unstable above pH 6.0.

- control of storage conditions (time and temperature of storage) to minimize the formation of glycerol;
- control of glycerol in bread by acting on the age of the flour, amount of yeast, and the length of fermentation before baking. Addition of glycerol as humectants should be evaluated before usage;
- control of natural lipase present in certain food ingredients (e.g., spices) that may release 3-MCPD from 3-MCPD esters present in oils.

Table 7.2 lists possible control measures for reducing 3-MCPD in selected product categories.

7.2.3 Furan

Furan is a carcinogen in laboratory animals, e.g., rodents, probably by genotoxic mechanisms; the liver and kidney are the main target organs. There are, however, no data from human studies. The IARC has classified furan as possible human carcinogen (Group 2B carcinogen) (32). Preliminary data on exposure to humans suggest that this may occur at doses that are close to the amounts that produce carcinogenic effects in experimental animals. Therefore, furan is considered to be a potential concern although today no firm conclusion can be made (see Chapter 2.4).

Furan is formed during heat treatment of products by the degradation of organic compounds, in particular carbohydrates. Studies by Perez-Locas and Yaylayan indicate ascorbic acid, sugars, or mixtures of amino acids and sugars as possible precursors (33). Furan can also be formed by the oxidation of polyunsaturated fatty acids (PUFAs) at elevated temperatures (e.g., above 80°C) (34, 35).

Consequently, furan occurs in a wide range of food products. Based on an analysis of large number of foods, Zoller and coworkers (36) report that furan can occur mainly in:

- moist foods that underwent a heat treatment in a sealed container, e.g., canned and jarred products containing meat and various vegetables;
- crusty and dry products, i.e., typically those that are baked and roasted, including coffee.

In most cases, foods processed in jars undergo a heat treatment such as sterilization (approx. 121°C). When consuming such foods, the reheating process in an open heating pan may reduce the furan content, although inhalation of vapor may expose the food handler to furan.

As in the case of chloropropanols, the formation pathway of furan is variable and different ingredients can contribute to its formation. Nevertheless, to minimize the formation of furan, it is possible to act on:

TABLE 7.2 Control measures for reducing 3-MCPD formation in selected foods and ingredients.[a]

Food/ingredient	Control measures
Malts	Control of roasting process, maximum temperature of 200 °C
Biscuits	Control of yeast fermentation process, e.g., minimize fermentation time as far as feasible
	Control of time–temperature process
Breakfast cereal	Product formulation: minimum amount of vegetable protein
	Selection of vegetable oil with minimum amount of 3-MCPD esters
Breads and pastry products	Control of the baking process, i.e., time and temperature designed to avoid excessive browning
	Control of glycerol and consider chloride (salt) content: limit storage time of flour
	Reduce yeast fermentation time. Note that this may have an adverse effect on acrylamide formation
Smoked products	Addition of calcium carbonate to the wood to reduce MCPD formation in the smoke
	Controlling the smoke process, e.g., minimizing smoke time
	Brine solutions instead of dry curing to reduce salt content in the product
Cooked meat and cheese products	Minimize cooking or microwave time
	Minimize addition of glycerol
Dry savory products	Inactivate lipase in spices before usage
	Product reformulation, i.e., use of spice extract
Dehydrated culinary products (during storage)	Recipe formulation, e.g., minimize fat
	Minimize glycerol formation through, for instance, yeast fermentation, lipid hydrolysis by enzyme or microbial activity
	Control of storage conditions (time and temperature)
Soy sauce and acid-HVPs and related products	Removal of 1,3-DCP by distillation of hydrolysate with overheated vapor at a reduced pressure
	Mitigation measures for monochloropropanols: controlling the hydrolysis process (time and temperature) and subsequent neutralization
	elimination of formed 3-MCPD by an alkali treatment process with potassium hydroxide, sodium hydroxide, ammonium hydroxide, or sodium carbonate to pH 8–9
	use of countercurrent liquid/liquid extracting process
	substitute hydrochloric acid with, for example, sulfuric acid
Soy sauce	Process reformulation: prevention of 3-MCPD formation by use of enzymatic hydrolysis during manufacturing process

[a]References 30 and 31.

- product formulation, e.g., considering the impact of ascorbic acid and PUFAs;
- minimizing the oxidation of PUFAs over the product shelf life;
- the heat treatment process, i.e., optimizing the heat treatment process so as to achieve the quality of product while minimizing the formation of furan;
- reheating of canned/jarred food.

7.2.4 PAHs

PAHs comprise a large group of carcinogenic substances, exhibiting a similar degree of toxicity (see Chapter 2.8). Among these, the most notable ones are dibenz[*a,h*]anthracene, benzo[*a*]pyrene, and dibenzo[*a,h*]pyrene. They are highly lipophilic compounds, and consequently they are found in the lipid tissues or fatty components of foods. They are formed as a result of incomplete combustion or pyrolysis of organic materials. They are also considered environmental contaminants as they may be present in the environment subsequent to industrial activity and pollution (soot, smoke). In food, over and above their occurrence due to contamination from the environment, they are formed during different processes such as grilling, smoking, or drying, particularly when fat in food or in the form of a drip is exposed to very high temperatures. The amount of PAHs in food is related to the distance from the heat source, the time of grilling, and whether melted fat is allowed to drop on the heat source. Highest levels of PAHs are found in charcoal-broiled meat, whereas liquid smoke flavoring generally contains less PAHs.

As mentioned earlier, other sources may be seafood, fruits and vegetables, and vegetable oils due to environmental pollution. In vegetable oils, deodorization can remove the light fraction. For the heavy fraction, active carbon treatment should be applied.

7.2.5 Toxic Nitrogen Compounds

During processing, a number of nitrogenous compounds may be formed, which are considered potentially harmful. These are, for example, ethylcarbamate and substances belonging to the class of biogenic amines and nitrosamines.

7.2.5.1 Biogenic Amines In foods and beverages, biogenic amines are formed by the decarboxylation of amino acids (Chapter 3.2). The reaction is catalyzed by the corresponding decarboxylase enzyme present in the food or resulting from the metabolic activity of certain bacteria, notably the Enterobacteriaceae (e.g., *Morganella morganii* and *Klebsiella pneumoniae*) or Lactobacillus. The principal biogenic amines are histamine, tyramine, cadaverine, putrescine, serotonin, and tryptamine.

Biogenic amines can virtually be formed in all food where the following conditions are fulfilled (37):

- availability of free amino acids;
- presence of decarboxylase-positive microorganisms or enzyme in the food;
- conditions allowing growth of decarboxylase-positive microorganisms and activity.

Such conditions can be encountered during fermentation or food spoilage. Therefore, in non-fermented foods, the presence of biogenic amines can be indicative of undesired microbial activity. Other factors influencing the formation of biogenic amines are pH, increase in temperature, aerobic conditions, and presence of antibiotics or preservatives.

Although many biogenic amines have an important metabolic function in humans and animals, exposure to high amounts of these can be toxic. The most notorious example is histamine, which can cause histamine poisoning. The presence of other biogenic amines, e.g., cadaverine or putrescine, may also enhance the effect of histamine. Individuals who have been subject to histamine poisoning can manifest a wide range of symptoms, e.g., rash, flushed skin, hot flash, sweating, facial swelling, nausea, vomiting, diarrhea, heart palpitation, burning throat, peppery taste in the mouth, dizziness, and headache. Most cases of histamine poisoning have been associated with consumption of certain fishes and in some cases cheese.

7.2.5.2 Nitrosamines Nitrosamines result from the interaction of nitrites with amines in food or *in vivo*, under acidic conditions. During fermentation, under microbial activity, nitrate can also convert to nitrites, precursors for nitrosamines.

Nitrosamines are carcinogens. They can lead to cancer of a variety of organs: liver, pancreas, stomach, lower gastrointestinal tract, and bladder. Nitrites and nitrates may occur naturally in water or in some foods such as leafy vegetables at high levels (e.g., 3000 mg/kg), due to fertilization. They may also be added to food as additive, for instance in curing meat products. This is done principally for preventing growth of the foodborne pathogen *C. botulinum*. Additionally, it gives a specific color and flavor to the food that consumers value. High levels of amines are found in protein foods, in particular fish, and may increase with the frozen storage. The nitrosation reaction can be mitigated by adding an adequate amount of ascorbic acid.

Human exposure to nitrosamines may also be due to natural occurrence in the food, although at low amounts, as well as other sources, such as tobacco. For further details, refer to Chapter 4.1.

7.2.5.3 Ethyl Carbamate Ethyl carbamate is of concern due to its mutagenic and carcinogenic properties. It results from the esterification of carbamide (HN=C=O, also termed cyanic acid) by ethanol during yeast fermentation.

Ethyl carbamate is found in a number of fermented foods, e.g., bread, soy sauce as well as wine and spirits. The main factors affecting the formation of ethyl carbamate are low pH and increase in temperature. In wine, it has been shown that the holding temperature influences ethyl carbamate formation. For instance, increasing the temperature by 10 °C can increase the amount of ethyl carbamate threefold. In some countries, cysteine-*N*-carbamide is used as a chemical agent to enhance the preparation of dough in bread making, though it is forbidden in the European Union (EU). For further details, refer to Chapter 3.1.

7.3 HACCP: THE BACKBONE OF FOOD SAFETY ASSURANCE

Today's management of food safety is to a great extent based on the application of the HACCP system. Originally, the system was introduced to ensure the microbiological safety of food products. Later on, its use was extended to all types of foodborne hazards, including chemical hazards. Although very different in nature, processing contaminants are similar to microbiological hazards in the sense that they are controlled through product formulation and processing steps. Therefore, the HACCP system can, with some adaptation, be readily used for the management of processing contaminants.

The following section aims to explain how to use the HACCP approach to minimize the formation of processing contaminants and ensure the safety of products in this respect. The chapter is preceded by a short summary on the HACCP system and its benefits. For more details on the subject, the reader is referred to other sources (38).

7.3.1 HACCP System: History and Benefits

The HACCP system as defined by the Codex Alimentarius Commission (CAC) is a system that identifies, evaluates, and controls hazards, which are significant for food safety. The value of the HACCP system lies in the fact that it is a scientific, rational, and systematic approach to the identification, assessment, and control of hazards during production, processing, manufacturing, preparation, and use of food to ensure that food is safe when consumed. Scientific and technical knowledge must be used proactively to manage food safety.

HACCP was originally designed by the Pillsbury Company, together with the National Aeronautics and Space Administration (NASA) and the US Army Laboratories at Natick who developed this system to ensure the safety of the astronauts' food. For many years after its conception, the system was promoted by international organizations such as the WHO and was applied on a voluntary basis in certain food industries. In 1993, the CAC recognized the HACCP system as a powerful tool to improve food safety and established the Codex guidelines for the application of the HACCP system (39). This has had major implications in the widespread implementation of the HACCP system. With the establishment of the World Trade Organization in 1995 and the coming into force of the Agreement on *Sanitary and Phytosanitary Measures* (WTO/SPS), the work of Codex, i.e., its standards, guidelines, and recommendations (including the Codex document on HACCP system and guidelines for its application) became the international reference for food safety. This implies that WTO member states needed to take the work of the Codex Alimentarius into consideration and adapt their national legislation to the provisions of the CAC. Indirectly, this meant that the application of HACCP became an international requirement for food safety assurance. Today, the principles of HACCP are integrated in the national legislation of many countries as well as in the ISO 22000 standards, i.e., the standards defining the requirements for management of food safety (40).

The introduction of HACCP brought many benefits to the food safety assurance system. The HACCP system:

- is a proactive approach to food safety management; this means it allows conceivable and reasonably expected hazards to be identified, even when failures have not previously been experienced. It is particularly useful for new operations.
- is flexible, i.e., necessary control measures can be adapted to changes in operations, such as change in equipment design, in processing procedures, and technological development. This is particularly important in the context of processing contaminants, as our knowledge about their health risks and mechanism of formation is continuously progressing.
- helps to target resources to the most critical part of the food operations.
- is applicable to the entire food chain, from the raw material to the end product, i.e., growing, harvesting, processing/manufacturing, transport and distribution, and preparation and consumption. As can be noted, for many processing contaminants, the control measures can be multiple and at different stages of the food chain.
- overcomes many of the limitations of the traditional approaches to food safety control, generally based on:
  - snapshot inspection, which is a rather ineffective approach in foreseeing potential problems, particularly for hazards associated with processing or product formulation.
  - end-product testing, which would entail high costs for analysis and which would lead to identifying problems without understanding their cause.

7.3.2 Principles of the HACCP System

As stipulated in the Codex guidelines on HACCP, the HACCP system is comprised of seven principles. These are outlined as follows.

Principle 1:	Conducting a hazard analysis
Principle 2:	Determining the critical control points (CCPs)
Principle 3:	Establishing critical limits
Principle 4:	Establishing a system to monitor control of the CCP
Principle 5:	Establishing the corrective action to be taken when monitoring indicates that a particular CCP is not under control
Principle 6:	Establishing procedures for verification to confirm that the HACCP system is working effectively
Principle 7:	Establishing documentation concerning all procedures and records appropriate to these principles and their application

To ensure the most effective outcome, the application of the HACCP system is carried out following a number of steps. The codex guidelines outline 12 steps for conducting a HACCP study and establishing a HACCP plan. To this should be added the training of different operators, the implementation and maintenance of the plan as well as a number of prerequisite activities (see Section 7.4.1). With regard to the HACCP application, the importance of validating the elements of the HACCP system needs to be highlighted. This means that at every step in the development of the HACCP study, it is important to ensure that decisions taken in relation to the different elements of the HACCP system are valid, i.e., that they are established based on a scientific and technical evidence. In particular, the control measures must be effective and achieve the expected outcome (e.g., regulatory or industry limits, performance or food safety objectives). As such, validation is the assurance *in the food safety assurance system.* As the Codex Guidelines for HACCP are not explicit on the subject of validation, separate guidelines on validation of control measures have been established by the Codex Alimentarius (41). For a more general overview of the food safety management system during food processing and manufacturing, the reader is referred elsewhere (42).

7.4 APPLICATION OF HACCP TO PROCESSING CONTAMINANTS

HACCP is a food safety management tool. As such, it can be used in different ways to support the management of processing contaminants.

1. Processing contaminants can be integrated in the overall HACCP study and HACCP plan of a product.
2. A generic HACCP study may be conducted on a product category. Based on the results, a code of practice or other forms of guidance can be developed providing advice on required control measures, limits, monitoring, procedures, etc. Such an approach may be taken by trade associations or governmental organizations to advise small food

operations, or to promote industry-wide actions (18, 43). It may also be used during product development to design the product so as to minimize the formation of processing contaminants.

3. The results of a generic HACCP study on a product category can also be used in health education interventions. They provide guidance on the recommendations that need to be given to domestic or professional food handlers preparing food in households or in food service establishments.

7.4.1 Prerequisites to the Application of HACCP

For the purpose of this text, the term "prerequisite" refers to all the measures and activities that need to be in place in order to support the application of the HACCP system, whether these are done in industry or by governments.

The prerequisite activities, as expressed by the term itself, are *sine qua non* conditions for the successful implementation of the HACCP system and management of processing contaminants. Depending on the nature of the product and the hazards considered, prerequisite activities may entail different types of measures. For processing contaminants, the following prerequisites need to be in place:

1. *Management Commitment.* Raising awareness of high-level decision makers on the health significance and importance of processing contaminants is essential, as the outcome of the study may indicate the need for modifying the manufacturing process (formulation, equipment, etc.), with a possible impact on product quality/sensorial properties and/or on the nutritional profile.

> ***Example:***
> Following the concern raised by 3-MCPD formation in the acid-HVP production processes, the food industry changed the production method and moved toward an enzymatic hydrolysis process. This meant millions of dollars of investment in terms of Research and Development as well as changes in the food production facilities.

2. *General Good Manufacturing Practices (*GMPs*).* These are basically a set of rules procedures and practices that have been established as a result of previous experiences to ensure that products are safe and fit for human consumption. In practice, they refer to generic control measures that apply to a given sector of the food chain, regardless of its specific conditions (e.g., environment, ingredients, product formulation, production, and processing). However, this does not preclude the fact that these practices may also be identified as control measures to specifically control a hazard in a HACCP study.

Example:
As a GMP in manufacturing confectionary products, material that has been burnt or overbaked, e.g., biscuits and wafers, should be discarded and not reworked. This measure *per se* can already contribute to reducing the level of 3-MCPD.

Examples:
The rotation of stocks in the warehouse and observation of the principle of "first in, first out" (FIFO) can also reduce the formation of certain processing contaminants. For instance, limiting the storage of potatoes can help to minimize the formation of acrylamide in French fries. Old flour may contain higher amounts of glycerol. Limiting its storage through application of the FIFO principle can contribute to minimizing the 3-MCPD amount in bread. 3-MCPD may also be released during storage due to lipase activity. Thus, limiting the storage of certain finished products and applying the FIFO principle can also maintain 3-MCPD at low levels.

It is to be noted that ISO 22000 distinguishes between the terms "prerequisites" and "operational prerequisites." The latter refers to control measures identified in the HACCP studies (Principle 1 on Hazard Analysis) that are necessary for controlling as a hazard. Since in most cases, processing contaminants will not be considered as a significant hazard at the levels present, it is the authors' understanding that control measures for processing contaminants would fall under such a definition. In other words, measures taken at the control points (CPs) mentioned later in this text would be part of an "operational prerequisite program." Due to the complexity that such terminology presents, these terms are not used in this text.

ISO 22000 defines the term "prerequisite" as follows [40]:

Prerequisite program
Basic conditions and activities that are necessary to maintain a hygienic environment throughout the food chain suitable for the production, handling, and provision of safe end products and safe food for human consumption

Operational prerequisite program (Operational PRP)
These are identified by the hazard analysis as essential in order to control the likelihood of introducing food safety hazards to and/or the contamination or proliferation of food safety hazards in the product(s) or in the processing environment

3. *Scientific Research.* Scientific and technical data and know-how are fundamental to any proactive and science-based food safety management system such as HACCP. The type of scientific research that is needed includes:
   - *Toxicological and epidemiological research.* HACCP is a risk-based system. As such, toxicological and epidemiological data are required to evaluate the health significance of the different compounds, the degree of risk they present, and their sources. In principle, such guidance should be provided by public health authorities. For some processing contaminants, such data are available. However, for a large number of processing contaminants, appropriate data are lacking, making the application of hazard analysis difficult.
   - *Mechanism of formation.* Research on the mechanism of formation is essential in order to design control measures or decide on the parameters that need to be controlled. As explained in different chapters of this book, the mechanism of formation of processing contaminants is very complex since it can follow different pathways; moreover, a multitude of factors, such as different precursors or intermediate compounds, catalysts and processing parameters can influence their formation, as exemplified for acrylamide in Fig. 7.1. Thus, to be able to effectively mitigate the formation of processing contaminants, there is a need for an in-depth understanding of the mechanism of formation and the factors that influence it.

> ***Example***:
> The discovery that acrylamide is formed from the reaction of asparagine with reducing sugars has paved the way for a wide range of possible control measures. Based on this insight, taking French fries as an example, a number of control measures such as selection of potato cultivars with low sugar contents, storage of potatoes above 6 °C, blanching and optimization of heat treatment conditions have been recommended.

   - *Validated analytical methods.* As in the expression "what is not measured is not managed," validated, robust analytical methods are needed in order to measure the processing contaminants in food products for the purpose of exposure assessment, risk management, and evaluation of efficacy of control measures. Sensitivity of the methods is also important as some processing contaminants can occur in low amounts, but still be important in terms of human exposure. The analysis of some contaminants, e.g., acrylamide, presently requires state-of-the-art methods practiced in dedicated laboratories. Methods simple enough for use in the control laboratory and capable of providing timely results still need to be developed.

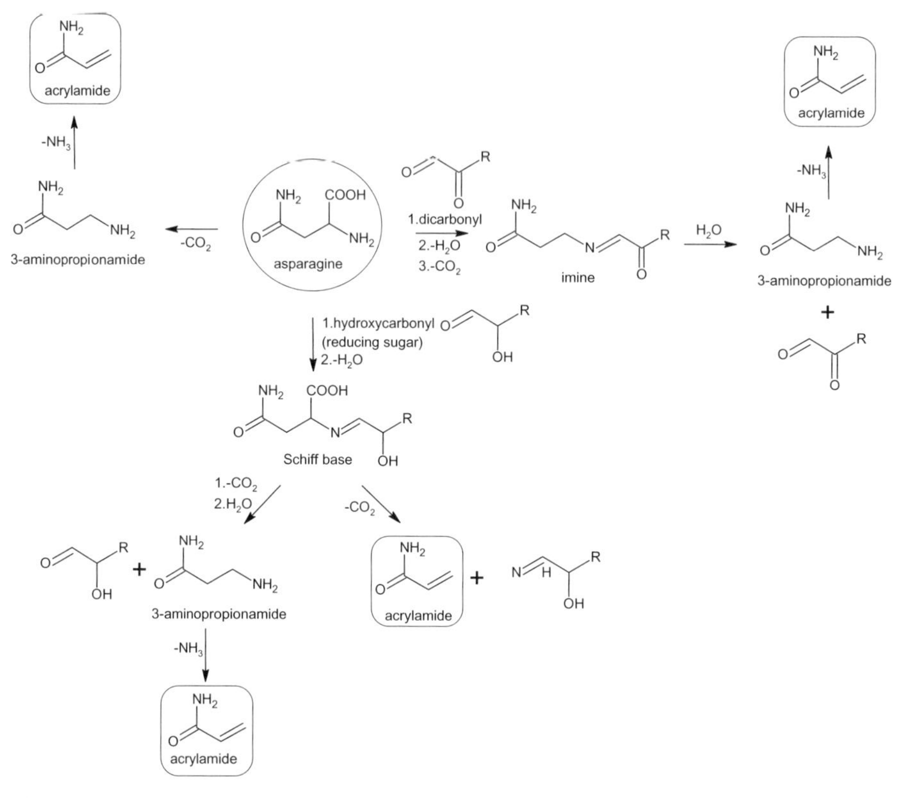

Figure 7.1 Possible pathways and intermediates leading to the formation of acrylamide in food.

4. *Data on the likelihood of occurrence.* The first principle of HACCP on hazard analysis calls for an evaluation of the risk, including the likelihood of occurrence. Data on the occurrence of processing contaminants in various food products are central to such an evaluation.
5. *Determination of acceptable level.* In managing a hazard, in particular when the occurrence of the agent cannot be fully prevented as is in the case of many processing contaminants, there is a need to know to what level the hazard in question must be controlled. This is also underpinned in Principle 1 of HACCP which, as part of hazard analysis, calls for the establishment of control measures.

A control measure is defined as "any action and activity that can be used to prevent or eliminate a food safety hazard or reduce it to an acceptable level." This brings in the concept of "acceptable level." Thus, to be able to manage processing contaminants, it is imperative to establish a limit of acceptability for the contaminant in question.

Establishing an acceptable limit is often a complex task. On the one hand, it has *in principle* to ensure the safety of products; as such, it should be based on the evaluation of the toxicological data. Where adequate toxicological data are not available and a risk assessment of some kind is not possible, it is difficult to determine a safe dietary intake. For instance, despite extensive studies that have been undertaken since the discovery of acrylamide, so far scientists have not been able to establish such a level. It is also doubtful that, in all cases, it may be technically feasible to prevent the formation of processing contaminants to levels that are considered safe from a toxicological point of view, without significantly altering the textural and organoleptic properties of the product. On the other hand, according to the present approach to food safety based on the risk analysis process, in establishing a standard, risk managers also need to take into account other factors such as feasibility, consumer preferences, and nutritional implications. Processing contaminants present multiple challenges in this respect:

i. *Risk/risk evaluation.* Several processing contaminants may be associated with the same processing step but their formation pathways differ. In fact, a processing step may promote the formation of a contaminant while it may have an adverse effect on another.

> ***Examples:***
> In preventing the PAH formation in grilled meat products by reducing grilling time, one may increase the microbiological risks. In baking bread, a high pH may inhibit 3-MCPD formation but increase the acrylamide formation (Fig. 7.2). Similarly, limiting the time of yeast fermentation may decrease the amount of 3-MPCD formation but increase the amount of acrylamide.

Thus, any mitigation measures targeted at a reduction of thermal input must consider the possible associated chemical and microbiological risks. Many outbreaks of *Escherichia coli* O157 associated with hamburger patties have been associated with undercooked hamburgers. This is not to say that in these cases, undercooking was done with the intention of preventing processing contaminants. However, care should be taken that in deciding control measures, whether they take place in the food processing and manufacturing industry or in homes, the various types of hazards are considered, the process is optimized, and limits are decided accordingly.

ii. *Role of precursors and internal dose.* Certain compounds may be formed *in vivo* and even if the product as such does not contain significant amounts of the contaminant itself, the contaminant may be released *in vivo* after consumption of the product. Quantification of the exposure

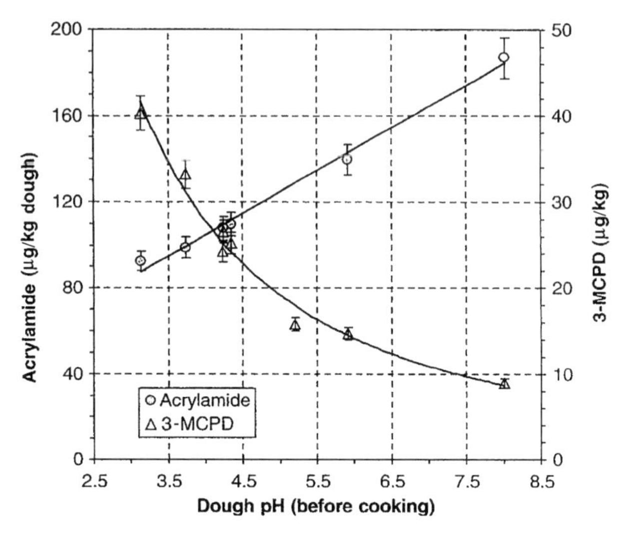

Figure 7.2 Impact of dough pH on the formation of 3-MCPD and acrylamide (reproduced with permission of Koning *et al.* (66)).

to the contaminant via the use of biomarkers may provide more reliable estimation of the exposure and would take into account the impact of precursors.

> ***Examples:***
> 1. 3-MCPD may be formed from the corresponding esters in the gastrointestinal tract through lipase activity.
> 2. Nitrosamines may be formed from nitrites and nitrate and amines *in vivo*, under acidic conditions of the stomach.

iii. *Risk/benefit considerations.* Mitigation measures may lead to the loss of certain positive health benefits of foods, such as their inherent nutritional value. These need to be taken into account.

> ***Example***:
> The use of white flour may lead to a lesser amount of acrylamide formation in bread. However, the nutritional benefits of wholemeal flour outweigh the apparent benefit of white flour.

Reducing processing contaminants may also affect the texture and organoleptic quality of products and undermine consumer acceptance of the product.

For instance, research on coffee and acrylamide has revealed that a decrease in acrylamide content can significantly affect the product quality.

A range of other factors may also interfere with this risk management decision. So to set a limit for a given contaminant, a holistic approach should be considered, taking into account all types of risks as well as benefits associated with a given product or processing and manufacturing process. Hence, the acceptable limits of a hazard for a food may not be strictly associated with the risk of the contaminant under consideration, but from the perspective of society, the limit is viewed here as the "safety standard."

At times, where the hazard in question cannot be prevented to acceptable levels (e.g., case of carcinogenic and genotoxic compounds for which no toxicological safe level can be established), or data are lacking to conduct a risk assessment, public health authorities may decide to apply the concept of "as low as reasonably achievable" (ALARA), or develop a code of practice (CoP) to promote best practices for minimizing the formation of processing contaminants. For instance, short of being able to decide a safety standard for acrylamide, the CAC decided to develop a Code of Practice to consolidate the know-how for minimizing acrylamide and promoting best practices (43).

In determining a limit of acceptability, further complexity arises from the variability in the occurrence and content of processing contaminants. For instance, data reported on acrylamide indicate a large variability within a specific food product, i.e., batch-to-batch. Part of this variability may come from the analytical method; another source of variation may be due to the raw material and its composition, seasonal and environmental influences (45), and variation in the conditions and method of processing, which may not be identical from one facility to another.

Limits of acceptability for a contaminant are usually established in the framework of regulatory measures, and must be complied with. Alternatively, where a regulatory limit is not possible, for instance due to the lack of adequate toxicological data and risk assessment, public health authorities may decide to establish a "guidance value" to guide industry in its efforts to manage the contaminants. Legally, such a guidance value is not binding but in practice, it is like a "legal limit" as all food manufacturers who care for the reputation of their company or brand will try to comply with such a limit. Such a guidance value can also be established internally by a food industry or a food sector. An example of such a guidance value is the German "signal value" (SV) established by the German authorities for acrylamide, based on an approach similar to the ALARA concept, but referred to as the minimization concept (Table 7.3). The SV is defined as the lowest value of those 10% of products of each commodity group that have the highest acrylamide content (46, 47). Such an approach is often a dynamic process and as scientific knowledge improves and it becomes possible to reduce the content of acrylamide further, the SVs are updated. For instance, in 2008, the German authorities updated their original SV set in 2003. It is to be noted that such an

TABLE 7.3 German signal values (SVs), 6th and 7th calculations (acrylamide, µg/kg).[a]

Product group	SV 2007	SV 2008
Crispbread	496	496
Breakfast cereals	180	80[b]
Fine bakery products, short pastry	300	260
Cakes and biscuits for diabetics	545	545
Children's biscuits (for infants and small children)	197	197
Gingerbread and bakery wares containing gingerbread	1000	1000
Spekulatius (thin almond biscuits)	416	416
Potato crisps	1000	1000
French fries, prepared	530	530
Potato fritter, prepared	1000	872[b]
Coffee, roasted	277	277
Coffee extract (soluble coffee)	969	937[b]
Coffee substitute	801	801

[a]References 46 and 47.
[b]Products that show decreased SVs from 2007 to 2008.

approach may give a competitive advantage to larger food industries that are supported by a strong research and development infrastructure. Such an approach would also present a disadvantage if several processing contaminants occur in the same product and a limit is established for a given contaminant without considering implications for others. Additionally, even for a single category of products, there may be large variations on the content of processing contaminants, due to factors beyond the control of manufacturers. This makes compliance with the present German SV difficult for some products. For instance, for breakfast cereal, over and above the composition of the product, seasonal variation may play an important role on the content of free asparagine in the cereals used, and thus on the amount of acrylamide in the final product.

A second reason for which an "acceptable limit" is required is for the purpose of hazard analysis (Principle 1): it consists of the evaluation of the significance of a hazard and its consequence for the consumer, if the step at which it is formed is not controlled. Such an evaluation helps to decide the stringency and frequency with which control and monitoring must be exercised at the given step. To carry out an evaluation, the questions that are usually raised are whether it is likely that an *unacceptable amount* of a processing contaminant is formed and if this is the case, whether it is likely that this will lead to injury or harm of consumers. Thus, the notion of limit of acceptance is raised again. *A priori* different types of limits can be considered.

Acute reference dose (ARfD): is defined as "an estimate of the amount of a substance in food or drinking water, normally expressed on a body weight basis, that can be ingested in a period of 24 h or less without appreciable health risks to the consumer on the basis of all known facts at the time of the evaluation" (78).

Reference dose (RfD): an estimate of the daily exposure dose that is likely to be without appreciable health effect even if continued exposure occurs over a lifetime.

7.4.1.1 ARfD Perhaps this would be the most practical value when evaluating the significance of a hazard in the context of hazard analysis and deciding the need for having a so-called "critical control point." This would mean evaluating if exposure to a hazard as a result of consumption of a specific product would cause injury or harm (as a product that would be contaminated with a foodborne pathogen). However, for most processing contaminants, such a value is not available. Also, if a hazard complies with this value but exceeds the regulatory standard or RfD, the product, even though *per se* not causing immediate harm to consumers, may be viewed by the regulatory authorities or by public opinion as unsafe. In 2006, a food manufacturer selling a product in Switzerland was forced to recall his biscuits due to a relatively higher content of acrylamide than that of other manufacturers. In this case, there was no ARfD, but if there had been one, it would have been unlikely that the product would have exceeded it.

7.4.1.2 RfD For many but not all chemical hazards, public health authorities have established an RfD, such as the tolerable daily intake (TDI). If a hazard is likely to occur at amounts exceeding this limit, one may, for purely management purposes, consider that there is a significant risk, although in reality the health risk is only upon long-term exposure. A public recall that occurred in 2006 illustrates this point, even though the contaminant in question was not a processing contaminant. In this case, bakery products were recalled in Hungary and some other European countries due to amounts of coumarin exceeding the TDI. In this example, although the temporary exposure of consumers to such a product would not have had an appreciable impact on health, the sale of the product was prohibited for food safety reasons. For the purpose of hazard analysis, such a value can be considered when there is no regulatory standard.

7.4.1.3 Regulatory Standards As mentioned earlier, a regulatory standard or a norm is a value that public health authorities determine as a limit of acceptability for a given contaminant, taking into consideration risks and other factors such as feasibility. Even though they may not always be strictly based on or related to safety assessment, they are viewed as the safety standard. Exceeding this limit on the short term does not necessarily mean that the consumption of the product will lead to injury of consumers; however, as this is the standard set for food safety purposes and food manufacturers are to comply with this limit, it is practical to consider this value as the limit of acceptability in hazard analysis. Table 7.4 gives examples of regulatory standards or norms for selected processing contaminants.

In summary, the management of processing contaminants follows the existing paradigm of risk analysis encompassing valuable input from fundamental research up to the risk management decision. Consideration of these elements is central to ensuring effective and sustainable food safety assurance programs (Fig. 7.3). As shown, for managing processing contaminants, a concerted effort of all stakeholders (academic, governmental, and industry sectors) is required.

7.4.2 Considerations in the Application of the HACCP System to Processing Contaminants

Application of HACCP principles is preceded by a number of activities including assembling a team, describing the product, identifying intended use, and constructing the flow diagram together with on-site confirmation.

All these activities are important with regard to the management of processing contaminants, and general guidance on the application of these as well as the implementation of the seven HACCP principles is given in the Codex Guidelines on HACCP (39). In the following section, necessary adaptations and considerations specific to processing contaminants are explained (38):

7.4.2.1 Assembling the Team For processing contaminants, the team should include individuals with expertise on:

- the product, its ingredients, and its process of production and manufacturing
- the processing contaminants and the mechanism of its formation
- regulatory requirements
- operations
- analytical methods
- food engineering
- quality management
- primary production, where applicable.

Depending on the case, individuals with expertise in microbiology, thermal processing and other quality aspects of the product should also be part of the team, in particular if the product will undergo fermentation or sterilization processes.

TABLE 7.4 Regulatory limits or tolerances/guidance levels for selected processing contaminants (non-exhaustive list).

Processing contaminants	Commodity	Limits/tolerances/guidance, mg/kg	Reference no.
3-MCPD	Acid-HVP in soy sauces	0.02[a]	EC (53)
	Acid-HVP	0.05	United States (54)
Dichloropropanols	Savory sauces	0.05	CH (55)
Benzo[*a*]pyrene	Smoked meats and smoked meat products	0.005	EC (56)
	Smoke flavoring used in foodstuffs	0.03	EC (56)
	Foods for infants and young children	0.001	EC (56)
Benzene	Bottled water	0.005	United States (57)
Histamine	Fresh fish	n = 9 c = 2 m = 100 M = 200	EC (59)
	Fish products	n = 9 c = 2 m = 200 M = 400	
N-Nitrosamines	Beer	0.0005[b]	CH (73)
Nitrites	Traditionally cured meat products	175[c]	EC (74)
Ethyl carbamate	Fruit brandies	0.8	EFSA (75)
	Wines	0.03	Canada (76)

[a] 40% solids.
[b] Sum of volatile *N*-nitrosamines.
[c] Expressed as $NaNO_2$.

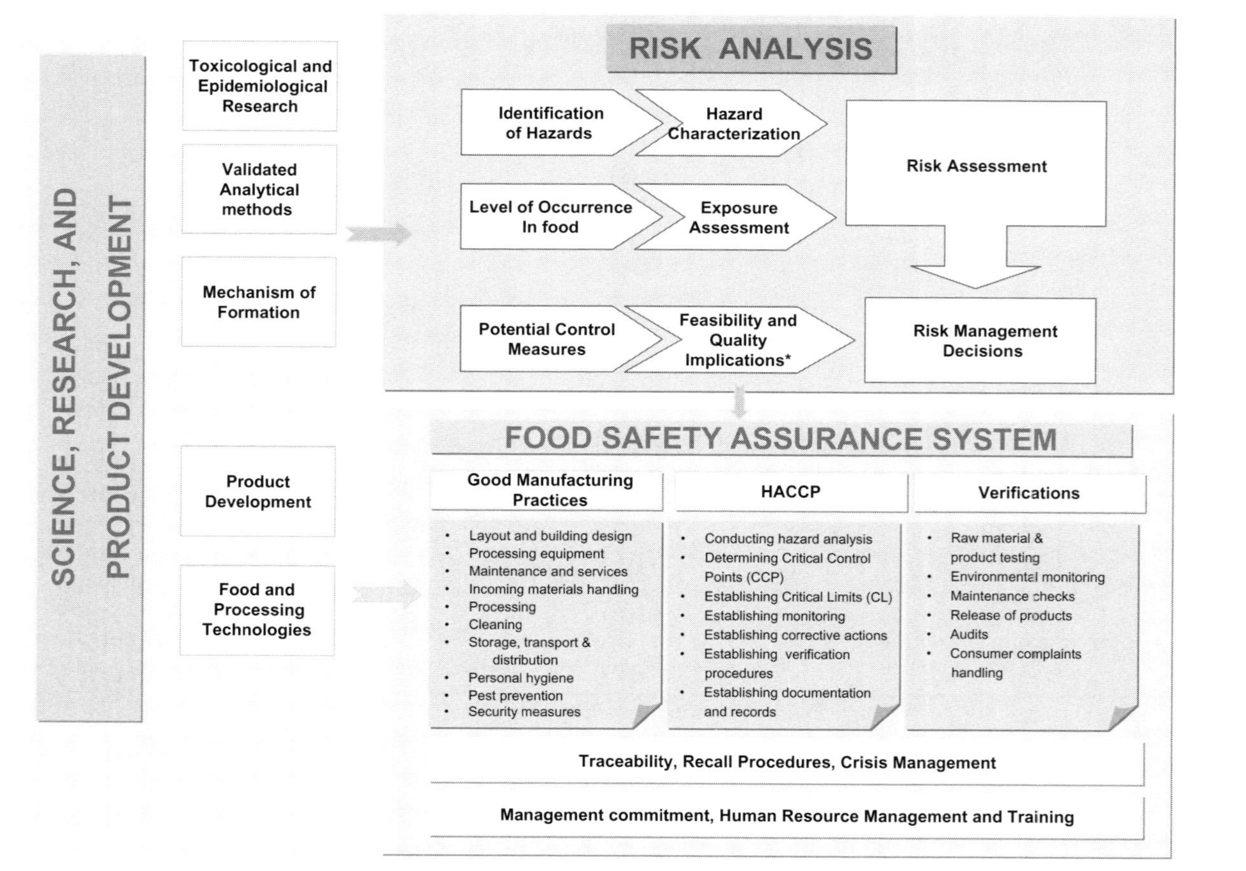

* Include nutritional, microbiological, chemical safety as well as cost and consumer acceptance

Figure 7.3 Illustration of the management of processing contaminants from fundamental research to the risk management decision and integration in the industry food safety assurance programs (adapted from Reference 42) See color insert

7.4.2.2 Description of the Product A full description of the product should be prepared, including conditions for raw material storage, transport and distribution, and preparation by end users. Composition of the food and the physicochemical properties are also essential. As mentioned in Section 7.2, the formation of many of processing contaminants is regulated by these factors. The more detailed the description, the less likely the risk of overlooking a factor which may impact the formation of processing contaminants.

7.4.2.3 Identification of Intended Use The intended use should be based on the expected use of the product by the end user or consumer, including the country where the product will be sold. This information is important to ensure that the safety of the product is designed according to the needs or requirements of the target consumers. For instance, if the product is to be sold in another country or for specific consumer groups, e.g., children, the limits applicable for the country in question or target consumers must be taken into account.

7.4.2.4 Construct the Flow Diagram The flow diagram should cover all the steps in the operation and conditions for these. As can be seen for the example, sometimes factors influencing the formation of processing contaminants occur at stages preceding or following the specified operation (e.g., storage). Therefore, the flow diagram should as far as possible provide appropriate details throughout the entire food chain.

7.4.2.5 On-site Confirmation of Flow Diagram This step in the development of HACCP is intended to verify that the flow diagram reflects the true manufacturing situation and that no important consideration is omitted. This step may not be feasible during the early step of product design where hazard analysis is carried out on unit operations or on a theoretical case.

7.4.2.6 Principle 1: Hazard Analysis This principle consists in listing all potential hazards associated with each step, evaluating their significance taking into account their likelihood of occurrence and their health consequences, and considering measures to control identified hazards. Where the hazard in question is already identified, e.g., decision to control acrylamide in French fries, at each step of the operations, precursors or factors that can lead to its formation are considered and measures for control are determined. More than one control measure may be required to control a hazard. In deciding on the control measures, it is fundamental to understand what are the parameters that characterize the control measures and to have full understanding of the mechanisms of formation, including precursors, catalysts, and processing conditions. If one takes French fries as an example, then a broad range of control measures can be identified as contributing to the reduction of acrylamide (Table 7.5). In addition, a number of other potential control measures can be explored further, albeit not all tested at the industrial scale, such as (i) addition of the enzyme asparaginase, (ii) pretreatment of potatoes with amino acids (glycine or glutamine, possibly in combination with citric acid), and (iii) addition of di- or trivalent cations (e.g., $CaCl_2$).

TABLE 7.5 Application of HACCP principles to the management of acrylamide in French fries.

Steps	Control measures	Limits/targets	Monitoring procedures	Corrective actions
Raw material	Selection of cultivars with reduced sugar content or cultivars that are less prone than others to low temperature sweetening	Specific cultivars with sugar levels lower than 0.3% fresh wt	Check the potato cultivars Test for sugar content or apply the fry test	Reject potatoes
Potato storage	Control conditions, i.e., temperature of storage	Storage temp. between 6 and 8 °C, in darkness	Monitor the temperature during storage and transport	Reconditioning of potatoes over a few weeks at higher temperatures (12–25 °C)
Processing	Coarse cut	14 × 14 mm/surface to volume ratio of 3.3 cm^{-1}		
	Blanching potato strips to lower sugar levels	Blanching in hot water at 85 °C for 3.5 min	Monitor the time and temperature of blanching	In line color sorting to remove dark fries
	Optimized heat treatment process (frying)	120 °C until golden yellow color	Observe the color and monitor the temperature of frying	
Preparation in the home	Optimized reheating process, no overcooking by consumers. Instruction for preparation by manufacturer and observation of instruction by end users	According to on-pack instructions (pan-frying until golden yellow color)	Observe the on-pack instructions and the final color	Discard dark French fries

7.4.2.7 Principle 2: Determine the CCPs In general, the CCP refers to the step in the operations at which control[2] is essential to eliminate, reduce, or maintain a hazard at an acceptable level; in other words, a step at which, if it is not controlled, the product may be considered unsafe[3,4]. The designation of a step as a CCP usually has many implications. Over and above the need for setting up a very strict monitoring procedure (see Section 7.4.2.9), it also implies that all the records pertaining to the implementation of control measures at this step need to be reviewed and verified before the product is released. This may sometimes delay the release of the product, for instance, if the monitoring is based on testing. Also, if the critical limits (see Section 7.4.2.8) at the CCP have been violated and the product has accidentally reached the market, consideration must be given to product recall. Additionally, a condition for a step to be considered a CCP is the fact that it should be possible to effectively monitor the step. Therefore, strictly speaking, the concept of CCP is reserved to critical steps.

In evaluating many processing contaminants, very few may actually occur at a level that will present a significant health risk if a noncomplying product incidentally reaches the consumer and is consumed. For instance, the harm caused by an occasional consumption of a portion of French fries, even with the highest level of acrylamide observed for this product category, may not be a significant health risk in a short-term exposure, although in the long run, it may constitute a health concern. Therefore, when conducting hazard analysis, at the amounts at which many processing contaminants occur, risks of consuming a product that incidentally has high contents of a processing contaminant may not be such as to warrant a CCP in the same way it is needed for a hazard such as salmonella in milk or a metal piece in ice cream. This should not be interpreted as the golden rule, and there are of course exceptions; for example, the occurrence of high amounts of histamine may lead to histamine poisoning. Thus, an evaluation of risk must be conducted on a case-by-case basis.

7.4.2.8 Principle 3: Establishing Critical Limits Critical limits are basically limits of acceptability or unacceptability of control parameters. As several parameters may be important for controlling one or more steps, there may be different critical limits and these have to be identified carefully. In HACCP, the term "critical limit" is associated with the CCPs. However, even if the HACCP study does not result in specific CCPs, acceptable limits for control parameters need to be defined. Note that where the formation of a contaminant is controlled through processing or product formulation, the critical limits

[2]There may be more than one CCP at which control is applied to address the same hazard.

[3]The determination of a CCP can be facilitated by the application of a decision tree, which indicates a logical reasoning approach.

[4]Note that if a hazard is identified at a step where control is necessary for safety and no possible control measure exists at that step or any other, then the product or process must be modified at that step or at any earlier or later stage, to include a control measure.

refer to the limit of acceptability of the process (temperature, time) or product parameters (pH, water activity, ingredient composition) and not to the contaminant itself. Sometimes, when the control measure takes place in homes or in food service establishments where tools for physical or chemical measurements are not available, indirect measures such as color or texture can be used as a monitoring parameter. For instance, in controlling the 3-MCPD formation in the bread-baking process, the golden color as target can be used as guidance.

7.4.2.9 Principle 4. Establishing a Monitoring System for Each CCP According to Codex Alimentarius, in the context of HACCP, monitoring is the scheduled measurement or observation of a CCP, relative to its critical limits. The monitoring procedure must be able to detect loss of control at the CCP. The frequency of monitoring should be set so as to enable the necessary adjustments, which will ensure control of the process and prevent violation of the critical limits. Where possible, process adjustment should be made when monitoring results indicate a trend toward loss of control at a CCP. From this, it foregoes that very rigorous monitoring procedures must be established at the CCP.

As mentioned earlier, for many processing contaminants, it may be viewed that the risk of a noncomplying product will not be such that it will present a significant health risk for consumers, and thus a CCP will not be warranted.

Nevertheless, even if a step is not considered a CCP, this should not be interpreted as that step not being important or preclude monitoring of the control measure(s) at that step. The step may remain crucial from the perspective of controlling the processing contaminants in question and/or meeting regulatory requirement. Therefore, a monitoring procedure may need to be established at the steps that play a decisive role for controlling the contaminant. The steps at which these control measures take place are referred to as CPs. For instance, in the French fries example, one could consider the conditions of storage of the raw potato and blanching as essential control measures for maintaining the acrylamide within given limits (Table 7.5). In this case, depending on the degree of variation expected, one or both steps may be considered CPs and may be monitored. To differentiate these two types of monitoring, for the purpose of this text, we will refer to CCP monitoring versus CP monitoring.

7.4.2.10 Principle 5. Establishing Corrective Actions When applying the HACCP system, specific corrective actions must be considered for each CCP so that any deviation in control parameters is corrected. The corrective actions must ensure that the CCP(s) has/have been brought under control before the product is put on the market. This can also include proper disposition of the affected product. In the case of a processing contaminant, such a principle can also be applied to a CP, i.e., a step that is not considered CCP but important for controlling processing contaminants. However, whether the product should be disposed of or released needs to be evaluated on a case-by-case basis, and in collaboration with public health authorities if the regulatory limit is exceeded.

7.4.2.11 Principle 6: Establishing Procedures for Verification Verification refers to all methods, procedures, and tests to determine if the HACCP system is working correctly. As such, verifications may include a variety of activities and collection of data to confirm that the HACCP plan is valid and is well implemented. It can include:

- review of the records and monitoring data confirming that the process parameters are kept under control and within established limits;
- product survey to ensure compliance with established limits for the contaminants;
- audit of manufacturing operations for confirming the implementation of prerequisite programs (e.g., GMPs) and of the HACCP system.

Principle 6 of HACCP also includes the concept of validation. This has been briefly addressed earlier (see Section 7.3.2).

7.4.2.12 Principle 7: Establishing Documentation and Record Keeping In this context, documentation and records include:

- HACCP study, including hazard analysis, determination of control measures, and process/control parameters and acceptable limits;
- validation studies showing that control measures achieve the regulatory or internally set limits for processing contaminants;
- records of monitoring procedures and corrective actions in case of deviations;
- reports of audits confirming compliance with prerequisites.

7.4.2.13 Maintenance of the HACCP Plan Although not explicitly presented as a principle of the HACCP system, maintenance of the HACCP plan is as important as its elaboration itself. To this end, various types of data and information need to be monitored and reviewed. Where necessary, changes in the HACCP study/plan or improvements in their implementation must be made. Section 7.5.4.7 provides example of data that would need to be monitored and reviewed for managing processing contaminants.

7.5 A CASE STUDY ON THE APPLICATION OF THE HACCP APPROACH FOR THE MANAGEMENT OF PROCESSING CONTAMINANTS: EXTRUDED CEREAL SNACK

This is a hypothetical HACCP study for an extruded cereal snack product developed solely for the purpose of illustrating the application of the HACCP approach for the management of processing contaminants. Although many of the ingredients and unit operations are valid for most products of this type, the product as described here does not represent a real product.

7.5.1 Scope of the Study

A broad range of contaminants including microbiological hazards (e.g., *Salmonella* spp., *Bacillus cereus*), chemical hazards (e.g., pesticide residues), allergens, physical hazards, vitamin overdosage, and choking risk, can be considered for a typical cereal-based snack.

However, for the purpose of this chapter, the scope of the HACCP study is limited to processing contaminants, in particular acrylamide, 3-MCPD, and PAHs, as cereal grain foods constitute an important dietary source of these contaminants. Therefore, this study does not cover all food safety aspects of the product. Also, this hypothetical study does not consider the impact of any change in product/process formulation carried out to minimize processing contaminants on other types of hazards. Such a consideration would be crucial when considering the overall safety of a product.

The study focuses on the operations from the perspective of processing and manufacturing. As far as possible, it will also highlight measures that are required up- or downstream during agriculture, distribution, and final preparation by consumers to ensure minimum exposure of consumers to processing contaminants and to meet regulatory and/or potential customer requirements.

In addition to acrylamide and 3-MCPD, which are the focus of this study, formation of furan is also a potential concern with cereal products. However, due to lack of sufficient information on its level of occurrence, mechanism of formation, and possible control measures, it is not possible to fully consider it in this HACCP study. Nevertheless, where possible, based on the limited available data, some consideration is given to the possible formation of furan.

7.5.2 Product Description

It is assumed that the product is a cereal-type snack, which will be marketed in the countries of the EU. For the purpose of this study, we consider that the product is to comply with the standards presented in Table 7.6.

The raw materials include whole wheat flour, rye flour, oat flour, vitamin mix, ferric phosphate, sucrose, honey, water, oil (palm oil), and antioxidant tocopherol.

A flow diagram describing the unit operations for the cereal product is presented in Fig. 7.4. As illustrated, all the dry ingredients except the sucrose are mixed in a mixer. The water, oil, and the dry mix are then dosed separately in the extruder. The mixture is cooked at 140–160 °C for 30–40 s at a throughput of 300 kg/h. Then it is dried on a belt drier (160–175 °C for 5 min) and

TABLE 7.6 Standards selected for processing contaminants possibly applicable to the cereal snack considered in the HACCP study.

Food/ingredient	Processing contaminant	Limits	Source/rationale
Vegetable oil	Benzo[*a*]pyrene[a]	2 μg/kg (fresh wt)	(68)
Honey	HMF	40 mg/kg	(77)
Extruded cereal snack	3-MCPD	<10 μg/kg	Nestlé unpublished monitoring data (2002–2005)
Extruded cereal snack	Acrylamide	80 μg/kg[b]	Based on German SV for 2008 for breakfast cereals (47)
Extruded cereal snack	Benzo[*a*]pyrene, as a marker for PAHs[a]	1 μg/kg	(56)
Extruded cereal snack	HMF	As low as practically achievable	Due diligence

[a]The source of benzo[*a*]pyrene or other PAHs in cereal-based products may be environment and/or processing, e.g., exposure of seeds with combustion gases may lead to contamination of vegetable oil.

[b]This value is presently not attainable for all categories of cereal products. It is to be considered tentative and for the purpose of this case study only.

cooled to 40–45 °C. A sugar/honey syrup slurry is then sprayed (coated) onto the extruded material in a tumbler and subsequently dried in a band drier at an air temperature of 128–130 °C for 12–15 min.

The product is filled into a bag-in-box packaging consisting of a non-barrier material, e.g., polyolefin and a carton box. The final product has a moisture content of about 2%, which corresponds to a water activity <0.3.

7.5.3 Distribution and Intended Use

The product is intended for consumption by the general population (from the age of 3 onwards) as a cereal snack.

Finished products will be distributed via approved food transporters, in trucks dedicated to finished packed products. The products will be stored at ambient temperature in warehouses, retailers, and households.

The product is considered a ready-to-eat (RTE) product, i.e., it is expected to be consumed without further treatment, and with a shelf life of 12 months.

7.5.4 HACCP Study

As mentioned before in this chapter, for most processing contaminants, occasional or incidental consumption of food products in amounts generally reported for cereal foods will not present a significant risk to health. Therefore,

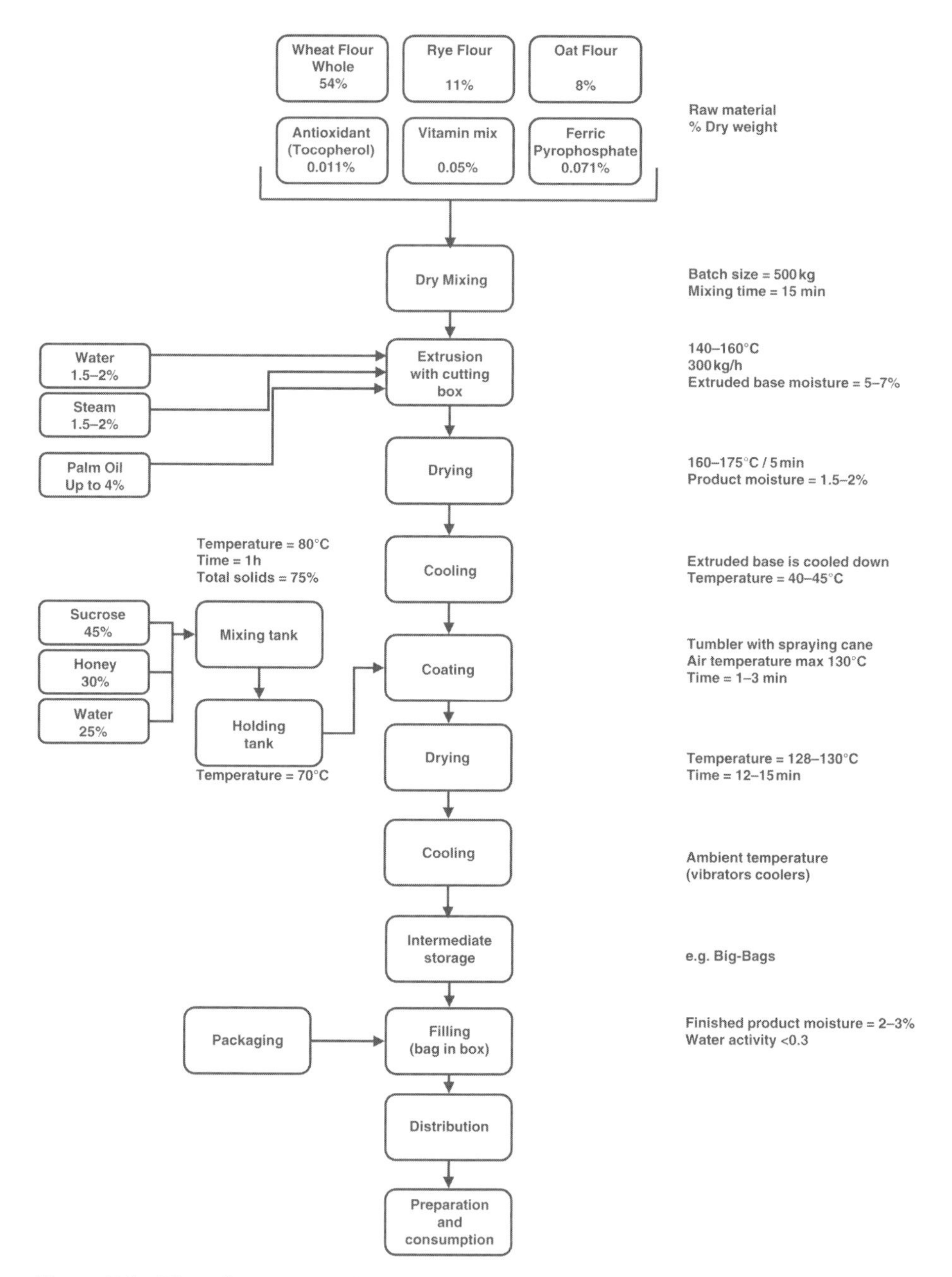

Figure 7.4 Flow diagram of unit operations in the extruded cereal snack HACCP case study.

the risk posed by the occurrence of these contaminants in the products is not considered to be significant to a degree that will warrant a CCP in the strict sense of the term. However, this does not undermine their importance for health and preclude the need for taking measures to reduce their amounts, while maintaining nutritional and organoleptic quality of products. Therefore, the intent of this study is to identify possible control measures and required monitoring to manufacture products that consistently have a low content of processing contaminants and that comply with regulatory requirements, where these exist.

Table 7.7 presents the outcome of the HACCP study and a plan for the management of processing contaminants for the cereal snack.

7.5.4.1 Hazard Analysis

7.5.4.1.1 Cereals Free asparagine, the precursor of acrylamide, is present in the major ingredients of the product, namely wheat, rye, and oat flour. In cereal-based products, free asparagine has been shown to be the limiting and the determining factor in the formation of acrylamide (60). The variety of cereals used and their growing conditions can have an influence on the content of free asparagine (18, 61). Thus, selecting a specific variety of cereals can be considered a potential control measure for reducing free asparagine. However, year-to-year variations or other factors (mineral deficiencies, drought, salt, toxic metals, and plant diseases) may also influence the content of free asparagine (45, 62). Therefore, such a control measure would necessitate monitoring of the amount of free asparagine in the incoming raw material or other types of indicators (e.g., rheological properties of dough) (44), short of which a consistent reduction in acrylamide content cannot be guaranteed.

However, a segregation of cereals based on the content of asparagine and their rejection in case of exceedingly high amounts may presently not be a practical and feasible control measure at industrial scale. Additionally, selection and possibly periodic change of supplier to ensure a low content of free asparagine in the raw material may lead to logistic problems and can create opportunities for errors in managing other hazards. Therefore, at this stage, selection of cereal varieties, although theoretically possible, is not a viable option for all situations. On the other hand, control of quality of fertilizers can contribute to a reduction of free asparagine in certain cereals. Several studies have shown that fertilizers poor in sulfur can lead to higher contents of asparagine in wheat, and a higher amount of acrylamide in the final product (58, 61, 63, 64). This finding needs to be further studied and also validated for other types of cereals.

Several studies have reported on the amount of free asparagine according to cereal type (61, 65, 66). Contents of free asparagine are found to vary considerably according to the cereal type. Rice has consistently been reported to contain relatively low amounts of asparagine. For wheat, rye, and oat, there

TABLE 7.7 A plan for the management of processing contaminants in line with the HACCP approach.

Ingredients/ steps	Hazards	Control measures	Limits	Monitoring	Corrective actions	Verification*
Cereals (wheat, rye, and oat)	High amount of free Asn	Advise suppliers on agronomic practices (adequate level of sulfate fertilization)	Adequate amount of sulfate in the fertilizers used: e.g., 40 kg/Ha S. Flour N/S ratio <17	Optional: testing for the N/S ratio of flour	Follow up with supplier possibly rejection of the raw material	Verify fertilization practices Periodic verification of N/S ratio in flour
	Free Asn varies according to the type/ variety of cereal and % of whole grain flour versus refined flour)	Option: use of refined and sifted cereal flour	Material quality conforming to internal specification	Check at reception	Rejection of the consignment	Ad hoc testing of flour quality to confirm compliance with specification
		If and where feasible, specify grains and varieties that are low in free Asn	Material quality conforming to internal specification	Check at reception	Rejection of the raw material if not conform to specifications	Supplier audits (grain segregation, traceability)
Cereals (wheat, rye, and oat)	Presence of PAHs above norms	(i) Sourcing from regions remote from urban areas (ii) Specification to suppliers	Material quality conforming to internal specification	Certificate of analysis (CoA)	Rejection of the consignment	(i) Periodic testing according to risk level (ii) Audit of the supplier (sourcing, traceability)
	High amounts of glycerol (precursor of 3-MCPD)	(i) Ensure freshly produced flour and short storage time (ii) Stock rotation at the supplier	Production date/best-before date	Check production/ best-before date at reception	Rejection of the consignment	Audit of the suppliers warehouse Ensure stock rotation and respect storage time

Vitamin premix	Vitamin C can be a precursor of furan	Correct % of vitamin C in the premix	As per chosen composition in the recipe	CoA	Rejection of the consignment	Periodic verification of vitamin premix composition at supplier
Veg. oil	PAH	Selection of oil treated with activated carbon at the supplier level	Material quality as per internal specification (PAHs <2 μg/kg)	CoA	Rejection of the consignment	Audit of oil suppliers to verify the deodorization process and active carbon treatment Periodic testing of oil for PAHs to confirm suppliers' compliance with the requirements
	3-MCPD esters	Selection of oil according to the amount of 3-MCPD esters.	Material quality as per internal specification (e.g., MCPD esters <1 mg/kg)	CoA	Rejection of the consignment	Periodic monitoring for 3-MCPD esters to confirm suppliers compliance with the requirements
Honey	Heat treatment in the presence of free Asn	Control of storage conditions of honey at the supplier's (time and temperature of storage)	Temperature of storage: max 25 °C. Level of HMF according to specification (<40 mg/kg)	CoA	Rejection of the consignment	Periodic test of honey for HMF (?); suppliers storage conditions
Storage of honey	HMF	Store honey in a cool place; storage time and stock rotation	Maximum temp. 25 °C Total storage time prior to manufacturing 3 months	Monitor temperature and time of storage	Block batch, restore storage temperature or relocate goods	Periodic inspection of the warehouse and verification of practices (stock rotation)

TABLE 7.7 ***Continued***

Ingredients/ steps	Hazards	Control measures	Limits	Monitoring	Corrective actions	Verification*
Storage of wheat, rye, and oat flour	Glycerol liberation in aged flour	Limit storage period of flour Apply stock rotation	According to the expiry date of flour	Check the best-before date of flour before use	Inform supplier; rejection of the consignment	Periodic inspection of the warehouse (stock rotation) and verification of records
Recipe	Concomitant high amounts of reducing sugars and cereals containing free Asn	Formulation to minimize amounts of AA precursors	Following product formulation	Recipe	Block batch, inform QA and production	Free Asn in selected batches
Dry mixing	Ingredient mismatch	Process control; training of operators; mass balance check of ingredients	Recipe	Dosing system	Block batch, inform QA and production	Free Asn levels in selected batches
Extrusion	AA 3-MCPD	Optimized extrusion conditions	Temperature: 140 °C, moisture content of extrudate: 7%	Extrusion parameters	Block batch, check contaminants and discard product if unacceptable high level of AA or 3-MCPD	Verify the implementation of the monitoring
Drying (belt)	AA 3-MCPD	Optimized drying conditions	Temp.150 °C Time: 6 min	Drying parameters, i.e., time and temp.	Block batch, discard product if unacceptable high level of AA or 3MCPD	Periodic audit of the manufacturing site, and verification of practices and records

Heating of honey syrup mixture	HMF	[illegible] requires validation of the heat treatment to ensure absence of microbiological risks.				
Coating	HMF	The process (130 °C, 1–3 min) will contribute to additional formation of HMF				
	AA	Despite only a short heating period, consider formation of AA at the surface of the product				
Drying (band drier)	AA	Temperature of >120 °C may contribute to AA formation				
	HMF	Consider possibility of formation of HMF under given conditions				
Filling and Packaging	None identified					
Distribution and retail	HMF	Controlled storage conditions (time and temperature) during distribution and retail; stock rotation in the warehouse to minimize storage time, respecting expiry date	Storage at maximum 25 °C for the period of shelf life (<12 months). Storage time is to be determined according to sales turnover	Monitor temperature and shelf life of the product on daily basis	Restore temperature of storage. Discard expired products	Periodic audit of distributors by the manufacturer t o verify the implementation of control measures, including monitoring temperature of storage, observation of expiration date, and stock rotation
Consumer: storage, preparation	HMF	Instruction should be provided by the manufacturer on the pack for ideal storage conditions Keeping the product in cool and dry conditions and respecting the expiry date.	Best-before date considering a shelf life of 12 months Storage temp. max. 25 °C	Consumer adhere to best-before date	Discard	N/A

*Additional verifications

- Periodic testing of the final product at the end of the shelf life to ensure compliance with set standards for AA, 3-MCPD, PAHs
- Data on any noncompliance reported by external bodies, e.g., customers and regulatory authorities.

Asn = asparagine; HMF = hydroxymethylfurfural; AA = acrylamide; N/A = not applicable.
Additional verification measures are listed in Section 7.5.4.5.

are conflicting reports and no firm conclusion can be drawn at this stage. Furthermore, certain cereals present important and specific nutritional benefits. For instance, oat or oat bran are reported to have hypocholesterolemic effects. Therefore, in principle it is possible to minimize the amount of acrylamide by selecting cereals that contain a reduced amount of free asparagine. However, from a nutritional perspective, this may not always be recommended, particularly for products that are an important part of the diet, because of the benefits they can bring. By the same token, regulatory norms, which do not take into account the composition of the product and its impact on the risk and benefits of the products, may force manufacturers to lower the content of the contaminant to the detriment of potentially long-term nutritional benefits.

Flour made from whole grains, in particular containing bran, may lead to higher amounts of acrylamide, as bran and germ contain higher contents of asparagine (61, 67). The lowest values of asparagine have been associated with sifted wheat flour (67). Thus, to minimize the formation of acrylamide, the following measures or combination of measures can be considered:

- Selection of cereal varieties with the lowest content of free asparagine, where feasible.
- Controlling conditions of agriculture, i.e., ensuring adequate amounts of sulfate fertilizers (40 kg/Ha S) (63). The group of Granvogl (44) showed that the nitrogen and sulfur content of wheat will be influenced by the fertilizers; thus, the ratio of N/S can be an indicator for a S deficiency. A flour with the optimum amount of sulfur should have a N/S ratio <17. This may be used as a means for either monitoring the quality of the flour or verifying agronomical conditions of wheat cultures.
- Composition of selected cereals: reduction of the proportion of flour made with whole grain, or selection of ingredients that lead to low amounts of acrylamide, e.g., rice and corn.

It is to be noted that the influence of the type and variety of cereal on the final content of acrylamide is not absolute and depends on the other ingredients and design of processing steps (extrusion, toasting, and drying). For instance, by designing an extrusion-puff process with low thermal input, a manufacturer has been able to produce products with low contents of acrylamide, yet containing up to 25% whole grain (66). As mentioned before, climatic and agronomical conditions under which cereals are produced can also significantly influence the amount of acrylamide in a product and cause year-to-year variations. The nutrient supply to the plants (quality of fertilizers), fungal and bacterial attacks, toxic metals, drought, and salt stress can all influence the content of acrylamide (64). However, there is no information on the amount of other processing contaminants (e.g., 3-MCPD).

Cereals can also be a source of PAH and amounts reported in the literature are in the order of 0.04–0.1 μg/kg, and up to 5 μg/kg for bran (68). Although the source of PAH in cereal grains is the environment where cereals are cultivated to ensure that the final product presents minimum amounts of PAH, it is important to source the cereal as far as possible from areas remote from urban and industrial activities, where environmental pollution appears less significant.

7.5.4.1.2 Vegetable Oil Becalski and colleagues (65) also point to a certain impact of the variety of vegetable oil on the acrylamide formation in fried potato slices; olive oil led to higher value of acrylamide in the final product than did corn oil. The CIAA (17) reviewed the impact of oil on acrylamide formation. Although they reported an influence of certain types of oil (e.g., a cold pressed wheat germ vegetable oil labeled biological), based on the review of all data they could not conclude on a definite impact of oil variety on acrylamide formation. Therefore, it is assumed that in the case of the cereal snack, the influence of the type of oil will be marginal, if any.

On the other hand, the selection of dietary fats and oils can have an influence on the formation of furan, as oxidation of PUFAs is considered a viable route. After coconut, palm oil—chosen as an ingredient in this study—has the lowest amount of PUFA and hence presents lesser concern in terms of formation of furan.

Additionally, recent studies have shown that the nature or quality of oil plays a central role in the formation of 3-MCPD. On the one hand, oil may provide glycerol, a known precursor of 3-MCPD. On the other hand, it has recently been demonstrated that oils, depending on their type, quality, and processing technique may contain different concentrations of 3-MCPD esters, ranging from 970 to 2435 μg/kg of oil, which under the action of lipase in food or *in vivo* may hydrolyze to furnish 3-MCPD (29). Further research is needed to better understand the impact of oil refinery on the formation of 3-MCPD. In the meantime, it is advisable to possibly select oils with a relatively lower amount of 3-MCPD esters.

Oil may also be a source of PAHs, which is considered both an environmental and a processing contaminant. As mentioned in Section 7.2.4, PAHs in oil can be reduced by the deodorization process and active carbon treatment.

7.5.4.1.3 Vitamin Premix One of the mechanisms proposed for the formation of furan is the degradation of vitamin C upon heat treatment. As such, to ensure stability of vitamin C and to prevent potential formation of furan, it would be important to add vitamin C, if necessary, after the heat treatment step. Another vitamin that has been studied for its potential role in formation of processing contaminants is vitamin E. The influence of this vitamin is described later (Section 7.5.4.1.5).

7.5.4.1.4 Honey Honey is composed of a large proportion of reducing sugars, i.e., fructose and glucose. The proportion of these sugars may vary and is in the range of 27–44% for fructose and 22–41% for glucose (69). Other sources report values of 46% for fructose and 37% for glucose (70). Hence, honey may be an important ingredient in terms of favoring acrylamide formation. Therefore, the processes of production of the cereal snack should be designed in such a way that it circumvents the heating of honey in the presence of free asparagine.

Heating honey can also lead to the formation of another processing contaminant, i.e., HMF[5] (see Chapter 2.5). Upon long-term storage, in particular at high ambient temperature, the amount of HMF may also increase. Sanz *et al.* (71) studied the effect of storage temperature on the increase of HMF. HMF increased only slightly during storage at 25 °C but increased noticeably at 35 °C. Gidamis *et al.* (72) also reported that the amount of HMF is far below the maximum acceptable content of 40 mg/kg as recommended by the CAC if stored for less than 6 months at ambient storage. Therefore, to minimize the formation and/or increase of HMF, it would be important to use freshly produced honey or control its storage conditions. A total of 3 months at maximum 25 °C was chosen for the storage of honey along the supply chain (supplier, transport, and manufacturing sites).

It is also to be noted that honey has a low pH and this may be favorable for the formation of 3-MCPD. The influence of honey on the formation of 3-MCPD, including during the storage period, needs further study.

Additionally, depending on its glucose concentration, honey can crystallize during storage. Such a phenomenon can change the water activity, which in turn can influence the rate of chemical reactions or impact on the quality of the product (see Section 7.5.4.1.12) (30).

7.5.4.1.5 Antioxidant Tocopherol (vitamin E) is used as antioxidant. Studies of Tareke (31) found that addition of antioxidants such as vitamin E can enhance the formation of acrylamide. Some other antioxidants such as rosemary extracts have led to a decrease in acrylamide in potato products. A recent study has shown that antioxidants from bamboo leaves or green tea can lead to substantial reduction of acrylamide in some products when tested at the laboratory scale (48). In a review of the subject, Taeymans and coworkers (17) concluded that the current data do not allow researchers to draw firm conclusions. One possibility is the interaction of oxidized polyphenols (quinones) with the intermediate 3-aminopropionamide, but clearly further research is needed to elucidate the underlying mechanisms. For the purpose of this study, no specific control measure is considered for tocopherol. However, further investigation is recommended on how tocopherol may lead to enhanced (or reduced) acrylamide in certain food matrices.

[5] 5-(Hydroxymethyl)-2-furfural (HMF) is formed from the Maillard reaction in many food products, e.g., honey, processed milk, apple juice, and breakfast cereal. It can also be formed by the caramelization of sugars in conditions of high temperature and low moisture content. At very high concentration, HMF is cytotoxic, causing irritation to the eyes, upper respiratory tract, skin, and mucous membranes. However, at concentrations occurring in food, its health effects, if any, are not clear as *in vitro* studies on genotoxicity/mutagenicity have given controversial results.

7.5.4.1.6 Ferric Pyrophosphate There is no known effect of ferric pyrophosphate on the formation of processing contaminants.

7.5.4.1.7 Water does not present a concern *per se*, except that the moisture content of the product, or more accurately the water vapor pressure that it exercises, expressed as the *water activity* of the product, at different stages of processing is a determining factor for the rate of chemical reactions, e.g., the Maillard reaction. The rate of the Maillard reaction increases at a water activity of 0.2 and is highest at a water activity of 0.6 (49).

7.5.4.1.8 Reception of Raw Materials With reference to the importance of the variety of vegetable oils and cereals selected, it is important to verify the authenticity of the raw material at this step.

7.5.4.1.9 Storage of Raw Materials As mentioned before, "aged" flour may contain higher amounts of glycerol, a possible precursor for 3-MCPD. 3-MCPD can also be released by action of lipase during storage. Therefore, it is important to adhere to appropriate storage conditions (temperature) and limit storage time by applying the FIFO principle.

7.5.4.1.10 Dry Mixing At this stage, all the dry ingredients, i.e., cereals, tocopherol, and ferric pyrophosphate, are mixed for about 10 min at ambient temperature. Although no specific hazard is associated with the dry mixing process itself, should product formulation, i.e., type and proportion of cereal used, be considered as a measure to minimize acrylamide, the control measure will need to be implemented at this step and monitored.

7.5.4.1.11 Extrusion Water, oil, and the dry mix are dosed separately and extruded at a temperature of 140–160 °C for 30–40 s. The moisture content of the extruded product is about 5–7%, which corresponds to a water activity of 0.4 at 25 °C. There is no consistent predictive model to estimate how such high extrusion temperature influences the water activity during the process. However, considering the heterogeneity of temperature and moisture content, it can be expected that at some specific point of the extrusion process, the water activity reaches values of 0.6–0.7 that are critical for Maillard reaction and may favor the formation of acrylamide and some other processing contaminants.

Most literature studies report on moisture content rather than on water activity. Accordingly, high moisture content and low temperature lead to the lowest amount of acrylamide. Konings *et al.* report formation of a measurable amount of acrylamide (50–100 μg/kg) at a moisture content ranging from 10% to <30% (66). Considering that a slight decrease in moisture content, e.g., 1–2%, can result in significant increase in acrylamide formation, it can be anticipated that under the present condition of moisture content, a substantial amount of acrylamide is formed. On the other hand, the temperature of the extrusion process is at the lower end of temperature usually reported for extrusion processes. Temperatures >170 °C can lead to a significant increase in acrylamide content in the extrudate (61). Note that the lowest content of acrylamide in rye extrudate has been observed at a temperature of 150 °C and 19% moisture content. Thus, for the present study, the temperature of 140 °C at 7% moisture content in extrudate is provisionally retained as limits of extrusion parameters. However, further studies need to be carried out to explore the possibility of increasing the moisture content in the extruded product and to optimize the time–temperature conditions of the extrusion process.

7.5.4.1.12 Drying The extruded product is dried on a belt drier at 160–175 °C for 5 min and cooled to 40–45 °C.

Based on the present state of knowledge, it is most likely that during this step, a major increase in acrylamide takes place as the moisture content is low and the temperature of drying is relatively high. Studies on the kinetics of the drying process have shown that this may also impact the amount of acrylamide. For instance, in cornflakes, toasting for a short time at high temperature would lead to lower contents of acrylamide than the reverse would (61). More research is needed to have a better understanding on the impact of kinetics of drying on acrylamide formation and the optimum conditions for drying or toasting.

Another contaminant that is also of concern at this step is 3-MCPD, which can also be formed at temperatures of 150 °C or higher (Fig. 7.5). In this case, the formation of both processing contaminants is to be considered. As reported by Konings *et al.* (66), the optimal moisture contents and temperature to minimize the formation of both acrylamide and 3-MCPD are 20% to 25% and about 150 °C, respectively. The formation of acrylamide is more sensitive to a decrease in moisture content, whereas in the case of 3-MCPD, an increase in temperature would be more important. However, there is limited knowledge on the impact of drying and toasting on the formation of these processing contaminants.

Heat applied at this step and at the extrusion step may inactivate lipase and in this way minimize 3-MCPD formation and thereby contribute to the stability of the product during storage and over shelf life.

Considering this, options to reduce the formation of processing contaminants at this step are to:

- toast at a high temperature but shortest possible time. This, in principle, should minimize the formation of acrylamide but may lead to a higher content of 3MCPD.
- toast at the lowest possible temperature (e.g., 150 °C) for a longer time. This would minimize the formation of 3-MCPD; however, the longer time required to dry the product to the required end moisture content may lead to a higher content of acrylamide.

The second option, possibly combined with the addition of asparaginase, is proposed in this case study. However, due to the nature of the process and

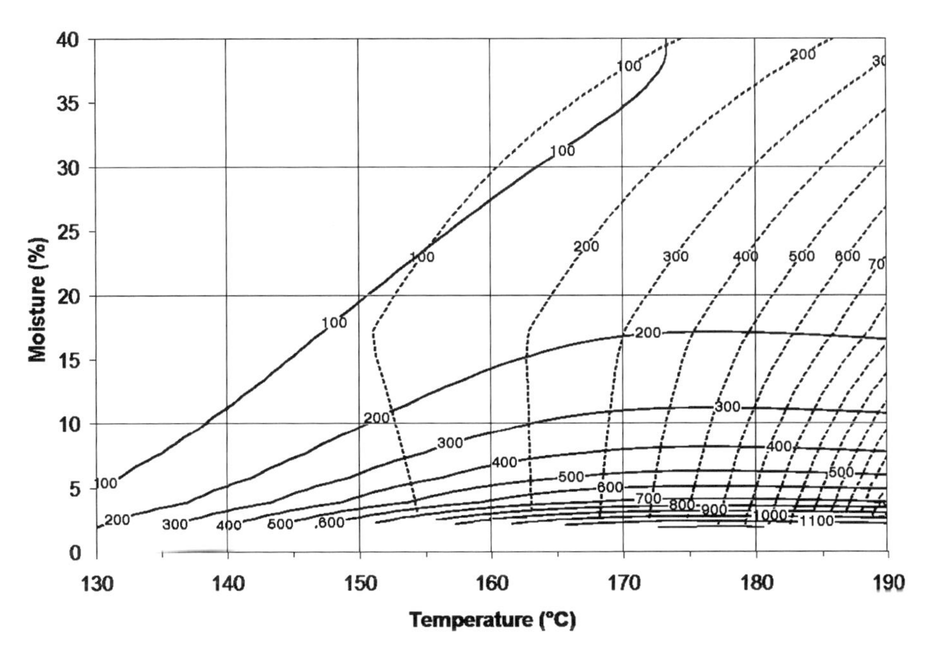

Figure 7.5 Graph showing the effects of moisture and temperature on acrylamide (solid line) and 3-MCPD (dashed line) generated in wheat dough (reproduced with permission of Koning *et al.* (66)).

need to extrude at low moisture, the enzyme may not be active and necessitate the addition of a minimum 20% water (estimate based on comparable studies). This modification will impact the texture of the product and final moisture content.

7.5.4.1.13 Coating A sugar/honey syrup mixed at about 70 °C is sprayed onto the extruded material in a tumbler (130 °C). Although the formation of 3-MCPD at this step remains low, high amounts of acrylamide may be formed from the reaction of free asparagine and the abundant amount of fructose and glucose present in the honey. However, as the coating of syrup will not diffuse into the dense extruded cereal mixture, the reactivity may only occur at the surface of the coated material. Addition of asparaginase as mentioned earlier can prevent the formation of acrylamide. Moreover, heating honey can lead to the formation of HMF.

Another factor that should be considered in relation to the coating is a possible phase transition and crystallization of sugars, i.e., glucose in honey and added sucrose, during the coating process or later during storage. Even if the honey is partly crystallized before mixing, heating the syrup will guarantee the full dissolution of dextrose and sucrose crystals. As a consequence, the sugars may be in an amorphous state with the potential to crystallize at a later stage, e.g., during drying or storage. As stated before, such a crystallization can impact water activity and the reactivity of different chemical components.

7.5.4.1.14 Drying The product is dried on a band drier at 128–130 °C for 12–15 min. This step can further lead to the formation of acrylamide should the semifinished product be rich in free asparagine. As explained earlier, sugar may take a crystalline or amorphous form, depending on the rate of drying. A rapid drying process tends to create an amorphous state, with the risk that the sugar will crystallize during storage for instance, and increase the water activity.

The heat treatment of honey can also increase the amount of HMF in the product. Based on data presented in Fig. 7.5, no substantial amount of 3-MCPD is expected to be formed at this stage.

7.5.4.1.15 Filling and Packaging The product is filled into bag-in-box packages. This step will have no impact on the formation of processing contaminants.

7.5.4.1.16 Distribution The finished product should be distributed via approved food transporters. At this stage, the processing contaminant that needs to be considered is HMF. To control this, the product should be kept under appropriate time–temperature conditions, i.e., optimally at 25 °C, and consumers should respect the shelf life of 12 months.

7.5.4.1.17 Storage, Preparation, and Consumption Similar to distribution, consumers should be advised to keep the product in cool and dry conditions and respect its best-before date.

7.5.4.2 Possible Modifications in the Product Formulation or Processing The present HACCP study demonstrates that the product as originally planned would most probably lead to amounts of processing contaminants above those stipulated in this hypothetical study as maximum contents for the category, in particular acrylamide and HMF. To minimize the formation of acrylamide and also to meet requirements set for the product (Table 7.6), specific control measures, including modification to product formulation and/or processing need to be considered.

Based on present available scientific studies, these measures include:

- Increasing moisture content in the extruded product and optimizing the time–temperature conditions of the extrusion process and belt drier. Such a change in the process may lead to a product with very different organoleptic quality that may not meet customer satisfaction.
- Investigating the use of the enzyme asparaginase in a preconditioning step, illustrated in Fig. 7.6, where this is legally possible (see Chapter 2.1). Such a modification would probably present the best option since elimination of free asparagine would prevent the formation of acrylamide. Under these conditions, the drying and toasting process could be designed with consideration of 3-MCPD alone, which would simplify the design of the product/process with regard to processing contaminants.

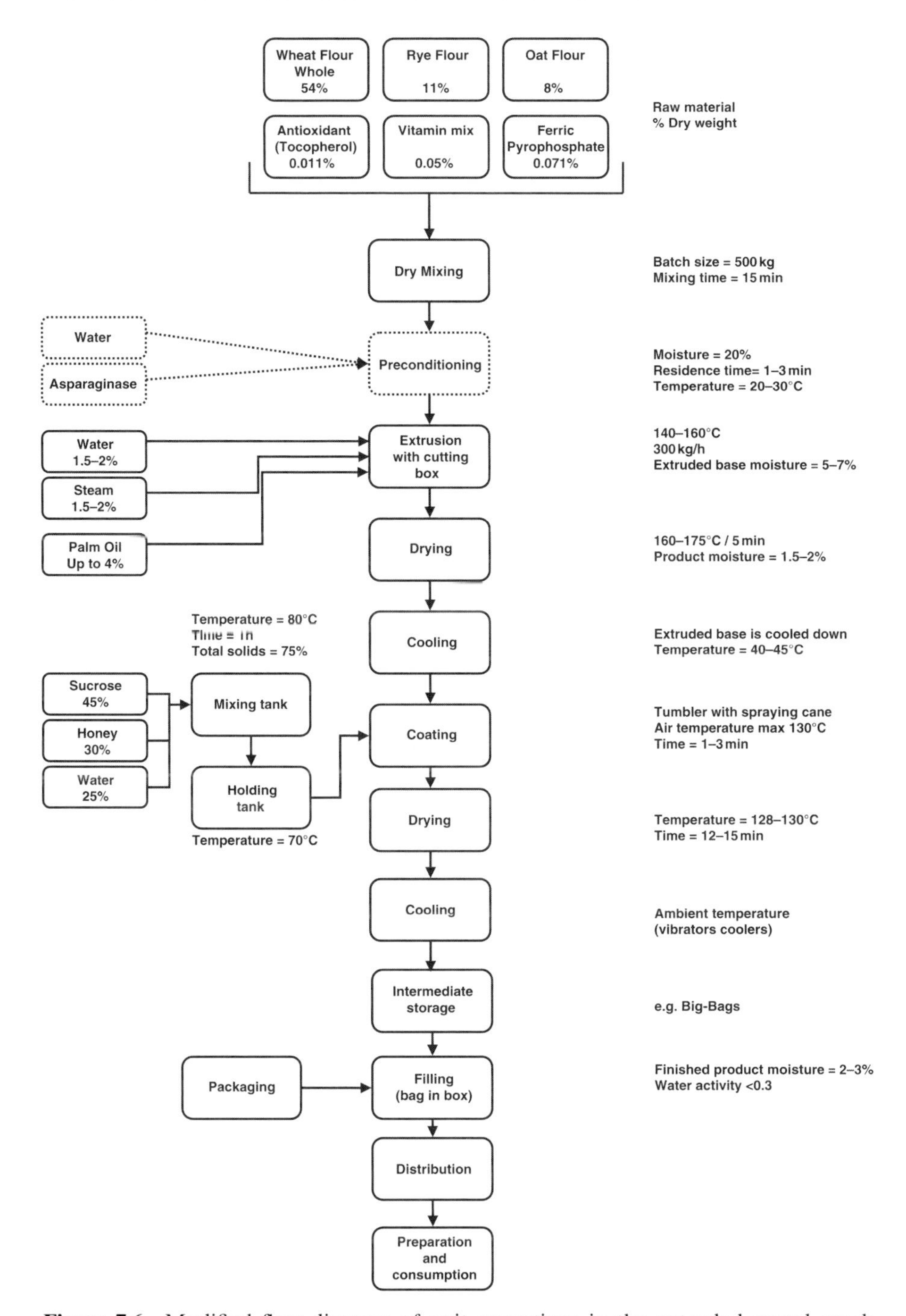

Figure 7.6 Modified flow diagram of unit operations in the extruded cereal snack HACCP case study.

In the future, with better knowledge about the agronomical conditions influencing free asparagine content of cereals and with the development of analytical tools, it may perhaps be feasible to reduce the amount of acrylamide by selecting specific varieties of cereals. This could be an alternative to enzyme application.

By selecting honey low in HMF and ensuring strict control during storage of the raw material and final product, the concentration of HMF can be minimized. However, honey is heated appreciably during the process, which favors HMF formation. Depending on the guidance value chosen for the contaminant/product combination, certain changes in the process may need to be considered.

7.5.4.3 Corrective Actions Although the study did not identify a CCP as such, this does not preclude that any deviation from the set limits or procedures should prompt the following corrective actions:

- blockage of the batch followed by an investigation;
- rejection of raw material, where necessary;
- follow-up with the suppliers and/or distributors;
- re-establishment of the control parameters and where applicable, revisiting operator training;
- rejection of products that show signs of severe thermal exposure.

7.5.4.4 Additional Validation Studies All along the HACCP study, scientific and technical data validating the hazard analysis and processing conditions required to reduce processing contaminants were provided. Additionally, the following validations need to be considered:

- Agronomical conditions influencing the amount of asparagine in different cereal grains, in particular the impact of sulfate-rich fertilizers on the free asparagine content of wheat, rye, and oat.
- Change in the quality of flour during storage (glycerol content) and maximum storage time.
- Optimum moisture content/water activity of the product and its rate of change, at different stages of processing, in particular the drying, heat treatment, and storage steps.
- Stability of the product during its storage, i.e., ensuring that at the end of the shelf-life period of the product, the 3-MCPD is not increased to unacceptable levels. In other words, the shelf life must be determined and validated taking into consideration possible changes in the amount of processing contaminants.

- Prelaunch test to confirm that the product as designed complies with the standards set for the product (acrylamide, 3-MCPD, and PAHs), and compared with products in the same category, presents low amounts of processing contaminants.

Additionally, aspects in relation to the implementation of the plan need also to be validated. Examples are:

- training of the operators to ensure that they understand the importance of control measures, implement the necessary monitoring, and respect the set standards;
- methods of analysis and competency of analytical laboratories;
- equipment design allowing implementation of control measures, e.g., twin-screw extruders can be operated at higher moisture content.

7.5.4.5 Verifications Over and above monitoring activities at the extruder and belt driers, a number of other tests and monitoring are required as "verification" to confirm the adequate implementation of the HACCP study:

- periodic tests of the raw material to verify and confirm compliance of suppliers, e.g., PAHs in oil and HMF in honey;
- periodic testing of the final product at the end of the shelf life to ensure compliance with set standards for acrylamide, 3-MCPD, and PAHs;
- audit of manufacturing site, warehouses as well as suppliers (oil, honey, cereal producers) and retailers to ensure that identified measures and relevant monitoring are well implemented and followed up.

7.5.4.6 Records Based on the present study, the following documents and records are recommended:

- the HACCP study and the HACCP plan;
- validation studies, including literature studies, relevant legislation, and customer specifications;
- specification to suppliers or communications to distributors and retailers;
- reports of audits confirming compliance of suppliers, distributors, and food processing and manufacturing operations with control measures identified in the HACCP study;
- data on monitoring activities;
- reports of laboratories on analytical results;
- customer complaints and other verification data;
- records on follow-up or corrective actions (e.g., minutes of meetings);

- records of training of personnel (e.g., subject of training and yearly training plan);
- standard operating procedures (SOPs) and other prerequisites.

7.5.5 Implementation of the HACCP Study

The outcome of the HACCP study is summarized in the HACCP plan (Table 7.7). The implementation consists of:

- carrying out the recommended modifications in product formulation or processing as identified in the HACCP study;
- providing specifications and guidance to suppliers and distributors in handling the raw material and finished products;
- training of personnel involved at the different steps in the production, from the receipt of the raw material to processing and manufacturing, and warehouse management;
- training of auditors, laboratory and other personnel engaged in verification measures;
- monitoring control measures as identified in the HACCP plan;
- implementing the aforementioned verification measures[6].

7.5.6 Maintenance of HACCP Plan and Continuous Improvement

The HACCP system is a dynamic system. This means that to ensure that products meet up-to-date standards for safety at all times, the HACCP study/plan and its implementation need to be continuously reviewed and where necessary, decisions revised or the implementation improved. In case of processing contaminants, the following types of data need to be considered in the review process:

- developments in scientific and technical know-how, e.g., increased understanding on risks, mechanisms of formation, or mitigation measures as well as emergence of a new processing contaminant;
- report of any complaints (e.g., customers or regulatory authorities);
- changes in regulatory requirements;
- changes in customer requirements;
- changes in factors that impact on the amounts of processing contaminants, e.g., change in supplier and quality of raw material, innovation or renovations leading to change in product formulation or processing conditions;
- benchmarking products against other products in the same category;

[6]These measures can be carried out as part of the overall quality management.

- review of monitoring and verification data (e.g., audit reports and product surveys) and potential gaps in the implementation of the HACCP plan (see Section 7.5.4.5).

7.5.7 Further Research Based on the Outcome of the Case Study

Despite the abundant number of research articles on processing contaminants, in particular on acrylamide, there are still many gaps in our knowledge to allow us to control and reduce processing contaminants in an effective and consistent manner, while respecting other aspects of product quality. The complex manner in which different precursors, ingredients, or their conditions of production or processing interact leads to an intertwined cascade of reactions, even more so if multiple processing contaminants are considered.

Based on the present HACCP study, to minimize or further reduce the amount of processing contaminants in the extruded cereal snack, the following research work is recommended:

- impact of agricultural conditions (crop variety, climatic conditions, agricultural practices such as the nature of fertilizers) on the content of free asparagine in different types of cereals (e.g., wheat, rye, and oat),
- factors influencing the amount of 3-MCPD esters in oils;
- influence of honey and its low pH on formation of 3-MCPD;
- the mechanism by which certain antioxidants may impact the formation of acrylamide;
- influence of water activity on acrylamide at different unit operations;
- impact of phase transitions such as glass transition or crystallization of amorphous fractions of sugars on water activity, molecular mobility, and reactivity leading to Maillard reaction and potential formation of a contaminant;
- potential changes in the product during its shelf life and optimum water activity to minimize chemical interactions over shelf life, e.g., potential formation of HMF;
- kinetics of heat treatment on the formation of acrylamide;
- exploring the impact of other mitigation measures such as use of citric acid, or adjusting the pH on product quality and consumer acceptance;
- extent of HMF formation, its risks, and defining possible measures for its prevention.

7.6 CONCLUSIONS

Processing, be it in industrial, catering, restaurant settings, or in the home, generates numerous chemical reactions in food, leading to a plethora of chemical compounds. Although such reactions in food have occurred since the beginning of the history of food processing and preparation as far back as the fire age, the state of knowledge on potential health effects of these chemical compounds is still in its infancy. The Heatox project (50) in Europe has identified a significant number of chemical compounds for which little is known about their toxicity, their adverse or beneficial health effects. As the formation of these compounds follows different pathways, care should be taken that when driving the level of one contaminant down, this does not lead to the increase of a more harmful one or to deterioration of the nutritional quality or microbial stability.

HACCP is a system that provides a logical, systematic, and rational approach to identifying and assessing hazards, and to determining control measures at different stages of the food chain. Although the system was first designed for managing microbiological hazards, it can be applied to processing contaminants, with some adaptation. The application of the HACCP system will facilitate:

- a systematic analysis of potential sources of hazards, in this case, precursors of processing contaminants or possible reactions between food components; and
- the determination and implementation of those control measures that are most effective in achieving the food safety objectives set.

In this chapter, the application of HACCP principles for the management of processing contaminants has been demonstrated by the example of an extruded snack cereal product. Through this case study, a systematic analysis of hazards, identification of possible control measures (including modification), monitoring, or other verifications at the different steps of food operations are illustrated. The identified control measures and required monitoring can then be applied as such in the framework of monitoring plans or be the basis of a code of practice for a given food sector. As illustrated in this case study, and as opposed to the general misperception, the application of the HACCP system does not necessarily mean identification and management of a contaminant through a CCP.

The application of the HACCP system also provides the opportunity to raise questions about various factors that may influence the formation of processing contaminants, to consider changes in design of the product or its processing (i.e., to look for alternative ingredients or processing). In this way, it can be instrumental in identifying areas where further research is needed.

In addition, the study identified:

- necessary communications to suppliers (e.g., specifications), distributors (guidance for storage and shelf life), and customers or consumers;
- the role of auditors when auditing suppliers, distributors, and manufacturing operations;

- the tasks of operators in terms of implementing control measures;
- the necessary dialogue with regulators with regard to what is achievable and necessary norms to ensure fair trade;
- the appropriateness of a product as such and further consideration of processing contaminants in research and product development.

Finally, it is to be stressed that the potential of HACCP as a tool for designing and managing the safety of products relies on the expertise that is deployed in the team. In the extruded cereal snack example described here, advice of scientists with solid knowledge in moisture isotherms and impact of water activity on Maillard and other chemical reactions was later found to be important.

ACKNOWLEDGMENT

The authors thank Drs Peter Ashby (Cereal Partners Worldwide) and Gilles Vuataz (Nestlé Research Centre) for their valuable contribution in reviewing this chapter.

REFERENCES

1. Kestens, K. (1999). *L'Alimentation et le Droit- Introduction Historique et Juridique au Droit de l'Alimentation*, S.A. La Charte, Brugge.
2. IARC (1994). *IARC Monographs on the Evaluation of Carcinogenic Risks to Humans*, Vol. 60, International Agency for Research on Cancer, Lyon, France.
3. Baum, M. (2007). Thermal processing of food: potential health benefits and risks. An Ex-vivo approach to assess low dose effects of acrylamide, in *Deutsche Forschungsgemeinschaft* (ed. Eisenbrand, G.), Wiley-VCH Verlag GmbH, pp. 90–103.
4. Hogervorst, J.G., Schouten, L.J., Konings, E.J., Goldbohm, R.A., van den Brandt, P.A. (2007). A prospective study of dietary acrylamide intake and the risk of endometrial, ovarian and breast cancer. *Cancer Epidemiology Biomarkers & Prevention*, *16*, 2304–2313.
5. Hogervorst, J.G., Schouten, L.J., Konings, E.J., Goldbohm, R.A., van den Brandt, P.A. (2008). Dietary acrylamide intake and the risk of renal cell, bladder, and prostate cancer. *American Journal of Clinical Nutrition*, *87* (**5**), 1428–1438.
6. Olesen, P.T., Olsen, A., Frandsen, H., Frederiksen, K., Overvad, K., Tjønneland, A. (2008). Acrylamide exposure and incidence of breast cancer among postmenopausal women in the Danish diet. Cancer and health study. *International Journal of Cancer*, *122*, 2094–2100.
7. Pelucchi, C., Galeone, C., Levi, F. *et al.* (2006). Dietary acrylamide and human cancer. *International Journal of Cancer*, *118*, 467–471.
8. Mucci, L.A., Dickman, P.W., Steineck, G., Adami, H-O, Augustsson, K. (2003). Dietary acrylamide and cancer of the large bowel, kidney, and bladder: absence of an association in a population-based study in Sweden. *British Journal of Cancer*, *88*, 84–89.

9. Mucci, L.A., Lindblad, P., Steineck, G., Adami, H.O. (2004). Dietary acrylamide and risk of renal cell cancer. *International Journal of Cancer*, *109*, 774–776.
10. Mucci, L.A., Sandin, S., Balter, K., Adami, H.O., Magnusson, C., Weiderpass, E. (2005). Acrylamide intake and breast cancer risk in Swedish women. *The Journal of American Medical Association*, *293*, 1326–1327.
11. Mucci, L.A., Adami, H.O., Wolk, A. (2006). Prospective study of dietary acrylamide and risk of colorectal cancer among women. *International Journal of Cancer*, *118*, 169–173.
12. Tristcher, A.M. (2004). Human health risk assessment of processing related compounds in food. *Toxicology Letters*, *149* (**1–3**), 177–186.
13. JECFA (Joint FAO/WHO Expert Committee on Food Additives) (2007). Report of the 68th meeting of the Joint FAO/WHO Expert Committee on Food Additives. World Health Organization, Geneva.
14. Tareke, E., Rydberg, P., Karlsson, P., Eriksson, S., Törnqvist, M. (2002). Analysis of acrylamide, a carcinogen formed in heated foodstuffs. *Journal of Agricultural and Food Chemistry*, *50* (**17**), 4998–5006.
15. Pedreschi, F., Kaack, K., Granby, K. (2004). Reduction of acrylamide formation in potato slices during frying. *Lebensmittel-Wissenschaft und-Technologie*, *37*, 679–685, September.
16. Pedreschi, F., Kaack, K., Granby, K., Troncoso, E. (2007). Acrylamide reduction under different pre-treatments in French fries. *Journal of Food Engineering*, *79* (**4**), 1287–1294.
17. Taeymans, D., Wood, J., Ashby, P., Blank, I., Studer, A., Stadler, R.H., Gondé, P., Van Eijck, P., Lalljie, S., Lingnert, H., Lindblom, M., Matissek, R., Müller, D., Tallmadge, D., O'Brien, J., Thompson, S., Silvani, D., Whitmore, T. (2004). A review of acrylamide: an industry perspective on research, analysis, formation and control. *Critical Reviews in Food Science and Nutrition*, *44*, 323–347.
18. CIAA. The CIAA Acrylamide "Toolbox" (2007). Confederation des Industries Agroalimentaires, Paris. http://ec.europa.eu/food/food/chemicalsafety/contaminants/ciaa_acrylamide_toolbox.pdf (accessed 10 April 2008).
19. Kuilman, M., Wilmsa, L. (2007). Safety of the enzyme asparaginase, a means of reduction of acrylamide in food. *Toxicology Letters*, *172* (**S1**), 196–197.
20. JECFA (Joint FAO/WHO Expert Committee on Food Additives) (1993). Report of 41st meeting of the Joint FAO/WHO Expert Committee on Food Additives. World Health Organization, Geneva.
21. Velisek, J., Davidek, J., Hajslova, J., Kubelka, V., Janicek, G., Mankova, B. (1978). Chlorohydrins in protein hydrolysates. *Zeitschrift fur Lebensmittel-Untersuchung und-Forschung*, *167*, 241–244.
22. Breitling-Utzmann, C.M., Hrenn, H., Haase, N.U., Unbehend, G.M. (2005). Influence of dough ingredients on 3-chloropropane-1,2 diol (3-MCPD) formation in toast. *Food Additives and Contaminants*, *22* (**2**), 97–103.
23. Collier, P.D., Cromie, D.D.O., Davies, A.P. (1991). Mechanism of formation of chloropropanols present in protein hydrolysates. *Journal of the American Oil Chemists Society*, *68*, 785–790.
24. Crews, C., Hough, P., Brereton, P., Harvey, H., MacArthur, R., Matthews, W. (2002). Survey of 3-monochloropropane, 1-2 diol in selected food groups 1999–2000. *Food Additives and Contaminants*, *19* (**1**), 22–27.

25. Hamlet, C.G., Sadd, P.A., Crews, C., Velisek, J., Baxter, D.E. (2002). Occurrence of 3-chloro-propane-1,2-diol (3-MCPD) and related compounds in foods: a review. *Food Additives and Contaminants*, *19*, 619–631.
26. Hamlet, C.G., Sadd, P.A. (2005). Effects of yeast stress and pH on 3-monochloropropanediol (3-MCPD)-producing reactions in model dough systems. *Food Additives and Contaminants*, *22* (**7**), 616–623.
27. CAC (2005). Discussion Paper on Acid HVP Containing products and other products containing chloropropanols. CX/FAC/06/38/33, November 2005, FAO/WHO Food Standard Programme, Food and Agriculture Organization.
28. Stadler, R.H., Theurillat, V., Studer, A., Scanlan, F., Seefelder, W. (2007). The formation of 3-monochloropropane-1,2-diol (3-MCPD) in food and potential measures of control, in *Thermal Processing of Food: Potential Health Benefits & Risks*, Deutsche Forschungsgemenischaft, Wiley-VCH Verlag GmbH, Germany, pp. 141–154.
29. Seefelder, W., Varga, N., Studer, A., Williamson, G., Scanlan, F.P., Stadler, R.H. (2008). Esters of 3-chloro-1,2-propanediol (3-MCPD) in vegetable oils: significance in the formation of 3-MCPD. *Food Additives and Contaminants*, *25* (**1**), 1–10.
30. Tosi, E.A., Ré, E., Lucero, H., Bulacio, L. (2004). Effect of honey high-temperature short-time heating on parameters related to quality, crystallisation phenomena and fungal inhibition. *Lebensmittel-Wissenschaft und-Technologie*, *37*, 669–678.
31. Tareke, E. (2003). Identification and origin of potential background carcinogens. Endogenous isoprene and oxiranes, dietary acrylamide. PhD Thesis, Department of Environmental Chemistry. Stockholm University.
32. IARC (1995). *IARC Monographs on the Evaluation of Carcinogenic Risks to Humans*, Vol. 63, International Agency for Research on Cancer, Lyon, France.
33. Perez-Locas, C., Yaylayan, V.A. (2004). Origin and mechanistic pathways of formation of the parent Furan-A food toxicant. *Journal of Agricultural and Food Chemistry*, *52* (**22**), 6830–6836.
34. Becalski, A., Seaman, S. (2005). Furan precursors in food: a model study and development of a simple headspace method for determination of Furan. *Journal of AOAC International*, *88*, 102–106.
35. Märk, J., Pollien, P., Lindinger, C., Blank, I., Märk, T. (2006). Quantitation of Furan and methylfuran formed in different precursors by proton transfer reaction mass spectroscopy. *Journal of Agricultural and Food Chemistry*, *54*, 2786–2793.
36. Zoller, O., Sager, F., Reinhard, H. (2007). Furan in food: head space method and product survey. *Food Additives and Contaminants*, *24* (**S1**), 91–107.
37. Silla-Santos, M.H. (2001). Toxic nitrogen compounds produced during processing: biogenic amines, ethyl carbamides, nitrosamines, in *Fermentation and Food Safety* (eds M.R. Adams, M.J.R. Nout), Aspen Publishers, Gaithersburg, MD, pp. 119–140.
38. Motarjemi, Y. (2001). An introduction to the Hazard Analysis and Critical Control Points (HACCP) system and its application to fermented foods, in *Fermentation and Food Safety* (eds M.R. Adams, M.J.R. Nout), Aspen Publishers, Inc., Gaithersburg, MD, pp. 53–70.
39. CAC Hazard Analysis and Critical Control Point (HACCP) (2003). System and Guidelines for its Application. Annex to Recommended International Code of Practice General Principles of Food Hygiene to CAC/RCP 1-1969 (Rev. 4—2003); FAO/WHO Secretariat of the Codex Alimentarius Commission, Rome.

40. ISO 22 000 (2005). Food Safety Management Systems—Requirements for Any Organisation in the Food Chain, International Organisation for Standardisation. Geneva.
41. CAC (2008). Guidelines for the Validation of Food Safety Control Measures, FAO/WHO Secretariat of the Codex Alimentarius Commission, Rome.
42. Motarjemi, Y. (2008). Management of food safety in the industrial setting. In: *Medical Sciences*, eds. Verhasselt, Y.L.G., Mansourian, B.P., Wojtczak, A.M., Sayers, B.M., Szczerban, J., Aluwihare, A.P.R., Napalkov, N.P., Brauer, G.W., Davies, A.M., Mahfouz, S.M., Manciaux, M.R.G., Arata, A.A., Pellegrini, A., Jablensky, A., Kitney, R., Kazanjian, A., Turmen, T., Leidl, R., Sorour, K., in *Encyclopedia of Life Support Systems*, developed under the auspices of UNESCO, EOLSS, Oxford, UK. http://www.eolss.net. (accessed 1 October 2008).
43. CAC (2008). Proposed Draft Code of Practice for the Reduction of Acrylamide in Food, Appendix V, Alinorm 08/31/41, pages 56–69. FAO/WHO Secretariat of Codex Alimentarius Commission, Rome.
44. Granvogl, M., Wieser, H., Koehler, P., Von Tucher, S., Schieberle, P. (2007). Influence of sulfur fertilization on the amounts of free amino acids in wheat. Correlation with baking properties as well as with 3-aminopropionamide and acrylamide generation during baking. *Journal of Agricultural and Food Chemistry*, *55* (**10**), 4271–4277.
45. Halford, N.G., Muttucumaru, N., Curtis, T.Y., Parry, M.A.J. (2007). Genetic and agronomic approaches to decreasing acrylamide precursors in crop plants. *Food Additives and Contaminants*, *24* (**S1**), 26–36.
46. Göbel, A., Kliemant, A. (2007). The German minimization concept for Acrylamide. *Food Additives and Contaminants*, *24* (**S1**), 82–90.
47. BVL (2008). Acrylamide-Minimierungskonzept-7. Berechnung der signalwerte. Bundesamt für Verbraucherschutz und Lebensmittelsicherheit. http://www.bvl.bund.de (accessed 1 October 2008).
48. Zhang, Y., Ying, T.J., Zhang, Y. (2008). Reduction of acrylamide and its kinetics by addition of bamboo leaves (AOB) and extract of green tea (EGT) in asparagine-glucose microwave heating system. *Journal of Food Science*, *73*, C60–C66.
49. Taylor and Francis Group (2008). *Fennema's Food Chemistry* (eds S. Damodara, K.L. Parkin, O.R. Fennema), CRC Press, New York.
50. Final report (HEATOX Project) (2007). http://www.slv.se/upload/heatox/documents/D62_final_project_leaflet_pdf (accessed 16 January 2008); http://www.slv.se/upload/heatox/documents/Heatox_Final%20_report.pdf (accessed 1 October 2008).
51. CAC (2008). Code of Practice for the reduction of 3-Monochloropropane, -1,2-diol (3MCPD) during the production of acid-hydrolysed vegetable proteins (Acid-HVPs) and products that contain acid-HVPs. FAO/WHO Secretariat of Codex Alimentarius Commission, Rome.
52. UK FSA. Draft Guidelines for the reduction of 3-MPCD in foods, UK Food Standard Agency. (In preparation).
53. European Commission (2001). Commission Regulation (EC) No 466/2001 of 8 March 2001 setting maximum levels for certain contaminants in foodstuffs (OJ L 77 16.3.2001, p12).
54. Committee on Food Chemicals Codex (1996). *Food Chemicals Codex: First Supplement to Fourth Edition*, Institute of Medicine of the National Academies, Washington, DC. http://www.iom.edu/report.asp?id=4590 (accessed 23 January 2007).

55. Département fédéral de l'intérieur (DFI) (2006). Ordonnance du DFI du 26 juin 1995 sur les substances étrangères et les composants dans les denrées alimentaires, RS 817.021.23 (DFI schedule on foreign substances and components in foodstuffs). Berne, Switzerland: DFI. http://www.admin.ch/ch/f/rs/c817_021_23.html (accessed 24 April 2008).
56. European Commission. (2005). Commission Regulation (EC) No. 2008/2005 of 4th February 2005 amending regulation (EC) No. 466/2001 as regards polycyclic aromatic hydrocarbons.
57. US /FDA (2007). Beverages. U. S. Code of Federal Regulations, Part 165, Section 165. 110, Title 21.
58. Granvogl, M., Schieberle, P. (2006). Thermally generated 3-aminopropionamide as a transient intermediate in the formation of acrylamide. *Journal of Agricultural and Food Chemistry*, *54* (**16**), 5933–5938.
59. European Commission. European Council Directive (1991). 91/493/EEC The health conditions for the production and the placing on the market of fishery products.
60. Weber, E.A., Graeff, S., Koller, W.D., Hermann, W., Merkt, N., Claupein, W. (2008). Impact of nitrogen amount and timing on the potential of acrylamide formation in winter wheat (*Triticum aestivum* L.). *Field Crops Research*, *106* (**1**), 44–52.
61. BLL (Research Association of the Food Industry) (2005). Development of new technologies to minimise acrylamide in food. http://www.dil-ev.de/staticsite/staticsite.php?menuid=363&topmenu=363&keepmenu=inactive (accessed 24 April 2008).
62. Lea, P.J., Sodek, L., Parry, M.A.J., Shewry, P.R., Halford, N.G. (2007). Asparagine in plants. *Annals of Applied Biology*, *150* (**1**), 1–26.
63. Muttucumaru, N., Halford, N.G., Elmore, J.S., Dodson, A.T., Parry, M., Shewry, P.R., Mottram, D.S. (2006). Formation of high levels of Acrylamide during the processing of flour derived from sulfate-deprived wheat. *Journal of Agricultural and Food Chemistry*, *54* (**23**), 8951–8955.
64. Claus, A., Schreiter, P., Weber, A., Graeff, S., Herrman, W., Claupein, W., Schieber, A., Carle, R. (2006). Influence of agronomic factors and extraction rate on the acrylamide contents in yeast-leavened breads. *Journal of Agricultural and Food Chemistry*, *54* (**23**), 8968–8976.
65. Becalski, A., Benjamin, P., Lewis, D., Seaman, S.W. (2003). Acrylamide in foods: occurrence, sources, and modeling. *Journal of Agricultural and Food Chemistry*, *51*, 802–808.
66. Konings, E.J.M., Ashby, P., Hamlet, C.G., Thompson, G.A.K. (2007). Acrylamide in cereal and cereal products: a review on progress in level reduction. *Food Additives and Contaminants*, *24* (**1**), 47–59.
67. Fredriksson, H., Tallving, J., Rosén, J., Åman, P. (2004). Fermentation reduces free asparagine in dough and acrylamide content in bread. *Cereal Chemistry*, *81* (**5**), 650–653.
68. SCF (2002). Scientific Committee on Food. Opinion of the Scientific Committee on Food on the risks to human health of Polycyclic Aromatic Hydrocarbons in food (expressed on 4 December 2002). SCF/CS/CNTM/PAH/29 Final. 4 December 2002.

69. Doner, L. (1977). The sugars of honey—a review. *Journal of Science and Food Agriculture*, *28*, 443–456.
70. Szczêsna, T. (2007). Study on the sugar composition of honeybee-collected pollen. *Journal of Apicultural Science*, *51* (**1**), 15–22.
71. Sanz, M-L., del Castillo, M.D., Corzo, N., Olano, A. (2003). 2-Furoylmethyl amino acids and hydroxymethylfurfural as indicators of honey quality. *Journal of Agricultural and Food Chemistry*, *51* (**15**), 4278–4283.
72. Gidamis, A.B., Chove, B.E., Shayo, N.B., Nnko, S.A., Bangu, N.T. (2004). Quality evaluation of honey harvested from selected areas in Tanzania with special emphasis on hydroxymethyl furfural (HMF) levels. *Plant Foods for Human Nutrition*, *59* (**3**), 129–132.
73. Département fédéral de l'intérieur (DFI) (1995). Ordonnance du DFI sur les substances étrangères et les composants dans les denrées alimentaires (817.021.23). http://www.admin.ch/ch/f/rs/c817_021_23.html (accessed 24 April 2008).
74. European Commission (2006). EC Directive 2006/52/EC DIRECTIVE 2006/52/EC of the European Parliament and of the Council of 5 July 2006.
75. EFSA (2007). Ethyl carbamate and hydrocyanic acid in food and beverages—Opinion of the scientific panel on contaminants in the food chain. *The EFSA Journal*, *551*, 1–44.
76. Health Canada (2007) Canadian Standards ("Maximum Limits") for Various Chemical Contaminants in Foods. http://hc-sc.gc.ca/fn-an/securit/chem-chim/contaminants-guidelines-directives_e.html (accessed 7 September 2007).
77. CAC (2001). Codes Standards for Honey 12-1981, FAO/WHO Secretariat of the Codex Alimentarius Commission, Rome.
78. IPCS. Guidance on setting of acute reference dose (ARfD) for pesticides. The International Programme on Chemical Safety, World Health Organsiation, Geneva. http://www.who.int/ipcs/food/jmpr/arfd/en/index.html (accessed 2 October 2008).

8

EMERGING FOOD TECHNOLOGIES

Fanbin Kong and R. Paul Singh
Department of Biological and Agricultural Engineering, University of California, Davis, CA 95616, USA

8.1 INTRODUCTION

Increasing demands from consumers for high-quality foods that are fresh-tasting and nutritious have created considerable interest in the development of new food-processing techniques. Traditional food-processing technologies such as freezing, canning, and drying rely on heating or cooling operations. Although these technologies have helped to ensure a high level of food safety, the heating and cooling of foods contribute to the degradation of various food quality attributes. The color, flavor, and texture of foods processed solely by heating may be irreversibly altered. To ameliorate the undesirable thermal effects on foods, considerable effort has been made in commercial and academic circles to develop technologies that could inactivate microorganisms while significantly reducing or completely eliminating the amount of heat required. These technologies are classified into two categories: nonthermal and thermal technologies, depending on the mechanisms of microbial inactivation. Nonthermal technologies being actively studied include pulsed electric field (PEF), pulsed ultraviolet (UV) light, and ultrasound processing. Thermal technologies include dielectric heating including microwave (MW) and radio-frequency (RF) heating, ohmic heating (OH), and infrared (IR) heating. This chapter provides a technical description of each of these technologies, along with a discussion of their applications in food processing.

Process-Induced Food Toxicants: Occurrence, Formation, Mitigation, and Health Risks,
Edited by Richard H. Stadler and David R. Lineback

8.2 PULSED ELECTRIC FIELDS

8.2.1 Fundamentals

Pulsed electric field processing involves fluid foods (e.g., milk and juice) placed between two electrodes and subjected to pulses of high voltage (typically 20–80 kV/cm). Inactivation of microbial cells is achieved by electroporation of their membranes. Foods are treated for very short time periods (a few microseconds); therefore, temperature rise is minimal, minimizing chemical or physical changes in the food material (1, 2).

The beneficial effects of PEFs on inactivating microbial levels in foods have been known for several decades. The pasteurizing effects of electric fields in foods were first observed in the early 1900s. According to Beattie and Lewis (3), in electrically treated milk supplied to the city of Liverpool in England, the bactericidal effects of the treatment were due not only to heat that was generated, but also to the electric field itself. In the 1920s, a process called *electropure* was introduced in the United States to pasteurize milk using electricity (4). This process did not use pulsed electricity. Pasteurization was mostly due to thermal effects. About 50 plants were still using this technology until the 1950s, but they were shut down as a result of competition from other methods of heating and increasing costs of electricity. The use of intensive electric field pulses on inactivation of microorganisms in foods was patented by Maxwell laboratories in the mid-1980s (2, 5). Pilot scale PEF equipment is available in processing apple juice, cider, and orange juice to produce foods with extended shelf life with fresh-like quality.

Electroporation is thought to be the major mechanism causing microbial inactivation, which implies the phenomenon that pores are created in cell membranes when the cell is placed in an electric field with a short pulse (1–100 μs), causing cells to swell and rupture (1). A natural pressure gradient exists across cellular membranes, so that when a cell is placed in an electric field, the transmembrane potential increases. If the applied electrical field is more than a certain critical value, then the cell wall ruptures. While the complete mechanism of cell wall breakdown in an electric field is not clearly understood, this observed phenomenon is used regularly in creating pores in a cell membrane. The technique has been used in the field of biotechnology to introduce foreign DNA into a cell. A field strength above 15 kV/cm is required for destruction of microbial membranes (6). In addition, reactive free radicals or electrolysis products are generated during PEF process that may also contribute to the microbial inactivation.

8.2.2 Technical Design

A food material contains ions that make it a good conductor of electricity. When a large flux of electrical current flows through a food material, a high-voltage PEF is generated within the food. The electric current is allowed to flow through the food object for a very short period of time, on the order of microseconds. Therefore, a capacitor is needed to generate pulses. The capacitor slowly charges and then quickly discharges its stored electrical energy.

The microbial inactivation in foods due to an imposed electrical field depends on the length of time the field is applied and the number of pulses. It is recommended that high electric field and short time pulses be used to minimize heat generation due to Joule heating effect. Otherwise, the benefits of PEF are lost due to thermal degradation of the treated food. There are operational problems with the use of monopolar pulses (7). Because many constituents of a food material, such as electrolytes, protein, and living cells, have a net electric charge, they tend to accumulate on the charged electrode surfaces. A shielding layer is therefore created on the electrodes that make the electric field nonuniform. The undesirable shielding layers are prevented when bipolar pulses are used (7).

Several technical issues that are important in the industrial application of PEF have been noted (7, 15). These include:

- determining the optimum electric field strength for inactivating bacteria;
- provision to cool the food material that heats up due to Joule heating effect;
- dielectric breakdown in foods;
- proper selection of power and flow rates;
- operational safety issues.

A simplified schematic design of a PEF system is shown in Fig. 8.1. The main components are the high-voltage generator, switch, capacitor, and electrodes. A cooling system with recirculating cold water is often used to dissipate the heat generated from electric currents and keep treated foods at low temperature (8).

Two types of pulses have been considered for PEF applications, namely exponential decay and square pulses. In square pulses, the voltage increases instantaneously to a peak value, where it is held for some time before decreas-

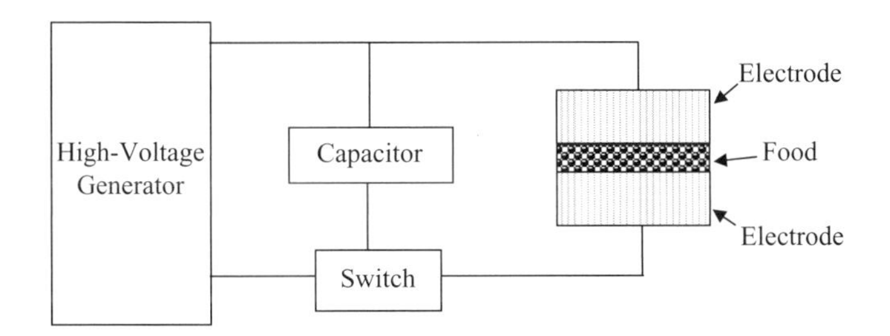

Figure 8.1 A simplified general design of a PEF apparatus (15).

ing to zero almost instantly. With exponential pulses, the long tail section of the pulse is not effective in killing bacteria. On the other hand, it generates excess heat. Square pulses can maintain their peak voltage for a longer time than exponential pulses, and they generate less heat. Although the generation of square pulses needs more complex circuits, it is preferred for its advantages in food applications.

While PEF is desirable for microbial inactivation, it causes undesirable arcing or dielectric breakdown in a material. Arcing occurs when the applied field strength becomes equal to the dielectric strength of the material. When a liquid food is subjected to PEF, any presence of vapor bubbles causes arcing. Gases or vapors have a much lower dielectric strength than do pure liquids. Any roughness of electrode surface also causes dielectric breakdown of the food material. Zhang *et al.* (7) recommend considerations of the following points to avoid arcing:

- using electrodes that are smooth;
- carefully designed treatment chambers to provide uniform electric field strength;
- degassing;
- pressurizing the liquid in the treatment chamber to prevent bubble formation.

Several different designs of PEF treatment chambers have been investigated (9). A static chamber used at Washington State University is shown in Fig. 8.2. The disk-shaped electrodes (area 27 cm^2) are made of stainless steel polished to mirror-like surface, with a gap that could be set at either 9.5 or 5.1 mm. Electric field strengths of up to 70 kV/cm could be used. Electrodes contain built-in jackets that allow circulation of water to maintain low temperatures. A modified version of this static cell has been used for continuous application, as shown in Fig. 8.3. To continuously pump a liquid food through the cell, baffled flow channels were added inside the treatment chamber. A pulse width of 2 to 15 μs with a repetition rate of 1 Hz has been tested, and the

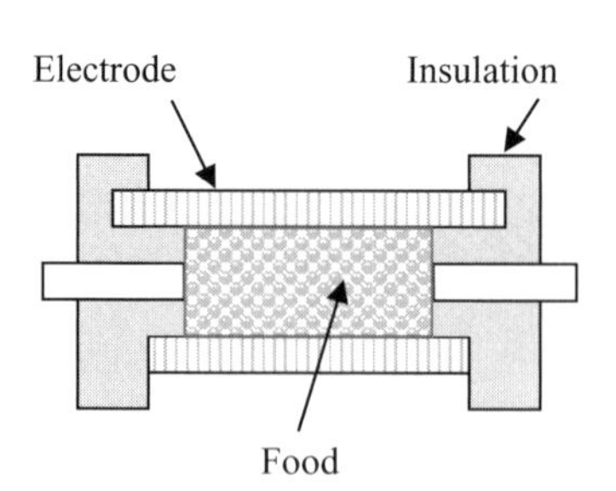

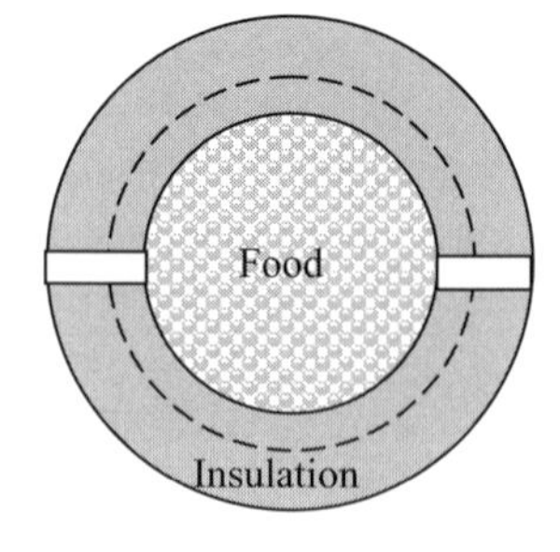

Figure 8.2 A static treatment chamber for PEFs developed at Washington State University (8).

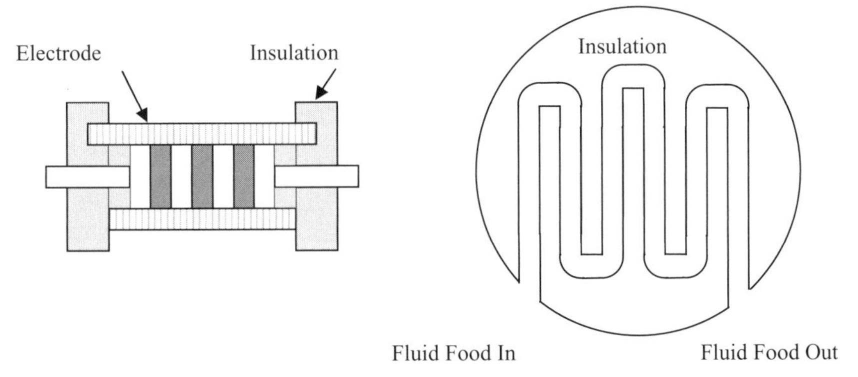

Figure 8.3 A schematic of a flow through treatment chamber. The fluid inside the chamber is baffled to avoid dead spaces (8).

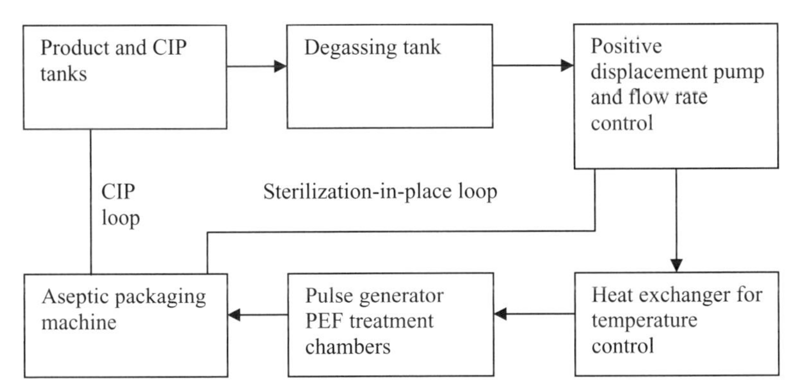

Figure 8.4 A simplified flow diagram for an integrated PEF and aseptic system for juice processing (11, 15).

flow rate of the test liquid food through this cell was reported to be either 1200 or 600 cm^3/min.

A PEF unit for treatment of fresh orange juice at a pilot scale was described by Qiu *et al.* (10). This system involved a continuous pilot-scale PEF unit integrated with an aseptic packaging machine (Fig. 8.4). The investigators used a 40,000 V/17 MWp high-voltage pulse generator with a multiple-state co-field PEF treatment chamber. The aseptic packaging machine was used to package PEF-treated food under either nitrogen or sterile air headspace. The pumping system (Moyno pump) was used to transport juice at a uniform rate from 75 to 200 L/h. The pulse generator had a 40-kV command charging power supply. For switching, they used a 50 kV/5 kA hollow anode thyratron. The maximum repetition rate of the pulse generator was set at 1000 Hz. The network could be changed to generate different pulse shapes, namely, square wave, exponential decay wave, and an under-damped RIC (resistance-inductance-capacitance) waveform. They used a set of co-field tubular treatment chambers with cooling capabilities. The diameter of the treatment zone was 0.48 cm, and the separation between the electrodes was set at 0.48 cm. The system was operated at 30 °C, and the feed flowed through 12 PEF treatment chambers. A system flow rate of 75 L/h was obtained, with an average of 3.3 pulses delivered to the feed stream in each cell.

8.2.3 Research Status and Applications

Most research on the applicability of PEF in food processing has focused on microbial inactivation in different liquid food products such as fruit juice and milk, whereby to extend the shelf life of foods by minimizing spoilage caused by microbial growth. In these studies, the rate of microbial growth is a key parameter that is compared between foods treated with PEF or traditional technologies. PEF treatment of fresh orange juice inactivated 99.9% of microbial flora, with the square waves being the most effective (10). Compared with heat pasteurization, the PEF-treated orange juice retained more vitamin C and flavor. Some illustrative examples of foods treated with PEF are shown in Table 8.1.

TABLE 8.1 Examples of foods processed using PEFs and change in their quality attributes other than microbial.

Product	Process and quality attributes	Reference no.
Apple juice, fresh and reconstituted	Pasteurization. No change in solids concentration, pH, and vitamin C. Loss of calcium, magnesium, sodium, and potassium. No sensory differences between processed and untreated juices.	8
Commercial cheese sauce, reformulated	Preservation. Better flavor and appearance than comparable products.	12
Green pea soup	Cooking. No difference in sensory properties after 4 weeks' storage at 4 °C.	8
Liquid whole egg	Pasteurization. Prevention of coagulation, superior quality.	9, 12, 13
Orange juice	Preservation at pilot-scale. Less than 6% flavor loss, negligible vitamin C, and color change.	10, 12, 13
Orange juice, fresh-squeezed	Pasteurization. Minimal loss of flavor compounds, color, and vitamin C.	12
Salsa	Preservation. Better flavor and appearance than comparable products.	12
Spaghetti sauce	Aseptic processing. Acceptable after 2 years and 80 °F storage.	12

PEF treatment has only limited effect on enzyme activity. It cannot inactivate bacterial spores. Therefore, it has to be combined with other "hurdles" to enable sufficient commercial shelf life, such as low storage temperature, high pressure, and ultrasonication (13). Significant shelf-life extensions have then been obtained with a minimum of quality loss. These applications are still under development stage. Information on the commercial application of PEF is very limited. Considerable more research is necessary to obtain data on the effects of PEF treatments on the sensory properties as well as nutritional content of foods (14, 15).

PEF is also used as pretreatment in food extraction, pressing, or drying (16, 17). When plant membranes are exposed to high-enough PEF (0.3–1 kV/cm) for a short time (0.025–0.1 s), an irreversible rupture (electropermeabilization) can be achieved, causing an increase in mass transfer coefficients that result in a faster transport of water to product surface. Therefore, a PEF pretreatment may lead to significant saving of energy and better utilization of production capacities in food dehydration and separation processes.

8.3 PULSED UV LIGHT

8.3.1 Fundamentals

Pulsed UV light processing involves the use of short-time high-frequency pulses of broad-spectrum light ranging from 100 to 1100 nm (from UV to the near infrared region) to kill microorganisms. Food is exposed to at least one pulse of light with a duration range from 1 μs to 0.1 s, having an energy density in the range between 0.01 and 50 J/cm^2 at the food surface (18). This technology has been widely applied in microbial inactivation of water, air, food, packages, and medical devices, as well as food packaging and processing equipment.

The mechanisms for using pulsed UV light to inactivate microorganisms include photochemical and photothermal effects (19). The highly conjugated double carbon bonds in proteins and nucleic acids of the microbes exposed to light at UV wavelengths absorb energy that disrupts cellular metabolism. DNA/RNA absorption of the UV light (UVC 200–280 nm) resulted in mutations in microorganisms leading to bacterial death. A significant increase in surface temperature (50–100 °C) contributes to thermal inactivation. Because the temperature rise is limited to outer surface of the foods, the overall thermal effect is minimal on the food quality. Therefore, it is considered to be a nonthermal technology capable of producing shelf-stable food products with premium nutritional quality.

8.3.2 Technical Design

Basic electrical components include pulse generator, associated switching, and control circuitry. A high-current pulse of high voltage is applied to inert gas lamps to illuminate the desired treatment area. An intense pulse of light with

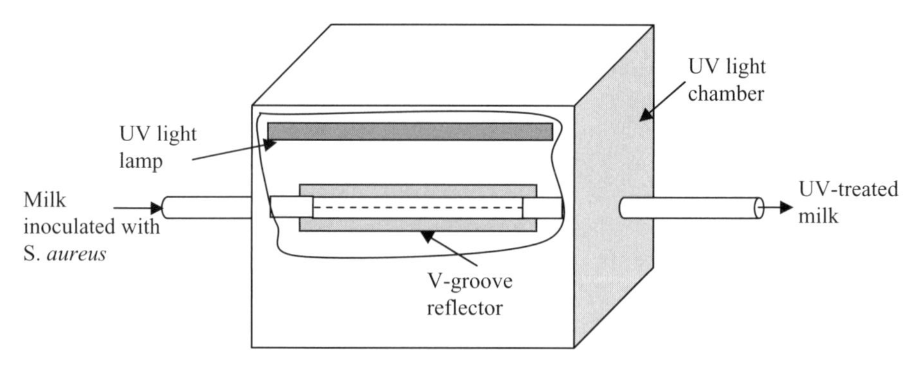

Figure 8.5 Schematic diagram of a continuous milk treatment system (20).

short duration (a few hundred microseconds) is emitted when the high current passes through the gas in the lamp. The system is designed in a way that the frequency of flashing, number of lamps, and flashing configuration can be adjusted to suit different processing requirements.

Critical process factors influencing the microbiocidal effect include characteristics of light such as wavelength, intensity, duration and number of the pulses, and packaging and food attributes (type, transparency, and color) (18). The number of lamps, their position, orientation, and design have a direct relevance to the energy delivered and dose effectiveness, and they determine the extent of lethality (20). The effect of pulsed UV light is also influenced by the distance of sample from the light source, treatment time, opacity of the liquid, and presence of particulate materials. For heat-sensitive products, such as minimally processed fruits and vegetables, a cooling system is usually used to reduce the thermal effect due to light absorption by the product (19).

Figure 8.5 shows a continuous milk treatment system using pulsed UV light treatment (20). An input voltage of 3800V is required for the pulsed light sterilization system to produce polychromatic radiation in the wavelength range of 100 to 1100nm, with 54% of the energy in the UV light region. The light generated is of three pulses per second and 1.27 J/cm^2 per pulse. A peristaltic pump is used to pump milk through a quartz tube (1.14cm i.d., 1.475cm o.d.) exposed to pulsed UV light. The distance between the quartz window and the central axis of the lamp is 5.8cm. The total exposure length of the quartz tube is 28cm. A V-groove reflector setup is used to hold the quartz tube and change the distance between the quartz tube and the UV light source. This reflector has a polished surface reflecting the energy back to the quartz that enhances the energy absorption by milk in the quartz tube.

8.3.3 Research Status and Applications

The US Food and Drug Administration (FDA) has already approved pulsed UV light for use on food materials and packaging (21). The major present application for pulsed UV light processing is in the decontamination of pharmaceuticals, water, air, and food packaging as well as baked goods. Pulsed UV light processing effectively inactivates vegetative bacteria, as well as bacterial and mold spores (18, 20). It is particularly effective in decontaminating dry and smooth surfaces, such as shell eggs and packaging material (22). It can also significantly reduce the microorganisms in foods with rough and opaque surface, such as packaged bread and cakes, meats, fish, chicken, shrimp, and tomato, with their shelf life extended from several days to 2 weeks (2). The effect of pulsed UV light is significantly influenced by the thicknesses of treated materials, treatment time, and energy density. Recently, the use of pulsed UV light has been extended to the pasteurization of fruit juices using UV light, which involves design of treatment chambers that utilize turbulent flow to form a continuously renewed surface, thus enabling microbial inactivation (23).

8.4 ULTRASOUND

8.4.1 Fundamentals

Ultrasound is generated by sound waves of 20,000 or more vibrations per second. It is able to travel through gas, liquid, and solid materials. It can be classified into two categories: high-frequency low-energy ultrasound and low-frequency high-energy ultrasound. High-frequency low-energy ultrasound covers frequencies higher than 100 kHz and intensities lower than $1\,W/cm^2$; it is capable of traveling through a medium without altering the material, allowing nondestructive measurements in foods. It has been successfully used for characterizing physicochemical properties of food materials. Low-frequency high-energy ultrasound is characterized by high power levels ($10–1000\,W/cm^2$) and relatively low frequencies (<0.1 MHz). Low-energy ultrasound applications in food processing include crystallization, drying, degassing, extraction, filtration, homogenization, meat tenderization, emulsification, drying, and freezing (24, 25). There has been a growing interest in the use of high-intensity ultrasound as a preservation method, including surface sanitation, microbial inactivation, and modification of enzyme activity.

The major mechanism for ultrasound microbial inactivation is cavitation. When ultrasound waves travel through liquids, bubbles or cavities are formed. When bubbles collapse, local shock waves are created that instantaneously increase the local temperature and pressure up to 5000 K and 100 MPa, respectively. The sudden changes in temperature and pressure that occur during cavitation are considered to be the main reasons for cell membrane damage and therefore microbial inactivation (26). Other mechanisms may include formation of free radicals and hydrogen peroxide, both of which have bactericidal properties.

8.4.2 Technical Design and Consideration

Due to the importance of cavitation in the bactericidal effect of ultrasound, critical processing factors are those that influence cavitation. The ultrasonic frequency used must be under 2.5 MHz; if it is greater than that, cavitation will not occur (27). Temperature is inversely proportional to the minimum oscillation of pressure that is required to produce cavitation. Other factors include frequency and amplitude of the ultrasonic waves, dissolved gas, the exposure or contact time with the microorganisms, hydrostatic pressure, specific heat of the liquid and the gas in the bubble, and the volume and composition of food to be processed (18, 26). The effect varies with the type of microorganisms.

Figure 8.6 shows a continuous-flow ultrasonic system designed by Villamiel and de Jong (27) using a 450 Sonifier II ultrasonic cell disruptor (Branson Ultrasonic Corporation, Stamford, CT). This sonicator has a tip of 12.7 mm in diameter and works at a fixed frequency (20 kHz). In the sonic converter, high-frequency electrical energy is transformed to mechanical vibration, which is

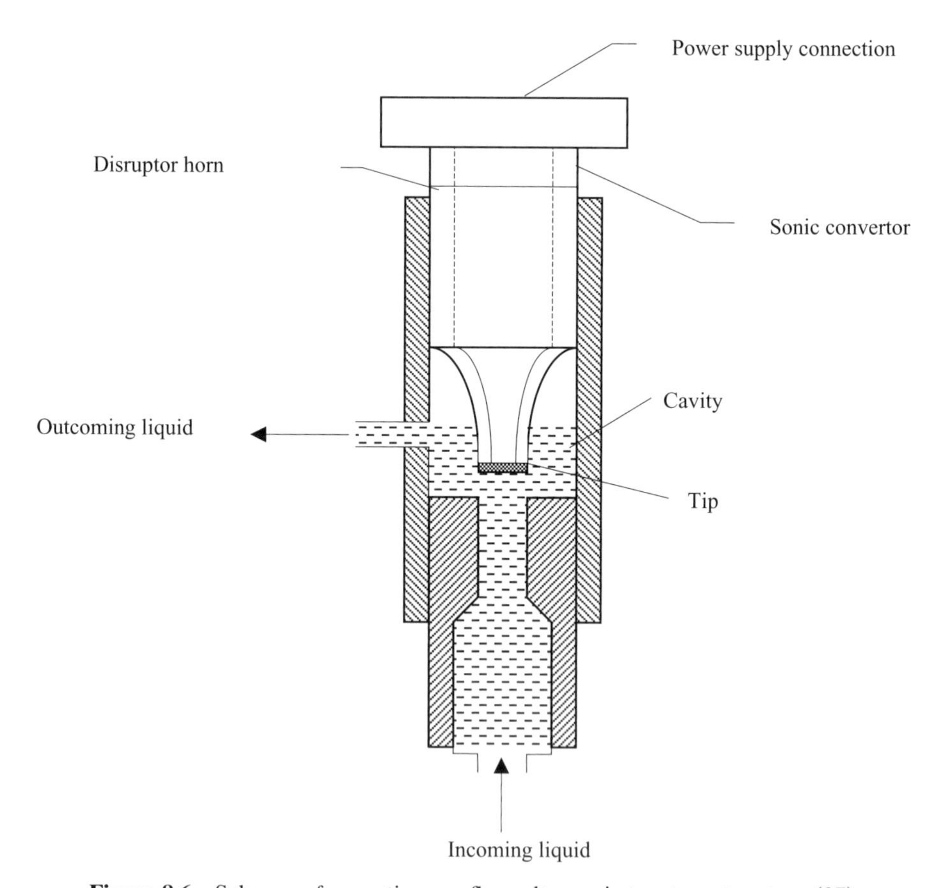

Figure 8.6 Scheme of a continuous-flow ultrasonic treatment system (27).

further transmitted to a titanium alloy disruptor horn. The horn amplifies the energy and delivers it into the tip immersed in the sample. Samples are pumped and flow through the insulated ultrasonic cavity. The total volume of the cavity is 18.76 mL. The process conditions can be changed by varying the output intensity level and the flow rate (11–50 mL/min) to adjust the theoretical residence time inside this cavity. In addition to microbial inactivation, the continuous-flow ultrasonic treatment enhances homogenization of the treated foods. It has lower energy consumption when compared with a batch ultrasonic system (27).

8.4.3 Research Status

High-energy ultrasound has been widely used for food applications including degassing of liquid foods, induction of oxidation/reduction reactions, enhancing extraction of enzymes and proteins, enhancing enzyme inactivation, and induction of nucleation and crystallization (24, 28). Latest development involves application of high-energy ultrasound in food freezing with a view to shorten the freezing process and obtain product of improved quality (25).

The potential of using ultrasound technology in food preservation has been demonstrated in the research literature. For example, *Salmonella* spp. showed a 4-log reduction in viable cell count when subjected to ultrasound of 160 kHz at a power of 100 W for 10 min in peptone water (29). However, using ultrasound alone cannot always achieve satisfactory bactericidal effect. Efficiency of ultrasound microbial inactivation depends on the microorganism itself as well as on the medium properties and operating conditions. Kim *et al.* (30) reported that ultrasound can reduce the bacterial count in leafy lettuce with a lower maximum reduction of 90%, compared with 99% by chlorinated (100–200 mg/L) water, and 99.95% by ozonated (0.1–1.5 mg/L) water. At present, the applications of ultrasound in food preservation process are not commercially feasible.

Synergistic effect could be achieved when ultrasound technology is combined with other preservation processes. Ultrasound has been used in conjunction with pressure treatment (manosonication), heat treatment (thermosonication), or both (manothermosonication). An enhanced mechanical disruption of cells has been observed that improves the efficiency in microbial killing. For example, the *D*-value for *Saccharomyces cerevisiae* was 739 min when subjected to 45 °C heat treatment alone; it decreased to 22.3 min when 20-kHz ultrasound was combined with heat treatment (31). When using ultrasonic treatment (20 kHz and amplitude of 117 μm) alone at ambient temperature, *Listeria monocytogenes* has a *D*-value of 4.3 min; by combining with pressure of 200 kPa (manosonication), the *D*-value decreased to 1.5 min. A further increase in pressure to 400 kPa reduced the *D*-value to 1.0 min (32). Ultrasound and temperature have been combined for inactivation of enzymes and microorganisms in raw milk (27). Sonication also improves the sterilizing effect of the chlorine solution in reducing microbial populations on broiler breast skin (33).

8.5 MICROWAVE AND RADIO-FREQUENCY PROCESSING

8.5.1 Fundamentals

Research on the use of microwave and radio frequency started during World War II as a by-product of the wide applications of radar technology. The domestic microwave oven is now a common appliance in households. In food applications, microwave heating is accomplished using frequencies of 2450 or 915 MHz, corresponding to 12 or 34 cm in wavelength. Domestic ovens operate at 2450 MHz. RF heating uses frequencies of 13.56, 27.12, and 40.68 MHz. Dielectric and ionic heating are the major mechanisms involved in heating foods with MW or RF. Dielectric heating results from the movement of the polar molecules trying to align themselves to the rapidly changing direction of the electric field that creates frictional heat. The dominant polar materials are water molecules; therefore, the water content of the food is an important factor for the dielectric heating performance of foods. Other food components such as salt, protein, and carbohydrates are also dipolar ingredients. As these are volumetrically distributed within food material, dielectric heating results in volumetric heating, which is fast and more uniform when compared with conventional heating. At higher temperatures, the electric resistance heating from the dissolved ions also plays a role in the heating mechanisms.

The dielectric properties of foods are the key parameters determining the coupling and distribution of electromagnetic energy, thus deciding the heating effectiveness of MW and RF. Dielectric properties are normally described by dielectric constant and loss factor. Dielectric constant describes the ability of a material to store energy in response to an applied electric field. The loss factor describes the ability of a material to dissipate energy in response to an applied electric field, i.e., the ability to generate heat. Dielectric properties depend on chemical compositions of the foods, moisture content, and bulk density. They are also highly dependent on temperature and the frequency of applied electric field. Dielectric property data for various materials are widely dispersed in the technical literature (34, 35).

Runaway heating, a potential problem associated with MW and RF heating, implies that the loss factor increases with the increase in temperature that leads to uneven heating in which regions of food with higher temperature will absorb more supplied energy. It has a higher probability to occur in RF than MW heating, due to the more rapid increase in loss factor with increasing temperature associated with lower frequencies. For example, research has shown that dielectric loss factors of whey protein products, and macaroni and cheese increased sharply at 27 and 40 MHz with increasing temperature, but only slightly increased at 915 and 1800 MHz (34).

Penetration depth is another important property for dielectric heating, which defines the distance an incident electromagnetic wave can penetrate beneath the surface of a material as the power decreases to 1/e of its power at the surface. It is determined by the dielectric constant and the loss factor of the food. The penetration depth is approximately 1 to 2 cm at 2450 MHz for a normal moist food, and decreases at higher temperatures. The penetration depths at RF range are about four times as deep as for microwave frequencies, which allow RF energy to penetrate dielectric material more deeply with more uniform heating along the depth of a food package than MW energy. Therefore, MW heating is suitable for packages with relatively smaller thickness (e.g., 1 to 2 cm), while RF heating can be applied for packages and trays with large institutional sizes up to 4 to 8 cm in depth (34, 36).

MW and RF heating for food pasteurization and sterilization purposes have been widely explored (18). Compared with conventional heating, they require less time to come up to the desired process temperature, which enable high-temperature, short-time processing for solid and semisolid foods that maximizes retention of desired nutrients. As a volumetric heating process, it is also possible to obtain greater uniformity in heating than conventional heating. Furthermore, it is also more convenient to control as the heating can be turned on and off instantaneously. In many applications, MW and RF processing systems can be more energy-efficient.

8.5.2 Technical Design

Microwave equipment mainly consists of a microwave generator (magnetron) that converts electric energy into microwaves, a metal cavity where foods are heated, and waveguides made of aluminum tubes. For a continuous operation, the cavity may be substituted with a tunnel fitted with a conveyor belt. Figure 8.7 shows a schematic diagram of a 915-MHz pilot-scale microwave system

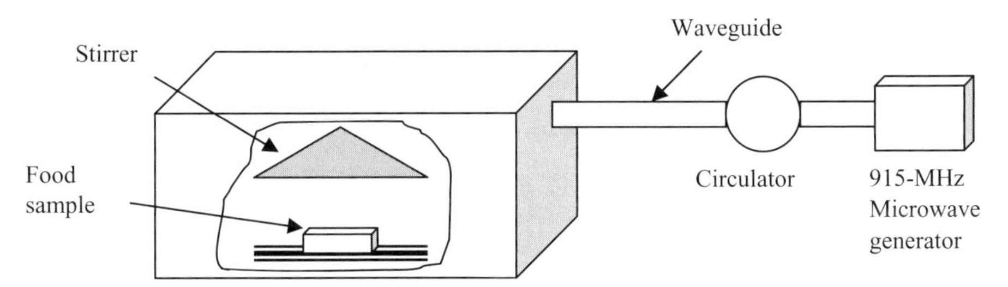

Figure 8.7 Schematic diagram of a 915-MHz pilot-scale microwave system (0–5 kW) (37).

designed for sterilization and pasteurization at Washington State University (37). It consists of a 5-kW power generator and a multimode stainless steel microwave cavity (1.07×1.22×1.47 m). The power unit transforms a 480-V power supply to 7500 V that is provided to the magnetron. The generator provides microwave power from 0.2 to 5 kW, which is controlled by the feedback of a 4- to 20-mA control signal from the magnetron anode current. Microwaves are generated by the magnetron and directed to the cavity via a circulator and rectangular waveguides. Power meters are used to monitor the microwave power input and reflected power from the cavity. The reflected waves from the cavity could damage the magnetron; therefore, they are directed to a matched water load by the circulator. A stirrer (0.87 m in diameter, 15 rpm) is used to enhance the uniformity of the microwaves in the cavity.

Heating uniformity remains a major challenge in the development of MW heating technologies. This nonuniform heating problem results from the discontinuous dielectric properties between foods and the surrounding medium (e.g., air or water) as well as from the variation in the dielectric properties of different food constituents (38). Composition and geometry of food and packaging materials and design of the MW cavity affect the heating uniformity. Various techniques have been attempted to improve the uniformity of heating. Studies have shown that 915-MHz microwaves combined with water immersion technique have a relatively uniform heat distribution within certain food products packaged in pouches and trays yielding high-quality product (36, 38). Food package is often rotated or oscillated to achieve uniform heating, which is used in a domestic microwave oven. Other techniques include cycling the power, variable frequency microwave processing, phase control microwave processing, and combination of microwaves with conventional heating for a surface treatment (18).

Unlike MW in which microwave source can be separated from the applicator, the RF source and applicator normally need to be designed and built together due to the high-impedance nature of RF coupling. RF power is generated in an oscillating circuit, in which food itself is part of it (Fig. 8.8). An important part of RF heating equipment is the design of the electrodes to

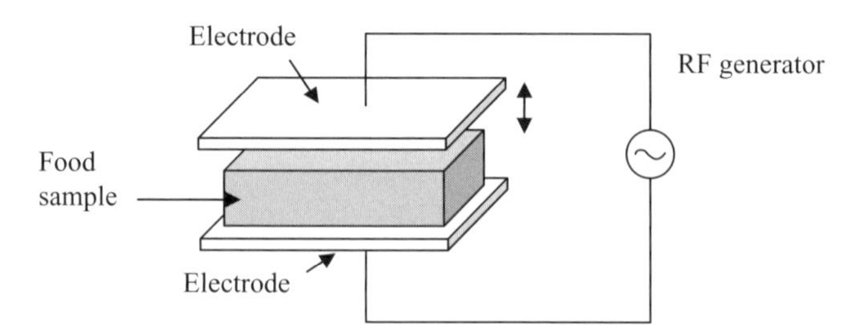

Figure 8.8 Schematic diagram for RF treatment on foods.

create uniform electric field patterns and thus uniform heating patterns, which is influenced by the required field strength, and food geometry and composition.

8.5.3 Research Status and Applications

MW and RF heating have found many applications in the food processing industry, including tempering of frozen foods for further processing, precooking of bacon, and finish drying of pasta, biscuits, and other cereal products. These systems are preferred in treating prepackaged food products because plastic packaging materials are transparent to microwaves. As a promising technique for food pasteurization and sterilization, MW and RF energy allow a quick heating process, thus a short overall processing time that reduces quality changes in the treated foods compared with conventional heating. Their use for pasteurization and sterilization of foods has been studied for more than 30 years (34, 38). The destruction kinetics of some microorganisms such as *S. cerevisiae, Lactobacillus plantarum*, and *Escherichia coli*, as well as inactivation of enzymes under the continuous microwave heating have been reported (18). The commercial application of MW processing in food pasteurization has been successfully accomplished in Europe and Japan, where MW heating is used to pasteurize and sterilize pre-packed foods including milk, yogurt, or pouch-packed meals (39).

In the United States, the MW sterilization process is still at a pilot scale and it has not been approved by the FDA. The major concern is heating uniformity that must be met in order to fulfill the quality advantages. To meet the food safety requirements, it is critical to know and control the lowest temperatures within the product (cold spot), where the microorganism destruction has the slowest rate of heating. For food that is heated mainly by conduction, the cold spot is usually located in the geometric center of the package. However, for foods heated by MW and RF energies, the cold spot varies with the food type, shape, temperature, and configuration of the applicator, imposing a difficulty to monitor and control the heating process. At Washington State University, a consortium was formed in 2001 with memberships from food processors, equipment manufacturers, ingredient suppliers, academia (food science, engineering, microbiology, and economics), US army, and government, with the aim to develop MW sterilization system for commercial use at industrial level and to develop a wide variety of shelf-stable products. Considerable effort has been made in developing predictive methods to quickly and reliably determine the location of cold spots. Chemical marker, digital imaging, and computer simulations have been used to generate 3D heating patterns of food. Fiber-optic sensors are used to monitor the rapid temperature rise that would occur in the packaged foods during heating. Experiments conducted to evaluate the effectiveness of MW sterilization on various food types included mashed potato, beef, rice, fish, and vegetables, packaged in plastic tray or pouch (34, 36, 37). Results from these studies have demonstrated the benefits of microwave sterilization in terms of improved sensory and nutritional quality compared with conventional retorting. These results indicate that microwave sterilization of foods should be a viable industrial process pending approval by the FDA.

8.6 OHMIC HEATING

8.6.1 Fundamentals

Ohmic heating is also called electrical resistance heating. It is a direct type of heating where food itself is a conductor of electricity and heat is generated with the passage of alternating electric currents, with frequency of 50 Hz in Europe and 60 Hz in the United States. The heat is generated by the electrical resistance of the food according to Ohm's law. A similarity between OH and MW heating is noted by the fact that the electric energy is converted into thermal energy volumetrically, while the difference is that food subjected to OH is in contact with electrodes. The applicability of OH depends on the electrical conductivity of the material to be heated, which makes foods a good candidate since most foods contain considerable moisture content and dissolved ionic salts. The application of OH in food processing started in the nineteenth century; however, active research on OH applications in foods was mainly conducted in the last two decades (40).

Thermal effect is the principal mechanism for OH to inactivate microorganisms. A mild electroporation may also occur during ohmic heating that contributes to bactericidal effect (18). Volumetric heating makes OH rapid in heating rate and relatively uniform in heating pattern, which is expected to improve food quality as compared with conventional heating using conduction, convection, or radiation. A special use of OH is to heat fluids containing particulates. Because the heating rate depends on the electrical conductivity of the material, the heating rates in the two phases (fluid and particles) can be adjusted by formulating the ionic contents of the fluid and particulate phase to ensure the appropriate levels of electrical conductivity. Heat can be generated faster in the particulate than in the liquid. This is advantageous when compared with conventional heating where heat has to be transferred to particulates from fluid, resulting in a significantly slower heating rate in the particulates than in the surrounding fluid and thus overcooking of the surrounding liquid. Therefore, OH is often considered to be a promising technology for continuous sterilization of liquid–particulate mixtures. Other advantages include the limited contact surface between electrodes and food that reduces the risk of fouling (compared with conventional heating), high energy efficiency because 90% of the electrical energy is converted into heat, and the ease of process control with instant switch-on and shutdown. The major disadvantage, however, is that OH heating is strongly influenced by the electrical heterogeneity of the food to be heated. There is a complex relationship between temperature, electrical field distributions, and the shape and orientation of

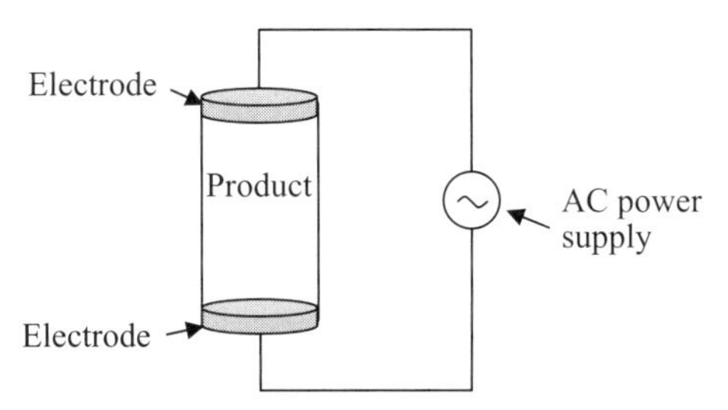

Figure 8.9 Schematic diagram showing the principle of OH.

particulates in the fluid, which causes nonuniformity in temperature distribution and increases the difficulty in process control and monitoring (41).

8.6.2 Technical Design

An ohmic heater consists of a pair of electrodes, a container for the food to be processed, and an alternating power supply (Fig. 8.9). The electrode materials need to be inert and must not release metal ions into foods. Currently, both batch and continuous processes are being considered. For continuous processing, the electric field can be either perpendicular to the flow (cross-field) or parallel to flow (in-field).

The most important parameter in OH is the electrical conductivity of the food, which increases with the increase in ion content of a food. Food products can be formulated with salts to achieve uniform ohmic heating. The electric field strength can be adjusted by changing the electrode gap and the applied voltage. Particle orientation and geometry directly influence the heating rate of the different constituents of a liquid food (40). Similar to dielectric heating, ohmic heating also has runaway heating problems because the electrical conductivity of most foods increases with temperature (18).

8.6.3 Research Status and Applications

Research conducted on OH applications in fruits, vegetables, meat products, and surimi has demonstrated its ability to improve heating uniformity and food quality with minimal structural, nutritional, or sensory changes (42). Food applications of ohmic heating include blanching, evaporation, dehydration, fermentation, and extraction. Most industrial applications are in Japan, the United Kingdom, and the United States (41). Studies investigating the effect of pasteurization and sterilization of ohmic heating on a number of food products have indicated that OH is effective in inactivating bacteria, spores, yeast, and mold (18). Furthermore, since ohmic heating system does not involve mechanical agitation that is often present in a conventional heat exchanger, particle integrity is maintained for particles up to 2 cm in diameter, improving attractive appearance and textural properties of treated foods (43).

Despite considerable potential of the ohmic heating process to produce high-value-added and shelf-stable products with an improved quality compared with current sterilization techniques, the FDA has not yet approved processes involving continuous OH for microbial inactivation of foods. Similar to the MF process, a major issue for using OH processing in microbial inactivation is to identify cold spot of the treated foods. This is necessary to effectively control the temperature profile of food materials, which is critical for the heating process to be sufficient for inactivating target pathogens to the desired extent. This is relatively simple for a homogeneous fluid medium where the cold spot is usually the fastest moving region. For the processing of solid–liquid mixtures, it is complicated as the cold spot may shift considerably during processing (18). Therefore, specific analysis must be conducted for different equipment designs and for individual products. This involves developing effective approaches for identification, control, and validation of all the critical control points as well as precise mapping of temperatures in the foods submitted to OH (44). Complete models need to be developed to characterize the flow of the food components when being processed in an ohmic heater. Heating distribution patterns in the fluid flow must take into account differences in electrical conductivity between the liquid and solid phases and the responses of the two phases to temperature changes. OH sterilization is expected to minimize the thermal degradation of desirable product attributes yet maintain a safe product; however, its commercialization will not occur until the ability of OH to uniformly render the product commercially sterile is proven and the adequate control of the rate of heating is achieved (18).

8.7 INFRARED HEATING

8.7.1 Fundamentals

On the electromagnetic spectrum, infrared waves are identified with a frequency greater than the visible light. Generally, IR can be divided into three regions based on the spectral ranges, including near-infrared (NIR, 0.75 to 1.4 μm), mid-infrared (MIR, 1.4 to 3 μm), and far-infrared (FIR, 3 to 1000 μm). When IR waves are incident on a material, they are either reflected, transmitted, or absorbed. Absorbed waves are transformed into heat, leading to an increase in the temperature of the material. In general, FIR radiation is preferred for food processing since most food components absorb radiative energy in the FIR region (45). IR has limited penetrating depth, the decay in IR power reaching 90% within a thin layer of 40 μm in bacterial suspension (46); thus, IR radiation can be considered surface treatment and is attractive primarily for surface-heating applications. The mechanisms of IR heating in inactivating microorganisms may include thermal effect and DNA damage similar to UV light (47). Commercial FIR heaters with high emissivity are now available. Interest in the application of FIR radiation in the food processing industry has increased during the past few years due to its characteristic of rapid and contactless heating.

8.7.2 Technical Design

The main component of an IR oven is a radiator, either a gas-fired heater or an electric heater. The latter is used for higher temperatures. IR temperatures used for heating foods are usually within 650–1200 °C to prevent charring of products. It is possible to selectively heat food components such as protein and carbohydrates because they absorb different wavelengths of infrared radiation. This feature can be also used to selectively inactivate microorganisms while minimizing changes in food quality (48). Figure 8.10 shows a novel FIR heating system designed for selective heating of soy protein and glucose (49). The FIR heating system consists of six ceramic lamps, an optical band pass filter selected to emit IR radiation in the spectral range corresponding to the dominant absorbance band of soy protein (6 to 11 μm), and a cone-shaped aluminum waveguide to direct IR radiation from the source to the sample with maximum efficiency. Without the presence of the filter, the temperature of glucose was higher than that of soy protein; when filter was used, the soy protein was heated by about 6 °C higher than glucose after 5 min of heating.

8.7.3 Research Status and Applications

IR heating has been widely used in drying low-moisture foods, such as breadcrumbs, cocoa, flour, grains, pasta, malt, and tea. It is also used as an initial heating stage in baking, roasting, and frying processes to allow a rapid increase

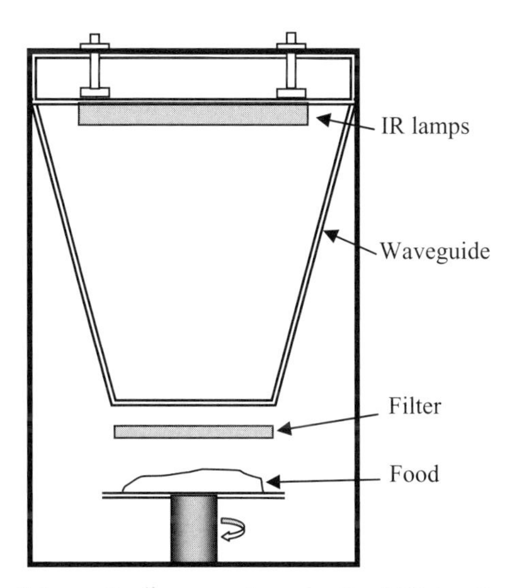

Figure 8.10 Schematic diagram of a selective FIR heating system (49).

in the surface temperature. Due to the low penetration depth, IR heating is often combined with other techniques, such as conventional convective heating, to achieve a synergistic effect and reduce processing time. For example, a combination of IR heating with freeze-drying in sweet potatoes can reduce the processing time by more than half (50).

IR heating is capable of inactivating bacteria, spores, yeast, and mold in both liquid and solid foods, which makes it a potential technique for pasteurization and sterilization of food products (48). IR is easy to control and relatively inexpensive with a high efficiency of energy utilization. It is faster than conventional heating. Studies conducted with honey showed that 3 to 4 min IR heat treatment was adequate for commercially acceptable products with the desired amount of yeast reduction and reduced thermal damage in quality (51). Due to the limited penetration depth, the effect of IR radiation on the microbial inactivation diminishes as the sample thickness increases; therefore, it is particularly suitable for surface pasteurization of pathogens. It was successfully used in pasteurization of strawberries, with a surface temperature high enough to effectively inactivate microorganisms and an internal temperature of the strawberries under 50 °C, thus avoiding deterioration in the quality attributes (52). In general, spores are more resistant than vegetative cells. Hamanaka *et al.* (47) proposed intermittent IR treatment to increase inactivation efficacy in spores, in which the first irradiation aims to activate spores into vegetative cells, followed by the second irradiation that effectively inactivates vegetative cells.

Increase in the power of IR heating source produces more energy, leading to higher level of microbial inactivation. Inactivation efficiency is also associated with the radiation spectrum: the total energy decreases as the peak wavelength increases. Other factors affecting efficacy of microbial inactivation include moisture content, physiological phase of microorganisms (exponential or stationary phase), and types of food materials (48).

REFERENCES

1. Señorans, F., Ibáñez, E., Cifuentes, A. (2003). New trends in food processing. *Critical Reviews in Food Science and Nutrition*, *43*, 507–526.
2. Ohlsson, T., Bengtsson, N. (2002). Minimal processing of foods with non-thermal methods, in *Minimal Processing Technologies in the Food Industry* (eds T. Ohlsson, N. Bengtsson), Woodhead Publishing, Cambridge, pp. 34–60.
3. Beattie, J.M., Lewis, F.C. (1924). The electric current (apart from the heat generated). A bacteriological agent in the sterilization of milk and other fluids. *Journal of Hygiene*, *24*, 123–137.
4. Palaniappan, S., Sastry, S.K. (1990). Effects of electricity in micro-organisms: a review. *Journal of Food Process Preservation*, *14*, 393–414.
5. Dunn, J.E., Pearlman, J.S. (1987). Methods and apparatus for extending the shelf-life of fluid food products, Maxwell Laboratories, Inc. US Patent 4.695.472.

6. Knorr, D. (1999). Novel approaches in food-processing technology: new technologies for preserving foods and modifying function. *Current Opinion in Biotechnology*, *10*, 485–491.
7. Zhang, Q., Barbosa-Canovas, G.V., Swanson, B.G. (1995). Engineering aspects of pulsed electric field pasteurization. *Journal of Food Engineering*, *25*, 261–281.
8. Barbosa-Canovas, G.V., Pothakamury, U.R., Palou, E., Swanson, B.G. (1998). *Nonthermal Preservation of Foods*, Marcel Dekker, Inc., New York.
9. Qin, B.L., Pothakamury, U.R., Barbosa-Canovas, G.V., Swanson, B.G. (1996). Nonthermal pasteurization of liquid foods using high intensity pulsed electric fields. *Critical Reviews in Food Science and Nutrition*, *36*, 603–627.
10. Qiu, X., Sharna, S., Tuhela, L., Jia, M., Zhang, Q.H. (1998). An integrated PEF pilot plant for continuous nonthermal pasteurization of fresh orange juice. Unpublished Report. Food Science and Technology Department, Ohio State University, Columbus, OH.
11. Knorr, D. (1998). Advantages, possibilities, and challenges of high pressure applications in food processing, in *The Properties of Water in Foods* (ed. D.S. Reid), Blackie Academic and Professional, London, pp. 419–437.
12. Mermelstein, N. (1998). Processing papers cover wide range of topics. *Food Technology*, *52* (7), 50–53.
13. Deeth, H., Datta, N., Ross, I.V.A., Dam, T.X. (2006). Pulsed electric field technology: effect on milk and fruit juices, in *Advances in Thermal and Non-Thermal Food Preservation* (eds G. Tewari, V. Juneja), Blackwell Publishing, Ames, IA, pp. 241–269.
14. Barbosa-Canovas, G.V., Gould, G.W. (2000). *Innovations in Food Processing*, CRC Press, Boca Raton, FL.
15. Singh, R.P. and Yousef, A.E. (2001). Technical elements of new and emerging nonthermal food technologies. Food and Agriculture Organization of the United Nations, http://www.fao.org/Ag/ags/Agsi/Nonthermal/nonthermal_1.htm, accessed 19 September 2008.
16. Praporscic, I., Ghnimi, S., Vorobiev, E. (2005). Enhancement of pressing of sugar beet cuts by combined ohmic heating and pulsed electric field treatment. *Journal of Food Processing and Preservation*, *29*, 378–389.
17. Toepfl, S., Mathys, A., Heinz, V., Knorr, D. (2006). Review: potential of high hydrostatic pressure and pulsed electric fields for energy efficient and environmentally friendly food processing. *Food Reviews International*, *22*, 405–423.
18. FDA (2000). *Kinetics of Microbial Inactivation for Alternative Food Processing Technologies*, U.S. Food and Drug Administration, Center for Food Safety and Applied Nutrition, June 2, 2000. http://www.cfsan.fda.gov/~comm/ift-toc.html (accessed 17 March 2008).
19. Gómez-López, M., Devlieghere, F., Bonduelle, V., Debevere, J. (2005). Factors affecting the inactivation of micro-organisms by intense light pulses. *Journal of Applied Microbiology*, *99* (3), 460–470.
20. Krishnamurthy, K., Demirci, A., Irudayaraj, J.M. (2007). Inactivation of *Staphylococcus aureus* in milk using flow-through pulsed UV-light treatment system. *Journal of Food Science*, *72* (7), M233–239.
21. FDA (1996). Code of Federal Regulations. 21CFR179.41. http://www.access.gpo.gov/cgi-bin/cfrassemble.cgi?title=200321 (accessed 17 March 2008).

22. Dunn, J., Clark, W., Ott, T. (1995). Pulsed-light treatment of food and packaging. *Food Technology*, *49* (9), 95–98.
23. Sizer, C.E., Balasubramaniam, V.M. (1999). New intervention processes for minimally processed juices. *Food Technology*, *53* (10), 64–67.
24. Knorr, D., Zenker, M., Heinz, V., Lee, D. (2004). Applications and potential of ultrasonics in food processing. *Trends in Food Science & Technology*, *15*, 261–266.
25. Zheng, L., Sun, D. (2006). Innovative applications of power ultrasound during food freezing processes—a review. *Trends in Food Science & Technology*, *17* (1), 16–23.
26. Piyasena, P., Mohareb, E., McKellar, R.C. (2003). Inactivation of microbes using ultrasound: a review. *International Journal of Food Microbiology*, *87*, 207–216.
27. Villamiel, M., de Jong, P. (2000). Inactivation of *Pseudomonas fluorescens* and *Streptococcus thermophilus* in trypticase soy broth and total bacteria in milk by continuous-flow ultrasonic treatment and conventional heating. *Journal of Food Engineering*, *45*, 171–179.
28. Vilkhu, K., Mawson, R., Simons, L., Bates, D. (2007). Applications and opportunities for ultrasound assisted extraction in the food industry—a review. *Innovative Food Science & Emerging Technologies*, *9(2)*, 161–169.
29. Lee, B.H., Kermasha, S., Baker, B.E. (1989). Thermal, ultrasonic and ultraviolet inactivation of *Salmonella* in thin films of aqueous media and chocolate. *Food Microbiology*, *6*, 143–152.
30. Kim, B., Kim, O., Kim, D., Kim, G. (1999). Development of a surface sterilization system combined with a washing process technology for leafy lettuce. Proceedings of the International Symposium on Quality of Fresh and Fermented Vegetables, Seoul, Korea Republic, 27–30 October 1997 (eds J.M. Lee, K.C. Gross, A.E. Watada and S.K. Lee), pp. 311–317.
31. Lopez-Malo, A., Guerrero, S., Alzamora, S.M. (1999). *Saccharomyces cerevisiae*, thermal inactivation kinetics combined with ultrasound. *Journal of Food Protection*, *62*, 1215–1217.
32. Pagan, R., Manas, P., Alvarez, I., Condon, S. (1999). Resistance of *Listeria monocytogenes* to ultrasonic waves under pressure at sublethal (manosonication) and lethal (manothermosonication) temperatures. *Food Microbiology*, *16*, 139–148.
33. Lillard, H.S. (1994). Decontamination of poultry skin by sonication. *Food Technology*, *48*, 72–73.
34. Wang, Y., Wig, T.D., Tang, J., Hallberg, L.M. (2003). Dielectric properties of food relevant to RF and microwave pasteurization and sterilization. *Journal of Food Engineering*, *57*, 257–268.
35. Wang, Y., Tang, J., Rasco, B., Kong, F., Wang, S. (2008). Dielectric properties of salmon fillets as a function of temperature and composition. *Journal of Food Engineering*, *87*, 236–246.
36. Guan, D., Cheng, M., Wang, Y., Tang, J. (2004). Dielectric properties of mashed potatoes relevant to microwave and radio-frequency pasteurization and sterilization processes. *Journal of Food Science*, *69* (1), FEP30–FEP37.
37. Lau, M.H., Tang, J. (2002). Pasteurization of pickled asparagus using 915 MHz microwaves. *Journal of Food Engineering*, *51* (4), 283–290.

38. Guan, D., Gray, P., Kang, D.H., Tang, J., Shafer, B., Ito, K., Younce, F., Yang, T.C.S. (2003). Microbiological validation of microwave-circulated water combination heating technology by inoculated pack studies. *Journal of Food Science*, *68* (4), 1428–1432.
39. Tewari, G. (2006). Microwave and radio-frequency heating, in *Advances in Thermal and Non-Thermal Food Preservation* (eds G. Tewari, V. Juneja), Blackwell Publishing, Ames, IA, pp. 91–98.
40. Vicente, A., Castro, I.A. (2006). Novel thermal processing technologies, in *Advances in Thermal and Non-Thermal Food Preservation* (eds G. Tewari, V. Juneja), Blackwell Publishing, Ames, IA, pp. 99–130.
41. Ohlsson, T., Bengtsson, N. (2002b). Minimal processing of foods with thermal methods, in *Minimal Processing Technologies in the Food Industry* (eds T. Ohlsson, N. Bengtsson), Woodhead Publishing, Cambridge, pp. 4–33.
42. McKenna, B.M., Lyng, J., Brunton, N., Shirsat, N. (2006). Advances in radio frequency and ohmic heating of meats. *Journal of Food Engineering*, *77*, 215–229.
43. Eliot-Godéreaux, S., Zuber, F., Goullieux, A. (2001). Processing and stabilisation of cauliflower by ohmic heating technology. *Innovative Food Science and Emerging Technologies*, *2*, 279–287.
44. Salengke, S., Sastry, S.K. (2007). Models for ohmic heating of solid liquid mixtures under worst-case heating scenarios. *Journal of Food Engineering*, *83*, 337–355.
45. Sandu, C. (1986). Infrared radiative drying in food engineering: a process analysis. *Biotechnology Progress*, *2*, 109–119.
46. Hashimoto, A., Sawai, J., Igarashi, H., Shimizu, M. (1991). Effect of far-infrared radiation on pasteurization of bacteria suspended in phosphate-buffered saline. *Kagaku Kogaku Ronbunshu*, *17*, 627–633.
47. Hamanaka, D., Dokan, S., Yasunaga, E., Kuroki, S., Uchino, T., Akimoto, K. (2000). The sterilization effects on infrared ray of the agricultural products spoilage microorganisms (part 1). An ASAE Meeting Presentation, Milwaukee, WI, July 9–12, No. 00 6090.
48. Krishnamurthy, K., Khurana, H.K., Jun, S., Irudayaraj, J., Demirc, A. (2007b). Infrared heating in food processing: an overview. *Comprehensive Reviews in Food Science and Food Safety*, *7* (1), 2–13.
49. Jun, S., Irudayaraj, J. (2003). Selective far infrared heating system—design and evaluation I. *Drying Technology*, *21* (1), 51–67.
50. Lin, Y.P., Tsen, J.H., King An-Erl, V. (2005). Effects of far-infrared radiation on the freeze-drying of sweet potato. *Journal of Food Engineering*, *68*, 249–255.
51. Hebbar, H.U., Nandini, K.E., Lakshmi, M.C., Subramanian, R. (2003). Microwave and infrared heat processing of honey and its quality. *Food Science and Technology Research*, *9*, 49–53.
52. Tanaka, F., Verboven, P., Scheerlinck, N., Morita, K., Iwasaki, K., Nicolaï, B. (2007). Investigation of far infrared radiation heating as an alternative technique for surface decontamination of strawberry. *Journal of Food Engineering*, *79*, 445–452.

9

FOOD PROCESSING AND NUTRITIONAL ASPECTS

JOSEF BURRI,[1] CONSTANTIN BERTOLI,[2] AND RICHARD H. STADLER[1]

[1] *Nestlé Product Technology Centre Orbe, CH-1350 Orbe, Switzerland*
[2] *Nestlé Product Technology Centre Konolfingen, 3510 Konolfingen, Switzerland*

9.1 INTRODUCTION

Food processing can be considered a set of practices—using defined technologies and techniques either single or in combination—to transform raw foods/commodities or intermediate products into food ready for consumption by humans or animals. Some basic techniques of food processing, also frequently encountered in the home are, for example, cutting/mincing/macerating, liquefication and emulsification, heat treatments (such as pasteurization and sterilization), canning, freezing, drying/dehydration (e.g., lyophilization and spray drying), addition of chemical preservatives (e.g., organic acids and nitrites), fermentation, salting, smoking, atmosphere modification, and air entrainment (e.g., gasification of beverages). In fact, salting and drying are two of the earliest methods of treating foods to help preserve freshness and improve flavor (1, 2). The practice of cooking is defined by numerous techniques such as boiling, broiling, baking, frying (e.g., vacuum frying, sautéing, and deep-frying), steaming, smoking, microwaving, and roasting (e.g., toasting, barbecuing, or grilling).

Many of these processes are practiced either in a domestic environment, i.e., in our everyday preparation of food in the home, in catering services, restaurants, and by the food manufacturing industry, with the goal toward the

Process-Induced Food Toxicants: Occurrence, Formation, Mitigation, and Health Risks,
Edited by Richard H. Stadler and David R. Lineback

production of foods that are nutritional, safe, tasty, consistently high quality, and that can be successfully marketed through their functionality, convenience, and affordability. In fact, more and more consumers indicate their preference for "health-enhancing" foods or "functional foods" containing active ingredients that impart a certain health-linked benefit, e.g., n-3 long-chain fatty acids in milk to reduce cardiovascular disease (CVD) risks, probiotics in yogurt to improve the growth of beneficial intestinal flora, and phytosterols in margarine to reduce cholesterol uptake. Such products are today taking increasingly more space on supermarket shelves.

The art of converting food raw materials into an edible state probably dates back to the advent of anthropogenic fire, which revolutionized hominid life in terms of hunting, cooking, nutrition, subsistence, and residence patterns. However, there are debates as to when hominids domesticated fires. Some authors hypothesize an early date (about 1.9 million years ago) (3), while others (4, 5) propose the domestication of fires some 400,000–300,000 years ago.

The invention of processing food with fire probably had the largest effect with respect to modern human traits (6). It widened the availability of food, making certain plant species edible (for example, through the breakdown of inherent plant toxins), and improved the overall quality of the diet. The softening of the food polymers (e.g., cellulose) through the cooking practice lessened the time spent for chewing, and may be one causative factor of reduced tooth size and changes to the masticatory system. Some authors propose cooking may have in fact impacted direct and indirect changes to the overall body frame, the gastrointestinal system as well as to cultural, social, and sexual behaviors (6).

9.2 FOOD PROCESSING AND MAJOR BENEFITS

In today's world, the benefits of food processing are multiple and clearly undisputed. For example, thermal treatment has several benefits from a nutritional point of view (Table 9.1).

Cooking enhances food digestibility through changes of the physicochemical structure of macronutrients, e.g., starches and proteins, increasing the caloric intake per meal. Heat input initiates changes in the plant cell wall structure and consequently the food matrix, and may enhance the bioavailability of bioactive food components.

The chemical reactions that result from food processing are usually complex, leading to the formation of a plethora of new compounds. One of the most prominent and valued consequences of processing of foods is the generation of taste, aroma, and color, reflected in our daily diet in foods and beverages such as bread/bread crust, soy sauce, breakfast cereals, roasted meat, malted beverages, and coffee. A chemical process that significantly contributes toward these attributes is the Maillard reaction. It describes a series of chemical reac-

TABLE 9.1 The major nutritional benefits of thermal treatment of food.

Formation of	• aroma and taste-active compounds • (novel) antioxidants • chemoprotective compounds
Improvement of	• digestibility • bioavailability of nutrients
Reduction or elimination of	• microbial load • natural toxins • enzyme inhibitors

tions initiated by the condensation of a reducing carbohydrate with an amino compound, usually an amino acid or protein moiety (7). Depending on the nature of the food material and severity of processing conditions (mainly thermal input), this complex cascade is the source of literally thousands of individual chemical compounds, termed collectively "Maillard reaction products" (MRPs).

The sequence of changes that comprise the Maillard pathway can be broadly discerned into the "early" and "advanced" stages. The former constitutes the formation of the Amadori rearrangement product (frequently abbreviated ARP), and highlights the initial stage of the Maillard reaction. The advanced stage involves the degradation of the Amadori compound via different routes as outlined by Hodge (8), the main pathway leading to unstable deoxysones by elimination of water, which in turn undergo secondary reactions leading to advanced MRPs.

The various pathways depend on the pH of the reaction milieux, the basicity of the amine attached to the sugar, and the thermal input. Different pathways involve numerous steps, which lead to the formation of a broad spectrum of volatile or soluble substances. Contributing to the final stage of the Maillard reaction and responsible for the brown color are polyfunctional macromolecules called melanoidins, with masses up to 100,000 Daltons (9). Their formation during thermal processing has been studied in model experiments and is apparently free radical driven through the formation of a transient Maillard intermediate named "Crosspy," which in fact represents a protein-bound pyrazinium radical cation that rapidly polymerizes to finally afford melanoidins (10).

There is a paucity of knowledge of the physiological relevance of MRPs at the typical concentrations at which they occur in our diet. In fact, many reports on biologically active MRPs have been published in the past two to three decades, and several of the studies have shown beneficial chemoprotective properties of the compounds either *in vitro* or *in vivo*, for example, by modulating the activity of detoxifying enzymes in the body (11, 12). For the food industry, a main goal of research is to elucidate the key pathways that lead to bioactive Maillard-related substances, and to modulate these in terms of providing foods with additional benefits such as improved aroma, taste, and textural properties.

9.2.1 Nutrients

9.2.1.1 Proteins and Amino Acids Proteins are complex molecules with a high molecular weight and any processing is likely to modify their physical and chemical properties. One obvious aim of food processing is to increase the palatability of the raw material. At the same time, processing will also affect the nutritional value. With respect to proteins, both positive and negative effects are observed: improved digestibility and availability is obtained by denaturing the original protein structure, but also by destroying antinutritional substances that might be present, such as trypsin inhibitors. On the other hand, processing can also induce the reduction of the protein quality. The most common case is of course the well-known Maillard reaction that involves reducing sugars and amino acids as described earlier in this chapter.

MRPs have a number of desired properties. In cereal-based foods, they confer the golden color and flavor of biscuits, and some intermediate MRPs have antioxidant properties that may increase the shelf life of foods. Bread crust is a good example of positive health benefits of MRPs. A study has shown that extracts of rye bread crust exhibited antioxidative properties in *in vitro* test systems, correlating positively with the degree of browning (13). The active structures involved are MRPs and more precisely condensates of the carbohydrate degradation product acetylformoin and *N*-α-acetyl-carboxymethyl-L-lysine, the product being termed simplistically by the authors as "pronyl-L-lysine" (13). This substance exhibits an antioxidative capacity fivefold higher than that of ascorbic acid, and its concentration is related to the bread-baking time. On the other hand, research has shown that bread crust contains higher amounts of undesired compounds such as acrylamide and 3-MCPD (see Chapters 2.1 and 2.6, respectively).

The sources of reducing sugars in the Maillard reaction are manifold: milk, maltodextrins, malt extract, fruits, honey, and so on, but the sugars can also be formed during the process, e.g., the fermentation of bread dough. The reaction partner is typically the cereal protein, or any other protein source from an ingredient. In cereal products, lysine is usually the amino acid limiting the protein quality. Whereas milk protein contains about 8g lysine/16gN, this value is in the range of 2–2.5 for wheat and corn, and up to 3.5 for rice or oats. A standard method to measure the protein quality is the PER (protein efficiency ratio), based on rat feeding trials. For cereal products, however, it is much easier to measure the reactive (available) lysine, as there is a good relationship between this value and the PER (14).

The Maillard reaction is fastest at a water activity (A_w) of 0.70–0.50, i.e., not during cooking but rather during the drying step. Typically, most cereal products (e.g., breakfast or infant cereals, snacks) are first submitted to a heat treatment well above an A_w of 0.8, and are then dried down to an A_w in the range of 0.10 to 0.40, passing thus through the critical region. The degree of protein damage is then determined mainly by the drying technology and the drying conditions. Freeze-drying and spray drying under mild conditions usually result in foods with little formation of MRPs (lysine blockage of 0–10%), whereas technologies such as oven baking and roller drying may afford products with substantial browning, resulting in lysine blockage of 30–70% (15, 16). Extrusion cooking can be conducted in such a way that the Maillard reaction is less pronounced than, for instance, in roller drying. In a set of trials, infant cereals based on wheat or rice flour and soy flour as main protein source have been either roller-dried or extruded. The wheat flour was partly hydrolyzed with alpha-amylase, to obtain increasing amounts of reducing sugars and thus vary the severity of the Maillard reaction. As illustrated in Fig. 9.1, reactive lysine and PER are well correlated.

In certain raw foods, protein availability is reduced due to the presence of antinutritional factors such as trypsin inhibitors or phytates. Hydrothermal treatments are very efficient to reduce the activity of protease inhibitors, and processes such as fermentation, germination, and soaking may reduce the phytate level. Chitra *et al* (17) have shown that germination reduced the phytic acid content of a variety of beans and peas by 40–60%, whereas fermentation reduced phytic acid contents by 26–39%. Both germination and fermentation greatly increased the *in vitro* protein digestibility. Lectins (haemagglutinins) are proteins or glycoproteins that may induce a hormonal imbalance and

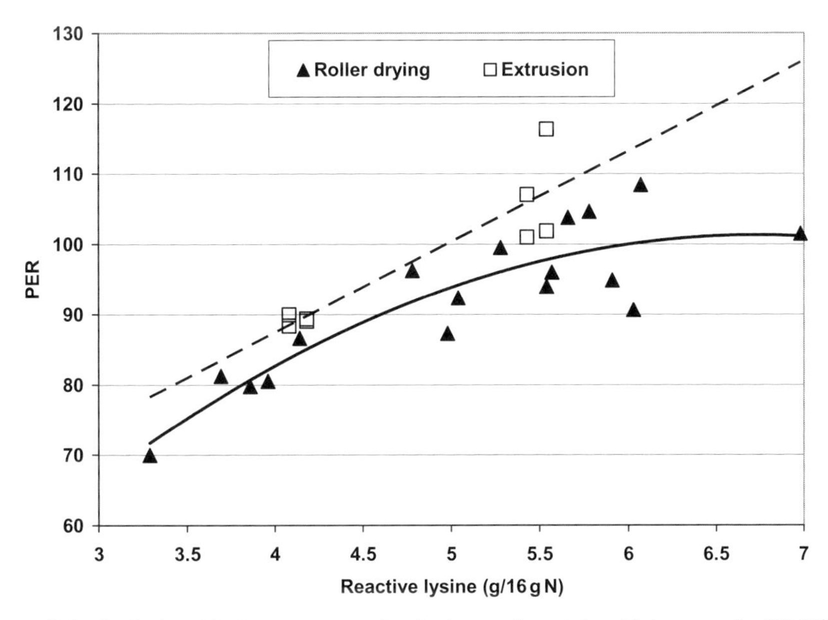

Figure 9.1 Relationship between reactive lysine and protein efficiency ratio (PER) in a soy cereal product.

influence the gut metabolism. They are found in soy and lentils and are also inactivated by heat treatment (18).

9.2.1.2 Carbohydrates Carbohydrates are the main constituents of cereals, but also fruits and to some extent of vegetables. They account for about 55% of the energy intake in a balanced diet, and as much as 80% in some countries. Carbohydrates need to be broken down to simple sugars for the body to be able to absorb them. Many carbohydrates, and in particular simple sugars, are well digested in their raw state without any processing as the digestive enzymes are capable to reduce them to monosaccharides. Native starch from most sources is quite easily digested. Raw potato starch, however, is indigestible because it is encapsulated within the starch granules, which hinders the accessibility of digestive enzymes (19). When potatoes are cooked, the starch granules are gelatinized and the starch becomes digestible. It has been shown (20) that with any of the standard processing means (boiling, mashing, baking, frying), less than 7% of the starch resisted digestion. On the other hand, processing of starch and subsequent retrogradation is likely to form resistant starch. Resistant starch is starch that "resists" digestion in the small intestine. In fact, resistant starches have been defined as "the sum of starch and products of starch digestion not absorbed in the small intestine of healthy individuals" (21). In many processing systems, a small amount of resistant starch is formed involuntarily. However, as it is considered dietary fiber, companies have developed processes to manufacture products with a high content of resistant starch (22).

The primary goal of food processing is to achieve optimal physical and organoleptic properties. With respect to starch, this can mean much more than simple gelatinization, and depends very much on the process. Cooking by steam injection and subsequent roller drying will result in the full development of viscosity and is the process of choice for the production of infant cereals. In order to obtain a pap with a well-defined nutrient and caloric density, the starch can be broken down by the use of amylolytic enzymes (alpha-amylase and amyloglucosidase). This will reduce the viscosity of the pap and enhance the sweetness without using added sugars, and also assures easy digestion. The roller-drying process is carried out at dry matter levels of 20% to 50%, in which starch is therefore fully hydrated. Extruders are typically run at much higher dry matter levels of up to 85% or more, and in addition, high pressure and shear are exerted. Starch therefore undergoes rather a melting than a gelatinization transition. Also, the starch molecules are disintegrated and broken down by the high shear. The resulting carbohydrate profile depends on the extruder configuration and the extrusion conditions, in particular the water content, pressure, and temperature. This will then also affect the product properties, such as viscosity, solubility, and so on. This is illustrated in Fig. 9.2, which shows typical carbohydrate profiles of extruded, roller-dried, and partly hydrolyzed wheat flours: the roller-dried flour contains mainly amylose and amylopectin, as would native wheat, whereas in the extruded flour, most of

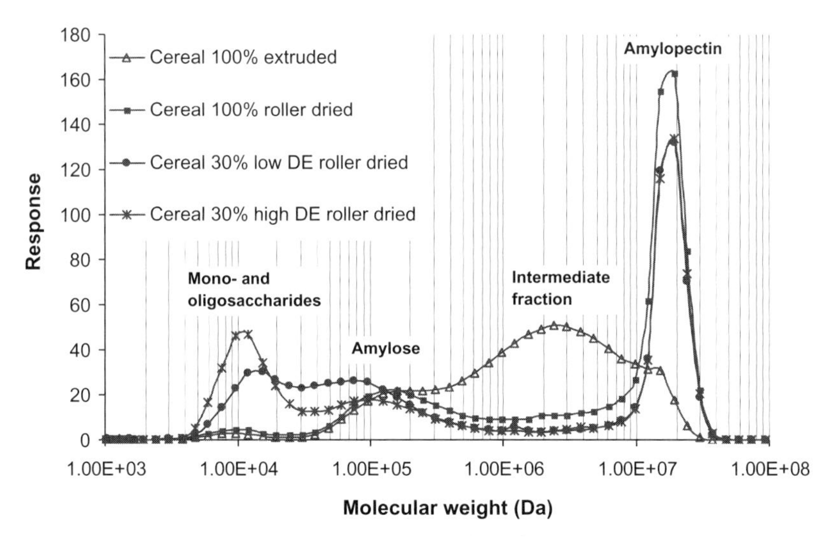

Figure 9.2 Carbohydrate profile of extruded and roller-dried cereals. Response expressed as arbitrary units.

the amylopectin is degraded and we find a huge peak with molecular weights between amylose and amylopectin. In the roller-dried cereal with a low dextrose equivalent (DE), produced only with alpha-amylase, a considerable part of the amylopectin is hydrolyzed, but very little oligosaccharides are formed. Finally, in the product where 30% of the cereal is hydrolyzed with both alpha-amylase and amyloglucosidase, and mixed with 70% of non-hydrolyzed flour, we observe a large peak for the low-molecular-weight oligosaccharides.

Most dietary fibers are carbohydrates. They are relatively little affected by standard processing operations. In the extrusion process, some insoluble fibers may be transformed into soluble ones, the amount of total dietary fiber remaining similar. In any process that implies enzymatic hydrolysis, there is a chance that fibers are also partly broken down. Commercial amylase preparations may contain side activities of cellulases and glucanases, or endogeneous hydrolases may become active (23). In processes such as germination and malting, both starch and fibers will be hydrolyzed (24).

Some leguminous seeds contain substantial amounts of flatulent sugars, mainly raffinose and stachyose. Although they do not give rise to any health concern, digestive comfort is reduced. Classically, the preparation of such food includes a soaking step where most of these sugars are washed out.

9.2.1.3 Fat

9.2.1.3.1 Nutritional Relevance Fats and oils are part of a healthy diet. However, the type of fat, and also the amount that is ingested are important. Both can negatively affect cardiovascular health by elevating blood lipid levels. Fat and oils are not only the most energy-dense nutrient group providing

9 kcal/g, they also provide two essential fatty acids (linoleic acid of the *n*-6 or ω6 and α-linolenic acid of the *n*-3 or ω3 family) and serve as a carrier of the four fat soluble vitamins (A, D, E, and K) and carotenoids. Health authorities worldwide recommend that the consumption of saturated fatty acids is restricted, and to favor a diet well balanced in *n*-3 and *n*-6 oils. The main responsibility of the food producer, therefore, lies in the choice of the lipid type and quantity. The main areas for the development of nutritionally improved products are:

- Reducing the total amount of fat: 20% to maximum 35% of daily total energy for adults (25, 26).
- Reducing or eliminating *trans* fatty acids (TFAs) as much as possible. In most cases, this means that partially hydrogenated fats have to be eliminated.
- Reducing the amount of saturated fatty acids: maximum 10% of daily total energy (25, 26).
- Adding oils with recognized health benefits.

In all of these cases, this may require a substantial reformulation of the product. It is a real challenge to maintain the original physical and sensory properties of the product, and in particular maintain the shelf-life performance.

9.2.1.3.2 Manufacture of Fats and Oils Fats and oils are derived from both animal and plant materials. The latter can today be considered the principal source for fats and oils. In 2005/2006, the global production of vegetable oils was reported at 118.2 mio. metric tons (27). Most crude fats and oils are not directly suitable for human consumption, as they contain many impurities that may affect the organoleptic properties, the shelf life, and product safety. Typical impurities include residual amounts of extraction solvents, free fatty acids, phospholipids, environmental contaminants such as pesticides, dioxins, and PCB-like dioxins, heavy metals, mycotoxins, and polycyclic aromatic hydrocarbons (PAHs). These impurities are reduced to an acceptable level for human consumption through refining. Defined as a multistep purification process, refining renders a crude oil edible, and is comprised of the following four steps:

1. *Degumming*: mainly for the removal of phospholipids and oxidation-enhancing metals.
2. *Neutralization*: for the removal of free fatty acids by reaction with caustic soda, resulting in the formation of oil insoluble soaps and the reduction of colored substances and certain pesticides. Chemically refined oils employ a chemical neutralization, whereas physically refined oils are neutralized during deodorization (steam stripping).
3. *Bleaching*: for the removal of pigments, heavy metals and PAHs, as well as the destruction of (hydro)peroxides.

4. *Deodorization*: for the removal of free fatty acids in case of a physical refining process, off-taste and undesired odorous components, and pesticides. This final refining step promotes *trans*-isomerization of polyunsaturated fatty acids (PUFAs), in particular if oils high in PUFAs are deodorized above 230 °C (28). Fully refined oils with less than 1% TFAs are now commonly available. However, the formation of TFAs decreases the nutritional quality of the oil.

Cold-pressed, extra virgin, and virgin oils, e.g., olive, pumpkin seed, nut oils, and cocoa butter, can be consumed without further purification. It should be noted that this is only possible if the oil has been obtained from specially selected sources, ensuring product safety with respect to the aforementioned environmental pollutants, and under mild extraction conditions without using solvents.

9.2.1.3.3 Storage, Handling, and Incorporation of Fats and Oils into Food Products After refining, the oil is relatively stable and bland in flavor. Precautions must to be taken during storage, handling, transport, and food production to avoid damage to the oil, which may result in the loss of quality, or even render the oil unfit for human consumption. The causes of damage can be several, for example, due to (i) oxidative degradation (formation of rancid compounds), (ii) contamination with substances from the environment, (iii) unclean transport equipment/containers, and (iv) the deliberate adulteration with other oils.

Lipid oxidation leads to chemical changes and consequently sensorial modification, with the ultimate result of food spoilage. Together with microbial spoilage and browning reactions, lipid oxidation is one of the three most important causes of food spoilage. The intrinsic nature of fats and oils that causes lipid oxidation cannot be changed, but degradation can be significantly slowed down by applying the following measures during handling, storing, transport, and incorporation into food products:

- Minimize contact with and absorption of air (oxygen).
- Avoid excessive moisture (risk of hydrolysis, free fatty acids are more prone to oxidation, than corresponding triglycerides).
- Avoid contact with, or at least minimize the content of pro-oxidants (mainly the oxidation catalysts iron and copper).
- Keep fats and oils in the dark (UV light may trigger photo-oxidation).
- Keep storage time and temperature to a minimum.
- Avoid excessive exposure to elevated temperatures.
- Avoid unnecessary agitation.
- Select an oil suitable for the application.
- Add antioxidant(s).

To provide the best possible protection against oxidative degradation, the antioxidant must be added to the oil at the end of the refining process by the oil manufacturer. The type of antioxidant or mixture of antioxidants to be used depends on the oil type, the application, and also the legislation of the country where the product will be sold. Further information on oil storage, handling, and transport is given by List *et al.* (29).

9.2.1.3.4 Lipid Oxidation Oxidative damage of lipids is caused by one of the most complex reaction chains occurring in food products. It occurs in the presence of catalysts, such as heat, light, metals, enzymes, metalloproteins, and microorganisms. Besides the development of off-flavors, which is the main reason for food spoilage, lipid oxidation can also cause the loss of other essential nutrients, such as amino acids, fat-soluble vitamins, and other bioactives.

The following deterioration reactions occur in lipids: (i) autoxidation, (ii) photo-oxidation, (iii) thermal oxidation, (iv) enzymatic oxidation, and (v) polymerization. Autoxidation, an autocatalytic reaction, is the most common oxidative deterioration. Autoxidation is initiated and propagated by free radical reactions and occurs via a series of chain reactions (see References 30 and 31 for details). If oxidation occurs, only a small amount of the lipid phase is typically affected. Lipid oxidation is classically described as a three-step chain reaction:

1. *Initiation*: the formation of free radicals.
2. *Propagation*: the reaction of free radicals with reactive oxygen species (singlet oxygen), and the formation of further free radicals, which will also react with oxygen. Hydroperoxides have been identified as primary products of autoxidation.
3. *Termination*: the formation of non-radical products, and the decomposition of primary into high sensory impact secondary oxidation products, such as aldehydes, ketones, alcohols, volatile organic acids, and epoxy compounds. This decomposition phase involves a large number of interrelated reactions of intermediates.

Due to the intrinsic nature of oxidative deterioration reactions, lipid oxidation products initially develop slowly. The reactions then accelerate during storage, rendering the food product inedible due to the presence of rancidity (30).

Primary oxidation products in oils and in complex food matrices are difficult to assess as they cannot be detected by sensory analysis. They can only be measured by chemical analytical techniques. The peroxide value (iodometric titration) is commonly accepted as a method to assess the initial state of oxidation in bulk oils. Many different analytical methods have been proposed for measuring the advanced state of oxidation in oils, and even more in food products (see References 30 and 31 for details). Secondary oxidation products are detectable by smell and taste at very low concentrations, often at part-per-billion levels (30). Hence, their impact on product sensory and final product acceptance is extremely high. Sensory analysis is still the most powerful and rapid method to determine rancidity in foods. Analytical values for secondary oxidation products are helpful in understanding the oxidation process. However, they do not provide direct information on the sensory acceptance of a product, a key product release criterion in commercial manufacture. To use analytical oxidation values for quality control in a meaningful way, a correlation between the analytical value and the corresponding sensory profile of the food product must first be established. This can be a very laborious and difficult task as it requires extensive sensory work, involving a trained sensory panel that assesses the sensory attributes of the food product in an objective way (product profiling).

With the exception of frying and deep-fat frying processes, which can cause severe oil damage, the influence of food processing on the overall oxidative status of fats and oils is generally very limited. This is a prerequisite for products with a good shelf-life performance. The fresh food product ready to be sold must be devoid of any sign of oxidation. Oxidative damage, which is already detectable in a freshly prepared food product, can often be traced back to an inadequate manufacturing process degrading the oil. The recipe itself can also negatively affect oil quality as certain ingredients, e.g., iron or copper salts, can catalyze oxidation reactions. To ensure high-quality food products, it is strongly recommended to taste oils prior to use in production, and to also determine the peroxide value. Food oxidation can be effectively reduced by choosing packaging that adequately protects the product against light. Packing the food product under an inert gas (e.g., nitrogen) is a possibility. It requires a gastight packaging material such as tins, glass jars, aluminum pouches, or plastics with an aluminum barrier.

9.2.1.3.5 Deep-Fat Frying and Oil Deterioration Deep-fat frying has become one of the most popular food preparation techniques. Vegetable oils fulfill two functions in this process. First, the oils are used as a heat-exchange medium to cook the food product and thus render the food edible. Second, the oils contribute to the quality of the food product (flavor, texture, overall sensorial properties, and nutritional value) as the oil is partly absorbed (up to 40%). Depending on the food product, frying temperatures range from 135 to 190 °C. Fats and oils undergo many chemical reactions during frying. These include hydrolysis, thermal oxidation, polymerization, steam distillation, flavor changes, and darkening. Some of these changes are desirable, whereas others are detrimental to the quality of the oil and the fried food. Deep-fat frying is one of the most complex, and least understood processes of the food industry.

Water released from the food product into the hot oil is transformed into steam, causing oil hydrolysis. Free fatty acids are more prone to oxidation than triglycerides, and decrease the smoke point of the oil. The steam also has positive effects on frying oil quality, as it blankets the oil. This limits oxygen availability from the air, which reduces the oxidation rate. Steam also acts as a stripping medium for volatiles, which in turn delays deterioration of the frying oil.

Some of the lipids from the food product are released into the frying medium. Food particles can lead to the darkening of the frying oil. The composition and stability of the frying medium are thus affected. Protcin and carbohydrates react with the frying oil forming desirable, but also undesirable flavors. Trace metals released from the food product, or even from the fryer when inadequate equipment is being used, also accelerate the degradation of the oil.

Fats and oils are degraded when exposed to high temperatures, and all oxidation reactions are accelerated. Heating alone of the frying shortening without any food preparation, will cause the oils to deteriorate over time. Eventually, the oil becomes unacceptable for human consumption.

Dimers and oligomers are the most predominant compounds in the group of nonvolatile molecules formed during frying. These molecules are mainly formed in phase three of the autoxidation reaction when nonpolar (C–C bonds) and polar compounds (C–O–C ether bonds) are formed. The formation of these high-molecular-weight compounds increases the viscosity of the oil, which leads to a higher oil uptake by the food.

The turnover rate is very important for the control of the frying process. It refers to the amount of oil added to the fryer to make up for the amount removed by the food during frying. A high turnover of the oil is important to minimize thermal deterioration of the oil, and to maintain the quality of the fried food products.

The decision as to when to discard a frying oil is one of the most difficult quality control problems of frying operations. The sensory properties of the food being fried should be the principle quality index for deep-fat frying (32). Different countries have established various guidelines for the evaluation of frying media. It is now generally accepted that frying media with a total polar material content of 24–28%, and a polymeric materials content of 12–15% are thermally abused, and must be discarded. A maximum content of 2% free fatty acids has also been suggested. Many different rapid tests have been developed to control oil quality during frying. The test suitability depends on the frying operation used (equipment, frying medium, frying process, food product). Test results must be correlated with the results of typical quality control methods and the sensorial properties of the fried foods.

Frying operations require high-stability fats and oils. This requires an α-linolenic acid content of preferably less than 2%, but which should not exceed 3%. Partially hydrogenated seed oils match these requirements very well. Until recently, seed oils were most commonly used as frying media. Due to a high to very high TFA content (up to 50%), seed oils have been replaced with unhardened vegetables oils, such as palm olein or high oleic sunflower oil, in many frying operations. High oleic oils do not always provide the desired fried flavor, as the linoleic acid content is too low. However, an increase of the linoleic acid content would cause faster oil degradation. As the oxidative stability of these high oleic oils is reduced compared with the hardened fats, the desired shelf life can no longer be reached. Gupta (33) and Frankel (34) discuss the various aspects of deep-fat frying in detail.

9.2.1.4 Vitamins and Minerals Food processing can also lower the nutritional value of some foods. In general, fresh food that has not been processed other than by washing and simple kitchen preparation, may be expected to contain a higher proportion of naturally occurring vitamins and minerals than the equivalent product processed in an industrial environment. For example, heat treatment may lead to a decrease in essential nutrients and consequently reduces the nutritional value of certain foods. In this context, some water-soluble vitamins are heat sensitive, e.g., vitamins C, B1, B2, B6, and folic acid, while lipid-soluble vitamins are not (35). The Maillard reaction itself may also lead to the loss of vitamins and proteins due to the transformations involved in the Maillard reaction.

However, a first and often very important loss of vitamins and minerals already occurs prior to any heat treatment, when the raw materials are physically prepared. This may occur by the practice of peeling fruits or vegetables, but the most striking example is probably the cereal milling process. Vitamins and minerals are found in high proportions in the germ and bran (and aleuron layer in the wheat grain), and only in smaller amounts in the endosperm. As the bran and germ are discarded in white flours, their respective vitamin and mineral contents are consequently much lower (Tables 9.2 and 9.3). The current dietary recommendations favoring whole grain products are thus an excellent means to increase exposure of consumers to trace elements.

Vitamins react very differently to factors linked to processing and storage, as shown in Fig. 9.3. Low and neutral pH have little effect, whereas alkaline treatments can be detrimental. Processes implying strong hydrothermal treatments have variable effects and are difficult to predict. In extruded rice cereals, vitamins A, C, and B1 were severely damaged, as less than 30% were retained

TABLE 9.2 Vitamin content of wholemeal and white wheat flours (average values).[a]

Nutrient	Wholemeal flour	White flour
Vitamin E (mg)	1.40	0.30
Thiamin (mg)	0.46	0.10
Riboflavin (mg)	0.27	0.00
Niacin (mg)	5.70	1.70
Vitamin B6 (mg)	0.50	0.15
Folate (µg)	57.00	22.00
Pantothenic acid (mg)	0.80	0.30
Biotin (µg)	7.00	1.00

[a]Nestlé internal data.

TABLE 9.3 Mineral content of wholemeal and white wheat flours (average values).[a]

Nutrient	Wholemeal flour	White flour
Sodium (mg)	3	3
Potassium (mg)	340	150
Calcium (mg)	38	15
Magnesium (mg)	120	20
Phosphorous (mg)	320	110
Iron (mg)	3.9	1.5
Copper (mg)	0.45	0.15
Zinc (mg)	2.9	0.6
Manganese (mg)	3.1	0.6
Selenium (μg)	53	4

[a]Nestlé internal data.

VITAMIN	ACID (LOW pH)	ALKALI (HIGH pH)	HEAT (>70°C)	LIGHT	OXYGEN
C	☺	☹	☹	☹	☹
B6	☺	☺	😐	☹	☺
FOLIC ACID	☺	☺	☹	😐	☺
A	☺	☺	☺	☹	☹
D	☺	😐	☺	😐	☹
E	☺	☺	☺	😐	☹

☺ **No effect** 😐 **sensitive** ☹ **Very sensitive**

Figure 9.3 Impact of different physical and chemical parameters on selected vitamins.

in the best case, whereas vitamins D and B5 were retained to 65–75% on an average (Fig. 9.4, minimum and maximum retention values). In roller-dried products, vitamin C was the most vulnerable with a retention of 75% (Fig. 9.4, RD). All other vitamins were retained at more than 80%.

There are a number of strategies to obtain processed products with an adequate content of vitamins, and some are exemplified as follows:

- Selection of raw materials with high initial vitamin content (e.g., wholemeal flours).

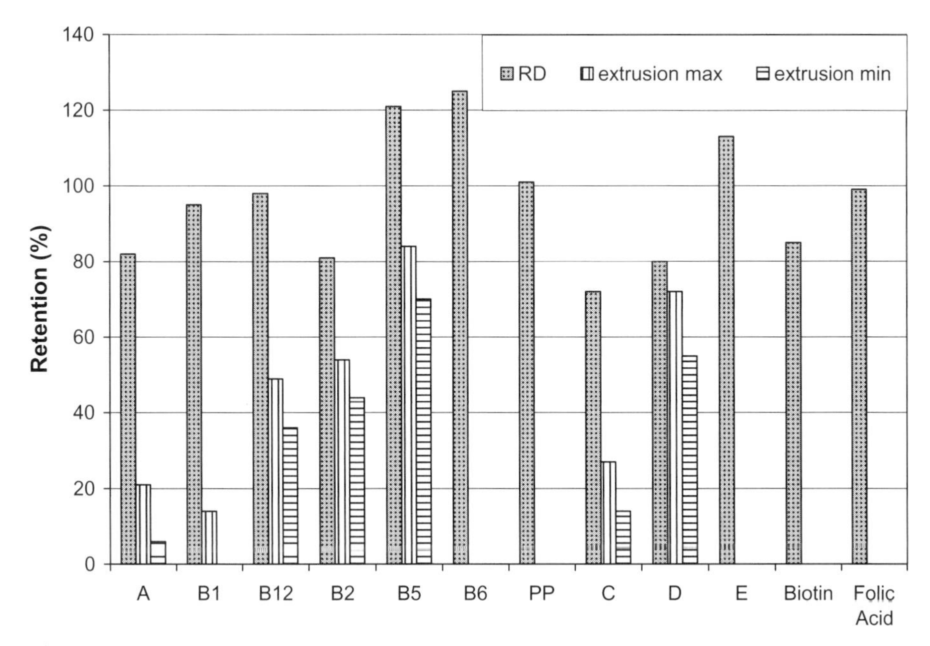

Figure 9.4 Vitamin retention during roller drying (RD) and extrusion of cereal flours.

- Preference to processes that have little impact on vitamin stability (e.g., freeze-drying or spray drying).
- Optimization of process parameters as feasible within given limits.
- For vitamin-enriched products, consider the method of vitamin addition as close as possible to the last manufacturing step, for example:
  —dry addition of a vitamin premix for dry powders such as milk or infant cereals
  —addition to the coating mix in products that are coated (sugar fruit or chocolate).

The amount of minerals in foods is not much affected by processing, except when this includes discarding certain constituents, as mentioned earlier for the case of flour milling. Unit operations such as cooking, drying, extrusion, and so on have little effect on the bioavailability of minerals. The latter is, however, influenced by the overall chemical composition of the food. The most important example is the bioavailability of iron. Iron absorption is substantially enhanced in the presence of vitamin C. A molar ratio of 2:1 for vitamin C:iron has been established as optimal, corresponding to a weight ratio of about 6.3:1 (36). On the other hand, iron absorption is strongly reduced by phytic acid, which is present in many cereals, vegetables, and pulses. A process has been developed to degrade phytic acid enzymatically (37), thus increasing iron utilization by the body. Further details on the effect of processing and home cooking on iron bioavailability are described by Hurrell *et al.* (38). In summary, the iron bioavailability can be improved in several ways such as:

- Reduction of the inhibiting effect of phytic acid.
- Addition of vitamin C (in pure form or as an ingredient) in appropriate proportions.
- Choosing an iron source that has a good bioavailability, e.g., ferrous fumarate in infant cereals (39).

9.2.2 Phytochemicals

Phytochemicals are secondary plant metabolites that are present usually in small and varying quantities and that have specific functions in the plants' natural development, growth, and resistance to pathogens and predators.

A plethora of different chemicals and structures have been defined in the past century in food plants, and it would be out of the scope of this chapter to elaborate in any great detail on these. In particular the "dietary phytochemicals" have attracted much attention as to their chemoprotective properties, with mounting evidence indicating protection against several chronic disease states such as diabetes type 2, certain cancers, and CVD. In this section, phytochemicals originating from unprocessed and processed cocoa, tea, coffee, and whole grain will be discussed.

9.2.2.1 Cocoa Cocoa and cocoa-related products have in recent years attained widespread recognition as an important source of dietary polyphenols. Cocoa is manufactured from cocoa beans, which are in essence the seeds of the cocoa tree *Theobroma cacao*. Whereas coffee and tea are extracts of plant parts, cocoa is eaten as a whole. Cocoa is in fact the nonfat component of cocoa liquor (the finely ground cocoa beans), which is used in chocolate making, or as cocoa powder, commonly with 12% fat and used in beverages such as chocolate drinks and in the confectionery and baking industries. The chemical constituents of cocoa have been subject of intense study in the past two decades, and focus has been on the polyphenolic fraction that comprises about 6.2% of the cocoa nib (the dried and roasted kernel of the cocoa bean that remains after the cocoa bean husk has been removed) (40). The most important subclass of particular interest to the scientific community is the flavanols (also termed flavan-3-ols or catechins). The main catechins in cocoa are (−)-epicatechin and (+)-catechin, as well as catechin dimers or oligomers/polymers made of up to 10 units and commonly referred to as procyanidins (Fig. 9.5) (41).

9.2.2.1.1 Chemical Changes during Processing The production of the different cocoa fractions encompasses several steps. The chocolate and cocoa industry use a large assortment of different equipment, techniques, and condi-

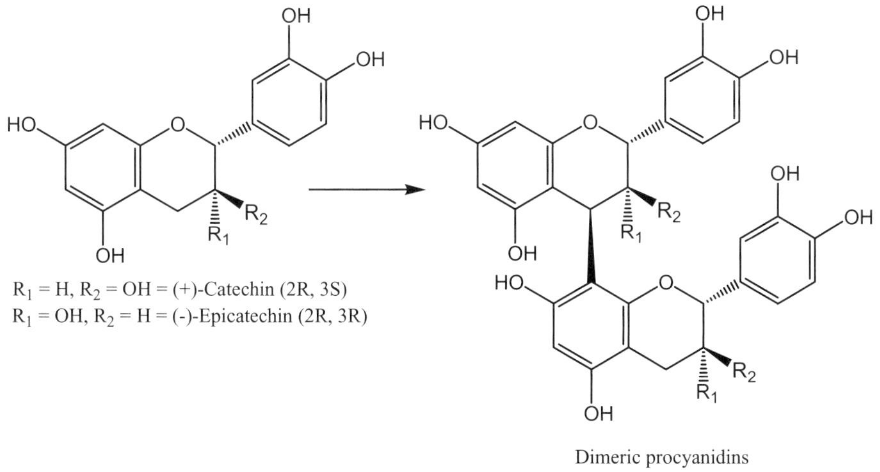

R_1 = H, R_2 = OH = (+)-Catechin (2R, 3S)
R_1 = OH, R_2 = H = (-)-Epicatechin (2R, 3R)

Dimeric procyanidins

Figure 9.5 Formation of procyanidins during cocoa fermentation and processing.

tions of processing, and it is out of the scope of this chapter to elaborate on any of the unit operations. Typically, it begins with the slicing of the cocoa pods (the fruit), and withdrawal of the pulp and beans, the latter subsequently fermented, which can take 3–7 days. After fermentation, the seeds are usually sun-dried for about 6 days, down to a moisture content of below 8%. The beans during this stage turn darker color due to the oxidation of phenolic compounds. In brief, the next steps of cocoa processing are more complex and comprise a roasting step typically at temperatures ranging from 110 to 140 °C for 20 to 40 min (42). This is the most important step for flavor development (43). The roasting time depends on the further use of the cocoa, and cocoa powder is usually made from darker roasted beans, whereas cocoa butter is made from slightly roasted beans (40). After bean roasting, the nibs are milled and the resulting cocoa mass (also termed "cocoa liquor") pressed to separate the butter fraction. The residual cake is milled resulting in cocoa powder that can also be treated with alkali (potassium or sodium carbonate) to increase color intensity and make the product less acidic. The alkalization treatment is also termed "Dutching," since it was developed by Van Houten toward the end of the nineteenth century in Holland.

All these processes have a marked effect on the polyphenolic constituents of cocoa. During fermentation, the contents of catechins are significantly lowered, especially in the early stage of the process (44). Overall, approximately 90% of the (–)-epicatechins may be converted to more complex polyphenols during fermentation, which lowers the typical astringency of the unfermented beans (40). Roasting impacts the contents of constituents such as hydroxy acids, polyphenols (e.g., epimerization of (–)-epicatechin), and of course consumes free amino acids and reducing sugars in condensation reactions, the severity depending on the conditions of roast (45, 46). Alkalization favors conversion of polyphenols to the corresponding quinones, as well as promoting epimerization reactions (47) and expectedly leads to a further reduction of (−)-epicatechin, by >20%.

For the majority of steps from fermented bean to finished product, there is very little change in total polyphenol content. In fact, selecting the cocoa source is the most important key factor in making a high-polyphenol chocolate. Further, the polyphenol content of cocoa solids can be maximized during processing, and ways to reduce (−)-epicatechin loss include (i) being selective about bean origin (epicatechin content can vary by ~20% based on varietal/clone), (ii) reduction of bean fermentation time and severity, (iii) gentle drying of the beans, and (iv) gentle nib roasting.

9.2.2.1.2 Cocoa Polyphenols and Health Epidemiological evidence supports the concept that diets rich in fruits, berries, and vegetables can promote health and attenuate, or delay, the onset of various diseases. The first *in vitro* studies on potentially beneficial antioxidant properties of cocoa polyphenol fractions were published more than a decade ago, demonstrating better activity of cocoa than red wine polyphenols using the suppression of low density lipoprotein (LDL) oxidation as the end point (48). Since then, several similar studies have appeared that compare the antioxidant capacity of different polyphenol-rich beverages on a common comparative scale (49, 50). Soon after the early *in vitro* study, Richelle and coworkers conducted the first bioavailability trial administering dark chocolate to human subjects and subsequently measuring epicatechin and its main metabolites in the plasma (51). After the intake of 80 g of chocolate, plasma epicatechin rose to 203 ng (0.7 μM), but clearance was very fast (2.8 h). The maximum concentration measured in the plasma was proportional to the dose of chocolate ingested and was not affected by the food matrix. The relative bioavailability of epicatechin from cocoa is low (around 2%) (49), and another factor that may impact bioavailability is the chirality of the polyphenols. As mentioned earlier, cocoa is rich in (+)-catechin, which is better absorbed than the (−)-form that apparently predominates in chocolate (52).

To date, more than 25 human intervention studies with cocoa have been published, measuring in most cases more than one end point per study and affording at least one positive outcome (53). The end points measured include blood pressure, platelet function, endothelial function, plasma antioxidant status, insulin sensitivity and resistance, and plasma lipids (levels and oxidation). In fact, most of the research are focused on the vascular system and indicate an association with reduced risk of CVD, by possibly acting through one or more of the mechanisms listed.

Polyphenolic procyanidins have also attracted much attention in terms of potential beneficial health effects (50, 54, 55). Most procyanidins are stable to the stomach environment and not degraded (56), and consequently reach the small intestine intact where they are available for absorption or further metabolism, viz. degraded to phenolic acid metabolites (57). The degradation products of procyanidins have not yet been fully characterized, and more work on the possible biological effects of the metabolites is warranted. Research so far indicates that the concentration of epicatechin in cocoa is the major factor to explain the association of cocoa and health (53).

9.2.2.2 Tea Tea (*Camellia sinensis*) is one of the most popular beverages consumed worldwide. Green and black teas can be considered the major categories, with the latter representing >75% of the world production and popularity dominating among North American and European consumers (58). However, production today is also diversifying into speciality teas such as white tea, fruit, herbal and scented teas, and other specific tea blends.

9.2.2.2.1 Chemical Changes during Processing The fundamental steps of tea processing are relatively straightforward (Fig. 9.6). Green tea is prepared by steaming (at about 100 °C) the tea leaves or pan-fired (as high as 230 °C) to produce Japanese green tea or Chinese green tea, respectively. These thermal treatments inactivate enzymes (polyphenol oxidases), avoiding further biochemical reactions (condensation/polymerization) in the tea leaves, with concomitant change in the chemical pattern. After heat treatment, the tea leaves are subjected to subsequent rolling and drying processes to afford a dry product. White tea is produced from the buds of young leaves, steamed and dried.

The composition of the tea leaf is characterized by a high amount of polyphenols (mainly flavonoids), with an average content of 28% flavonoids on a dry weight basis (58). Importantly, the tea classes differ in their relative flavonoid composition, and these include catechins, theaflavins, thearubigins, and flavonol glycosides.

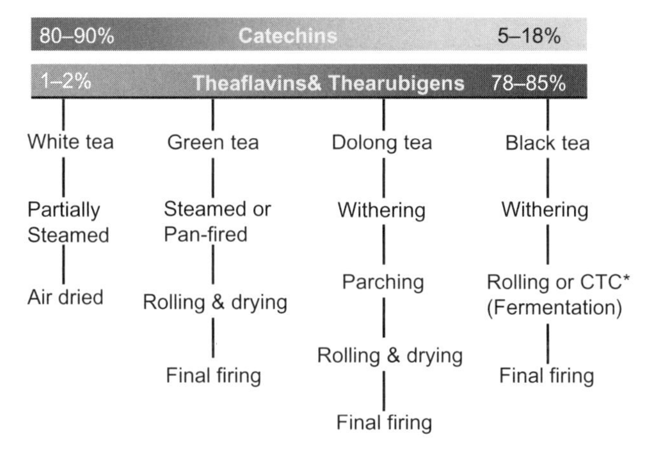

Figure 9.6 Overview of traditional tea manufacture and its impact on polyphenol content (*CTC = crushing/tearing/curling). See color insert.

The tea leaf contains a polyphenol oxidase enzyme that catalyzes the aerobic oxidation of catechins upon disruption of the leaf cell structure. This is a key step in the manufacture of black tea and is in fact a drying period after rolling or crushing/tearing/curling (CTC) to initiate the enzymic reaction, and is clearly distinct from the yeast-mediated alcoholic fermentation (59). The different quinones produced by enzymatic oxidation undergo condensation reactions resulting in a series of compounds, including bisflavanols, theaflavins, and thearubigens, which contribute to the major organoleptic properties of black tea. The steps of black tea manufacture are designed to achieve oxidation of the tea catechins and produce a product of optimal quality flavor and color (60). Thus, the pattern of polyphenols in tea depends on the degree of oxidation (61). Oolong tea, on the other hand, is less oxidized, and prepared by firing the leaves after rolling to terminate the oxidative processes, thereby resembling more closely green tea (Fig. 9.6).

Catechins are the major flavonoids in fresh unprocessed green tea leaves, the principal compounds being (–)-epigallocatechin (EGC), (–)-epicatechin gallate (ECG), and (–)-epigallocatechin gallate (EGCG). These are subjected to oxidative process during tea manufacture, procuring the corresponding quinones by action of polyphenol oxidase. These reactive compounds can readily dimerize or polymerize via radical-driven phenol-coupling reactions to form polyphenolic compounds as typically found in oolong and black teas. The major condensation compounds are theaflavins and the polymeric thearubigens, the former giving tea the characteristic bright orange-red color of black tea. In fact, theaflavins account for 1–2% dry weight of the water-extractable fraction, and in percentage of total phenolics represents >50% (59) (Fig. 9.7). Thearubigens, which can reach concentrations of 10–20% on a dry matter basis, represent the major oxidation products of catechins during fermentation, increasing with fermentation time with a concomitant loss of catechins such

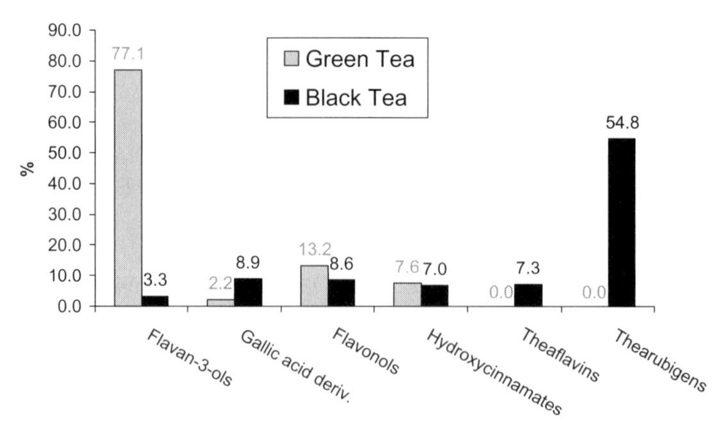

Figure 9.7 Composition (%) of the different classes of phenolics in black and green tea.

as EGC and EGCG (62). Total polyphenols in green and black tea do not differ significantly (16–17 g/100 g), as there is no net loss of chemicals but rather a conversion to more complex polyphenolic structures (61). However, the ranges given in the literature are only indicative, as actually measured amounts of polyphenols in commercial teas can vary widely and certain compounds in black tea such as the theaflavins may oxidize during storage (63).

Flavor is the most important factor that determines the quality of tea, and involves both taste and aroma. The balance of astringency, bitterness, and brothy taste is important to the characteristic taste of tea; catechins and caffeine are counted as the major contributors to astringency and bitterness. Notably, many variables may impact the final product composition of an aqueous tea infusion, such as the field conditions, treatments, and manufacturing conditions, which may differ from one factory to another (64). Hence, commercial teas show a high variation in polyphenol content. Furthermore, during a few minutes infusion, catechins from tea leaves are not fully extracted. The preparation method (e.g., brew time, water temperature, and degree of agitation) will affect the extraction efficacy and consequently the content of polyphenols in the cup, with averages ranging from 0.28 to 1.3 g/L (64). Thus, although efforts are made to optimize catechin content in the final product, consumer behavior will determine to a great extent the dietary intake of tea polyphenols.

9.2.2.2.2 Tea and Health Benefits Several reviews on the possible health benefits of flavonoids from tea have been published in the past decades (65–67), and similarly to the research on cocoa polyphenols, a main focus has been on the antioxidant capacity and chemoprotective properties of the key tea constituents. Especially EGCG, theaflavins, thearubigens, and flavonol glycosides are major components that contribute to the antioxidant properties of tea. However, comparable to studies on cocoa polyphenols, the bioavailability of tea catechins, measured as plasma concentration in rodent studies, is low (<5%) (68).

Animal studies indicate that green tea and its components in the diet provide some protection against degenerative diseases and act as antitumorigenic/chemoprotective agents (69–71). A review on the health effects of intake of green tea/extracts and catechins based on *in vivo* animal models at doses in the range 0.01–2.5% and measuring various biological end points is given by Crespy and Williamson (72). There is also evidence to suggest that tea may have beneficial effects on certain immune parameters (73), which has implications for improving gut health and resistance against infections (74). Green tea catechins exhibit hypolipidemic activity, attenuating diet-induced body fat accretion in mice (75). It is most likely that the hypolipidemic activity is mediated by inhibition of absorption of cholesterol and triglycerides (76). Tea could have a beneficial effect against lipid and glucose metabolism disorders implicated in type 2 diabetes (77), and could also reduce the risk of coronary disease (78).

Although the evidence on the purported beneficial effects of tea and tea polyphenols is promising, future *in vivo* studies are warranted at realistic dose levels, considering the very large variability in dietary exposures. Further, well-designed chronic studies measuring appropriate metabolic end points and biomarkers should be designed to demonstrate the effectiveness of the active compounds and their potential role in disease prevention.

9.2.2.3 Coffee There are two coffee species of commercial importance, namely *Coffea arabica* (Arabica) and *Coffea canephora* (Robusta). Arabica accounts for some 64% while Robusta accounts for about 35% of the world's coffee production. Regarding the chemical composition, both species are characterized by different contents of minerals, volatile substances, chlorogenic acids, and caffeine (79). The chlorogenic acids are a group of *trans*-cinnamic acids esterified to quinic acid (Fig. 9.8). This class of compounds accounts for 6% to 10% of the weight of green coffee (80). One of the most abundant chlorogenic acids is 5-caffeoyl quinic acid that represents 4% to 5% of the total green coffee weight. As depicted in Fig. 9.8, mono- and dicaffeoyl quinic acids have been identified in coffee with substitution at the 3-, 4-, and 5-position of quinic acid.

Green coffee is usually not consumed as such, although certain coffee products containing a portion of green coffee extract are gaining popularity. Robusta coffee contains relatively higher amounts of chlorogenic acids compared with Arabica beans. However, a large part of the chlorogenic acids is progressively decomposed and transformed during roasting—a process that takes place at temperatures typically in the range of 220–250 °C over a time period of 6 to 15 min, with losses of around 8–10% for every 1% loss in dry matter (80). The roasting process and blend of green beans affords the typical organoleptic pattern of the final product and encompasses the formation of literally hundreds of new compounds such as catechol, 4-ethyl catechol, vinyl catechol, phenol, and guiacol. The polyphenols formed during the roasting of coffee are reactive and enter into the formation of the brown polymeric materials termed melanoidins, which are characteristic of roasted coffee.

9.2.2.3.1 Coffee and Health Benefits Coffee is a food plant with a long history of safe use and, in contrast to most other traditional foods, has been the subject of extensive scientific research addressing its potential impact on human health (81, 82). Early research pertaining to the health impact of coffee was more focused on the potentially deleterious effects of the beverage and its constituents. These concerns were raised based upon the observation of mutagenic activity of coffee in bacterial and mammalian test systems using various end points (83, 84). Similar *in vitro* cytotoxic effects were observed in wine and tea (85), supporting the hypothesis that hydrogen peroxide and oxygen radicals are involved in cell toxicity (86). Further investigations provided insight into the mechanisms of formation of hydrogen peroxide in polyphenol-rich beverages (87). These studies also highlighted the problems

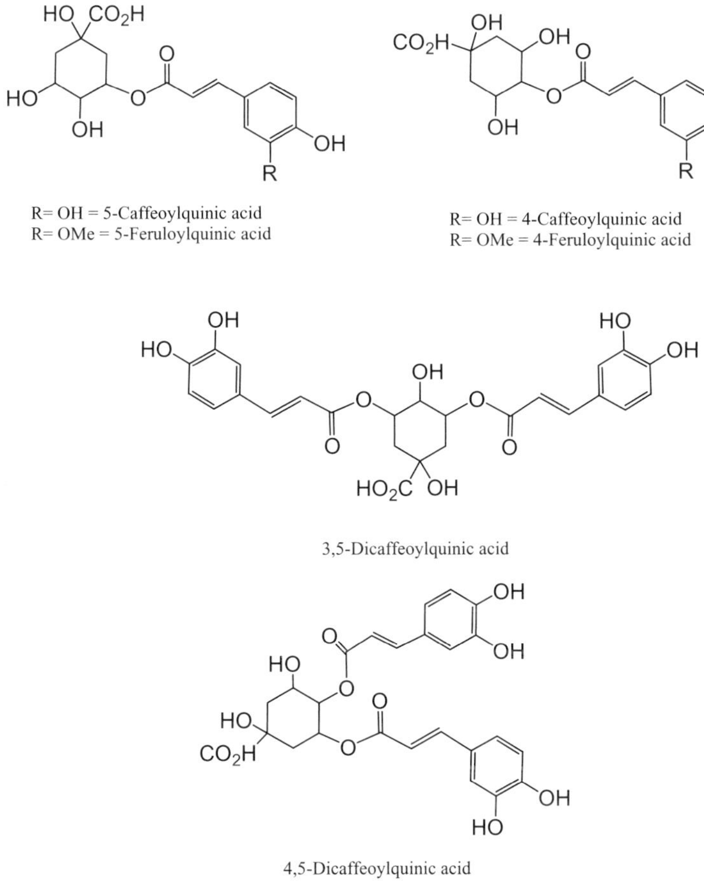

R= OH = 5-Caffeoylquinic acid
R= OMe = 5-Feruloylquinic acid

R= OH = 4-Caffeoylquinic acid
R= OMe = 4-Feruloylquinic acid

3,5-Dicaffeoylquinic acid

4,5-Dicaffeoylquinic acid

Figure 9.8 Chemical structures of chlorogenic acids commonly found in coffee.

of chemical and biological *in vitro* assays that may give conflicting results and cannot be easily extrapolated to assess the impact of coffee on human health and safety.

Coffee represents one of the richest dietary sources of polyphenols. The chlorogenic acid content of an average 200-mL cup of Robusta coffee has been reported in the range of 70–350 mg (80). In fact, the antioxidant capacity of coffee has been attributed to the natural polyphenols already present in the green coffee bean as well as the melanoidins generated during the roasting of coffee (88–91). It is well known that during the typical temperatures of roasting, the loss of antioxidant activity due to the thermal degradation of natural antioxidants—mainly polyphenols such as chlorogenic acids—can be compensated or even enhanced by the formation of active MRPs such as melanoidins (92), which have been proven to be important contributors to the radical scavenging properties of coffee (93).

In fact, the radical scavenging capacity of coffee has been reported to be dependent on the roasting degree of coffee. Different *in vitro* test systems have been employed, and some authors have observed a higher antioxidant activity for light- and medium-roasted coffees than for green or dark coffee, when using the $ABTS^{\bullet+}$ [2,2′-azinobis(3-ethylbenzothiazoline-6-sulfonic acid)] method (94). Also Borrelli and coworkers (91) observed that the free radical scavenging properties of melanoidins determined by $ABTS^{\bullet+}$ and $DMPD^{\bullet+}$ (1,1-dipheny-2-picrylhydrazyl) decreased as the intensity of roasting increased.

In an *in vivo* study performed with rats, it was observed that feeding rats with coffee brew resulted in an increase of the total antioxidant capacity of the plasma (95). A major contributor to the antioxidant activity was identified as *N*-methylpyridinium, a recently discovered alkaloid present in the roasted bean at concentrations of up to 0.25% on a dry weight basis (96). *N*-methylpyridinium is formed by the decarboxylation of trigonelline during roasting, and trigonelline is known to also furnish nicotinic acid (demethylation reaction), providing 1–3 mg per standard cup.

The importance of coffee in delivering a major part of the dietary polyphenols is illustrated in a survey of dietary contribution of antioxidants in Norway. In this study, coffee accounted for 64% of total intakes, followed by fruits, berries, tea, wine, cereals, and vegetables (97). Pellegrini *et al.* reported similar results from their study in Italy that compared total antioxidant capacity of plant foods, beverages, and oils (98). A study in Finland reports on the content of phenolic acids in the most common fruits and beverages and a wide range of berries (99). Coffee was the best source among the beverages with 97 mg/100 g, while tea contained less than half that amount (30–36 mg/100 g). A study on the relative antioxidant capacity of commonly consumed phenolic-rich beverages, estimated on a cup serving basis, revealed that soluble coffee was more potent than cocoa or tea (100).

Epidemiological studies strongly suggest that coffee consumption may help reduce the risk of several chronic diseases, including type 2 diabetes (101), Parkinson's disease (102), and liver disease (103). Coffee consumption has also been associated with reduced incidences of several types of cancer (104, 105). A recent cohort study concluded that the consumption of coffee may inhibit inflammation and thereby the risk of cardiovascular and other inflammatory diseases in postmenopausal women, attributed to the antioxidants present in the brew (106).

In summary, the evidence supporting a direct link between coffee intake and adverse health effects is very limited and inconsistent. Studies on the safety of individual chemicals administered to experimental animals at hugely excessive concentrations *vis-a-vis* human dietary intake levels will not provide a clear answer as to the possible risks for human health. In this respect, a more "holistic" approach in toxicity studies taking the beverage (or food) as whole may provide a more balanced picture.

9.2.2.4 Whole Grains Over the last decades, numerous epidemiological studies have demonstrated the health benefits of whole grains. Although many of the benefits may have a strong link to dietary fiber, it becomes evident that whole grain means more than just higher levels of fiber, vitamins, and minerals. Whole grains also contain substantial amounts of phytonutrients such as polyphenols, phytate, lignan, and so on According to Adom and Liu (107), the phytochemical contents in whole grains have been underestimated in the literature, as bound phytochemicals were not included. Corn had the highest total phenolic content, with ferulic acid being the major phenol compound. Corn also had the highest total antioxidant activity, followed by wheat and oats. About 60% to 90% of phytochemicals were bound and could thus potentially survive stomach and intestinal digestion to reach the colon. This may partly explain the mechanism of grain consumption in the prevention of colon and other digestive cancers.

Health claims in several countries link whole grain consumption and heart health. In a meta-analysis, it was shown that individuals with the highest intake of whole grains had a significantly reduced risk for CVD (108). The protective effect appears not simply to be due to cholesterol lowering or the fiber component. It is likely that whole grains are exerting their effects via other components such as the antioxidants or their effects on glucose/insulin homeostasis.

Strong epidemiological evidence also links consumption of whole grains to a reduced risk of type 2 diabetes (109, 110) and a reduced risk of cancer (111). Emerging evidence suggests that whole grain foods may be helpful in weight management (112).

Analytical methods to quantify the different phytonutrients are scarce and further development and validation is needed. However, it can be speculated that most processing will have a much lower effect on phytonutrient levels in cereals, compared with coffee, as the processing conditions applied are in general much milder.

9.3 CONCLUSIONS AND FUTURE RESEARCH NEEDS

The main objective of food processing is to obtain food that is safe, tasty, nutritional, appealing, affordable, and convenient to use and transport. In ancient times, food "processing" was exclusively done in a domestic environment, but nowadays the food manufacturing industry plays an important role in supplying the consumers with complete meals or prepared food items that will be integrated into the diet. Producing food with high nutritional quality means selecting the right ingredients, choosing adequate manufacturing processes and conditions, and taking all steps to assure that the nutritional quality is maintained over the entire shelf life.

From a nutritional point of view, processing enhances food digestibility, increases the bioavailability of many nutrients, and reduces or eliminates certain antinutritional factors. Processing even creates bioactive substances such as antioxidants. On the other hand, processing can also lead to reduced nutritional quality in a number of ways, for example, by (i) loss of nutritionally important components such as bran in cereals, (ii) reduced protein quality (Maillard reaction), (iii) loss of vitamins, or (iv) the formation of undesired substances such as *trans* fatty acids in the fat hydrogenation process. It is then up to the manufacturer to take appropriate measures to restore the food's nutritional quality.

While the major nutrients, vitamins, and minerals have been studied in quite some depth, the large segment of the "phytonutrients" or "phytochemicals" warrants further research in several areas. These include:

- Development and validation of analytical methods to reliably quantify the various classes of chemicals and their corresponding metabolites in both foods and biological materials (plasma, tissues).
- Further studies on the beneficial health effects, such as antioxidant capacity, and possibly identify other mechanisms of action.
- Assess the impact of food constituents, chemical/physical properties and food processing on the bioavailability of phytochemicals.
- Establish Recommended Dietary Allowance (RDA) values and determine tolerable upper intake levels (UL). Are there segments of the population (e.g., children, individuals with specific health problems) that require special attention, i.e., specific RDAs or ULs?

REFERENCES

1. Connor, J.M., Schiek, W.A. (1997). *Food Processing. An Industrial Powerhouse in Transition*, John Wiley & Sons, Inc., New York, pp. 8–22.
2. Pariza, M.W. (1997). CAST Issue Paper Number 8, November, Examination of dietary recommendations for salt-cured, smoked, and nitrite-preserved foods.
3. Wrangham, R.W., Jones, J.H., Laden, G., Pilbeam, D., Conklin-Brittain, N.L. (1999). Cooking and the ecology of human origins. *Current Anthropology*, *40*, 567–616.
4. Brace Loring, C. (1995). *The Stages of Human Evolution*, 5th edn, Prentice-Hall, Englewood Cliffs, NJ, p. 371.
5. Straus, L.G. (1989). On early hominid use of fire. *Current Anthropology*, *30*, 488–491.
6. Laden, G. (2007). Symposium Proceedings of the DFG: thermal processing of food: potential health benefits and risks, Wiley-VCH Verlag GmbH, Weinheim, pp. 208–223.
7. Maillard, L.C. (1912). Action des acides amine sur les sucres: formation des melanoidines par voie methodique. *Comptes Rendus*, *154*, 66–68.
8. Hodge, J.E. (1953). Browning reactions in model systems. *Journal of Agricultural and Food Chemistry*, *1*, 928–943.

9. Soldo, T., Blank, I., Hofmann, T. (2003). (+)-(S)-alapyridaine—a general taste enhancer? *Chemical Senses*, *28*, 371.
10. Hofmann, T., Bors, W., Stettmeier, K. (2002). A radical intermediate of melanoidin formation in roasted coffee, in *Free Radicals in Food—Chemistry, Nutrition and Health Effects* (eds M.J. Morello, F. Sahidi, C.T. Ho), ACS symposium Series *807*, American Chemical Society, Washington, DC, p. 49.
11. Marko, D. (2007). Symposium Proceedings of the DFG. Thermal processing of food: potential health benefits and risks, Wiley-VCH Verlag GmbH, Weinheim, pp. 66–74.
12. Yilmaz, Y., Toledo, R. (2005). Antioxidant activity of water-soluble Maillard reaction products. *Food Chemistry*, *93*, 273–278.
13. Lindenmeier, M., Faist, V., Hofmann, T. (2002). Structural and functional characterization of pronyl-lysine, a novel protein modification in bread crust melanoidins showing in vitro antioxidative and phase I/II enzyme modulating activity. *Journal of Agricultural and Food Chemistry*, *50*, 6997–7006.
14. Finot, P.A. (1997). Effect of processing and storage on the nutritional value of food proteins, in *Food Proteins and their Applications* (ed. S. Damodaran, A. Paraf), Marcel Dekker Inc., pp. 551–577.
15. Finot, P.A., Deutsch, R., Bujard, E. (1981). The extent of the Maillard reaction during the processing of milk. *Progress in Food & Nutrition Science*, *5*, 345–355.
16. Hurrell, R., Finot, P.A. (1983). Food processing and storage as a determinant of protein and amino acid availability; in nutritional adequacy, nutrient availability and needs, in *Experientia Supplementum 44*, Nestlé Research Symposium, Birkhäuser, pp. 135–156.
17. Chitra, U., Sing, U., Venkateswara Rao, P. (1996). Phytic acid, in vitro protein digestibility, dietary fiber, and minerals of pulses as influenced by processing methods. *Plant Foods for Human Nutrition*, *49*, 307–316.
18. Carvalho, M.R.B., Sgarbieri, V.C. (1997). Heat treatment and inactivation of trypsin-chymotrypsin inhibitors and lectins from beans. *Journal of Food Biochemistry*, *21*, 219–233.
19. Gallant, D.J., Bouchet, B., Buléon, A., Pérez, S. (1992). Physical characteristics of starch granules and susceptibility to enzymatic degradation. *European Journal of Clinical Nutrition*, *46* (Suppl. 2), 3–16.
20. Garcia-Alonso, A., Goni, I. (2000). Effect of processing on potato starch: in vitro availability and glycaemic index. *Nahrung*, *44*, 19–22.
21. Asp, N-G. (1997). Resistant starch—an update on its physiological effects. *Advances in Experimental Medicine and Biology*, *427*, 201–210.
22. Shi, Y.C., Cui, W.Y., Birket, A.M., Thatcher, M.G. (Inventors) (National Starch) (2003). European Patent EP1362869, Filing date: 05/14/2003. Resistant starch prepared by isoamylase debranching of low amylose starch.
23. Mühlenchemie GmbH & Co. KG (2008). Documentation on products for flour improvement, Ahrensburg, Germany.
24. Pyler, R.E., Thomas, D.A. (1991). Malted cereals: their production and use, in *Handbook of Cereal Science and Technology* (eds K. Kulp, J.G. Ponte), Marcel Dekker, New York, pp. 685–696.
25. U.S. Department of Health and Human Services (2005). Dietary Guidelines for Americans, pp. 40–45.

26. FSA Nutrient and Food Based Guidelines for UK Institutions (2006). http://www.food.gov.uk/multimedia/pdfs/nutguideuk.pdf (accessed 28 April 2008).
27. World Agricultural Supply and Demand Estimates (WASDE-455 Report) (2008). World Agricultural Outlook Board of the U.S. Department of Agriculture, February 8, 2008. http://future.aae.wisc.edu/outlook/wasde455.pdf (accessed 28 April 2008).
28. Bertoli, C., Bellini, A., Delvecchio, A., Durand, P., Gumy, D., Stancanelli, M. (1998). Formation of trans fatty acids during deodorization of low erucic acid rapeseed oil. In Proceedings of the World Conference on Oilseed and Edible Oil Processing, Vol. 2 (eds S.S. Koseoglu, K.C. Rhee, R.F. Wilson), AOCS Press, Champaign, pp. 67–71.
29. List, G.R., Wang, T., Shukla, V.K.S. (2005). Storage, handling, and transport of oils and fats, in *Bailey's Industrial Oil and Fat Products*, Vol. 5, 6th edn (ed. F. Shahidi), John Wiley & Sons, Inc., New York, pp. 191–221.
30. Frankel, E.N. (2005). Methods to determine extent of oxidation, in *Lipid Oxidation*, 2nd edn, The Oily Press, Bridgwater, England, pp. 99–103.
31. Schaich, K.M. (2005). Lipid oxidation: theoretical aspects, in *Bailey's Industrial Oil and Fat Products*, Vol. 1, 6th edn (ed. F. Shahidi), John Wiley & Sons, Inc., New York, 269–356.
32. 4th International Symposium on Deep-Fat Frying: 11–13 January 2004, Hagen/Westphalia, Germany. http://www.dgfett.de/material/recomm.htm (accessed 29 April 2008).
33. Gupta, M.K. (2005). Frying oils and frying of food and snack food production, in *Bailey's Industrial Oil and Fat Products*, Vol. 4, 6th edn (ed. F. Shahidi), John Wiley & Sons, Inc., New York, pp. 1–31; pp. 269–315.
34. Frankel, E.N. (2005). Frying fats, in *Lipid Oxidation*, 2nd edn, The Oily Press, Bridgwater, England, pp. 355–389.
35. Rechkemmer, G. (2007). Symposium Proceedings of the DFG: thermal processing of food: potential health benefits and risks, John Wiley & Sons, Ltd, Weinheim, pp. 50–65.
36. WHO/FAO, Allen, L., De Benoist, B., Dary, O., Hurrell, R. (2006). Guidelines on Food Fortification with Micronutrients. http://www.who.int/nutrition/publications/guide_food_fortification_micronutrients.pdf (accessed 28 April 2008).
37. Hurrell, R.F., Reddy, M.B., Juillerat, M.-A., Cook, J.D. (2003). Degradation of phytic acid in cereal porridges improved iron absorption in humans. *American Journal of Clinical Nutrition*, *77*, 1213–1219.
38. Hurrell, R.F., Reddy, M.B., Burri, J., Cook, J.D. (2002). Effect of industrial processing and home cooking on iron absorption from cereal-based foods. *British Journal of Nutrition*, *76*, 165–171.
39. Hurrell, R.F., Furniss, D.E., Burri, J., Whittaker, P., Lynch, S.R., Cook, J.D. (1989). Iron fortification of infant cereals: a proposal for the use of ferrous fumarate or ferrous succinate. *American Journal of Clinical Nutrition*, *49*, 1274–1282.
40. Schieberle, P. (2000). The chemistry and technology of cocoa, in *Caffeinated Beverages: Health Benefits, Physiological Effects, and Chemistry* (eds T.H. Parliment, C.T. Ho, P. Schieberle), ACS Symp. Ser., *754*, American Chemical Society, Washington, DC, pp. 262–275.

41. Wollgast, J., Anklam, E. (2000). Review on polyphenols in *Theobroma cacao*: changes in composition during the manufacture of chocolate and methodology for identification and quantification. *Food Research International*, *33*, 423–447.
42. Hashim, L. (2000). Flavour development of cocoa during roasting, in *Caffeinated Beverages: Health Benefits, Physiological Effects, and Chemistry* (eds T.H. Parliment, C.T. Ho, P. Schieberle), ACS Symp. Ser., *754*, American Chemical Society, Washington, DC, pp. 276–285.
43. Ramli, N., Hassan, O., Said, M., Samsudin, W., Idris, N.A. (2006). Influence of roasting conditions on volatile flavor of roasted Malaysian cocoa beans. *Journal of Food Processing and Preservation*, *30*, 280–298.
44. Kim, H., Keeney, P.G. (1984). Epicatechin content in fermented and unfermented cocoa beans. *Journal of Food Science*, *49*, 1090–1092.
45. Caligiani, A., Cirlini, M., Palla, G., Ravaglia, R., Arlorio, M. (2007). GC-MS detection of chiral markers in cocoa beans of different quality and geographic origin. *Chirality*, *19*, 329–334.
46. Nazaruddin, R., Seng, L.K., Hassan, O., Said, M. (2006). Effect of pulp preconditioning on the content of polyphenols in cocoa beans (*Theobroma cacao*) during fermentation. *Industrial Crops and Products*, *24*, 87–94.
47. Kofink, M., Papagiannopoulos, M., Galensa, R. (2007). Catechin in cocoa and chocolate: occurrence and analysis of an atypical flavan-3-ol enantiomer. *Molecules*, *12*, 1274–1288.
48. Waterhouse, A.L., Shirley, J.R., Donovan, J.L. (1996). Antioxidants in chocolate. *Lancet*, *348*, 834.
49. Richelle, M., Huynh-Ba, T., Tavazzi, I., Mooser, V., Enslen, M., Offord, E.A. (2000). 2000). Antioxidant capacity and epicatechin bioavailability of polyphenolic-rich beverages (cocoa and teas, in *Caffeinated Beverages: Health Benefits, Physiological Effects, and Chemistry, ACS Symp. Ser* (eds T.H. Parliment, C.T. Ho, P. Schieberle), ACS Symp. Ser., *754*, American Chemical Society, Washington, DC, pp. 102–110.
50. Bearden, M.M., Pearson, D.A., Rein, D., Chevaux, K.A., Carpenter, D.R., Keen, C.L., Schmitz, H. (2000). Potential cardiovascular health benefits of procyanidins present in chocolate and cocoa, in *Caffeinated Beverages: Health Benefits, Physiological Effects, and Chemistry* (eds T.H. Parliment, C.T. Ho, P. Schieberle), ACS Symp. Ser., *754*, American Chemical Society, Washington, DC, pp. 177–186.
51. Richelle, M., Tavazzi, I., Enslen, M., Offord, E.A. (1999). Plasma kinetics in man of epicatechin from black chocolate. *European Journal of Clinical Nutrition*, *53*, 22–26.
52. Donovan, J.L., Crespy, V., Oliveira, M., Cooper, K.A., Gibson, B.B., Williamson, G. (2006). (+)-Catechin is more bioavailable than (–)-catechin: relevance to the bioavailability of catechin from cocoa. *Free Radical Research*, *40*, 1029–1034.
53. Cooper, K.A., Donovan, J.L., Waterhouse, A.L., Williamson, G., (2008). Cocoa and health: a decade of research. *British Journal of Nutrition*, *99*, 1–11.
54. Miller, K.B., Stuart, D.A., Smith, N.L., Lee, C.Y., McHale, N.L., Flanagan, J.A., Ou, B., Hurst, W.J. (2006). Antioxidant activity and polyphenol and procyanidin contents of selected commercially available cocoa-containing and chocolate products in the United States. *Journal of Agricultural and Food Chemistry*, *54*, 4062–4068.

55. Murphy, K.J., Chronopoulos, A.K., Singh, I., Francis, M.A., Moriarty, H., Pike, M.J., Turner, A.H., Mann, N.J., Sinclair, A.J. (2003). Dietary flavanols and procyanidin oligomers from cocoa (*Theobroma cacao*) inhibit platelet function. *American Journal of Clinical Nutrition*, *77*, 1466–1473.

56. Scalbert, A., Rios, L.Y., Gonthier, M.-P., Manach, C., Morand, C., Remesy, C. (2002). The specificity of cocoa polyphenols recent advances in their bioavailability. *Polyphenols Actualites*, *22*, 14–18.

57. Rios, L.Y., Bennett, R.N., Lazarus, S.A., Remesy, C., Scalbert, A., Williamson, G. (2002). Cocoa procyanidins are stable during gastric transit in humans. *American Journal of Clinical Nutrition*, *76*, 1106–1110.

58. Ho, C.-T., Zhu, N. (2000). *The Chemistry of Tea* (ed. T.H. Parliament), ACS Symposium Series, *754* (Caffeinated Beverages; Health Benefits, Physiological Effects, and Chemistry), American Chemical Society, Washington, DC, pp. 316–326, April 2000.

59. Del Rio, D., Stewart, A.J., Mullen, W., Burns, J., Lean, M.E.J., Brighenti, F., Crozier, A. (2004). HPLC-MSn analysis of phenolic compounds and purine alkaloids in green and black tea. *Journal of Agricultural and Food Chemistry*, *52*, 2807–2815.

60. Temple, S.J., Temple, C.M., Van Boxtel, A.J.B., Clifford, M.N. (2001). The effect of drying on black tea quality. *Journal of the Science of Food and Agriculture*, *81*, 764–772.

61. Ho, C.-T. (1999). The chemistry of tea. Book of Abstracts, 217th ACS National Meeting, Anaheim, CA, 21–25 March.

62. Muthumani, T., Kumar, R.S.S. (2006). Influence of fermentation time on the development of compounds responsible for quality in black tea. *Food Chemistry*, *101*, 98–102.

63. Yao, L.H., Jiang, Y.M., Caffin, N., D'Arcy, B., Datta, N., Liu, X., Singanusong, R., Xu, Y. (2006). Phenolic compounds in tea from Australian supermarkets. *Food Chemistry*, *96*, 614–620.

64. Astill, C., Birch, M.R., Dacombe, C., Humphrey, P.G., Martin, P.T. (2001). Factors affecting the caffeine and polyphenol contents of black and green tea infusions. *Journal of Agricultural and Food Chemistry*, *49*, 5340–5347.

65. Wolfram, S. (2007). Effects of green tea and EGCG on cardiovascular and metabolic health. *Journal of the American College of Nutrition*, *26*, 373–388.

66. Zhao, B. (2006). The health effects of tea polyphenols and their antioxidant mechanism. *Journal of Clinical Biochemistry and Nutrition*, *38*, 59–68.

67. Weisburger, John H. (2006). Tea is a health-promoting beverage in lowering the risk of premature killing chronic diseases: a review, in *Protective Effects of Tea on Human Health* (eds N.K. Jain, M. Siddiqi, J. Weisburger), CAB International, Wallingford, pp. 1–5.

68. Catterall, F., King, L.J., Clifford, M.N., Ioannides, C. (2003). Bioavailability of dietary doses of 3H-labelled tea antioxidants (+)-catechin and (–)-epicatechin in rat. *Xenobiotica*, *33*, 743–753.

69. Hirose, M., Nishikawa, A., Shibutani, M., Imai, T., Shirai, T. (2002). Chemoprevention of heterocyclic amine-induced mammary carcinogenesis in rats. *Environmental and Molecular Mutagenesis*, *39*, 271–278.

70. Clark, J., You, M. (2006). Chemoprevention of lung cancer by tea. *Molecular Nutrition & Food Research*, *50*, 144–151.
71. Carter, O., Wang, R., Orner, G.A., Fischer, K.A., Lohr, C.V., Pereira, C.B., Bailey, G.S., Williams, D.E., Dashwood, R.H. (2007). Comparison of white tea, green tea, epigallocatechin-3-gallate, and caffeine as inhibitors of PhIP-induced colonic aberrant crypts. *Nutrition & Cancer*, *58*, 60–65.
72. Crespy, V., Williamson, G. (2004). A review of the health effects of green tea catechins in in vivo animal models. *Journal of Nutrition*, *134*, 3431S–3440S.
73. Singh, J., Qazi, G.N. (2006). Immunomodulatory activity of tea, in *Protective Effects of Tea on Human Health* (eds N.K. Jain, M. Siddiqi, J. Weisburger), CAB International, Wallingford, pp. 34–44.
74. Hamer, M. (2007). The beneficial effects of tea on immune function and inflammation: a review of evidence from in vitro, animal, and human research. *Nutrition Research*, *27*, 373–379.
75. Klaus, S., Pueltz, S., Thoene-Reineke, C., Wolfram, S. (2005). Epigallocatechin gallate attenuates diet-induced obesity in mice by decreasing energy absorption and increasing fat oxidation. *International Journal of Obesity*, *29*, 615–623.
76. Chen, Z.-Y, Fong, W.P. (2000). *Hypolipidemic Activity of Green Tea Epicatechins*, ACS Symposium Series, *754* (Caffeinated Beverages), American Chemical Society, Washington, DC, pp. 156–164
77. Kao, Y.-H., Chang, H.-H., Lee, M.-J., Chen, C.-L. (2006). Tea, obesity, and diabetes. *Molecular Nutrition & Food Research*, *50*, 188–210.
78. Weber, P. (2004). Teavigo—a powerful constituent of green tea to benefit human health. *Agro Food Industry Hi-Tech*, *15*, 20–22.
79. Rubayiza, A.B., Meurens, M. (2005). Chemical discrimination of Arabica and Robusta coffees by Fourier transform Raman spectroscopy. *Journal of Agricultural and Food Chemistry*, *53*, 4654–4659.
80. Clifford, M.N. (1999). Chlorogenic acids and other cinnamates—nature, occurrence and dietary burden. *Journal of the Science of Food and Agriculture*, *79*, 362–372.
81. Schilter, B., Cavin, C., Tritscher, A., Constable, A. (2001). Health effects and safety considerations, in *Coffee: Recent Developments* (eds R.J. Clarke, O.G. Vitzthum), Blackwell Science, London, p. 165.
82. Higdon, J.V., Frei, B. (2006). Coffee and health: a review of recent human research. *Critical Reviews in Food Science and Nutrition*, *46*, 101–123.
83. Nakasato, F., Nakayasu, M., Fujita, Y., Nagao, M., Terada, M., Sugimura, T. (1984). Mutagenicity of instant coffee on cultured Chinese hamster lung cells. *Mutation Research*, *141*, 109–112.
84. Nagao, M., Fujita, Y.M. Wakabayashi, K., Nukaya, H., Kosuge, T., Sugimura, T. (1986). Mutagens in coffee and other beverages. *Environmental Health Perspectives*, *67*, 89–91.
85. Ariza, R.R., Dorado, G., Barbancho, M., Pueyo, C. (1988). Study on the causes of direct acting mutagenicity in coffee and tea using the Ara test in *S. Typhimurium*. *Mutation Research*, *201*, 89–96.
86. Itagaki, S.K., Kobayashi, T., Kitagawa, Y., Iwata, S., Nukaya, H., Tsuji, K. (1992). Cytotoxicity of coffee in human intestinal cells in vitro and its inhibition by peroxidise. *Toxicology in Vitro*, *6*, 417–421.

87. Stadler, R.H., Turesky, R.J., Mueller, O., Markovic, J., Leong-Morgenthaler, P.-M. (1994). The inhibitory effects of coffee on radical-mediated oxidation and mutagenicity. *Mutation Research*, *308*, 177–190.
88. Yen, W.J., Wang, B.S., Chang, L.W., Duh, P.D. (2005). Antioxidant properties of roasted coffee residues. *Journal of Agricultural and Food Chemistry*, *53*, 2658–2663.
89. Anese, M., Nicoli, M.C. (2003). Antioxidant properties of ready-to-drink coffee brews. *Journal of Agricultural and Food Chemistry*, *51*, 942–946.
90. Delgado-Andrade, C., Morales, F.J. (2005). Unraveling the contribution of melanoidins to the antioxidant activity of coffee brews. *Journal of Agricultural and Food Chemistry*, *53*, 1403–1407.
91. Borrelli, R.C., Visconti, A., Mennella, C., Anese, M., Fogliano, V. (2002). Chemical characterisation and antioxidants properties of coffee melanoidins. *Journal of Agricultural and Food Chemistry*, *50*, 6527–6533.
92. López-Galilea, I., Andueza, S., di Leonardo, I., Paz de Peña, M., Cid, C. (2004). Influence of torrefacto roast on antioxidant and pro-oxidant activity of coffee. *Food Chemistry*, *94*, 75–80.
93. Hofmann, T., Bors, W., Stettmeier, K. (1999). On the radical-assisted melanoidin formation during thermal processing of foods as well as under physiological conditions. *Journal of Agricultural and Food Chemistry*, *47*, 391–396.
94. Del Castillo, M.D., Ames, J.M., Gordon, M.H. (2002). Effects of roasting on the antioxidant activity of coffee brews. *Journal of Agricultural and Food Chemistry*, *50*, 3698–3703.
95. Somoza, V., Lindenmeier, M., Wenzel, E., Frank, O., Ersbersdobler, H.F., Hofmann, T. (2003). Activity-guided identification of a chemopreventive compound in coffee beverage using in vitro and in vivo techniques. *Journal of Agricultural and Food Chemistry*, *51*, 6861–6869.
96. Stadler, R.H., Varga, N., Milo, C., Schilter, B., Arce Vera, F., Welti, D.H. (2002). Alkylpyridiniums. 2. Isolation and quantification in roasted and ground coffees. *Journal of Agricultural and Food Chemistry*, *50*, 1200–1206.
97. Svilaas, A., Sakhi, A.K., Andersen, L.F., Svilaas, T., Ström, E.C., Jacobs, D.R., Ose, L., Blomhoff, R. (2004). Intakes of antioxidants in coffee, wine and vegetables are correlated with plasma carotenoids in humans. *Journal of Nutrition*, *134*, 562–567.
98. Pellegrini, N., Serafini, M., Colombi, B., Del Rio, D., Salvatore, S., Bianchi, M., Brighenti, F. (2003). Total antioxidant capacity of plant foods, beverages and oils consumed in Italy assessed by three in-vitro assays. *Journal of Nutrition*, *133*, 2812–2819.
99. Mattila, P., Hellstroem, J., Toerroenen, R. (2006). Phenolic acids in berries, fruits, and beverages. *Journal of Agricultural and Food Chemistry*, *54*, 7193–7199.
100. Richelle, M., Tavazzi, I., Offord, E. (2001). Comparison of the antioxidant activity of commonly consumed polyphenolic beverages (coffee, cocoa, and tea) prepared per cup serving. *Journal of Agricultural and Food Chemistry*, *49*, 3438–3442.
101. Van Dam, R.M., Feskens, E.J. (2002). Coffee consumption and risk of type 2 diabetes mellitus. *Lancet*, *360*, 1477–1478.
102. Ascherio, A., Zhang, S.M., Hernan, M.A., Kawachi, I., Colditz, G.A., Speizer, F.E., Willet, W.C. (2001). Prospective study of caffeine consumption and risk of Parkinson's disease in men and women. *Annals of Neurology*, *50*, 56–63.

103. Tverdal, A., Skurtveit, S. (2003). Coffee intake and mortality from liver cirrhosis. *Annals of Epidemiology*, *13*, 419–423.
104. Tavani, A., La Vecchia, C. (2000). Coffee and cancer: a review of epidemiological studies, 1990–1999. *European Journal of Cancer Prevention*, *9*, 241–256.
105. Leitzmann, M.F., Stampler, M.J., Willet, W.C., Spiegelman, D., Colditz, G.A., Giovannucci, E.L. (2002). Coffee intake is associated with lower risk of symptomatic gallstone disease in women. *Gastroenterology*, *123*, 1823–1830.
106. Frost Andersen, L., Jacobs, D.R., Carlsen, M.H., Blomhoff, R. (2006). Consumption of coffee is associated with reduced risk of death attributed to inflammatory and cardiovascular diseases in the Iowa Women's Health Study. *American Journal of Clinical Nutrition*, *83*, 1039–1046.
107. Adom, K.K., Liu, R.H. (2002). Antioxidant activity of grains. *Journal of Agricultural and Food Chemistry*, *9*, 6182–6187.
108. Anderson, J.W. (2000). Whole grain foods and heart disease risk. *Journal of the American College of Nutrition*, *19* (Suppl. 3), 291S–299S.
109. Murtaugh, M.A., Jacobs, D.R., Jacob, B., Steffen, L.M., Marquart, L. (2003). Epidemiological support for the protection of whole grain against diabetes. *Proceedings of the Nutrition Society*, *62*, 143–149.
110. Liu, S. (2003). Whole grain foods, dietary fibre, and type 2 diabetes: searching for a kernel of truth. *American Journal of Clinical Nutrition*, *77*, 527–529.
111. Jacobs, D.R., Marquart, L., Slavin, J.L., Kushi, L.H. (1998). Whole grain intake and cancer: an expanded review and meta-analysis. *Nutrition and Cancer*, *30*, 85–96.
112. Liu, S., Willett, W.C., Manson, J.E. (2003). Relation between intakes of dietary fibre and grain products and changes in weight and development of obesity among middle-aged women. *American Journal of Clinical Nutrition*, *78*, 920–927.

10

RISK COMMUNICATION

David Schmidt and Danielle Schor

International Food Information Council, 1100 Connecticut Ave. NW, Suite 430, Washington, DC 20036, USA

10.1 INTRODUCTION

Risk communication is an important component of the risk analysis framework. Together, risk assessment, risk management, and risk communication provide a framework for determining the risks in food, managing those risks through public policies and education, and communicating this information to a variety of stakeholders. Ideally, risk communication should be integral to the entire risk analysis process rather than an activity that is carried out after the risk assessment and risk management steps are completed. According to the National Academy of Sciences (NAS), risk communication is "an interactive process of exchange of information and opinion among individuals, groups, and institutions. It involves multiple messages about the nature of risk and other messages, not strictly about risk, that express concerns, opinions, or reactions to risk messages or to legal and institutional arrangements for risk management." The NAS further noted .... "even though good risk communication cannot always be expected to improve a situation, poor risk communication will nearly always make it worse" (1).

Risk communication has its roots in risk perception. The study of risk perception helps to explain the ways we subconsciously decide what to be afraid of and how afraid to be. Our responses to risk are not rational risk analyses but involve emotions, values, and instincts. Fears often do not match the facts. Successful risk communicators provide information in a way that recognizes

Process-Induced Food Toxicants: Occurrence, Formation, Mitigation, and Health Risks,
Edited by Richard H. Stadler and David R. Lineback

these complex interactions (2). Risk communication about process-induced food toxicants and health risks is a relatively new challenge that received particular attention in April 2002, when researchers at the Swedish National Food Administration and Stockholm University announced they had discovered the presence of acrylamide in plant-based foods—particularly potato and wheat products—that are prepared by heating at high temperatures. Although most epidemiological studies to date appear to indicate that dietary acrylamide is not associated with an increased risk of any of several types of cancers studied, acrylamide has been identified as a toxic, cancer-causing industrial chemical based on high-dose animal studies. Exposure assessments identified fried potato products and breakfast cereals as the most significant sources of dietary acrylamide in the US diet.

The acrylamide issue provides a case study in the challenges of risk communication for several reasons. First, the issue emerged with little warning. While Swedish researchers had long been interested in studying the effects of acrylamide, a press event carried out by the Swedish National Food Administration and Stockholm University in 2002 suddenly made acrylamide an international issue (3). Second, it provided the opportunity for sensational headlines such as "Can French Fries Give You Cancer?" (4). Third, numerous gaps in knowledge concerning acrylamide existed, and this limited the amount of definitive information about the chemical's risk available to human health scientists and policy officials to communicate.

This contrasts in at least two ways with bovine spongiform encephalopathy (BSE) in the United States, where a thorough risk assessment was conducted by a highly respected academic organization before the first case of BSE was detected domestically (5). The risk assessment enabled risk communicators to more easily develop credible messages about risk. In addition, specific regulatory steps were quickly taken by the various regulatory agencies with oversight for animal health and food safety to reduce risks (6). While acrylamide is specifically used in this chapter for the purposes of discussing risk communication, heat-generated food toxicants in general, such as heterocyclic amines, are also relevant to the discussion.

Through this and other case studies, those involved in risk analysis, including risk communication, can gain insight into the necessary elements of a good risk communication process and can use knowledge gained in future efforts.

10.2 ATTITUDES ABOUT FOOD-RELATED RISKS

Public reactions to risk and in particular those related to food are affected by numerous factors. First, the public is more likely to be concerned about unfamiliar risks. Chemicals with long names, for example, install fear even though the foods we eat are composed of chemicals with names largely unknown to the public. Second, the lack of personal control over the risk enhances concern. This explains why many individuals have difficulty letting another person drive and often become "backseat drivers." Third, a risk that might produce delayed effects is of more concern than an immediate risk. Fourth, if children are involved, and future generations might be affected, the concern multiplies. Fifth, if the risk has occurred before, the public is less understanding about it happening again. Sixth, if there is lack of trust in relevant institutions, such as regulatory agencies, and decision processes carried out by these institutions are not understood, the concern is magnified. Transparency is important in reducing this type of concern. Lastly, if there are unclear benefits, the public is less willing to accept the risk. Regardless of whether a substance is safe, there usually must be a tangible benefit—whether it is taste, convenience, or variety—in order for the public to accept it. In essence, consumers perform their own risk–benefit analyses. These are just some of the many factors that affect attitudes about food-related risks (7).

An additional factor, which is directly related to process-induced toxicants, is that risks caused by humans, rather than by nature, generate more concern. This may be the reason why the acrylamide "outrage" factor was not as high as might have been expected. The substances that cause the formation of acrylamide during processing are present in the raw materials. Thus, acrylamide is seen as somewhat of a "natural" product. Or at least it is perceived as somewhere in between a natural and man-made risk.

Qualitative research conducted by Cogent Research on behalf of the International Food Information Council (IFIC) in 2003 illustrates this reaction (8). The overarching goal of the research was to understand consumer response to then-recent media reports about the formation of the compound acrylamide during the food cooking process. To achieve this objective, IFIC investigated consumers' knowledge of the issue and any changes in habits as a result of concerns about acrylamide. To broaden understanding of the motivating factors behind behavioral shifts, IFIC also sought to uncover the types of food or health issues and information that are the root cause for consumers to make sudden dietary changes.

Cogent Research conducted a total of six focus groups in three markets. In each market, it conducted two groups—one among those who were unaware of acrylamide and one among those who had heard at least a little about the issue. All participants were recruited to provide a demographically diverse mix of general consumers who met the acrylamide awareness requirement. The groups were conducted on April 24, 2003 (Merriam, KS), April 29, 2003 (Cambridge, MA), and April 30, 2003 (Marina Del Ray, CA). Each session was approximately 2 h in length.

While the findings of focus groups provide useful insights into a particular subject and allow researchers to unearth unanticipated key issues, it is important to note that they are not statistically valid. One cannot project the findings of a single group, or for that matter, multiple focus groups to the entire population.

In terms of behavioral shifts toward food ingredients, avoidance of specific foods or ingredients is universal. All participants had removed or limited their intake of at least one item (although items mentioned were generally a group or type of food such as carbohydrates or fatty foods). The most commonly mentioned items that participants are avoiding are those that have received intensive and long-standing media coverage that has been validated by other sources (and/or common sense). The ability of new or isolated "health scares" to create an impact seems limited unless the story relates to a specific and current threat.

In terms of motivating factors, in general, the reduction or avoidance of any food is mainly the result of (or threat of) a definite and immediate personal negative outcome. In other words, avoidance is warranted in cases where people have a disease (e.g., diabetes), where nutrition plays a role in overall health or disease management, or where people are adversely affected by a food substance (e.g., allergies, general malaise, and weight gain).

Food avoidance is much less frequently the result of anxiety over the potential for longer-term health problems. A few participants with a family proclivity toward certain diseases were attempting to lessen their risk through dietary means; however, some with a genetic predisposition to a health problem admitted they do not always eat the way they should to lessen the risks. One notable exception to this is parents. Although not concerned for themselves, they have a strong desire to protect their children. Thus, many of the participants with children were restricting their children's intake of foods they felt have little nutritional value or can negatively impact long-term health (e.g., fried foods and fast foods). To the extent they avoid the foods themselves, it is mainly to set a good example for their children.

Regardless of their awareness of past media coverage on acrylamide, most consumers are not concerned about this issue, despite the fact that cancer is a serious concern for many participants. This lack of concern remained low among most participants even after extensive discussions and education (9).

Indeed, the vast majority of consumers say they had not changed their eating behavior as a result of hearing about acrylamide. The key exception to this seemed to be pregnant women or nursing mothers who have a heightened awareness and concern about most food/health issues. However, even if they admit that once they have stopped nursing, they will most likely resume consumption of foods associated with acrylamide. Parents and grandparents also indicated elevated levels of concern for their children and grandchildren, although they were not concerned for themselves. It is worth noting that a small number of consumers had changed their eating habits due to acrylamide, but a few of these had not been able to sustain their behavior.

There are some key reasons for the overall lack of concern, which results in a lack of motivation to change eating habits. They include the following:

1. Most consumers have become highly skeptical of research studies in general given past experience with contradictory studies. There is a strong negative opinion about the fact that these types of reports are

continually being changed (e.g., one week something is good, the next it is bad), which limits the ability of any new "health scare" to have a significant impact.

2. In particular, until something passes the test of time (i.e., not contradicted over time, but in fact, confirmed by other studies), many will not take note. Participants felt it is too early to be concerned about this issue, since the current studies were inconclusive. That said, a few in each group said they would keep their eye on the issue to see what further studies find. There was some backlash against the ever-constant stream of new food/health threats, with some participants indicating there is always something new to be concerned about, and if they cut out all of the foods that contained potential health risks, there would be nothing left to eat. In this case in particular, "aware" consumers feel they would have heard more about acrylamide if it were a serious risk.
3. The sensationalism that surrounds this type of story undermines credibility. In this case in particular, but in other cases as well, the way the media hype stories in order to get and retain viewers limits the credibility of the research and the real impact on health.
4. Many have already changed their behavior and reduced consumption of those foods that have been most closely associated with acrylamide (e.g., French fries and snack foods) for other health reasons; thus, they see no additional reason to change. Even when expanded to other foods such as cereal and bread, most participants indicated they would be unlikely to make a shift. Only a couple of aware participants had made any changes to their diet as a specific reaction to acrylamide.

Misconceptions involving acrylamide were pronounced in that consumers (especially those in the "aware" groups) often confuse acrylamide with other health issues; in particular, the trans fat issue that resulted from widespread media coverage about fast-food outlets changing their oil in order to make French fries more healthful. Even after reading IFIC's acrylamide Q&A, consumers often took away very different information or confused it with trans fat, oil in general, cooking techniques, or other issues. While many consumers (knowledgeable and unknowledgeable) did understand that acrylamide is a naturally occurring compound that is the result of cooking or heating, there were still a number of participants who felt it is only found in processed foods, not in whole foods made in the home.

A few consumers were fairly vocal about their feelings that government agencies are not always entirely forthcoming about the health risks associated with specific issues because of special interest groups and lobbyists. Interestingly, it seems the only way to assure some consumers that the government is unbiased is to definitively indicate that a food is "bad." Only in this way did some participants indicate they were willing to believe that the government is an unbiased party. However, other consumers express confidence in government agencies, and defend their role in ensuring thorough research and evaluation.

Medical professionals also face some criticism on health/nutrition-related issues, with some participants indicating doctors are not up-to-date on all the newest issues. That said, many participants indicated they would avoid foods if their doctor told them to. It is interesting to note that some medical associations (American Medical Association [AMA] in particular) also face a certain level of distrust, with participants indicating they are also subject to outside influences and lobbying. Indeed, money is often seen as the key motivator in many of these types of studies. It fuels the process throughout the entire chain. Scientists are seen to make decisions about what to study based on their need for funding. Charitable organizations and industry alike are believed to support studies that deliver results they want to hear. And the government is oftentimes believed to be withholding information from the public in order to alleviate any negative economic effect.

10.3 RANKING RISKS: PLACING RISKS IN CONTEXT

Comparing various risks is a difficult task for scientists. There are many uncertainties in risk estimates due to many factors, including the relationship between risk factors and human disease and the fact that risk assessment methods are often criticized as not being quantitatively valid or predictable.

Beyond these challenges, the process risk scientists go through to compare risks is extremely different from the process used by the public. While scientists tend to base these rankings on criteria that involve giving priority to the greatest risk to the greatest number, the public evaluates risk using a different set of criteria, where context plays a major role. For example, the public tends to place a higher value on the well-being of children as opposed to adults (10). Risk perception, in addition to risk numbers, must be considered by the risk communicator.

10.4 EFFECTIVE COMMUNICATION TECHNIQUES: EXAMPLES

Risk communication fails if it tells people only what the communicators want them to know in order to get them to behave "rationally," with rationally being defined as the viewpoint of the communicator. Scientists often assume that ignorance is at the root of conflict over science. The facts are assumed to speak for themselves (11). Risk communication is more likely to succeed if it sets the more realistic goal of helping people to understand the facts, in ways that are relevant to their own lives, feeling, and values so they can put the risk in perspective and make more informed choices.

"Framing" messages to connect with diverse audiences should not be viewed as being untrue to the science. Scientists may argue that this approach permits an irrational basis for policy decisions. One way to ensure that policy decisions are science-based is to conduct careful and rigorous risk analyses that can support policy decisions. However, a carefully crafted policy decision based on a risk assessment is difficult to implement with consumer opposition in a democratic society where public opinion matters. It is in the best interests of all stakeholders to craft a risk assessment that addresses public questions and concerns so that the information from the risk assessment that is communicated resonates with the public. Bringing the public along throughout the process results in the risk assessment having more impact than if the public is simply fed the information at the end with the expectation that it will be accepted without question.

Numerous lists providing risk communication best practices can be found from a variety of sources. The following were also articulated by the NRC-NAS in its 1989 report on Improving Risk Communication (12):

- Risk communication is successful only to the extent that it raises the level of understanding of relevant issues or actions and satisfies those involved that they are adequately informed within the limits of available knowledge.
- Risk communication is a component of risk management. Successful risk communication does not guarantee that risk management decisions will maximize general welfare; it only ensures that decision makers will understand what is known about the implications for welfare and the available options.
- A risk communication process that disseminates accurate information is not successful unless the potential recipients achieve a sufficient understanding. The recipient of the information must be able to achieve a complete understanding of the information he/she desires.
- Risk communication is more than one-way transmission of expert knowledge to the uninformed. Messages about expert knowledge are necessary to the risk communication process; they are not sufficient, however, for the process to be successful.
- Although consensus on controversial issues is often the criterion of success for producers of risk messages, it may not be appropriate for risk communication process in a democracy. Successful risk communication need not result in consensus about controversial issues.

The following "Seven Cardinal Rules of Risk Communication" are provided by the U.S. Environmental Protection Agency (13).

10.4.1 The EPA's Seven Cardinal Rules of Risk Communication

10.4.1.1 Rule 1. Accept and Involve the Public as a Legitimate Partner

Two basic tenets of risk communication in a democracy are generally understood and accepted. First, people and communities have a right to participate in decisions that affect their lives, their property, and the things they value. Second, the goal of risk communication should not be to diffuse public concerns or avoid action. The goal should be to produce an informed public that is involved, interested, reasonable, thoughtful, solution-oriented, and collaborative.

Guidelines: Demonstrate respect for the public by involving the community early, before important decisions are made. Clarify that decisions about risks will be based not only on the magnitude of the risk but on factors of concern to the public. Involve all parties that have an interest or a stake in the particular risk in question. Adhere to highest moral and ethical standards; recognize that people hold you accountable.

10.4.1.2 Rule 2. Listen to the Audience People are often more concerned about issues such as trust, credibility, control, benefits, competence, voluntariness, fairness, empathy, caring, courtesy, and compassion than about mortality statistics and the details of quantitative risk assessment. If people feel or perceive that they are not being heard, they cannot be expected to listen. Effective risk communication is a two-way activity.

Guidelines: Do not make assumptions about what people know, think, or want done about risks. Take the time to find out what people are thinking; use techniques such as interviews, facilitated discussion groups, advisory groups, toll-free numbers, and surveys. Let all parties that have an interest or a stake in the issue be heard. Identify with your audience and try to put yourself in their place. Recognize people's emotions. Let people know that what they said has been understood, addressing their concerns as well as yours. Recognize the "hidden agendas," symbolic meanings, and broader social, cultural, economic, or political considerations that often underlie and complicate the task of risk communication.

10.4.1.3 Rule 3. Be Honest, Frank, and Open Before a risk communication can be accepted, the messenger must be perceived as trustworthy and credible. Therefore, the first goal of risk communication is to establish trust and credibility. Trust and credibility judgments are resistant to change once made. Short-term judgments of trust and credibility are based largely on verbal and nonverbal communications. Long-term judgments of trust and credibility are based largely on actions and performance. In communicating risk information, trust and credibility are a spokesperson's most precious assets. Trust and credibility are difficult to obtain. Once lost, they are almost impossible to regain.

Guidelines: State credentials; but do not ask or expect to be trusted by the public. If an answer is unknown or uncertain, express willingness to get back to the questioner with answers. Make corrections if errors are made. Disclose risk information as soon as possible (emphasizing appropriate reservations about reliability). Do not minimize or exaggerate the level of risk. Speculate only with great caution. If in doubt, lean toward sharing more information, not less—or people may think something significant is being hidden. Discuss data uncertainties, strengths, and weaknesses—including the ones identified by other credible sources. Identify worst-case estimates as such, and cite ranges of risk estimates when appropriate.

10.4.1.4 Rule 4. Coordinate and Collaborate with Other Credible Sources Allies can be effective in helping communicate risk information. Few things make risk communication more difficult than conflicts or public disagreements with other credible sources.

Guidelines: Take time to coordinate all interorganizational and intraorganizational communications. Devote effort and resources to the slow, hard work of building bridges, partnerships, and alliances with other organizations. Use credible and authoritative intermediaries. Consult with others to determine who is best able to answer questions about risk. Try to issue communications jointly with other trustworthy sources such as credible university scientists, physicians, citizen advisory groups, trusted local officials, and national or local opinion leaders.

10.4.1.5 Rule 5. Meet the Needs of the Media The media are a prime transmitter of information on risks. They play a critical role in setting agendas and in determining outcomes. The media are generally more interested in politics than in risk; more interested in simplicity than in complexity; and more interested in wrongdoing, blame, and danger than in safety.

Guidelines: Be open with and accessible to reporters. Respect their deadlines. Provide information tailored to the needs of each type of media, such as sound bites, graphics, and other visual aids for television. Agree with the reporter in advance about the specific topic of the interview; stick to the topic in the interview. Prepare a limited number of positive key messages in advance and repeat the messages several times during the interview. Provide background material on complex risk issues. Do not speculate. Say only those things that you are willing to have repeated: everything you say in an interview is on the record. Keep interviews short. Follow up on stories with praise or criticism, as warranted. Try to establish long-term relationships of trust with specific editors and reporters.

10.4.1.6 Rule 6. Speak Clearly and with Compassion Technical language and jargon are useful as professional shorthand. But they are barriers to successful communication with the public. In low-trust, high-concern situations, empathy and caring often carry more weight than numbers and technical facts.

Guidelines: Use clear, nontechnical language. Be sensitive to local norms, such as speech and dress. Strive for brevity, but respect people's information needs and offer to provide more information. Use graphics and other pictorial material to clarify messages. Personalize risk data; use stories, examples, and anecdotes that make technical data come alive. Avoid distant, abstract, unfeeling language about deaths, injuries, and illnesses. Acknowledge and respond (both in words and with actions) to emotions that people express, such as anxiety, fear, anger, outrage, and helplessness. Acknowledge and respond to the distinctions that the public views as important in evaluating risks. Use risk comparisons to help put risks in perspective, but avoid comparisons that ignore distinctions that people consider important. Always try to include a discussion of actions that are under way or can be taken. Promise only that which can be delivered, and follow through. Acknowledge, and say, that any illness, injury, or death is a tragedy and to be avoided.

10.4.1.7 Rule 7. Plan Carefully and Evaluate Performance Different goals, audiences, and media require different risk communication strategies. Risk communication will be successful only if carefully planned and evaluated.

Guidelines: Begin with clear, explicit objectives—such as providing information to the public, providing reassurance, encouraging protective action and behavior change, stimulating emergency response, or involving stakeholders in dialogue and joint problem solving. Evaluate technical information about risks and know its strengths and weaknesses. Identify important stakeholders and subgroups within the audience. Aim communications at specific stakeholders and subgroups in the audience. Recruit spokespersons with effective presentation and human interaction skills. Train staff—including technical staff—in communication skills; recognize and reward outstanding performance. Pretest messages. Carefully evaluate efforts and learn from mistakes.

10.4.2 Empowering the Public

The National Center for Food Protection and Defense of the University of Minnesota has prepared best practices that include the importance of empowering consumers to reduce risk (14). This is not always possible, but to the extent that it is, it should be emphasized since the more consumers have control over a risk, the less anxious they are about that risk. A good example is food preparation in the home. While proper food preparation is not a substitute for good industry quality control programs or for good regulatory policy, it does contribute to minimizing risks from certain foods and should be emphasized.

For acrylamide, research has identified home food preparation methods consumers can use to reduce the amount of acrylamide formed during cooking. During the Second Joint Institute for Food Safety and Applied Nutrition (JIFSAN) Acrylamide Workshop, held in Chicago, Illinois, in April 2004, Dr. Lauren Jackson of the FDA Center for Food Safety and Applied Nutrition, National Center for Food Safety & Technology in Summit-Argo, Illinois, noted that most of the acrylamide formed during the toasting of bread was found in the darker portions of the toast. When this portion of the toast was scraped off the bread, the toast scrapings had significantly more acrylamide than the remainder of the toast. Dr. Jackson also reported the effects of frying times and temperatures on acrylamide formation in French fries. Because the amount of acrylamide and the degree of browning both increase with cooking times and temperatures, the degree of browning may be used as an indicator of the amount of acrylamide formed in some foods during cooking or processing. Additionally, to reduce acrylamide when frying fresh potatoes, storage of the potatoes in a cool (but not refrigerated) dry, dark room is important. This will prevent accumulation of sugars as well as sprout formation.

She also noted that some washing treatments, including soaking and rinsing the sliced or cut potatoes in plain water, are effective in reducing acrylamide in the final cooked potatoes (15).

An example of risk communication provided on acrylamide was contained in the May/June 2004 *Food Insight* article published by the International Food Information Council Foundation in answer to the question, "Can I Reduce Acrylamide Levels in Food?" *Food Insight* provided these recommendations (16):

> Until more is known about acrylamide in food, experts from FDA recommend that consumers focus not on acrylamide, but on eating a balanced diet, choosing a variety of foods that are low in trans fat and saturated fat, and rich in high-fiber grains, fruits and vegetables. However, if you are still interested in reducing the trace levels of acrylamide that may be present in food, here are a few tips for home cooking:
>
> - Avoid overcooking or using extremely high temperatures in cooking food. (NOTE: Undercooking some foods may result in foodborne illness.)
> - Fry foods to a LIGHT, rather than dark, golden brown.
> - Scrape the darker crumbs off toast and other baked items before consuming.
> - Store potatoes at room temperature in a dry location, then soak and rinse cut or sliced potatoes, before frying or baking. It is important to drain the potato slices before frying to prevent pan fires.
> - Enjoy a moderate amount of a wide variety of foods to stay healthy! (p. 3)

One of the reasons the Swedish National Food Administration was criticized for its handling of its 2002 press event on acrylamide was that the risks associated with eating certain foods were amplified, and no concrete advice was given on how to minimize that risk (3).

10.4.3 Credible Information Sources

Effective communication also involves working with credible information sources. In many cases, knowledge of what foods to avoid comes from commonly asserted knowledge or personal observation rather than from outside sources. Almost no one questions "common sense" information on nutrition and health, nor do they seem to question information that contains personally verifiable knowledge (e.g., weight loss from carbohydrate reduction).

Believable outside information sources are most often a physician, health magazine, or family member/friend. In terms of providing sound nutritional advice, physicians and family members have a distinct advantage over media sources or literature. Health magazines are generally regarded as a competent source whereas there is less enthusiasm for books, which are very divisive. While some participants (especially those reducing foods for weight management reasons) cite books as a competent source because they "have to tell the truth to get published," others clearly discount that idea.

News reports (especially television news) are mentioned less frequently as a source of information, and when cited, tend to be mentioned in conjunction with another source. That said, participants are very aware of health and food-related news reports (e.g., trans fat and Alar), and these do make an impact in consumers' minds even though it does not always result in changes to their consumption behavior. This is most likely because news reports are not the most trusted source, unless they confirm "common sense" or other long-standing health/nutrition associations, or involve a current (and short-term) health warning. News stories heralding new health information often pass by consumers due not only to the high number of new "health risks" but also to the number of conflicting stories. Many participants made note of the fact that the continuous changing of information on health and food makes them less likely to believe the stories they hear, if they even notice them in the first place.

While news reports and friends/family are most often cited as the original source for new health-related information, almost all participants indicated they would use the Internet if they wanted to get more information about acrylamide (or other new health "stories"). The Internet is considered the key place to get information, although most consumers recognize that the information found is subject to the web site's biases, and needs to be taken with caution if it is not validated in multiple places (8).

10.4.4 Timing

In addition to using "best practices" and relaying messages through credible information sources, the timing of messages must be considered. The risk analysis process should ideally be viewed as interactive, where the public has input into the risk assessment questions that are asked by scientists and into the risk management decisions that are made. Unfortunately, risk communication is often left to the end of the risk analysis process, where the result of the risk assessment and risk management strategies are simply communicated to the public.

10.5 CHALLENGES AND BARRIERS

Many challenges exist to seeing good risk communication practices implemented in the real world. While it is easy to discuss these practices theoretically, they are much more difficult to implement for many reasons.

First, scientific hubris is not easily overcome. In order to effectively communicate risks, the risk communicator must be willing to listen and empathize and frame the issue so that it is relevant to the audience.

Second, the ability to detect smaller and smaller amounts of substances that may pose risks is leading to increasing announcements about possible new hazards. We have gone from detecting parts per thousand in the 1950s and '60s to parts per billion in this decade. This phenomenon, often referred to as "chasing zero," poses challenge in communicating when the presence of a substance is indeed a risk or not. As analytical tools have improved, we are more likely to find things in food, but our tools for determining the significance of these findings have not kept up. Thus, we are faced with new findings that cannot be put into context, leading to erosion in consumer confidence.

Third, conflicting messages will always exist on heated issues, but similar organizations need to do a better job in developing coordinated messages to make it less confusing for the public to examine the issues. For example, regulatory agencies often find it difficult to coordinate messages, and this inability greatly affects public confidence. If regulatory agencies are not on the same page, the public begins to wonder what the real story is. Industry organizations also benefit from coordinating messages even when competing interests exist.

Fourth, the uncertainty, complexity, and incompleteness of data provide a challenge to good risk communication. Risk assessments are designed to provide information on the potential harm posed by threats to health, safety, or the environment. Unfortunately, large gaps remain in our understanding of risk.

A fifth obstacle is selective reporting by the news media. The media are critical to the delivery of risk information to the general public, but journalists are highly selective in reporting about risk. They tend to focus their attention on issues that play to the same "outrage factors" that the public uses in evaluating risks. In addition, many media stories about risk contain substantial omissions, or present oversimplified, distorted, inaccurate information. Studies have revealed, for example, that media reports on cancer risks often fail to provide adequate statistics on general cancer rates for purposes of comparison (17).

And sixth, risk communication works best when the public has a choice or a role. Situations where consumers can choose or avoid certain products or where the consumer is empowered to take steps to reduce risks are most successful in terms of risk communication.

10.6 FUTURE PROSPECTS

We often underestimate the consuming public's willingness to accept some level of risk when it comes to their food supply. Consumers are more likely to accept the fact that there are at least low levels of risks in foods that we consume when they understand what those risks are and what measures they can take to reduce them if they are concerned. More research is needed to better understand ways to communicate with consumers about risks in a manner that they may find more compelling. The old adage that "the dose makes the poison" is an accurate one but not a phrase that many consumers find reassuring (D. Schmidt and D. Schor, personal communication).

For the future, success will depend on the ability of those who control the message—whether they are scientists, policy makers, or influential stakeholders—to recognize that risk communication is more than good public relations. Good risk communication requires a willingness to learn, a willingness to partner, and a willingness to frame the message based on the audience. It is also necessary to recognize that public buy-in is not automatic but must be earned. Having good science is not enough—communicating the science well is at least half the task.

REFERENCES

1. National Research Council, National Academy of Sciences (NRC-NAS) (1989). *Improving Risk Communication*, National Academy Press, Washington, DC, p. 21.
2. Gray, G., Ropeik, D. (2002). Dealing with the dangers of fear: the role of risk communication. *Health Affairs*, *21*, 106–116.
3. Lofstedt, R. (2003). Science communication and the Swedish acrylamide "alarm". *Journal of Health Communication*, *8*, 407–432.
4. Hellmich, N. (2002). Can French fries give you cancer? USA Today, October 7, 2002.
5. Harvard Center for Risk Analysis, Harvard School of Public Health, and Center for Computational Epidemiology, College of Veterinary Medicine and Tuskegee University (2001). Evaluation of the Potential for Bovine Spongiform Encephalopathy in the United States.
6. FSIS/USDA. (2004). Prohibition on the Use of Specified Risk Materials for Human Food and Requirements for Disposition of Non-Ambulatory Disabled Cattle (FSIS Docket No. 03-025IF; 69 FR 1861), 12 January 2004.
7. National Research Council, National Academy of Sciences (NRC-NAS) (1989). *Improving Risk Communication*, National Academy Press, Washington, DC, p. 51.
8. IFIC. (2003). *Consumer Behavioural Shifts: Understanding Consumer Response to Acrylamide & Other Food/Health Issues*, International Food Information Council. http://www.ific.org/research/acrylamideres.cfm (accessed 12 February 2008).
9. IFIC. (2005). *Questions and Answers about Acrylamide*, International Food Information Council. http://www.ific.org/publications/qa/acrylamideqa.cfm (accessed 13 February 2008).
10. Silbergeld, E. (1995). The risks of comparing risks. *New York University Environmental Law Journal*, *3*, 405.
11. Nisbet, M.C., Scheufele, D.A. (2007). The future of public engagement. *The Scientist*, *21* (**10**), 38.
12. National Research Council, National Academy of Sciences (NRC-NAS) (1989). *Improving Risk Communication*, National Academy Press, Washington, DC, pp. 1–29.

13. Environmental Protection Agency (1988). Seven Cardinal Rules of Risk Communication. http://www.epa.gov/care/library/7_cardinal_rules.pdf (accessed 12 February 2008).
14. A homeland security center of excellence, in *National Center for Food Protection and Defense*, University of Minnesota. http://www.ncfpd.umn.edu/index.cfm (accessed 12 February 2008).
15. University of Maryland (2004). 2004 Acrylamide in Food Workshop: Update—Scientific Issues, Uncertainties, and Research Strategies, in *Joint Institute for Food Safety and Applied Nutrition*, University of Maryland. http://www.jifsan.umd.edu/acrylamide2004.htm (accessed 11 February 2008).
16. International Information Food Council (2004). Acrylamide: putting the current findings into perspective, in *May/June 2004 Food Insight*, International Information Food Council. http://www.ific.org/foodinsight/2004/mj/acrylamidefi304.cfm (accessed 11 February 2008).
17. Covello, V., Sandman, M. (2001). *Solutions to an Environment in Peril*, Johns Hopkins University Press, pp. 164–178.

11

RISK/RISK AND RISK/BENEFIT CONSIDERATIONS

LEIF BUSK
Department of Research and Development, National Food Administration, Box 622, Uppsala 75126, Sweden

11.1 INTRODUCTION

Food is a foundation for human existence. The access to food has been connected, in most cultures, to religious beliefs and ceremonies. Food is a sacred gift from nature, providing health and life. Since not everything in nature can be eaten, the knowledge of what is edible or not, had to be passed on between generations by oral tradition, often linked to and supported by religious doctrines.

Cooking of food has evolved over thousands of years, primarily to improve texture, taste, and palatability. As natural science has evolved, more and more has been learned about the benefits of cooking, not least in microbiological terms. The concept that cooking could introduce risks is relatively new and also to some extent, in contrast to the inherited notion that food is something nature has given, i.e., sound and healthy. This adds on to problems in communicating risks associated with cooking.

11.1.1 Risks and Benefits of Cooked Foods

Smoked foods were among the first to be identified as problematic. Over the last 30 years, additional knowledge has emerged indicating that hazardous

Process-Induced Food Toxicants: Occurrence, Formation, Mitigation, and Health Risks,
Edited by Richard H. Stadler and David R. Lineback

compounds can be formed as a result of heat treatment. Mitigation strategies have been suggested to reduce these risks. Shorter cooking times and lower temperatures have been central elements in the strategies.

However, concerns have been raised that the mitigation could introduce new problems, i.e., loss of protective effects or beneficial compounds or generation of other unwanted and hazardous compounds.

Out of a societal perspective, it is important that advice from authorities take this into account in order to avoid Pyrrhic victories that could result in decreased health and increased costs for medical care. It is also vital for maintaining public trust in authorities and in the safety of foods on the market. Consequently, there is a need for quantitative risk/risk or risk/benefit assessments before mitigation strategies are introduced, both in domestic cooking and in industrial production of foods.

In order to perform a transparent and understandable quantitative assessment of this type, there is a need for a common or at least comparable denominator of risks and benefits. This is where a major challenge lies today.

11.1.2 Guidelines for the Assessment of Risks

In the present risk analysis concept and guidelines adopted by regulatory bodies in Europe and elsewhere, there is ample guidance on how to perform risk assessments of chemicals in foods. However, current methods and available data usually generate point estimates, such as the no observed effect level (NOEL), which is the highest dose in an experiment that produces no toxicity of concern. The NOEL is then used to derive an accepted daily intake (ADI) for humans, by introducing a safety factor, usually 100 for well-conducted animal experiments. The ADI value is regarded as an exposure that carries no appreciable risk. NOEL and ADI are only suitable for toxic effects that have a threshold. For toxicity generally regarded as having no threshold (e.g., cancer), where every dose increment over zero corresponds to an increased risk, there are no mutually internationally accepted ways of quantifying the risk.

In addition, there are no guidelines when it comes to comparing different types of risks. This is an area where science cannot make assessments on its own. Since the dread factor can vary considerably between different types of effects, e.g., from skin rash, over liver damage and malformations to cancer, cardiovascular diseases (CVDs), and death, it is a societal decision how to quantify them in a way that is acceptable to most people. This is a task that has to be addressed in the near future. A good starting point is the work done using the DALY concept where different types of health defects are given weight factors and then multiplied by duration of the defect. DALY stands for Disability Adjusted Life Years. It has been used to assess the relative importance of risk factors in the Dutch diet by van Kreijl *et al.* (1). Another approach is to put different types of toxic effects or health defects in a small number of categories (3–5 have been proposed), for example, by an expert panel appointed by the International Life Science Institute (ILSI) (2).

11.1.3 Guidelines for the Assessment of Benefits

When it comes to benefit assessments, the situation is even more difficult. They have previously often been seen as something that could be added after the completion of the scientific assessment, sometimes as a part of risk management. There are, at present, no specific guidelines for benefit assessment of food or food components comparable to those available for assessment of risk. For pharmaceutical drugs, there are guidelines for assessment of efficacy that could be regarded as assessments of benefits. They are, however, quite general and do not include guidelines on risk/benefit assessments.

In order to assess health benefits of a diet or a specific dietary factor, one would need to:

- Define what the actual health outcome is
  - disease or disease risk marker
  - biochemical or clinical changes
  - nutrient balance or nutrient status indicator
- Establish causality
  - establish dose response in order to quantify
- Estimate benefit for a given intake
  - prevalence, magnitude of effect
- Develop a suitable measure of the benefit that in a meaningful way could be compared with some established measure of risk.

In the case of micronutrients, a model has been established for requirements. This model includes the use of average requirement (AR), defined as the daily intake that meets the requirements of 50% of a population and recommended daily allowance (RDA), defined as the intake of a nutrient that is adequate to meet the requirements of practically all healthy persons. RDA = AR + 2 standard deviations and covers the need of 97–98% of a population, assuming a normal distribution of requirement and coefficient of variation 10–15%.

This model has been further developed in a structured approach by Renwick *et al.* (3) to cover risk/benefit analysis of micronutrients. By using default coefficients of variation of 15% for benefits and 45% for toxicity (based on analyses of human variability in the effects of therapeutic drugs), it is possible to model recommended ranges of intake based on a balance between risk of deficiency (or lack of benefit) and risk of toxicity.

The model demands data of a magnitude and quality that are usually not available for compounds of interest when assessing risks and benefits with mitigation strategies in cooked foods. The model works with distributions of both benefits and risks and requires a very close collaboration between assessors and managers, since no point estimates are used but rather a continuum of both medical and societal parameters that requires value-based judgments.

No general cutoffs can be established. It is a very good illustration of the problems one faces in attempting to perform risk/benefit assessments of food components.

11.1.4 Risk/Benefit Assessment of Fish Consumption

The most elaborate work on risks and benefits in the food area has been done on fish. The risks are represented by the occurrence of methyl mercury and chlorinated contaminants such as dioxins and PCBs. The benefits are a reduced risk for CVD, at least partly attributed to the presence of omega-3 (n3) polyunsaturated fatty acids (PUFAs) (for examples, see References 4–6). Mozzaffarian and Rimm (4) conducted meta-analysis of a great number of studies relating fish or fish oil consumption and CVD, methyl mercury and early neurodevelopment, methyl mercury and CVD or neurological outcomes in adults, and health risks of dioxins and PCBs from fish. Their conclusions were that for major health outcomes among adults, based on both the strength of the evidence and the potential magnitudes of effect, the benefits of fish intake exceed the potential risks. For women of childbearing age, benefits of modest fish intake, except for a few selected species, also outweighed risks. These conclusions are supported by the work of Alexander (5).

With the available data it was possible to conduct a risk/benefit assessment. However, one must conclude that the amount and quality of data put into this assessment are not available for any other risk/benefit considerations within the food area.

11.2 MITIGATION AND POTENTIAL RISK OF OTHER POSSIBLE FOOD SAFETY-RELATED ISSUES

11.2.1 Acrylamide

Mitigation of acrylamide (AA) in the most relevant food commodities has been proposed along the following lines (see References 7 and 8).

- For potato products
  - use varieties with low concentrations of reducing sugars
  - store potatoes at higher temperature
  - blanching of potato slices and cuts before frying
  - Vacuum frying of potato crisps
- For cereal-based products
  - leavening of bread with yeast that utilizes asparagine more effectively
  - prolonged leavening of bread
  - exchange ammonium hydrogen carbonate for sodium hydrogen carbonate in dry biscuits

- For both potato and cereal-based products
  - lower cooking temperature
  - shorter cooking time
  - addition of an enzyme (asparaginase) in certain dough-based products before cooking
- For coffee
  - prolonged storage of more than 6 months of roasted coffee. It is important to note that this probably also affects the quality of the product and hence is of limited value.

11.2.1.1 Effects on Other Maillard Reaction Products In all proposals the aim is to reduce the amounts of AA. Since AA is formed by the Maillard reaction, one can suspect that the amounts of other Maillard reaction products (MRPs) are influenced as well, unknown if the concentrations increase or decrease. This has caused some concern, for two reasons. First, is it possible that the concentrations of other MRPs, as toxic as AA, increase and thereby create a bigger problem than initially? Whether this is the case or not is not known. There are, however, some indications of possible MRPs to investigate further. Chaudhry *et al.* (9) compiled all MRPs, as part of the HEATOX project, known to be present in food and performed molecular modeling to predict mutagenicity and carcinogenicity. The database contains 740 compounds that may be formed in the Maillard reaction or through lipid oxidation. The probabilities of being mutagenic and/or carcinogenic were calculated using a toxicity expert system. Fifty-two compounds were predicted to be both genotoxic and carcinogenic (see Tables 11.1 and 11.2). All lipid oxidation products have been found in cooked foods whereas six of the MRPs have not yet been identified.

Unfortunately, there are no systematic investigations on the concentrations of these suspected mutagens/carcinogens in cooked foods where the amounts of AA have been decreased after applying one or more of the mitigation strategies mentioned. A way forward could be to test the compounds in Tables 11.1 and 11.2 for genotoxicity and then develop a multi-method for detection of the compounds that turned out to be genotoxic. The effects of different mitigation strategies directed toward AA could then be studied to ensure that concentrations of mutagenic MRPs do not increase.

11.2.1.2 Effects on Hydroxymethylfurfural From the data available it seems especially interesting to study 5-(hydroxymethyl)-2-furancarboxaldehyde, trivial name hydroxymethylfurfural (HMF), present in high concentrations in normal Western diets.

HMF has not been tested so far for carcinogenicity in experimental animals. In addition, there are no epidemiological data available. The genotoxic activity is rather weak in a number of test systems, but this might be due to low activity of sulfotransferases necessary to activate HMF to its sulfonated DNA-reactive

TABLE 11.1 Lipid oxidation products suspected to be mutagenic and carcinogenic.[a]

3-penten-2-one	1-octen-3-one
[E]-2-butenal	[Z]-2-butenal
[E]-2-nonenal	[Z]-2-nonenal
trans-4,5-epoxy-[E]-2-decenal	[E,E]-2,4-undecadienal
[E]-2-butenoic acid	4-hydroxynonanoic acid lactone
4-hydroxydecanoic acid lactone	5-hydroxydecanoic acid lactone
4-hydroxy-2-decenoic acid lactone	4-hydroxy-2-nonenoic acid lactone

[a]Compiled from Chaudhry *et al.* (9).

metabolite 5-sulfoxymethylfurfural (SMF). The amounts of sulfotransferases, for example SULT-1, are low in tissues of experimental animals but high in many human tissues (10). By introducing the genes for human SULTs into Chinese hamster cells, Glatt *et al.* (11) noticed a dramatic increase in the genotoxicity of HMF. A reasonable hypothesis is that current genotoxicity tests underestimate the potency of HMF, which can be present in concentrations three orders of magnitude higher than AA.

No systematic investigations have been made so far on the concentrations of AA and HMF under different cooking conditions in relevant foods. However, Rufián-Henares *et al.* (12) found no correlation between AA content and the concentration of HMF in cereals on the Spanish market, although the amounts of AA varied between 62 and 803 μg/g.

From the data presently available, it is not possible to draw any general conclusions on the risk for elevated concentrations of other MRPs or lipid oxidation products when altering conditions, in such a way that cooking results in a lower AA content.

11.2.1.3 Effects on Antioxidants The second concern when lowering the concentrations of AA by changing the conditions for the Maillard reaction is that the amount of beneficial compounds formed would decrease. The discussions have mainly focused on antioxidants formed via the Maillard reaction (13, 14).

Ehling and Shibamoto (15) studied the relationship between AA formation and antioxidant activity in an asparagine/d-glucose browning model system. The formation of AA was closely correlated to the browning, but no correlation was found between the formation of AA and antioxidants.

Summa *et al.* (16) investigated the correlation of AA content and the antioxidative activity of self-prepared model cookies. The antioxidative effect was measured as a reduction of Fremy's salt radicals. They found a direct correlation between the concentration of AA and the antioxidative activity. However, a simple model for the prediction of AA contents and antioxidative activity of samples from baking time, color, protein content, or moisture of the samples was not found. No identification of the antioxidants was undertaken.

TABLE 11.2 MRPs suspected to be mutagenic and carcinogenic.[a]

Main functional group	Compound
Pyrazine	Quinoxalin
Pyrazine	2-Methylquinoxaline
Pyrazine	2-(2-furyl)pyrazine
Pyrazine	1-Methylpyrrolo[1,2-*a*]-pyrazine
Thiophene	Thieno(3,4*b*)thiophene
Thiazole	2-Formylthiazole[b]
Thiazole	2-(2-furyl)thiazole
Pyrrole	1,2-Dimethyl-1H-pyrrole
Pyrrole	1-Ethyl-1H-pyrrole-2-carboxaldehyde
Pyrrole	1-(2-furfuryl)pyrrole-2-carboxaldehyde
Pyrrole	2-Cyano-5-methylpyrrole[b]
Pyrrole	4-(2,5-dimethyl-1-pyrrolyl)butanoic acid[b]
Furan	5-(Hydroxymethyl)-2-furancarboxaldehyde
Furan	Benzofuran
Furan	3,4-Dimethyl-2(5*H*)-furanone
Furan	4,5-Dihydro-2-methyl-3(2*H*)-furanone
Furan	2-Methylbenzofuran
Furan	4-Hydroxy-5-methyl-3(2*H*)-furanone
Furan	1-(2-furyl)-1,2-ethanediol
Furan	4,6-Dimethyl-3(2*H*)-benzofuranone
Furan	3-Hydroxy-2-methyltetrahydrofuran
Furan	1-(5-hydroxymethyl-2-furyl)-1-ethanone[b]
Furan	5-Formyl-2-furfuryl methanoate[b]
Pyridine	3-Aminopyridine
Pyridine	2-Methylfuro[2,3-*c*]pyridine[b]
Oxazole	Benzoxazole
Hydrocarbon	Styrene
Oxygen-containing	2,4-Pentanedione
Oxygen-containing	2,3-Butanedione
Oxygen-containing	1-(acetyloxy)-2-propanone
Oxygen-containing	3-Penten-2-one
Oxygen-containing	3-Methyl-3-buten-2-one
Oxygen-containing	2,3-Dimethyl-2-cyclopenten-1-one
Oxygen-containing	3-Methyl-2H-1-benzopyran-2-one
Oxygen-containing	2-Butenal
Oxygen-containing	2-Hydroxy-3-ethyl-2-cyclopenten-1-one
Oxygen-containing	2,3-Dihydro-3,5-dihydroxy-6-methyl-4H-pyran-4-one
Oxygen-containing	2,3-Dihydro-5-hydroxy-6-methyl-4H-pyran-4-one

[a]Compiled from Chaudhry *et al.* (9).
[b]Not yet detected in cooked foods.

Generally speaking, it is difficult to predict biological consequences from measurements of antioxidative or oxidative effects of food or food components. Whether a compound exerts antioxidant or oxidative effects can be dependent both on dose and the matrix where measurements take place. In addition, the antioxidants must be taken up by the organism and transported to the site of action within the body in order to have a biological effect.

Antioxidants have been suggested to have a number of different effects in humans (for a review, see Reference 13), including allergenic, antiallergenic, antibiotic, carcinogenic, anticarcinogenic, mutagenic, clastogenic, cytotoxic, antimutagenic, oxidative, and antioxidative effects. Somoza (14) has pointed to the possible relation between high-molecular-weight MRP, melanoidins, and their ability to promote glycation reactions *in vivo*. Such reactions are suggested to be involved in the progression of several diseases, e.g., diabetes mellitus, cardiovascular complications, and Alzheimer's disease. If the concentrations of these melanoidins follow the concentrations of AA, mitigation of the latter might have additional beneficial effects. The suggested beneficial effects are derived from both *in vitro* studies and studies in animals and humans.

11.2.1.4 Effects on Secondary Metabolites from Colonic Bacteria MRPs might also have effects that are not connected to their antioxidative capacity, mediated by microorganisms in the gut. The Maillard reaction produces modified proteins that are not digested, as most proteins are, in the small intestine. These modified proteins thus become available for fermentation by microorganisms in the colon. Not very much is known about the fate of the modified proteins (17), but it has been suggested that they can alter the composition of the colonic flora and hence produce secondary effects. Examples of this could be effects on energy utilization, on the immune system, hypertension, and certain mental disorders, mediated by different indoles, amines, and phenols resulting from amino acid fermentation (17). Though this might be speculative, it illustrates the complexity of the possible interactions between MRP and human well-being.

11.2.1.5 Effects on Advanced Glycation Products Recently, another class of MRPs, advanced glycation end products (AGEs) has attracted the interest of both chemists and toxicologists. AGEs are low-molecular-weight compounds that are generated in the later stages of the Maillard reaction, after the formation of a Schiff base and the Amadori rearrangement. The highest amounts have been reported in fat-rich foods, such as butter, olive oil, and almonds, somewhat lower concentrations in meat products and drastically lower concentrations in cereal-based foods. This means that the concentrations are low in food commodities usually high in AA and vice versa. It has been shown that increased exposure to AGEs is related to vascular and kidney tissue damage in experimental animals. Observations in animals suggest that AGEs have the same effects in humans. Restriction of AGEs in the diet reduces several immune defects, insulin resistance, and diabetic complications. It is proposed that AGEs increase oxidant stress, thereby impairing innate immune defense that eventually leads to inappropriate inflammatory responses. AGEs also cross-link protein fragments, decreasing their digestibility and their nutritional value (for an overview, see Reference 18). No information is available on the formation of AGEs using the suggested mitigation strategies for AA. One should also keep in mind that the highest concentrations of AGEs are found in food commodities usually low in AA content and hence not subjected to AA mitigation. Further research on the correlation between formation of AGEs and AA, together with increased knowledge about the health hazards of AGE exposure, is needed.

In summary, comparing data on all the aforementioned possible effects on human health of different MRPs with the knowledge available for PUFAs in the fish risk/benefit example and the data needed in order to assess health benefits of a diet or a specific dietary factor, as mentioned earlier, one must conclude that available scientific information on health benefits of MRP is sparse. It will be very difficult to perform meaningful risk/benefit assessments unless the database is expanded considerably.

Given the plethora of both possible beneficial and detrimental effects and their high relevance to human well-being, it is very important to investigate further both the individual and the combined relationships between AA and MRP.

11.2.2 Chloropropanols

3-Monochloropropane-1,2-diol (3-MCPD) and its esters has received most attention among the chloropropanols found in foods. Assessments made by international bodies such as the European Food Safety Authority (EFSA) or the FAO/WHO Joint Expert Committee on Food Additives (JECFA) conclude that 3-MCPD is a non-genotoxic carcinogen. JECFA and the former Scientific Committee on Foods of the European Commission hence allocated 3-MCPD an ADI value of 2 μg/kg bw using a safety factor of 500 from the dose that was without any carcinogenic effect in animals. There are no reasons to believe that humans are more sensitive although this has not been studied in detail. The European Commission has then set a regulatory limit of 20 μg/kg soy sauce. This means that unless the relevant exposure of 3-MCPD increases with a factor of 500, it is not meaningful to make a risk/risk assessment with AA since the latter is regarded as a genotoxic compound without any safe exposure at all. In the eyes of a regulator, AA probably represents a higher risk, even at very low doses obtained after mitigation procedures. It is of course of interest to investigate if mitigation strategies for AA could result in such a drastic increase in bioavailable 3-MCPD concentrations. Whether mitigation of 3-MCPD itself gives rise to unwanted compounds or loss of beneficial factors has not been studied.

11.2.3 Polycyclic Aromatic Hydrocarbons (PAHs)

The relevant exposure to PAH via food comes from smoked products or foods grilled over open fire. PAH is formed under incomplete combustion of organic material and is carried with smoke or heat conversion to the food that is smoked or grilled. No appreciable amounts of PAH are formed in the food as such. There is consequently no substantial link to formation of AA. Mitigation of PAH formation, for example, by washing the smoke or making sure that no flames are visible while grilling, cannot be expected to influence the concentrations of AA. Drying of cereals and roasting of coffee by direct contact with combustion exhausts also increase the formation of PAHs. Whether AA has been measured under such conditions is not known.

Dennis *et al.* (19) found a good correlation between PAH formation and *N*-nitrosamines in Icelandic smoked cured mutton. This was somewhat unexpected since *N*-nitroso compounds are formed under heat in the food itself from secondary or tertiary amines and nitrite, while PAH is formed by burning the fuel that generates smoke.

11.2.4 Biogenic Amines

These compounds are mainly the result of fermentation processes. Mitigation often consists of good hygiene, which is beneficial also from other viewpoints. Formation of biogenic amines in wine is mitigated by the use of bentonite and edible oils can sometimes be cleaned by bentonite. In some instances, bentonite has been shown to be naturally contaminated by dioxins (20).

11.2.5 Ethyl Carbamate

The formation of ethyl carbamate or urethane in foods is usually connected to fermentation processes. A number of pathways have been described involving diethyl pyrocarbonate and ammonia, carbamoyl phosphate and ethanol, urea and ethanol, cyanide, cyanate, and alcohol (21). Mitigation strategies have been proposed involving controlled fertilization of vineyards to reduce nitrogen contents of grapes, the use of yeast strains producing low concentrations of urea, selection of yeasts that produce low concentrations of citrullin in the case of malolactic fermentation, addition of urease, and minimizing heat exposure during storage and transport (22). There is no information on the formation of other hazardous compounds or loss of beneficial factors as a consequence of these proposals.

11.2.6 Nitrosamines

Nitrosamines are considered one of the most potent groups of carcinogens. Approximately 300 of these compounds have been tested, and about 90% of them have been found to be carcinogenic in laboratory animals. Special emphasis has been put on *N*-nitroso-dimethylamine, the most potent carcinogen in this group found in foods. The exposure to nitrosamines via food comes primarily from smoked cured meat, beer, and in some countries fish (23). Nitrosamines in cured meat are formed from nitrite and secondary amines. Mitigation strategies involve addition of antioxidants such as sodium ascorbate and/or α-tocopherol. This has proved very efficient in products such as bacon. Generally, the mitigation effect is more pronounced in foods with a high fat content (24, 25). There are no data indicating formation of other toxic substances or loss of beneficial factors after adding these antioxidants to reduce the concentration of nitrosamines.

Nitrite is added to cured meat products for several reasons: one being that nitrite inhibits the growth and toxin production of *Clostridium botulinum*. In addition to adding antioxidants, there have been discussions on the necessity of adding nitrite to cured meat products to prevent the growth of *C. botulinum*. A lower addition of nitrite would result in lower production of nitrosamines. In 2003, EFSA's Scientific Panel on Biological Hazards evaluated the need for nitrites/nitrates to ensure microbiological safety of meat products (26). Their conclusion was that it is necessary to add 50–150 mg/kg of nitrate to inhibit the growth of *C. botulinum* in some meat products, especially cured products with a low salt content and a prolonged shelf life. No risk/benefit assessment was performed, taking nitrosamine formation into account. Walker (23) recommends nitrite addition to cured meat should not be higher than necessary to ensure microbiological safety. There have been no suggestions to abolish nitrite addition to lower the exposure of nitrosamines and to accept a higher risk for growth of microorganisms.

11.2.7 Irradiation

Food irradiation has been considered a safe processing technology for improving food safety by preservation, eliminating efficiently bacterial pathogens, parasites, and insects. However, there have also been concerns raised about the safety of irradiated food. Generation of long-lived radicals and loss of vitamins have been put forward in the debate but also stable organic compounds like 2-alkylcyclobutanones (2-ACBs), radiolytic derivatives of triglycerides, formed uniquely upon irradiation of fat-containing food (27). The toxicity of various synthetic 2-ACBs has so far only been studied *in vitro*, revealing cytotoxic and DNA-damaging potential. Further studies are necessary to clarify mechanisms of action and potential relevance for human health before any risk/benefit assessment can be performed. The relevance of vitamin losses for consumers has not been determined.

11.2.8 Benzene

Ascorbic acid and benzoates can cause slight to moderate increases of benzene in soft drinks of various types. No information is available on other toxic or beneficial compounds after mitigation by lowering the addition of ascorbic acid and benzoates (28).

11.2.9 Trans Fatty Acids

The undesirable effects of trans fatty acids (TFAs) on lipoproteins, important in CVD, were confirmed around 1990. This started considerable mitigation efforts by industry in order to reduce consumer exposure to TFAs. There has been a tendency in low-trans margarines to replace TFAs with palmitic acid. This has caused some concern since palmitic acid, which has a potent hypercholesterolemic effect, then became the dominating saturated fatty acid in this type of product. It has been suggested that at least a part of the TFAs could be replaced with stearic acid instead of palmitic acid. Stearic acid has a lower hypercholesterolemic effect than palmitic acid (29).

When partially hydrogenated vegetable fats are to be replaced in some processed foods, e.g., croissants and wafers, unsaturated vegetable oils are not an alternative. Out of technological reasons, coconut oil and other palm kernel oils are used. They contain high amounts of saturated fatty acids and low amounts of polyunsaturated fatty acids. When TFAs are reduced in industrial fats, this can result in an unfavorable exposure to higher amounts of saturated fatty acids and lower concentrations of polyunsaturated fatty acids.

No systematic studies have been identified that, in a quantitative way, compare the risks of different alternatives when TFAs are to be replaced in different types of fats.

11.3 RESEARCH NEEDS

In order to facilitate risk/risk and risk/benefit assessments in the future, there is a need to develop:

- concepts and methods to enable comparison between different types of toxic effects
- concepts and methods to enable quantitative assessments of benefits
- a common denominator for risks and benefits to enable comparisons.

Furthermore, as indicated by the examples given, there is a general need for much more data on both risks and benefits within many areas. However, it seems less likely that there will be huge datasets available on hard end points derived from humans, both with regard to risks and benefits, in more than a few instances in the future. Consequently, there is a need for validated surrogate end points that are easier to obtain.

As evident from the text, there are a great number of specific questions regarding risks and benefits that have been raised. One could argue that there is a need to validate risks and benefits with cooking using a more holistic approach. By breaking down cooking into narrow, specific questions, there is a risk that one will not see the forest for all trees. Hence there is a need to develop aggregated measures of risks and benefits of different cooking methods.

11.4 CHALLENGES AND FUTURE PROSPECTS

Risk/benefit assessments contain value-based judgments that should not be the sole responsibility of scientists. An obvious example is the quantification of dread caused by different types of toxic effects. There is a challenge in getting public acceptance for such procedures and also to get acceptance for the common denominators of risks and benefits that need to be developed.

One might also expect pressure for consumers to be involved, not only in the final weighing and management of risks and benefits but also in formulating questions to the scientists before the assessments start.

Another challenge is to get scientists from various disciplines to cooperate. In order to investigate if and how cooking methods need to be revised to ensure human health, scientists from a number of fields, including chemistry, food science, toxicology, microbiology, nutrition, and medicine, need to work together closely. It is the task of grant-providing bodies to facilitate such research programs.

REFERENCES

1. van Kreijl, C.F., Knaap, A.G.A.C., van Raaij, J.M.A. (2006). *Our Food, Our Health*, National Institute for Public Health and the Environment, Bilthoven.
2. Owens, J.W. (2002). Chemical toxicity indicators for human health: case study for classification of chronic noncancer chemical hazards in life-cycle assessment. *Environmental Toxicology and Chemistry*, *21* (1), 207–225.
3. Renwick, A.G., Flynn, A., Fletcher, R.J., Muller, D.J.G., Tuijtelaars, S., Verhagen, H. (2004). Risk-benefit analysis of micronutrients. *Food and Chemical Toxicology*, *42*, 1903–1922.
4. Mozzaffarian, D., Rimm, E.B. (2006). Fish intake, contaminants, and human health: evaluating the risks and the benefits. *The Journal of the American Medical Association*, *296*, 1885–1899.
5. Alexander, J., Frøyland, L., Hemre, G.-I., Jacobsen, B.K., Lund, E., Meltzer, H.M., Skåre, J.U. (Chair). (2006) Benefits and risks of fish consumption in Norway, in *The EFSA's 6th Scientific Colloquium Report—Risk-Benefit Analysis of Foods: Methods and Approaches*, European Food Safety Authority, Parma, pp. 101–117. http://www.efsa.europa.eu/cs/BlobServer/Scientific_Document/comm_colloque_6_en.pdf?ssbinary=true (accessed 3 October 2008).
6. Gochfeld, M., Burger, J. (2005). Good fish/bad fish: a composite benefit-risk by dose curve. *NeuroToxicology*, *26*, 511–521.

7. CIAA. (2006). CIAA "Acrylamide Toolbox." http://www.ciaa.eu/documents/brochures/toolbox%20rev11%20nov%202007final.pdf (accessed 8 March 2008).
8. *Final Report* (2007). HEATOX DG Sanco 6th Framework project Contract No. FOOD-CT-2003-506820. http://www.slv.se/upload/heatox/documents/Heatox_Final%20_report.pdf (accessed 28 April 2008).
9. Chaudhry, Q., Cotterill, J., Watkins, R. (2006). A molecular modelling approach to predict the toxicity of compounds generated during heat treatment of food, in *Acrylamide and Other Health Hazardous Compounds in Heat-Treated Foods* (eds K. Skog, J. Alexander), Woodhead Publishing, pp. 132–160.
10. Glatt, H.R. (2005). Activation and inactivation of carcinogens by human sulfotransferases, in *Human Sulphotransferases* (eds G.M. Pacifici, M.W.H. Coughtrie), Taylor & Francis, London, pp. 281–306.
11. Glatt, H.R., Schneider, H., Liu, Y.G. (2005) V79-hCYP2E1-hSULT1A1, a cell line for the sensitive detection of genotoxic effects induced by carbohydrate pyrolysis products and other food-borne chemicals. *Mutation Research*, *580*, 41–52.
12. Rufián-Henares, J.A., Delgado-Andrade, C., Morales, F.J. (2006). Relationship between acrylamide and thermal-processing indexes in commercial breakfast cereals: a survey of Spanish breakfast cereals. *Molecular Nutrition & Food Research*, *50*, 756–762.
13. Friedman, M. (2005). Biological effects of Maillard browning products that may affect acrylamide safety in food: biological effects of Maillard products. *Advances in Experimental Medicine and Biology*, *561*, 135–156.
14. Somoza, V. (2005). Five years of research on health risks and benefits of Maillard reaction products: an update. *Molecular Nutrition in Food Research*, *49*, 663–672.
15. Ehling, S., Shibamoto, T. (2005). Correlation of acrylamide generation in thermally processed model systems of asparagine and glucose with colour formation, amounts of pyrazines formed, and antioxidative properties of extracts. *Journal of Agriculture and Food Chemistry*, *53*, 4813–4819.
16. Summa, C., Wenzl, T., Brohee, M., de la Calle, B., Anklam, E. (2006). Investigation on the correlation of the acrylamide content and antioxidant activity of model cookies. *Journal of Agriculture and Food Chemistry*, *54*, 853–859.
17. Tuohy, K.M., Hinton, D.J.S., Davies, S.J., Crabbe, M.J.C., Gibson, G.R., Ames, J.M. (2006). Metabolism of Maillard reaction products by the human gut microbiota—implications for health. *Molecular Nutrition in Food Research*, *50*, 847–857.
18. van Nguyen, C. (2006). Toxicity of the AGEs generated from the Maillard reaction: on the relationship of food-AGEs and biological-AGEs. *Molecular Nutrition in Food Research*, *50*, 1140–1149.
19. Dennis, M.J., Cripps, G.S., Tricker, A.R., Massey, R.C., McWeeny, D.J. (1984). N-nitroso compounds and polycyclic aromatic hydrocarbons in Icelandic smoked cured mutton. *Food and Chemical Toxicology*, *22*, 305–306.
20. Wright, C., Davenport, E.J., Kan-King-Yu, D., Jefferies, D., Cubberly, R., Lalljie, S.P. (2005). Inter-laboratory comparison of polychlorinated dibenzo-p-dioxins (PCDDs) and dibenzofurans (PCDFs) in bleaching earth used in the refinement of edible oils. *Food Additives and Contaminants*, *22*, 716–725.
21. Uggla, A., Busk, L. (1992). Ethyl carbamate (urethane) in alcoholic beverages and foodstuffs—a nordic view, in *Nordiske Seminar- og Arbejdsrapporter*, Vol. *570*, Nordic Council of Ministers, Copenhagen, pp. 10–12.

22. U.S. Food and Drug Administration (1997) Centre for Food Safety and Applied Nutrition, Ethyl Carbamate Preventative Action Manual. http://www.cfsan.fda.gov/~frf/ecaction.html.
23. Walker, R. (1990). Nitrates, nitrites and N-nitrosocompounds: a review of the occurrence in food and diet and the toxicological implications. *Food Additives and Contaminants*, *7*, 717–768.
24. Lathia, D., Blum, A. (1989). Role of vitamin E as nitrite scavenger and N-nitrosamine inhibitor: a review. *International Journal of Vitamin and Nutrition Research*, *59*, 430–438.
25. Fiddler, W., Pensabene, J.W., Piotrowski, E.G., Phillips, J.G., Keating, J., Mergens, W.J., Newmark, H.L. (1978). Inhibition of formation of volatile nitrosamines in fried bacon by the use of cure-solubilized a-tocopherol. *Journal of Agriculture and Food Chemistry*, *26* (3), 653–656.
26. EFSA Scientific Panel on Biological Hazards (2003). Opinion of the scientific panel on biological hazards on the request from the commission related to the effects of nitrites/nitrates on the microbiological safety of meat products. *The EFSA Journal*, *14*, 1–31.
27. Hartwig, A., Pelzer, A., Burnouf, D., Titéca, H., Delincée, H., Briviba, K., Soika, C., Hodapp, C., Raul, F., Miesch, M., Werner, D., Horvatovich, P., Marchioni, E. (2007). Toxicological potential of 2-alkylcyclobutanones—specific radiolytic products in irradiated fat-containing food—in bacteria and human cell lines. *Food and Chemical Toxicology*, *45*, 2581–2591.
28. McNeal, T.P., Nyman, P.J., Diachenko, G.W., Hollifield, H.C. (1993). Survey of benzene in foods by using headspace concentration techniques and capillary gas chromatography. *Journal of AOAC International*, *76*, 1213–1219.
29. Pedersen, J.I. (2006). Composition of "post trans" margarines (abstract). Fats and health—update on dietary phytosterols, trans fatty acids and conjugated linoleic acids. DGF Meeting, October 2006, Frankfurt, pp. 19–20.

INDEX

Accepted Daily Intake (ADI) 696
Acenaphthene 245
Acenaphthylene 245
Acetaldehyde 60
Acetol (3-hydroxyacetone) 200
Acid-Hydrolysed Vegetable Proteins (acid-HVP) 177, 203, 539, 541, 567
Acrolein 51
  acceptable daily intake (ADI) 65
  acute dermal LD_{50} 62
  acute oral LD_{50} 62
  ambient air 53, 64
  analysis methods 55–59
  chemical and physical properties 56
  emission 51, 53
  Environmental Protection Agency (EPA) 51
  formation 51
  herbicide 51
  inhalation toxicity 62
  mechanisms of formation 59–62
  occurrence in foods 53–55
  precursors 53
  toxicity 51, 62–65
  use 51
Acrylaldehyde, *see* Acrolein
Acrylamide 10, 23, 566, 582, 698
  agronomical factors 32, 33
  analysis 30
  brochures 39, 41
  CIAA Acrylamide Toolbox 16, 31, 39, 40, *see also* CIAA
  Codex Code of Practice 41
  exposure estimates 35, 37
  FAO/WHO consultation 33
  haemoglobin adduct 25, 39
  health risks 37–39
  JECFA evaluation 38
  JIFSAN database 40
  mechanisms of formation 26–30
  occurrence in foods 24–25
  polymer 23
  railway tunnel accident 23
  reduction in food 31–33, 41
Acrylaway 32
Acrylic acid 59, 61
Acute Reference Dose (ARfD) 585
Adiabatic heating 451
Adrenal Medulla lesions 129

Process-Induced Food Toxicants: Occurrence, Formation, Mitigation, and Health Risks, Edited by Richard H. Stadler and David R. Lineback

Advanced Glycation Endproducts (AGEs) 151, 215, 702
analysis 223–227
exposure 228–229
formation 218–223
health risks 229–232
historical perspective 215–218
mitigation 227–228
Agency for Toxic Substances and Disease Registry 65
AGE-RAGE interactions 230
Agmatine 323
Alcoholic beverages 140–143, 297, 373, *see also* Beer; Wine
Aldehyde
α,β-unsaturated 158
reactive 155, 701
Strecker, *see* Strecker aldehyde
Alfalfa seeds 513
2-Alkylcyclobutanone 393
2-Alkylcyclobutanones 705
Allyl alcohol 196
Almonds, roasted 138
Alzheimer's disease 340, 702
Amadori compound 27, 220, 221, 511, 512, 515, *see also* Amadori rearrangement product
Amadori compound, protein bound 221
Amadori rearrangement 218
Amadori rearrangement product (ARP) 150–151, 647
American Beverage Association 432
Ames *Salmonella typhimurium* mutagenesis assay 82, 88, 122, 131, 157, 288, 405
Amino acids 648
cross-linked 222, 475, 480
D-isomers 475, 482
Amino acid decarboxylase activity 324
β-Aminoalanine 480
2-Amino-1,4-dimethyl-5H-pyrido [4,3-b]indole (Trp-P-1) 77
2-Amino-1-dimethyl-5H-pyrido [4,3-b]indole (Trp-P-2) 77
2-Amino-1-methyl-6-phenylimidazo [4,5-b]pyridine (PhIP) 78
2-Amino-3,4,8-trimethylimidazo [4,5-f]quinoxaline (4,8-DiMeIQx) 78
2-Amino-3,8-dimethylimidazo[4,5-f]quinoxaline (8-MeIQx) 78
2-Amino-3-methyl-9H-pyrido[2,3-b]indole (MeAαC) 77
2-Amino-3-methylimidazo[4,5-f]quinoline (IQ) 78
2-Amino-3-methylimidazo[4,5-f]quinoxaline (IQx) 78
2-Amino-9H-pyrido[2,3-b]indole (AαC) 77
Anthracene 245
Antioxidants 32, 87, 604, 700
Arrhenius equation 153
As Low as Reasonably Achievable (ALARA) 14, 33, 436, 583
As Low as Reasonably Practicable (ALARP) 14
Ascorbic acid 5, 428–429
Asparaginase 32, 607
Asparagine 27–28
Aspergillus oryzae 542
Automated headspace sampling 119
Azodicarbonamide 307
Azomethine ylide 29

Baby food 26, 119–120, 140, 490, *see also* Infant formula
Bacillus cereus 594
Bacteria starter cultures 327
Bacterial mutagenesis assays 88–89, *see also* Ames *Salmonella typhimurium* mutagenesis assay
Bacterial spores 446, 461
Bakery products 222, 568
Bamboo leaves 604
Barbecuing 84–85, *see also* Cooking
Barley 199, 487
Beef 328, *see also* Meat and meat products
Beer 331, 332, 373, 374, 520
Benchmark Dose Lower Confidence Interval (BMDL) 11, 38, 268, 301, 379
Benefits
guidelines for assessment 697
Benz[a]anthracene 245
Benzaldehyde 429
Benzene 413, 705
analysis 415–421
exposure 433–434
formation 426–431
health risk 434–435

mitigation 431–432
occurrence 421–426
risk management 435–436
Benzo[a]pyrene (B[a]P) 245
Benzo[b]fluoranthene 245
Benzo[g,h,i]perylene 245
Benzo[k]fluoranthene 245
Benzoate 415
and ascorbic acid interaction 428
Benzoic acid 424, *see also* Benzoate
Beta-carbolines 80
Beverage industry 431
Beverages 26, 256
alcoholic 297, 299, 309
carbonated 424
non-carbonated 422
Biogenic Amine Index (BAI) 327
beer 331
Biogenic amines 321, 572, 704
analysis 333–337
exposure 340–342
formation 337–339
health risks 342–343
historical perspective 321
mitigation 339–340
occurrence 324–333
risk management 343
Biomarkers, of exposure 259
Biosensors 337
Biscuits 568
Boronic acid derivatives, of MCPD 183
Bouillon cubes 86
Bovine Spongiform Encephalopathy (BSE) 680
BRAFO project 15
Bread 138, 140, 191, 195, 198, 224, 300, 308, 513, 568, *see also* Biscuits; Tortillas
crumb 140, 193, 224
crust 190, 195, 224
Bread machines, domestic 198
Breakfast cereals 26, 140, 568
Breast milk 181
Brewed coffee 119–120, *see also* Coffee
3-Bromopropane-1,2-diol 194
2-Bromopropane-1,3-diol 194
2,3-Butanedione 701
2-Butenal 700
Butylated hydroxyanisole (BHA) 87
Butylated hydroxytoluene (BHT) 87
CAC General Standard for Irradiated Food 394
Cadaverine 323
Caffeic acid 87
Calcium, *see* Divalent cations
California 42
Camembert cheese 396
Cancer
colon and colorectal 90, 154, 260
leukaemia 434
mammary 38, 90
ovarian 39
pancreatic 261
prostate 90
Canned food products 120
Canned soups 119
Canning 448
Cannizzaro reaction 429
Capillary Electrophoresis (CE) 83, 147
Caramelization 148–150, 218
Carbamic acid 306
Carbamyl phosphate 306
Carbamylation 306
Carbohydrate 454, 650
Carboxymethyllysine (CML) 221, 224
Carcinogenicity Potency Database (CPDB) 11
Cardiovascular disease (CVD) 646, 697
Casein 481, 485
Catechins 87, 661
Celery tuber 463
Cereal products 26, 191, 197, 300, 477, 486
Cereal-based foods 140
3-MCPD and acrylamide formation 201, 582
Cereals 199, 256, 300
CHARRED database 260
Cheese 191, 326, 375, 513
Cheese, mozzarella 488
Chelation of metal ions 432, 495
Chicken 87, 191, 327, *see also* Meat and meat products; Poultry
Chicory 138, 254
Chicory-coffee mixtures 139
Cost Action 927 17
Chinese Restaurant Syndrome 518

Chloroesters (e.g., 3-MCPD-esters) 177–178, 180, 189, 198, *see also* Chloropropanols
  and human breast milk 181
  and human gut metabolism 180
Chlorogenic acids 666–667
2-Chloropropane-1,3-diol (2-MCPD) 175–177, 539, *see also* Chloropropanols
3-Chloropropane-1,2-diol (3-MCPD) 175–176, 640, *see also* Chloropropanols
Chloropropanols 175, 539, 567, 580, 703
  analysis 181–189, 546
  chemical structures 177–178, 540
  exposure 201–203
  formation and precursors 195–200, 546–552
  health risks 179–181
  historical perspective 175–178
  maximum limits 203
  mitigation 200–201, 552–557
  occurrence 190–195
  protein hydrolysates 541–546
  risk management 203–204
Christmas pudding 298
Chrysene 245
CIAA (European Food & Drink Federation) 16, 31–32
Cigarette smoke 75, 78, 433
Cis-2-butene-1,4-dial 121, 130
Citric acid 119
Clostridium botulinum 5, 389, 448, 461, 565, 573, 705
Cocoa 227, 375, 660
  epimerisation, of (−)-epicatechin 661
  human intervention studies 662
  polyphenol content 663
  processing and changes 600
Cocoa powder 26
Cocoa products 26
Codex Alimentarius Acrylamide Code of Practice 41
Coffee 138, 191, 194, 222, 224, 229, 256, 520, 666
  Arabica 666
  decaffeinated 138, 195
  epidemiological studies 668
  green coffee 422, 427, 666
  health benefits 666
  instant 138, 195, 375
  mutagenic activity 666
  polyphenol content 662–668
  processing 666
  Robusta 667
  torrefacto 139
Coffee beans 422, *see also* Coffee
Coffee filters 193
Colorimetric methods 145
Comet assay 402–403, 406
Consumer information 41–42
Cooked beef 80, 84
Cooked meats 78, 80
Cooking
  barbecuing 250
  charbroiling 250
  deep fat frying 53
  extrusion 602, 606
  frying 53, 86, 249
  grilling 87, 248, 250, 431
  high pressure pasteurization 226
  microwave 86
  pan-residue scrapings 85
  risks and benefits of 695, *see also* Risk/benefit considerations; Risk/risk considerations
  roasting 248, 252, 422
  steaming 249
Cooking oils 53–54
Copper 305, 428, 430, 431, 495
Corn meal 487, *see also* Maize
Cornflakes 606
Cranberries 424
Cranberry juice 424
Creatine 80
Crispbread 568
Crude oil 247
Curing salts 371
Cyanic acid 306
Cyclobutanones 393
Cytochrome P-450 monooxygenases (CYP450) 90–92, 94–95 121–122, 131, 261, 288, 289, 378

Dairy products 250, 325, 488, 513, *see also* Milk; Yogurt
D-Amino acids 509
  Analysis 511

D-alanine 522
D-arginine 523
D-aspartic acid 523
D-cysteine 523
D-cystine 524
D-histidine 524
digestion 521–527
D-lysine 524
D-methionine 525
D-phenylalanine 525
D-proline 525
D-serine 526
D-threonine 526
D-tryptophan 526
D-tyrosine 527
D-valine 527
occurrence 513–520
racemization mechanisms 511–512
Deep fat frying, *see* Cooking
Dehydroalanine 479, 485
Dehydrochlorination 549
3-Deoxyglucosulose-lysine-dimer (DOLD) 222
Deterministic models 35
DFG-Senate Commission for Food Safety (SKLM) 453
Diabetes 217, 233
Diabetic nephropathy 231
Dialysis patients 231
Dibenz[a,h]anthracene 245
Dicarbonyl compounds 28, 30, 150, 151
Dichloropropanol esters 545
Dichloropropanols 179–182, *see also* Chloropropanols
1,3-Dichloropropane-2-ol (1,3-DCP) 543
1,3-Dichloropropanol (1,3-DCP) 175–177, 193, *see also* Dichloropropanols
2,3-Dichloropropane-1-ol (2,3-DCP) 543
2,3-Dichloropropanol (2,3-DCP) 175–177, *see also* Dichloropropanols
Diels-Alder rearrangement 258, 463
Dietary staples 37
Diethyl dicarbonate 287
Digestibility 646
2,5-Dimethylpyrazine 80
Dioxolane/Dioxane derivatives 187
Direct fire drying 257
Disability Adjusted Life Years (DALY) 15, 696
Dissociation reactions 457
Divalent cations 32, 153
DNA adducts 90, 130, 261
DNA comet assay 395
Dried fruit 138
Dried herbs 389
Drinking water 178, 426
Drying 249, 257
DSM Food Specialties 32
Duplicate diet study 34
D-Value 448
Dynamic Purge and Trap (P&T) HS analysis 415

EC DG-SANCO 16
EC SCOOP task 192
Edible oils, *see* Fats and oils
EDTA 419, 429
Eggs 396, 434, 489, 518
Electron Spin Resonance (ESR) spectroscopy 394
Electrophiles 93, 121, 377
Emulsifier 197
1,2-Enolisation 148, 150
Enteral formula 142–143, 226
(−)-Epicatechin gallate (ECG) 87, 664, *see also* Catechins
(−)-Epigallocatechin (EGC) 87, 664, *see also* Catechins
(−)-Epigallocatechin gallate (EGCG) 664, *see also* Catechins
Enzyme Linked Immunosorben Assay (ELISA) 83, 223, 336
Epichlorhydrin 180, 191, 193
Epidemiology studies 39, 89–96, 157, 260
Epoxide Hydrolase (EH) 261
ERA Net Scheme 17
Erythorbic acid 423
Escherichia coli 86, 635
Escherichia coli O157 581
Ethanedial, *see* Glyoxal
Ethyl carbamate (urethane) 285, 573, 704
analysis 289–297
chemical and physical properties 290
formation 303–308
historical perspective 285–287
legislation 309 310

Ethyl carbamate (urethane) (*cont'd*)
  mitigation 308
  occurrence and exposure 297–301
  risk assessment 301–302
  toxicology 287–289
European Committee for Standardization (CEN) 394
European Food Safety Authority (EFSA) 15, 298, 397, 703
European standards for the detection of irradiated foodstuffs 395
European Technology Platform (ETP) "Food for life" 17
European Union (EU) Scientific Committee on Food (SCF) 404
Expert Committee on the Wholesomeness of Irradiated Food (JECFI) 398
Exposure estimates 33–35
Extruded cereal snack 593
Extrusion cooking 649, *see also* Cooking

Fats and oils 256
  bleaching 662
  degumming 662
  deodorization 663
  free fatty acids 642
  neutralization 662
  refining 652, 653
  storage, handling 663
Fenton's reaction 428
Fermentation 55, 325
Fertilisers 32, 602
Fish 84, 191, 328, 489, *see also* Marine products; Pickled herring
  smoked 329, 372
Flavonoids 80
Flour 307
Fluoranthene 245
Fluorene 245
Food additives 142
Food chemical contaminants 7
  routes of contamination 8
Food consumption surveys 34, 36, 301
Food contact materials 178, 191
Food flavourings 178, 192
Food ingredients 142, 192
Food irradiation 119–120, 340, 387, 415, 423, 427, 705
  applications 388–390
  benzene formation 407
  carbohydrates 391
  detection methods 394–397
  lipids 391, 393
  proteins 391
  radiation chemistry 390–394
  toxicology 397–407
Food preservation 4–5
  liquid smoke flavouring 5
  modern means 4
  nitrite 5
  nitrosamines 5
  PAHs 5
Food processing 3–6
  and food safety hazards 6–8
  and nutritional aspects; see Chapter 9
  canning 4
  early types of 4
  history 3–6
  major benefits; see Chapter 9
  minimal food processing 4
  pathogens 4,6
  physical hazards 7
  role 3–6
  techniques 5
  technological developments 9
Food spoilage 639
Food supplements 397
Formaldehyde 60
Formic acid 152
Free radical mechanisms 61
Free radicals 60, 87
French fries 194, 568, 579, 590
Fresh foods 55
Fruit 422, *see also* Dried fruit; Fruits and vegetables; Stone fruit spirits; *names of specific fruits*
Fruit drinks 424
Fruit juice 139–140, 518
Fruits and vegetables 26, 513
Functional foods 646
Furan 117, 570
  analysis119
  exposure assessment 117, 120
  metabolism 121–122
  mitigation 121

occurrence 139, 119–120
toxicology 122–130
Furfural 139, 148, 151–152
Furoylmethyl amino acids (fMAAs) 224
F-Value 448

Gamma irradiation, *see* Food irradiation
Garlic 192, 196
Gas Chromatography/Electron Capture Detection (GC/ECD) 59
Gas Chromatography/Mass Spectrometry (GC/MS) 31, 56, 59, 83, 119, 181, 183, 257, 290, 415, 479
Gelatin 477
Genetic factors 33
Genotoxicity 38, 127, 154, 179
German minimisation concept 41
German "Signal Value" (SV) 583
Gibbs function 450
Glutathione 121, 159, 228
Glutathione *S*-transferase (GST) 94, 159–160
Glycation compounds
bioavailability 231
intake 229
Glycerol 196–197, 547, 555
Glycerol dehydration 60, 62
Glycidamide 37–38
Glycidol 180, 197, 200, 549
Glycosylamines 150
Glycotoxins 218, 230
Glyoxal 30, 150, 227
GOLD (gyloxal-lysine-dimer) 222
Good manufacturing practice (GMP) 7, 553, *see also* Hazard analysis critical control point, prerequisites
Grapes 139
Green tea, *see* Tea
Ground beef, 590 396
Guidance limits, for process toxicants 586

Ham 327
Hamburger patties 86
Hazard analysis critical control point (HACCP) 7, 12, 565
application to process toxicants 576
benefits 574
case study cereal snack 593
control point (CP) 578
critical control point (CCP) 576
history 574
operational prerequisites 578
prerequisites 577
principles 575
Hazards
characterization 14
identification 13
Headspace Solid Phase Microextraction (SPME) 420
Headspace/SPME/Gas Chromatography/Mass Spectrometry (HS/SPME/GC/MS) 59
Health Canada 41
HEATOX project 10, 40, 699
Hepatocellular neoplasms 128
Heptaflourobutyryl (HFB) ester derivative 181
Herbs 397, *see also* Dried herbs
Heterocyclic Aromatic Amines (HAAs) 75
aminoimidazoarenes (AIAs) 78
analytical methods 82–84
antioxidants as inhibitors of 87
bioavailability 89
chemical structures 77
endogenous formation 80–82
health risks 96–97
mechanisms of formation 77–80
metabolism 89–96
mutagenic potency 88–89
occurrence 84–86
pyrolytic HAAs 77
reduction in cooked meat 86–88
Heyns products 150, 218
High Performance Liquid Chromatography (HPLC) 55, 257, 335
High-Pressure Processing (HPP) 340, 445
allergenic potential 462–464
basic principles 452
chemical and matrix effects 453–460
microbial effects 460–462
HighQ RTE project 9
Histamine 323

Histamine intoxication 341
Histidinoalanine 480
Histidinomethylalanine 479
Honey 137, 227, 518, 603
Honey, Manuka 227, 229
Human Immunodeficiency Virus type 1 (HIV-1) 527
Hydrogen peroxide 60
Hydroxyl radical 427
3-Hydroxy-B[a]P 259
5-Hydroxymethylfurfural (HMF) 135, 227, 610, 699
  analysis 143–148
  chemical and physical properties 137
  decomposition 151
  exposure 160
  formation 148–151
  HMF value 140
  mitigation 152
  occurrence in foods 137–143
  quality indicator, storage 139
  toxicology 154–160
Hydroxymethyloxirane 556
Hypoallergenic infant formula 220
Hypochlorous acid 196

In vivo glycation 230
Indeno[1,2,3-cd]pyrene 245
Infant cereals 140
Infant foods 26, 119–120, *see also* Baby food
Infant formula 140, 375, 488, 518
Inflammatory markers 231
Infrared heating 638
  applications in drying 639
  honey 640
  pasteurization 640
  penetration depth 640
Institute for Reference Materials and Measurements (IRMM) 25
International Agency for Research on Cancer (IARC) 122, 245, 289, 566
International Council of Beverages Associations 431
International Food Information Council (IFIC) 681
Iron 429, 430
Irradiation, *see* Food irradiation
ISO 22000 578

Jam 139, 424
Jarred products 572

Kidney toxicity 493, 526, *see also* Renal failure
Koji 542

Lactalbumin 482
Lactic Acid Bacteria (LAB) 304, 307, 327
Lanthionine 480
Lanthionine (LAN) isomers 524
Levulinic acid 152
Linoleic acid 656
Linolenic acid 652
Lipase 198–199, *see also* Chloroesters
Lipid oxidation 60–61, 221, 227, 453, 700
Lipids 54, 453, *see also* Fats and oils
Liquid Chromatography/Mass Spectrometry (LC/MS) 31, 59, 83, 223, 292
Liquid smoke flavoring (LSF) 248, 415, 422
Liquid egg 396
Liver Cholangiocarcinomas 127, 129
Lowest Observed Adverse Effect Level (LOAEL) 129
Lowest Observed Effect Level (LOEL) 179
Lysine blockage 649
Lysine modification 229
Lysinoalanine 475
  analysis 479–486
  digestibility and nutritional quality 491–498
  formation 477–479
  kidney toxicity 493–495
  mitigation 482–483
Lysinoalanine (LAL) isomers 524
Lysinomethylalanine 480
Lysophospholipids 197

Maggi, Julius 541
Magnesium, *see* Divalent cations
Maillard cross-link 222
Maillard reaction 10, 27, 29, 53, 80, 148–151, 217, 219, 223

Maillard Reaction Products (MRPs) 498, 647, 699, 701, 702
Maize (corn grain) 518
MALDI/TOF/MS 225
Malt and malt products 192–193, 199, 571
Mammalian cell assays 88
Mango 396
Margin of Exposure (MoE) 11, 12, 38, 203, 269, 301, 379
Marine products 250, *see also* Fish
Meat and meat products 84–85, 191, 192, 224, 248–249, 250, 325, 371–372, 489–490, *see also* Sausage
Meat extract 86, 193
Meat marinades 87
Meat, charcoal broiled 249, 572
Melanoidins 151, 232, 647, 666–668, 702
Metal chelating agents 428
Metalloenzymes 491
Methional 59–60
Methylfurfural 148
Methylglyoxal 220, 226–227
Methylglyoxal-lysine-dimer (MOLD) 222
Michael addition 157–158
Microbial flora, inactivation 455, 460
Micronutrients 37
Microwave and radio-frequency processing 632
  applications 633, 635
  pre-packed foods 635
Microwave cooking 86, *see also* Cooking
Milk 140, 222, 459, 516, 520, *see also* Dairy products
  powders 223–224, 375, 477
  quality 325
  sterilized 224
  UHT 224, 477, 488
  UV treatment 628
Mineral content
  in white wheat flours 658
  in wholemeal wheat flours 658
Mineral water 431
Mitigation strategies, *see names of individual substances*
Modified starch 192–193
Molecular Imprinted Polymers (MIP) 188
Monoamine oxidase (MAO) 338
Monochloropropanol esters 545, *see* Chloroesters
3-Monochloropropane-1,2-diol-dipalmitate 189
3-Monochloropropane-1,2-diol (3-MCPD) 703, *see* Chloropropanols
Monosodium glutamate (MSG) 516
Monte Carlo simulation 120

N-(2-furoylmethyl) amino acids (FMAAs) 225
Namiki pathway 151
Naphthalene 245
N-ε-carboxymethyllysine (CML) 221, 230
N-ε-fructosyllysine 229
Nervous system effects 38
Nitrite 325, 369–370, 705
Nitrogen monoxide 371
Nitrogen oxides 370
Nitrosamines 573, *see also* N-nitrosamines
Nitrosation
  acid catalysed 368
  formaldehyde catalysed 368
  secondary amine mediated 368
N-Methylhydrazine 58
N-Methylpyridinium 668
N-Nitrosamides 367
N-Nitrosamines 365, 573, 704
  analysis 365
  exposure 375–377, 380
  formation 367–369
  non-volatile 367, 370
  occurrence in foods 366, 375
  toxicology 377–380
  volatile 361, 370
N-Nitrosodimethylamine (NDMA) 370–377
N-Nitroso-oxazolidinones 379
No Observed Adverse Effect Level (NOAEL) 38, 65, 129
No Observed Effect Level (NOEL) 179, 696
2-Nonenal 405, 700
Non-enzymatic browning, *see* Maillard reaction

Noodles 477
Novozymes 32
NTP 2-Year rodent study 126

Ohmic heating 636
  applications 637
  cold spot 638
  microbial inactivation 637
Oil refining, *see* Refined edible oils
Oil seeds 476
Olive oil 54, 452
Olives 26
Ornithinoalanine 480
Oxidative damage 64, 526
3-Oxopropanamide 28
Oyster sauce 546

P-450, *see* Cytochrome P-450 monooxygenases
Packaging material 464
Packaging, vacuum 432
Palm oil 603
Palm olein 642, 656
Papaya 396
Paper, tea bag 193
Parkinson's disease 340, 341
Pasteurization 513, 515
Peanuts 476
Pentosidine 222, 232
Peroxide value 654, 655
Persistent Organic Pollutants (POPs) 7
Phase II enzymes 94
Phenanthrene 245
Phenylboronic acid 183
Phenylethylaminoalanine 480
Phosphatidylglycerols 197
Phospholipids 547, 551
Photostimulated Luminescence (PSL) 396
Phthalates 464
Physiological-Based Pharmacokinetic (PBPK) model 121
Phytic acid 649, 659–660, 669
Phytochemicals 646, 660, 669, 670
Pickled herring 194
Plasticizers 464
Polychlorinated Biphenyls (PCBs) 698
Polycyclic Aromatic Hydrocarbons (PAHs) 243, 572, 704
  analysis 257–258
  biomonitoring 259–260
  carcinogenicity 245, 265–268
  exposure 259–260
  health risks/effects 260–268
  mechanisms of formation 258
  metabolic activation 262
  mitigation 258
  occurrence 246–257
  risk management 268–270
Polyphenol oxidase 664
Polyphenols 660–669
Polyunsaturated Fatty Acids (PUFAs) 570, 653, 698
Pork fat 396, *see also* Meat and meat products
Pork juice 87, *see also* Meat and meat products
Potato
  acrylamide 26, 42
  crisps 36
  storage 32, 590
Poultry 80, 84
Prebiotic effect 232
Preventase 32
Probabilistic models 35
Probable Daily Intake (PDI) for children 434
Process toxicants
  anthropogenic sources 247
  definition 8
  interdisciplinary efforts 16
  research needs 16
  risk assessment 12
  setting priorities 9–13
Processed flavourings 86
Procyanidins 660, 661, 662
Pronyl-lysine 222, 224
Protein Efficiency Ratio (PER) 399, 492, 648
Protein modifications 217, *see also* Advanced Glycation Endproducts (AGEs)
Protein tertiary structure 455
Proteins 454, 648
  enzymatic hydrolysates 541
Provisional Maximum Tolerable Daily Intake (PMTDI) 179, 202

Pulsed Electric Field (PEF) processing 622
impact on enzyme activity 627
impact on quality attributes 626
juice processing 625
microbial inactivation 624
Pulsed ultraviolet (UV) light 627
applications 627
microbial inactivation 627
milk treatment 628
Putrescine 323
Pyrene 245
Pyrraline 222, 228, 232

Quality Adjusted Life Years (QUALYs) 15
Quantitative Structure-Activity Relationship (QSAR) models 11
Quercitin 87

Radiation, *see* Food irradiation
RAGE (AGE receptors) 230
Raltech studies 398
Rat hepatocytes 64
Raw milk 177, *see also* Milk
Reactive nitrogen species 92
Reactive oxygen species 261–262
Recommended Daily Allowance (RDA) 697
Reference Dose (RfD) 585
Refined edible oils 194–195
Renal failure patients 217, 230, *see also* Kidney toxicity
Restaurant hamburgers 86, *see also* Meat and meat products
Rice, hypoallergenic 463
Risk analysis 12
Risk assessment 12, 13, 696
guidelines 696
Risk communication 12, 679
attitudes about risks 680–684
credibility 689
effective communication 684
media 687, 691
ranking risks 684
rules 685–688
Risk management 12, 39
Risk perception 679
Risk quantification 696
Risk strategies 13
Risk/benefit considerations 407, 695, 705
and fish consumption 694
Risk/risk considerations 581, 695
Roasted almonds 138
Roasted cereals 199
Roller drying 649, 650, 658, 659
Rosmarinic acid 87

SAFE Consortium 17
Safety factor 12
Salami 195, 325, *see also* Meat and meat products
Salmonella 594
Sauerkraut 300, 331–333
Sausage 325, 327, 371
casings 193
smoke flavored 192
Schiff base 27, 80, 150
with asparagine 29
Scombrotoxicosis 328, 341
Seasonings 389
Selenomethionine isomers 525
Shelf-life 388
Smoke flavorings 192, 248
Smoked fish 329
Smoked products 247, 270, 571, *see also* Meat and meat products
SN2-reaction 548
Sodium caseinate 477
Soft drinks 425
Soil 247, 256
Solid Phase Extraction (SPE) 82–83, 333
Solid Phase Microextraction (SPME) 59, 188, 290, 295
Soups, gravies 86
Soy, alkali treated 492
Soy protein 475, 477, 481, 483, 484, 486, 487, 491, 493, 494, 497, 498, 512, 513, 516, 517, 519
Soy sauce 179, 300, 540, 542, 545, 554, 571
Soybean meal 547, 554
Spectrophotometric methods 146
Spermidine 323
Spermine 323
Spices 389, 397

Spirit drinks, *see* Stone fruit spirits
Staphylococcus aureus 565
Starter cultures 325
Static headspace (HS) analysis 418
Steak, charcoal grilled 249
Steam stripping 554
Stone fruit spirits 297
Strecker aldehyde 28, 150–151
Strecker degradation 28, 59, 60, 61, 80
Styrene 700–701
Sulfotransferases 158
5-Sulfoxymethylfurfural 157–159
Sunflower oil 54
Surface water 247
Swiss Federal Office of Public Health 34, 42
Swedish National Food Administration 25

Tandem SPE technique 83
Tea 87, 249, 255, 663
  black tea 663
  chemoprotective effects 87, 665
  green tea 87, 604, 663
  polyphenols 87, 663
  processing 663
  white tea 663
Tea bag paper 193
Thaumatin 334, 486
Theaflavins 334, 663, 665
Thearubigens 663, 665
Thermal Energy Analyzer (TEA) 366
Thermodynamics 448
Thermoluminescence (TL) 396
Thiazolidine-4-carboxylic acid 370
Thin Layer Chromatography (TLC) 56, 336, 479
Threshold of Toxicological Concern (TTC) 14
Toasted bread 191, 195, 300, 569, *see also* Bread
Tobacco smoke, *see* Cigarette smoke
Tolerable Daily Intake (TDI) 161, 434, 544
Tomato products 141, 516
Tomato sauce 459, 466
Tortillas 487
Total Diet Studies (TDS) 421
Toxicity prediction systems 699
  Derek 11
  Topkat 11
Trans fats, *see* Trans Fatty Acids
Trans Fatty Acids (TFAs) 652, 656, 670, 690
  industrial fats 706
  margarines 706
Transgene animals 88
Transgenic maize 518
Transition metals 430, *see also* Copper; Iron
Triglycerides 61, 180, 549, *see also* Lipids
Trigonelline 668
Trypsin inhibitors 648, 649
Type 2 diabetes 231, 665, 668, 669, 702
Tyramine 323

UK Food Standards Agency (FSA) 41
Ultrasound 629
  applications 629
  D-value 631
  microbial inactivation 629
  preservation 629
  raw milk 631
United States Environmental Protection Agency (US EPA) 64–65, 244
United States Food and Drug Administration (US FDA) 25
Urea 305, 307
Uremic patients 218, 230, 233, *see also* Renal failure patients
Uremic toxins 230, *see also* Advanced Glycation Endproducts (AGEs)
Urethane, *see* Ethyl carbamate
Urine metabolites 93, 155
US FDA Draft Action Plan 39–40

Vacuum atmosphere 340
Vacuum Distillation (VD) 420
Vacuum sealed packaging 432
Vegetables 256, 422, 487, 518, *see also* Fruits and vegetables; Sauerkraut; *names of specific vegetables*
Vegetables, fermented 376
Vegetative cells, *see* Microbial flora
Vinegar 138, 141–142

Viruses 462
Vitamin C, *see* Ascorbic acid
Vitamin E 604
Vitamin premix 599, 603
Vitamins
  fat soluble 453, 652
  impact of processing on 658, 659
  water-soluble 453, 657
  in white wheat flours 658
  in wholemeal wheat flours 657
Volatile Organic Compound (VOC) 425

Water 456, 457, 458, 460
Water activity 605, 669
Wet strength resins 178, 200
Wheat gluten 481, 484, 547
Whey powder 223
Whisky 141
Whole grain 602, 657, 669
Wine 138, 140–141, 227, 297, 307, 330
Wood smoke 200
Wool 490

Yeast fermentation 307
Yeast extract 300
Yeast proteins 490
Yogurt 227

z-Value 448